ANIMAL PHYSIOLOGY

The S203 Course Team

Chairman and General Editor
Michael Stewart

Academic Editors
Caroline M. Pond (Book 1)
Norman Cohen (Book 2)
Michael Stewart (Book 3)
Irene Ridge (Book 4)
Brian Goodwin (Book 5)

Authors
Mary Bell (Book 4)
Sarah Bullock (Book 2)
Norman Cohen (Book 2)
Anna Furth (Book 2)
Brian Goodwin (Book 5)
Tim Halliday (Book 3)
Robin Harding (Book 5)
Stephen Hurry (Book 2)
Judith Metcalfe (Book 5)
Pat Murphy (Book 4)
Phil Parker (Book 4)
Brian Pearce (Book 3)
Caroline M. Pond (Book 1)
Irene Ridge (Book 4)
David Robinson (Book 3)
Michael Stewart (Book 3)
Margaret Swithenby (Book 2)
Frederick Toates (Book 3)
Peggy Varley (Book 3)
Colin Walker (Book 2)

Course Managers
Phil Parker
Colin Walker

Editors
Perry Morley
Carol Russell
Dick Sharp
Margaret Swithenby

Design Group
Diane Mole (Designer)
Pam Owen (Graphic Artist)
Ros Porter (Design Group Coordinator)

External Course Assessor
Professor John Currey (University of York)

General Course Consultant
Peggy Varley

Course Secretaries
Valerie Shadbolt
Christine Randall

ANIMAL PHYSIOLOGY

Edited by Michael Stewart

BIOLOGY: FORM AND FUNCTION

Hodder & Stoughton   The Open University

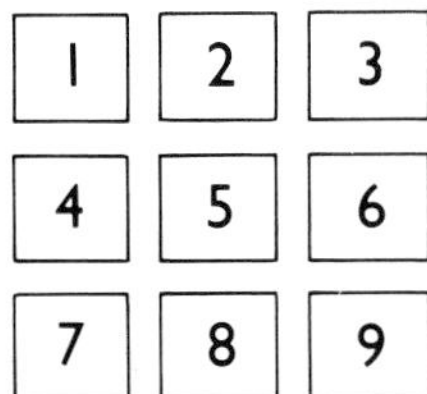

1. Mammalian liver section stained with haematoxylin and eosin, showing liver parenchyma and (in the centre of the picture) a portal triad of vein, artery and bile duct (× 175).

2. Chordae tendineae in the left ventricle of the heart of a large polar bear.

3. Mammalian bladder section stained with Masson trichrome showing the transitional epithelium stained red (× 500).

4. Adult parasitic helminth worms (*Schistosoma* sp.) in small blood vessels in the gut of a laboratory hamster.

5. Inner surface of stomach of a polar bear.

6. Sea anemone (phylum Cnidaria, class Anthozoa).

7. Section of skeletal muscle of a mammal impregnated with gold to shown nerve fibres and motor end plates (× 400).

8. Section of the trachea of a mammal; it is lined by ciliated columnar epithelium (× 400).

9. Intestine of adult mute swan showing fat-filled mesenteries and pancreas.

Back cover: Computer-generated pseudocolour image of an autoradiogram showing the binding of a radiolabelled GABA ligand, [^{3}H]muscimol to cerebellum of one-day-old chick brain. The highest levels of activity are coded red.

British Library Cataloguing in Publication Data
Animal physiology.
1. Animals. Physiology
I. Stewart, Michael II. Series
591.1

ISBN 0-340-53187-8

First published 1991.

Designed by the Graphic Design Group of the Open University.

The text forms part of an Open University course. Further information on Open University courses may be obtained from the Admissions Office, The Open University, P.O. Box 48, Walton Hall, Milton Keynes, MK7 6AB.

Typeset by Wearside Tradespools, Fulwell, Sunderland, printed in The United Kingdom by Thomson Litho Ltd, East Kilbride for the educational division of Hodder and Stoughton Ltd, Mill Road, Dunton Green, Sevenoaks, Kent TN13 2YA, in association with the Open University, Walton Hall, Milton Keynes, MK7 6AB.
1.1

CONTENTS

PREFACE

Animal Physiology is the third in a series of five volumes that provide a general introduction to biology. It is designed so that it can be read on its own (like any other textbook) or studied as part of S203, *Biology: Form and Function*, a second-level course for Open University students. As well as the five textbooks, the course consists of five associated study texts, 30 television programmes, several audiocassettes and a series of home experiments. As is the case with other Open University courses, students of S203 are required both to complete written assignments during the year and to sit an examination at the end of the course.

In this book, each subject is introduced in a way that makes it readily accessible to readers without any specific knowledge of that area. The major learning objectives are listed at the end of each chapter, and there are questions (with answers given at the end of the book) that allow readers to assess how well they have achieved these objectives. Key words are identified in **bold** type both in the text where they are explained and also in the index, for ease of reference. A 'further reading' list is included for those who wish to pursue certain topics beyond the limits of this book.

INTRODUCTION

Physiology is the study of the function of organisms and their constituent parts. This book is concerned with the mechanisms that operate in animals and with the different solutions to problems posed that enable them to live successfully in their environments. Both vertebrates and invertebrates are studied, with particular emphasis on mammalian physiology.

Over the past hundred years or so biology has moved from the stage of cataloguing and description to that of analysis and experiment. Under the general heading of physiology are included specific fields of interest, such as cell physiology, neurophysiology, functional anatomy, environmental physiology, and invertebrate and vertebrate physiology. However, one of the important modern developments has been the breaking down of some of the barriers that have been built around these specialities. Physiology today is more often concerned with appreciating the major general principles and how these relate to those of other major areas of biology, for example, anatomy, ecology, behaviour and biochemistry.

As you read the chapters of this book, bear in mind two points. First, to appreciate how animals function we must be aware of their external form and internal structure. In this sense, the term 'structure' includes the chemical structure of the cell constituents, the ultrastructure of cells as revealed by the electron microscope, the organization of cells into tissues as seen in the light microscope, and the gross anatomy of the animal as revealed by dissection. The second point is that an animal's physiology should be related to what we know of its natural environment. It is unwise to extrapolate too readily our findings in the artificial conditions of the laboratory to the much more complex conditions in the real world. We should remember the saying that what is true for *E. coli* is not necessarily true for *E. lephant.* There is a very wide range of natural environments—deserts, cold water and mountains, for example—that different animals have to face. Even in one particular environment, say a hot desert, the ways in which different animals cope with the heat are likely to be different, depending on a number of factors such as the animal's size, its diet and its basic organization. All this means that physiologists are often concerned with the variety of physiological adaptations shown by different species of animals in different environments. This is very much the field of *comparative physiologists*, whereas *cellular physiologists* tend to be interested in the physiological characteristics that living animals have in common, which are often more obvious at the cellular or biochemical level. These two approaches are complementary and often reflect approaches to common problems at different levels; both views are represented in the chapters that follow. In writing this book we have assumed that the reader has a basic level of knowledge of animal diversity, and of cell biology and biochemistry such as will be found in Books 1 and 2 of this series.*

*Caroline M. Pond (ed.) (1991) *Diversity of Organisms*, and Norman Cohen (ed.) (1991) *Cell Structure, Function and Metabolism*, Hodder and Stoughton Ltd, in association with The Open University (S203, *Biology: Form and Function*).

By way of introduction let us highlight some of the key issues discussed in the chapters and describe the logical links between them. One of the important concepts in physiology is *regulation and control* of physiological processes. The functions of organisms are controlled and regulated by physiological communication systems: nervous and hormonal. Most biologists consider the nervous system to have the primary coordinating role, and indeed nerves have the great advantage of conveying information quickly to precise target organs. Chapters 1–3 introduce some fundamental aspects of the nervous system, in particular the structure and function of the basic unit, the neuron or nerve cell. Signals pass along nerves as electrical impulses that are generated by changes in the permeability of membranes; hormones are generally carried in the blood system. Both nervous and hormonal systems interact with receptors on specific target cells through the release of chemical messengers. The ensuing events that lead to the final responses depends on what the target cell is differentiated to do (e.g. muscle fibre, secretory cell or nerve cell). Chapter 4 picks up the theme of regulation and control and examines it further by taking a 'systems' overview of the biological phenomena, encouraging students to think about the broader issues of how animals function. It considers the principles by which animals regulate their essential physiological variables, such as body temperature and water content: this is the principle of *homeostasis*.

Chapters 5–7 are concerned with the circulatory and respiratory systems of animals, which are closely linked. A circulatory system is a necessity for most animals over a certain size, especially homoiothermic animals which have a high metabolic rate. The circulatory systems of a range of organisms are studied and whilst there are major differences between species, some common principles are clear. Relationships between structure and function are emphasized, not only for the gross components of the systems—hearts and blood vessels—but also for the micro-circulation, in which the exchange of oxygen, nutrients and waste materials occurs. The themes of regulation and control are particularly relevant in the last two sections of Chapter 5, in which the human circulatory system is analysed. In Chapters 6 and 7, mechanisms of respiration are contrasted in fishes, insects, birds and mammals. The processes of transporting oxygen and carbon dioxide in the blood are considered, in particular the function of respiratory pigments. Environmental physiology is an important theme, and a study is made of how vertebrate animals are adapted physiologically to the very different problems posed by high altitude, or deep water, or the stress of exercise.

In Chapter 8 the dietary requirements of animals (the types and amounts of chemicals that different animals require) are outlined as an introduction to feeding mechanisms. The relationship between the type of food ingested and the method used to obtain it is stressed. The mechanisms that animals use to gather and process food are very diverse, but we can more easily organize our knowledge by recognizing general categories that depend on the size and composition of the food taken in. For example, animals that feed on relatively small particles often display common features, even though they may be unrelated taxonomically. Keep in mind that such a system of classification, however good, is unnatural and sometimes arbitrary. Note too that the conclusions we draw are *generalizations*, not scientific laws, and physiologists frequently come across animals that provide exceptions to the rules. The chapter ends by examining digestion and absorption of food, primarily in mammals, together with the coordination (control) of digestive processes.

Chapter 9 returns to the theme of regulation and control by considering the subject of blood sugar regulation. Glucose is the most important 'fuel' for mammals; cells take it up from the blood and it is essential that there should always be an adequate supply available. Too high or low a concentration leads to deleterious effects, so there is a need for homeostasis of blood sugar level. Mammals regulate this within strict limits through a number of control systems. The final section shows how errors in human blood sugar regulation result in two main types of *diabetes mellitus*, and how these can be controlled by medication and attention to diet.

In Chapter 10, control mechanisms are again emphasized, this time in the context of reproduction. Many mammals have breeding cycles, usually correlated with the seasons, and the importance of the role of hormones and of environmental cues (such as light, temperature and the presence of other individuals) in the control of reproductive processes is considered. From this emerges an understanding of the correlation between the environment and the nervous and hormonal control systems. Knowledge of hormonal and nervous control of reproductive cycles has important implications in the field of contraception and human reproduction, and this is given due consideration. Because the chapter concentrates on mammals, which have been studied more extensively than other groups (largely for clinical and economic reasons), it ends by considering the diversity in reproductive controlling mechanisms in other animals including birds, reptiles and amphibians.

In Chapters 11 and 12 the theme of the relationship between an animal's physiology and its environment is stressed. Animal tissues are bathed with fluids that may differ in ionic content and water content from the external medium, so salts and water may be gained by the body fluids or lost to the environment. Animals may either tolerate variations in composition of their body fluids, or much more commonly, maintain constant levels of ions and other solutes in the fluids around and within cells, that is they show homeostasis of their body fluids. The problems posed by different environments (seawater, freshwater and land) are discussed, in particular the osmotic and ionic stresses encountered by the animals living there. The strategies adopted by invertebrates and vertebrates in these three environments are considered, and the particular problems of moving between the environments are discussed. Although external body surfaces play a key role in control of ion and water movements, in many animals such as birds and mammals the principle role is that played by the renal organs. Their importance in osmoregulation and excretion is examined, and the physiology of the mammalian kidney and the insect Malpighian tubule is studied in detail.

An important feature that you will find throughout this text is the presence of a summary of the main points either at the end of chapter sections, or at the end of each chapter: these should be helpful in pointing to key facts and ideas.

1.1 INTERCELLULAR COMMUNICATION

An extremely important feature of all living organisms is their ability to respond to changes in both their external environment (e.g. fluctuations in temperature, pressure, and light), and internal environment (e.g. changes in blood glucose after a meal, the state of contraction or relaxation of a muscle). Multicellular animals also need to control the activity of various organs and tissues, such as muscles, blood vessels and reproductive organs. Such control must be carefully coordinated if the animal is to function properly. It is therefore necessary to have a specialized means of communication between cells and tissues.

For centuries the nervous system has been recognized as being involved in regulatory processes. The Roman physiologist Galen thought that the brain converted 'vital spirits' in the blood into 'animal spirits', which were then carried along the nerves to all parts of the body, including movement in muscles and 'sensations' in body organs. By the nineteenth century, it was known that communication along nerve fibres was via the rapid transmission of electrical 'impulses'. In the early part of the twentieth century, evidence was beginning to accumulate that suggested that these electrical impulses stimulated the release of chemical messenger molecules from nerve endings. These messenger molecules are called **neurotransmitters**, and they act to alter the activity of other nerve cells, organs or tissues. At the same time, another communication and control system was discovered, which also involved the release of chemical messenger molecules, but this time into the bloodstream. These substances are called **hormones**, and they are generally released from specialized glandular tissues called endocrine glands. The bloodstream acts as a transport system for the hormones, so an endocrine gland in one part of the organism can influence the activity of cells in another part. Thus animals possess two communication systems, a nervous system and a hormonal or endocrine system.

In general, nervous communication is extremely rapid and short in duration, and the response is precisely located; hormonal effects are usually slower in onset, longer in duration and often involve widely scattered cells. Neurotransmitters usually act over very short distances (15–100 nm), whereas hormones act over much longer distances (from millimetres to metres). The nervous system is ideally suited for coordinating activities that involve rapid or continuous adjustments (e.g. breathing or running), whereas the hormonal system may be considered to be best suited for relatively long-term regulation (e.g. the control of reproductive cycles—see Chapter 10; or the maintenance of blood glucose levels—see Chapter 9).

The nervous and hormonal systems are not, however, entirely separate modes of intercellular communication. They are highly interdependent, since much of the endocrine system is controlled by the nervous system. There are also striking similarities between the two systems: for example, some nerves release hormones, and several chemical messenger molecules can act as both neurotransmitters and hormones. Moreover, intercellular communication in

RECEPTOR
A molecule or molecule complex capable of recognizing and interacting with a specific hormone or neurotransmitter.

SECOND MESSENGERS
Messenger molecules formed within cells in response to the interaction of a hormone or neurotransmitter and its receptor. Their role is to amplify this signal & convert it into an appropriate cellular response.

both systems involves the interaction of the messenger molecule and specific recognition molecules, which are part of the target cell or tissue. These recognition sites are called **receptors** and they play a pivotal role in the receipt and translation of the information carried by the messenger molecule. The specificity of these receptors ensures that target cells and tissues respond only to those hormones and neurotransmitters for which they have receptors (Figure 1.1).

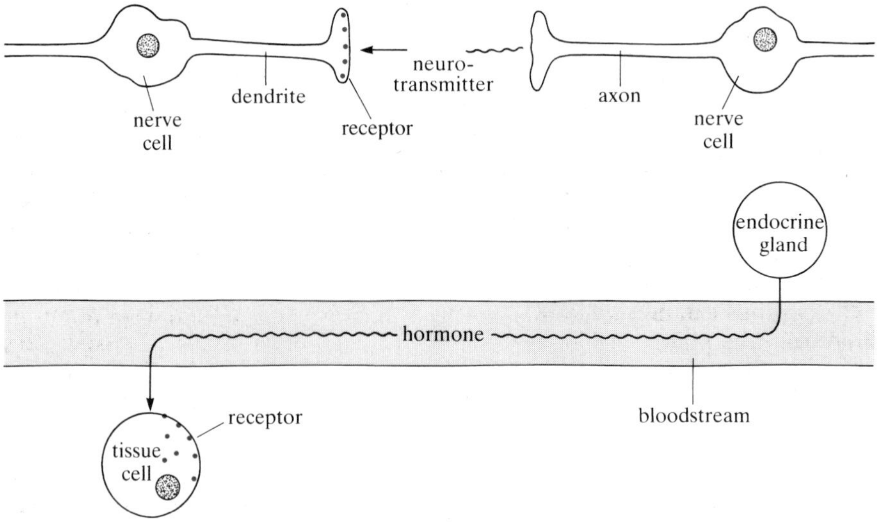

Figure 1.1 Chemical communication in the nervous system and the hormonal (endocrine) system. Both systems use signalling substances (messenger molecules) but act over different distances.

SIMILARITIES BETWEEN NEUROTRANSMITTERS & HORMONES
- both have receptors
- both generate second messengers to amplify

Another area of similarity is the way in which cells process the information carried by neurotransmitters and hormones. In many cases, the interaction of a messenger molecule and its receptor leads to the generation of a second set of chemical messengers, which act entirely within the cell. These are called **second messengers** and they serve to amplify the original signal and then initiate the appropriate cellular response to that signal.

In the remainder of this chapter we shall deal specifically with communication in the nervous system. We shall examine the cells that comprise it and their interrelationships, the mechanisms underlying the transmission of signals along nerve fibres, and the way in which nerves are formed into circuits capable of processing information into coordinated and controlled responses. In the following chapters we will discuss hormonal communication (Chapter 2) and communication at the intracellular level (Chapter 3).

1.2 THE NERVOUS SYSTEM: CELLS

The nervous system is directly involved in all the processes that regulate and control an animal's physiology. The complexity of the system depends, to a great extent, upon what the animal needs to do. In some animals, the nervous system may consist of just a few nerve cells and their fibres, grouped together to form a simple brain. In contrast, the human brain is composed of more than 10^{12} nerve cells with more than a thousand times that number of interconnections. Despite the differences in complexity, the basic mechanisms underlying communication in the nervous systems are the same in the nerves of a squid, for example, as they are in the nerves of human beings. Before discussing the way in which the nervous system works, we should consider the cells that comprise it.

The nervous system is composed of two types of cell, **neurons** and **glia**, but within these broad divisions there are subtypes of cell that play specialized roles. We shall deal with glial cells and their relationships with neurons later in this chapter, but first we shall concentrate on neurons because these are the cells responsible for the transmission of signals throughout the nervous system.

1.2.1 Neurons

The internal structure of a neuron is very similar to that of many other types of cell. The neuron cell body contains a nucleus, extensive endoplasmic reticulum, Golgi apparatus and mitochondria. Not surprisingly, neurons perform the basic functions of all cells, synthesizing macromolecules, transporting some and breaking down others. A feature of neurons is that they possess organized outgrowths from their cell bodies called **axons** and **dendrites**. It is via these outgrowths, or processes as they are often termed, that neurons communicate with other neurons and tissues. Processes from adjacent neurons can link up to form a network such that cell-to-cell communication can take place over long distances. The three-dimensional organization of these processes is extremely varied, and neurons can be classified morphologically and, to some extent, functionally according to the arrangement of these processes.

Most neurons have a dendritic zone and a single axon, which leaves the cell body at the **axon hillock**. Figure 1.2 shows how the structure of a typical neuron is related to its functions, although this can be difficult to see in some

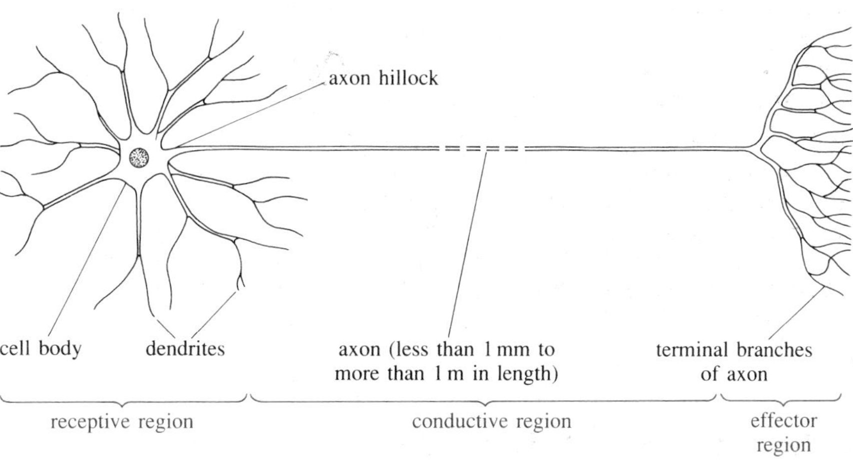

Figure 1.2 The structural and functional organization of a neuron.

neurons. As the derivation suggests (from the Greek for 'a tree'), dendrites are generally 'tree-like', although their shape varies in different neurons (Figure 1.3). The branching of dendrites may be simple, as in Figure 1.3a, or very complex as on the neuron in Figure 1.3d; on many of the branches there are short projections called dendritic spines. The dendrites form part of the receptive region of the neuron, receiving inputs via the axons of other neurons. The axon conducts signals, in the form of electrical impulses, away from the cell body towards the effector region, which is at the end of the axon furthest from the cell body. The axon may either branch to form a number of swellings at its tips or develop such swellings along its length. These swellings form part of the **synapse** (Figure 1.4a), the contact region between a tip of the

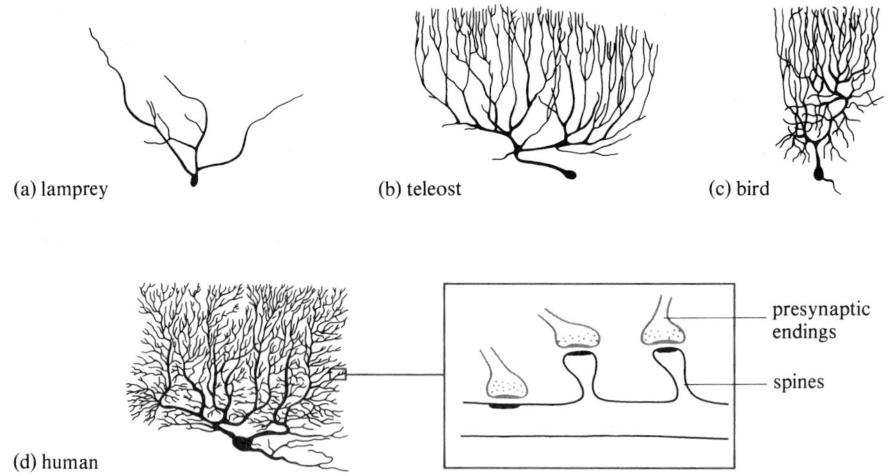

Figure 1.3 Shapes of dendritic 'trees' of a type of neuron (Purkinje cell) from the cerebellum of different vertebrates. One dendrite is shown enlarged, and dendritic spines can be seen with incoming presynaptic endings.

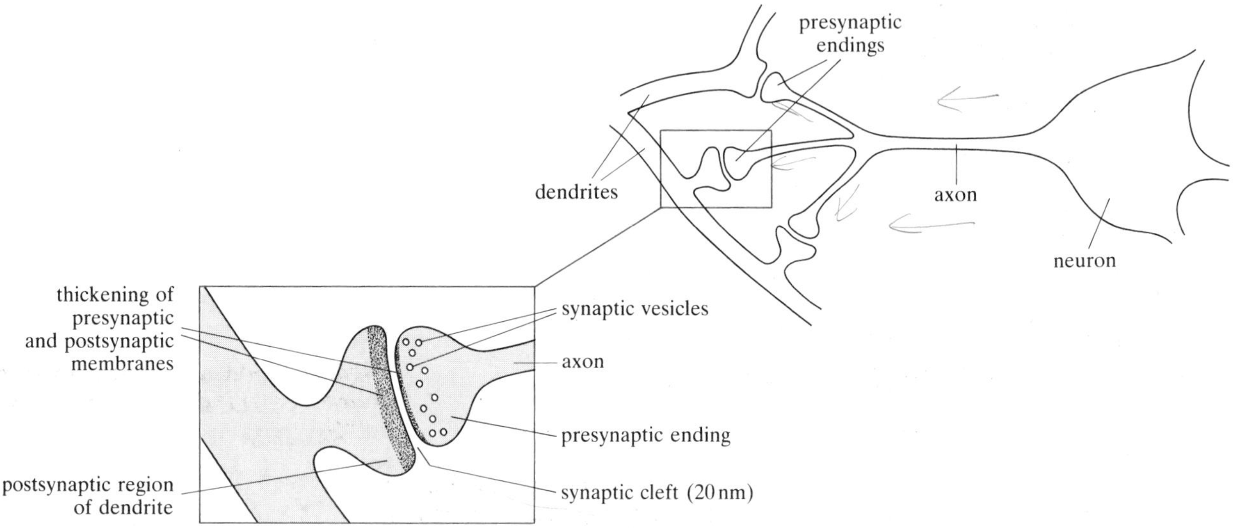

(a)

Figure 1.4 (a) The major components of a synapse, and their relationship to axons and dendrites of nerve cells. (b) A synapse as seen in the electron microscope (magnification of approximately 50 000 times).

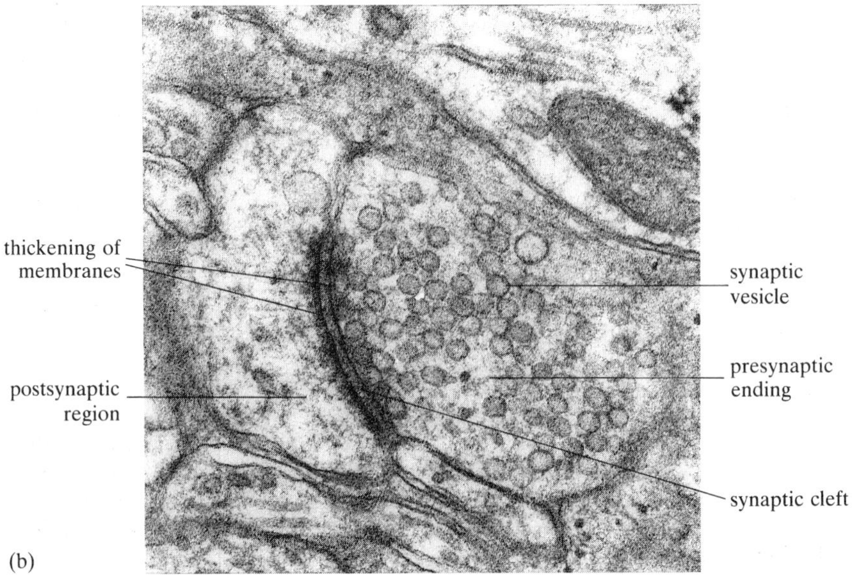

(b)

axon (the presynaptic ending or terminal) and a postsynaptic target area, which may be a dendrite or a dendritic spine, a neuron cell body, a muscle fibre or an organ of some kind. The small space between the pre- and postsynaptic regions is called the synaptic cleft. Figure 1.4b is an electron micrograph of a synapse: it shows that the presynaptic terminal contains a number of spherical vesicles and that both the presynaptic and postsynaptic membranes are somewhat thickened. The significance of some of these structures will become clear later in this chapter. Figure 1.5 gives some idea of the multiplicity of synaptic inputs that can be achieved on dendrites (some large neurons can receive as many as 10^8 synaptic inputs).

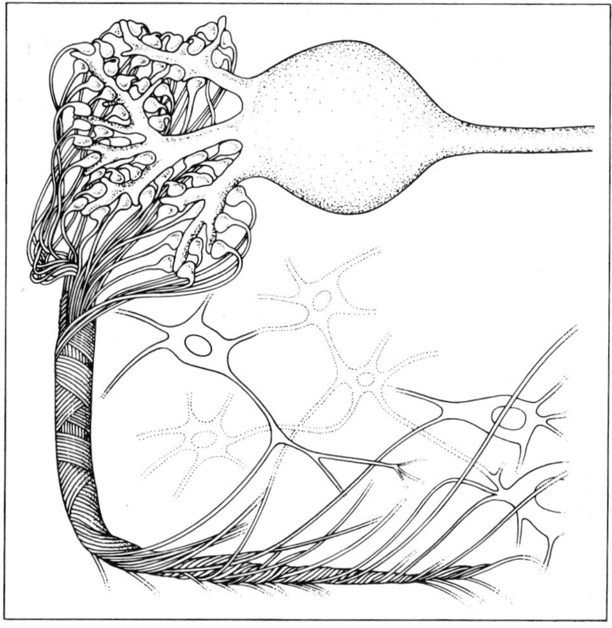

Figure 1.5 An artist's impression of the density of synaptic inputs to a single neuron.

In summary, the neuron can be thought of as having receptive, conductive and effector regions. The relative extent of these regions varies, depending upon the type of neuron, and may imply something about the function of that particular neuron. The dendritic field (the extent of dendritic branching) can be enormous. Axons can run for several metres and eventually split into a network of terminals forming synapses with the target cells or tissue. The final result is somewhat similar in arrangement to a huge, interconnected, telephone network (although there the analogy ends).

◇ What could be the function of extensive dendritic branching and of dendritic spines?

◆ Dendrites receive inputs from axons forming synapses on them. The large surface area provided by branching and by the spines on them, increases the membrane area available for synaptic inputs.

Now attempt Question 1, on p. 40.

The basis of communication in the nervous system is the generation and conduction of electrical impulses along axons. These impulses are brought about by momentary changes in the permeability of axonal membranes to various ions and, as these changes in ion movements are fairly complex, we shall examine this process step-by-step.

MEMBRANE POTENTIAL E_m

Electrical potential difference (usually inside negative) across a membrane. Determined by the distribution of ions across the membrane & the membrane's permeability to those ions.

EQUILIBRIUM POTENTIAL

Equ. pot. for a particular ion can be calculated using Nernst equation, and is the electrical potential across a membrane at which there is no movement of that ion across the membrane.

1.3 SIGNALLING ALONG AXONS

1.3.1 The membrane potential

Neurons, like many other types of cell, have a small difference in electrical potential across their membranes. This is due to the maintenance of unequal concentrations of ions, particularly Na^+ and K^+ ions, on either side of the membrane and the membrane's permeability to these ions. If the membrane is penetrated with a microelectrode connected to a voltmeter, this potential difference can be measured: it is called the **membrane potential** (E_m). It is rapid changes in the E_m, caused by momentary changes in the membrane's permeability to ions, that underlie the generation and propagation of electrical impulses in axons. Before considering how these electrical impulses are generated, we need to know how the movement of ions across the membrane influences the E_m.

Table 1.1 lists the concentrations of Na^+, K^+ and Cl^- ions on the inside and the outside of some nerve and muscle cell membranes. Notice that there are unequal distributions of these ions across the membrane. The movement of an ion across the membrane is determined by its electrochemical gradient, which comprises both the electrical and concentration gradients of that ion across the membrane. If we take the squid giant axon as an example (see Table 1.1) and assume that the membrane is permeable to K^+ ions only, we would expect K^+ to move out of the axon down its concentration gradient. However, this movement of K^+ is eventually opposed by the build-up of positive charge on the outside of the membrane which repels the movement of any more K^+ ions. When these effects balance each other such that there is no net movement of K^+ ions across the membrane, an equilibrium point is reached. This situation can be described mathematically using the **Nernst equation**, which allows the calculation of the **equilibrium potential** for any particular ion.

Table 1.1 Concentrations (mmol l^{-1}) of certain ions inside and outside various cells, and the resting membrane potentials E_m recorded when these cells are penetrated by a microelectrode.

		Squid giant axon	Mammalian motor neuron	Frog muscle
K^+	inside	400	150	140
	outside	20	2.5	2.5
Na^+	inside	50	15	9
	outside	440	145	120
Cl^-	inside	40–100	9	3
	outside	560	101	120
	E_m/mV	−60	−70	−90

The Nernst equation can be written as:

$$E = \frac{RT}{ZF} \times 2.303 \log_{10} \frac{c_o}{c_i} \text{ volts}$$

where R is the gas constant (8.314 J K^{-1} mol^{-1}), T is absolute temperature (291 K), Z is the valency of the ion (+1 for a positively charged monovalent ion), F is the Faraday constant (96 500 coulombs mol^{-1}) and c_0 and c_i are the ion concentrations outside and inside the membrane, respectively.

Substituting the data for the squid giant axon (Table 1.1) into this equation, we can calculate the equilibrium potential for K^+ (E_{K^+}):

$$E_{K^+} = \frac{8.314 \times 291}{+1 \times 96\,500} \times 2.303 \log_{10} \frac{20}{400} \text{ volts}$$

$$E_{K^+} = 0.025 \times -3.0 = -0.075 \text{ volts } (-75 \text{ millivolts})$$

◇ Using the Nernst equation and the Na^+ ion concentrations either side of the squid axon membrane shown in Table 1.1, calculate the equilibrium potential for Na^+ (E_{Na^+}).

◆ The answer is +55 mV.

When a squid giant axon is penetrated with a microelectrode, the E_m recorded across the membrane is usually around −60 mV. The resting membrane is thus permanently polarized with the inside of the membrane negatively charged. Notice that the recorded E_m is similar to E_{K^+} (−75 mV) but is not identical with it.

◇ Can you suggest a reason for the difference between the recorded E_m and the calculated E_{K^+}?

◆ The difference between these values suggests that the axonal membrane is permeable to other ions as well as K^+.

Axonal membranes are, in fact, permeable to Na^+, Cl^- and some other ions, such as Ca^{2+}. However, the membrane is selectively permeable, for example it is about 80 times more permeable to K^+ than it is to Na^+. Nevertheless, the small influx of Na^+ ions (remember the concentration gradient) is sufficient to raise the equilibrium E_m from the E_{K^+} of −75 mV to its recorded level of approximately −60 mV. Any change in the permeability to either ion will immediately change the E_m, and it is the degree of permeability to K^+ and Na^+ that is the key to the generation and propagation of signals along axons.

1.3.2 Local and action potentials

To start with, think of an axon as an electric cable. The axon contains axoplasm, a solution of ionic salts and proteins, within a lipid–protein membrane. An electric cable consists of a metal wire contained within a plastic covering. But how does an axon compare with an electric cable in terms of its electrical properties? The ions of the axoplasm are charged and therefore their movement generates a current, just like the movement of electrons in the wire. Axoplasm however, is a poor conductor of electricity compared with the metal wire because the number of charge carriers is smaller and their mobility is less. Moreover, because the membrane around the axoplasm is permeable to ions it is not a perfect insulator, and ions (i.e. current) leak out of the membrane. This would not seem to be a propitious start for a communication system that relies on the generation and propagation of electrical impulses.

Figure 1.6 is a sketch of an experiment to illustrate the poor electrical properties of an axon. A squid giant axon is impaled by three microelectrodes. Microelectrode P is used to pass a current (in a sense, to inject ions) across the membrane. The direction the current flows (which depends upon whether anions or cations are injected) will either displace the E_m (−60 mV) towards zero (i.e. make the inside of the membrane more positive) or make the E_m more negative (i.e. increase the negative charge inside the mem-

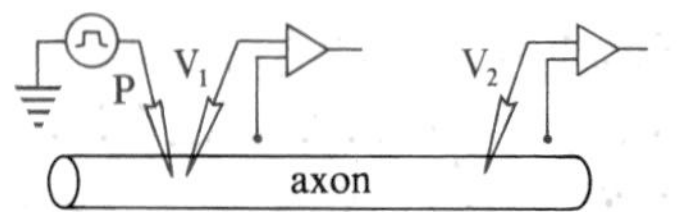

Figure 1.6 Diagram of a set-up to measure local and action potentials in a large axon.

HYPERPOLARIZATION
In contrast to depolarization where the membrane potential (E_m) is displaced towards zero, during hyperpolarization the E_m is made more negative.
DEPOLARIZATION
The displacement of the membrane potential (E_m) from its resting value (approx −60 to −70 mV) towards zero.
LOCAL POTENTIAL
A change in membrane potential (E_m) which tends to decay with distance along a nerve fibre. If the local potential is sufficiently large (ie above threshold) it may trigger an action potential which is faithfully reproduced along the length of the nerve fibre.

brane). The terms **depolarize** and **hyperpolarize** are usually used to describe these respective changes in the membrane potential. The other two electrodes in Figure 1.6 (V_1 and V_2) record the E_m at different distances along the axon from P.

A current is passed through the electrode P. The membrane is hyperpolarized at time A, and a little later at time B it is depolarized, as indicated at the top of Figure 1.7a. Look at the E_m value recorded at V_1. The membrane is hyperpolarized and then depolarized as current flows from P to V_1. Now look at the recording at V_2, a short distance away from V_1. The E_m changes at V_2 as current is passed at P but the changes are much smaller than those recorded at V_1.

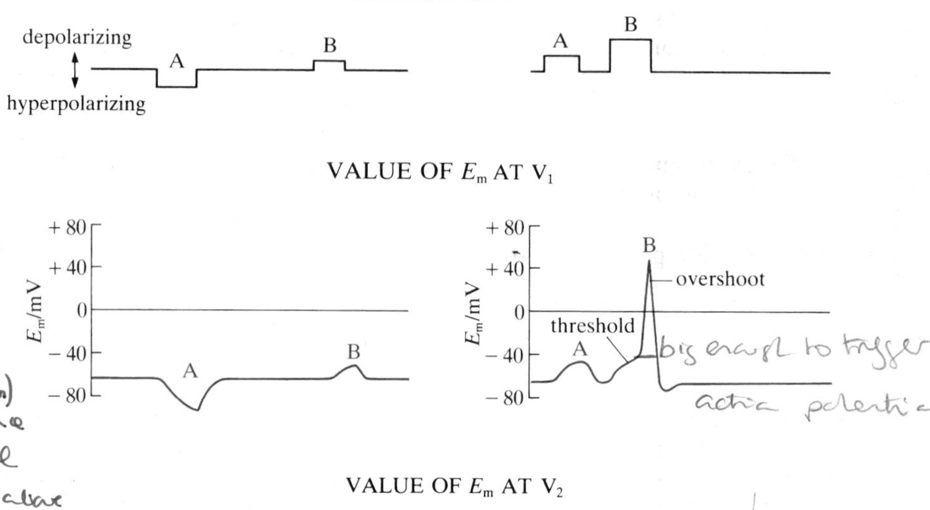

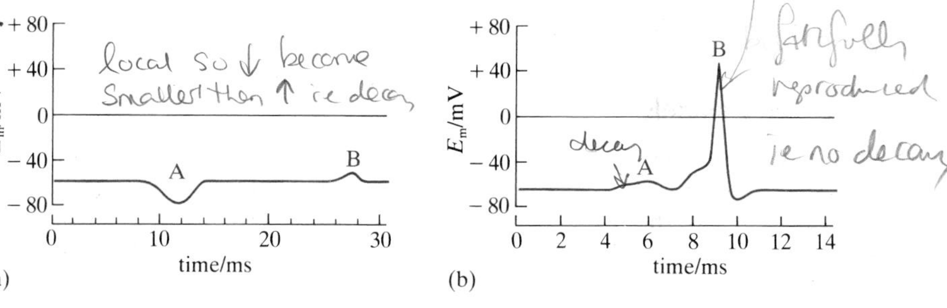

Figure 1.7 (a) Local potentials and their effect on the value of E_m. Notice how the effect is less at V_2 than at V_1. (b) The recording of an action potential. Notice that the effect is the same at V_2 as at V_1.

ACTION POTENTIAL
Momentary localized change in elec. potential across a nerve membrane, caused by a rapid change in the membrane's permeability to Na^+. Transmission of info across nerve fibres takes place by means of trains of action potentials, the nature of the info. conveyed being determined by the frequency & duration of action potentials; their size is constant.

The reduced potential recorded at V_2 is caused by the poor electrical properties of the axon. Consequently even if the signal did manage to reach the target, it would bear no resemblance to the signal originally transmitted. This experiment shows that changes in E_m decay over distance (and with time) along the axon because of its poor electrical properties—these changes in E_m are called **local potentials**. Obviously, if the axon of a nerve cell is very short (the distance between P and V_1 in Figure 1.6), representative signals can be transmitted. Such local potentials may be very important in small neurons with short axons.

Now look at Figure 1.7b. This time the membrane is depolarized at both times A and B at P, and the E_m is again recorded at V_1 and V_2. The first small depolarization (A) decays between V_1 and V_2, as in Figure 1.7a, but the second depolarization (B), which is larger, produces a massive change in the E_m, labelled an overshoot. This overshoot is faithfully reproduced at V_2 a few milliseconds later: it is termed an **action potential**, and it lasts for about 1.5 ms before the E_m is restored to the resting value. If a depolarizing current even larger than B were passed through P another action potential would be generated, but it would be exactly the same size as the previous one. Thus, once the E_m has been pushed beyond a threshold value, an action potential of fixed size and duration results.

If we were to attempt to generate a second action potential just 0.1 ms after the first, there would be no response. This is because there is a period of enforced 'silence', called a **refractory period**, at the end of an action potential. During this refractory period the action potential travels away from the point of stimulation, because the membrane adjacent to the site of the stimulus is in an electrically receptive state, whereas the membrane at the site of the stimulus is in an electrically inactive state. We shall discuss the refractory period in more detail in the next Section.

In the experiment shown in Figure 1.6 the change in E_m is generated near the middle of the axon. Normally action potentials are generated at the point where the axon leaves the neuron cell body (the axon hillock—the most excitable part of the axon) and will, therefore, pass down the axon towards the synapse. If the microelectrode is moved from V_2 to a position 1 metre down the axon, and a suitably large depolarizing current is passed at P, we should still record an action potential at this distant point.

◇ How do action potentials differ from local potentials?

◆ Local potentials decay over distance whereas an action potential will be accurately reproduced all the way along the axon.

Local potentials displace the E_m above the threshold value, and generate an action potential. This action potential then acts as a local potential and generates another action potential in the next patch of membrane. The action potential is neither attenuated nor distorted because it is recreated in each successive patch of axonal membrane. The action potential is the key to signalling along axons, in the next Section we shall consider what produces it.

1.3.3 Ion channels

Ions cross cell membranes via channels formed by specialized proteins that constitute part of the membrane structure. These **ion channels** are transient: in other words, they are opened or closed in response to an appropriate stimulus. Some are operated by signal molecules such as hormones and neurotransmitters, and are termed 'ligand-dependent ion channels'. Others are activated by changes in the membrane potential, and are termed 'voltage-dependent ion channels'.

In the nervous system, neurotransmitters open or close ligand-dependent ion channels that either form part of or are closely associated with the receptors for that neurotransmitter (see also Section 1.4). The resulting change in the membrane's permeability to particular ions causes a redistribution of ions across the membrane thus changing the E_m; this change in E_m may be sufficient to activate voltage-dependent ion channels that, in turn, further displace the E_m thus opening more channels, and so on. In a sense, the opening of ligand-dependent channels acts as a 'primer' to shift the E_m to a threshold position, whereupon the voltage-dependent channels are activated. Once these are opened, the result is an immediate change in the membrane's permeability to ions and, consequently, a rapid change in the E_m.

During an action potential the inside of the axon becomes positively charged with respect to the outside (cf. Figure 1.7b). This must mean that there has been a redistribution of ions across the membrane, caused by either an inflow of cations or an outflow of anions, or both. In fact, the depolarizing phase of an action potential is caused by a rapid change in the membrane's permeability to Na^+ ions due to the opening of voltage-dependent Na^+ channels. This can be shown by determining the ionic conductance (G) of the membrane to different ions during the action potential (see Figure 1.8). A change in the E_m, from −60 mV to −50 mV, increases the sodium conductance across the axonal membrane by some eight times (Figure 1.8b). This in turn leads to the massive change in the E_m of 100 mV (−60 to +40 mV) as more of the voltage-dependent Na^+ channels are opened. At the peak of the action potential the E_m is +40 mV, a value close to that for E_{Na^+} (+55 mV).

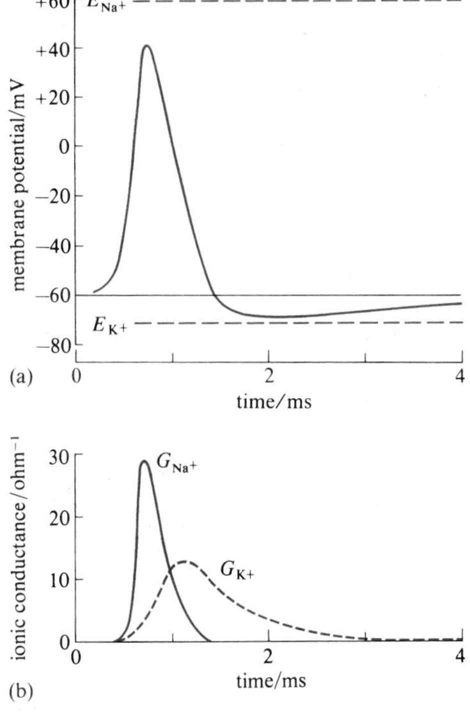

Figure 1.8 Changes in (a) membrane potential and (b) ionic conductance during an action potential. Notice that the increase in sodium conductance (G_{Na^+}) precedes the action potential by a fraction of a millisecond.

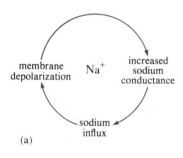

◇ How does Na^+ enter the axon?

◆ Na^+ moves passively down an electrochemical gradient through voltage-dependent channels.

◇ What sort of feedback regulation is illustrated by the activation of voltage-dependent Na^+ channels?

◆ This is an example of positive feedback: as the membrane's permeability to Na^+ increases (i.e. the sodium conductance increases) the E_m approaches zero; this in turn further increases the membrane's permeability to Na^+ and the E_m is displaced even further (Figure 1.9a).

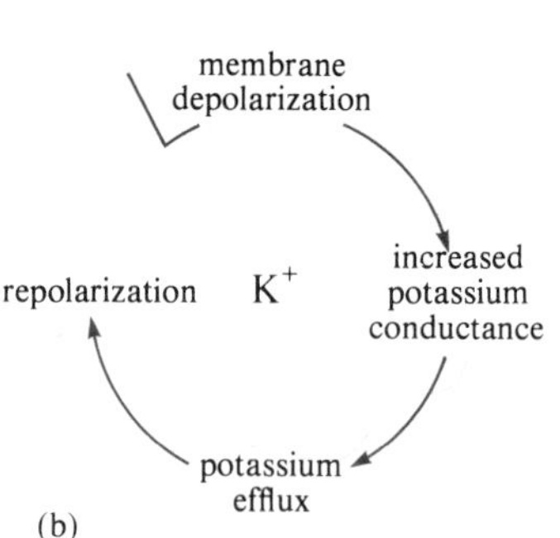

Figure 1.9 Changes in the conductance of (a) sodium and (b) potassium. In the case of sodium, positive feedback is operating, whereas for potassium negative feedback occurs.

Sodium ions are therefore responsible for the depolarizing phase of the action potential. However, this increase in Na^+ permeability is short-lived. The end-point (or peak) is reached when the Na^+ channels close and the K^+ channels are opened to their greatest extent (Figure 1.8b). Many, but not all, nerve cell membranes possess voltage-dependent K^+ channels. Like the Na^+ channels, these are also opened in response to a change in the membrane potential; however, they open very slowly in comparison with the Na^+ channels. By opening just as the Na^+ channels are closing, the K^+ channels help the membrane to return to its resting state.

◇ If the membrane's permeability to K^+ were increased, in which direction would K^+ move?

◆ K^+ would tend to move out of the axon down its concentration gradient.

This outflow of K^+ restores the E_m to the resting condition; that is, it repolarizes the membrane. Note that this is an example of *negative* feedback, as indicated in Figure 1.9b. Note also that the resting value for E_m closely approaches E_{K^+} (−75 mV).

These sequential changes in permeability to specific cations during the action potential have been substantiated by measurements of Na^+ and K^+ fluxes in response to externally imposed changes of membrane potential and by the use of tetrodotoxin and tetraethylammonium ions, which block voltage-dependent Na^+ and K^+ channels, respectively. It is important to realize that the action potential results from the purely passive redistribution of ions. The consequence is that some Na^+ are gained by the axon and some K^+ lost; however, relatively few ions traverse each channel during a single impulse. The action potential is thus very economical.

After a period of continuous firing of action potentials, the Na^+–K^+ exchange pumps in the axonal membrane restore the concentrations of external Na^+ and internal K^+. This pumping is an active transport process and quite distinct from the passive processes involved in generating action potentials.

The action potential is localized but passes along the axon as a wave of depolarization moving at a speed of some hundred metres per second. Propagation occurs because of the amplification effect of the voltage-dependent ion channels. The direction of propagation is away from the cell body because the ion channels beyond the point of stimulation are ready to be opened whereas those nearer the cell body cannot be opened. You will recall that, following the generation of an action potential, there is a refractory period during which another action potential cannot be initiated. This is caused by the conformation of the voltage-dependent Na^+ ion channels at this point. The channels are formed by complex proteins that undergo changes in conformation in response to changes in the membrane potential. It was stated earlier that at the peak of the action potential, the Na^+ channels close. In fact, they are inactivated such that the channel is in a different conformation to that which exists when the channel is closed. A closed channel has the capability to be opened (by a change in membrane potential) but an inactivated channel does not. Inactivated channels require a period of time to recover before reverting to the closed conformation, hence the refractory period.

So far, we have considered the role of Na^+ and K^+ ions in the process of signalling along axons, but other ions may also be involved. Ca^{2+} ions play a key role in a number of processes, and it has long been known that a reduction in the level of extracellular Ca^{2+} lowers the threshold for the initiation of impulses in both nerves and muscles. Extracellular Ca^{2+} appears to influence the relation between the membrane potential and its permeability to Na^+ and K^+, tending to stabilize the membrane by increasing the amount of depolarization needed to push the E_m beyond threshold. In some situations, Ca^{2+} entry through voltage-dependent Ca^{2+} channels can contribute to the generation of action potentials, for example in the muscle fibres of the heart and in some invertebrate neurons.

1.3.4 Fibre size, insulation and the speed of transmission

In Section 1.3.2 we compared the electrical properties of an axon with those of an electric cable. You will recall that the axoplasm is a poor conductor of electric current in comparison with the metal wire. In addition, the axonal membrane is a poor insulator, allowing current (ions) to leak across the membrane, unlike the electric cable, which is usually insulated by a plastic covering. The result is that signal intensity diminishes along the length of the axon.

We have suggested that this can be overcome by the generation of action potentials, which faithfully reproduce the signal along the axon. However, an important facet of the way in which the nervous system functions is the speed at which these signals can be transmitted and the distance they must travel along the axon. Nerves involved in reflex responses, for example the flip of a lobster's tail (an escape response), conduct impulses at a speed of over 100 metres per second and in some large animals such nerves may be many metres long.

In Section 1.3.2 we saw how local potentials displace the E_m above a threshold to generate an action potential which, in turn, acts as a local potential for the generation of another action potential and so on along the axon. However, because of the poor electrical properties of the axon, these local potentials tend to attenuate over distance such that they are unable to push the E_m above the threshold. When this happens no action potential can be generated. So the question is, how are signals transmitted accurately along the entire length of the nerve?

The size of the axon is important in this regard. The larger the axon the less the leakage of ions and thus the further the local potential will spread without attenuation. Consequently, the action potential proceeds rapidly along the axon. Many invertebrate nerves are of large diameter, for example the squid axons are approximately 0.5 mm in diameter. However, most vertebrate nerve fibres are only a few micrometres in diameter yet they may have to transmit signals over considerable distances.

◇ What effect would fibres of small diameter have on the spread of local potentials in vertebrates?

◆ Local potentials will attenuate rapidly over distance. This in turn influences the speed with which the change in local potential affects the adjacent patch of axonal membrane and generates an action potential.

Vertebrate nerve fibres can also achieve high speeds of impulse transmission because many of them are insulated with a lipid-rich substance called **myelin**. This is provided by specialized glial cells (see also Section 1.5) called Schwann cells. These cells wrap themselves around the axon to give concentric layers of tightly apposed myelin membranes (Figure 1.10). The myelin sheath acts as

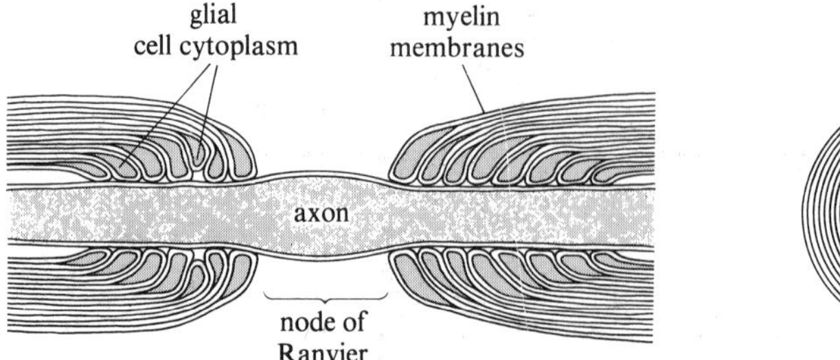

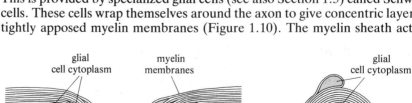

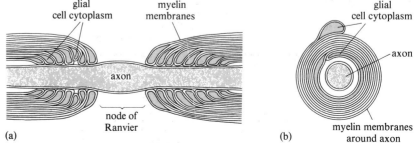

Figure 1.10 (a) Longitudinal section and (b) cross-section of the myelination of an axon by a glial cell. At intervals (called nodes of Ranvier) the axon is bare of myelin.

an insulator preventing leakage of ions (current) across the axonal membrane. At intervals of about 1 mm, the sheath is interrupted, exposing a patch of naked axonal membrane called the **node of Ranvier** (after its discoverer). As ions cannot flow into or out of the myelinated (insulated) part of the nerve, during an impulse, current flows along the axon to the next node where an action potential is produced, and so on. Consequently, only a small proportion of the axonal membrane actually changes its permeability to ions and, not surprisingly, the majority of voltage-dependent ion channels are situated at these nodes. From Figure 1.11 you will note that the impulse travels rapidly in jumps from node to node—a mode called **saltatory conduction** (from the Latin for 'to jump').

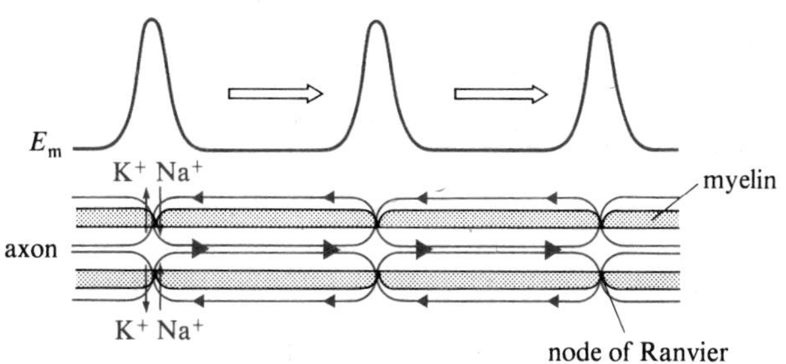

Figure 1.11 Saltatory conduction. Current flows between the nodes of Ranvier in a myelinated nerve: action potentials are generated at the nodes. Open arrows show the direction of movement of the action potentials.

Myelination confers a distinct advantage on vertebrates as it allows a considerable reduction in the diameter of fibres needed to attain a given speed of conduction. Such small diameter fibres are also advantageous in lines of communication because more fibres can be packed into a given space.

Summary of Section 1.3

1 The membrane potential (E_m) of an axonal membrane is dependent upon the distribution of ions across the membrane and its permeability to these ions.

2 The basis of communication in the nervous system involves the generation and conduction of action potentials along axons. This can best be explained by considering the points 1–10 illustrated in Figure 1.12.

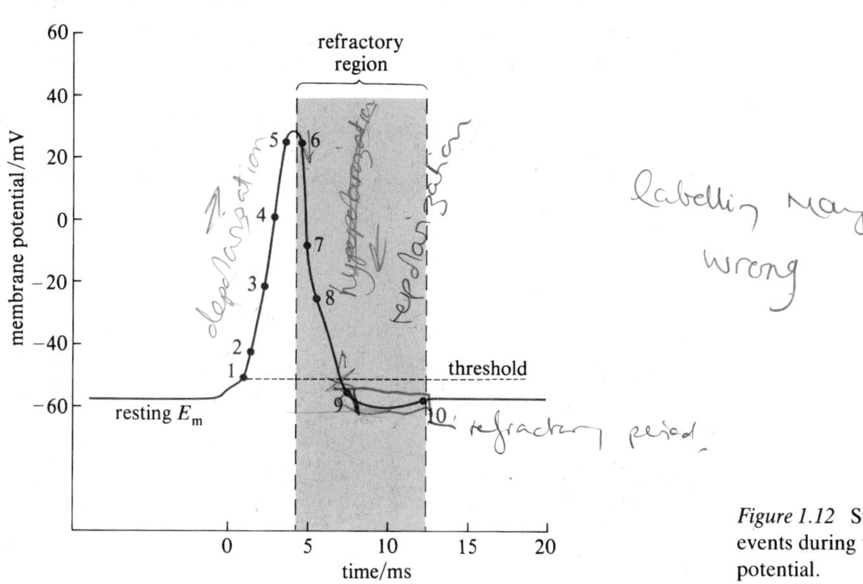

Figure 1.12 Summary of the sequence of events during the generation of an action potential.

3 Local potentials can displace the E_m towards a threshold (1), whereupon voltage-dependent Na^+ ion channels open (2).

4 By positive feedback, more and more Na^+ channels are opened (2–5) and the E_m is driven rapidly towards E_{Na^+}. The membrane is depolarized and an action potential generated.

5 As the action potential reaches its peak, Na^+ channels are inactivated and K^+ channels begin to open (6). The E_m inverts and moves towards E_{K^+} (7–9).

6 During the hyperpolarizing phase (6–10) the patch of axonal membrane is refractory and cannot be depolarized; however, the change in E_m stimulates the membrane just in front to depolarize. In this way the action potential is propagated along the membrane.

7 Subthreshold changes in E_m decay rapidly with distance along the axonal membrane.

8 The speed of conduction is determined by the diameter of the axon and/or the presence of an insulating myelin sheath. Action potentials travel along myelinated axons by a process called saltatory conduction.

Now attempt Questions 2–4, on pp. 40–41.

1.4 TRANSMISSION AT SYNAPSES

ACETYLCHOLINE
A neurotransmitter released from certain synapses and neuromuscular junctions. The nerves that release acetylcholine, and the synapses at which it acts, are termed cholinergic.

In the previous Section we saw how signals are generated and transmitted along axons. However, between the ending of a nerve fibre and its target (another nerve cell or a tissue such as a muscle) there is a gap in the line of communication: in effect a break in the circuit. It is now accepted that the means of communication across this gap (the synaptic cleft) is chemical, although some synaptic clefts are so narrow that direct electrical transmission is possible.

At the turn of the century, there was heated debate between physiologists about the means of transmission at the synapse. The idea of chemical transmission was largely based upon anatomical and electrical observations but was not popular. Synaptic transmission was known to take place in a small fraction of a second, and to many it seemed more likely that direct electrical contact was involved. In 1921, Otto Loewi reversed this thinking dramatically. Loewi perfused the heart of a frog and electrically stimulated the vagus nerve, which slows the heartbeat. He then transferred samples of the perfusate from this heart to a second one and noted that the beat of this heart also slowed. 'Something' released by the vagus nerve, he termed it 'vagustoff', produced a slowing of the heart. Loewi went on to show that vagustoff was in fact **acetylcholine**. He explained later that the idea for this experiment came to him in a dream and he wrote it down in the middle of the night. Unfortunately, the next morning he could not read his own writing, but when the dream returned, he rushed immediately to his laboratory to perform the experiment.

It is now known that acetylcholine is one of dozens of chemicals (neurotransmitters) known to be involved in transmission at synapses.

In the remainder of this Section we shall consider how electrical impulses trigger the release of neurotransmitters, how these subsequently elicit electrical activity in the target (postsynaptic) structures and how their effects are terminated.

1.4.1 Presynaptic events

Many of the pioneering experiments in this area were carried out on preparations such as the frog neuromuscular junction (the synapses between nerves and muscle fibres, see Figure 1.13). The fact that acetylcholine is known to be the neurotransmitter released from these nerves, together with the large size of the synapse (compared with those in the brain, for example), make it an ideal preparation for studying the presynaptic release of, and postsynaptic response to, this neurotransmitter.

If an impulse is passed down the axon to the neuromuscular junction, there is a delay of about 0.5 ms from its arrival at the synapse before a response can be recorded in the muscle fibre. This delay is sensitive to temperature: if the preparation is cooled to 2 °C, then the delay is increased to 7 ms. If acetylcholine were simply diffusing across the synaptic cleft, the delay need only be 50 μs. Thus some kind of metabolic event appears to be involved in synaptic transmission.

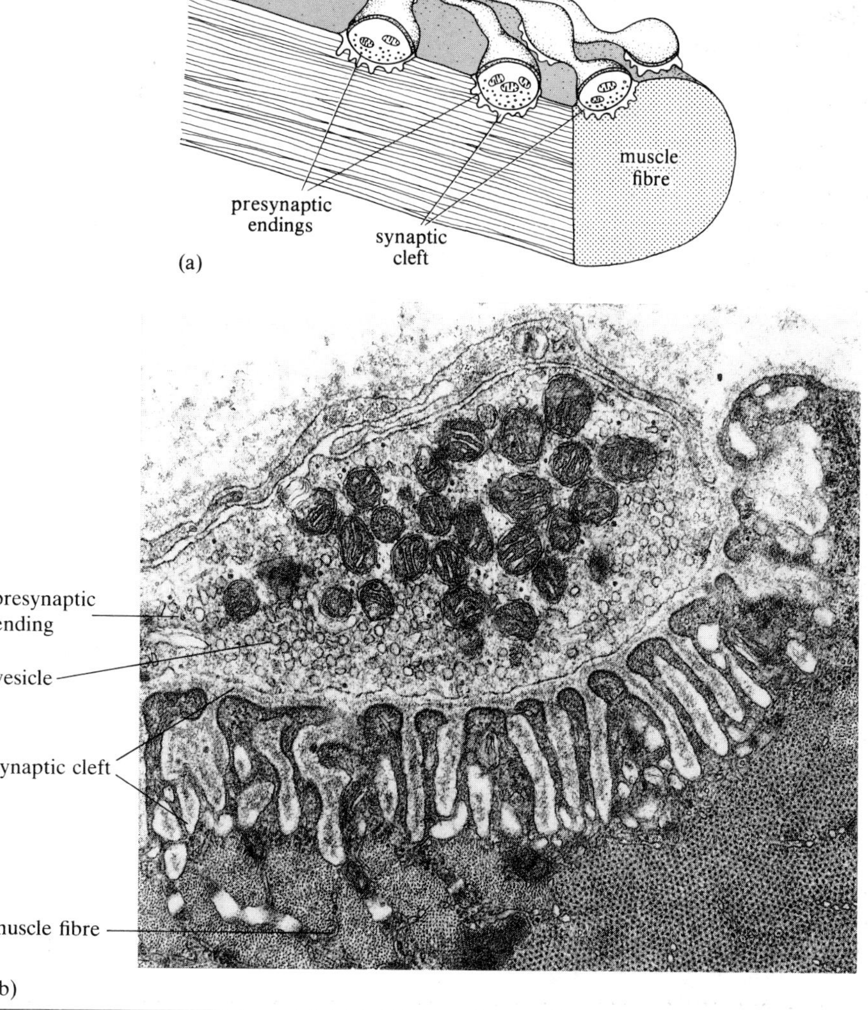

Figure 1.13 (a) The neuromuscular junction: the axon divides into a number of specialized endings that form synapses with a muscle fibre. (b) One such synapse seen in the electron microscope (magnification approximately 18 000).

You will recall from Section 1.3 that Ca^{2+} is involved in signalling along axons; it is also involved in synaptic transmission. It has been been found that if the concentration of Ca^{2+} in the fluid around the synapse is lowered, the release of acetylcholine is also reduced. It has now been established that Ca^{2+} is involved not only in the release of neurotransmitters at all synapses, but also in other processes, such as the release of hormones.

The link between Ca^{2+} and neurotransmitter release was elegantly demonstrated by Bernard Katz and Ricardo Miledi working in London. They found that by artificially holding the membrane potential of the presynaptic terminal at the equilibrium potential for Ca^{2+} ($E_{Ca^{2+}}$) they could inhibit acetylcholine release at the synapse.

◇ Can you explain why this should occur?

◆ You should recall from Section 1.3 that at the equilibrium potential for a particular ion there is no movement of that ion across the membrane. Thus at $E_{Ca^{2+}}$, there is no influx of Ca^{2+} into the presynaptic terminal in response to the arrival of a nerve impulse.

So the sequence of events under normal conditions appears to be: depolarization of the presynaptic terminal by the arrival of an action potential, the entry of Ca^{2+} and the release of a neurotransmitter. But how is the neurotransmitter released?

Two findings suggest that the release of acetylcholine occurs in quanta (small packages of fixed size). First, even if no action potentials are travelling down the axon to the synapse, small, spontaneous 'miniature' potentials (of about 1 mV) can be recorded at the postsynaptic membrane. Second, if the presynaptic membrane is gradually depolarized, the miniature potentials recorded at the postsynaptic membrane increase in size proportionally (see Figure 1.14). In the resting state, the spontaneous release of quanta gives rise to the miniature potentials. When the presynaptic terminal is gradually depolarized, as a result of nerve stimulation, more and more quanta are released. This quantal release theory of neurotransmission is now generally accepted, but the mechanism of release and exactly how the neurotransmitter is packaged into quanta are both hotly disputed.

When examined by electron microscopy the presynaptic terminal can be seen to contain a number of vesicle-like structures that would appear to be prime candidates for storing the neurotransmitter (cf. Figure 1.4b). These vesicles (40–200 nm in diameter) can be separated from the terminal and do indeed contain neurotransmitter. They can also be stained in situ with agents that are known to react with particular neurotransmitters. If a nerve is stimulated

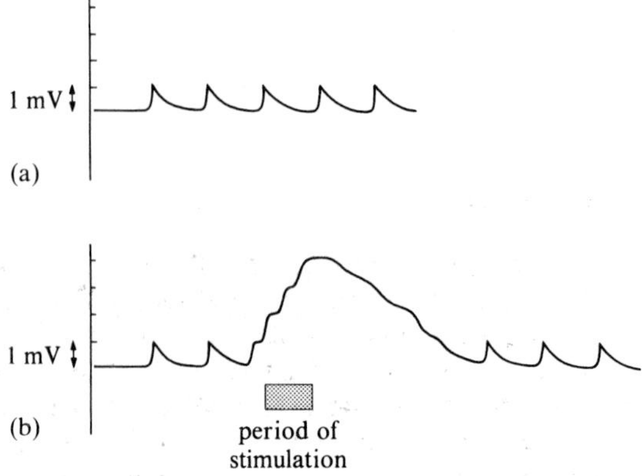

Figure 1.14 The quantal nature of neurotransmitter release. (a) In the resting state, miniature potentials of fixed size and regular frequency are recorded in the muscle fibre. (b) Stimulation results in a potential that is the sum of small spontaneous potentials, occurring with increased frequency.

excessively, so that a large amount of neurotransmitter is released, the number of vesicles decreases. Conversely, when the nerve is recovering from such treatment, the number of vesicles increases again. The release of neurotransmitter is believed to be by exocytosis, that is to say, the vesicles fuse with the presynaptic membrane and empty their contents into the synaptic cleft (Figure 1.15). Ca^{2+} is somehow involved in the fusion of vesicles to the synaptic membrane, and depolarization of the membrane simply increases the frequency of this process.

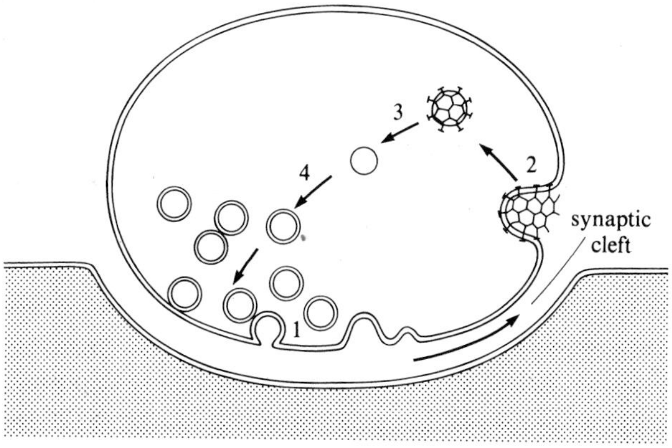

Figure 1.15 The vesicular hypothesis of neurotransmitter release.

(1) Vesicles fuse with the presynaptic membrane and release their contents by exocytosis. (2) Endocytosis of the membrane pinches off new vesicles. (3) New vesicles are formed. (4) Vesicles are ready to release a neurotransmitter.

This hypothesis of the mechanism of neurotransmitter release is attractive, and is generally accepted. However, microscopic evidence for exocytosis is scarce, and it is possible that the vesicles simply act as reservoirs for the neurotransmitter and that release is via specialized channels in the presynaptic membrane.

Figure 1.16 summarizes these presynaptic events. An action potential arrives at the presynaptic terminal and depolarizes the terminal membrane. This promotes an influx of Ca^{2+}, which in turn triggers the release of the stored neurotransmitter. The neurotransmitter is secreted in a regulated, quantal form. Figure 1.16 also shows some of the events that take place at the postsynaptic membrane; we shall consider these in more detail in the next Section.

POSTSYNAPTIC POTENTIAL (PSP)
Interaction of a neurotransmitter and its receptor on the postsynaptic membrane leads to the opening of ligand-dependent ion channels. Resultant change in the membrane's permeability to ions causes a change in the membrane potential (PSP). PSP may be either inhibitory (IPSP) or excitatory (EPSP) depending upon the nature of the neurotran. and the ion channels activated.

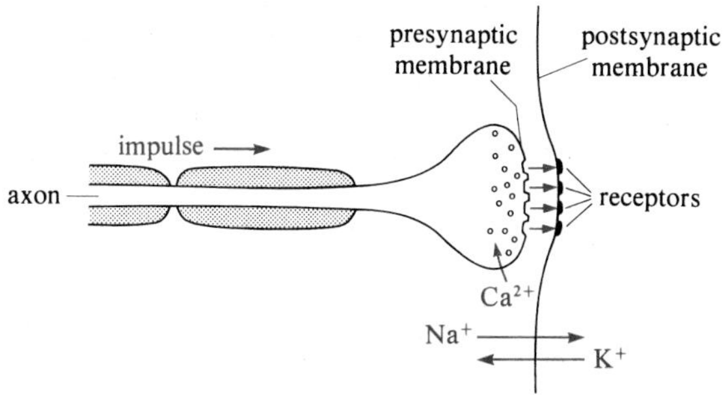

Figure 1.16 Summary of the events that occur during synaptic transmission.

1.4.2 Postsynaptic events

Once released from the presynaptic terminal, neurotransmitter molecules diffuse across the narrow synaptic cleft and interact with receptors situated in the postsynaptic membrane. This interaction leads to a local change in the postsynaptic membrane potential owing to a change in the membrane's permeability to ions. The **postsynaptic potential (PSP)** is generated by the opening of ligand-dependent ion channels. You will recall from Section 1.3.3

that ligand-dependent ion channels form part of, or are closely associated with, certain neurotransmitter receptors and are entirely different from the voltage-dependent channels responsible for the generation of action potentials. The extent to which the PSP is changed depends upon how much neurotransmitter is released into the synaptic cleft and how long it stays there.

Ligand-dependent ion channels have two important features: (i) owing to the specificity of receptor–neurotransmitter interactions, these ion channels will be opened only by the neurotransmitter released at that synapse and (ii) the channels are selective for particular ions or types of ion (e.g. cations rather than anions).

Acetylcholine is the neurotransmitter at frog neuromuscular junctions and, as a result of nerve stimulation, it has been found that the postsynaptic membrane becomes more permeable to Na^+, K^+ and Ca^{2+}, but not to Cl^-.

◇ What can you deduce about the selectivity of the ion channels linked to acetylcholine receptors at the frog neuromuscular junction?

◆ These channels are selective for cations rather than anions.

It should be noted that these changes in permeability are not the same as those that underlie the action potential. During an action potential, permeability of the axonal membrane to Na^+ increases first followed by an increase in permeability to K^+. In the postsynaptic membrane, the changes in permeability to ions are simultaneous but permeability may be differential. In other words, the postsynaptic membrane may become relatively more permeable to one ion than to another.

◇ At the frog neuromuscular junction the resting postsynaptic membrane potential (E_m) is about -90 mV. Given that E_{Na^+} is $+45$ mV and E_{K^+} is -74 mV for this membrane, predict very roughly a value for E_m if there is an equal and simultaneous increase in the postsynaptic membrane's permeability to Na^+ and K^+.

◆ If the increase in permeability is the same for both ions, the E_m takes an intermediate value between E_{Na^+} and E_{K^+}, that is about -15 to -20 mV.

◇ When measured, the postsynaptic membrane potential is actually close to zero. What does this suggest?

◆ That the permeability to Na^+ is increased more than the permeability to K^+ because the E_m (about zero) is closer to E_{Na^+} (+45 mV) than it is to E_{K^+} (-74 mV).

The differential permeability to Na^+ and K^+ varies at different synapses. In the example used so far (frog neuromuscular junction), acetylcholine opens Na^+ channels in the postsynaptic membrane such that the membrane is depolarized (the E_m approaches zero). This change in E_m may then be sufficient (above threshold) to initiate an action potential in the muscle fibre. Such changes in postsynaptic membrane potential are termed 'excitatory postsynaptic potentials' (EPSPs). However, at synapses on the mammalian heart, acetylcholine is inhibitory, that is, it generates 'inhibitory postsynaptic potentials' (IPSPs). This is achieved by increasing the postsynaptic membrane's permeability to K^+, which hyperpolarizes the membrane. Acetylcholine can thus produce either an EPSP or an IPSP depending upon the particular synapse or, more correctly, depending upon the relationship between the acetylcholine receptors and the ion channels to which they are linked.

The neuromuscular junctions of crustaceans and some insects (e.g. the locust) have both excitatory and inhibitory synapses, their postsynaptic effects being mediated by changes in the permeability to Na^+ and Cl^-, respectively. The neurotransmitters released at these synapses are amino acids, glutamic acid being released at the excitatory and γ-aminobutyric acid (GABA) at the inhibitory synapses.

Evidence now indicates that glutamic acid and GABA serve a similar role in the vertebrate brain. Indeed it would appear that they are the predominant excitatory and inhibitory neurotransmitters in the central nervous system. Some have proposed that amino acids are entirely responsible for rapid point-to-point transmission, whereas other neurotransmitters, such as acetylcholine, noradrenalin, dopamine, serotonin and peptides, play more subsidiary modulatory roles.

Figure 1.17 depicts a neuron receiving both excitatory and inhibitory inputs from other neurons. The neurotransmitter (glutamic acid) released at the excitatory synapse increases the postsynaptic membrane's permeability to Na^+, while at the inhibitory synapse, GABA increases its permeability to Cl^-. If the excitatory input is activated the influx of Na^+ depolarizes the membrane (E_{Na^+} is more positive than the resting E_m) and pushes the E_m towards the threshold at which an action potential can be initiated at the axon hillock of the target neuron. If the inhibitory input is activated the reverse occurs, that is, the membrane is hyperpolarized, and the E_m moves away from the threshold. Because the E_m is now more negative (see Figure 1.17b), it becomes more difficult for the excitatory input to depolarize the membrane sufficiently to initiate another action potential.

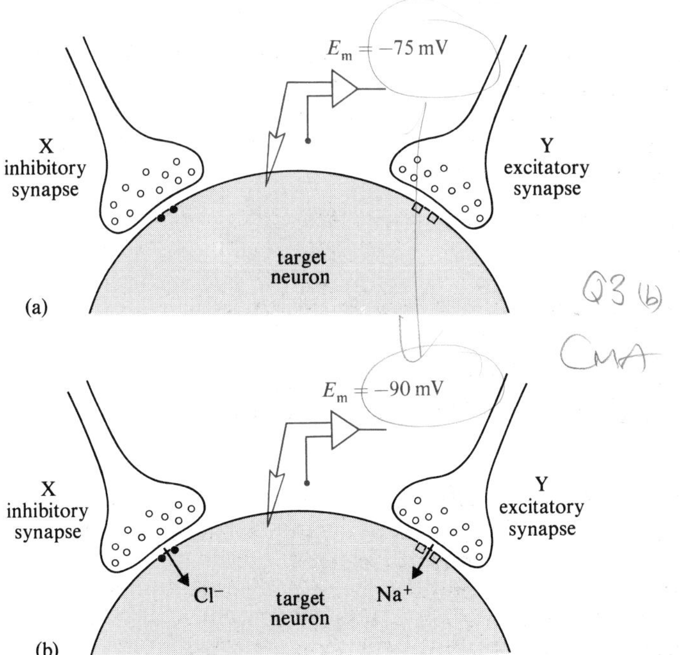

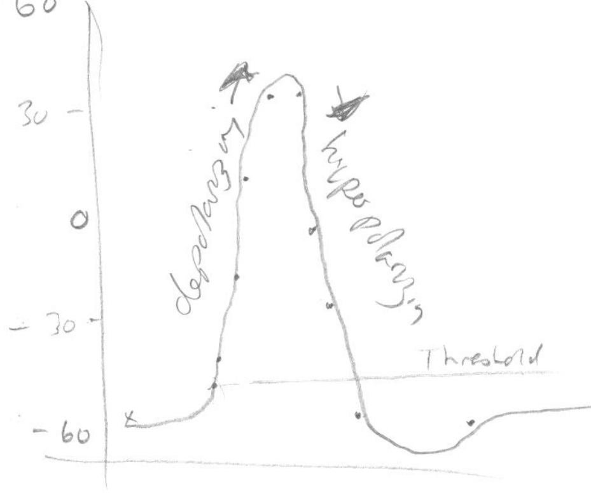

Figure 1.17 (a) A neuron receives both inhibitory (X) and excitatory (Y) synapses. (b) Input from X pushes the E_m away from the threshold and so it is harder for any subsequent input from Y to depolarize the neuron sufficiently.

A single neuron can receive many thousands of inputs, some inhibitory and some excitatory. The subsequent initiation of an action potential in the neuron depends upon the integration of conflicting inhibitory (hyperpolarizing) and excitatory (depolarizing) inputs, over both space and time. If the net result is a local potential of sufficient size to reach the threshold, then an action potential will be generated in the axon of the target neuron. If inhibitory inputs prevail, no action potential will be generated. This is important in the processing of information in the nervous system, because the synapses act as 'relay stations' in the line of communication, summing up converging inputs and then either passing on the signals, or blocking them. We shall return to this subject in Section 1.6.

1.4.3 Regulation of synaptic transmission

ACETYLCHOLINESTERASE

Enzyme responsible for terminating the action of acetylcholine. Present in the synaptic cleft, where it breaks down acetylcholine to its constituent parts, choline & acetate, which are then reused to synthesize more acetylcholine in the presynaptic terminal.

The duration of the postsynaptic potential is short because mechanisms exist that either deactivate the neurotransmitter or regulate its release from the presynaptic terminal. Acetylcholine is rapidly degraded by an enzyme, **acetylcholinesterase**, that is present in the synaptic cleft. The enzyme splits the acetylcholine molecule into its constituent parts, choline and acetate, which are then taken up into the presynaptic terminal where they are reused to synthesize more acetylcholine. Other neurotransmitters, such as amino acids, are rapidly transported back into the presynaptic terminal to be metabolized or are taken up by glial cells, which surround the synapse. These are active uptake mechanisms and the uptake sites often have a large capacity for the neurotransmitter.

At some synapses there are more complex mechanisms for terminating synaptic transmission. Specific receptors for the neurotransmitter are present on both the presynaptic and postsynaptic membranes. Once a critical concentration of neurotransmitter in the synaptic cleft is reached (a concentration that must be sufficient to ensure that the postsynaptic receptors are stimulated), the presynaptic receptors are activated and further neurotransmitter release is halted. These receptors are termed 'presynaptic autoreceptors'. In essence, the neurotransmitter regulates its own release and hence duration of action. Similarly the release of one neurotransmitter can be regulated by another neurotransmitter released in parallel from the same presynaptic terminal or from an axon synapsing onto that presynaptic terminal. Thus the postsynaptic effect of one neurotransmitter can be modified by the effect of another.

The mechanisms that regulate events at the synapse are important because (i) they ensure that the postsynaptic cell is not overstimulated and (ii) they allow the correct information to be conveyed to the target cell. This latter point is particularly important in complex neural circuits where a single neuron may receive many thousands of synaptic inputs, some excitatory and some inhibitory, at different positions on the cell and at different times (see also Section 1.6).

Regulation of synaptic transmission can also be achieved with the use of pharmacological agents. In fact the synapse, because of its critical role in the line of communication between neurons, is the site of action of a number of commonly used drugs. Some block the reuptake of neurotransmitters thus prolonging their presence in the synaptic cleft and hence their duration of action. The barbiturates and tranquillizers of the benzodiazepine family (the most common being valium) are two classes of psychoactive drugs which are commonly prescribed, or perhaps over-prescribed, for the treatment of anxiety. Their site of action is at the postsynaptic membrane where they modify the effect of the neurotransmitter GABA. As described earlier, GABA exerts its effect on target cells by opening Cl^- channels and thus hyperpolarizing the postsynaptic membrane. Both these drugs enhance the inhibitory effects of GABA by binding to sites associated with the GABA receptors and the Cl^- ion channels to which they are linked. Recently, with the advent of sophisticated techniques that allow the analysis of events at individual ion channels in membranes, it has been found that the barbiturates enhance the effect of GABA by prolonging the duration of Cl^- channel opening, and that the benzodiazepines increase the frequency with which the channels open. The overall effect of these drugs is, therefore, to augment the inhibitory postsynaptic potential at GABA synapses which, in turn, leads to muscle relaxation and sedation.

1.5 GLIAL CELLS

Most neurons are surrounded by cells that are collectively termed glia or neuroglia (meaning 'nerve-glue'). Although glial cells outnumber neurons in the mammalian brain by 10 to 1, their role has in the past been dismissed somewhat summarily. It was thought that these cells played only a supporting role (both structurally, as a kind of connective tissue, and metabolically, as a provider of nutrients) for neurons. However, more recent evidence suggests that there are many very complex interactions between neurons and glial cells.

In Section 1.3.4 you saw how one type of glial cell, the Schwann cell, is involved in myelinating (insulating) axons in some areas of the vertebrate nervous system. In the vertebrate brain another type of glial cell, the oligodendroglia (from the Greek for 'glia with few processes'), performs a similar role to Schwann cells, spinning myelin membranes around axons. Glial cells also play an important role in the general regulation of the ionic environment around neurons. Some glia (called astroglia or astrocytes) form connections with the blood vessels that pervade the brain, supplying it with glucose and oxygen. These connections form a selectively permeable barrier between the blood and the cells of the brain. The blood–brain barrier (see Figure 1.18), as it is called, is important in maintaining the environment around nerve cells: it protects them against toxins and metabolites in the blood and buffers the brain against wide fluctuations in blood constituents. During development, glial cells provide substances that regulate neural growth and differentiation and have been implicated in the guidance of developing axons to their final target areas.

Apart from forming connections with blood vessels, astroglia surround synapses and even invade the synaptic cleft with their processes. As mentioned in Section 1.4.3, these cells function as regulators of synaptic transmission because they are able to take up (or deactivate) neurotransmitters released into the synapse. If a glial cell is penetrated with a microelectrode, a large, stable E_m of -70 to -90 mV can be recorded. Changes in this E_m can

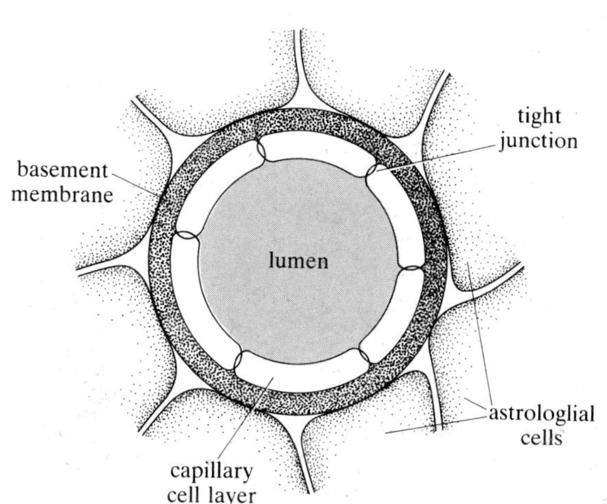

Figure 1.18 Elements of the blood–brain barrier. A brain capillary is formed by endothelial cells, which are connected by continuous belts of tight junctions, thus restricting diffusion between cells. The basement membrane and astroglial cell layer add to the barrier properties.

be recorded either when adjacent neurons are stimulated or when exogenous neurotransmitters are applied to them but, although they possess voltage-dependent ion channels, action potentials have never been recorded in glial cells. They do however possess membrane receptors for a number of neurotransmitters and peptide hormones, suggesting that they might be targets for signals released from neurons. Indeed there have been a number of observations of axons forming synapses on astroglial cells. Glial cells are also capable of releasing neuroactive substances that might influence the activity of adjacent neurons. In all, it seems that there is a high degree of cooperation between neurons and glia in the functioning of the nervous system.

Summary of Sections 1.4 and 1.5

1 The arrival of an action potential at the presynaptic terminal depolarizes the presynaptic membrane, triggering the release of a neurotransmitter into the synaptic cleft.

2 The neurotransmitter diffuses across the synaptic cleft and interacts with specific receptor sites on the postsynaptic membrane.

3 The receptors are coupled to ligand-dependent ion channels, which allow the movement of particular ions across the postsynaptic membrane. Neurotransmitters are termed excitatory or inhibitory, depending upon which ion channels they open.

4 Changes in the postsynaptic membrane's permeability to ions lead to an alteration of the membrane potential (E_m). If the membrane is depolarized, the E_m may be displaced sufficiently towards the threshold to allow the initiation of an action potential in the cell. Hyperpolarization of the membrane shifts the E_m further from the threshold making it more difficult for an action potential to be generated.

5 The overall change in the E_m of the target cell membrane represents the sum of all synaptic inputs. The generation of an action potential in the receiving cell depends upon the type, number and time of arrival of inputs to the cell.

6 There are mechanisms at the synapse that serve to terminate or regulate both the release and duration of action of neurotransmitters.

7 Glial cells form intimate and functional relationships with neurons. They are both targets for, and sources of neuroactive substances; they deactivate neurotransmitters by removing them from the synaptic cleft; they insulate axons and form part of the blood–brain barrier.

Now attempt Questions 5–7, on p. 41.

1.6 CODING AND TRANSMITTING IN CIRCUITS

In Sections 1.3 and 1.4 we looked at the way that two neurons, or a neuron and a muscle fibre, communicate with one another. In this Section we shall investigate how nervous impulses are 'coded' and integrated with electrical activity in other pathways in the nervous system, and at how cells are organized into circuits and circuits into systems.

As an example of how information, in the form of electrical impulses, can be coded, we shall look at the way in which sensory neurons detect changes in the animal's environment and transmit that information to the rest of the nervous system.

1.6.1 Generating signals in sensory neurons

An organism's perception of its environment, and of its place within that environment, depends upon specialized sensory neurons that detect changes either from within the organism (e.g. the contracted or relaxed state of a muscle) or outside in its surroundings (e.g. pressure on the skin or a change in light intensity). A sensory process involves translating the signal from one form (e.g. the contraction of a muscle) into another (e.g. the electrical events in the sensory neuron and subsequently the spinal cord). This conversion process is termed **transduction**.

TRANSDUCTION
The process by which sensory information (eg. pressure on the skin, light acting on the retina, muscle relaxation/contraction) is converted into elec. signals in sensory nerve fibres.

Transduction always results in an action potential being initiated in the sensory cell axon, but the mechanism of transduction varies according to the nature of the stimulus (light, pressure, temperature etc.).

Figure 1.19 shows a sensory receptor cell and the axon of a sensory neuron within a specialized receptive region of the skin of a fish. This represents one of thousands of such cells in this region, and they respond to the force and direction of water currents along the body. If a microelectrode is positioned close to the sensory axon, electrical activity can be recorded, and as the hairs on the cell are bent in one direction, the frequency of impulses travelling down the axon increases. When the hairs are bent in the opposite direction the frequency decreases.

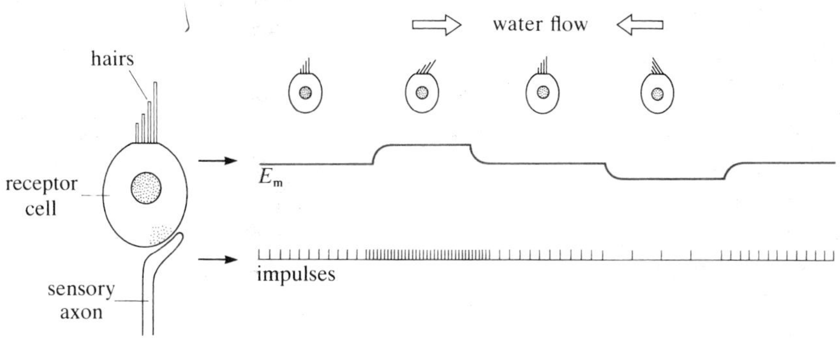

Figure 1.19 A sensory hair cell and the axon of a sensory neuron in the skin of a fish. When the hair is bent there is a change in the E_m in the sensory cell. The lower line shows the frequency of nerve impulses that are recorded in the axon.

◇ Why is there change in the frequency of action potentials rather than a change in their amplitude?

◆ Remember that the action potential has a fixed size. Once the threshold E_m is reached (the stimulus being, in this example, the deformation of the hairs on the cell), the action potential initiated is of fixed size, irrespective of any further increase in the intensity of the stimulus.

Any change in the stimulus (e.g. in intensity, or 'direction' in this case) is thus coded for by a change in the frequency of action potentials in the axon. The coding of information in the nervous system is, and can only be, by changes in the frequency of action potentials, sometimes known as the rate of 'bursting'.

1.6.2 Coding and integrating information

Communication in the nervous system is achieved through the movement of ions across nerve membranes, but because of the 'all or none' properties of the action potential it would appear that communication must be limited. A receiving neuron will respond to the arrival of an action potential but will not be 'aware' of the intensity of the stimulus that generated that signal—nor indeed the nature of the original stimulus. The amount of 'information' in one action potential is thus very limited. Information in the nervous system is coded, and because of the fixed size of the action potential, the only possible means of coding is according to the frequency of action potentials. Where the inputs occur (i.e. where messages arise) is indicated by the way neurons connect with one another to form pathways and circuits. Finally, information is processed in the nervous system. To store information, and modify it in the light of all the other information available, there must be sites in nervous pathways where information is amplified or overruled and integrated with information arriving from other places. The sites that carry out integration are the synapses between neurons: a system of nerves without synapses could only perform stop-go functions.

In Section 1.6.1 you saw how information about water flow is coded in the skin of a fish. Now look again at Figure 1.19.

◇ What is the electrical behaviour in the sensory cell when it is not being stimulated?

◆ Action potentials fire at a constant rate.

◇ What coding characteristic does the sensory cell show when the water flows from left to right?

◆ There is an increase in the frequency of action potentials in the axon.

◇ How is the reversed water flow coded?

◆ The frequency of action potentials is decreased.

The receptor cell provides useful information about the stimulus because the frequency of firing in the associated sensory axon varies above and below the baseline for spontaneous firing. Spontaneous firing is a property of many sensory cells and presumably results from the 'leakiness' of the cell membrane. As ions leak in and out, the potential reaches threshold and an action potential is generated. A stimulus simply increases the leakiness, and the result is either a depolarization or a hyperpolarization.

◇ What will indicate whether the membrane is depolarized or hyperpolarized?

◆ The nature of the ions crossing the membrane. If Na^+ is involved, the membrane will be depolarized. If K^+ or Cl^- is involved, the membrane will be hyperpolarized.

Sensory neurons transduce environmental information and this is coded in a number of ways. The onset and end of a stimulus is coded by changes in the rate of firing; the strength by the frequency (bursting); and the duration of the stimulus by the duration of bursting.

The dendrites of a neuron can receive multiple inputs. If simultaneous excitatory and inhibitory inputs arrive, the response of the neuron will depend upon the size of the resulting change in membrane potential, which in turn depends upon the summation of inputs. If the local potential reaches the threshold it will fire an action potential; if below the threshold, it will not. However, if a second excitatory input arrives a millisecond later than the first, the potential may already be slightly displaced towards the threshold and the additional input may be sufficient to push the potential above the threshold. In this case the efficiency of the synaptic input depends upon its timing; but it also depends upon where on the receiving neuron it arrives.

As Figure 1.20 shows, inputs can be on the axon (A), on the cell body (B) and on the dendrites (C and D). Some inputs are on the dendrite close to the cell body (C) whereas others may be some distance from the cell body on the tips of the dendrite (D). Because current spreads passively and attenuates rapidly with distance, what chance does a signal arriving at a synapse on a remote part of the dendrite have of being able to effect a change in the membrane potential at the cell body? Synapses at the cell body will obviously exert a stronger influence than others because of their immediate access to the axon hillock. Very often these synapses are inhibitory, and no matter how many excitatory inputs are received on the dendrites, the inhibitory input to the cell body prevails thus blocking transmission of the excitatory signals. The position of the synapses can therefore contribute to the integrative function of the neuron. When this is taken with the fluctuations in threshold potentials, the different diameters of the dendrites, the timing of inputs and whether the inputs are excitatory or inhibitory, it is obvious that the neuron is a very sophisticated 'analyser' of information. In fact, the integration of information at synapses is very complex and outside the scope of this chapter. The point to realize is that impulses are not simply passed on through synapses, but decay, or are reinforced, depending upon activity in other circuits that also have inputs to those synapses.

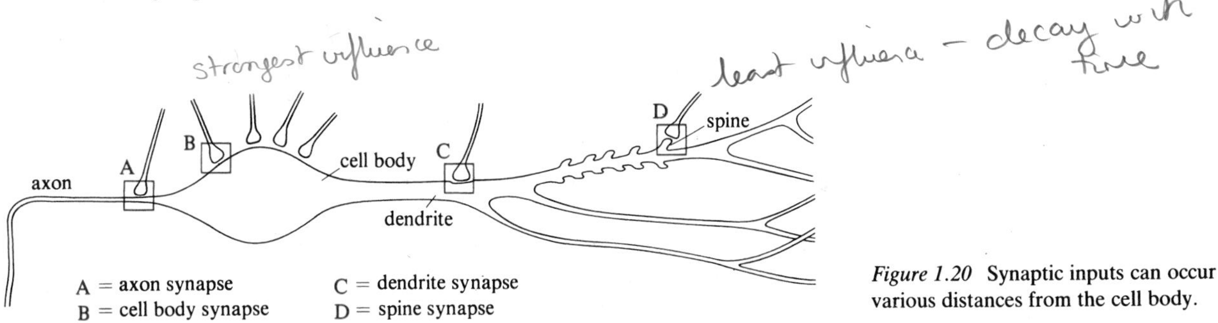

Figure 1.20 Synaptic inputs can occur at various distances from the cell body.

1.6.3 Cells, circuits and systems

To explain the principles of communication in the nervous system, we have concentrated on the ways in which neurons connect with one another. The neurons we have considered are largely typical of the central nervous system (brain and spinal cord) and peripheral sensory systems. This has certainly provided us with a basis on which to build up a view of signalling, but what happens in a system such as that of our own brain, which contains some 10^{12} neurons? As you can imagine, the millions of connections involved make the system very complex.

To show how neurons become arranged into more and more complex circuits, we shall take a spinal cord reflex. The knee-jerk reflex is something most people have experienced. When one leg is crossed over the other, the muscle that extends the knee joint is released. A sharp tap on the tendon of the knee pulls the muscle, stretching it, and instantaneously the muscle contracts and the knee jerks.

Figure 1.21 illustrates the neuronal circuitry involved in this reflex. The tap on the knee stretches the muscle fibres, which stimulates stretch receptors located deep within the muscle. These send a stream of action potentials via the sensory (afferent) nerve to the spinal cord, where an excitatory synapse conveys the information to a large motorneuron. This initiates an action potential that proceeds to the muscle via a motor (efferent) nerve, ending in a nerve–muscle junction where acetylcholine is released causing muscular

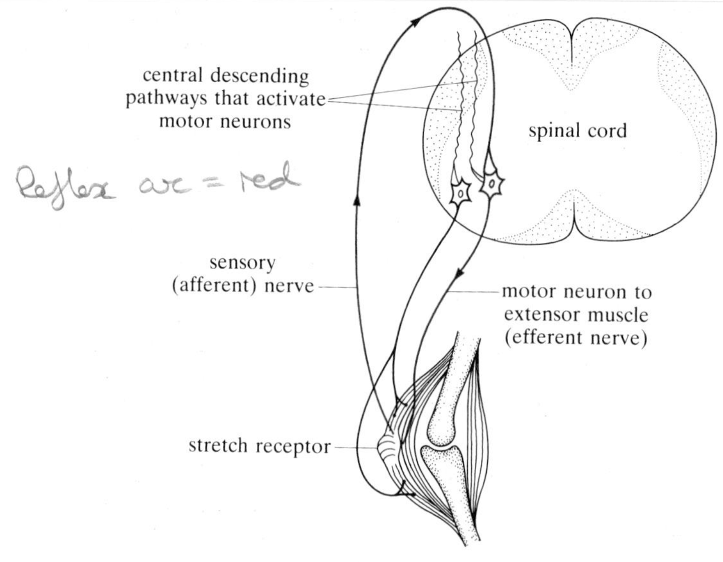

Figure 1.21 Nerve circuits involved in the knee-jerk reflex (a monosynaptic reflex). The reflex arc is shown in red. Note the possible involvement of central descending pathways from the brain (in black).

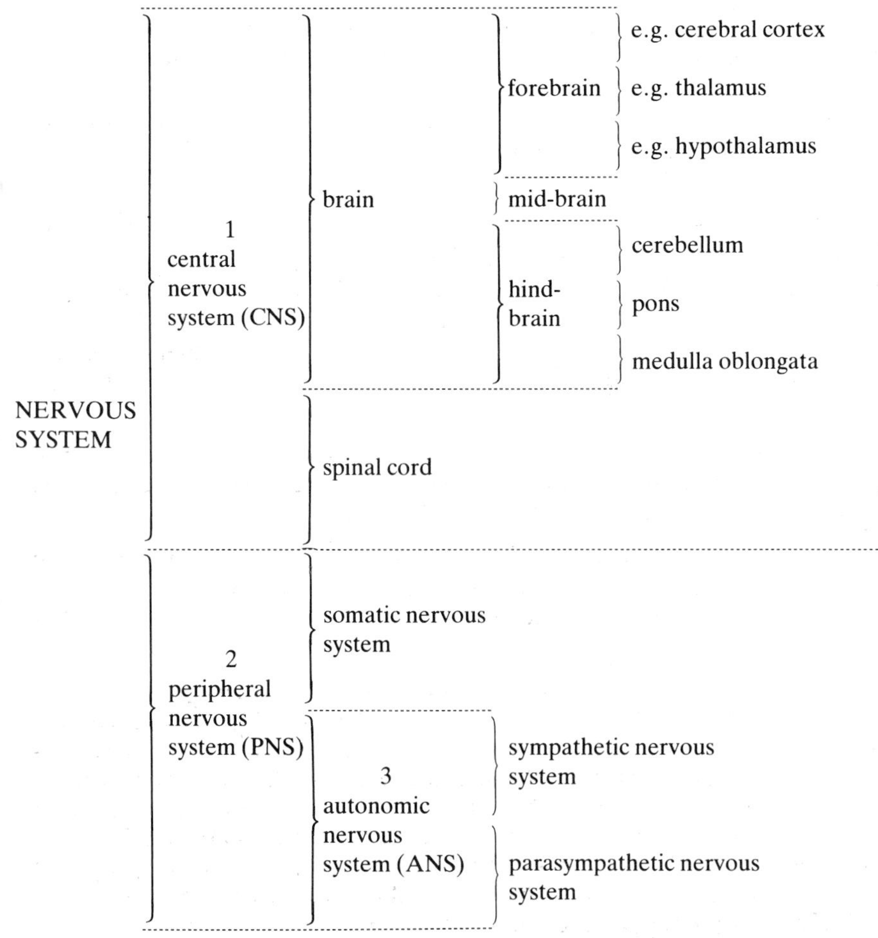

Figure 1.22 The organization of the human nervous system.

contraction. This circuit of neurons is called a reflex arc, and the knee-jerk—a 'mono-synaptic' (i.e. involving only one synapse) reflex—is probably the simplest reflex of all.

Now take a more complex system of neurons. Consider what happens when you touch a hot object. Muscles contract to remove the finger from the object, but often the body moves back at the same time. The reflex action coordinates movement, this time in a number of muscles, which must involve more neurons and therefore more synapses. The circuit must involve motorneurons at different places in the spinal cord because a number of sets of muscles are contracted. The other major part of the circuit must involve the brain because heat, and often pain, is experienced. So, the afferent sensory fibre must have access to relays that send information up the spinal cord to the brain and down the cord to various motorneurons. In this way neuronal circuits are built up, and the more complex the behaviour, the more neurons are brought into the circuit.

The complexity of nervous systems varies in different animal groups. In cnidarians, a true nervous system is present with neurons and synapses organized into a net of nerves. With most bilaterally symmetric animals, neurons and synapses become concentrated into groups or ganglia (singular, ganglion) with the preponderance of ganglia in the head.

In vertebrates, the nervous system can be divided into three parts (Figure 1.22).

1 The *central nervous system* (CNS), which comprises the brain and spinal cord.

2 The *peripheral nervous system* (PNS), which includes the sensory neurons and their axons, together with motor fibres that innervate the main muscles of the body.

3 The *autonomic nervous system* (ANS), which is concerned with the innervation of visceral organs and glands.

The autonomic nervous system is well developed in mammals and birds and consists of two parts, separated both anatomically and functionally. These are called the sympathetic and parasympathetic systems (Figure 1.23).

The nerve fibres of the sympathetic system arise from neurons in the trunk part of the spinal cord and communicate with neurons located in ganglia alongside the spinal cord. Efferent nerves from these ganglia innervate organs and glands at synapses that usually, but not always, use noradrenalin as the neurotransmitter.

The parasympathetic part of the autonomic nervous system arises from in front of and behind the sympathetic system. These neurons use acetylcholine as the neurotransmitter at synapses on organs and glands.

Both sympathetic and parasympathetic systems mostly innervate the same structures but exert different effects. Take, for example, the heart. The parasympathetic system slows the heart rate, the sympathetic system speeds it up.

The major control systems for the autonomic nervous system lie in the brain. These control regions (e.g. the hypothalamus, Figure 1.24) are considered in more detail in Chapter 2 because it is in these areas that the various communication systems principally interact, and from these areas much of the endocrine system is regulated. It is not possible to consider brain structure and function in detail here, but remember that it functions essentially as we have already described: signalling from neuron to neuron via synapses. The complexity of the brain is founded in the millions of connections both within the brain and to other parts of the nervous system.

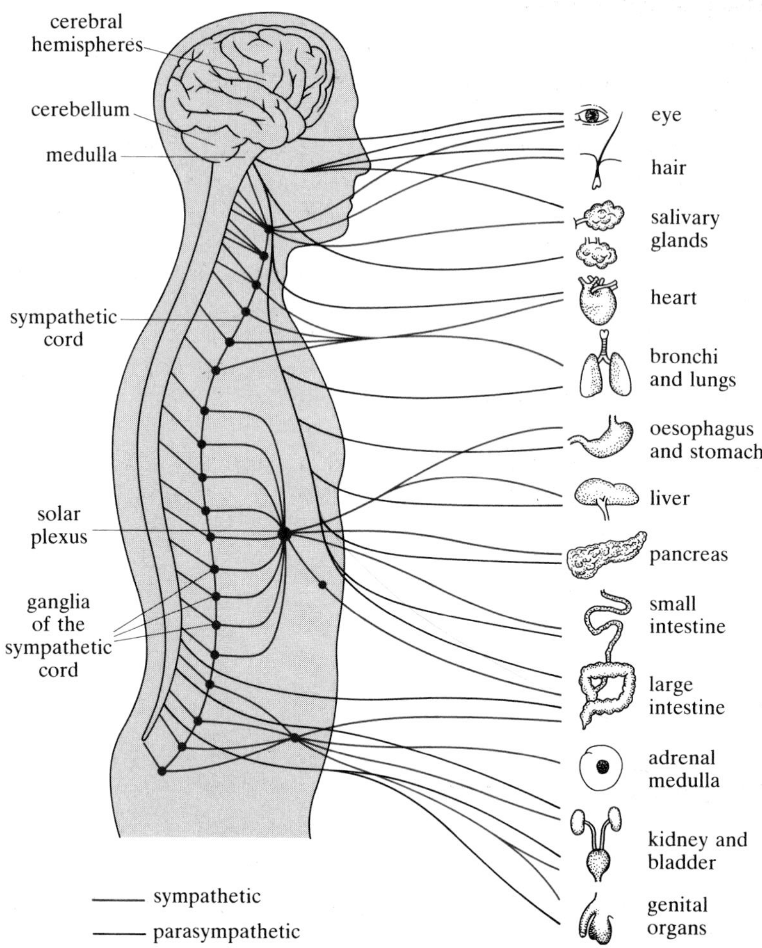

Figure 1.23 A schematic view of the autonomic nervous system in a man.

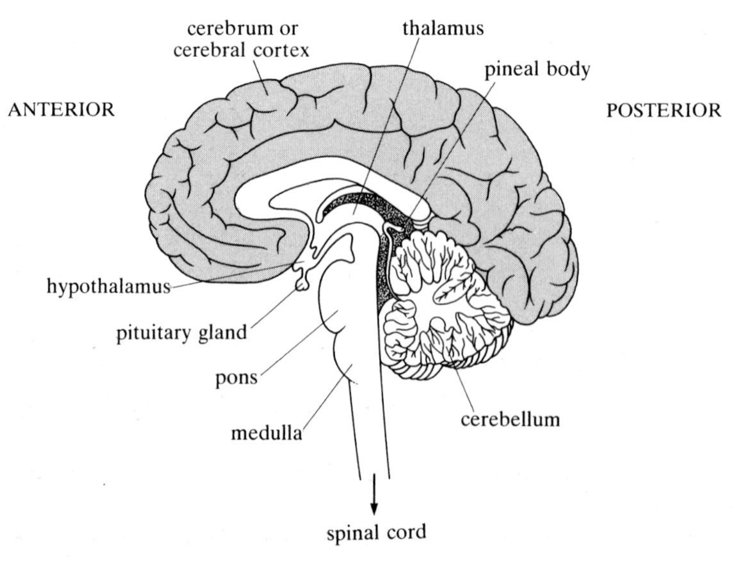

Figure 1.24 A section through the human brain to show the major areas discussed in this chapter and other chapters in this book. The cerebral cortex is shown as grey.

Summary of Section 1.6

1 Sensory receptors transduce 'environmental' stimuli into action potentials in sensory nerve axons. The information is 'coded' in the frequency of action potentials and their duration.

2 Coded information is integrated at the synapses between neurons. The nature of the synapse (excitatory or inhibitory) and its position (dendrite, cell body, etc.) on the receiving neurons are important in determining how the information is passed on to other neurons.

3 The knee-jerk reflex is a monosynaptic pathway. More complex reflexes involve more neurons, and central control from the brain can dominate the expression of muscular activity (motor output).

4 Neuronal circuits build up into nervous systems, reaching great complexity in mammals in which there is a division into central, peripheral and autonomic nervous systems.

5 Control over these systems is predominantly exerted by the brain, whose functional basis lies in the transmission of impulses from neuron to neuron via synapses.

Now attempt Questions 8 and 9, on p. 41.

OBJECTIVES FOR CHAPTER 1

Now that you have read this chapter you should be able to:

1.1 Define and use, or recognise definitions and applications of each of the terms printed in **bold** in the text.

1.2 Illustrate with a simple diagram the structural and functional zones of a neuron. (*Question 1*)

1.3 Describe how the membrane potential of a neuron arises, in terms of the distribution of ions and the permeability of the membrane. (*Questions 2 and 3*)

1.4 Distinguish between an action potential and a local potential in terms of the permeability of membranes to ions. (*Questions 2 and 3*)

1.5 List the principal events in the course of an action potential. (*Questions 2 and 3*)

1.6 Describe how an action potential travels along an axon and how fibre size and insulation are important in this regard. (*Question 4*)

1.7 Describe the events between the arrival of an action potential at an axonal ending and the generation of another in a target neuron or muscle fibre. (*Question 5*)

1.8 Explain, in terms of changes in the permeability of membranes to ions, how some synapses are inhibitory and others excitatory. (*Question 5*)

1.9 Describe mechanisms by which synaptic transmission is regulated. (*Question 6*)

1.10 List at least three examples of the relationship between neurons and glial cells. (*Question 7*)

1.11 Describe how the nervous system codes the information that it transmits. (*Question 8*)

1.12 Demonstrate how the structure and function of synapses determine how inputs are integrated. (*Question 9*)

1.13 Show how circuits can be 'engaged' in the nervous system with reference to simple and more complex reflexes. (*Question 9*)

QUESTIONS FOR CHAPTER 1

Question 1 (*Objective 1.2*) Figure 1.25 shows three neurons communicating with one another. Study the diagram carefully and select three correct descriptions from (i)–(vii).

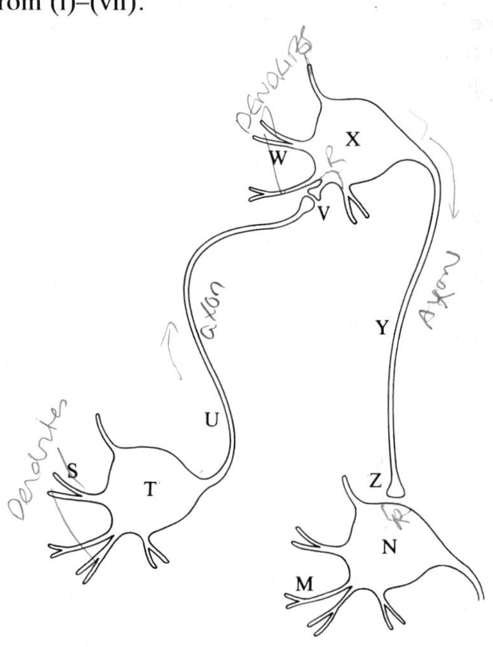

Figure 1.25 For use with Question 1.

(i) Signals pass from neuron N to neuron T via neuron X.
(ii) U is a dendrite of neuron T.
(iii) X communicates with N via a synapse on a dendrite.
(iv) Neurons T, X and N show distinct receptive and effector regions.
(v) T communicates with X via a synapse that includes a dendrite.
(vi) Y is a dendrite and likely to possess dendritic spines.
(vii) Signals pass from neuron T to neuron N via neuron X.

Question 2 (*Objectives 1.3–1.5*) The E_m of a squid axon is normally −60 mV. What would you expect to happen to the E_m immediately after each of (a)–(d) (i.e. would the membrane hyperpolarize or depolarize), and why?

(a) A sudden increase in the permeability to Na^+.
(b) A sudden increase in the permeability to Cl^-.
(c) A rupture of the membrane.
(d) A rise in the concentration of K^+ outside the axon membrane.

Question 3 (*Objectives 1.3–1.5*) What is the explanation for the fact that a local potential decays over distance along an axon whereas an action potential is faithfully reproduced all the way along an axon?

Question 4 (*Objective 1.6*) Table 1.2 shows the speed of conduction and the fibre size of selected axons from three species.

Table 1.2 The speed of conduction and the fibre size of selected axons from three species.

	Diameter of fibre/μm	Speed of conduction/m s^{-1}
species X	7	1.2
species Y	500	33
species Z	15	90

(a) Two of the three species are invertebrates. Which are they and why do you classify them in this way?

(b) What might explain why the speed of conduction in species Z is greater than in species Y?

(c) Why is the speed of conduction in species Y greater than in species X?

Question 5 (*Objectives 1.7 and 1.8*) At synapse A a neurotransmitter opens channels for Na^+ and K^+ in the postsynaptic membrane whereas at synapse B, the same neurotransmitter opens just Na^+ channels. At which synapse will there be the larger change in the E_m, and in which direction will this be (hyperpolarization or depolarization)?

Question 6 (*Objective 1.9*) Why is the effect of acetylcholine on the post-synaptic membrane short-lived?

Question 7 (*Objective 1.10*) Select the three correct statements from (i)–(vi).

(i) The blood–brain barrier prevents all blood-borne substances except gases from entering the brain.

(ii) Schwann cells and oligodendroglia form insulating layers of myelin around axons.

(iii) Glial cells possess membrane receptors for a number of neurotransmitters.

(iv) Glial cells act as regulators of the extracellular environment around neurons.

(v) Glial cells never receive synaptic inputs from neurons.

(vi) Action potentials are commonly recorded in glial cells.

Question 8 (*Objective 1.11*) Suggest three ways in which environmental information arriving at a sensory cell can be coded within the nervous system.

Question 9 (*Objectives 1.12 and 1.13*) Construct a flow diagram to show where the stimulus is generally communicated within areas of the nervous system when you inadvertently prick your finger.

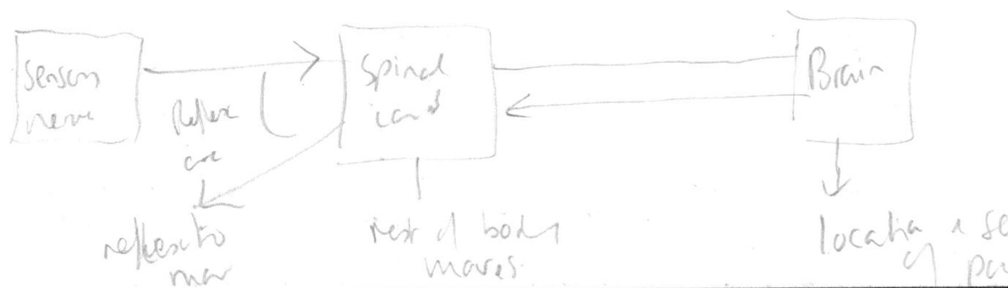

2.1 HORMONAL CONTROL

Castration of domestic animals has been practised from very early times. Aristotle (384–322 BC) described the effects of the castration of birds and, although he had no understanding of the biological mechanisms involved, his observations led him to believe that the testes are related to the sexual characteristics and reproductive capacity of the male. The first real indication that this was indeed the case came in 1848. Arnold Berthold castrated six young cockerels and then implanted one testicle into the body cavity of some of the birds. The castrated birds with testicular implants exhibited normal sexual behaviour and developed typical male features (i.e. large combs and wattles), whereas those without implants did not (Figure 2.1). Subsequent examination showed that the implants had re-established a connection with the blood but not with the nervous system. Berthold's interpretation was that some substance must be released from the testicular implant into the blood to maintain male behaviour and secondary sexual characters. Surprisingly, he did not carry these experiments further and their real significance was not realized until some 50 years later.

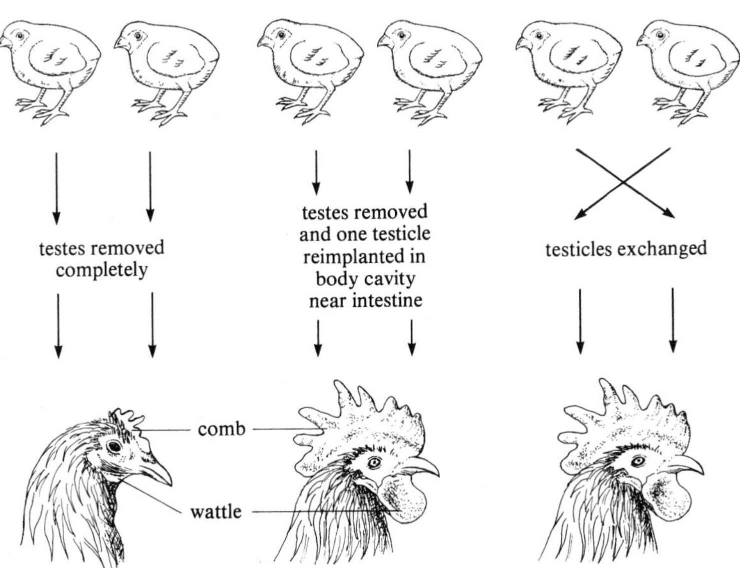

Figure 2.1 Details of Berthold's original experiment.

In a series of experiments between 1902 and 1905 two physiologists, Bayliss and Starling, demonstrated unequivocally that coordination of the functions of organs could occur without the intervention of the nervous system. They showed that a chemical substance was secreted by cells in the small intestine when acidified food was emptied from the stomach (Figure 2.2). They named

PANCREAS

A gland situated close to the duodenum, which secretes digestive juices via a duct into the intestine. It also contains specialized groups of cells (islets of Langerhans) which secrete insulin and glucagon into the bloodstream according to the level of blood sugar.

ENDOCRINOLOGY

Overall term for the study of hormones, their effects and the glands that synthesize and secrete them.

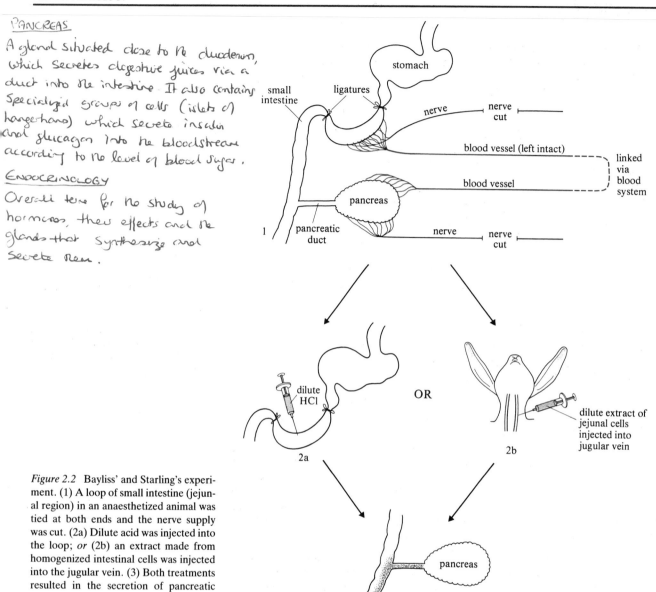

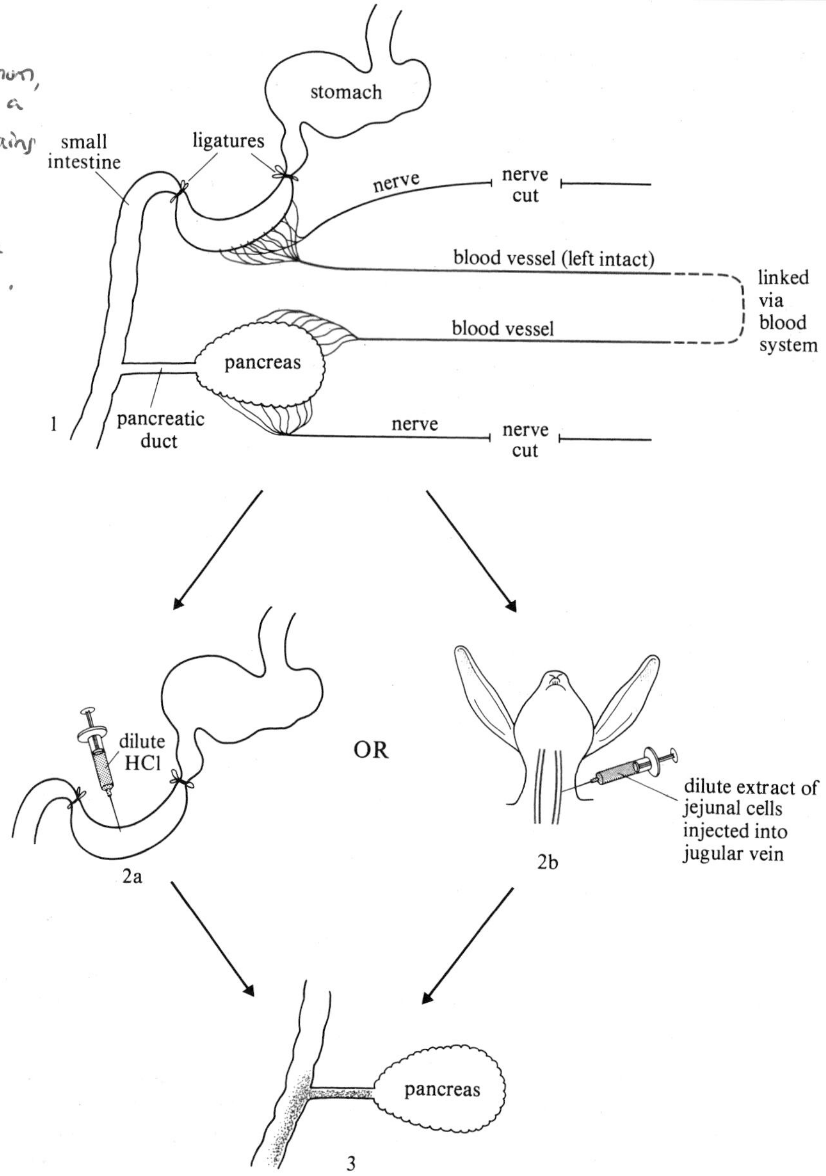

Figure 2.2 Bayliss' and Starling's experiment. (1) A loop of small intestine (jejunal region) in an anaesthetized animal was tied at both ends and the nerve supply was cut. (2a) Dilute acid was injected into the loop; *or* (2b) an extract made from homogenized intestinal cells was injected into the jugular vein. (3) Both treatments resulted in the secretion of pancreatic juice.

this chemical secretin and demonstrated that it was conveyed in the blood to the **pancreas**. Here, it stimulated some of the pancreatic cells to produce an alkaline juice that was then discharged through a duct into the intestine. Secretin is extremely potent and tiny amounts are able to produce very marked effects. Starling suggested the name hormone (from the Greek for 'to excite') as a general term to describe substances that are secreted into the bloodstream in minute amounts from one tissue and influence the activity of a distant target tissue. If taken literally, this definition of a hormone is somewhat misleading as it is now known that hormones may be inhibitory as well as excitatory.

In the remainder of this chapter we shall look more closely at hormones, the glands that produce them and the target organs that respond to them. The overall term for this branch of physiology is **endocrinology**.

2.2 HORMONE STRUCTURE

Two main groups of hormones are known. There are those synthesized from fatty acid precursors and those produced from amino acids or closely related compounds (Figure 2.3).

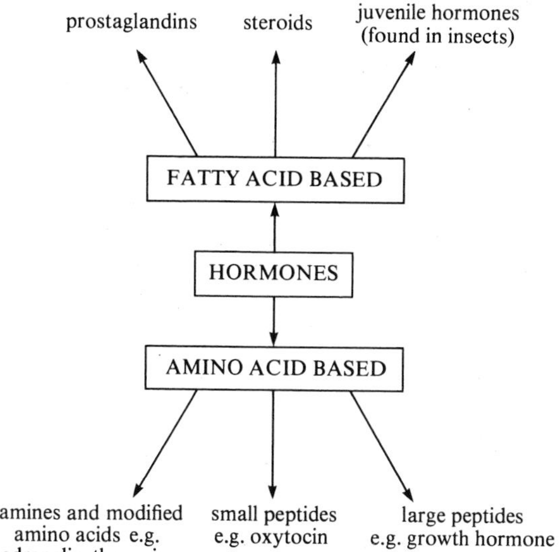

Figure 2.3 Various groups of hormones and their structural relationships. Most neurotransmitters are amines or modified amino acids.

Hormones based on fatty acids, such as the **steroid hormones**, are relatively small in size. They are fairly constant in structure because each is generated by small modifications of a basic molecule. In this case they are all based on a molecular structure that is synthesized from cholesterol and consists of four rings of carbon atoms (Figure 2.4). In vertebrates, the important groups of steroids are the sex hormones (see Chapter 10), oestrogens (based on 18 carbon atoms: C_{18}), androgens (C_{19}) and the progestins (C_{21}) and the corticosteroids (C_{21}). Within each of these groups different hormones are synthesized by the addition (or removal) of oxygen and/or hydrogen to (from) the basic steroid nucleus (Figure 2.5).

In contrast, **peptide hormones**, which are based on amino acids, vary enormously in both size and structure. The hormone thyroxin, which is involved in the control of tissue metabolism, is composed of just two modified amino acids; whereas growth hormone, which is concerned with the regulation of growth, is composed of 190 amino acids and has a complex three-dimensional structure. Distinct 'families' of peptide hormones containing similar amino acid sequences can be identified; for example, oxytocin and antidiuretic hormone are both small peptide hormones that differ by only two amino acid substitutions (Figure 2.6), yet they produce quite different effects. On the other hand, a small number of differences in the amino acid sequence may not change the effect of the hormone to any great extent. For example, slightly different versions of oxytocin are found in different groups of vertebrates (Figure 2.7), but the effects they exert are similar. Such polymorphism (from the Greek for 'many forms') is quite common among the larger peptide hormones.

Figure 2.4 The relationship between (a) the parent cholesterol molecule and (b) the four-ringed steroid nucleus. All naturally occurring steroid hormones contain the 17 carbon atoms that make up the rings. Steroids differ from each other in terms of the number of carbon, hydrogen and oxygen atoms added onto the basic nucleus.

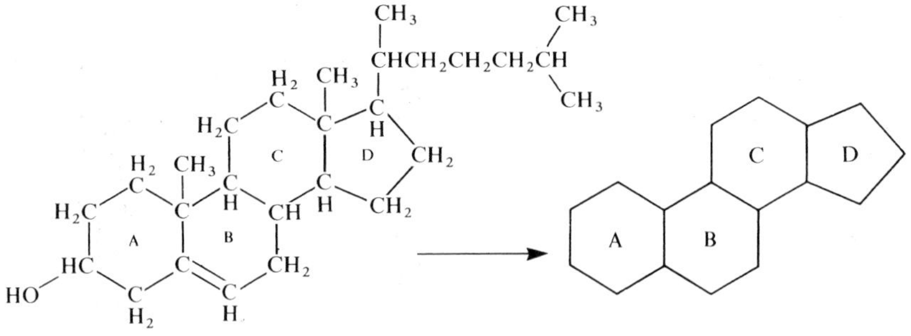

Figure 2.5 Some C_{21} steroids. Although the overall molecule is complex, the differences between the hormones are very simple. Three corticosteroids are shown. Different corticosteroids are synthesized by the addition (or removal) of a hydrogen (or oxygen) atom at key points on the molecule. (Compare the groups shown in red.) The progestins, for example progesterone, are also C_{21} steroids. However, their basic structure is different from that of corticosteroids. (Compare the positions indicated by red arrows with the same positions in the corticosteroids.)

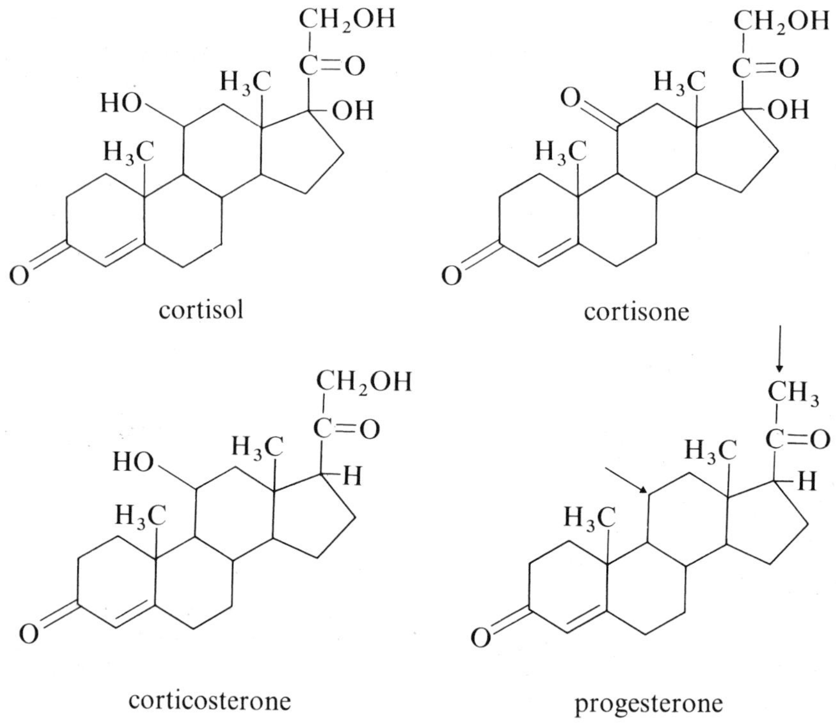

Figure 2.6 The amino acid sequence of (a) oxytocin and (b) antidiuretic hormone. The two cysteine molecules are joined by a disulphide bridge (S—S bond), so part of the molecule has a ring structure. The differences are shaded pink.

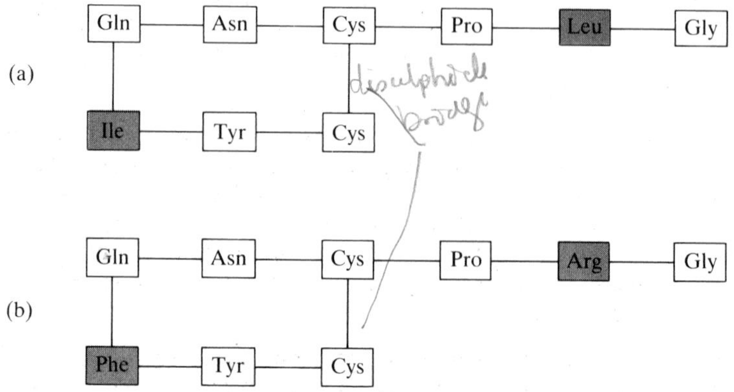

	POSITION 1	2	3	4	5	6	7	8	9
OXYTOCIN (mammals)	Cys	Tyr	Ile	Gln	Asn	Cys	Pro	Leu	Gly
MESOTOCIN (birds, reptiles, amphibians)	Cys	Tyr	Ile	Gln	Asn	Cys	Pro	Ile	Gly
ISOTOCIN (bony fishes)	Cys	Tyr	Ile	Ser	Asn	Cys	Pro	Ile	Gly

Figure 2.7 Oxytocin is an example of a polymorphic hormone. Different groups of vertebrates have slightly different oxytocins. Mesotocin has one amino acid changed (isoleucine at position 8), and isotocin has two changes (serine at position 4 and isoleucine at position 8). The differences are shown in pink.

2.2.1 Hormone synthesis and storage

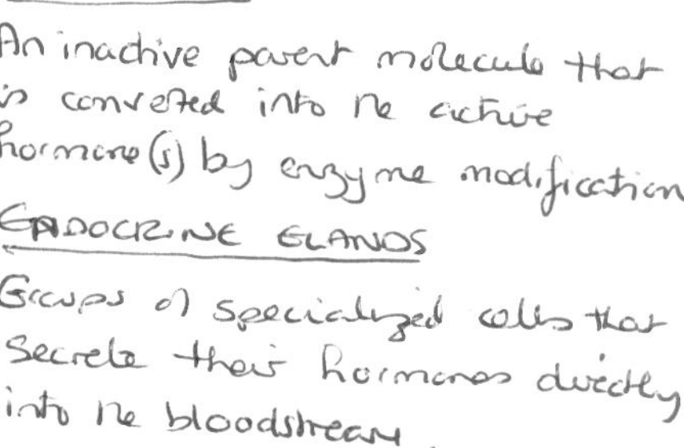

PROHORMONE
An inactive parent molecule that is converted into the active hormone(s) by enzyme modification
ENDOCRINE GLANDS
Groups of specialized cells that secrete their hormones directly into the bloodstream.

Many peptide hormones are synthesized as larger molecules, which subsequently undergo modification by enzymes to yield the smaller, active hormone. The parent molecule, or **prohormone**, may be a polymer of the active hormone and often has little or no biological activity compared with the actual hormone. Postsynthetic processing of the prohormone may occur before or after secretion. Sometimes several different hormones may be synthesized as a result of enzymes cutting up the parent molecule in different ways (Figure 2.8).

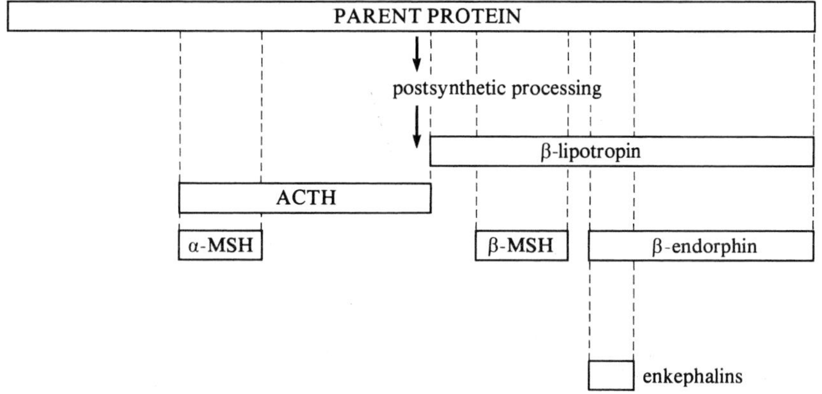

Figure 2.8 At least five different hormones can be produced from the same parent protein (prohormone) in certain endocrine cells of the pituitary gland. The hormones are cut out of the parent protein by specific enzymes. Some of these cells secrete several of these hormones, others secrete predominantly one hormone. The relationship of the hormones to the parent protein is indicated in the diagram.

Peptide hormones are usually stored in membrane-bound vesicles within the cytoplasm of hormone-producing cells. These, together with the rough endoplasmic reticulum and prominent Golgi apparatus, are the major characteristics of these cells.

In vertebrates, steroid hormones are synthesized by a common sequence of reactions. Cholesterol is always the precursor molecule, and the enzymes responsible for its conversion are distributed between the endoplasmic reticulum and the mitochondria. The presence or absence of particular enzymes determines the ability of the steroid-producing cell to synthesize particular steroid hormones.

Steroid-synthesizing cells share a characteristic fine structure that is quite different from those cells that produce peptide hormones. These cells have extensive smooth endoplasmic reticulum, the Golgi apparatus is often prominent and the mitochondrial cristae have a tubular or vesicular appearance. Steroid hormones, unlike peptides, are not usually stored prior to release; the cytoplasmic lipid droplets that are commonly seen in these cells contain precursor materials.

The cells that produce hormones are often gathered together to form distinct tissues called **endocrine glands**. In vertebrates, these organs are richly supplied with networks of small blood vessels (capillaries) that facilitate the secretion of hormones directly into the bloodstream. Endocrine tissues are often referred to as 'ductless' glands to distinguish them from other glandular tissues, such as sweat and salivary glands. The latter are termed **exocrine glands** and release their fluid secretions via ducts.

The sites of the major mammalian endocrine glands are shown in Figure 2.9. Such glands seem to be less common in invertebrates, although they are known to be present in cephalopod molluscs, crustaceans, echinoderms and insects.

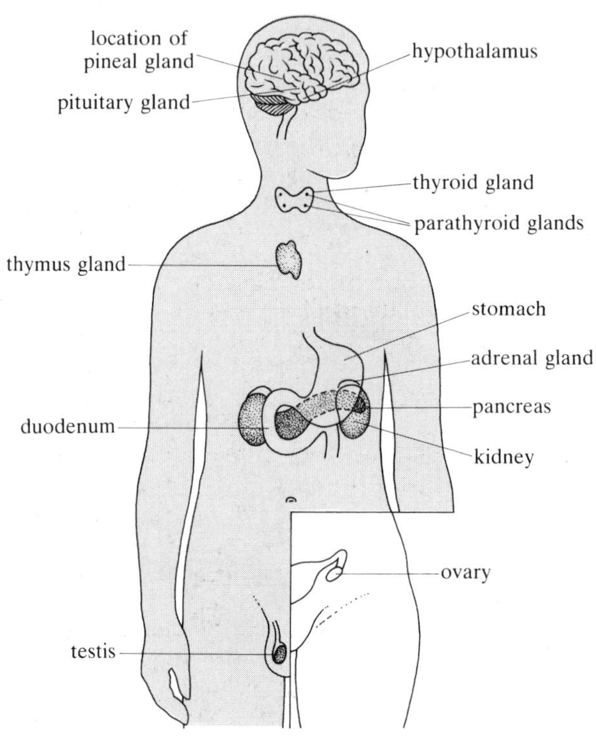

Figure 2.9 Major sites of hormone production in mammals.

Endocrine glands are often identified experimentally by the ablation (removal) of the gland followed by reimplantation or an injection of the hormone. Removal of the putative endocrine gland should generate deficiency symptoms, and reimplantation should relieve these symptoms if the blood supply to the gland is resumed. (Note the similarity of this to Berthold's original experiments.) The injection of extracts containing the 'active' product (hormone) secreted by the gland should also relieve symptoms of deficiency.

These methods have been used extensively, but they are not always appropriate, for the following reasons:

1 Some endocrine cells are found scattered among other cells that form part of a larger organ. For example, the pancreas is a mixture of endocrine and exocrine cells.

2 Some endocrine tissues contain a variety of cell types producing different hormones. For example, the mammalian adrenal gland (Figure 2.10) contains two distinct hormone-producing regions: an inner medulla that secretes noradrenalin and adrenalin, and an outer cortex that secretes mainly corticosteroids. The cortex can be further subdivided into three distinct regions, each of which produces different groups of C_{21} steroid hormones.

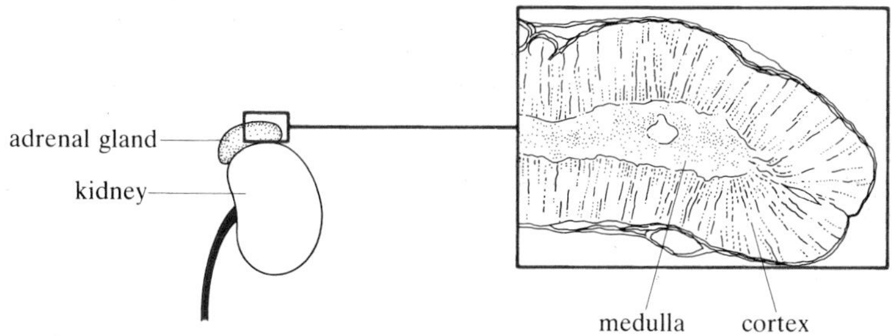

Figure 2.10 The location and structure of the mammalian adrenal gland. The enlargement is a drawing of a low-power histological section. Note the single large vein in the centre of the medulla.

3 A gland that serves a single function may have different anatomical appearances in different groups of animals. For example, the adrenal cortex and medulla are quite separate structures in fish.

4 Any surgical operation will obviously affect a whole range of physiological processes for a short while. However, this can be allowed for to some extent by doing control experiments in which sham operations are performed; for example, a piece of muscle may be implanted rather than an endocrine gland.

The removal–reimplantation or removal–replacement approach to identifying endocrine glands is therefore not ideal and can give very misleading results. The removal of a complex, multifunctional organ produces a wide range of physiological changes that are not necessarily related to particular hormonal effects. Consequently, modern endocrinologists often make more use of ultrastructural and biochemical features that are known to be unique to endocrine cells as a means of identification. These features generally relate to the type of hormone produced by the cell. Many vertebrate endocrine glands contain complex blood systems that permit short-distance communication between different regions of the gland. For example, the relative amounts of noradrenalin and adrenalin secreted by the adrenal medulla are controlled by corticosteroids from the adrenal cortex. Corticosteroids stimulate the cells of the medulla to produce an enzyme that converts noradrenalin into adrenalin, so corticosteroids increase the amount of adrenalin secreted. This particular effect is important in modifying the organism's response to prolonged stress. However, the cells of the medulla respond only to high levels of corticosteroids—levels far higher than those found in the bloodstream. How are these levels achieved? The gland is fed by several arteries, but all the blood leaving the gland drains into a single vein that runs through the centre of the medulla, and it is in this vein that the concentration of corticosteroids reaches a high level.

2.2.2 Hormone release and duration of action

The discharge of hormones from endocrine cells is controlled by nervous, hormonal or metabolic stimuli, which serve to alter the rate of release into the bloodstream. As with the release of neurotransmitters, Ca^{2+} seems to play an important part in the secretion process. In the adrenal medulla hormones are stored in granules, and Ca^{2+} stimulates the movement of these granules towards, and their fusion with, the plasma membrane. The release of hormones is not random; in general it is controlled by precise and specific stimuli but, under conditions of stress, non-specific stimuli of sufficient intensity and duration can elicit hormone release. Under normal conditions hormone release is often intermittent or cyclical (Figure 2.11). These cycles may be circadian (i.e. they follow an approximate 24-hour cycle) or seasonal. Other periodicities are also known, for example in the hormones controlling

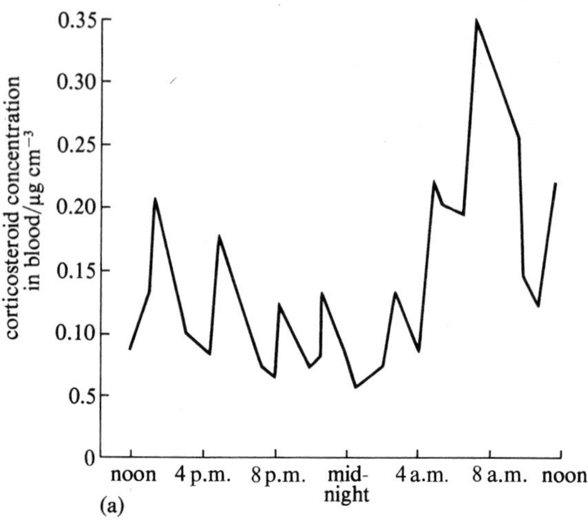

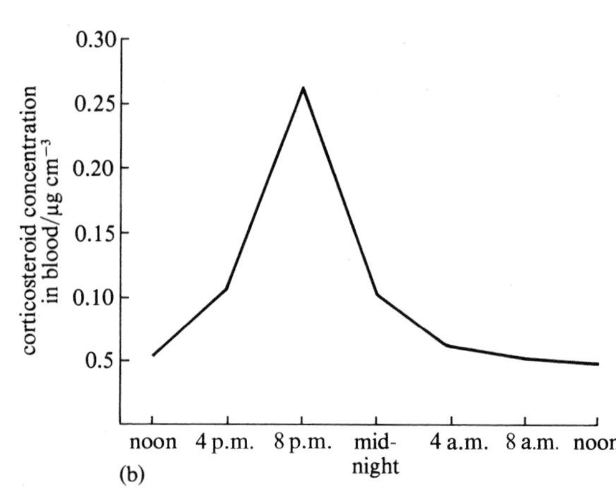

Figure 2.11 (a) Changes in human corticosteroid levels over a 24-hour period. Levels are low at night and higher during the day. (b) Corticosteroid levels in a rat. Here levels are highest at night because the rat is nocturnal.

the human menstrual cycle (see Chapter 10). In contrast, some hormones are secreted more or less continuously, such as thyroxin and triiodothyronine from the **thyroid gland**.

The active life of a hormone in the bloodstream varies from a few seconds to almost a week, depending on the hormone. The activity of the hormone is terminated by enzymes that break down the hormone in the liver and target tissue. In addition, small hormones may be excreted in the urine as blood passes through the kidney.

Many small or insoluble hormones form complexes with large proteins that are found in the blood or are produced by the parent endocrine cell. These **hormone-binding proteins** perform several functions: they prevent the rapid excretion of the hormone when the blood is filtered by the kidney (the hormone–protein complex is too large to pass through the filter); they prolong the hormone's active life by preventing its degradation by enzymes; they provide a source of hormone for slow release.

Slow release of hormone is possible because the amount of hormone bound to these plasma proteins is always in equilibrium with a small amount of 'free' (unbound) hormone. Receptors on the target tissue have a higher affinity for the hormone than do the binding proteins, so the 'free' hormone tends to bind to the receptors (Figure 2.12). When this happens the equilibrium between protein-bound hormone, free hormone and the receptor-bound hormone shifts, and the hormone dissociates from the hormone-binding protein.

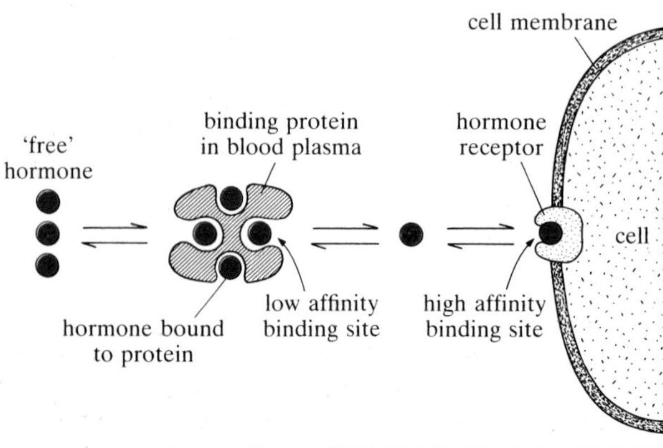

Figure 2.12 Equilibrium between 'free' hormone, protein-bound hormone in the plasma and receptor-bound hormone.

2.2.3 Hormone concentrations

Figure 2.13 summarizes the various factors that govern hormone levels in the bloodstream. The concentrations of particular hormones in the bloodstream vary widely and are related to the volume of blood in the system (which is a function of the animal's size), physiological state, metabolic rate and the hormone-binding capacity of the proteins in the blood. However, concentrations of hormones are always very low compared with those of other constituents of the blood.

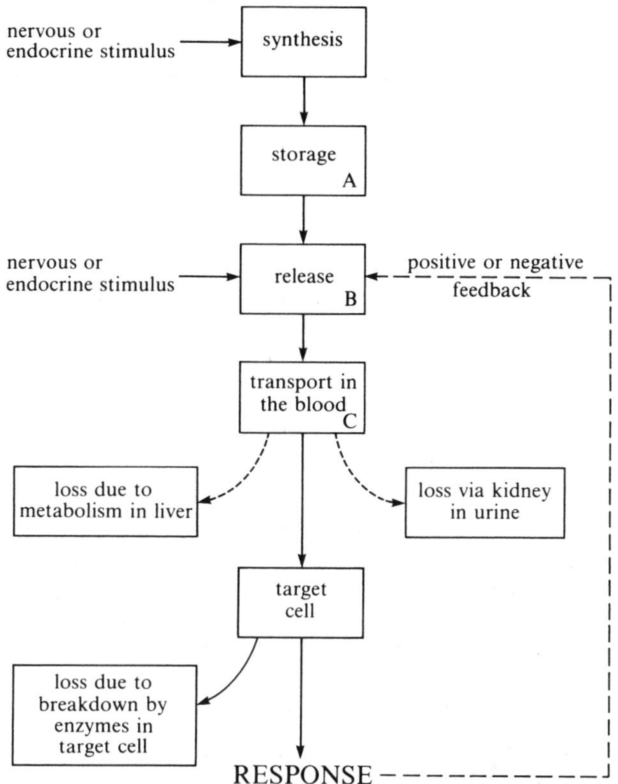

Figure 2.13 The life of a hormone. (A) *Storage.* The hormone may be stored bound to a larger polypeptide, or be present as part of a larger parent polypeptide. This probably keeps osmotic pressure in the vesicle low. (B) *Release.* The hormone may be released intermittently, continuously, or in distinct pulses. (C) *Transport in the blood.* The hormone may be bound to proteins in the plasma. This prevents loss in the kidney or liver. The location of the gland in relation to the blood system may also be important.

Hormones such as glucagon and insulin, which are concerned with the regulation of blood sugar levels (see Chapter 9), are normally present at a concentration of 10^{-9} mol l^{-1} whereas sex hormone concentrations can vary between 10^{-7} mol l^{-1} and 10^{-10} mol l^{-1} because of their rhythmic release.

Detecting such low concentrations of hormone is about equivalent to tasting a single spoonful of sugar in a cup of coffee the size of a conventional swimming pool. Consequently, the isolation of sufficient amounts of hormone for chemical analysis is something of a herculean task. Over 4 tons of sows' ovaries were processed to produce the 12 mg of crystals that were eventually used to elucidate the structure of the female sex hormone oestradiol.

Two methods are generally used to determine the concentration of a hormone in a sample of blood or a tissue extract, these are termed **bioassay** and **radioimmunoassay**.

Bioassay is particularly useful in the analysis of hormones of uncertain nature or which are not available in pure form. The technique makes use of the biological effects produced by a particular hormone. For example, growth hormone (GH) stimulates the growth of cartilage in young mice. The thickness of the cartilage produced is proportional to the amount of hormone applied to it. By adopting an arbitrary standard (e.g. 1 unit of GH produces a cartilage thickness of 1 mm in 17 days; 2 units produce 2 mm etc.), it is possible to quantify unknown amounts of hormone in terms of these standard units. Bioassays can be devised for most hormones but they may not always be particularly accurate because tissues from individual animals may vary in their responsiveness to the same amount of hormone.

Radioimmunoassay is widely used in endocrinology. Its main advantage over other methods is that it allows the measurement of extremely low concentrations of hormone. Basically, this method allows radioactively labelled and unlabelled (i.e. the sample to be analysed) hormones to compete for a limited number of binding sites on a specific antibody to that hormone. Production of such specific antibodies relies on the fact that different vertebrates use slightly different versions of the same hormone (see Figure 2.7). So, if human prolactin (PRL) is injected into a rabbit its white blood cells classify it as 'foreign' because its amino acid sequence is different from that of rabbit PRL, thus the animal produces large amounts of antibodies that are highly specific for human PRL. The radioimmunoassay method is outlined in Figure 2.14. The reaction must take place in the presence of an excess of hormone, so that at equilibrium the ratio of bound radioactively-labelled hormone to bound unlabelled hormone equals the ratio of the initial concentrations of the labelled and unlabelled hormone. The unbound or 'free' hormone is then removed, usually by absorbing it onto dextran-coated charcoal, and the remainder, which contains only the hormone bound to the antibody, can be measured. The radioactivity in this fraction is thus a measure of the proportion of labelled hormone that is bound; it will decrease as the initial concentration of unlabelled hormone increases. By constructing a standard curve, as shown in Figure 2.15, using known amounts of hormone, the hormone content of an unknown sample can be assessed by measuring the ratio of the radioactivity bound to the total radioactivity and making reference to the standard curve.

Figure 2.14 The radioimmunoassay technique.

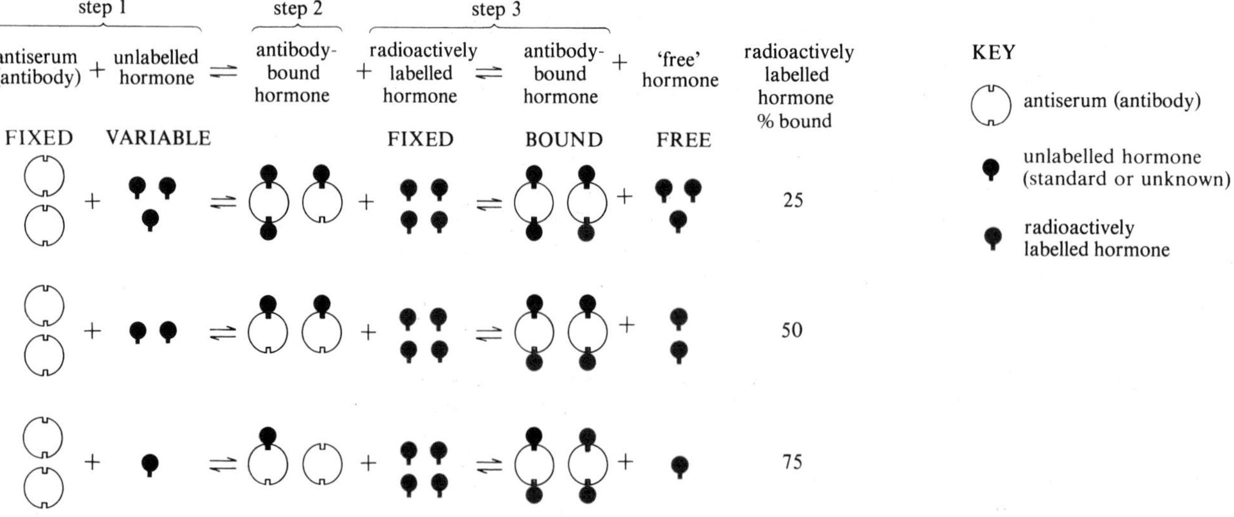

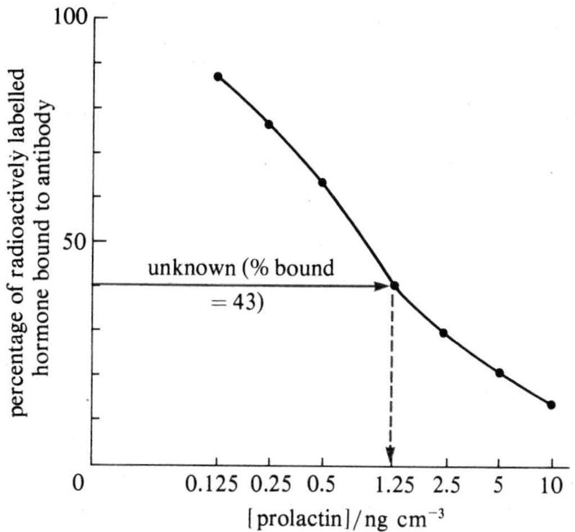

Figure 2.15 A standard graph showing the percentage of radioactively labelled hormone bound to antibody against the hormone concentration (measured in ng cm^{-3}). If the x axis is plotted on a linear scale, a curve is obtained. It is not very easy to read off unknown sample concentrations from such a graph, so it is more usual to plot hormone concentrations on a log scale, which produces a relatively straight line.

The application of radioimmunoassay has had such an impact on medical diagnosis that its originator Rosalind Yalow was awarded a Nobel Prize in 1977.

Summary of Section 2.2

1 Hormones can be broadly subdivided into those based on (a) fatty acids and (b) amino acids. Steroids are important examples of the first group and peptide hormones of the second.

2 Many hormones are synthesized as inactive parent molecules, which are subsequently converted into the active hormone by the action of specific enzymes.

3 Families of peptide hormones, with short sequences of amino acids in common, can be distinguished. For some hormones this can be explained on the basis of each one being derived by postsynthetic processing from a common parent molecule.

4 Hormone-producing cells have recognizable ultrastructural features that relate to the type of hormone they produce.

5 The cells that produce hormones are often situated together in a distinct tissue—an endocrine gland.

6 Hormones may be released (a) more or less continuously, (b) in short distinct pulses (say, over a 24 h period) or (c) in longer cycles.

7 The duration of a hormone's action depends on the longevity of the hormone. This is governed by (a) its rate of breakdown in the liver and at the target tissue and (b) its rate of excretion via the kidneys. The effective life of small hormones is often extended by interaction with binding proteins that are present in the blood.

8 The concentration of circulating hormones is usually very low—this means that techniques such as bioassay and radioimmunoassay are required to measure them.

Now attempt Questions 1–3 on pp. 63–64.

2.3 NEUROSECRETION

'It appears indeed as if there is a much closer relationship between nervous and glandular (endocrine) tissues than is commonly recognised'. (Comment made in a review by Ernst and Berta Scharrer published in 1940.) 'We have just heard some very interesting things—and also a great deal of nonsense'. (Comment addressed to Ernst Scharrer when he presented the concept of neurosecretion at a scientific conference in London in 1953.)

During the 1920s a German scientist, Ernst Scharrer, noticed that certain neurons in a particular region of the minnow brain had structural features similar to those of endocrine cells. The axons of these neurons passed out of a region of the brain called the **hypothalamus** and into the **pituitary gland**, which is situated beneath it (see also next Section). Instead of making synaptic contacts with other nerves or effector (target) cells, these axons ended in large bulbous swellings that were closely applied to certain of the numerous capillaries that pass through the pituitary gland. These findings led Scharrer to suggest that at least some of the hormones released from the pituitary gland were actually synthesized in neurons. If this were so, the pituitary might act merely as a storage and release depot rather than as a conventional endocrine gland. At this time, the idea was almost universally rejected by the scientific community because it cut right across the concept of distinct nervous and hormonal communication systems. However, Scharrer and his wife Berta persisted in their studies, and by the 1950s a considerable body of evidence supported their original hypothesis. The term **neurosecretion** was introduced to indicate that nerves secreted material into the bloodstream in much the same way as other endocrine cells. The secretory products were termed **neurohormones** to indicate their origin in the nervous system, and the organs involved in storage and release were termed **neurohaemal organs** because of the close association of nerve endings and blood capillaries within these structures.

In general, neurosecretory neurons differ from conventional neurons in the position and structure of the nerve endings, and in the size of the granules found within the cell. Neurosecretory granules are much larger (200–500 nm in diameter) than those found at standard nerve endings (usually 40–100 nm in diameter) and similar in size to those found in many endocrine cells. In comparison with conventional nerves, neurosecretory neurons have more nerve endings (Figure 2.16), which enables them to secrete relatively large amounts of neurohormones. This is an important structural adaptation because these cells secrete their contents into the bloodstream, which has an enormous volume in comparison with the synaptic cleft. Like those of conventional neurons, neurosecretory axons produce and conduct action potentials, but changes in potential develop more slowly and last longer. This seems to be another adaptation that increases the amount of neurohormone secreted per stimulation.

2.3.1 The pituitary and hypothalamus

The concept of neurosecretion not only provided new insights into the nervous system but also focused attention on the pituitary gland which, by this time, was known to be involved somehow in the regulation of various physiological processes. The pituitary gland is a small organ (the human pituitary weighs 0.5–0.8 g) situated in the floor of the skull just above the roof of the mouth (Figure 2.17). The gland was discovered by the sixteenth century anatomist Vesalius, who misguidedly thought it was responsible for the secretion of pituita (nasal fluid), hence the name pituitary. Because of its

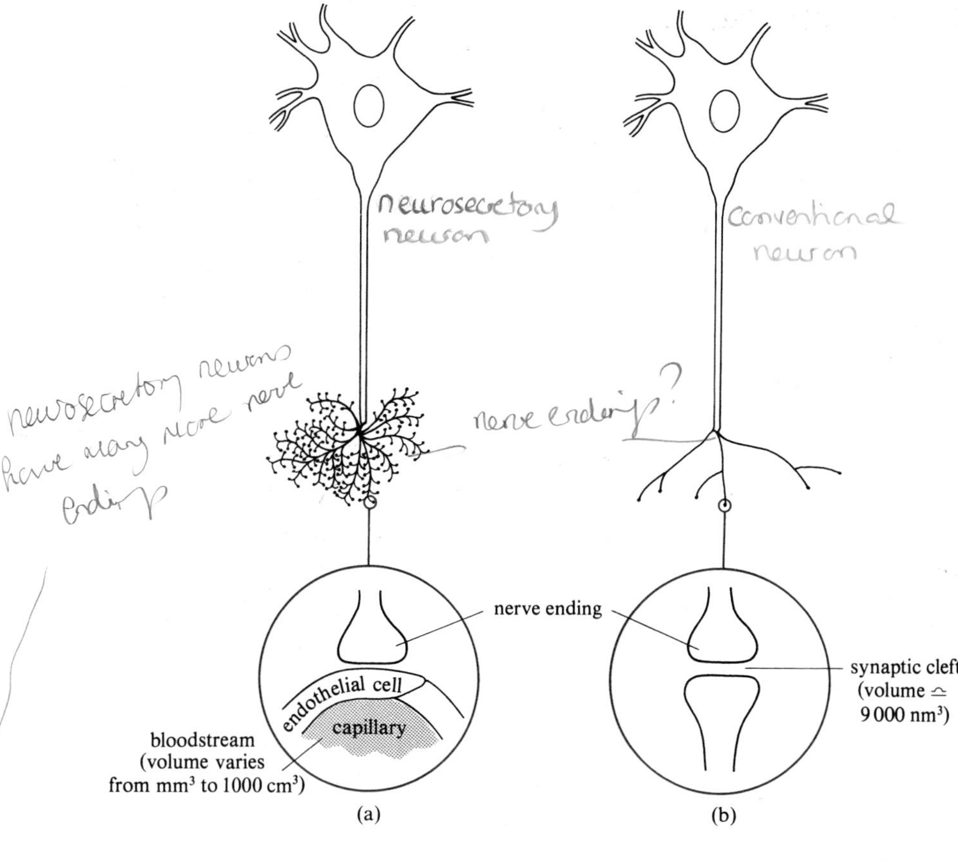

Figure 2.16 (a) A typical neurosecretory neuron. (b) A conventional neuron.

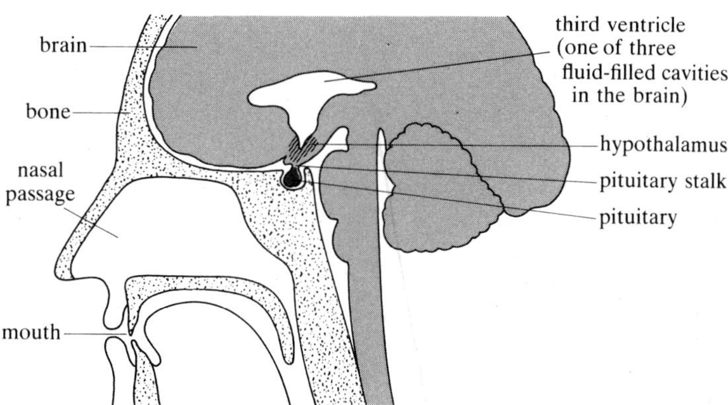

Figure 2.17 The location of the pituitary gland in humans.

position immediately below the hypothalamic region of the brain, the pituitary is also known as the hypophysis (from the Greek: hypo, under; physis, growth). The gland is composed of two embryologically distinct tissues: the adenohypophysis and the neurohypophysis. In mammals, these two sections form the anterior and posterior lobes of the pituitary, and we shall use these terms in this chapter. The pituitary is joined to the brain by a thin stalk of nervous tissue known as the pituitary stalk.

The posterior lobe of the pituitary (Figure 2.18) consists largely of neurosecretory nerve endings, most of whose cell bodies originate in two specific regions of the hypothalamus (these were the neurons originally investigated by Scharrer). Two hormones are synthesized in these hypothalamic cells: antidiuretic hormone (ADH) and oxytocin. The hormones are

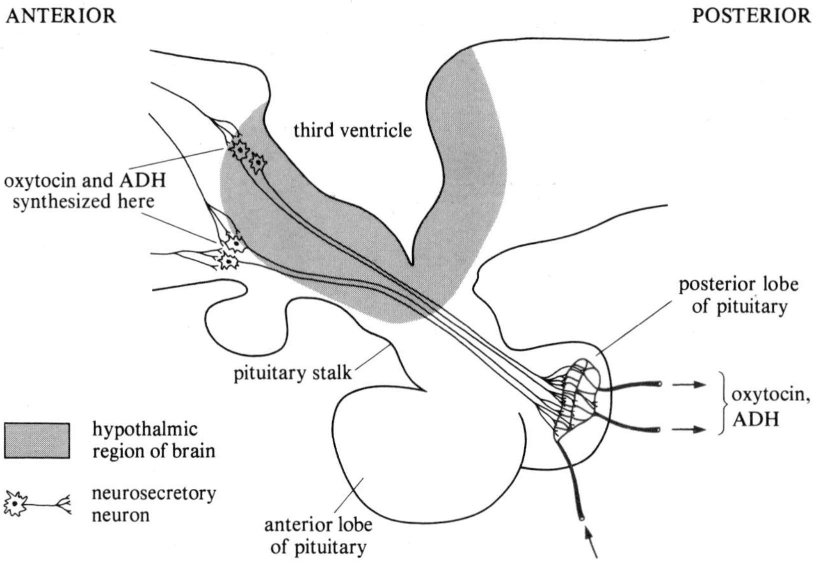

Figure 2.18 The posterior lobe of the pituitary and its associated neurosecretory neurons in the hypothalamus.

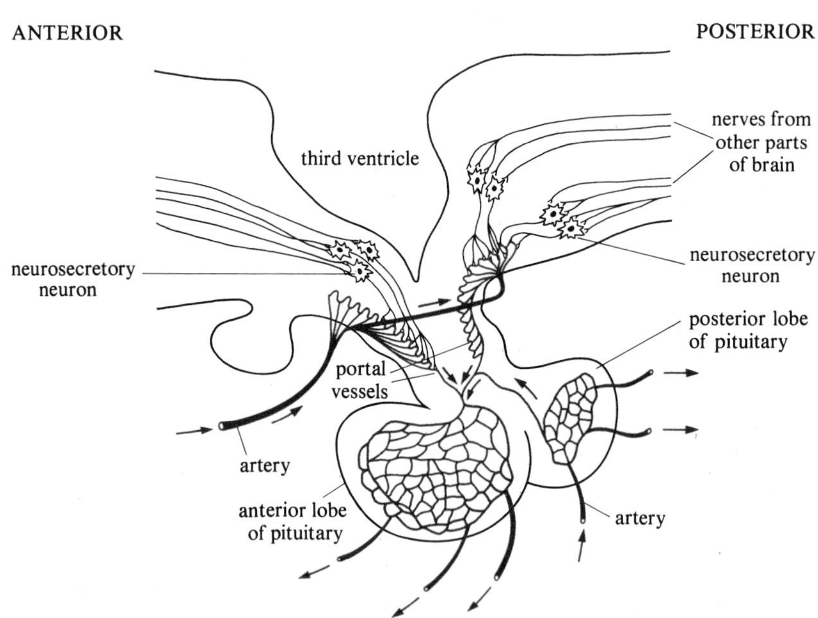

Figure 2.19 The blood supply to and from the pituitary gland.

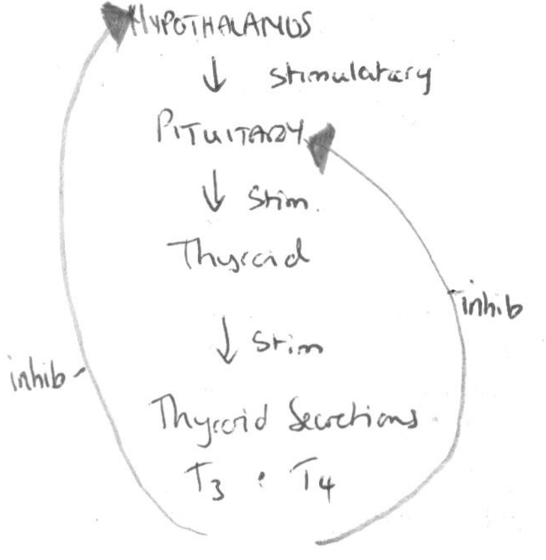

combined with larger proteins, then transported along the axons and subsequently stored and released from nerve endings in the posterior lobe. In contrast to those within the posterior lobe, cells contained within the anterior lobe of the pituitary are responsible for both the synthesis and the release of hormones.

Table 2.1 lists a number of the hormones produced by the anterior lobe of the pituitary. Most of these stimulate the secretion of other hormones from endocrine glands situated in different regions of the body; they are therefore said to exert a tropic action (from the Greek for 'to turn' or 'change'), and such hormones are often referred to as tropins. For example, thyroid-stimulating hormone stimulates the thyroid gland to secrete two thyroid hormones, thyroxin and triiodothyronine. Most of these hormones are relatively large peptides. Unlike the posterior lobe of the pituitary, the anterior lobe did not appear to have a nerve supply, so it was rather a surprise when experiments showed that electrical stimulation of certain parts of the hypothalamus dramatically increased the secretion of some anterior lobe hormones and decreased the secretion of others. A possible solution to this problem became evident when the blood system of the pituitary was investigated (Figure 2.19).

◇ Suggest how the hypothalamus could influence the release of hormones from the anterior lobe of the pituitary.

Table 2.1 Hormones produced by the anterior lobe of the pituitary (adenohypophysis)

Hormone (name used in this Book)	Abbreviation		Target cells	Action
thyroid-stimulating hormone	TSH		thyroid gland	stimulates the synthesis and secretion of thyroxin and triiodothyronine
follicle-stimulating hormone	FSH	these two hormones are collectively termed *gonadotropins*	testis and ovary	controls the development and maturation of germ cells, i.e. spermatozoa and ova
luteinizing hormone	LH		testis and ovary	controls the secretion of the steroid hormones responsible for the male and female sexual characteristics; also triggers ovulation
prolactin	PRL		mammary gland and corpus luteum	stimulates milk production
			liver	stimulates the production of a pheromone that controls maternal behaviour in some mammals
			fish gills	involved in the maintenance of salt balance and in osmoregulation
growth hormone	GH		liver	stimulates the liver to form somatomedins, which alter the metabolism of tissues (e.g. liver, muscle, adipose tissue)
adrenocorticotropic hormone	ACTH		adrenal cortex	stimulates the synthesis and release of glucocorticoids
β-lipotropin	β-LPH		adipose tissue	mobilizes lipids; but may be more important as a precursor (parent molecule) for endorphin and enkephalins
β-endorphin, enkephalins			neurons	alter the activity of certain neurons
melanocyte-stimulating hormone	MSH		melanocytes* (also neurons)	darkens the skin in lower vertebrates; controls hair colouration in some mammals; affects the activity of some neurons

*Cells containing melanin that are responsible for skin colour change. (Also cells with other colours present in many fishes and other vertebrates.) Other terms will be explained in later Chapters.

◆ You should have noticed from Figure 2.19 that (a) there is an elaborate capillary network that is arranged so that blood passing into the hypothalamus drains into the anterior lobe and (b) a number of neurosecretory nerve endings are closely associated with this capillary network. Given this anatomical arrangement it could be that neurohormones released into the bloodstream from hypothalamic neurosecretory nerves after electrical stimulation either stimulate or inhibit the secretion from the target cells in the anterior lobe.

◇ Suggest a method by which you might test the above.

◆ One way would be to disrupt (i.e. cut) the blood capillaries (or portal vessels as they are termed) that connect the hypothalamus to the anterior lobe of the pituitary, and then repeat the electrical stimulation experiments. Stimulating the appropriate regions of the hypothalamus should now have no effect on the secretion of hormones from the anterior lobe.

This hypothesis was originally put forward by Geoffrey Harris at Oxford University in the 1940s. Although his ideas have turned out to be correct, they were not easy to substantiate. To give the hypothesis credibility, the neurohormones (or **releasing and release-inhibiting factors**, as they are now called) had first to be isolated. They proved to be very elusive, so much so that many physiologists began to think that the hypothesis was wrong. However, in 1969 two different research groups announced that they had isolated and chemically identified a releasing factor. This was thyroid-stimulating hormone releasing factor (TRF), which stimulates certain cells of the anterior lobe of the pituitary to release thyroid-stimulating hormone. This was a major milestone in endocrinology and the leaders of the two groups, Roger Guillemin and Andrew Schally, were subsequently awarded Nobel Prizes in 1977.

Several other releasing, e.g. gonadotropin releasing hormone (GnRH, see Chapter 10), and release-inhibiting factors have now been isolated and they are all relatively simple peptides (Figure 2.20). This has practical implications because chemists can manufacture small peptides quite cheaply whereas the chemical synthesis of the larger peptide hormones secreted by cells of the anterior pituitary is prohibitively expensive. Various drug companies quickly realized the potential of synthetic versions of these factors, which is why so much effort was expended on their isolation. Synthetic versions of TRF are currently used in a very sensitive test for pituitary malfunction.

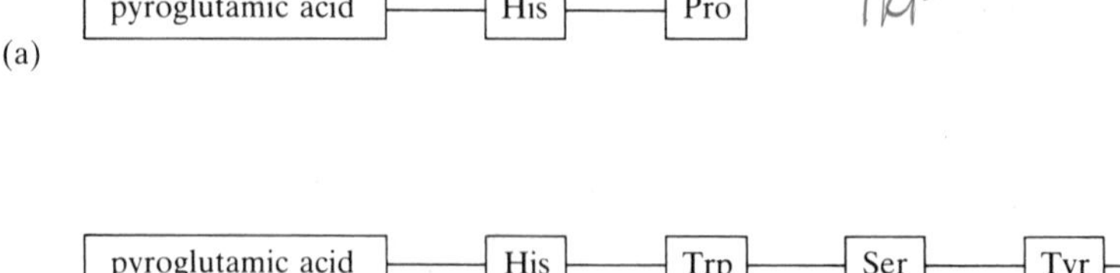

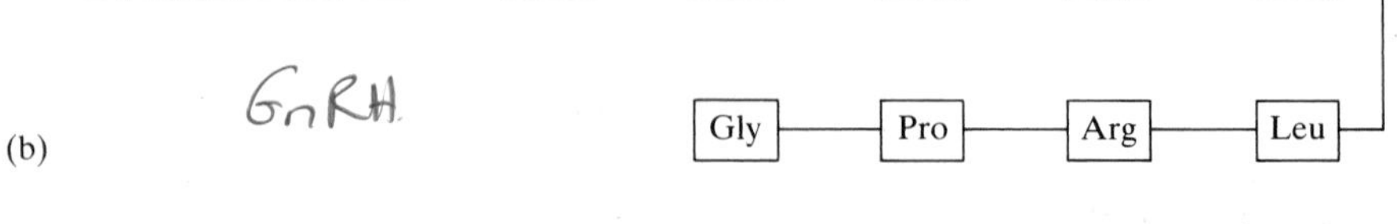

Figure 2.20 The structure of (a) thyroid-stimulating hormone releasing factor (TRF) and (b) gonadotropin releasing hormone (GnRH), so-called because this releasing factor controls the release of follicle-stimulating hormone (FSH), and luteinizing hormone (LH), which are known together as gonadotropins.

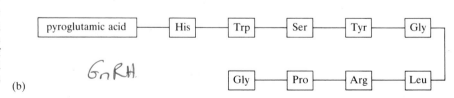

Figure 2.21 summarizes the interrelations of the hypothalamus and various regions of the pituitary. As you can see, the pituitary is involved in the control of a wide range of organs and physiological processes.

KEY
blood system supplies connecting link
neurosecretory neurons supply connecting link

hypothalamus
pituitary stalk
posterior lobe
ADH
oxytocin
kidney
uterus
mammary gland
releasing factors
inhibiting factors
anterior lobe
TSH
FSH
LH
PRL
GH
ACTH
β-lipotropin
β-endorphin enkephalins
MSH
thyroid gland
testes
ovary
mammary gland
liver
adrenal gland
adipose tissue?
nerves?
melanocytes, nerves?
thyroxin
testosterone
progesterone oestrogen
somatomedin
corticosteroids

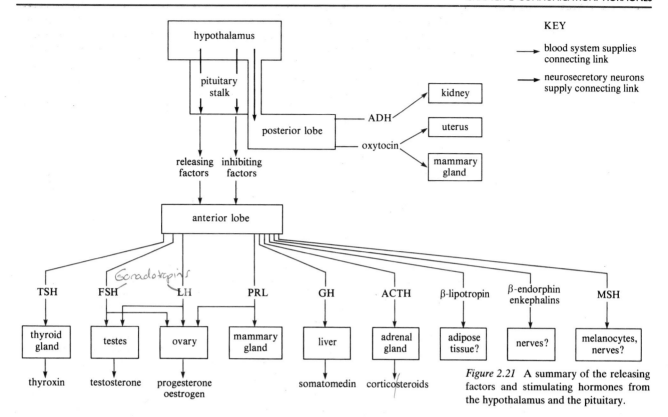

Figure 2.21 A summary of the releasing factors and stimulating hormones from the hypothalamus and the pituitary.

2.3.2 Feedback loops

The release of many pituitary hormones is a carefully controlled process involving **feedback loops**. If two or more variables are mutually interdependent, then feedback can be said to exist between them. For example, if the secretion of a hormone is decreased as a result of the response elicited by that hormone, the feedback is said to be negative. On the other hand, if the reverse occurs i.e. hormone secretion is increased, then the feedback is said to be positive. Hormone secretion from the posterior lobe of the pituitary is controlled by fairly simple feedback loops.

◇ Follow the sequence of events shown in Figure 2.22 (Steps 1–9). This illustrates the order of events in the regulation of body-fluid volume by

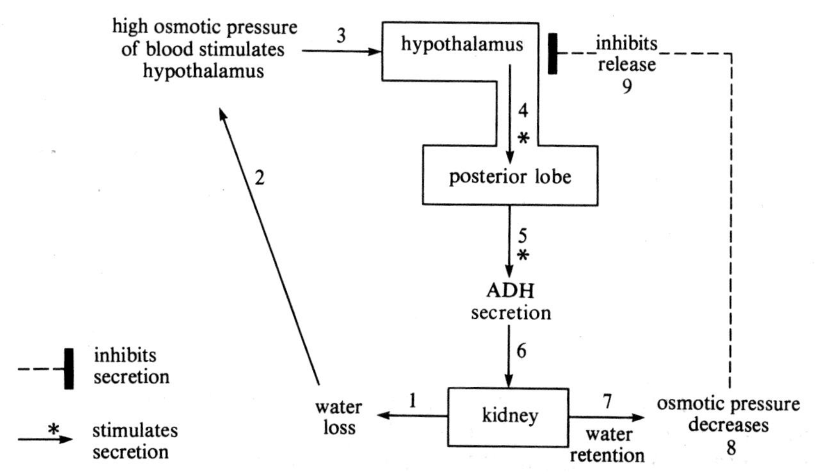

Figure 2.22 Feedback control of the secretion of antidiuretic hormone. The release of the hormone is regulated by an end-product of its own action (rather than by another hormone).

antidiuretic hormone (ADH), the hormone that controls loss of water from the body. Is the control of ADH secretion an example of positive or negative feedback?

◆ It is a negative feedback loop because an increase in the secretion of ADH leads to a change in osmotic pressure, which in turn leads to a decrease in ADH secretion.

The control of hormone secretion from the anterior pituitary is more complex because feedback control is often exercised at both the hypothalamus and the pituitary, so two feedback loops are involved (Figure 2.23). These are often termed long and short loops. For example, a high level of thyroxin in the blood inhibits the release of thyroid-stimulating hormone (TSH). Inhibition takes place at two sites, the hypothalamus and the anterior pituitary.

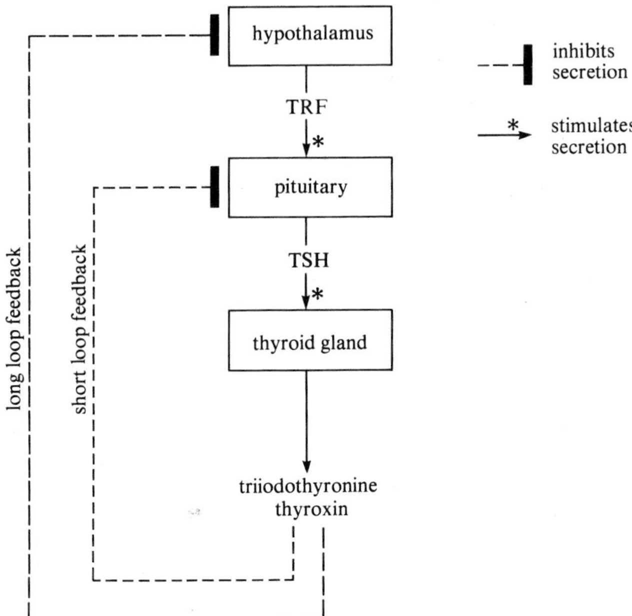

Figure 2.23 Feedback control of the secretion of tropic hormones. The thyroid hormone system is illustrated here.

Recent evidence indicates that different feedback loops can interact with each other. For example, high levels of corticosteroids inhibit the secretion of adrenocorticotropin (ACTH) by means of a standard, negative feedback loop because ACTH stimulates the secretion of corticosteroids. However, high levels of corticosteroids also make cells that secrete TSH more sensitive to the releasing factor TRF so that more TSH is released in response to a given amount of TRF. Corticosteroids therefore inhibit the secretion of ACTH and stimulate the secretion of TSH.

2.4 INTEGRATION OF NERVOUS AND ENDOCRINE SYSTEMS

By now you should appreciate that the pituitary and the hypothalamus are key centres of physiological control. Unlike the rest of the brain, the hypothalamus is not protected by the blood–brain barrier, so it is exposed to any physical and chemical variations that take place in the bloodstream. Various sensory devices in the hypothalamus are supplied with additional sensory information from various centres in the brain. In short, the hypotha-

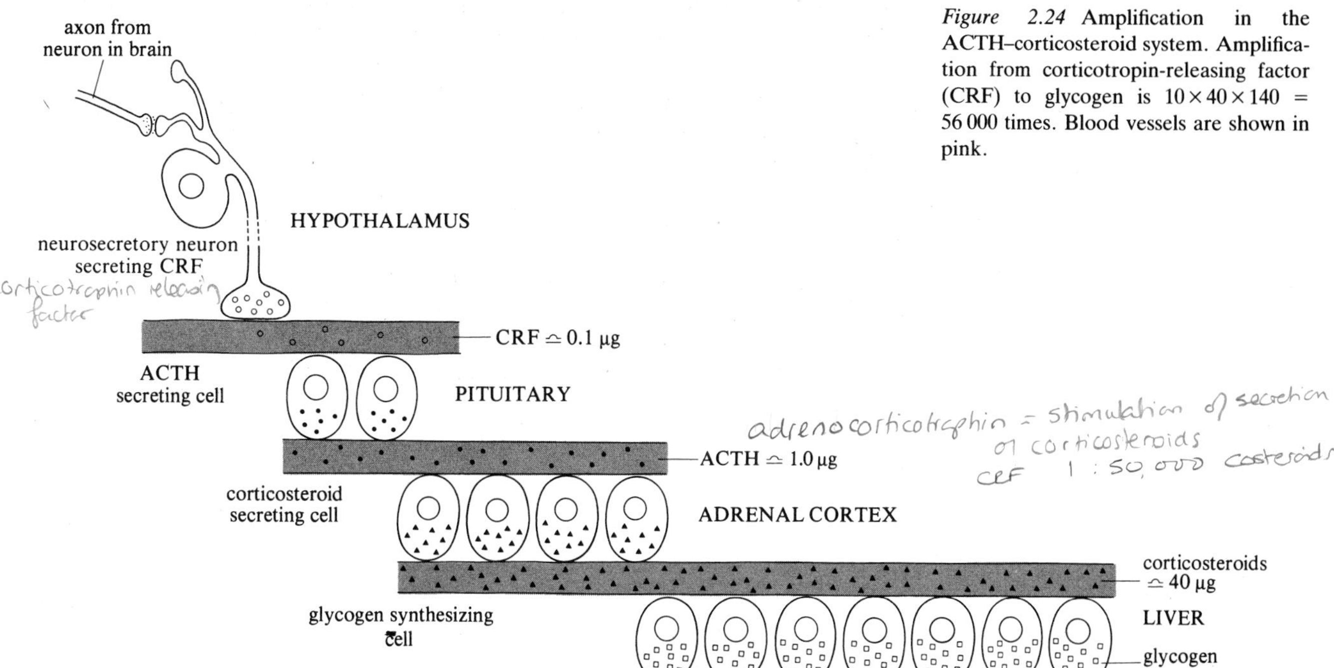

Figure 2.24 Amplification in the ACTH–corticosteroid system. Amplification from corticotropin-releasing factor (CRF) to glycogen is 10 × 40 × 140 = 56 000 times. Blood vessels are shown in pink.

lamus occupies the crossroads at which nervous and blood-borne information meet. It is thus uniquely placed to coordinate both sets of inputs and to activate or suppress pituitary secretions accordingly.

The arrangement, neurons ⟶ hypothalamic neurosecretory neuron ⟶ endocrine gland, possessed by vertebrates, enables rapidly evoked nervous activity to be translated into sustained hormonal stimulation that results in long-term physiological changes. The sequential system of hypothalamus ⟶ pituitary ⟶ endocrine gland enables an initially weak chemical signal to be amplified many times. For example, it has been determined experimentally that 0.1 μg of corticotropin-releasing factor can stimulate the deposition of 5 600 μg of glycogen in the liver via the adrenocorticotropic hormone–corticosteroid system (Figure 2.24).

At each stage in the system, the secretion of a small number of signal molecules results in the release of a much larger number of different signal molecules that operate the next stage. Such amplification is necessary if the nervous and endocrine systems are to be integrated, because a few neurons, releasing minute amounts of neurotransmitter, must influence large numbers of cells. The ACTH–corticosteroid system affects virtually every cell in the body across what is in effect a synaptic cleft (the blood system) of enormous volume.

It is not known whether this integration of nervous and endocrine systems is widespread among the invertebrates, largely because many invertebrate endocrine systems are only just being discovered. However, insects seem to have a similar sort of system.

2.4.1 Neuroendocrinology

We have seen that certain cells display both neuronal and endocrine features. You may now wonder if any endocrine glands display neuronal properties and whether hormones can influence the nervous system.

The endocrine cells of the pancreas, anterior pituitary and thyroid have all been shown to produce action potentials similar to those of nerves—so the answer to the first question is yes. Thyroxin, corticosteroids and sex hormones exert extremely important influences on the brain during development. Malfunction of the thyroid gland during foetal life results in irreversible brain damage, which can lead to a condition known as 'cretinism'. The role of steroid hormones in influencing behaviour is not restricted to the new-born animal. For example, the application of minute amounts of the steroid testosterone to one area of the rat hypothalamus stimulates male sexual activity whereas its application to a different area elicits maternal behaviour—so the answer to the second question is also yes.

The situation is further complicated by the fact that we can no longer assign hormones and neurotransmitters to defined locations in specific endocrine glands or brain structures. This is particularly true for peptide messenger molecules, which are now known to be secreted by both nerves and endocrine cells.

The development of modern radioimmunoassay and histochemical methods has allowed the detection and detailed mapping of these peptides or **neuropeptides** as they are now commonly called, in the central nervous system and in the peripheral autonomic system. In some cases neuropeptides are present in discrete nerve cells, and in others they coexist with other neurotransmitters. For example, noradrenalin may coexist with somatostatin in some sympathetic ganglion cells or with enkephalin in others. Interestingly, these neuropeptides can act in either the conventional way as a neurotransmitter or by modulating the effect of another neurotransmitter.

Summary of Sections 2.3 and 2.4

1 Specialized neurons (neurosecretory neurons) release substances (neurohormones) into the bloodstream, which transports them to target cells.

2 Neurosecretory neurons permit the activity of the nervous system to be coordinated with that of the endocrine system. Such interactions enable the short-term activity of a few neurons to be amplified into an action of relatively long duration that can influence a large number of cells.

3 The hypothalamus and pituitary are important centres of physiological regulation because within these two structures a large amount of nervous and hormonal information is integrated and used to regulate a wide range of body functions.

4 The pituitary has two parts, the anterior and posterior lobes. The latter acts as a releasing site (a neurohaemal organ) for neurohormones produced within the hypothalamus; the former is a complex tissue containing different types of endocrine cell. Anterior pituitary cells are controlled by neurosecretory neurons that originate in the hypothalamus and terminate in the pituitary stalk. These cells secrete neurohormones that are conveyed to the anterior pituitary by a network of blood vessels (a portal system). These neurohormones (releasing or release-inhibiting factors) either stimulate or inhibit the release of particular hormones from the anterior pituitary. Many of the anterior pituitary hormones have a tropic effect (i.e. they stimulate the release of hormones from other endocrine glands).

5 The neurosecretory cell ⟶ endocrine cell ⟶ endocrine cell arrangement exemplified by the vertebrate endocrine system produces a large amplification of the original signal. It also permits the generation of a range of feedback loops that enable the whole system to be finely regulated.

6 The endocrine and nervous systems interact: nerves control the activity of the endocrine system and the endocrine system regulates the activity of nerves.

7 There are many similarities between neurons and endocrine cells; some endocrine cells exhibit action potentials and secrete substances normally associated with neurons; similarly, some neurons secrete peptide hormones usually associated with endocrine cells.

Now attempt Questions 4–6, on p. 64.

OBJECTIVES FOR CHAPTER 2

Now that you have read this chapter, you should be able to:

2.1 Define and use, or recognize definitions and applications of each of the terms printed in **bold** in the text.

2.2 Describe the basic difference between hormones synthesized from fatty acids and those synthesized from amino acids. (*Question 1*)

2.3 Outline the various factors that govern the 'effective' life of a hormone and hence the duration of its action. (*Questions 1, 2 and 3*)

2.4 On the basis of its ultrastructure, distinguish a cell that produces peptide hormones from one that produces steroid hormones. (*Question 1*)

2.5 Explain how the concentration of hormones in the blood or other physiological fluids can be measured. (*Question 3*)

2.6 Outline how putative endocrine glands can be identified experimentally and the limitations of these methods.

2.7 Explain how neurosecretory neurons differ from conventional neurons. (*Question 4*)

2.8 Outline, giving specific examples, the different types of interaction that take place between the hypothalamus, the pituitary and peripherally sited endocrine glands. (*Question 5*)

2.9 Explain, giving examples, the ways in which hormone secretion is controlled by feedback loops. (*Question 6*)

2.10 Discuss the relationship between nerve cells and endocrine cells. (*Question 5*)

QUESTIONS FOR CHAPTER 2

Question 1 (*Objectives 2.2, 2.3 and 2.4*) Are the following statements true or false? Give reasons for your answers.

(a) The molecular structures of neurotransmitters and peptide hormones are more closely related than those of steroid hormones and peptide hormones, or those of carbohydrates and peptide hormones.

(b) Cross-reactivity is more likely to occur between progesterone and cortisol than between growth hormone and cortisol.

(c) Small hormones bind only to hormone receptors.

(d) The enzyme lactate dehydrogenase (LDH) has the same amino acid sequence irrespective of the animal or tissue from which it is isolated. Similarly, growth hormone isolated from different animals should also have identical sequences of amino acids. This is because the function of a protein depends on its entire amino acid sequence.

(e) Steroid-secreting cells have specialized mitochondria and a prominent smooth endoplasmic reticulum.

(f) Some endocrine cells secrete inactive forms of hormones that are later modified by enzymes in the blood or at the target tissue.

Question 2 (*Objective 2.3*) List four factors that determine which target tissue responds to a particular hormone, and the duration of the response.

Question 3 (*Objectives 2.3 and 2.5*) Doctors often ask hospital laboratories to analyse blood and/or urine samples for specific hormones in order to diagnose certain diseases.

(a) Explain why the doctor needs to know the approximate time of day the sample was taken in order to interpret the information provided by the laboratory.

(b) What sort of hormones are likely to appear in the urine?

(c) What kind of information could this analysis yield?

Question 4 (*Objective 2.7*) List four ways in which neurosecretory neurons differ from conventional neurons.

Question 5 (*Objectives 2.8 and 2.10*) The following kinds of statement are often written hurriedly during examinations. They contain some truth but not the whole truth and are therefore misleading. Explain briefly what the writer has omitted to say.

(a) Cells of the anterior lobe of the pituitary are connected to the hypothalamus by neurons.

(b) The secretion of hormones from the anterior lobe of the pituitary is controlled by releasing factors secreted by the brain.

(c) The secretion of corticosteroids is controlled by the pituitary.

(d) Nerves communicate via electrical impulses; in contrast, endocrine cells use hormones.

Question 6 (*Objective 2.9*) Organ transplant patients are often given massive amounts of corticosteroids. These are used because they suppress the body's natural defence mechanism (its immune system) and so help to prevent the rejection of the foreign tissue. From the information in this chapter, predict how this therapy will affect the blood levels of at least three other hormones.

3.1 THE ACTION OF NEUROTRANSMITTERS AND HORMONES AT THE CELLULAR LEVEL

In both the nervous and endocrine systems, intercellular communication is mediated by chemical messenger molecules (neurotransmitters and hormones) released into either the synaptic cleft or the bloodstream, respectively. The information that these messenger molecules convey must be received by the target cell or tissue then translated into an appropriate response. You will recall from the previous chapters that target cells and tissues possess specific receptors for neurotransmitters and/or hormones. It is these that determine the target cell's ability to respond to a particular messenger molecule; a cell will only respond to a messenger molecule if it possesses the appropriate receptors. In Chapter 1 we saw how some neurotransmitter receptors are coupled to specific ion channels. Receptor activation leads to the transient opening of these channels and a change in the postsynaptic membrane potential. However, many neurotransmitters and hormones activate receptors that are not coupled to ion channels. So how is the information that is carried by these messenger molecules converted into a cellular response?

In these cases the initiation of that response involves the formation of a second set of messenger molecules, which act entirely within the cell. These intracellular messengers, or second messengers as they are called, are formed as a result of receptor stimulation, and their role is threefold. They serve to convert the extracellular signal into an intelligible intracellular signal, they amplify the extracellular signal and they modulate the activity of the effector systems that control the observed cellular response. This process, summarized in Figure 3.1, involves three steps.

Figure 3.1 An intercellular communication system (a) compared with a more familiar communication system (b). Notice how similar the two are in terms of their basic components.

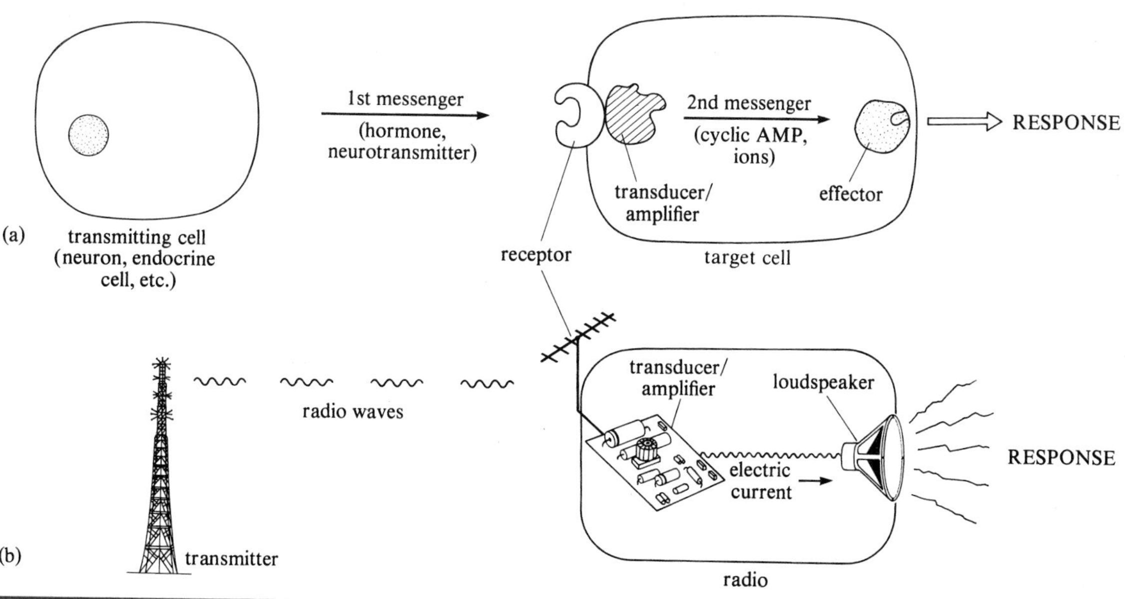

These three steps are:

1 An initial binding of the neurotransmitter or hormone (**first messenger**) to specific receptors situated on the target cell or tissue.

2 A transduction step in which the interaction of the first messenger and receptor triggers the production of the second messengers. Amplification of the initial signal usually occurs at this stage because the interaction of one first messenger molecule with an appropriate receptor can produce many molecules of the second messenger.

3 A final effector step in which the second messengers act directly (or indirectly) to activate (or inactivate) a specific, cellular effector system, which brings about the observed cellular response.

In the remainder of this chapter we will look more closely at the mechanisms involved in the receipt and translation of information at the cellular level.

3.2 RECEPTORS

The receptor is analogous to the aerial shown in Figure 3.1b and, like an aerial, it is tuned to receive signals on a particular molecular waveband. In reality each receptor has a unique, three-dimensional structure that incorporates a highly specific binding site, which recognizes certain facets of the molecular structure of a hormone or neurotransmitter. This specificity ensures that cells respond only to those first messengers for which they possess receptors.

◇ What class of biological molecules is most likely to be used in the construction of receptors?

◆ In many ways receptors are similar to enzymes and membrane transport molecules, thus the answer is proteins of various kinds. Not surprisingly, most receptors possess many of the properties exhibited by enzymes and transport molecules. The binding of a first messenger to its receptor is saturable and reversible and can be characterized by K_d, a measure of affinity of binding and B_{max}, a measure of the density of binding sites. These terms are the equivalents of K_t and J_{max} in membrane transport.

Table 3.1 contains some detailed information on three receptors.

During stimulation of the nervous or endocrine system, the concentration of neurotransmitter or hormone in the immediate vicinity of the target cell rises.

Table 3.1 The characteristics of receptor molecules for three different messengers

Receptor for	Chemical nature	Relative molecular mass	No. per cell	K_d value/mol l^{-1}
acetylcholine	glycolipoprotein arranged in several subunits	88 000–500 000	depends on the cell, but can reach values as high as 10^{11}	$\simeq 2 \times 10^{-8}$
thyroid-stimulating hormone (TSH)	glycoprotein arranged in several subunits	280 000	500 (thyroid cells)	1.9×10^{-9}
insulin	glycoprotein arranged in several subunits	300 000	75 000 (liver cells)	1.2×10^{-6}

This results in more and more of the receptor sites being occupied. Generally, but not invariably, the observed cellular response is proportional to the number of receptor sites occupied.

Many receptor molecules are situated on the outside of the cell membrane. This is not surprising when you consider that many neurotransmitters and hormones are proteins or amino acid derivatives. As such, they are hydrophilic and would not cross the cell membrane easily. On the other hand, steroids are derived from a lipid molecule (cholesterol) so you would expect steroids to cross membranes readily. Steroid receptors could therefore be sited on the outside of the membrane or in the cell cytoplasm. The majority of steroid receptors are indeed cytoplasmic, but recent evidence suggests that some cells may possess membrane-bound steroid receptors.

Two techniques are commonly used to confirm the presence or precise location of particular receptors. One is **autoradiography** where cells or tissues are exposed to radioactively labelled first messenger molecules that bind to their receptor sites. The tissue is then coated with a photographic emulsion, and the radioactive particles emitted from the tissue react with the emulsion to form grains of metallic silver. The receptor sites can then be seen, usually under a microscope, as black dots when the autoradiogram is developed. The other method is called **radioligand binding**, and is similar in many ways to the radioimmunoassay outlined in the preceding chapter.

In ligand binding assays, radioactively labelled hormones or neurotransmitters are allowed to bind to membrane or cytosolic fractions instead of antibodies.

By performing experiments similar to those described in Chapter 2, Figures 2.14 and 2.15, the presence or absence of particular receptors in those fractions can be determined. Both methods allow the determination of the density (B_{max}) and affinity (K_d) of the receptors.

3.2.1 Receptor agonists and antagonists

It should not be assumed that only the endogenous neurotransmitter or hormone can interact with its receptor. Other molecules, often with a similar structure, which either occur naturally or are specifically manufactured, can also bind to these sites. In some cases such molecules mimic the action of the normal first messenger—these are termed **agonists**. Others interact with the receptor in such a way that the natural first messenger, or an agonist, can no longer bind to the receptor. These molecules are called **antagonists** and their effect is to block receptor sites.

Figure 3.2 shows some examples of agonists and an antagonist together with the natural first messengers.

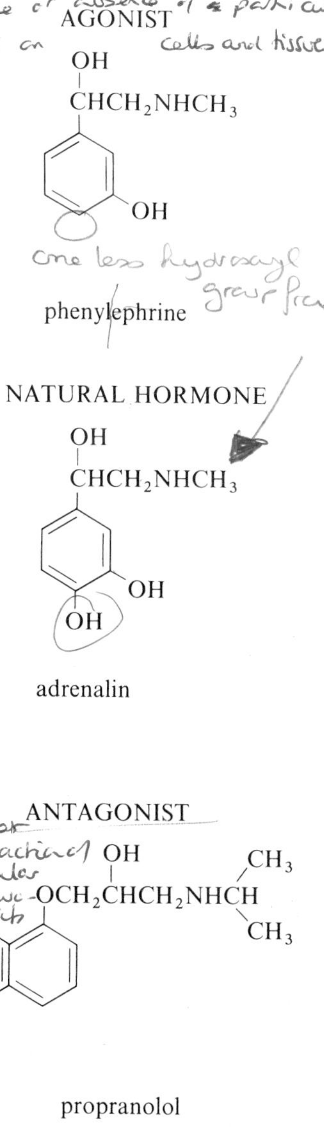
AGONIST
OH
CHCH2NHCH3
OH
phenylephrine
NATURAL HORMONE
OH
CHCH2NHCH3
OH
OH
adrenalin
ANTAGONIST
OH
CH3
OCH2CHCH2NHCH
CH3
propranolol

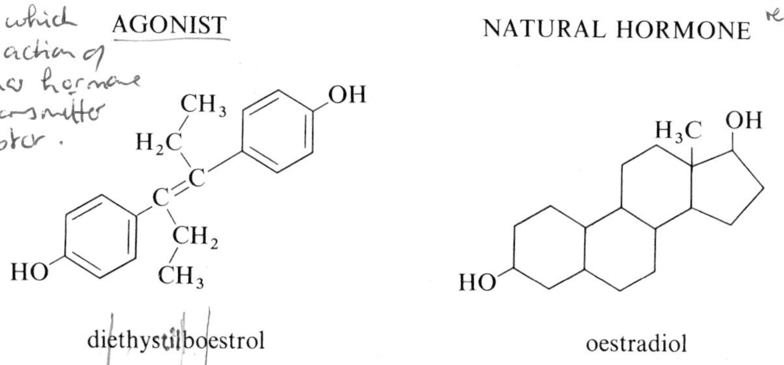
AGONIST
OH
CH3
H2C
C
C
CH2
CH3
HO
diethystilboestrol
NATURAL HORMONE
H3C
OH
HO
oestradiol

Figure 3.2 Some examples of synthetic agonists and an antagonist. (Above) Adrenalin and its agonist phenylephrine and antagonist propranolol. (Left) Oestradiol and its agonist diethylstilboestrol.

◇ By carefully examining Figure 3.2 you should be able to determine why phenylephrine and propranolol act at the same receptor as adrenalin.

◆ Clearly all three molecules are very similar in their chemical structure. This is particularly evident in the case of phenylephrine in that it differs from adrenalin in only one respect—it has one less hydroxyl group.

Receptor agonists and antagonists are widely used as research tools in studying the properties of specific receptors. This has led to the design and manufacture of vast numbers of such compounds, which are now commonly used in medicine and agriculture. For example, phenylephrine is used in various 'cold-relief' treatments because of its ability to prevent excessive nasal mucous secretion. Compounds closely related to propranolol constitute the active ingredient in treatments for certain cardiac disorders and asthma. In the former they block the effect of adrenalin and noradrenalin on the heart thus reducing the heart rate, and in the latter they prevent the constriction of the bronchi. Some receptor agonists and antagonists, many of which are naturally occurring, are extremely toxic; nevertheless they are commonly used. For example, the vegetable poison curare is often used as a general muscle relaxant in patients undergoing major surgery. Its effect is to block the action of the neurotransmitter acetylcholine at the neuromuscular junction. Some snake venoms act in a similar way, such that when injected into the prey animal it is rendered helpless.

With the aid of these compounds, research has established that receptors for particular first messengers can be divided into subtypes. Acetylcholine is released from cholinergic synapses on skeletal muscle and at parasympathetic synapses on various organs. Based on the relative potencies of the agonists nicotine and the fungal alkaloid muscarine, it has been found that the receptors at each target area are different. The former are more sensitive to nicotine and are termed nicotinic receptors. The latter are known as muscarinic receptors because muscarine is the more potent agonist.

3.2.2 Receptor regulation

So far, we have tacitly assumed that cells possess a fixed number of receptor sites for a particular neurotransmitter or hormone. However, this is not the case. Cells are capable of adjusting their receptor populations in response to the extracellular environment. This process provides a failsafe component in the communication system that can compensate to some extent for malfunctions of various sorts.

◇ Assume that a cell is exposed to a fixed concentration of a first messenger. Predict in general terms how (a) increasing and (b) decreasing the number of receptors would affect the cell's response to that first messenger.

◆ Increasing the number of receptor molecules while keeping the concentration of first messenger steady ought to result in the occupation of a greater number of receptor sites because this increases the chances of a first messenger molecule meeting a receptor. Decreasing the number of receptors should have the opposite effect. On this basis, the net effect of a cell's increasing its receptor population would be to increase its response to the first messenger, whereas decreasing the number of receptors would decrease the cell's response.

In reality, the extracellular concentration of a first messenger is not fixed, it varies depending upon the degree of stimulation of the nervous or endocrine systems. Under some conditions target cells may be exposed over long periods of time to higher than normal concentrations of a hormone or neurotransmitter. Often this results in the target cell or tissue becoming less sensitive to that first messenger so that higher concentrations of it are required to produce a response. This adaptive change in the target cell's sensitivity is generally achieved by a change in the number of its receptors available for stimulation, but in some cases a change in receptor affinity can bring about the same effect. This process is termed **receptor regulation**. In the instance outlined above, the sustained exposure to a high concentration of a first messenger would elicit a decrease in receptor numbers or affinity; this process is termed receptor down-regulation. It is thought that this may be one of the ways in which drug addiction occurs. Continuous drug use results in the receptors to which the drug binds becoming less sensitive. The addict then requires more drug more often to experience the effect of that drug. The opposite of this situation is also true, that is, receptor up-regulation can occur. If a target cell has not been stimulated by an appropriate first messenger for some time the cell's receptor population or affinity may be increased so that lower concentrations of the first messenger are required to evoke a response.

In some cases sensitivity to a particular first messenger may be modified by other messenger molecules—you will see an example of this in a subsequent chapter on the control of reproduction. Some target cells have different degrees of sensitivity at different stages of their lives.

The molecular mechanisms involved in receptor regulation are only now beginning to be fully understood, so detailed discussion of this point is inappropriate here. However, an indication of how this process might occur will be given in the following Section.

3.3 SECOND MESSENGERS

The interaction of a neurotransmitter or hormone with its receptor is often translated into intracellular action via an intracellular second messenger. The number of known second messengers appears to be surprisingly small, yet they are capable of regulating a wide variety of physiological and biochemical events.

The great majority of receptors for polypeptide hormones and neurotransmitters are coupled to one of two second messenger-generating systems. One involves the production of a cyclic nucleotide—**cyclic adenosine monophosphate (cyclic AMP)**, the other employs Ca^{2+} ions. In contrast, the interaction of steroid hormones and their receptors leads to the formation of messenger RNA molecules (mRNA), which then serve as second messengers. Some first messenger receptors are coupled to systems other than those indicated above. For example, stimulation of certain receptors leads to the formation of another cyclic nucleotide, cyclic guanosine monophosphate (cyclic GMP). However, such second messenger systems appear to be restricted to relatively few target cell types and are, as yet, poorly understood.

3.3.1 Cyclic AMP as a second messenger

The interaction of particular hormones and neurotransmitters with their receptors leads to the activation of an enzyme, **adenylate cyclase**, situated on the inner face of the cell membrane (Figure 3.3, left-hand side). Adenylate cyclase catalyses the conversion of ATP into cyclic AMP; a single activated adenylate cyclase molecule can produce some 1 000 cyclic AMP molecules per minute. It is in this way that the enzyme amplifies the initial interaction of the first messenger and receptor and converts it into an intracellular signal.

However, before this can occur, the receptor on the outer face and enzyme on the inner face of the membrane must become coupled in some way. In other words, the information conveyed by the external signal must be transmitted across the cell membrane. This transduction step is mediated by regulatory proteins called **G-proteins**, so-called because they must bind guanosine triphosphate (GTP) to become active. Basically, the binding of the first messenger and receptor causes a conformational change in the receptor. This is transmitted through the membrane causing the G-protein to bind GTP and undergo a change in conformation itself. It is in this form that the G-protein can interact with adenylate cyclase.

Two regulatory G-proteins are known to be associated with adenylate cyclase: one is stimulatory (G_s) the other inhibitory (G_i). This allows cells to possess receptors that are either positively or negatively coupled to adenylate cyclase; the former stimulate the production of cyclic AMP, and the latter inhibit it.

Increases in the intracellular concentration of cyclic AMP result in the initiation of a variety of cellular processes, which are in the main mediated by a **protein kinase**. Protein kinases that are specifically activated by cyclic AMP

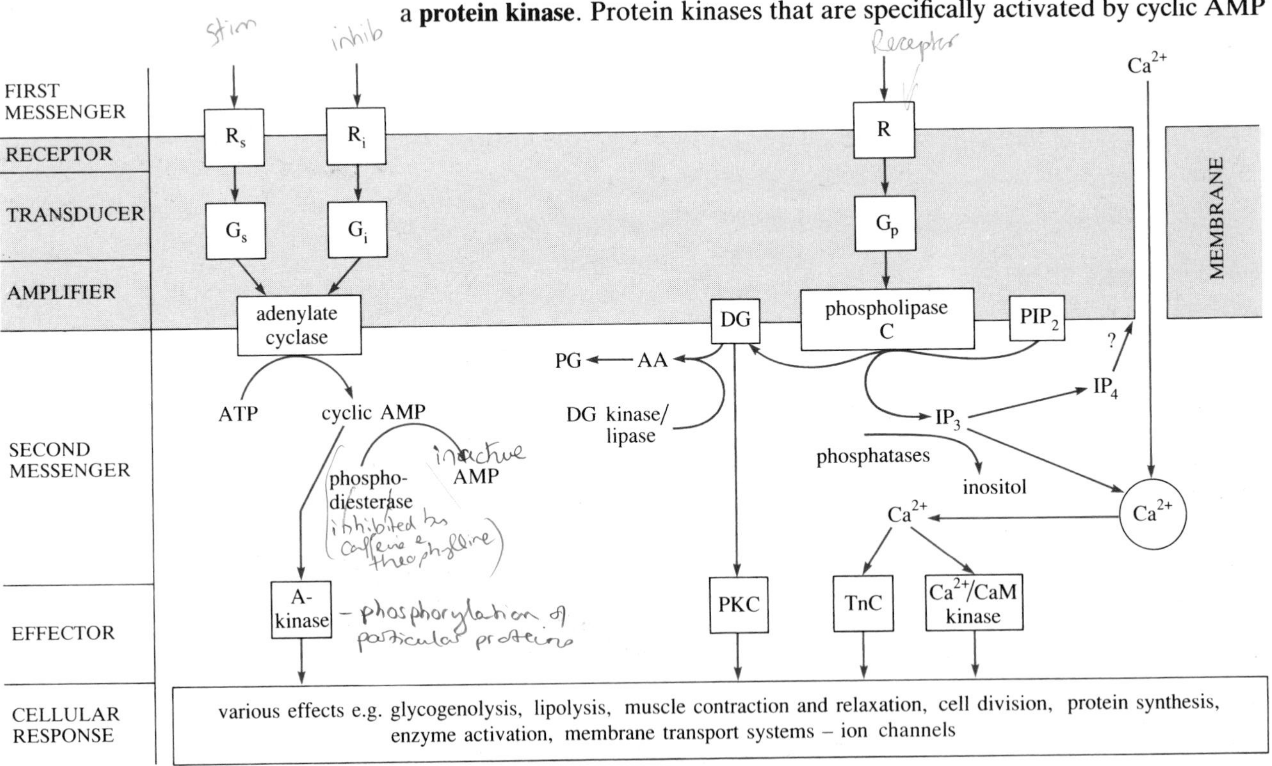

Figure 3.3 The details of intracellular second messenger systems. In the cyclic AMP pathway (left) signals can act at either stimulatory (R_s) or inhibitory (R_i) receptors to regulate the amplifier enzyme, adenylate cyclase. This is achieved via the agency of the appropriate regulatory G-proteins (G_s and G_i). Increases in the intracellular concentration of cyclic AMP activate specific protein kinases (A-kinase), which in turn regulate the cellular response.

In the inositol lipid pathway (right) signals acting at a receptor (R) activate phospholipase C via a G-protein (G_p), which then cleaves PIP_2 to yield inositol trisphosphate (IP_3) and diacylglycerol (DG). IP_3 mobilizes intracellular Ca^{2+}, and DG activates protein kinase C (PKC). The increased cytoplasmic Ca^{2+} concentration activates specific proteins [troponin C (TnC)] or protein kinases [Ca^{2+}–calmodulin kinase (Ca^{2+}/CaM kinase)]. IP_3 can be further phosphorylated to inositol tetrakisphosphate (IP_4), and DG can be metabolized by a specific lipase to yield arachidonic acid (AA) for prostaglandin (PG) synthesis.

are termed A-kinases and their effect is to phosphorylate (that is, to add a phosphate group to a particular amino acid) particular proteins. These substrate proteins may be enzymes, structural proteins such as ribosomes, nuclear proteins such as histones or membrane proteins. Once phosphorylated, these protein molecules may adopt a different shape. For enzymes this change in shape may underlie the activation or inhibition of catalytic properties, for membranes it may result in changes in permeability to ions (opening or closing of ion channels). Another group of membrane proteins that may be similarly phosphorylated are receptors. In this state the receptor may no longer be able to bind first messenger molecules, in effect the receptors become inactive. This seems to be one of the steps involved in the down-regulation of receptors (recall Section 3.2.2) and may be part of the process by which intercellular communication is terminated at the target cell.

Table 3.2 gives some examples of the responses initiated by increasing intracellular cyclic AMP in various cells.

Table 3.2 Responses of some target cells to particular first messengers, via the cyclic AMP pathway

First messenger	Target	Cellular response
thyroid-stimulating hormone	thyroid gland	thyroxin secretion
adrenalin	skeletal muscle	glycogen breakdown
adrenalin	fat cells	increased lipid breakdown
adrenalin	smooth muscle	relaxation

The effects of cyclic AMP are terminated by the action of a cytoplasmic enzyme called **cyclic AMP phosphodiesterase**, which degrades cyclic AMP to inactive AMP. Under resting conditions cyclic AMP is degraded as fast as it is produced. When production of cyclic AMP is stimulated, the capacity of the cell's phosphodiesterase is soon exceeded allowing the intracellular level of cyclic AMP to rise. Once adenylate cyclase is inactivated, either by the removal of the stimulus or by depletion of the enzyme, the level of cyclic AMP falls as the phosphodiesterase is once more able to metabolize the cyclic AMP molecules.

Our understanding of the cyclic AMP second messenger system has been aided by the fact that there are a number of agents that can be used to manipulate various parts of the pathway. Caffeine and theophylline, for example, inhibit the phosphodiesterase that degrades cyclic AMP. Other compounds, such as forskolin and the toxin from the cholera bacillus, stimulate cells to produce cyclic AMP in the absence of a first messenger. Forskolin, an organic molecule isolated from the root of an Indian herb, activates adenylate cyclase directly whereas cholera toxin activates the stimulatory G-protein (G_s) coupled to adenylate cyclase. It is suggested that this is the reason for the severe diarrhoea suffered by cholera victims as cyclic AMP stimulates fluid secretion in the intestine.

◇A particular first messenger stimulates the production of cyclic AMP in a target cell. What would happen to the intracellular concentration of cyclic AMP if the same cell is exposed to the same first messenger in the presence of caffeine?

◆ The effect of an exposure to the first messenger plus caffeine would be to increase the intracellular cyclic AMP concentration to a level above that found with the first messenger alone. Caffeine inhibits cyclic AMP phosphodiesterase thus blocking cyclic AMP breakdown. This amplifies the effect of the first messenger on intracellular cyclic AMP levels.

3.3.2 Calcium ions as a second messenger

It has been known for some time that Ca^{2+} regulates a variety of cellular processes. The difference between the intracellular (approximately 10^{-7} mol l^{-1}) and extracellular (approximately 10^{-3} mol l^{-1}) concentrations means that a stimulus could increase the cytoplasmic Ca^{2+} concentration simply by opening Ca^{2+} ion channels in the membrane and allowing Ca^{2+} to enter the cell down a concentration gradient. In some situations, for example the depolarization of a nerve terminal and the subsequent release of neurotransmitter, Ca^{2+} influx occurs through voltage-dependent ion channels (Chapter 1, Section 1.4). Ca^{2+} can also enter cells via ligand-dependent channels. Receptors for some first messenger molecules are closely coupled to membrane Ca^{2+} channels such that when the receptors are activated the channel is opened transiently allowing Ca^{2+} to enter the cell. Because the opening of such an ion channel results in the influx of many molecules of Ca^{2+}, the original interaction of first messenger and receptor is amplified. However, the interactions of many first messengers and their receptors lead to an increase in cellular Ca^{2+} concentration that is *not* the result of the opening of membrane Ca^{2+} channels. In these cases the Ca^{2+} used as an intracellular signal is released from stores within the target cell. Precisely how neurotransmitters and hormones release this stored Ca^{2+} is currently (1990) the subject of intense research. This is a rapidly expanding area of investigation, and so the second messenger pathway for Ca^{2+} shown on the right-hand side of Figure 3.3 should be regarded purely as a summary of the salient points as we currently understand them.

The remarkable feature of this second messenger system is that receptor stimulation results in the metabolism of one of the constituents of the cell membrane. This is a minor membrane phospholipid, which is unusual in that it contains an inositol sugar portion that carries three phosphate groups (PO_4^-). This lipid is called **phosphatidylinositol 4,5-bisphosphate** (PIP_2) and is one of a group of inositol-containing phospholipids called the phosphoinositides.

As in the cyclic AMP pathway, activated receptors on the cell surface are coupled to an enzyme on the inner face of the membrane by a G-protein. It is not known whether this G-protein (G_p) exists in both stimulatory and inhibitory forms, but it is clear that it is different from that which couples receptors to adenylate cyclase. The enzyme activated is termed **phospholipase C** and it cleaves PIP_2 to yield two molecules, **inositol trisphosphate** (IP_3) and **diacylglycerol** (DG) (Figure 3.4).

IP_3 is soluble in the cytoplasm. In contrast, diacylglycerol is a lipid molecule so it does not enter the cytoplasm but remains in the membrane.

IP_3 functions as the intracellular signal responsible for initiating the release of Ca^{2+} from the cell's internal stores. It does this by interacting with specific IP_3 receptors on the storage pool which, in most cells, is situated in the endoplasmic reticulum. The precise molecular mechanism involved in IP_3-stimulated Ca^{2+} release has yet to be fully elucidated, but evidence suggests the involvement of regulatory G-proteins in this process. The cytoplasmic

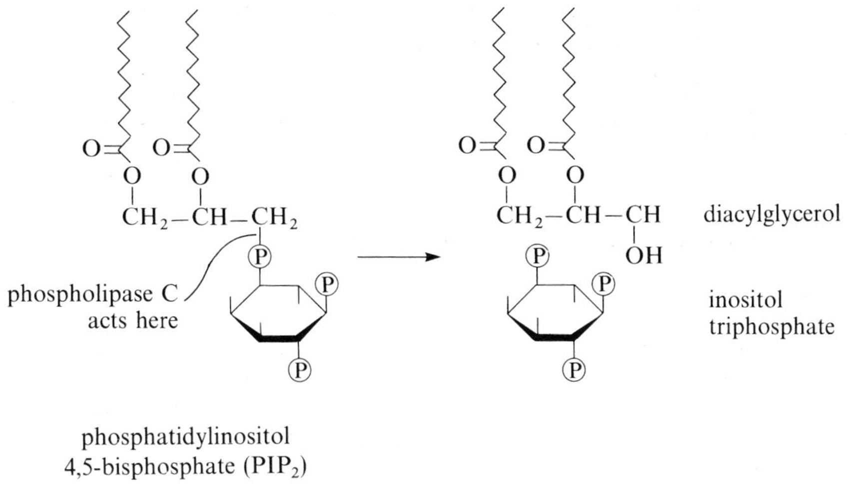

phospholipase C acts here

phosphatidylinositol 4,5-bisphosphate (PIP_2)

diacylglycerol

inositol triphosphate

Figure 3.4 The molecular structures of inositol trisphosphate (IP_3), diacylglycerol (DG) and their parent molecule phosphatidylinositol 4,5-bisphosphate (PIP_2).

Ca^{2+} concentration is raised from $10^{-7}\,mol\,l^{-1}$ in the resting state to approximately $10^{-6}\,mol\,l^{-1}$ during stimulation. This increase is transient, owing to the limited supply of stored Ca^{2+} available for release.

An increase in the cytoplasmic Ca^{2+} concentration can therefore be achieved either by releasing Ca^{2+} from stores within the cell or by opening Ca^{2+} channels in the cell membrane. This increase is the stimulus for the activation of specific protein kinases, such as Ca^{2+}–calmodulin kinase, which, like the A-kinases, are the effectors responsible for initiating the cellular response—some examples of which are given in Table 3.3. Some proteins, such as troponin C, which is involved in muscle contraction, are activated directly by Ca^{2+}.

Table 3.3 Responses of some target cells to particular first messengers, via the PIP_2 pathway

First messenger	Target	Cellular response
acetylcholine	smooth muscle	contraction
acetylcholine	pancreas	amylase secretion
growth factors	fibroblasts	DNA synthesis
thyrotropin-releasing hormone	anterior lobe of pituitary gland	prolactin secretion

The effect of IP_3 is terminated by the action of specific cytoplasmic phosphatases, which sequentially remove its phosphate groups to leave free inositol which is then reincorporated into the membrane phosphoinositides. Intracellular Ca^{2+} homeostasis is maintained by the removal of Ca^{2+} from the cell via membrane-bound pumps, and by its sequestration into organelles such as mitochondria.

A very recent and important discovery is that IP_3 can be further phosphorylated to inositol tetrakisphosphate (IP_4). This molecule also appears to act as an intracellular messenger although its precise role is still subject to debate. Some believe that it acts to open membrane Ca^{2+} channels. The resultant influx of extracellular Ca^{2+} could then replenish the intracellular Ca^{2+} pool and maintain cellular responses set in motion by the initial release of cytoplasmic Ca^{2+}. Others suggest that IP_4 may serve a similar function by redistributing Ca^{2+} within the cell.

As already stated, receptor-stimulated PIP_2 breakdown yields two molecules, IP_3 and diacylglycerol. The latter, being a lipid, remains in the cell membrane where it activates a specific protein kinase called protein kinase C. Protein kinase C is a multifunctional enzyme responsible for regulating a wide variety of cellular responses, such as glycogen metabolism in the liver and cell growth and differentiation. It also appears to be involved in receptor down-regulation where it acts, like the cyclic AMP-dependent protein kinases mentioned in Section 3.3.1, by phosphorylating receptors thus rendering them inactive.

Diacylglycerol is rapidly degraded by a specific kinase and recycled into the membrane phosphoinositides. However, in some cells it is metabolized via a lipase enzyme which liberates a certain portion of its structure. This portion is arachidonic acid, and it is the precursor of another group of intercellular messenger molecules called **prostaglandins**. These molecules are involved in the control of a number of cellular responses, such as inflammation, regulation of the vasculature and the reproductive cycle (see Chapter 10).

Clearly this second messenger system is more complex than that involving cyclic AMP. Nonetheless, it is open to manipulation with pharmacological agents. For example, the intracellular Ca^{2+} concentration can be elevated with the use of a substance, called an ionophore, that shields the electric charge of Ca^{2+} ions and carries them across the membrane. The oil extracted from the seed of a particular tree found in southeastern Asia contains a group of substances called phorbol esters, which directly stimulate protein kinase C in the same way as diacylglycerol.

With the use of these agents, research has established that in many cases the formation of both IP_3 and diacylglycerol is required to initiate a cellular response—in other words, they act synergistically. A notable example of this is the regulation of cell growth. Diacylglycerol activates protein kinase C, which in turn stimulates a membrane-bound pump that removes protons (H^+) from the cell. The resultant rise in cytoplasmic pH, in conjunction with the IP_3-induced increase in the cytoplasmic Ca^{2+} concentration, is thought to contribute to the synthesis of RNA which prepares the cell to synthesize DNA. Any deficit in this pathway could result in abnormal or uncontrolled cell growth, perhaps leading to the formation of a cancerous tumour.

3.3.3 Messenger RNA as second messenger

Steroid receptors are probably composed of two subunits, each of which binds one steroid hormone molecule (Figure 3.5). When both receptor sites R_A and R_B are occupied by the appropriate steroid, the hormone–receptor complex migrates to the cell nucleus. The ensuing events are most completely worked out for the progesterone receptor found in the chick oviduct cells, but other steroid hormones appear to have a similar mode of action. The two subunits of the progesterone receptor are not identical and have different functions, although they are similar in size. One subunit (B) ensures that the hormone–receptor complex binds to a specific acidic protein that forms part of chromatin. A and B dissociate and the A subunit then interacts directly with the adjacent region of the DNA molecule. This results in the transcription of this section of DNA and the subsequent production of many mRNA molecules. In this system, mRNA molecules constitute the second messengers and polysomes are the effectors responsible for the cellular response. The thyroid hormones seem to act in a similar manner except that their receptors are sited permanently in the cell nucleus.

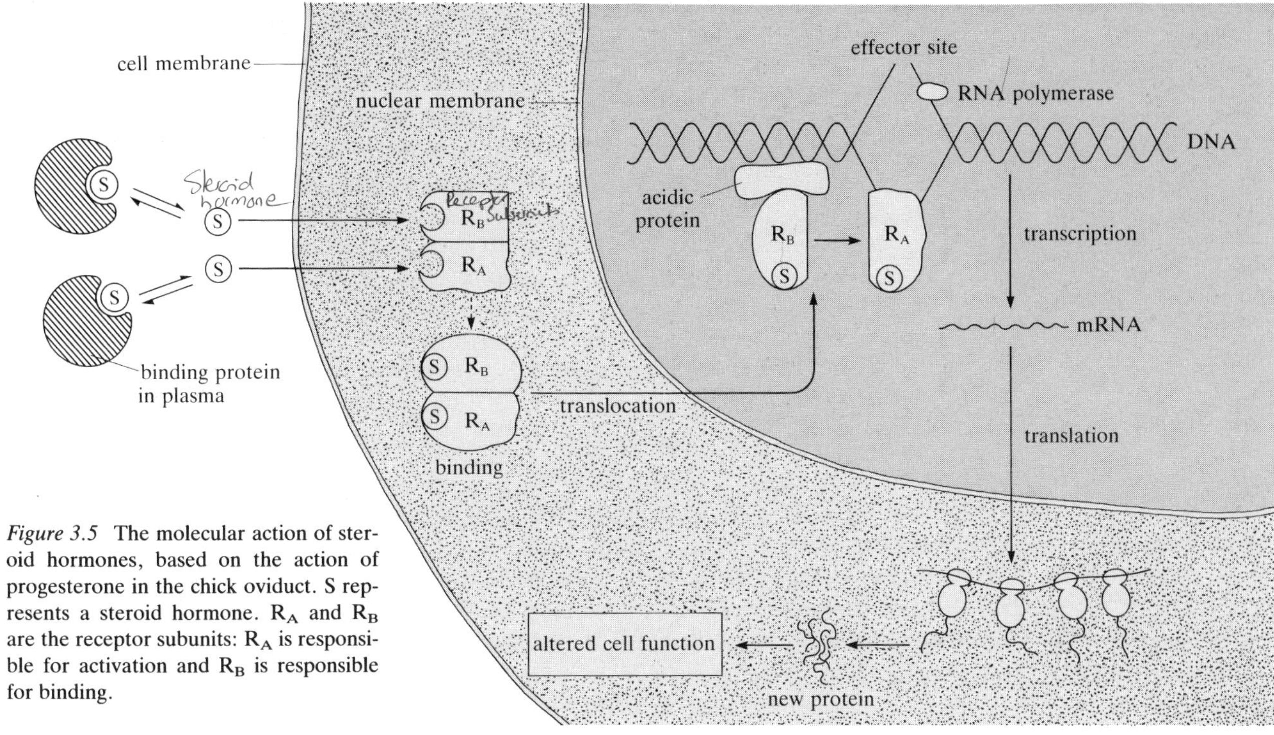

Figure 3.5 The molecular action of steroid hormones, based on the action of progesterone in the chick oviduct. S represents a steroid hormone. R_A and R_B are the receptor subunits: R_A is responsible for activation and R_B is responsible for binding.

3.4 INTERACTIONS BETWEEN SECOND MESSENGER SYSTEMS

It should not be assumed that a particular cell type possesses only one of the second messenger systems so far described, or that each system mediates a cellular response entirely independently of another. It is the interplay between second messenger systems that allows the fine control over biological processes.

There are, for example, many kinds of interaction between cyclic AMP and Ca^{2+}. In smooth muscle cells Ca^{2+} causes muscle contraction whereas cyclic AMP causes relaxation. Here the two systems are antagonistic, one initiates an effector process and the other terminates it. But equally their effects may be synergistic or sequential (the stimulation of one second messenger system is required before the other can be activated). The mechanisms underlying these interactions are complex and beyond the scope of this chapter. However, there is one simple example that will give you some idea of how these interactions take place. In some cells a rise in cytoplasmic Ca^{2+} leads to an inhibition of cyclic AMP production. This seems to occur because Ca^{2+} activates Ca^{2+}–calmodulin, which in turn activates cyclic AMP phosphodiesterase. By increasing the activity of the enzyme that degrades cyclic AMP, the Ca^{2+} signal inhibits the cyclic AMP signal.

SUMMARY OF CHAPTER 3

1 Changes in the activity of target cells in response to hormonal or nervous signals involve the binding of a hormone or neurotransmitter (first messenger) to specific receptors situated either in the cell membrane or within the cell. Receptors for neurotransmitters and polypeptide hormones are usually

situated on the outside of the cell membrane, whereas receptors for steroid and thyroid hormones are usually situated within the cell cytoplasm or nucleus. Binding of hormones and neurotransmitters to their receptors is characterized in the same way as the interactions of enzymes and their substrates. Consequently, the density of particular receptor sites (B_{max}) and their affinity (K_d) can be assessed.

2 The action of a particular hormone or neurotransmitter can often be either mimicked or blocked by structurally similar chemical compounds; such compounds are called agonists and antagonists, respectively.

3 The sensitivity of a target cell to a particular hormone or neurotransmitter can be modified by changes in the density of its receptor population. Such changes are regulated by a variety of factors (e.g. hormone levels in the blood, the duration of exposure to the first messenger, the developmental stage of the cell, etc.). The overall term for this process is receptor regulation and it can take two forms; an increase (up-regulation) or a decrease (down-regulation) in receptor numbers.

4 The interaction of a first messenger and its receptor is often translated into cellular action via the formation of intracellular signal molecules called second messengers. Second messengers serve three functions: (i) they convert the extracellular signal into an intelligible intracellular signal; (ii) they serve to amplify the extracellular signal; (iii) they modulate specific effector systems, which initiate the cellular response.

5 The majority of receptors for polypeptide hormones and neurotransmitters are coupled, via regulatory G-proteins, to one of two intracellular second messenger systems; one involves cyclic AMP and the other employs Ca^{2+}. In the former, receptors are coupled to an enzyme, adenylate cyclase, which converts ATP into cyclic AMP, which then acts as an intracellular second messenger. In the latter, the cellular responses initiated by a first messenger are mediated by an increase in the cytoplasmic concentration of Ca^{2+}.

Increased cytoplasmic Ca^{2+} can be achieved either by the opening of Ca^{2+} channels in the cell membrane, which allows Ca^{2+} to enter the cell down a concentration gradient, or by releasing Ca^{2+} from intracellular stores, a process that is mediated by the receptor-stimulated breakdown of a membrane phospholipid (PIP_2) to yield the second messenger IP_3. This process also yields another second messenger molecule, diacylglycerol, which activates a specific protein kinase called protein kinase C.

Steroid and thyroid hormones employ messenger RNA molecules as second messengers.

6 The overall response of a target cell to a particular first messenger depends upon the type of effector system possessed by the cell. These effector systems are often protein kinases which serve to modify the activity of, for example, other enzymes, or ion channels.

7 The effects of the natural first messenger on these second messenger systems can be mimicked by pharmacological agents which, for example, may raise or lower the levels of cyclic AMP and/or Ca^{2+} within the cell.

8 There are often complex interactions between second messengers in the control of biological processes.

Now attempt Questions 1–5, on pp. 77–78.

OBJECTIVES FOR CHAPTER 3

Now that you have read this chapter you should be able to:

3.1 Define and use, or recognize definitions and applications of each of the terms printed in **bold** in the text.

3.2 List the major steps between the release of a first messenger molecule and the response elicited in the target cell.

3.3 Interpret simple data relating to the kinetics of the binding of first messengers to their receptors on target cells. (*Questions 1 and 3*)

3.4 Design a simple experiment to test whether a drug is a receptor agonist or an antagonist. (*Questions 2 and 3*)

3.5 Describe with simple diagrams the steps involved in the generation of two intracellular second messengers. (*Question 4*)

3.6 Interpret simple data from an experiment designed to determine which second messengers are involved in regulating a particular cellular response. (*Question 5*)

QUESTIONS FOR CHAPTER 3

Question 1 (*Objective 3.3*) The binding of hormones and neurotransmitters to their receptors can be characterized in the same way as interactions of enzymes and substrates. The K_d for the binding of T_3 (triiodothyronine) to thyroid hormone receptors in the nucleus is 10^{-11} mol l^{-1} whereas for T_4 (thyroxin) the K_d is 10^{-10} mol l^{-1}. Explain what this difference indicates and suggest which of the two hormones would be the more effective.

Question 2 (*Objective 3.4*) Substance Z is a natural messenger that causes the synthesis of a particular protein in rat liver cells through the agency of cyclic AMP. You wish to produce, synthetically, a substance very like Z, with the same effect, and end up with four compounds (A–D) that are similar in structure to Z. How would you set about testing your four candidates?

Question 3 (*Objectives 3.3 and 3.4*) Carrying out the experiments in Question 2 gives the results in Table 3.4. Classify each of the compounds (A–D) as an agonist or an antagonist or neither.

Table 3.4 The effects of substances Z and A–D on the level of cyclic AMP and protein production in liver cells

Substance	Protein production (% increase)	Change in cyclic AMP (% increase)
Z alone	100	100
A alone	no change	no change
B alone	50	50
C alone	no change	no change
D alone	100	100
Z plus A	no change	no change
Z plus B	125	115
Z plus C	100	100
Z plus D	175	175

Question 4 (*Objective 3.5*) Which of the following statements are incorrect and why?

(i) Cyclic AMP and inositol trisphosphate (IP_3) are inactivated by phosphodiesterase and phosphatase enzymes, respectively.

(ii) Receptors for steroid hormones are invariably found on the surface of cell membranes.

(iii) The role of IP_3 is to increase the cytoplasmic concentration of Ca^{2+} by opening Ca^{2+} channels in the cell membrane.

(iv) Regulatory G-proteins are involved in the transduction of some signals across cell membranes.

(v) Adenylate cyclase acts to amplify external signals by producing large numbers of cyclic AMP molecules within the cell.

(vi) Protein kinase C is activated by IP_3.

Question 5 (*Objective 3.6*) You wish to determine which second messenger systems are involved in mediating three different responses (A, B and C) in a particular cell type. You are given some forskolin and a Ca^{2+} ionophore to manipulate the second messenger pathways, and your experiments give the results in Table 3.5 (100% = maximum response). From these data you should be able to conclude which second messenger systems are involved in mediating each response.

Table 3.5 The effect of forskolin and a Ca^{2+} ionophore on responses A, B and C. The values are percentages of the maximum effect

Agent	Response		
	A	B	C
forskolin alone	100	0	25
ionophore alone	0	100	25
forskolin plus ionophore	50	100	100

REGULATION AND CONTROL ◆ CHAPTER 4 ◆

4.1 INTRODUCTION

Such things as the rate at which urine is formed by the kidney, the heart rate, the concentration of sodium in the blood and the rate at which sweat is formed are generally termed physiological **variables**. A variable might suddenly change its value, as in the increase of heart rate upon performing vigorous exercise. However, another variable might show near constancy over lengthy periods of time, e.g. human body temperature. No matter how much actual variation is observed and under what conditions, the term 'variable' serves to direct our attention to both actual and *potential* changes in the thing being observed.

Other things are safely assumed to remain constant during the course of an observation. For instance, in describing physiological changes in the body during exercise, we might reasonably and safely assume that the bones forming the skeleton do not change in length. Similarly, in studying digestion of a meal, we would also assume that the gut remains a constant length. Such things as the length of a bone would be termed **parameters**. Clearly one cannot always draw a completely hard-and-fast distinction between variables and parameters; usage can change according to exactly what it is that we are trying to explain. However, for most practical purposes, a distinction can usually be drawn.

Now consider the state of your body over a period of time and what happens to some of its better-known physiological variables, such as temperature, urine flow-rate and fluid volume. Some variables show near constancy over long periods of time, in spite of the body engaging in widely different activities and being subjected to environmental stresses. By contrast, other variables fluctuate enormously. For example, your body temperature and the sodium concentration of your blood plasma are fairly constant, whereas your heart rate can show massive increases at times of physical exertion. Sweating rate varies enormously in response to changes in the environmental temperature that might be associated with only very slight changes in core body temperature. Similarly, the rate of urine production alters by large amounts according to our ingestion of fluids. A similarly wide variation can be seen in *behaviours* that have consequences for physiological variables. For instance, you might sit motionless on a hot day, minimizing heat generation, or engage in vigorous exercise to generate heat in a cold environment.

In general, the physiological processes and behaviours mentioned would be referred to as examples of **regulation** and **control**. These are terms in common use in such disciplines as engineering, economics and psychology, as well as in various branches of biology. However, the terms are often used interchangeably and sometimes with quite different meanings by different authors. In this chapter, we shall concentrate on physiological systems and propose a logical usage of the terms 'regulation' and 'control'. The principles that emerge from the chapter should, with suitable adaptation, be relevant to other areas of biology.

When you consider a number of the variables of the human body, you will appreciate that the near constancy of some of them is vital to survival. For example, the human body can only function optimally provided that body temperature remains close to 37 °C. A departure far from this in either direction can soon become lethal. Contrast this with, say, the rate at which the body produces sweat or urine. Both of these rates can vary very widely according to circumstances, and with no detrimental consequences. Indeed, their wide fluctuation is necessary in serving the constancy of body temperature and body-fluid volume, respectively. It is imperative that some physiological variables remain fairly constant at what is assumed to be an **optimum value** for efficient functioning of the body. Action is automatically taken when they depart far from this value, and the action serves to return them to normal. Such variables are classified as being *regulated.* Changes in other physiological variables (e.g. sweating) act in the service of maintaining the constancy of the regulated variable; these physiological variables are classified as being *controlled.*

◇ Classify (a) body-fluid volume, (b) drinking and (c) formation rate of urine as being either regulated or controlled variables.

◆ Body-fluid volume is classified as being a *regulated* variable. It is biologically imperative that this is maintained nearly constant, and there are mechanisms that serve this end. By the exertion of *control* over the physiological variables, drinking and formation rate of urine, body-fluid volume is regulated.

In the sections that follow, we shall examine how some physiological variables are regulated by the exertion of control over other physiological variables.

Summary of Section 4.1

1 It is vital to the efficient functioning of the body that some physiological variables are held at nearly constant values.

2 Regulated physiological variables (e.g. body temperature) are those that are held at or near constant values as a result of changes in other physiological variables.

3 Controlled physiological variables (e.g. sweating) are those that serve to maintain constancy of a regulated physiological variable (e.g. body temperature).

4 In serving regulation, controlled physiological variables can fluctuate widely depending upon circumstances.

4.2 PRINCIPLES OF FEEDBACK AND HOMEOSTASIS

There are often several processes involved in maintaining the constancy of a regulated physiological variable. One such is described by the term **feedback**. Most people have some idea of what this means but have great difficulty in defining exactly what the term signifies. For example, a distance-teaching university might ask for feedback from the students taking a course. Figure 4.1 illustrates this. There is a flow of information from the university to the students, e.g. books and television programmes sent out. This is termed the **forward path**. In a good teaching system, there is also a flow back from the

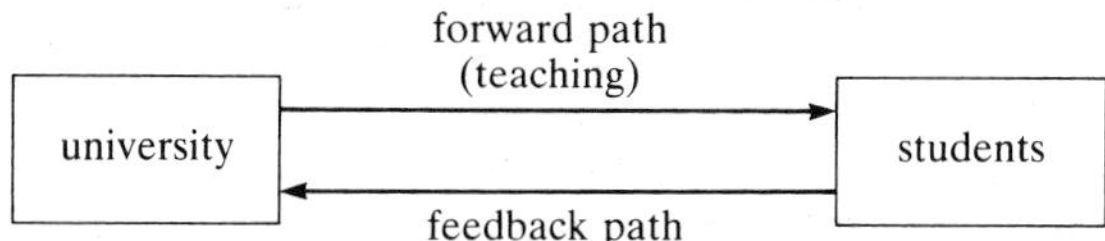

Figure 4.1 The principle of feedback. Information flows in one direction from the university teacher to the students. In the opposite direction, the feedback path, information travels from the students back to the teacher.

students to the university, e.g. surveys of the students' comprehension of each course. This is the **feedback path**. If information in the feedback path influences that in the forward path then a **feedback system** is created.

A feedback system is one in which there is not a simple, one-way sequence of events from cause to effect, but rather where the effect feeds back to influence the cause. For example, in a distance-teaching university the quality and content of courses would, one hopes, be modified according to reports obtained on the students' understanding of them. Figure 4.1 shows that a loop is formed in feedback systems, and another expression that is commonly used to refer to a system involving feedback is **closed loop**. The concept of feedback is employed in considering metabolic pathways. The principles underlying feedback are essentially the same in both biochemical and physiological domains. However, the terminology used here, deriving from engineering control principles, is perhaps more appropriate to the physiological level.

Systems involving feedback can be intimidatingly complex but, somewhat deceptively, they can also be very simple. The meaning of 'feedback' can be well illustrated by the simple examples in Figure 4.2. Two broad categories of feedback exist, and these are represented in this figure. Take the situation shown in part a. A bar rotates around a pivot. Suppose we displace this bar from its resting position, the vertical, by an angle θ. A force tending to return the bar to the vertical (labelled F) is thereby introduced. The feedback or closed loop aspect of this system is that (i) a displacement (D) from vertical creates a restoring force (F) (i.e. $D \longrightarrow F$) and (ii) the restoring force influences the displacement ($F \longrightarrow D$). Compare this with Figure 1.

The system of Figure 4.2a shows **negative feedback**. The qualifying word 'negative' is used to describe the fact that the deviation from normal creates a force that tends to *eliminate* the deviation. Similarly, one would hope that student feedback helps to correct deviations from the university's expectations concerning its teaching material. Now compare the system shown in Figure 4.2a with that shown in part b, where the bar is balanced precariously above the pivot and then a deviation from vertical occurs. Still the deviation creates a force ($D \longrightarrow F$). Still the force so created exerts an influence on the deviation ($F \longrightarrow D$). However, this time the deviation, rather than being eliminated, is strengthened. Force creates θ, which creates more force, which enlarges θ... Because of this self-reinforcing effect, in contrast to self-correction, the system property is termed **positive feedback**.

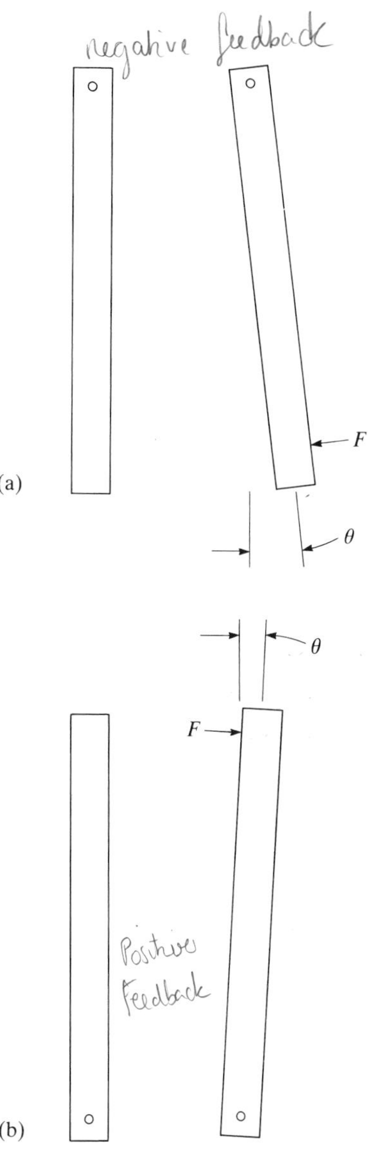

Figure 4.2 The two types of feedback. (a) Negative feedback: to the left, the bar is in its resting position; any disturbance D (as measured by θ, the angle of deviation of the bar from the vertical) creates a force F that corrects the disturbance. (b) Positive feedback: the bar is initially balanced in the vertical position; any disturbance D (angle θ) creates a force F that reinforces the deviation.

Feedback is evident then in simple and complex systems. This chapter is concerned with a particular type of system, in which a physiological variable is regulated by the exertion of control over other variables. Where negative feedback is involved in serving the function of regulation and control we speak of a **negative feedback control system**. For example, in humans, an elevation in body temperature causes sweating. The subsequent evaporation of the sweat cools the body and thereby corrects the deviation from normal temperature.

◇ A person ingests a litre of water and then, in response to the swelling of body fluids, forms a relatively large amount of urine. Is there negative or positive feedback present in this system?

◆ The deviation from normal causes action to be taken of a kind to eliminate the deviation, so this system shows the property of negative feedback.

As an example of both negative and positive feedback, we can consider the pumping effectiveness of the human heart following haemorrhage. Figure 4.3 shows this. The normal human heart pumps about 5 litres of blood per minute. Suppose there is a loss of 1 litre of blood. As shown, pumping effectiveness falls sharply. However, it recovers to normal in about 2 hours. This implies that negative feedback processes for maintaining arterial pressure and cardiac output have been actuated and are effective. Now contrast this with what happens when the blood loss is 2 litres. Pumping effectiveness falls over the 2 hour period and leads to death. This decline does not mean that negative feedback mechanisms were wholly ineffective. The fall in cardiac output would have been the trigger for negative feedback action just as when 1 litre was lost. It does mean that the negative feedback effects were overridden by other factors and these have the property of positive feedback. The positive feedback loop is as follows. As a result of loss of blood, arterial pressure falls. As a consequence of this, flow of blood to cardiac muscle through the coronary vessels falls. As a result of the reduction in blood supply to the heart, cardiac pumping efficiency declines, which further reduces the blood supply to heart muscle, and so on, until death. The term 'vicious circle' aptly summarizes the nature of positive feedback in a case such as this. We shall have rather little to say about positive feedback loops in this chapter, but it is important to recognize their existence. They often arise in situations where things go wrong, as in the example of Figure 4.3, but they also form part of some normally functioning biological systems. The remainder of the chapter concerns negative feedback.

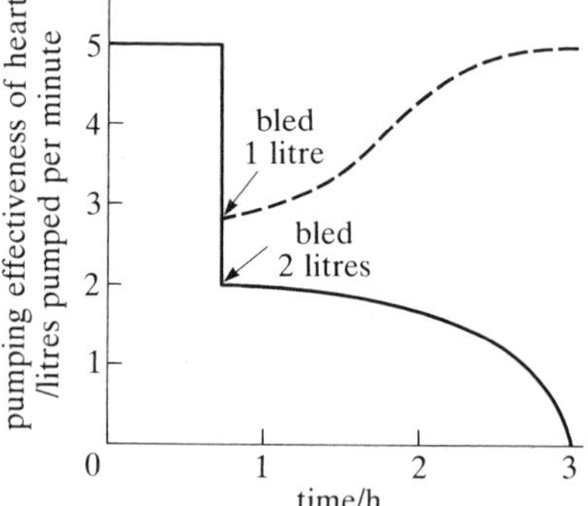

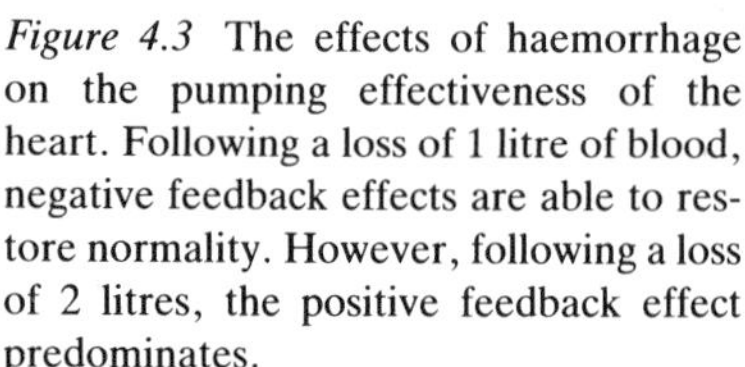
Figure 4.3 The effects of haemorrhage on the pumping effectiveness of the heart. Following a loss of 1 litre of blood, negative feedback effects are able to restore normality. However, following a loss of 2 litres, the positive feedback effect predominates.

In discussions of biological control systems, comparisons are frequently drawn between these systems and those in engineering that also involve negative feedback. Consider the system that regulates the temperature of a house. The temperature that the occupant sets on the dial of the thermostat is termed the **set-point** of the system. The system is built to compare the actual temperature of the house with the set-point temperature. When the actual temperature falls below the set-point, heating is activated and continues until the actual temperature equals the set-point value. In other words, the system might be said to be 'error activated' and, in negative feedback fashion, error is self-eliminating.

How close is the similarity between this system and negative feedback systems in biology? In terms of understanding the principles underlying the performance of the systems, the one is closely analogous to the other. However, if we examine the details of two biological systems, we shall see that caution is sometimes in order in drawing analogies. First, we shall consider a case where a set-point seems unambiguously to be involved. Consider a situation where a subject is asked to point one arm straight ahead. A command is sent from the brain to the muscles controlling the position of the arm. The command constitutes the set-point of the system. The actual position of the arm is compared with the set-point to determine muscular activity. The subject is able to see the arm and make fine adjustments so as to maintain the actual arm position in alignment with the set-point. Suppose that suddenly a weight is attached to the arm. The arm might be lowered slightly but this deviation from the set-point would be detected and would cause increased neuromuscular activity to restore the horizontal position. Suppose now the subject is plunged into darkness but asked to keep the arm horizontal. Although the subject is denied visual feedback on arm position, there are other sources of feedback that are utilized. The state of tension in the muscles is detected by specialized neurons, and information on this state is fed back and compared with the command. Although we might have some difficulty actually locating the set-point within the central nervous system, none-the-less there is good evidence to suppose that the system involves a set-point. Indeed, it is difficult to imagine how it could possibly work without one.

Now consider a description of a hypothetical system that shows negative feedback and that tends to maintain a variable within limits but for which we do not need to build a set-point into the description. The amount of a substance (we will call it substance ZZ) in the blood depends simply upon two factors, its excretion through the kidney and its absorption from the gut. It is always supplied in abundance in the diet and is neither stored nor broken down in the body. Suppose that the rate of excretion of substance ZZ through the kidney is related to its concentration in the blood, according to the graph shown in Figure 4.4a. The relationship between the rate of absorption of substance ZZ from the gut and its concentration in the blood is shown in Figure 4.4b. In Figure 4.4c the graphs of excretion and absorption rates are superimposed to give an estimate of the net rate of uptake of substance ZZ.

Now suppose that the concentration of substance ZZ in the blood is at value B. Excretion rate is equal to absorption rate, so there is no net rate of change. Some of the substance is then injected into the blood, so that concentration rises to value C. Now excretion rate is higher than absorption rate (by value y), and so there will be a net loss of the substance from the body. Concentration will fall until value B is reached. Suppose now that some of the substance is artificially removed from the blood so that concentration falls to value A. Absorption rate is higher than excretion rate (by value x) and so there will be a net gain of the substance. This will continue until concentration rises to value B. So the system behaves in such a way that any departure from value B is associated with variables taking values that tend to return concentration to value B. In this sense, value B acts like a set-point. However, unlike the system of the thermostatic heating of a room, there is no actual physically realizable set-point. Nowhere in describing the system was the operation of comparison of a variable with a set-point mentioned. The point is that systems that are different in their construction can sometimes behave in much the same way.

Should we call value B in Figure 4.4 a set-point? In the literature, there are at least three attitudes taken to the issue of set-points in such a system. Some argue that since there is no set-point having any physical embodiment in a system of this kind, the expression is misleading and should not be incorporated into formal models that purport to explain the biological phenomena. In contrast, others argue that it is the *system property* that is important; the set-point exists in the sense that the system takes action to return itself to this value. The system behaves *as if* a set-point exists. A compromise position on the usage of the expression set-point is that we should employ it where the

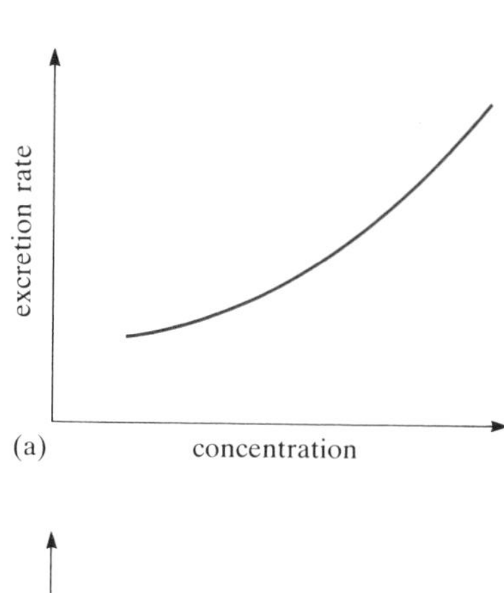

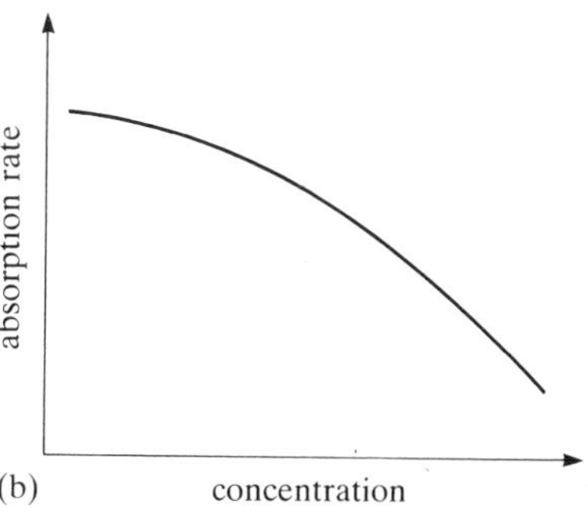

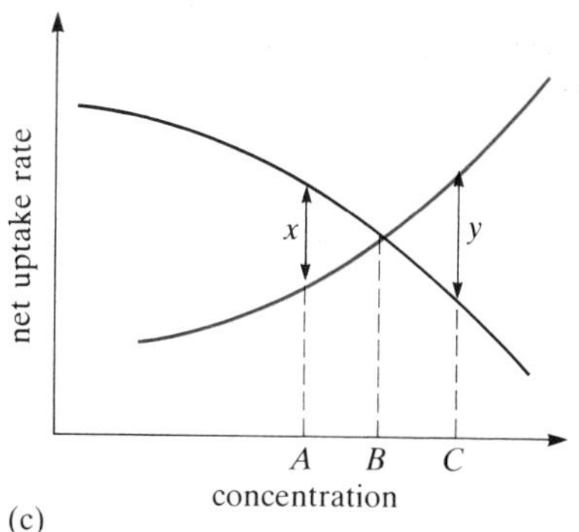

Figure 4.4 A hypothetical system that serves to regulate the concentration of a substance ZZ in the blood: (a) a graph showing excretion rate as a function of blood concentration; (b) a graph showing rate of absorption from the gut as a function of blood concentration; (c) the two curves superimposed to show the net rate of uptake or loss of substance ZZ.

term is felt to yield insight, provided that we qualify carefully the sense in which we are using it. In some cases, if we fail to do this, the expression can indeed be seriously misleading. Perhaps 'set-point' will go on being employed in descriptions of systems of the kind represented in Figure 4.4, but placed in inverted commas.

Stop and think about some biological systems (e.g. the system regulating body temperature, involving sweating), and you will see that the property of negative feedback is very widely built into them. Indeed, so ubiquitous are negative feedback loops that the notion of regulation and control is sometimes taken as almost synonymous with negative feedback control. However, there are ways in which control can be exerted over a regulated parameter other than by negative feedback, as you will shortly see.

The terms 'regulation' and 'control' are closely associated with the physiological concept of **homeostasis**. This term was coined by the American physiologist Walter Cannon, and refers to the constant, or near-constant, levels of some of the vital physiological parameters of the body. The word derives from two Greek words meaning 'similar to standing still'. As defined by Cannon, homeostasis refers to a property of physiological systems, i.e. their observed outcome, the attainment of near constancy. It does not specify the mechanism whereby this outcome is achieved. In developing the concept of homeostasis, Cannon had an intellectual precursor in the famous French physiologist Claude Bernard, who emphasized the importance of maintaining constancy in the internal environment (the *milieu intérieur*) of the body. Bernard coined one of the classic statements of physiological science:

> It is the fixity of the *milieu intérieur* which is the condition of free and independent life, and all the vital mechanisms, however varied they may be, have only one object, that of preserving constant the conditions of life in the internal environment.

Since the pioneering work of Bernard and Cannon, there has been a common tendency to treat the terms 'homeostatic' and 'negative feedback' as virtually synonymous. Thus the expression 'homeostatic behaviour' is commonly used to refer to behaviour that (a) is caused by a deviation from optimum in a physiological variable and (b) serves to correct the deviation. An obvious example would be the drinking of water following a period of dehydration. The existence of negative feedback loops ensures that deviations from optimum tend to be self-correcting. However, in addition to the negative feedback loops, there are other mechanisms for ensuring homeostasis. Uncritical equating of the terms homeostasis and negative feedback can detract from the role played by these other processes. One such process is termed **feedforward**. Feedforward is where action appropriate for correcting a deviation occurs in a sytem but the action is initiated *before* any deviation occurs. In so doing, the system can either pre-empt or at least minimize the deviation. To understand this, suppose, for example, that, on leaving the house, you look out of the window and see that there has been a sudden fall of snow. You will most likely put on a winter-coat before venturing out. This is not in *response to* bodily hypothermia (fall of body temperature), which would be feedback. Rather, by *anticipating* potential hypothermia, the system is able to take action that pre-empts it. This is feedforward. In the discussion that follows you will meet several examples of feedforward control.

◇ The expression 'demand led' is used in the description of the relationship between metabolic need and pumping activity by the heart, as will be discussed in Chapter 6. When the need of the body tissue for oxygen and energy increases, as a result of, say, sudden exertion, the pumping activity of the heart increases. Is this an example of feedback or feedforward?

- It is a feedback system. The heart is responding to the current state of the tissues, not to a future state of the tissue. (Later in this chapter we shall discuss how feedforward also exerts an influence on the circulatory system).

In the service of homeostasis use is made of both (a) *intrinsic* physiological mechanisms that do not involve behaviour and (b) *extrinsic* mechanisms involving the behaviour of the whole animal. The extent to which control resides with these two varies (i) between individual regulatory systems, (ii) from species to species and (iii) according to the current situation for a given animal. For example, both humans and lizards regulate body temperature. Humans do so very effectively by either intrinsic physiological (termed **autonomic**) means (e.g. shivering) or behavioural means (e.g. moving to California). However, a lizard has rather fewer means of exerting control, relying heavily upon behaviour. A lizard will shuttle back and forth between sunshine (where it absorbs heat) and shade (where it loses heat). Thereby, it can regulate body temperature within a range between the extremes that would be reached if it were to remain in either location. Of course, this mode of control would be unavailable on a cloudy day!

As an example of shifting between intrinsic and extrinsic controls, consider the reaction of a rat to an injection of hypertonic saline (a saline solution more concentrated than the body fluids). Both the kidney and the drinking systems will act so as to correct the deviation in body fluids. The kidney produces a urine that has a greater concentration of sodium chloride than the plasma of the blood (i.e. a hypertonic urine). The deviation from normal in body fluids will create the **motivation** for gaining water and drinking. For example, the animal will go to locations where, in the past, water was to be found. In the laboratory, it might perform 'tricks' that it has learned, such as pressing a lever that in the past has earned the animal water. By means of ingested water, the body fluids will be diluted. If drinking water is not available, a greater load will be placed upon the kidney; it will excrete a more strongly hypertonic urine.

Summary of Section 4.2

1 A feedback system is one in which an effect feeds back to influence a cause.

2 A negative feedback system is one in which displacements from a given state tend to be self-correcting.

3 A positive feedback system is one in which displacements from a given state are self-reinforcing.

4 A negative feedback control system is one in which a regulated variable is held close to a constant value by means of negative feedback.

5 Biological systems are analogous to engineering control systems involving a set-point. However, a set-point cannot always be identified in a system and yet the system might behave as if a set-point were built in.

6 The term homeostasis refers to the constancy of the vital physiological variables of the body, such as temperature.

7 Feedforward is action taken to correct a deviation in a system that is taken before the deviation occurs.

8 Both negative feedback and feedforward can play a role in maintaining homeostasis.

9 Both intrinsic physiological controls (e.g. panting) and behavioural controls (e.g. moving to a warmer location) are involved in maintaining homeostasis.

Now attempt Questions 1 to 5, on p. 122.

4.3 THE TECHNIQUES OF SYSTEMS MODELLING

For a considerable time now there has been a healthy cross-fertilization between the biological and engineering sciences. In the present context, the need to design and understand the performance of feedback control systems in engineering (e.g. the thermostatically controlled heating system of a house and automatic landing systems for aircraft) has provided a rich body of theory that can, with adaptation and suitable caution, be applied to biological systems. The input to biology from control systems theory has been particularly apparent in the physiological sciences, and the *American Journal of Physiology* has regular features on systems theory. As another indication of the influence of control systems theory in physiology, it is worth noting that what is now one of the most influential physiology textbooks, Arthur Guyton's *A Textbook of Medical Physiology*, devotes most of its first chapter to this topic.

In this section, the basic terms and nomenclature of control systems theory, are introduced and later applied to a number of biological systems. This chapter will employ the form of representation known as the **block diagram**. In this representation, the variables of the system are each represented by an arrow.

Figure 4.5 a shows an example of a block diagram. There are three variables, A, B and C, that are acted upon by a particular kind of 'parameter' (or, to use the standard language of representation of systems, an **operator**). The term 'parameter' here conveys the sense of constancy. Variables might fluctuate widely but the operation carried out on them is a constant one, in order to calculate the value of another variable. In Figure 4.5a the operator is a **summing junction**, which performs the addition of the variables A, B and C to give another variable, Y. The direction of information flow is shown by the arrows; Y depends upon A, B and C. For example, suppose we are interested in representing the total rate of secretion of a particular hormone. As Figure 4.5b shows, the total secretion rate of a hormone might be given by its rate of secretion from nerve endings plus its rate of secretion from a particular gland.

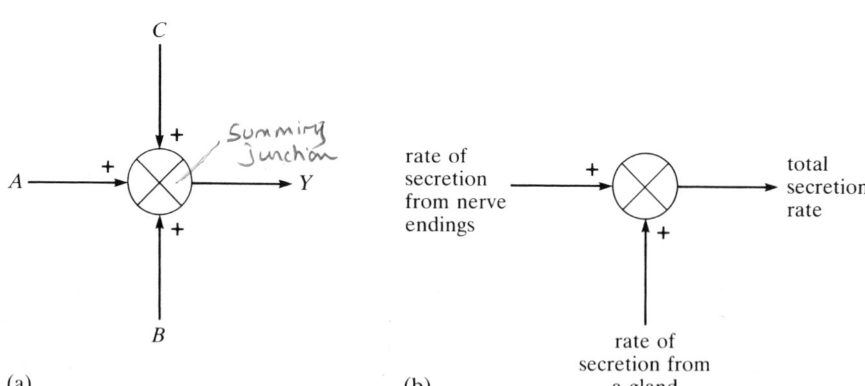

Figure 4.5 (a) General representation of a summing junction. The output Y is given by the sum of the input signals A, B and C. (b) Summing junction representing addition of the component secretions of a hormone to give the total secretion rate.

Figure 4.6a also shows a summing junction, but now, as well as addition, it represents subtraction of one variable from another ($Y = A + B - C$). The term 'algebraic addition' describes the operation shown in Figure 4.6a. Suppose we want to represent the calculation that the *net* rate at which a hormone is entering the blood is given by the rate at which it is being secreted from nerve endings *minus* the rate at which it is being destroyed by enzymes. Figure 4.6b represents this calculation.

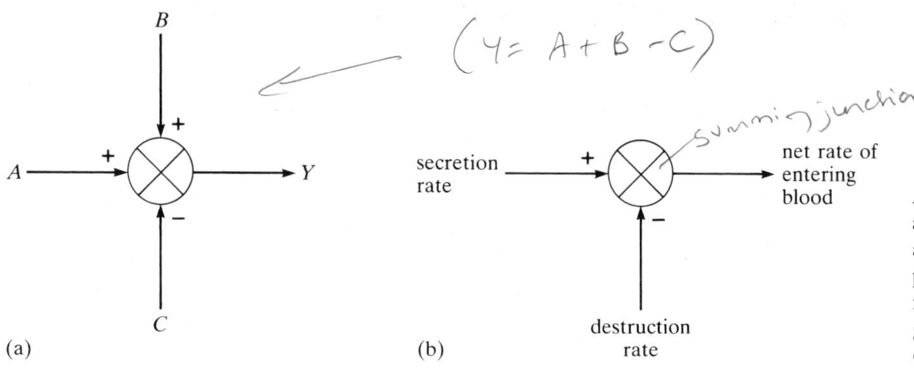

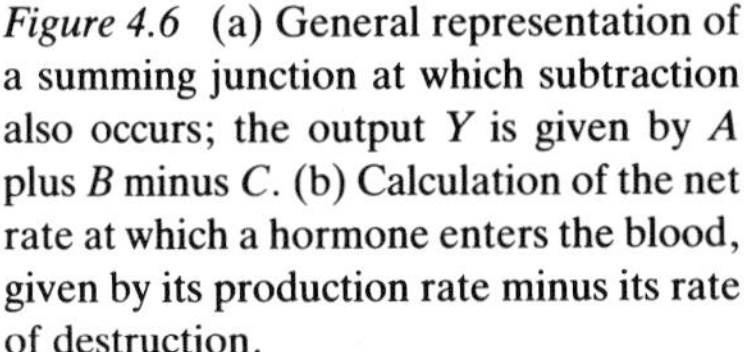

Figure 4.6 (a) General representation of a summing junction at which subtraction also occurs; the output Y is given by A plus B minus C. (b) Calculation of the net rate at which a hormone enters the blood, given by its production rate minus its rate of destruction.

Figure 4.7a represents the process of **integration** (Y is the integral of X). The meaning of this term is best illustrated by example. A bucket of water integrates the flow into it (litres min^{-1}) to give a volume (litres). The blood integrates the net *flow* of a hormone released into it, to give the *amount* of hormone in the blood, as shown in Figure 4.7b.

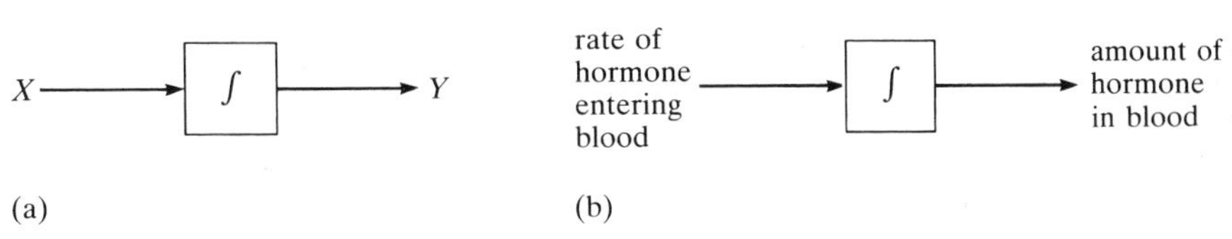

Figure 4.7 (a) General representation of integration; the output Y of this operator (called an integrator) is given by the integral of the input X. (b) An integrator that relates the amount of hormone in the blood to the net flow of the hormone into the blood.

The operation of division ($Y = A/B$) is represented in Figure 4.8a. For example, as Figure 4.8b shows, the concentration of a particular protein that is distributed throughout the blood is given by the amount of the protein divided by the blood volume.

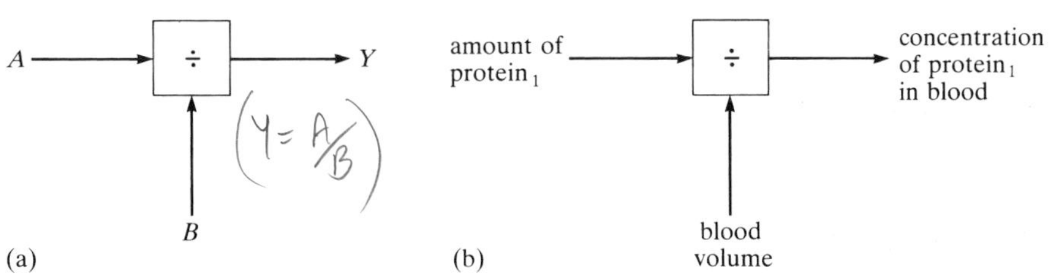

Figure 4.8 (a) General representation of the operation of division. (b) Calculation of the concentration of protein$_1$ by dividing the amount of protein$_1$ by blood volume.

The component operators that make up the representation of a physiological system are normally first identified in this way. They are then assembled to give a representation of the whole system under study. The performance of, and predictions from, the modelled system can then be tested and compared with the real biological system. The simple example shown in Figure 4.9 models the performance of a tank with an outlet at the bottom and a flow in. Figure 4.9a represents the fact that the net flow into the tank is given by flow in minus flow out. Figure 4.9b represents the relationship between net flow in and volume of fluid in the container; this relationship is given by the process of integration. Figure 4.9c represents the fact that the flow out of a container is directly proportional to the volume of fluid in it. The symbol K represents a constant term that defines the relationship between flow out and volume, i.e. flow out $= K \times$ volume. Figure 4.9d assembles these components to give the whole system. One can use the model to make predictions.

Figure 4.9 A water tank with flow in and flow out. (a) The net flow is given by flow in minus flow out; (b) the volume of water in the tank is given by the integral of the net flow; (c) the flow out is given by the volume multiplied by a constant term; (d) the components assembled to give a representation of the whole system. The diagram shows how volume depends upon flow into the container and how a negative feedback loop is implicit in the system.

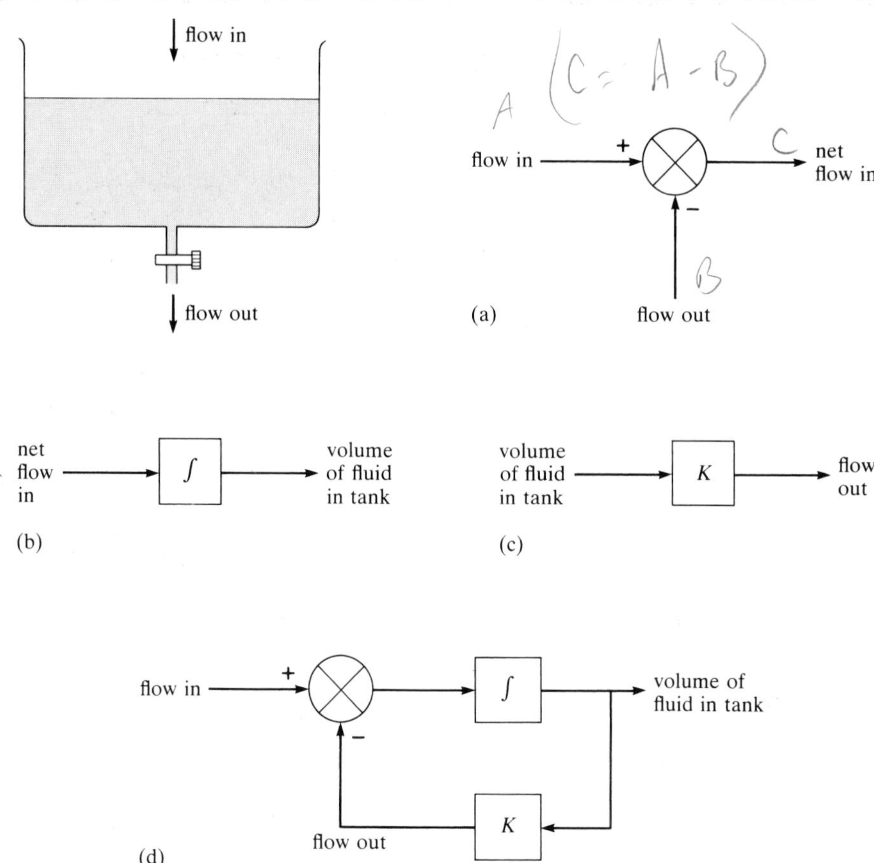

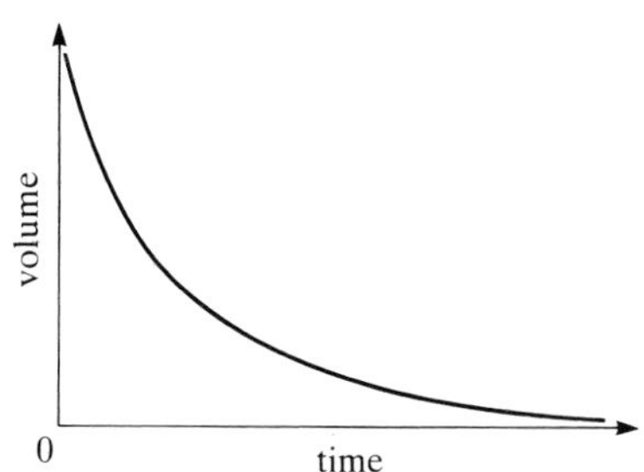

Figure 4.10 The predicted fall in volume if the tank is full and then the tap is opened at time zero.

◇ Suppose the tank is full, the flow in is zero and the outlet tap is closed. Then the outlet tap is opened. What does the model predict as far as the subsequent fluid volume is concerned?

◆ The flow out is in direct proportion to the fluid volume remaining in the tank. Therefore the tank empties as shown in Figure 4.10, the fluid volume falling rapidly at first and gradually more slowly.

◇ Suppose the tank is empty. The outlet tap is open. The inlet tap is opened and a constant flow enters the tank. What is the prediction of the model for fluid volume in the tank?

◆ Fluid volume will rise as shown in Figure 4.11, until flow out is equal to flow in. At this point, fluid volume will remain constant. Since flow out is given by volume multiplied by *K*, the volume rises until the pressure forcing fluid out is sufficient that flow out matches flow in.

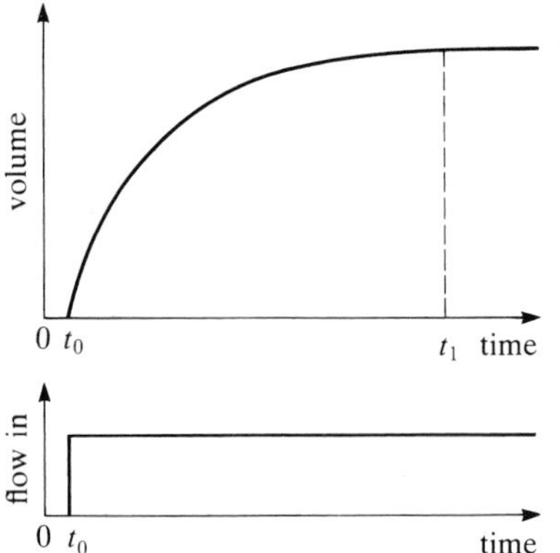

Figure 4.11 The predicted rise in volume following a change at time t_0 in the value of flow in from zero to some constant value.

When the variables of a system are each observed to be maintained at a constant value over a period of time, the system is described as being in a **steady state**. In Figure 4.11, beyond time t_1, steady state conditions prevail.

Figure 4.12 shows a model that represents the basics of the insulin control system for regulating glucose concentration in extracellular fluid, a system that will be discussed further in Chapter 9. The model is a simplification in that much has been left out, but it captures some of the important features of how this negative feedback system functions. An elevation in extracellular

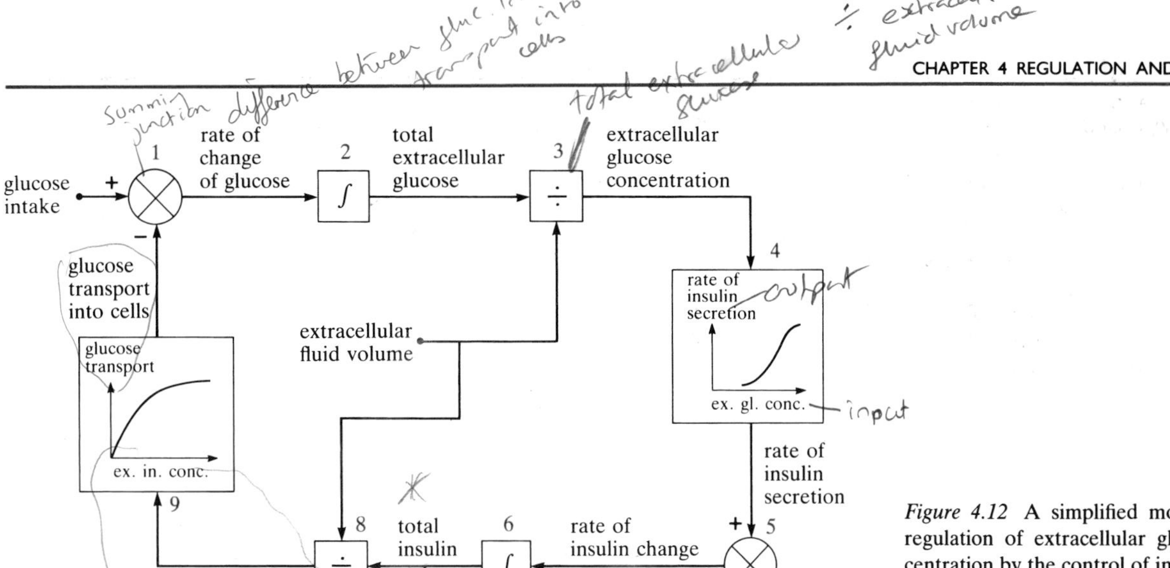

Figure 4.12 A simplified model of the regulation of extracellular glucose concentration by the control of insulin secretion rate. (Abbreviations: ex. in. conc. = extracellular insulin concentration; ex. gl. conc. = extracellular glucose concentration.)

glucose concentration causes an increased rate of insulin secretion, which in turn increases glucose transport into the cells. This lowers extracellular glucose concentration. In Figure 4.12, Block 1 is a summing junction at which the rate glucose enters the body ('glucose intake') has subtracted from it the rate of glucose transport into the cells. The difference is the net rate of glucose entry into the extracellular fluid, which is termed 'rate of change of glucose'. Block 2 integrates this rate to give the total amount of extracellular glucose. Block 3 divides this by extracellular fluid volume in order to give extracellular glucose concentration. Block 4 is a function box, the graph within which shows the relationship between the input, extracellular glucose concentration, and the output, rate of insulin secretion. Block 5 calculates the net rate of insulin entry into the extracellular fluid, which is termed 'rate of insulin change'. This is given by the difference between rate of insulin secretion and rate of insulin destruction. Block 6 integrates this net rate in order to give the total amount of insulin. The rate of insulin destruction is given by a constant K multiplied by the total insulin. Block 7 performs this calculation. Block 8 calculates extracellular insulin concentration, by dividing total insulin by extracellular fluid volume. Finally Block 9 completes the loop; the graph within this function box relates glucose transport into cells to the extracellular insulin concentration.

Guyton compared the predictions from this model with the performance of the real biological system in response to a constant infusion of glucose solution. Figure 4.13 shows the prediction of the model, which is close to what happens in the biological system. Extracellular glucose concentration at first rises and then falls to a level that is slightly above the starting level, but far below the level at which it would be, were it not for the negative feedback effect of the system.

◇ With reference to Figure 4.12, can you suggest a reason why glucose concentration should initially rise above its steady state level and then decline?

◆ There are time delays in the system. It will take time for total insulin to rise to a level at which glucose transport into cells is equal to glucose intake. Until these two are equal, total extracellular glucose will rise in response to the infusion.

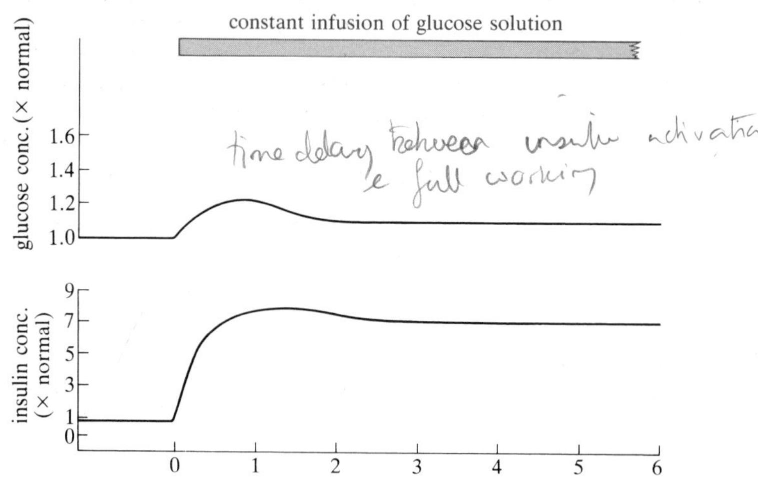

Figure 4.13 The response of the insulin–glucose system to an infusion of glucose.

Guyton's model, though a gross simplification of reality, vividly illustrates important features of how the physiological system functions. It enables us to understand such things as the rise and fall of glucose concentration, shown in Figure 4.13, in terms of the properties of the system.

This section has served as an introduction to the basics of systems modelling. In the three sections that follow, three physiological systems ((a) body temperature regulation, (b) drinking and regulation of body fluids and (c) the system relating feeding and body energy) will be discussed and the use of control theory models illustrated.

Summary of Section 4.3

1 In a block diagram, variables are each represented by an arrow, and parameters by an operator.

2 An operator acts upon one or more variables to give another variable.

3 Physiological systems can be represented by combining a number of operators and variables.

Now attempt Questions 6 to 8, on pp. 122–123.

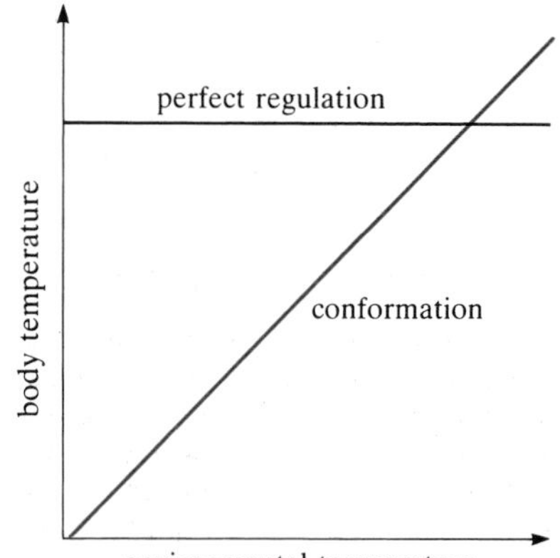

Figure 4.14 Two possible relationships between body temperature and environmental temperature: perfect regulation (black line) and conformation (red line).

4.4 REGULATION OF BODY TEMPERATURE

There is a very narrow range of core body temperatures within which a mammal or bird is able to function efficiently. The width of this range will vary to some extent with the species in question. The rates of biochemical reactions, upon which physiological processes depend, are a function of body temperature. For birds and mammals, life itself is usually only possible provided that body temperature remains within a certain narrow range, a 'range of tolerance'. For other animals, e.g. many reptiles, though the range of tolerance is wider, the physiological and biochemical processes of the body function best within a restricted range of temperature. For instance, in many species, muscles work ineffectively at low body temperatures.

The temperature of the environment in which an animal lives will often vary widely. To what extent does body temperature change as a function of environmental temperature? Figure 4.14 shows two extreme possibilities that might be found when an animal is confronted with different environmental temperatures: perfect regulation and **conformation**. In the case of perfect regulation, the control system would be very sensitive to the slightest deviation from normal body temperature and take corrective action. In the case of conformation, no control action is taken and body temperature conforms to environmental temperature. Between these two extremes there is a range of intermediate possibilities, where some regulation occurs but not enough to maintain a stable body temperature over the whole range of environmental temperature.

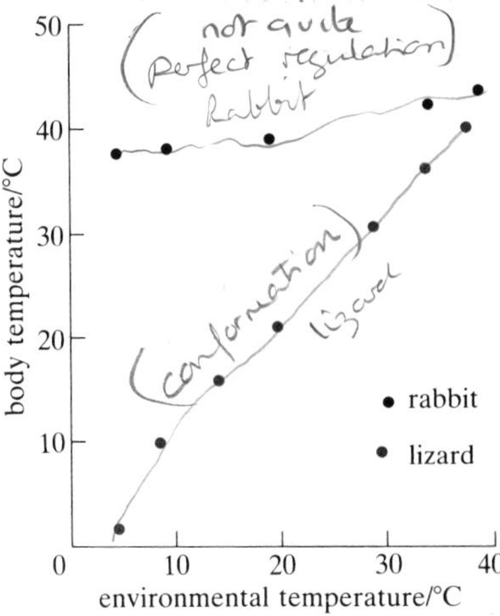

Figure 4.15 The relationship between body temperature and environmental temperature for two animals, a lizard (red dots) and a rabbit (black dots).

Figure 4.15 compares the body temperatures of a rabbit and a lizard when subjected to a range of external temperatures in a temperature-controlled chamber.

◇ Do these animals show perfect regulation, conformation or something between these extremes?

◆ The lizard shows conformation. The rabbit shows not quite perfect regulation.

From this result we could conclude that, in the particular environment employed in the experiment, either (a) the lizard does not exert control action in the service of temperature regulation or (b) any action it exerts is ineffective. However, compare this result with that shown in Figure 4.16. Note that, on this sunny day, the lizards' body temperature showed rather good regulation. By spending time basking and altering their orientation to the sun, the lizards are able to regulate body temperature. Whether they show regulation or conformation depends upon the opportunities for control action provided by the environment. However, some animals could be classed as **conformers**, irrespective of the nature of the environment in which they are placed, meaning that their body temperature invariably reflects that of the environment. As you might have imagined, many animals that spend all of their time living in water tend to be conformers.

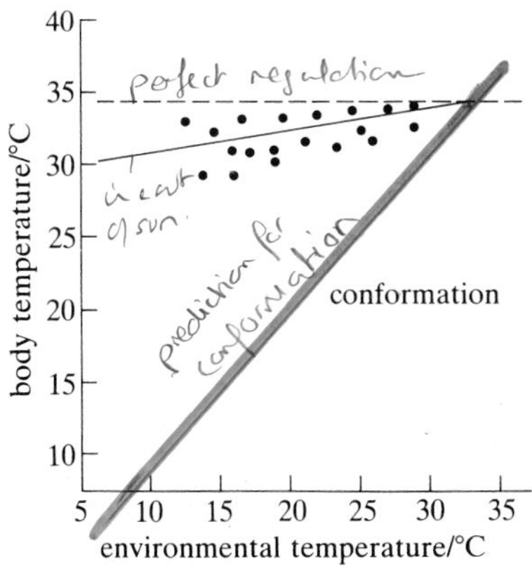

• actual data
--- perfect thermoregulation

Figure 4.16 The body temperature of a group of Spanish wall lizards (*Podacris hispanica*) on a sunny day in April, shown as a function of environmental temperature. The red line shows the prediction for conformation, and the dashed line the prediction for perfect regulation. The solid black line is the best fit to the individual data points. (Data collected by G. Patterson and P. Davies of Nottingham University.)

There are some simple principles of physics necessary for understanding temperature regulation. The body of an animal contains a certain amount of heat, and this determines the temperature of the body. As a result of metabolic processes (e.g. in muscle contraction), the body of any animal generates more heat. An animal will also gain heat from and lose heat to its environment: according to the circumstances, there might be either a net loss or a net gain. In control theory terms, the amount of heat in the body is given by the integral of the net rate of heat gain. When gain exceeds loss, body temperature rises. Conversely, when loss exceeds gain, body temperature falls. Regulatory physiology and behavioural physiology enter the picture when we consider whether the animal is able to take action to *alter* the rate of (a) heat generation and (b) heat exchange with the environment.

The classification of animals according to the nature of their systems of temperature regulation has had a somewhat confused and problematic history. The confusion is still evident in the terminology of the subject. For vertebrates, one popular classification is into warm-blooded (i.e. mammals and birds) and cold-blooded (i.e. fish, reptiles and amphibians). The technical terms are **homoiotherm** for warm-blooded animals and **poikilotherm** for the cold-blooded. However, whereas it is true that, for example, a lizard can tolerate a lower core body temperature than a bird, it by no means follows that the lizard's core body temperature would invariably be below that of the bird. The term 'homoiotherm' implies constancy of core body temperature, in distinction to the fluctuating nature of body temperature in the poikilotherm. Indeed, the fluctuations in body temperature tolerated by poikilotherms are in general greater than for homoiotherms. However, as Figure 4.16 shows, given the opportunity to exert control, the lizard exhibits rather effective regulation of body temperature.

Another traditional classification is in terms of **endotherms** (meaning heat is generated within) and **ectotherms** (meaning the heat source is outside). To some extent this corresponds to the classification of warm-blooded and cold-blooded, respectively. For example, the lizards described in Figure 4.16 are exhibiting ectothermic regulation, by utilizing the warmth of the sun. The rabbit whose data are presented in Figure 4.15 is showing endothermic regulation, by utilizing, for example, shivering. However, some qualification is needed before an animal is called an endotherm or an ectotherm. Some so-called cold-blooded animals have a significant endothermic capacity, for example the Indian python. Some teleosts, e.g. mackerel and tuna, can raise their body temperature by muscle activity from swimming. An animal classed as an endotherm, such as the rabbit, is able to exploit external sources of heat. Therefore we need some caution in categorizing animals as endotherms or ectotherms, though a division of animals along these lines is standard. It might have proved less confusing all round had zoologists referred to endothermic and ectothermic mechanisms or *strategies* for exerting control. Birds and mammals would be said to exploit both whereas, say, a lizard would be said to depend upon ectothermic strategies.

So to summarize the argument, both endotherms and ectotherms have an optimal level of body temperature. Both tend to take action that maintains their bodies near to this temperature, action being either intrinsic physiological activity not involving the animal's behaviour, termed autonomic (e.g. shivering), or extrinsic 'behavioural control' (e.g. moving to a new environment). The ectotherms are either wholly, or at least heavily, dependent upon behavioural control. The endotherms can utilize either autonomic or behavioural control (or, more usually, both) according to circumstances.

As far as endotherms studied in the laboratory are concerned, given a choice of environment and all other things being equal, they usually elect to remain in what is sometimes termed a **thermoneutral region**. This means a region in which minimum demand is placed upon the intrinsic physiological controls exerted over temperature regulation, such as shivering or sweating.

As far as terrestrial vertebrates are concerned, control actions exerted in the interests of temperature regulation can be divided into two categories: (a) endotherms can alter their rate of heat production, and (b) both endotherms and ectotherms are able to adjust the rate at which they exchange heat with the environment. As a first approximation, category b further sub-divides into (i) moving to a new environment and (ii) remaining in the same environment but changing the interface with this environment, such as huddling together with other animals, changing feather orientation to increase insulation, or curling into a ball.

Temperature regulation in mammals has long been subject to analysis in terms of control systems theory, and there exists an extensive literature on the application of these techniques to understanding the biological system. There is good evidence to show that the body contains a number of neurons (Chapter 1) whose electrical activity is sensitive to temperature and which play a role in the system of temperature regulation. (A caution is in order here, since a large number of neurons are sensitive to temperature and yet play no known role in temperature regulation.) When the local environment of the class of neurons with which we are concerned changes in temperature, the frequency with which they generate action potentials changes. The activity of some neurons, termed **warm neurons**, increases (i.e. higher frequency of action potential generation) with increases in temperature. Other neurons, termed **cold neurons**, decrease their activity as temperature increases. Figure 4.17 shows the performance of these two kinds of temperature-sensitive neuron. Some of these neurons are located at the skin, respond to peripheral temperature changes and, in effect, inform the central nervous system of the environmental temperature. Others are located deep inside the body, at the core, and respond to changes in core temperature.

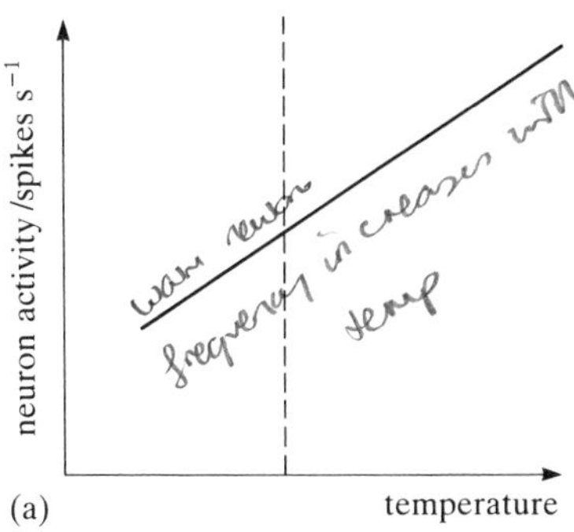

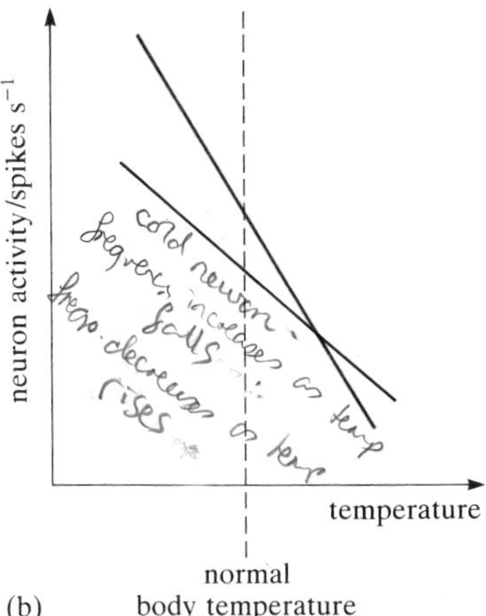

Figure 4.17 The response of temperature-sensitive neurons. (a) Warm neuron. (b) Cold neuron: the black line represents the normal response and the red line the response after the sensitivity of the cold neuron has been increased by, for example, a drug.

From the viewpoints of regulatory physiology, behavioural science and control systems analysis, our concern is how these two classes of neurons influence the control actions of the animal. However, before discussing the physiological aspects, it is useful to consider a fundamental principle of heat exchange. An argument commonly found in the literature is that, since the core exhibits the most stable temperature of any region in the body, it is here that one might expect the detectors of temperature to be located. The logic is somewhat suspect. It does not follow logically that, in a control system, the region showing the least fluctuation in a regulated variable is the site at which the variable is detected. For example, in the set-up illustrated in Figure 4.18, it is easy to demonstrate that the most stable temperature is at the centre of the thick-walled flask, and yet temperature is detected and control exerted at another location (the thermostat).

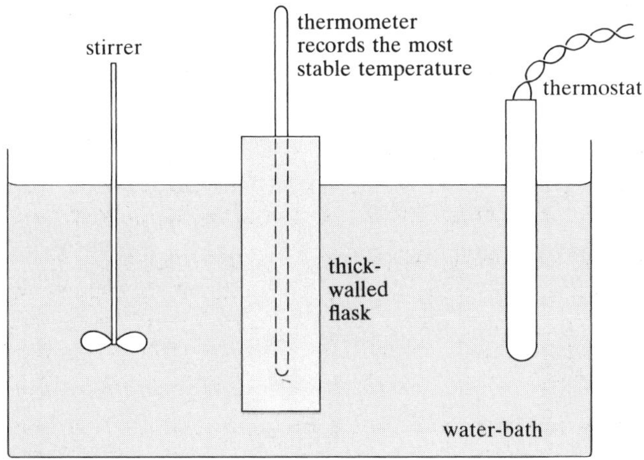

Figure 4.18 A water-bath containing a flask, a thermostat and a stirrer. When the temperature is recorded at various locations it is found that the most stable temperature is in the flask.

A second aspect of the fallacy soon becomes apparent when we take a control systems perspective. It could be dangerous to assume that regulation of core temperature could be best achieved by monitoring deep core temperature and basing control action upon this. A moment's reflection will show that, for temperature stresses that arise in the environment, deep core temperature will be the last to be influenced and will usually be the least affected. Therefore, to achieve rapid and sensitive control in response to environmental challenges, the system could not rely upon a detector located at the core. Rather, it might be supposed that a detector nearer the body surface would be more effective. However, deep-core temperature is obviously of great importance, and a temperature stress might arise there. We would therefore not expect the control system to be insensitive to what is happening at the core.

WARM NEURONS
A neuron whose frequency of generating action potentials is dependent upon its temp; the dependence is such that frequency increases with increases in temperature.
COLD NEURONS
A neuron whose frequency of generating action potentials is dependent on temperature, the dependence being of the form that as temperature falls, then frequency increases.

In some species, one can identify problems that could arise if the detector involved in the control action were at the skin. For example, consider the response of sweating in humans. Sweat cools the skin and thereby is able to reduce core temperature. It would be maladaptive if a detector located there were able to switch off prematurely the command to sweat, which is generated more centrally. As Benzinger (1964) so aptly expresses it:

> Temperature regulation in a home would not be improved if the warm sensor of the thermostat were sprinkled with water and cooled during periods of overheating.

From such considerations of how systems function we might expect the biological system not to depend exclusively upon either central or peripheral detectors in determining control action, and indeed this is what happens. Control action depends upon information from both sites. We might also expect important species differences in the weighting attached to central and peripheral detection, depending upon (a) the exact means of exerting control, (b) body size and (c) habitat.

◇ Why might we expect the size of the animal to be a significant factor in the evolution of a temperature regulation system?

◆ The smaller the body, the larger is the ratio of the body surface area to the body volume. The smaller the body, then the more rapidly temperature stresses arising in the environment are likely to be felt in the core of the body. One might expect the temperature regulation system of a small animal to place relatively more weight upon core temperature. On the other hand, it could well be maladaptive for the system of a very large animal to place weight upon core temperature, since this will be very well protected against temperature stresses arising in the environment.

Indeed, temperature detectors located deep in the brain (in the hypothalamus, to be more precise) play a larger part in control in smaller animals such as chipmunks, compared with, say, seals.

Figure 4.19 shows a model designed to account for our current understanding of temperature regulation. It is a simplification; for instance, detector

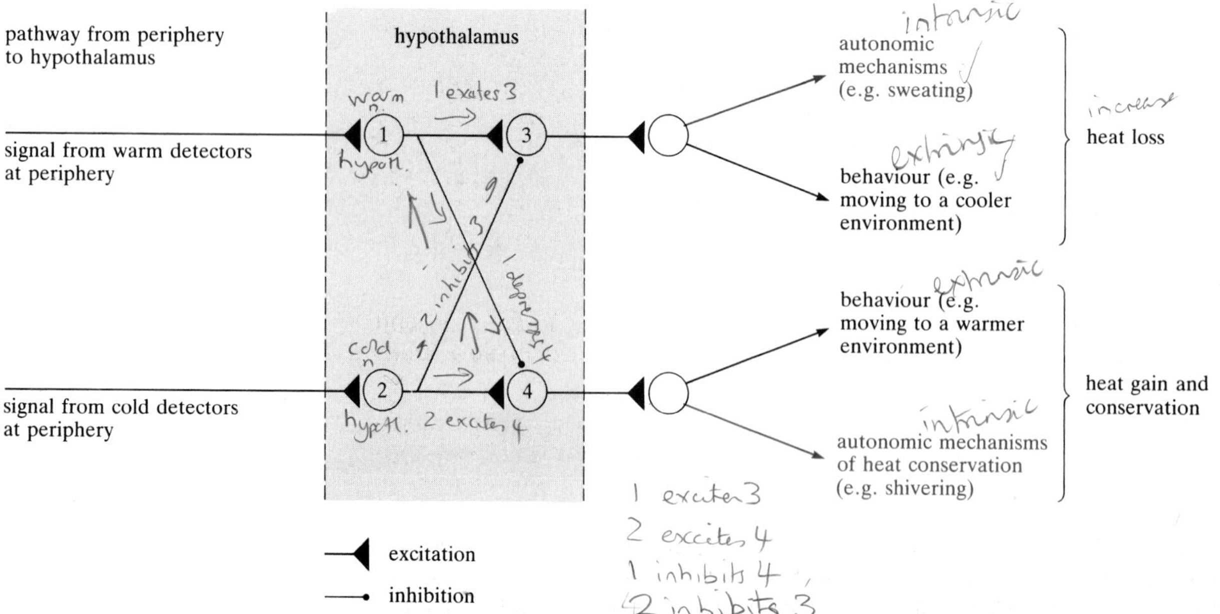

Figure 4.19 A model of the body temperature regulation system.

neurons might also be located at various other sites in the body, depending upon the species. Neuron 1 is a warm neuron, i.e. one that increases its firing rate as temperature increases. Neuron 1 is located in the hypothalamus. Warm detectors at the periphery provide an excitatory input to neuron 1. Neuron 1 excites neuron 3, which in turn activates autonomic controls of heat loss. It also motivates the animal to move to a cooler environment. Neuron 2 is a hypothalamic cold neuron. As brain temperature falls, so this neuron increases its firing. Peripheral cold detectors feed an excitatory input to neuron 2. Neuron 2 excites neuron 4, which causes the activation of mechanisms of heat production and conservation. The animal is motivated to move to a warmer environment. There are also *inhibitory* links between these pathways. Thus activity of neuron 2 inhibits the activity of neuron 3, and neuron 1 depresses the activity of neuron 4.

A thermoneutral region is one in which the demands for exerting control over either heat production or heat loss are minimal. In such an environment, there will be a certain background rate of metabolic heat generation. There will also be a certain exchange of heat with the environment. A background level of activity ('spontaneous activity') would be expected in the temperature-sensitive neurons (1 and 2). However, in such an environment, by means of the mutual inhibitory pathways, we would expect little or no net excitation of neurons 3 and 4. Therefore, there is no temperature control action taken. Suppose now that the environment suddenly cools and this cools the periphery of the body. This will excite 2, via the input from the peripheral cold detectors. Increased activity of 2 will increase the excitation of 4.

◇ What other influence on neuron 4 will change?

◆ The signal arising from the warm detectors will decrease. This will decrease the activity of neuron 1, which will in turn decrease the activity in the inhibitory pathway to neuron 4 and result in a further increase in the activity of neuron 4.

Strong activity in neuron 4 will therefore stimulate mechanisms of heat conservation and production, as well as possibly motivating the animal to move to a warmer environment.

Suppose that a human subject is in a thermoneutral region and body temperature is at 37 °C. Then a temperature stress is applied by placing the subject in a very hot environment. Body temperature will start to rise but, in response to this rise, autonomic effectors will be recruited, and these will restrain the shift of body temperature. One measure of the effectiveness of the control system is the ratio of how far temperature would shift if it were not for the negative feedback effects, divided by how far it does in fact shift. Some investigators term this ratio the **gain** of the control system, though this term has other uses in control theory. Guyton (1986) estimates that this ratio is something like 33 for the human temperature control system.

◇ Working on the basis of a gain of 33, a temperature stress that would shift the temperature of a physical object from 37 °C to 47 °C would move body temperature by how much?

◆ Divide the shift in temperature that the stress would cause (10 °C) by the gain (33). This gives about 0.3 °C. Therefore temperature would shift from 37 °C to only about 37.3 °C.

In Section 2, the concept of set-point was introduced. In discussions of biological control systems, comparisons are frequently drawn between the biological temperature regulation system and the system that thermostatically regulates the temperature of a house. Indeed, in the regulation system for the body temperature of humans, the magnitude of thermoregulatory effector action can be quantified as proportional to the deviation of actual body temperature from its 'set-point value' of 37 °C. As was discussed in Section 2, the house-heating system incorporates a set-point. Do we need to include a set-point in the model of biological temperature control? It is easy to see how the system represented in Figure 4.19 can account for the property of temperature regulation, the holding of body temperature near to 37 °C. However, nowhere in constructing Figure 4.19 have we explicitly incorporated a set-point in the model. We cannot point to a part of the nervous system and physically identify this as the embodiment of a set-point. The system behaves *as if* it incorporates a set-point. Whether the term is used, is somewhat a matter of taste.

A good example of how the set-point notion is used in discussions of biological control systems concerns the phenomenon of fever. In fever, body temperature is regulated at a higher than normal level and this helps to combat the infection. The higher level is defended against challenges. Many authors describe fever in terms of an elevation of set-point, caused by the presence of agents, termed 'pyrogens', that affect the neurons of the temperature system.

Physiologists have a good understanding of the basis of fever. For instance, it is known that thermosensitive neurons in the hypothalamus change their firing pattern following intravenous injection of pyrogens. There is evidence that cold-sensitive neurons are sensitized by pyrogens.

◇ Look at Figures 4.17 and 4.19 and suppose that the body is at 37 °C. What would be the predicted effect of a pyrogenically caused increase in the sensitivity of cold-sensitive neurons in the hypothalamus?

◆ There would be a net heat generation effect, which would continue until body temperature rose to a level at which the effects of warm-sensitive and cold-sensitive neurons again cancel. In other words, the system would be regulated at a higher temperature.

In response to pyrogens, not only do autonomic mechanisms act so as to raise body temperature, but so do behavioural mechanisms. Humans sometimes retreat to bed with a hot-water bottle!

Compared with some other biological systems (e.g. the control of feeding), temperature regulation is relatively easy to understand, and therefore is sometimes used as an example to illustrate general principles of control having a broader relevance. For example, Figure 4.20 shows a model that Eikelboom and Stewart (1982) use to illustrate an argument regarding the

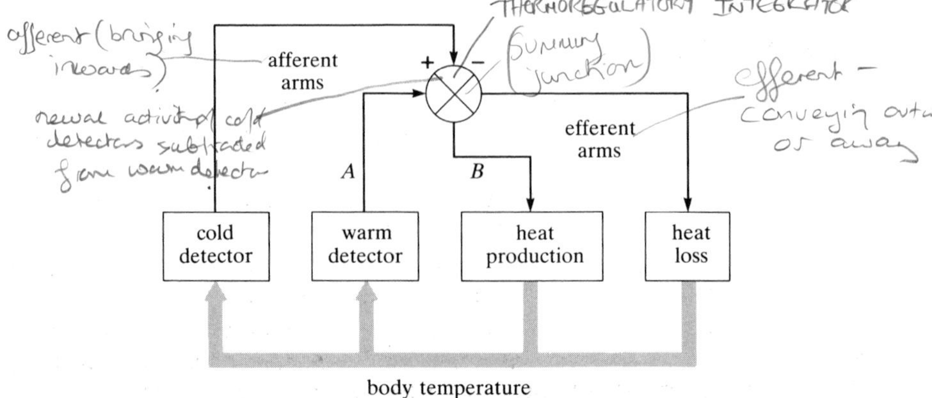

Figure 4.20 A model of body temperature regulation proposed by Eikelboom and Stewart (1982). Both cold detectors and warm detectors provide information to the central integrator. The signal from the cold detector is subtracted from that of the warm detector. If the outcome of this subtraction is positive, heat loss mechanisms are activated. If the outcome is negative, heat production is activated. Body temperature depends upon heat production and heat loss.

action of drugs on biological regulatory systems. Their paper has had a powerful influence, being widely employed by researchers in the areas of drug addiction and immunity to infections, amongst others. Eikelboom and Stewart use temperature regulation to make their argument explicit, and, though simplified, Figure 4.20 is based upon virtually identical assumptions to Figure 4.19.

In Figure 4.20, at the summing junction, the neural activity of the cold detector is subtracted from that of the warm detector. The inputs from these detectors are termed the 'afferent arm' of the system. If temperature is at 37 °C, the warm and cold signals cancel and the effectors are not activated. Suppose body temperature decreases. Activity in the cold detector increases and that in the warm detector decreases. This gives a signal that activates heat production at the 'efferent arm' of the system. Conversely, a rise in body temperature to above 37 °C will activate heat loss. To reiterate the earlier message, no set-point is explicitly incorporated in this model, though the system behaves *as if* it is being held at a set-point value. The summing junction, with its inputs and outputs, is known as a 'thermoregulatory integrator'.

Eikelboom and Stewart discuss drugs that alter body temperature and how their action can be better understood with the help of the model. Suppose a drug acts somewhere on the efferent side of the system; for example, a drug might act directly on the cells of the body to decrease their metabolic rate. This will result in a fall in body temperature, which will cause an increase in activity of the cold detector and a decrease in activity of the warm detector. The changes in activity of the detectors constitute inputs to the thermoregulatory integrator, which causes action in the efferent arms: (a) increased heat production and (b) behaviour to increase body temperature. For a drug that acts on the *efferent* side then, the model predicts that autonomic and behavioural actions are such as to oppose the effect of the drug. This is indeed what is observed in practice (Eikelboom and Stewart, 1982). Contrast this sequence of events with what happens to the temperature regulation system following injection of a small dose of morphine. There is an increase in body temperature as a result of increased physiological heat production, but also the animal's behaviour is such as to *increase* body temperature. This would only make sense in terms of Figure 4.20 if the site of action of morphine were at the *afferent* side of the integrator, since it would then be able to mobilize autonomic and behavioural control actions. For example, it might be that morphine changes the sensitivity of the cold detector neuron in the way shown in Figure 4.17.

Summary of Section 4.4

1 Confronted with variations in environmental temperature, an animal can exhibit conformation, perfect regulation or something between these extremes.

2 Action to regulate body temperature can be either endothermic (heat source within) or ectothermic (external heat source).

3 An animal such as a lizard relies upon ectothermic means of regulation. A bird or mammal can utilize ectothermic or endothermic means.

4 Neurons whose frequency of generating action potentials is temperature-dependent are implicated in the control system serving regulation of body temperature. There are so-called warm neurons and cold neurons.

5 The term 'set-point' can sometimes be descriptively useful in the context of body temperature. This need not imply that there is an anatomical embodiment of a standard against which actual temperature is compared.

6 Neurons implicated in the detection of body temperature are located both at the core and near to the periphery.

7 Control actions appear to arise from a balance between the outputs of cold neurons and warm neurons.

Now attempt Questions 9 to 11, on p. 123.

4.5 CONTROL THEORY AND DRINKING

In contrast to the relatively long history of application of control theory models to temperature regulation, these techniques have been applied to body-fluid regulation and drinking only since the late 1960s. The system involving the control of drinking and the regulation of body-fluid volume is one that has now been rather thoroughly analysed using a combination of physiological research and systems theory. First, it is useful to review the background physiology of the system.

The body of a rat, the favourite animal for research in this area, is about 69% water. Some of this water is in the various cells of the body (e.g. skin cells, brain cells) and collectively constitutes what is termed the **cellular water** (or 'intracellular water'). It is said to be contained in the 'intracellular compartment' (Figure 4.21). Although the water is distributed throughout the cells of the body, in some respects it can conveniently be treated as if it were a single entity. The remainder of body water, the water outside the cells, is termed the **extracellular water**. This is located in both the plasma of the blood and in the spaces between the cells, the interstitial water (see Figure 4.21). It is described as being contained in the 'extracellular compartment'.

Water is lost from the rat's body in urine, by evaporation from the lungs, and in the saliva spread by the rat over its body when it grooms itself. Some water is also lost in the faeces, and at times of high ambient temperature rats cool

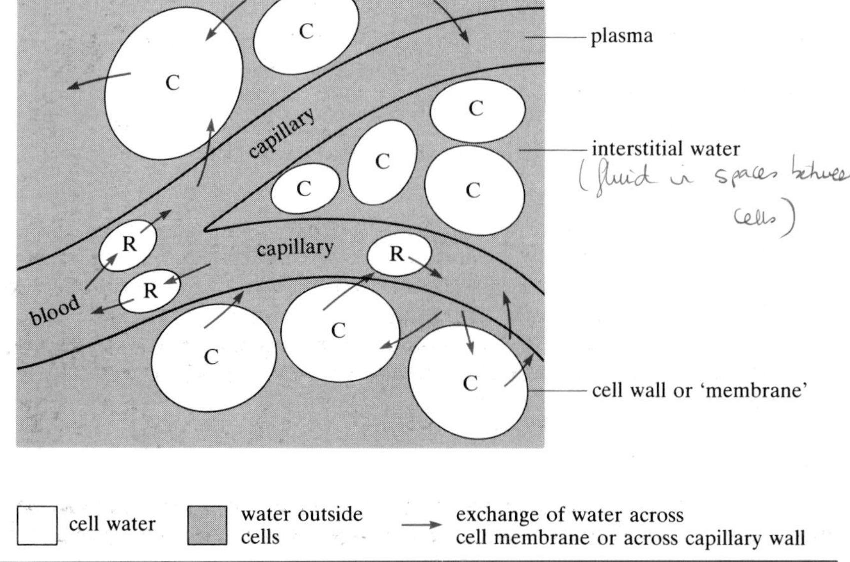

Figure 4.21 The relationship between intracellular and extracellular compartments, shown as a sample of body fluids. The cells labelled C are bathed by interstitial water; the cells labelled R are red blood cells. All the water within the cells collectively constitutes the intracellular water, and all that outside the cell is defined as extracellular water. The extracellular compartment is thus made up of the fluid part of the blood (the plasma) and the fluid in the spaces between the cells (the interstitial fluid).

their bodies by spreading relatively large amounts of saliva over the body surface. They do not sweat. Water is gained by drinking, from the water contained within some foods and by the metabolism of food.

Both the kidneys and the drinking motivation mechanisms serve to regulate body fluids. The kidneys filter the plasma of the blood, excreting some of what is filtered and reabsorbing the remainder back into the blood. This process of reabsorption is under the control of a number of hormones, the best-known being **antidiuretic hormone** (ADH). This will be described more fully in Chapter 11. When body-fluid volume falls, there is an increased rate of secretion of ADH, which stimulates reabsorption of water in the kidney. Hence, loss of water as urine is minimized. Conversely, if body-fluid volume increases to above normal, there is a suppression of the secretion of ADH. The rate of urine production rises, until the excess water has been excreted. The same detectors (of cellular and extracellular volumes) that are implicated in the drinking motivation system are also involved in determining the secretion of ADH. In this respect, there is a close parallel between the intrinsic and extrinsic control systems that serve to regulate body fluids.

Depriving a rat of water for a period of time causes a fall in the volume of body water, which arouses thirst. From this, we can conclude that loss of water from one or other, or both compartments, arouses this state. Indeed, we know that a loss of water from either body-fluid compartment can stimulate drinking. Removal of blood, an extracellular loss, is a potent stimulus to drinking. Drinking can also be powerfully aroused by injection of hypertonic saline. This stimulus is particularly revealing as far as the mechanisms of drinking are concerned. Injection of hypertonic saline very rapidly triggers drinking, and yet it does not cause any immediate loss of water from the body. Rather it causes a loss from the cellular compartment, as water moves from the cells into the extracellular compartment. Any loss from the cells is a gain to the extracellular compartment. In other words, the arousal of drinking by water deficiency in the cellular compartment is not cancelled by a surplus of water in the extracellular compartment.

It is reasonable to speak, in rather gross terms, of a deficit in fluid held in either compartment arousing thirst. However, when we examine the system more closely, it is necessary to consider exactly how and where the signals arise within each compartment. The consensus at the moment is that there are two detectors, one looking at a sample of the cellular compartment and the other at a sample of the extracellular compartment. The cellular detector is believed to be a neuron (or group of neurons) in the brain, whose state of shrinkage constitutes, in control theory terms, the error signal. Local application of minute amounts of hypertonic saline in the hypothalamus causes copious drinking. The detector in the extracellular compartment appears to be a stretch receptor in a blood vessel. So to summarize, a loss of water from either the cellular or the extracellular compartment, or both together, triggers drinking.

In rats and dogs, drinking is rapid. The animal soon ingests an amount commensurate with its deficit, and stops drinking. Water passes down the oesophagus, into the stomach and then into the intestine. It is absorbed across the wall of the stomach and intestine and into the blood and then distributed throughout the extracellular and cellular compartments. From physiological measurements, we know, for example, the speed with which ingested water leaves the stomach and enters the intestine, its rate of absorption across the wall of the intestine and the rate at which water appearing in the blood is distributed to the cells.

It is possible to build a model in control theory terms, using components of the kind shown in Figures 4.5–4.8, that assimilates all such physiological data

SIMULATION

A simulation of a real system is a dynamic model of the real system that performs in a manner similar in important respects to reality but in a domain that is different to that of the real system. A biological system might be simulated on a computer.

on water movements and volumes. This information can then be fed into a computer, and a **simulation** of the biological system run. The simulation can then give what is, in effect, a running commentary on what is happening throughout the body. Checks can be made to compare the simulation with the behaviour of real animals. For example, Figure 4.22 shows the response of a real rat and a simulated rat to a slow continuous infusion of hypertonic saline.

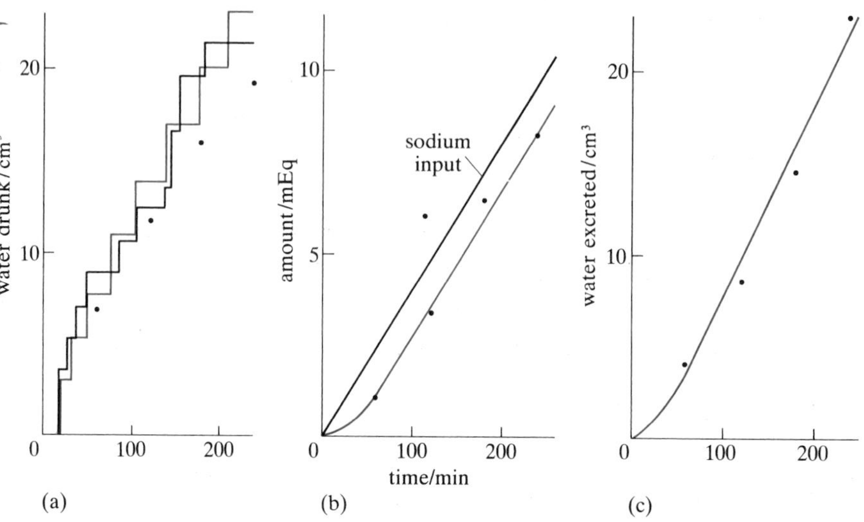

Figure 4.22 A comparison of the responses of real rats (black lines for a typical rat, black dots for the average) and a 'computer rat' (red lines) to a continuous intravenous infusion of hypertonic saline. (a) The amount of water drunk. (b) The amount of sodium excreted. The amount of sodium infused is also shown; this was the same for both real and simulated rats. (c) The amount of water excreted as urine.

SHORT-TERM FEEDBACK LOOPS

In the system that controls drinking, feedback occurs by ingested water ultimately restoring the body-water state to normal. However, by means of feedback from mouth to stomach, drinking is switched off before such restoration is complete. Such feedback termed 'short-term feedback'.

With the help of such a model, it is clear that, following, say, a single hypertonic saline injection or a period of water deprivation, restoration of cellular fluid has hardly begun at the time drinking is terminated. Correction of the depleted general physiological state of the cellular compartment is therefore not the factor that terminates drinking. The model suggests two possible (and non-mutually exclusive) explanations of what brings drinking to an end. Water might be taken up more quickly by brain tissue than by tissue elsewhere. There might be so-called **short-term feedback loops** arising from the mouth and stomach and inhibiting the drinking tendency. Figure 4.23 illustrates a model of drinking that incorporates such short-term feedback.

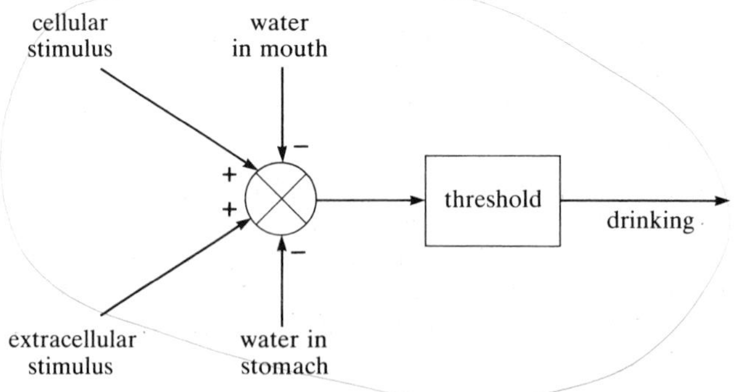

Figure 4.23 A model of drinking in the rat. Drinking tendency is excited by depletion of water in either the cellular or the extracellular compartment, or both. It is inhibited by water in the stomach and the memory of the passage of water through the mouth. At the summing junction, inhibitory tendencies are subtracted from excitatory tendencies. If the net signal is above a threshold level then the rat starts drinking and continues until the tendency is zero.

There are at least two ways in which one can override the normal homeostatic control of drinking and persuade a rat to ingest fluid much above the needs of regulation (such techniques are, of course, not unknown in human interaction!). Figure 4.24 shows the effect of replacing the rat's normal drinking fluid with a highly palatable saccharin solution. The other technique is to make the rat hungry by a period of food deprivation and then to feed it with small pellets of food at a rate of, say, one per minute. The animal is not deprived of water. Massive amounts of water are ingested during the course of such a test,

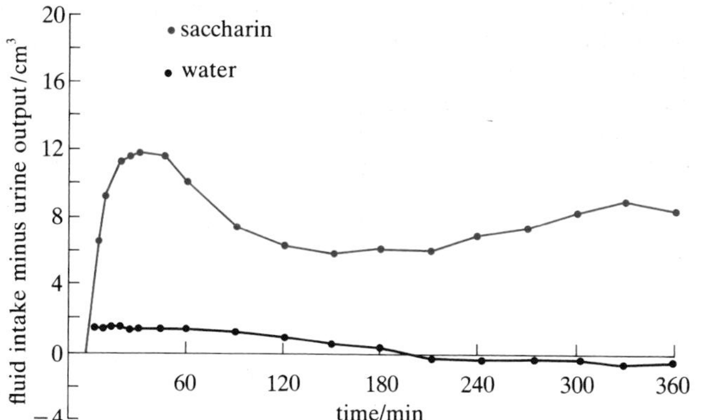

Figure 4.24 A comparison of a rat's intake of water (black line) and a dilute solution of saccharin (red line).

for reasons that are not at all clear. For our purposes, the relevance of this phenomenon, termed **schedule-induced polydipsia**, is what happens to the regulatory system. The excess water is detected by the ADH–kidney system and urination is increased. Figure 4.25 shows the result of an experiment of this kind. Note that excretion of urine starts off at a low rate. This is what would be expected; body-fluid volume increases to some extent and thereby suppresses the production of ADH.

◇ For the rat showing schedule-induced polydipsia, give an account of the variables of the system in the steady state.

◆ The rat is drinking at a steady rate. It is forming urine at a steady rate, which removes water from the body as fast as it enters. Body water will be at a level above normal but will be maintained constant. The elevated level of body water suppresses ADH secretion such that the kidney can lose water at a high rate.

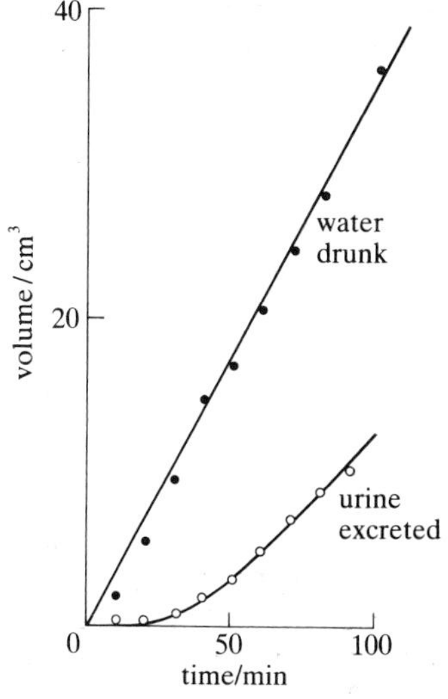

Figure 4.25 The phenomenon of schedule-induced polydipsia: the volume of water drunk and urine excreted by a group of hungry (but not water-deprived) rats receiving small food pellets at a rate of one every 30 seconds.

Another way in which the rate of urine formation can be greatly elevated is by damage to the system that secretes ADH. In fact, a disease condition known as **diabetes insipidus** is characterized by a failure to produce ADH (or a failure of the kidney to be responsive to the ADH that is produced) and hence by a much greater rate of urine formation.

◇ In terms of the regulatory system, both diabetes insipidus and schedule-induced polydipsia are characterized by abnormally high rates of drinking and urination. What is the fundamental difference as far as the control actions of the animal are concerned?

◆ In diabetes insipidus, the excessive rate of formation of urine is primary and the elevated drinking is a control action in response to loss of body fluids. By contrast, in schedule-induced polydipsia, the primary abnormality is in the drinking system, and the elevated urine formation is a control action in response to this. In Figure 4.25 the delay in the response of urine formation shows this.

Summary of Section 4.5

1 Body water is conveniently described as being composed of a cellular component (all of the water contained within the cells of the body) and an extracellular component (all of the water outside the cells of the body).

2 Depending upon the mammalian species, water can be lost from the body in evaporation from the lungs, in saliva spread over the body surface, and in

sweat, urine and faeces. It can be gained by drinking, from the water content of food and by the metabolism of food.

3 Both the behavioural mechanism of drinking and the intrinsic physiological mechanism of antidiuretic hormone/excretion serve to regulate body water.

4 The control of drinking depends upon the detection of both cellular and extracellular fluid states.

5 The body-fluid system can show abnormalities of function in disease conditions (e.g. diabetes insipidus) and in a behavioural treatment (e.g. schedule-induced polydipsia).

Now attempt Questions 12 to 14, on p. 123.

4.6 CONTROL THEORY AND FEEDING

The fuel that an animal needs for the cells of its body is obtained via the metabolism of its food. Whereas the needs of the cells might be fairly constant over long periods of time, energy supply from food might fluctuate widely as a function of the availability of food. This necessitates a storage mechanism to retain energy in the periods after feeding. For example, carbohydrates can be converted into fat and stored in this form. This can then be utilized for metabolism at times when food is not available. Feeding also serves needs other than the supply of energy. Vitamins, amino acids and minerals, essential for the functioning of the body, are obtained in the diet, and in some cases feeding behaviour can be directed towards specific dietary needs.

Some researchers have made impressive progress in understanding the determinants of feeding in invertebrates, a good example being the now-classic work of Dethier (1976) on the house fly. Bolles (1980) presented a control theory interpretation and model based upon Dethier's experimental evidence. Figure 4.26 shows this model. The chemical signal representing the presence of food is labelled g. An operator, the constant K_1, relates g to the excitation that food sets up in the fly's nervous system. The net feeding tendency is given by the difference between excitation and inhibition exerted upon the feeding controller. The output of the operator marked 'Feeding mechanism' is the rate of feeding. Integration of the rate of feeding gives the gut contents. Another operator, the constant K_2, relates gut contents to the strength of the inhibitory signal arising from gut contents. So long as excitation is above inhibition the fly feeds.

One way of looking at the functioning of feedback systems is to see what happens when the feedback loop is broken. Such a system is then said to be on **open loop**, a somewhat contradictory expression but one that is universal in control theory. In the feeding system of the house fly, it is possible to break

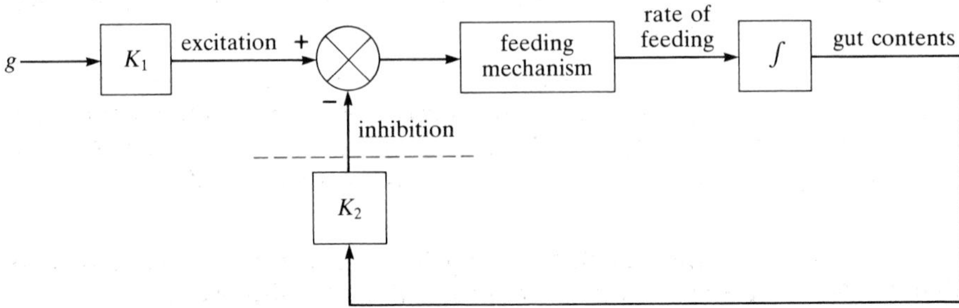

Figure 4.26 A model of feeding in the house fly.

the feedback loop by cutting the inhibitory neurons, at the point shown by the red dotted line in Figure 4.26. When this is done, the fly quite literally eats until it bursts.

An elegant series of experiments, carried out recently by Charles Lent and Michael Dickinson on the feeding control system of the leech (*Hirudo medicinalis*), has revealed important and fundamental similarities with the system of the house fly (Lent and Dickinson, 1988). Hungry leeches lie in wait, typically at the edge of a pond. A disturbance in the water alerts the leech, which then accurately orients itself towards the source of the disturbance and swims in that direction. On establishing contact with the target, which is localized by the warmth of its body, the leech bites. Blood is sucked and pumped into the crop. An amount of blood up to nine times the leech's initial body weight is ingested and then the meal is terminated. Now, in the satiated state, the leech seeks deeper water. In this state, rather than a warm surface eliciting the biting reflex, the leech actively avoids the potential prey. Satiation from a meal can sometimes last for up to a year.

The research of Lent and Dickinson was directed to understanding the precise neurobiology underlying the cycle of hunger and satiation. Such a search is aided by the relative simplicity of the leech's nervous system, containing relatively large, localizable and easily identifiable neurons. What exactly is the variable that underlies the switch from the hunger state to that of satiety? Lent and Dickinson speculated that distension of the body wall might constitute the inhibitory influence. To test this, during the course of feeding, leeches were artificially distended by injecting saline solution into the crop. The effect was an immediate cessation of feeding. Conversely, when ingested blood was removed from a leech during the course of feeding, then the bout of feeding lasted indefinitely. From a series of experiments of this kind, the conclusion was reached that physical distension constitutes the negative feedback influence.

The neurons that play a key role in controlling feeding in the leech secrete the chemical transmitter serotonin. Application of a warm stimulus specifically to the leech's lip evokes activity in a number of large and identifiable serotonergic neurons. This simulates the natural stimulus to feeding. Bathing a leech in fluid containing serotonin greatly amplifies activities associated with feeding, such as swimming in the direction of wave ripples, biting, and secretion of saliva, as well as increasing the amount of blood ingested. Serotonin arouses biting in otherwise satiated leeches. A similar spectrum of feeding or food-seeking activities could be elicited by artificial electrical stimulation of the effector neurons. Application of a neurotoxin with a specific effect upon serotonergic neurons caused a 'hungry' leech to display a pattern of responses characteristic of the satiety state.

◇ It was shown that electrical activity in a number of serotonergic effector neurons caused the pattern of food-seeking and feeding activities. It is postulated that the negative feedback effect arises from bodily distension. What would we expect the effect of distension to be as far as the animal's nervous system is concerned?

◆ Consider a neuron Z that is excited by distension. We might expect one of two possibilities: (a) that a set of neurons excited by the serotonergic neurons are *inhibited* by activity in neuron Z, or (b) that activity in the serotonergic neurons themselves is inhibited by activity in Z.

In fact, the results showed that activity in the serotonergic neurons, and thereby their release of serotonin, was suppressed by distension.

Such a simple system is very effective for the leech. Unlike the situation for some other species, the animal has no problem concerning diet choice in the interests of getting a balanced diet. For example, the necessary minerals would automatically be gained in the blood of the victim.

In more complex animals, such as the rat, the neurobiology is more difficult to tease apart. However, in studying them, we will encounter systems that, like those of the fly and the leech, are sensitive to (a) available food in the environment and (b) some measure of body energy level. This is reflected in the control of behaviour.

The history of the scientific study of feeding in mammals, such as the rat, dog and human, has consisted of various proposals as to where and how in the body the state of energy is detected and how this information is translated into a command to feed. Older theories include an empty stomach and a low blood glucose concentration as the internal triggers to feeding.

◇ As was discussed in Section 4.3, a system involving insulin is implicated in the regulation of blood glucose concentration. Can you suggest a problem that would be inherent in a control system relying upon blood glucose concentration as the trigger for feeding?

◆ Blood glucose concentration is regulated by intrinsic systems that are activated when the level falls. Therefore blood glucose is buffered against disturbances and would provide only an insensitive measure of body energy state. The argument that control action would be relatively ineffective if based upon a variable that is closely regulated by intrinsic physiological systems is similar to that concerning location of temperature-sensitive neurons in the deep core or nearer the body surface (Section 4.4).

An influential theory, and one that accords well with much of the experimental evidence, is that of Booth (1976). This contained the proposal that, for an animal such as a rat, on a balanced diet, feeding is under the control of the supply of energy to the liver. For example, at the end of a meal the animal experiences satiety and refrains from further feeding because the supply of nutrients from the gut is relatively high. When this supply falls, hunger is aroused. Suppose an animal is force-fed for a period of time. At the cessation of force-feeding, the animal will tend to refrain from feeding. According to Booth's argument, the breakdown of the inflated fat deposits in the animal is supplying sufficient energy to inhibit feeding.

Theories stated in words can sometimes sound intuitively appealing, and a number of experiments that lend support to the theory can be found, but still there can remain inherent contradictions within the theory. One way in which the assumptions within a theory can be made unambiguous and stated with greater rigour is in terms of a model of the system. The assumptions on which the theory rests are stated in exact terms and then the assumptions are assembled in a model and the model's predictions tested. Booth's theory was embodied in a mathematical model of rat feeding and simulated on a computer. Amongst other factors taken into account was the influence of environmental illumination. Wild rats are, of course, exposed to 24 hour cycles of light and dark. In the laboratory, such a cycle is produced by artificial illumination, normally 12 hours light followed by 12 hours darkness.

Under laboratory conditions and having food freely available, rats eat predominantly in the dark. Indeed, some refrain from eating at all in the 12 hours of light. However, the metabolic rate changes only slightly as a function of the phase of the cycle. During the 12 hours of light, a period during which metabolic rate is higher than energy intake, there is a net **lipolysis** (fat deposits are breaking down and supplying the rat's energy needs). During the dark phase, a period when energy intake exceeds metabolic rate, there is a net **lipogenesis** (energy intake exceeds requirements and there is a net investment into the fat deposits).

At the time that the model was being constructed, it was found experimentally that the rate at which the gut empties is considerably lower in the light than in the dark phase. The effect on the model of disrupting this normal rhythm was investigated. Slight differences had a large effect. This result was of considerable interest for a number of reasons, as follows.

If a part of the brain, known as the ventromedial hypothalamus (VMH), is destroyed, a particular pattern of abnormal feeding is seen. It might strike the reader as a curious way of investigating biological systems to destroy part of them and see what happens. For instance, economists do not go around blowing up banks to see what the effect of this is on the economy, and neither do they wish to do so. Curious as it might be, none-the-less, discussion of the brain-damaged rat occupies a large percentage of any textbook on neural biology or psychobiology. So it was felt relevant to investigate the effect of ventromedial hypothalamic damage as far as the predictions of the model were concerned. The effect of the damage on the rat's behaviour is unambiguous; it over-eats and becomes grossly obese. There then follows so-called *static obesity*, when weight is elevated but stable, and food intake is normal. Some of the older accounts of this phenomenon described the VMH as the brain's 'satiety centre', the logic being that its destruction left the 'hunger centre' in the neighbouring lateral hypothalamus unopposed. A rather different interpretation of this syndrome, arising directly from the simulation studies, is as follows (Duggan and Booth, 1986).

The brain-damaged rat takes meals abnormally frequently in the light phase, and feeding in the dark phase is nearly normal. The model predicted that speeding up gut emptying would lead to more frequent meals in the light phase and, in consequence, to obesity. Duggan and Booth observed that VMH lesions disrupt the control of stomach emptying by the autonomic nervous system; specifically, the slow daylight emptying is speeded up. If the normal parameters of gut emptying are replaced by the abnormal parameters measured in the VMH-lesioned rat, then over-eating and obesity very similar to those seen in real rats are predicted.

Drugs can also affect the rate of gastric emptying. For example, the drug fenfluramine is known to depress appetite and to slow gastric emptying. The simulation suggests that its mode of action on appetite could be, at least in part, via its action on gastric emptying. So lesions to the VMH might well have their effect as a result of changes in gastric emptying rather than by impairing a satiety centre.

The VMH-lesioned rat and the computer simulation serve as a warning to the theorist. Control theory concepts and terminology can be an invaluable guide to investigation and to understanding the data. However, they can also be fundamentally misleading. The VMH-lesioned rat might act *as if* a body-weight set-point has been elevated. It might act *as if* actual body weight is being compared with a set-point of body weight, and the comparison operation perceives a mismatch. However, the results give no evidence for such a regulation. Rather, they suggest that change in a rather distant parameter, gut emptying rate, can have a profound influence upon body weight. Sometimes the following analogy is used to illustrate this point. A village pond can show a near constancy of volume over many years. It might even look as if the volume is being measured and control action taken in response to a deviation from normal, but we know that it is not appropriate to

SENSORY-SPECIFIC SATIETY

A state of satiety, that is, to some extent, specific to the food that the animal has just eaten. The magnitude of the sensory-specific part of satiety can be measured by the degree to which appetite is revived by a different food.

TASTE-CONSEQUENCE LEARNING

Learning in which an animal forms an association between the taste of an ingested substance and the consequences (eg gastro-intestinal disorder or recovery from vitamin def.) that follow sometime after ingestion.

postulate such a measurement process. Rather the volume is simply the consequence of a balance between flows in and out. So much for this aspect of the model; we now turn to further complications.

A model based simply upon energy supply, though useful up to a point, cannot account for the phenomenon of diet choice: the rat's ability to select amongst nutrients so as to obtain a balance in its intake. The need for a capacity of this kind is apparent when we consider the ecology of the rat and contrast it with other species. For example, a leech or a carnivore would automatically obtain a range of proteins and minerals in its standard diet. By contrast, the rat normally exploits a wide range of habitats and needs to make use of various different foodstuffs of widely different chemical composition. It might be, by luck, that a particular foodstuff contains all the necessary components, but it might also be that the rat needs to sample from a variety of diets to obtain all of them.

In principle, one might imagine that the rat is equipped with an innate recognition system for what it needs. Deprive the rat of a vitamin or an essential amino acid, and it would be motivated specifically towards foods that contain the needed item. This is indeed how the system works for sodium deficiency. The sodium-depleted rat shows an immediate preference for sodium-rich foods, but sodium seems unique in the respect that there is a specific recognition system built into the rat's nervous system. As for the other dietary components, there are two phenomena demonstrated in the laboratory that appear relevant to how the rat in the wild would tackle this problem: **sensory-specific satiety** and **taste-consequence learning**.

When a rat stops feeding it is said to be satiated. However, an animal that is satiated on one particular diet will often start eating again if the diet is changed to one that tastes different. In other words, satiety is to some extent specific to the food that the rat has eaten, hence the term 'sensory-specific satiety'. To investigate the role of variety in the diet, we need to look at a rat's intake when choice is available compared with when only a single diet is present. Figure 4.27 shows the result of such an experiment. The columns represent the amount of food in grams (g) eaten by rats in two hours. In phase

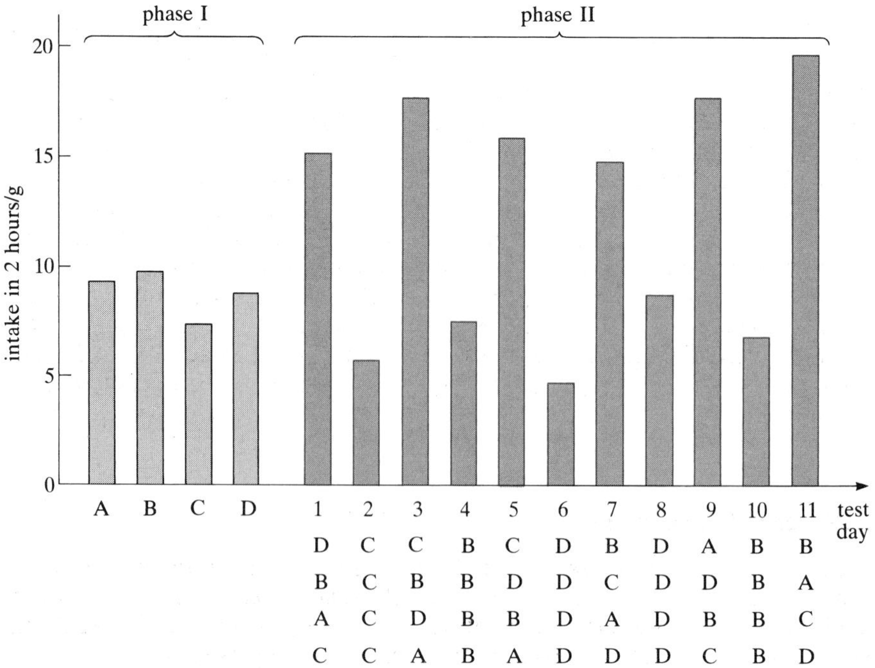

Figure 4.27 The intake of food by rats during a 2 hour feeding session. Phase I: on four consecutive days the rats were fed the four test diets, A, B, C and D; there was no significant difference in the amounts ingested. Phase II: on alternate days rats were allowed either variety or no variety in the diet; note the amount ingested on day 1, when they were allowed each of the four diets for 30 minutes, and compare this with day 2, when only diet C was available for the whole of the 2 hours. These results show that variety is a powerful stimulus to ingestion.

I of the experiment, over four successive days, the subjects were fed for two hours with one of four different foods, labelled A, B, C and D. Each food was of the same composition, with the sole exception that they were each made to taste slightly different by addition of an arbitrary flavour tag. For example, diet A would carry a lemon tag and diet B an almond tag. The diagram shows that there was no significant difference in the amount eaten. Now look at phase II of the experiment. On test day 1, rats were allowed 30 minutes of access to diet D, and then 30 min access to diet B. There then followed diets A and C, each presented for 30 min. Compare the intake over the two hours (15 g) with the intake on the following day when only diet C was available for the whole two hours. As is also shown on successive days of the experiment, variety is a powerful stimulus to ingestion (this doubtless fits your own experience of human feeding!). Translated to the natural habitat such a process would tend to lead to the rat sampling various available and palatable diets as opposed to reliance upon just one, and hence acquiring all the nutrients it needs.

The other mechanism mentioned, taste-consequence learning, gives the rat enormous flexibility in its behaviour. The rat learns the relationship between a particular taste and the consequences for the body of ingesting the food. These consequences can be revealed up to several hours after ingestion, and the animal can still form an association between the taste and the nutritional consequence. The best-known and most dramatic example of this is when a rat suffers gastrointestinal illness some hours after ingesting a food of a particular flavour and will avoid that particular food in the future. This effect is generally termed **taste-aversion learning**, but is also commonly called the **Garcia effect**, after its discover, John Garcia. Perhaps less well known is the positive aspect of the same process: that substances whose ingestion is followed by recovery from a deficiency of a vitamin or essential amino-acid will become preferred in the animal's diet choice. In this way, a rat suffering from, for example, an amino-acid deficiency, could sample various diets and form a positive association with the one supplying the missing dietary component. Such a learning process might well be more practical and effective than a series of genetically determined ('built-in') recognition systems each tuned to recognize particular dietary components.

At this point it is useful to summarize the description of the system that is emerging as far as the control of feeding is concerned. A detector of energy state is present in the system and this plays a role in arousing feeding motivation. However, there are also 'external' factors (e.g. food quality) that act in conjunction with the internal factor in determining whether food will be ingested. One such is the intrinsic palatability of the available food. Another is the variety of available food. A third factor is the associations that have been learned with the food. If, in the past, ingestion of the food has been followed by beneficial consequences, the food will tend to be ingested. If the learned association is with negative consequences, the food will be actively avoided. The learning characteristic means that we would describe the feeding system in terms of a special class of control system, an **adaptive control system**. In the language of control theory, this refers to a system that can modify the value of its operators (e.g. gain) in the light of experience.

Summary of Section 4.6

1 In the case of both the house fly and the leech (*Hirudo medicinalis*), feeding tendency is determined by the difference between excitation from food and inhibition arising from stomach distension. Breaking the feedback loop results in abnormally large amounts of food being ingested.

AD LIBITUM

A condition in which food or water, or both, are freely available at all times.

2 In a mammal such as the rat, feeding tendency is determined by (a) some measure of internal energy state (e.g. the supply of energy to the liver), (b) the availability and quality of food in the environment and (c) the associations that the animal has formed with the foodstuff.

3 A lesion of the ventromedial hypothalamus in the brain of the rat is followed by abnormally large intake of food. The rate at which the stomach empties is also altered and the possibility is raised that the change in stomach emptying might play an important role in the exaggerated food intake.

4 For mammals, the term 'sensory-specific satiety' refers to the fact that satiety depends to some extent upon the diet on which satiety was attained. Changing the diet can result in a revival of food intake in otherwise satiated rats. Variety in the diet is a powerful means of increasing intake.

5 The expression 'taste-consequence learning' refers to the association that an animal can form between the taste of a particular foodstuff and the consequences of its ingestion.

Now attempt Questions 15 to 17, on pp. 123–124.

4.7 INTERACTIONS BETWEEN ACTIVITIES

So far we have considered three distinct systems, by looking at physiological states of body temperature, body fluids, and energy, and at the control actions exerted over them by heat gain/loss, drinking/urination and feeding. We have not yet looked at interactions either between these systems or with other systems. A process of simplification, by considering a given system in isolation in this way, is necessary for gaining understanding. Thus, when we disturb a system, such as the one regulating body-fluid level, in order to see how controls over drinking and urination react, we deliberately attempt as best we can to keep all other disturbances to a minimum. For example, environmental temperature is held constant. If the experiment is of short duration, e.g. an hour, the animal's food might be taken away just for this short period, so that body fluids are not disturbed by the ingestion of food. For longer duration experiments on drinking, the animal might normally be allowed food freely (termed **ad libitum**) so that a hunger signal does not arise. In this way, we attempt to isolate the independent variable, the manipulated thirst stimulus, as the cause of any drinking. Such investigation is in the tradition of science. We simplify a complex system in order to try to understand it.

As necessary as such simplification often is, it is clear that, except under somewhat contrived conditions, systems do not function in isolation. Rather there are normally interactions between physiological systems. Indeed it might give insight to see how the animal copes with a variety of challenges, involving interactions between systems. First, we shall consider the mechanisms underlying interactions, and then we shall consider the interactions from a functional viewpoint.

◇ In the case of the rat, what are some examples of interactions between the systems of body fluids and body temperature?

◆ (a) When environmental temperature rises there will be an increased loss of water from the lungs. The rat protects itself against an increase in body temperature above normal (hyperthermia) by spreading saliva over its body surface. This will increase the loss of water from the body fluid. (b) Ingestion of water that has a temperature less than body temperature will lower body temperature.

As these examples show, activity on the part of one physiological system can place a load upon another system. Similarly, eating a high-protein diet places a load on the body-fluid system by elevating water loss through the kidney. Feeding and drinking provide a rich source of information on how control systems interact.

4.7.1 Drinking and feeding

Interactions between, on the one hand, feeding and energy, and, on the other, drinking and body-fluid state, can occur at a number of different levels in the two systems. In general, it is useful to distinguish a class of *central* interactions that directly involve the nervous system, and a class of *peripheral* interactions that arise in the body tissues outside the nervous system. We have already noted that eating a diet of high-protein content places a demand upon the body-fluid regulation system. This is a peripheral interaction, since it results from an elevated rate of urine formation. As another example of a peripheral interaction, consider what happens when a rat has its food supply restricted or stopped. As Figure 4.28 shows, the amount of water taken each day is reduced. However, there is no need to postulate inhibition from the energy-regulation system to the drinking control system, and there is no evidence for such a pathway. (It is difficult to see any useful purpose that could be served by such an interaction.) Rather, as food intake declines, then so the rat's need for water declines. The amount of water lost from the kidney falls.

In this regard, it is interesting to compare the rat and the rabbit. In contrast to the rat, the water intake of the rabbit increases substantially as a result of food deprivation. The reason for this behaviour is known, and the sequence

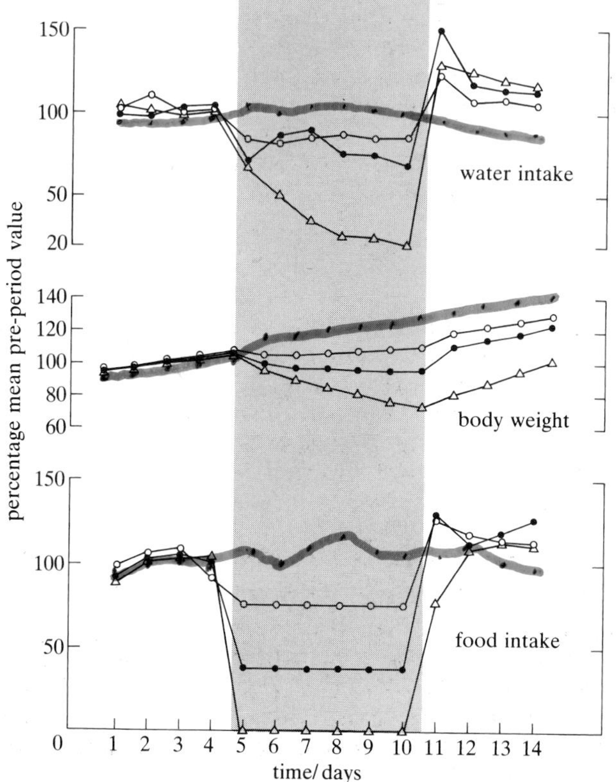

Figure 4.28 The effect of food deprivation or food restriction on water intake in rats allowed water ad libitum. Symbols: ●, food ad libitum; ○, two-thirds of normal intake; ●, one-third of normal intake; △, no food allowed. Food restriction was imposed only between days 5 and 10, as indicated by the shaded area; note the decline in water intake over this period.

of events is as follows. The kidney of the rabbit has a relatively poor ability to conserve sodium, and when the supply of sodium is halted the animal rapidly suffers sodium deficiency as a result of loss of sodium in the urine.

◇ The injection of hypertonic saline leads to a *rise* in plasma sodium concentration. In Section 4.5, we described the arousal of thirst by such injection. Can you suggest how a *fall* in plasma sodium concentration (in both rats and rabbits) is also able to lead to increased drinking?

◆ When plasma sodium concentration falls, by osmosis water migrates from the extracellular space into the cells. This results in a fall in extracellular volume, which triggers drinking.

The amount drunk by the food-deprived rabbit is brought to normal levels by replacing water as the available drinking fluid with a dilute saline solution. This enables body-fluid state to be returned to normal.

By contrast to such peripheral interactions, there are also well-recognized examples of central interactions between the feeding and drinking systems. These involve the central nervous system and can be characterized as a direct effect upon a motivational system. We shall now consider some examples of behaviour that implicate central interactions.

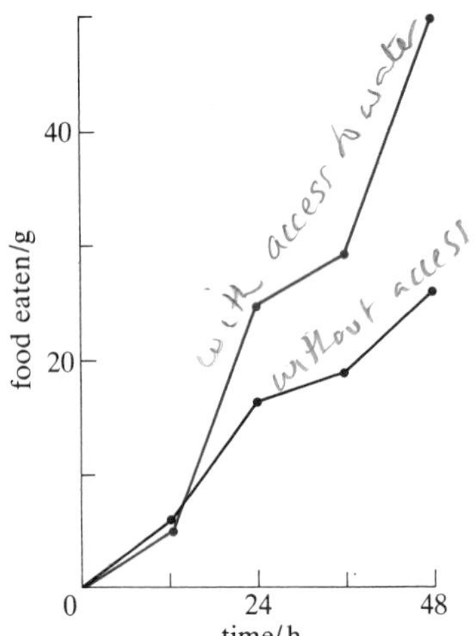

Figure 4.29 Food intake in rats with (red line) and without (black line) access to water.

Figure 4.29 shows the food ingestion of two groups of laboratory rats (*Rattus norvegicus*) over a period of time. One group is deprived of water and is compared with a group having water available ad libitum. Depriving the animal of water results in a reduction in food intake. As far as we know, the rat's energy needs do not fall during water deprivation. However, it is not difficult to appreciate the advantage to the animal of reducing food intake at a time of water shortage. Excretion of the breakdown products of metabolism leads to loss of water in the urine. Also food attracts water into the alimentary tract and this will have the effect of depleting the water in the extracellular fluid space. (The magnitude of this is shown in Figure 4.30.) It is suggested that there is a direct neural inhibition upon feeding, the signal in the inhibitory pathway arising from body-fluid depletion.

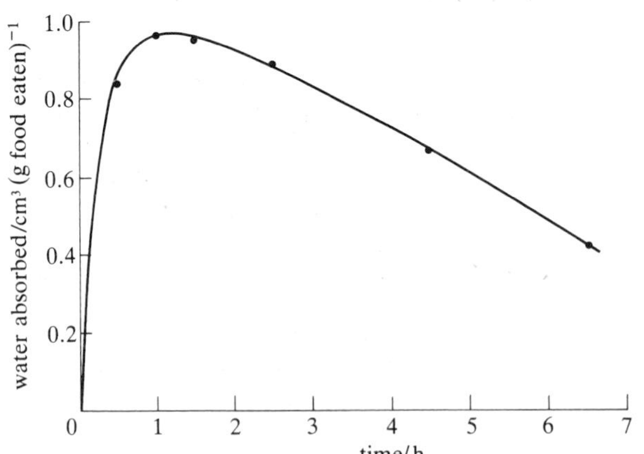

Figure 4.30 The amount of water absorbed into the stomach and small intestine of the rat as a function of the time after commencement of feeding at time zero. Water was not available for drinking during the experiment.

It might be supposed that a water-deprived rat cuts down on food intake simply because a dry mouth makes it harder to swallow food. Such a factor might play some role but it would seem not to be the whole explanation, as a number of pieces of experimental evidence show. For example, rats can first be trained to earn small food pellets by pressing a lever. The pellets of food can then be removed from the delivery apparatus, and the bar-pressing activity investigated. Such bar-pressing in the absence of food is assumed to provide an index of motivation that is divorced from the complications of swallowing food. Hungry rats were compared in the conditions (a) water deprived and (b) having water ad libitum. The water-deprived rats bar-press less energetically than those having water ad libitum. This suggests that the motivation to seek food has been lowered.

To understand another example of a central interaction, consider the time-course of the accumulation of water in the gut of a rat following a meal of standard laboratory chow, as shown in Figure 4.30. The rat had drinking water available except in the few minutes before the meal and during the meal. It can be seen that each gram of food pulls about 1 cm^3 of water into the gut*. Interestingly, in association with a meal of chow, rats drink something like 1 cm^3 of water per gram eaten. Logically, one might suppose that such meal- associated drinking can be explained by the dehydrating effect of feeding, an example of a peripheral interaction. That is to say, the loss of fluid from body fluids proper would arouse drinking by the processes described in Section 4.5. For a rat without access to drinking water, there is a rise in osmolality of the plasma following ingestion of a meal. This is caused by a move of water from the plasma of the blood to the gut. Undoubtedly, such dehydration would be a *sufficient* condition to arouse drinking. However, when we look closer at the time-course of dehydration, it is clear that, under laboratory conditions of ad libitum access to food and water, drinking normally occurs too soon after the meal to be explicable in these terms. Also, neither intragastric nor intravenous infusion of substantial amounts of water in association with a meal suppresses meal-associated drinking. Therefore we must conclude that, rather than drinking in response to systemic dehydration, the rat is able to pre-empt such dehydration.

The advantage of such timing is not difficult to see. Were the rat to work in the most simple negative feedback mode, it would be dehydrated by x cm^3 for each gram eaten. It would then need to drink x cm^3 to correct the deficit. Thereby $2 \times x$ cm^3 plus the food would be in its alimentary tract. According to the predictions of a computer simulation of the process, the gut would get very full. The result shown in Figure 4.30, amongst others, led researchers to use the expression **anticipatory drinking** to refer to meal-associated drinking. Such anticipatory drinking was seen as an example of *feedforward* rather than feedback.

So much for describing the mode of control; how do we explain what is going on? It has been proposed that by experience, i.e. a mechanism of *learning* (to be discussed in detail in Section 4.8), the rat comes to anticipate the forthcoming dehydration. A somewhat different interpretation of the evidence is as follows. A series of neuroendocrine changes occurs as food passes through the mouth, down the oesophagus and into the stomach. Eating provokes release of histamine from cells in the gastric mucosa. Histamine plays a vital role in the digestion of food. In addition, histamine has long been recognized as a dipsogenic (meaning 'thirst-provoking') substance. Blockage of histamine receptors with an antagonistic substance (see Chapter 2) inhibits more than 60% of meal-associated drinking.

◇ Can we unambiguously use this result, on its own, to support the conclusion that histamine is responsible for meal-associated drinking?

◆ It is compatible with this conclusion, but we need to be careful. Possibly the antagonist has a depressing effect on any behaviour. We need additional experiments to rule out this possibility.

* Note that the centimetre cubed (cm^3) is an SI unit of volume. For all measurements except those of high precision it can be taken as equal to the millilitre (ml).

In fact the antagonist fails to depress the drinking that is caused by water deprivation or salt injection; its anti-dipsogenic effect is specific to food-associated drinking.

So much for the rat; in spite of the concentration of research effort on rats, it can also be informative to look at the way other species solve their regulatory problems. Examine Figure 4.31. It shows when the Mongolian gerbil (*Meriones unguiculatus*) drinks in relation to when it eats. The gerbil is housed in a laboratory cage with ad libitum access to water and dry food.

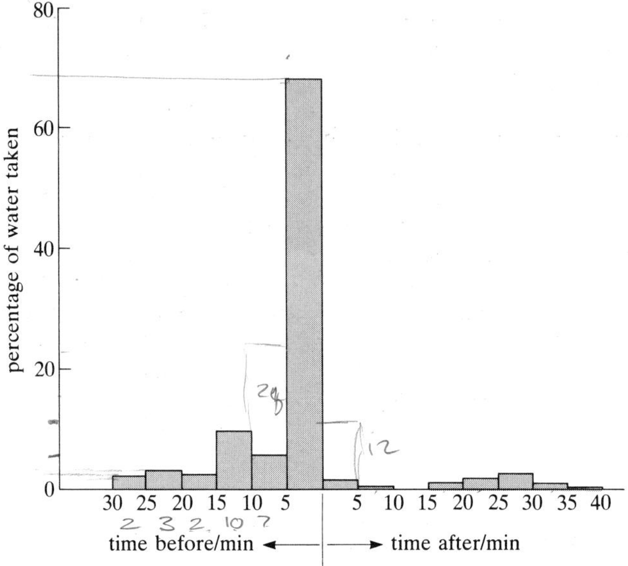

Figure 4.31 The timing of drinks in a group of gerbils in relation to meals.

◇ In what way does the gerbil's behaviour differ from that of the rat?

◆ The rat drinks predominantly in the period just after a meal, whereas most of the gerbil's daily water intake is in the 5 minutes just before a meal.

Before we try to explain this behaviour, it is important to note that a laboratory cage with dry food and free water is far removed from the ecological niche, the arid desert environment, in which the gerbil lives. Therefore, we need a considerable amount of caution. None-the-less, it is potentially useful to speculate on the animal's behaviour. In Figure 4.31, there is no evidence that feeding normally arouses drinking.

◇ Give two possible explanations for this.

◆ A mechanism of excitation from feeding to drinking might be absent in this species. In an environment with little or no free water, there would presumably be little adaptive value in an excitatory interaction from feeding to drinking. Alternatively, there might be such a mechanism present, but the effectiveness of drinking just before meals might be such as to provide all of the necessary water.

How do we explain the association in which drinking occurs before feeding? Suppose the gerbil simply wakes up once in a while and proceeds to take care of any needs it has. This would not amount to a true interaction between feeding/energy and drinking/body fluids, but might explain a correlation in time between meals and drinks. However, we can presumably dismiss this idea since one would expect, by chance, drinking to follow feeding as often as coming before it. As Figure 4.31 shows, drinking very rarely follows feeding.

◇ Given that, in both rats and gerbils, loss of body fluids has an inhibitory effect upon feeding, how might we explain the result shown in Figure 4.31?

◆ Ingested water would lift the inhibition upon feeding tendency, and thereby the gerbil's feeding tendency would increase in strength. The question as to what extent this process can explain the association needs further investigation.

Having looked at causal mechanisms of interaction between drinking/body fluids and feeding/energy, it is now appropriate to discuss their functional significance. To do so, it would be useful first to reconsider briefly the function of homeostasis. Looking at each individual system in isolation, it is clear that (a) it is in the animal's best interests to defend internal conditions at a certain level and (b) deviations from normal cause corrective action. However, looking at the interests of the *whole organism*, it can sometimes be to an animal's advantage to sacrifice partially the interests of one system in the interests of overall integrity of the body. For example, as Figure 4.29 shows, the rat that is water-deprived reduces its intake of food. This cannot be explained in terms of a reduction in energy need; metabolic rate does not fall. Indeed, the rat might need to be particularly active in the search for water. Acute arousal of thirst by hypertonic saline injection also depresses food intake, an effect that is immediately reversed by allowing the rat to drink water. The sacrifice of food intake is the result of inhibition upon feeding, and is in the interests of preserving body fluids. On a functional level, we might speculate that the ultimate arbiter of what is the optimum strength of such interactions is the overall fitness of the animal.

Summary of Section 4.7

1 In order to gain an understanding of individual systems (e.g. temperature regulation or drinking/body fluids), it is sometimes necessary either to eliminate or to hold constant the influence of other systems. At other times insight can be gained by exploring the interactions between biological systems.

2 A rat that is deprived of food or restricted in its access to food will cut down on water intake. This is because the demand for water is reduced.

3 A rat that is deprived of water or restricted in its access to water will cut down on its food intake. The evidence suggests the existence of an inhibitory pathway from the body-fluid system to the system of feeding/energy intake.

4 A rat drinks in association with a meal. Drinking is in anticipation of the loss of body fluids to the gut and thereby preempts such loss.

5 Whereas, in the laboratory with water and dry food available, rats tend to drink mainly in the period immediately after a meal, the Mongolian gerbil (*Meriones unguiculatus*) drinks mainly in the period just before a meal.

Now attempt Questions 18 and 19, on p. 124.

4.8 FEEDBACK AND FEEDFORWARD

In Section 4.2 the concepts of feedback and feedforward were introduced and an example of feedforward, drinking associated with feeding, was given in Section 4.7.1. In the present section, further examples of the exertion of control by feedback and feedforward are considered. The way in which feedforward control can arise as a result of learning mechanisms is investigated.

Just to reiterate, negative feedback control consists of responding to a disturbance in such a way that the response tends to correct the disturbance. Drinking *in response to* the rise in body-fluid osmolality caused by injection of hypertonic saline is an unambiguous case of negative feedback control. There is a disturbance in a regulated variable, and the disturbance instigates action, in this case drinking, that serves to correct the disturbance and return the system to its prior state. Feedforward is a process whereby action is taken *in anticipation of* a disturbance and thereby the disturbance is either pre-empted, or at least to some extent neutralized. There are a number of means by which feedforward control can be realized in a biological system, and, as we shall see, one of these involves a mechanism of conditioning. At this point, it is therefore necessary to make a short digression to consider the fundamentals of conditioning, as studied by the Russian physiologist Ivan Pavlov.

Pavlov's basic experiment is well known. Indeed the term 'conditioning' has entered the popular vocabulary, though it is often misused. Our understanding of conditioning has advanced greatly since the days of Pavlov. Though there are reports of conditioning going back at least to 1615, Pavlov is usually credited with the first scientific account of the phenomenon. Pavlov's discovery came whilst he was making a study of the factors causing the secretion of saliva and digestive juices in dogs. The secretion of digestive juices is a fixed, automatic (termed 'reflex') response to the presence of food in the stomach. Saliva is similarly secreted as a result of the stimulus of food in the mouth. Quite incidental to his mainstream research, Pavlov noticed that, in addition to the stimuli of food in the mouth and stomach, other stimuli were also able to provoke the secretion of saliva and digestive juices. Such things as the presentation of the food dish, which in the past had accompanied feeding, were able to elicit salivation. This led Pavlov to a formal series of experiments on the nature of the stimuli that could elicit salivation.

As is now well known, if a bell was rung (a **neutral stimulus**, NS) at the time of food presentation, then after a few pairings of bell and food, the sound of the bell presented on its own (i.e. in the absence of the food) was able to elicit salivation. The bell is termed 'neutral' because, on its own, it has no capacity to evoke salivation. Following pairing with food, the former neutral stimulus, the bell, is termed the **conditional stimulus** (CS), sometimes called the 'conditioned stimulus'. The meaning of this term is that the power of the bell to evoke salivation is *conditional* upon its earlier association with food; it has no intrinsic potency to elicit salivation. By contrast, food does have an intrinsic ability to evoke salivation; it does not require an earlier association. Therefore food would be described as an **unconditional stimulus** (UCS) to elicit secretion of saliva and gastric juices. The response elicited by the conditional stimulus is termed the **conditional response** (CR), and that elicited by the unconditional stimulus is termed the **unconditional response** (UCR). In this particular case, the CR is rather similar to the UCR. Both consist of the secretion of digestive juices, though the amount secreted by presentation of the CS is less than that caused by the UCS. However, in some conditioning situations the CR is very different from the UCR. For our purposes, suffice it to say that an otherwise neutral stimulus acquires a potency by virtue of its pairing with an unconditional stimulus. The procedure adopted by Pavlov and the kind of conditioning that results is termed **classical conditioning**.

Is the *pairing* of an NS (e.g. tone) and a UCS (e.g. food) important in order for the NS to acquire some of the capacity of the UCS? In order to conclude this unambiguously, a control group needs to be run. For example, dogs in an experimental group would be exposed to 20 presentations of tone followed by food. Dogs in a control group would be exposed to 20 presentations of a bell and food but at random times, not paired together. Only if the experimental group salivate more to the bell than do the control group would one accept that specifically pairing was responsible for giving the tone its capacity. When such control groups are run, they do show the power of pairing the NS and UCS.

Viewed in terms of the properties of biological systems, the function of classical conditioning is not difficult to appreciate. By acting in anticipation of the arrival of food in the mouth and stomach, the system is able to prepare the alimentary tract for the arrival of food. There is a rich variety of other biological systems that can be used to illustrate the general principle that it can often be to an animal's advantage to take anticipatory action. Such anticipatory action is made possible by a variety of Pavlovian conditioning. For instance, it is shown in the case of territorial defence by male Blue gouramis (*Trichogaster trichopterus*). Male fish were presented with a rival male. For some fish, the arrival of the new male was signalled by a red light. For others, it was unsignalled. Signalling gave the resident male a distinct advantage in the subsequent contest, the aggressive display being initiated sooner.

Though in psychology more recent studies of Pavlovian conditioning have now transformed the original interpretation, for our purposes the basis of this form of conditioning is the neural association formed between two *stimuli* to the nervous system. A previously neutral stimulus acquires some kind of potency by virtue of its association with a stimulus that has an unconditional physiological effect.

◇ What are some similarities between the processes of (a) classical conditioning and (b) feedforward?

◆ In both cases the word *anticipation* is appropriate to describe the process. (a) In the case of Pavlovian conditioning, by acting in response to the CS the system is prepared for the UCS. For instance, as far as juices of digestion are concerned, the mouth and stomach are in a condition prepared for the arrival of food. (b) In a feedforward system, by taking action in anticipation of a disturbance, the disturbance can to some extent be preempted.

Suppose that, rather than looking at simple reflexes, we stimulate a physiological negative-feedback system on a number of occasions and pair this stimulation with presentation of a neutral stimulus. It would then be interesting to see whether, by Pavlovian conditioning, the neutral stimulus has acquired any power to influence the negative feedback system. Has the neutral stimulus become a conditional stimulus? Over the years a large number of experiments of this kind have been performed, but up to quite recently the results were difficult to interpret (Eikelboom and Stewart, 1982). It was clear that neutral stimuli acquired a potency, i.e. became conditional stimuli, but it was very difficult to predict the direction of the conditioned effect. For example, in some cases, when a sound was paired with a physiological stimulus that caused hypothermia (a lowering of body temperature), the sound on its own would cause hypothermia. In other cases, it would cause hyperthermia (an increase in body temperature). Eikelboom and Stewart were able to bring a considerable amount of order to this chaos, by examining the negative feedback property of the systems under investigation.

Their analysis starts by noting that conditioning consists in forming an association between two stimuli to the nervous system (termed afferent stimuli). To understand the reaction to conditional stimuli, a knowledge of what the physiological challenge does to the nervous system is crucial. Let us return to considering Figure 4.20, a simple model of temperature regulation.

Suppose that we inject a drug that causes hypothermia by acting on the *afferent* arm, and pair a bell with this on a number of occasions. For instance, a drug might increase the sensitivity of warm neurons.

◇ What would be the predicted effect of sounding the bell on its own?

◆ One would expect a conditioned hypothermia.

Figure 4.32 represents this.

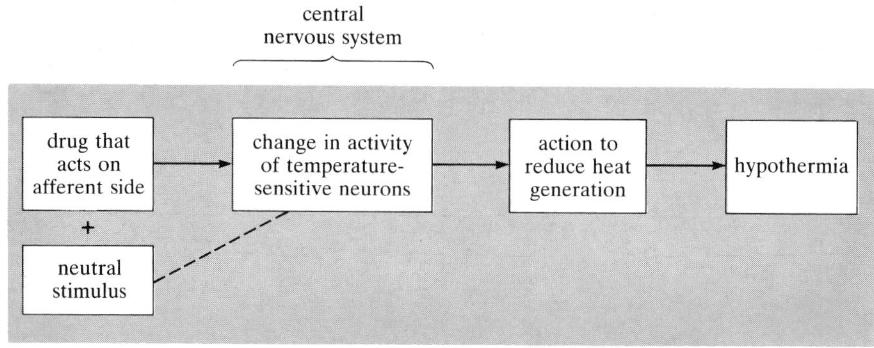

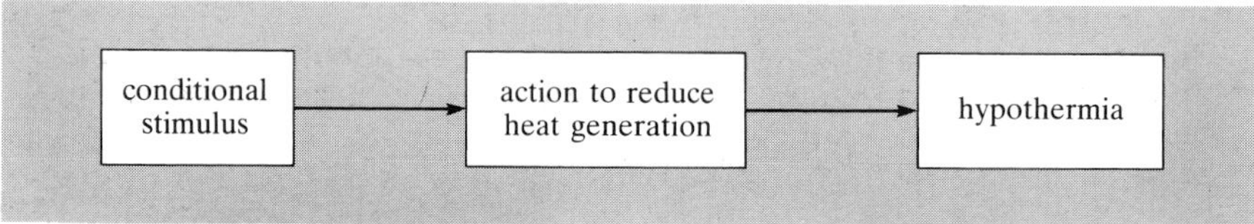

Figure 4.32 The sequence of events when a neutral stimulus is paired with a drug that acts on the afferent side of the temperature regulation system. During the conditioning phase, the neutral stimulus forms an association with the change in activity in the neurons of the central nervous system. Following conditioning, when the previously neutral stimulus is presented in the absence of the drug, it mimics the activity of the neurons and hence decreases heat generation. Hypothermia results. The neutral stimulus has become a conditional stimulus.

Suppose now that, rather than introducing a disturbance at the afferent side, we disturb the *efferent* side such as to result in hypothermia. For example, a drug might lower the metabolism of the cells of the body. This results in a changed neural input to the integrator arising from increased activity by cold neurons (afferent input) and, in consequence, action to generate heat. Suppose that we pair the sound of a bell with such a procedure, and the sound forms an association with the afferent input such that it is able to mimic this input. After a number of conditioning trials, we present the bell on its own.

◇ The physiological stimulus that the bell is paired with causes hypothermia. What would you predict the effect of the bell alone to be?

◆ The conditioned effect is one of hyperthermia, since the bell forms an association with the input to the integrator, which is such as to cause a heating effort.

Figure 4.33 represents this sequence of events.

The results do indeed fit the predictions based upon the known sites of action of drugs. The basic principle of the model can also be applied to exercise physiology and to the effects of drugs of abuse.

Muscular exertion in exercise is associated with an increase in heart rate, increased blood flow to the muscles concerned and an increase in energy

CONDITIONING PHASE

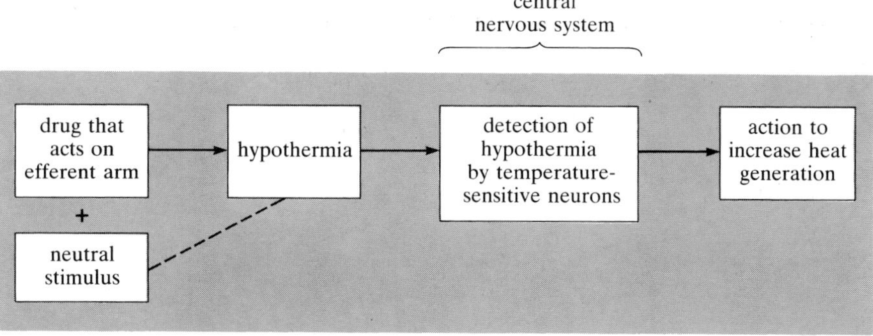

TESTING FOR CONDITIONING

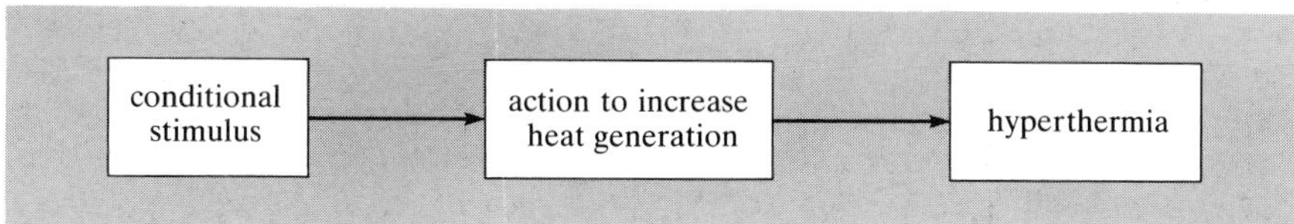

Figure 4.33 The sequence of events when a neutral stimulus is paired with a drug that acts on the efferent side of the temperature regulation system to induce hypothermia. In the conditioning phase, the neutral stimulus forms an association with the input to the central nervous system. Following conditioning, when the previously neutral stimulus is presented in the absence of the drug, it mimics the input to the central nervous system and hence causes increased heat generation and some hyperthermia. The neutral stimulus has become a conditional stimulus.

requirements (see Chapter 6). As early as 1940 an experiment was conducted that investigated the possibility of conditioning the various physiological responses associated with exercise. A human subject's exertion of effort on a bicycle ergometer was studied. Unconditional responses to this exertion included increased heart rate, blood pressure and respiration. An auditory stimulus was paired with the exercise and its conditional power then tested in the absence of physical exertion, i.e. the CS is given without the UCS. The subjects exhibited an increase in blood pressure in response to the CS.

Generally speaking, in Pavlovian conditioning, the strength of a CR depends upon the strength of the UCR with which it has been associated. For instance, if a bell is paired with a type of food that elicits much salivation, the bell will acquire a relatively large capacity to elicit salivation. In an experiment carried out in 1951, three different combinations of CS (tone) and UCS (muscular exertion) were tested on human subjects: (1) $tone_1$ + maximal exercise, (2) $tone_2$ + one-half maximal exercise and (3) $tone_3$ + one-sixth maximal exercise. It was found that the magnitude of the CR was proportional to the magnitude of the UCR produced by the various UCSs; $tone_1$ evoked the strongest CR and $tone_3$ the weakest. Similarly, a conditioned increase in blood pressure and heart rate is found when dogs with a history of exercising on a treadmill are later exposed only to stimuli paired earlier with exertion. Note that the heart is not responding to the *actual* need of the tissues for oxygen and glucose, but for the future need, a case of feedforward.

◇ In an evolutionary context, what might be the adaptive significance of the process revealed in these experiments?

◆ By increasing its heart rate and the blood flow to the muscles, and mobilizing energy reserves, the animal puts itself into a physiological state appropriate for exertion. In an evolutionary context, the exertion might be part of fleeing from danger, in which case there could be a distinct advantage in early preparation.

Note the direction of the conditioned effect associated with exercise; it is in the same direction as the physiological change with which it is paired. The withdrawal symptoms that joggers experience when they are not exercising might be due, at least in part, to their experiencing some of the cues that have been associated in the past with jogging. These cues start anticipatory changes in body physiology that are not realized in exercise. Siegel applies a similar logic to relapse in heroin addicts, the subject of the next discussion.

Suppose the body is subjected to extraneous stimulation from such substances as opiates (e.g. heroin) or benzodiazepines. The body responds to their presence by taking corrective negative feedback action that compensates for the disturbance to such processes as respiration. Now suppose that the substance is repeatedly administered in the presence of a set of stimuli, such as a particular room and use of a particular hypodermic syringe. Such stimuli might become conditional stimuli, in that they acquire some of the potency of the drug's action. Unfortunately, the situation is more complex than the drug actions we described earlier. It soon becomes apparent that opiates have multiple effects at various sites in the body. They induce changes in the central nervous system as well as peripheral changes. Therefore, there are at least two sets of effects with which, in principle, the CS could become associated: (a) the 'high' mood that the addict experiences as a result of the drug, and (b) the corrective action that the body takes to compensate for the disturbance caused by the drug. There is evidence that both reactions can become conditioned. Amongst other things, junkies sometimes report getting high by sham injections. However, not surprisingly from the viewpoint of medical science, most attention in the literature concerns conditioning of the compensatory mechanisms. What is of particular interest here is the relevance of conditioning to the phenomenon of **tolerance**.

If an animal, human or otherwise, is repeatedly injected with, say, morphine, then the effect of the injection decreases as the number of injections increases. In order for the drug to achieve a given effect, then larger and larger doses need to be given. This phenomenon is termed 'drug tolerance'. Its magnitude is enormous; for example, a morphine addict can take a dose many times higher than the dose that is lethal for someone experiencing morphine for the first time. With experience, changes in receptor number and binding characteristics (see Chapter 3) occur. However, in addition, conditioning seems to play an important part in the phenomenon of tolerance.

◇ If conditioning plays an important role in the development of drug tolerance, what would this imply about the circumstances under which the drug is administered?

◆ That there is a constant stimulus factor present that can form the CS. This might consist primarily in the injection procedure itself, but the room in which the injection was carried out might also form a CS.

◇ If the environment in which the drug is injected forms part, or all, of the CS for the conditioned compensation, what would be the effect of injecting the drug in an environment radically different from that in which the drug tolerance had been developed?

◆ The compensation would be weaker than normal and hence the drug would have a more potent effect.

In this context, it is interesting to note that sometimes drug addicts die from a dose that is relatively mild for a tolerant individual and low by their standards of experience. Of course, there could be a variety of factors implicated here, but conditioning seems to play a major role. In one survey, the majority of drug addicts who survived a heroin overdose reported that on the near-fatal occasion the drug was taken in an unusual environmental context. This suggests that the compensation that would normally be evoked was absent.

Only by appreciating the negative feedback principle that is at work in biological control systems, could we begin to make sense of these results. That is to say, a disturbance to a physiological variable causes action to be taken to correct the disturbance. Suppose we tried to understand the phenomena described here in terms of a simple chain of causality from conditional stimulus to conditional response, ignoring this feedback principle. This would leave us with the puzzle as to why sometimes the conditional response should be similar to the unconditional response (e.g. salivation to the food and to the bell) and sometimes diametrically opposite to the unconditional response (e.g. hypothermia to a peripherally acting drug but hyperthermia to a bell paired with drug injection). Therefore an understanding of feedforward requires an understanding of the nature of the feedback system with which it is associated.

Summary of Section 4.8

1 In the process of classical conditioning, a neutral stimulus to the central nervous system forms an association with an unconditional stimulus. Following conditioning, the neutral stimulus acquires some of the potency of the unconditional stimulus, and is then termed a conditional stimulus.

2 A neutral stimulus can be paired with an input that changes the state of a negative feedback system.

3 If a neutral stimulus is paired with a stimulus to the *afferent* side of a negative feedback system, the neutral stimulus will form an association with the afferent information. The neutral stimulus (now termed a conditional stimulus) will thereby acquire a capacity to act in the same direction as the stimulus with which it has been paired.

4 If a stimulus acts on the *efferent* side of a negative feedback system, it will cause, via feedback, an *afferent* signal to arise. As a result of this afferent information, action will occur that will counteract the effect of the stimulus to the efferent side. Pairing a neutral stimulus with such an efferent stimulus will result in an association between the two inputs to the central nervous system: the neutral stimulus (now termed a conditional stimulus) and the afferent information. Subsequent presentation of the conditional stimulus on its own will result in the regulated variable being changed in the opposite direction to the change with which it was paired.

Now attempt Question 20, on p. 124.

SUMMARY AND CONCLUSIONS

This chapter has discussed several examples of mechanisms that maintain constancy in the internal environment of an animal. The best-known such mechanism is that of negative feedback, the tendency of a deviation in a given variable to cause action that will eliminate the deviation. We described the regulation of physiological variables, e.g. body temperature and body-fluid volume. There are detectors of the variables that are regulated (e.g. temperature-sensitive neurons). As we have seen in the examples chosen, control is exerted in the service of regulation and this control can have both intrinsic physiological (e.g. sweating, excreting a urine concentrated in sodium chloride) and behavioural (e.g. moving to the shade, drinking, feeding) aspects. Controlled variables (e.g. urine flow-rate and sweat rate) are not measured by the body. Regulation is also achieved by the process of feedforward control acting together with feedback control; the animal can sometimes take action in the anticipation of a disturbance, and thereby pre-empt the disturbance. Even where the challenge is an abnormal one, such as intravenous drug injection, then both feedback and feedforward control act in such a way that physiological integrity is maintained. One means by which feedforward arises is through a process of Pavlovian conditioning, i.e. learning by experience.

In considering the systems of body-temperature regulation, drinking/body-fluids and feeding/body energy in mammals, a number of similarities and differences become apparent. The system of body-fluid regulation seems to function by monitoring the state of two representative parts of the body-fluid environment: the state of shrinkage of a neuron in the brain and the stretch of a blood vessel. Any disturbance to the body-fluid state would (a) normally arise rather gradually as a result of loss through, say, the kidney or sweating, and (b) soon be felt in either or both of the locations of the proposed detectors. A rather different system property is evident in the case of temperature and energy regulation. A temperature stress might often arise suddenly in the external environment, and considerable time might elapse before a temperature-sensitive neuron in the core of the body would perceive the disturbance. By then, serious hypothermia or hyperthermia might prevail elsewhere. In fact, as we have seen, the system regulating body temperature, though sensitive to core temperature, is also sensitive to temperature at, or near, the body surface.

In the case of regulation of energy state in mammals, some evidence points to the detector being located at the liver, a point that is sensitive to the energy supply. Some theorists have postulated the existence of a transducer that is sensitive to blood glucose level in an artery. However, the evidence for such a transducer playing a role in energy regulation is not good. Its ability to regulate energy state would be seriously hampered by the fact that intrinsic physiological mechanisms, e.g. the insulin–glucose system (Section 4.3; see also Chapter 10) exerts effective control over blood glucose level. Such a transducer would be buffered against disturbances, in a similar way to a core detector of body temperature in a large animal.

In one important respect for an animal such as a rat, the exertion of control over temperature and body fluids by behavioural means is rather simpler than the control of feeding. The commodity with which the animal needs to interact (i.e. heat and water, respectively) does not involve a complex recognition and decision mechanism. There are temperature-sensitive neurons at the periphery, and the mouth contains neurons whose frequency of firing is changed by coming into contact with water. Similarly, neurons that are sensitive, say, to the concentration of sodium ions in the fluids of the mouth could play a part in the control of ingestion of sodium chloride solutions at times of depletion. Ingestion of food by the rat has an area of uncertainty about it. The rat cannot recognize all potential poisons or needed dietary components by taste. Therefore it needs to form associations between whatever taste cues are available and the consequences of ingestion. Rats have the ability to form associations between taste and subsequent physiological events; this is an example of an adaptive control system.

OBJECTIVES FOR CHAPTER 4

Now that you have read this chapter, you should be able to:

4.1 Define and use, or recognize definitions and applications of, each of the terms printed in **bold** in the text.

4.2 Distinguish between the terms 'regulation' and 'control', and give examples of each. (*Question 1*)

4.3 Define what is meant by feedback, distinguishing it from feedforward. Distinguish between negative and positive feedback, and give examples of each in physiology. (*Questions 2–4 and 19*)

4.4 Explain what is meant by homeostasis, and give examples of two processes that serve to maintain homeostasis. (*Questions 1, 2 and 4*)

4.5 Distinguish between intrinsic physiological and behavioural controls over regulated physiological variables, and give an example of where both can be employed in the case of a given physiological variable. (*Questions 5 and 13*)

4.6 Describe some of the common operators that represent in block diagram form the components of a physiological system. (*Question 6*)

4.7 Show how the operators represented in block diagram form can be assembled to build up a system. (*Questions 6–8*)

4.8 Describe the performance of a relatively simple control system in terms of the block diagram components that make it up and the variables of the system. (*Questions 6–8, 10 and 11*)

4.9 Describe how human body temperature is regulated by autonomic and behavioural control. (*Questions 9–11*)

4.10 Explain how neurons, whose activity is sensitive to temperature, could form part of a system that regulates body temperature. (*Questions 9 and 10*)

4.11 Explain the meaning of the term 'set-point', and why it is commonly used in the literature of physiological regulation. Explain also why one needs to exercise caution in the use of the term.

4.12 In terms of the properties of a negative feedback system, give an account of our understanding of the modes of action of fever and drugs on temperature regulation.

4.13 Distinguish between the terms 'afferent' and 'efferent', and explain their relevance to the feedback control of body temperature. (*Question 11*)

4.14 Explain how drinking and the kidney jointly exert control over body fluids. (*Questions 12 and 13*)

4.15 Describe the bodily states that instigate drinking and those that serve to terminate it. (*Questions 12–14*)

4.16 Explain the function served by short-term feedback loops in the regulation of body fluids. (*Question 12*)

4.17 Contrast diabetes insipidus and schedule-induced polydipsia as far as the control actions serving regulation of body fluids are concerned.

4.18 Draw a simple block diagram to represent the feeding system of the house fly, and in the context of this diagram explain the meaning of the term 'open-loop'. (*Question 15*)

4.19 Give an account of the feeding system of the leech (*Hirudo medicinalis*), explaining how a knowledge of neurons and their interconnections is relevant to understanding how the system works. (*Question 15*)

4.20 Explain, with reference to the rat suffering damage to the ventromedial hypothalamus, why extreme caution is needed in the interpretation of experimental evidence.

4.21 Describe some processes other than detection of energy state that are involved in the control system of feeding in the rat, and in this context explain what is meant by the terms sensory-specific satiety, taste-consequence learning and adaptive control system. (*Questions 16 and 17*)

4.22 Explain why we sometimes need to study a given regulatory system in isolation from other systems, and at other times need to investigate the interactions between systems. (*Questions 18 and 19*)

4.23 Distinguish between interactions that arise (a) in the peripheral physiology of the energy and fluid regulation systems and (b) in the central nervous system processes underlying the motivation of feeding and drinking. (*Questions 18 and 19*)

4.24 Describe the possible adaptive significance of some of the interactions between feeding and drinking. (*Question 19*)

4.25 Describe what is meant by the expression 'classical conditioning' and how it relates to the expression 'anticipation'. (*Question 20*)

4.26 Explain how, in a feedback control system, the mechanism of classical conditioning can provide feedforward control. (*Question 20*)

4.27 Explain how, in terms of the function of feedback systems, we can understand that in some situations a conditional stimulus can have effects that are opposite to those of the physiological stimulus with which it has been associated.

QUESTIONS FOR CHAPTER 4

Question 1 (*Objectives 4.2 and 4.4*) Following an intravenous injection of hypertonic saline (meaning a saline solution more concentrated than the body fluids), the kidney of a rat is observed to excrete a strongly hypertonic urine. Write a sentence interpreting this in terms of regulation and control.

Question 2 (*Objectives 4.3 and 4.4*) Rats that have been depleted of body sodium exhibit a strong appetite for a solution of concentrated sodium chloride. Explain how the term 'feedback' could be applied to the system of which such salt appetite forms part and what kind of feedback it involves.

Question 3 (*Objective 4.3*) Chapter 1 describes the relationship between membrane potential and membrane permeability during the action potential in excitable cells. Is there any evidence for positive feedback in this relationship?

Question 4 (*Objectives 4.3 and 4.4*) Is drinking in response to blood loss (haemorrhage) an example of feedback control or feedforward control?

Question 5 (*Objectives 4.4 and 4.5*) Suppose a laboratory rat is deprived of drinking water for a period of time. Following the return of drinking water, the rat immediately starts drinking. Identify an intrinsic physiological control that, during the period of water deprivation, serves the same end as the behavioural control described here.

Question 6 (*Objectives 4.6, 4.7 and 4.8*) Represent in block diagram form the fact that the net rate of heat entry into a physiological system is given by the difference between (i) the rates of heat production and heat gain from the environment and (ii) the rate of heat loss to the environment. On a separate diagram, show in block diagram form that the amount of heat in the body of an animal is given by the integral of the net rate of heat flow. Finally combine the two diagrams to show the relationship between heat flows and the amount of heat in the body.

Question 7 (*Objectives 4.7 and 4.8*) With reference to Figure 4.12 and against a background of a steady rate of glucose intake, as far as the variable 'total insulin' is concerned, what would be the effect of a sudden injection of a large amount of insulin? Show a graph of total insulin against time.

Question 8 (*Objectives 4.7 and 4.8*) Now for a tough one, but have a go at it! With reference to Figure 4.12, consider that over a period of time the system exhibits a steady state, with glucose intake exactly matched to glucose transport into cells. What does this imply about the signal entering the integrator labelled block 6?

Question 9 (*Objectives 4.9 and 4.10*) What exactly does it mean to say that a neuron is 'temperature-sensitive'?

Question 10 (*Objectives 4.8, 4.9 and 4.10*) Refer to Figure 4.19. Suppose a drug is injected that has the effect of increasing the sensitivity of hypothalamic neuron 1. What would be the predicted effect of this as far as body temperature is concerned?

Question 11 (*Objectives 4.8, 4.9, 4.10 and 4.13*) With reference to Figure 4.20, consider the effect of the local application into the brain of a drug that causes a sharp increase in the firing rate of any neuron in the vicinity of the injection. The drug acts at point A. Contrast this effect with that of the drug when it acts at point B.

Question 12 (*Objectives 4.14, 4.15 and 4.16*) Suppose we were able to cut the short-term feedback loops in the drinking control system. What influence would this be expected to have upon the amount a rat drinks in response to a given deficit in body fluids?

Question 13 (*Objectives 4.5, 4.14 and 4.15*) A group of rats was injected with an amount of hypertonic sodium chloride solution. Drinking water was removed just before the injection. The drinking water was returned at a certain time after the injection. Within a time range of two hours, the amount that the rat drank decreased as a function of the length of time elapsing before water was returned. How might we explain this?

Question 14 (*Objective 4.15*) A rat has a choice of various concentrations of sodium chloride solution to drink. Following haemorrhage, what is the optimal drinking fluid for the quickest and most efficient restoration of the normal state of body fluids?

Question 15 (*Objectives 4.18 and 4.19*) Given the experimental set-up available to Lent and Dickinson, how might one open the feedback loop of the leech's feeding system?

Question 16 (*Objective 4.21*) In phase II of the experiment whose result is shown in Figure 4.27, the food dish and food were changed each 30 minutes even when the same food was available throughout a two-hour session (e.g. AAAA). Why did the experimenters consider it necessary to do this?

Question 17 (*Objective 4.21*) Contrast the underlying mechanisms responsible for the behavioural phenomena of (a) the sodium-deficient rat favouring a solution rich in sodium chloride and (b) a rat deficient in vitamin B favouring a diet containing vitamin B.

Question 18 (*Objectives 4.22, 4.23 and 4.24*) Two groups of rats are deprived of water for 24 h and then their drinking is measured over a 1 h period following the return of water. Group 1 had food ad libitum during the 24 h period, whereas group 2 is simultaneously deprived of food and water. In the 1 h drinking observation period, which group would be expected to drink more, and why?

Question 19 (*Objectives 4.3, 4.22, 4.23 and 4.24*) Why are the terms 'anticipatory' and 'feedforward' applicable to the drinking that a rat performs in association with a meal?

Question 20 (*Objectives 4.25 and 4.26*) In the experiment using a bicycle ergometer, described in Section 4.8, what kind of control group would one need to run in order to be sure that it was specifically *conditioning* by pairing that was revealed by the potency of the tone?

REFERENCES

Benzinger, T. H. (1964) The thermal homeostasis of man, in G. H. Hughes (ed.) *Homeostasis and Feedback Mechanisms*, Cambridge University Press, pp. 49–80.

Bolles, R. C. (1980) Some functionalistic thoughts about regulation, in F. M. Toates and T. R. Halliday (eds.) *Analysis of Motivational Processes*, Academic Press, London, pp. 63–75.

Booth, D. A. (1976) *Hunger Models*, Academic Press, London.

Dethier, V. G. (1976) *The Hungry Fly*, Harvard University Press, Cambridge, Mass.

Duggan, J. P. and Booth, D. A. (1986) Obesity, overeating and rapid gastric emptying in rats with ventromedial hypothalamic lesions, *Science*, 231, 609–611.

Eikelboom, R. and Stewart, J. (1982) Conditioning of drug-induced physiological responses, *Psychological Review*, 89, 507–528.

Guyton, A. C. (1986) *A Textbook of Medical Physiology*, W. B. Saunders, Philadelphia.

Lent, C. M. and Dickinson, M. H. (1988) The neurobiology of feeding in leeches, *Scientific American*, 258 (6), 78–83.

5.1 WHY HAVE A CIRCULATORY SYSTEM?

The metabolic processes of all cells require a continuous supply of oxygen and nutrients. These same processes generate waste products, which must be removed from the cells. In small animals the process of diffusion can take care of supply and disposal, but in larger animals diffusion alone cannot maintain an appropriate rate, and a circulatory system of some sort is present. This chapter is about the variety of circulatory systems found in animals, but much of the information about the mechanics is drawn from vertebrates, particularly mammals. It is not too surprising that this should be so, considering the extensive medical interest in the human heart. What is surprising, however, is that the knowledge that there is a circulatory system associated with the heart lay hidden until the early part of the seventeenth century, when William Harvey published his *Treatise on the motion of the heart and blood*. Prior to that date it was thought that the blood ebbed and flowed in the blood vessels as the heart contracted and relaxed. Harvey made an anatomical study of the human heart, noting the valves and deducing that they permitted flow in one direction only. He made some measurements on contracting hearts and calculated that in one hour the heart could pump into the arteries a weight of blood greater than the weight of the whole body. He did not himself observe that the arteries and veins were linked, but he did deduce that such links must exist since he could see that blood flowed through the heart and arteries in one direction only.

Flow is the underlying theme of this chapter. We shall be looking at the problems of obtaining an appropriate flow of blood through tissues, the means of controlling the flow, and the means by which flow is matched to demand in a dynamic system. The driving force behind the evolution of circulatory systems has been the need to maintain diffusive movement of molecules into, and out of, cells at the least energetic cost. Circulation is really an *additional* feature of effective diffusion, so it is appropriate that we should start by considering the mechanism of diffusion, and exchange between blood vessels and tissues.

5.1.1 Factors affecting diffusion

The main functions of the circulatory system are to transport nutrients, respiratory gases, hormones, waste products of metabolism, and cells.

For some animals, transport of these materials can be achieved largely by passive diffusion. In most unicellular organisms slow diffusion between the cell and the external aqueous environment is sufficient.

◇ Name two other factors that influence diffusion rate, in addition to the distance over which diffusion takes place.

◆ The surface area over which diffusion takes place, and the concentration difference between the ends of the diffusion path (i.e. the concentration gradient).

You may be familiar with a relationship formulated by Fick in 1855, which enables us to calculate the rate of diffusion across a cell membrane, if we know the properties of the membrane. **Fick's equation** for free diffusion is

$$J_{oi} = P_d A(C_o - C_i) \tag{5.1}$$

where

J_{oi} is the rate of diffusion from ***o***utside to ***i***nside
P_d is a diffusion constant combining a number of properties of the membrane and the absolute temperature
A is the surface area of the membrane
$(C_o - C_i)$ is the concentration gradient from ***o***utside to ***i***nside

Once a substance has crossed the membrane, the rate at which it diffuses through the cytoplasm is inversely proportional to the square of the distance over which diffusion takes place. So, any increase in the diffusion distance, for instance as a result of an increase in the size of an organism, will greatly decrease the rate at which diffusion can occur, and therefore lower the total fluxes into and out of the organism. If the metabolic demands of animals are calculated and compared with the rates of diffusion of gases through the cytoplasm, it becomes clear that diffusion alone is unlikely to satisfy the oxygen demands of organisms where the diffusion distances exceed about a millimetre.

This figure does not necessarily mean that animals larger than 1 mm cannot utilize diffusion. A flatworm such as *Dendrocoelum* is about 10 mm long and relies on diffusion for its oxygen supply. However, it is only around 1 mm thick, which means that diffusion distances remain small. Many sponges are much larger than flatworms and are often nearly spherical. They too rely on diffusion for gas exchange. Figure 5.1 shows a vertical section through a sponge.

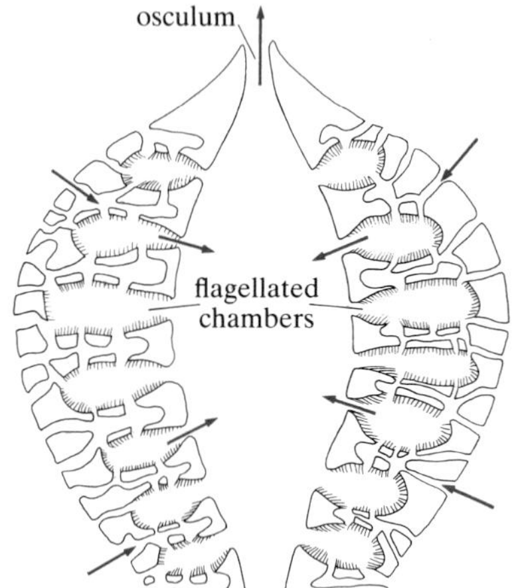

Figure 5.1 The flow of water through the chambers of a sponge. The arrows show the direction of water flow produced by the beating flagella.

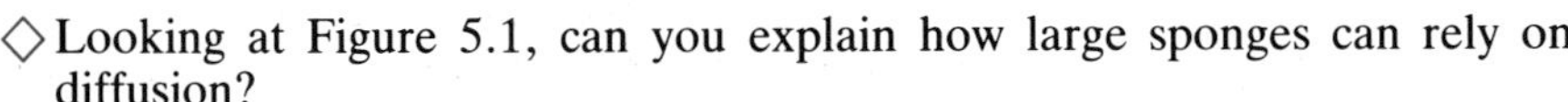
◇ Looking at Figure 5.1, can you explain how large sponges can rely on diffusion?

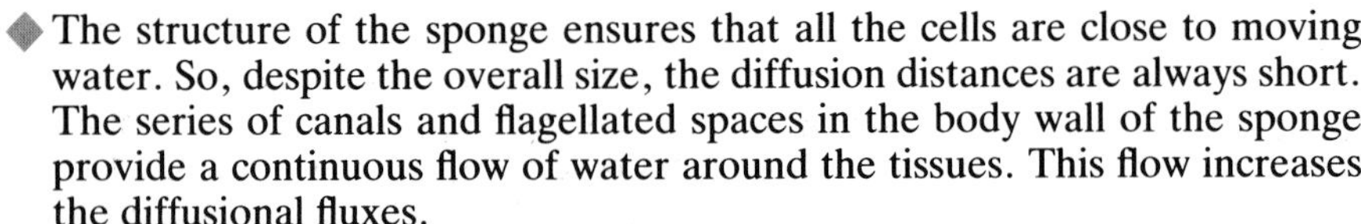
◆ The structure of the sponge ensures that all the cells are close to moving water. So, despite the overall size, the diffusion distances are always short. The series of canals and flagellated spaces in the body wall of the sponge provide a continuous flow of water around the tissues. This flow increases the diffusional fluxes.

However, this structure precludes the development of thick tissue beds or discrete organs of any size. If we look at all those animals that do not possess any form of circulatory system we find that, although they may be large, the tissues are not very thick. Echinoids (Figure 5.2) are spherical, and may be several inches across, but there is not a great deal of tissue inside their calcium carbonate shell (the shell is called a 'test') and they do not have a circulatory system. However, the water vascular system that links the tube-feet together almost certainly provides a pathway for oxygen into the tissues. The fluid in the vascular system is under pressure, which would tend to force water out through the walls of the tube-feet. This pressure is more than balanced by the osmotic difference between the fluid and the seawater outside, which means that there is a steady flow of water inwards. This flow brings in oxygen.

These systems serve some or all of the basic functions listed at the beginning of this section, but for larger animals these functions necessitate a circulatory system. To find examples, we need to look at other groups of animals, such as arthropods, molluscs and vertebrates.

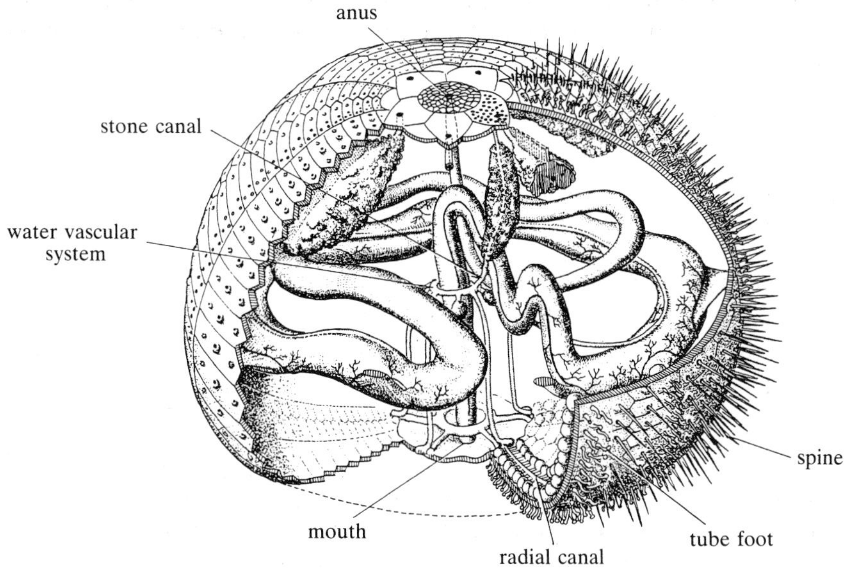

Figure 5.2 An echinoid (sea urchin).

5.1.2 Dynamic compartments

A circulatory system evolved in animals as tissues became differentiated to form discrete organs with a variety of functions. Inevitably there was an increase in size, and a point was reached at which diffusion alone could no longer supply nutrients and gases at the rate required by the cells. Most animals have a circulatory system in which fluid is pumped by one or more hearts. Two main types of circulation are distinguishable: open and closed.

In an open circulatory system (Figure 5.3), fluid is pumped through a series of tubes by the heart and then dumped into the body cavity. In insects the heart lies in a pericardial cavity. When the heart contracts, the pericardial cavity is enlarged and fluid is drawn in from the main body cavity—the perivisceral cavity—and enters the heart. The entry of fluid into the heart causes it to expand. On the next contraction the fluid is forced out of the heart back into the perivisceral cavity. Since the maximum size of the perivisceral cavity is determined by the rigid exoskeleton, the fluid pumped out by the heart raises the pressure in the perivisceral cavity, which forces fluid back into the pericardial cavity.

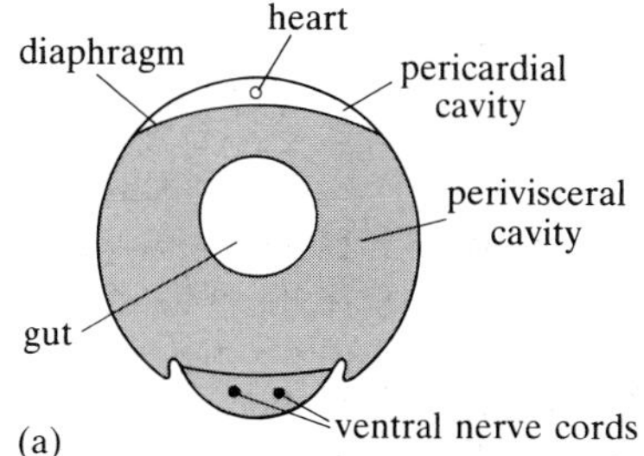

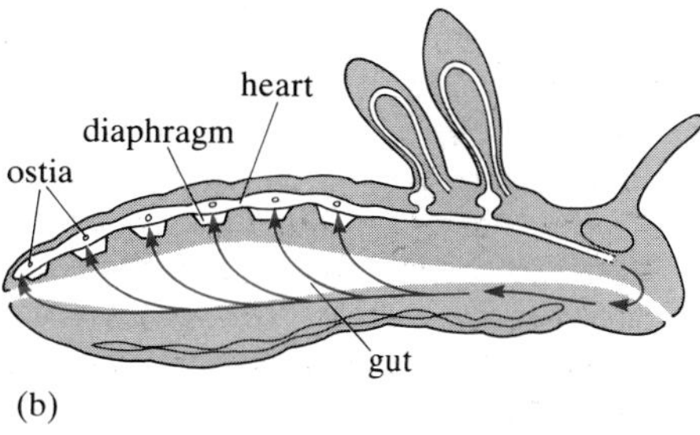

Figure 5.3 (a) Transverse section through the abdomen of a typical insect, showing the relationship between the heart and the two main body cavities. (b) Schematic longitudinal section of a typical insect, showing the direction of blood flow (red arrows).

A closed circulatory system is the one that will probably be more familiar to you since it is the type found in almost all vertebrates, as well as annelid worms and some molluscs. This type, as the name implies, is a closed system in which the circulating transport medium—the blood—is contained within tubular vessels that form a closed loop. It is important to realize that the circulatory system provides for the needs of the tissues, and that different tissues have different requirements. Tissues such as skin, muscle and liver all have different metabolic rates and require different flow rates of nutrients and gases. Also, each of these rates may vary with time. So the circulation is not a static entity: it is made up of a number of dynamic compartments.

For the present discussion it is useful to regard each tissue as a separate compartment. It will have varying requirements for oxygen depending upon its metabolic rate and hence varying requirements for blood flow, because the flow rate through the tissue sets limits on the rate of cellular metabolism. The blood vessels themselves are not static tubes; their diameter can alter. Finally the pump itself, the heart, does not have a fixed rate of contraction or a fixed output per contraction. It too is dynamic. All these dynamic compartments are linked and changes in one can affect the others, but in normal circumstances the tissues can be regarded as almost independent. When you suddenly accelerate to catch a bus, the leg muscles 'take' more blood through them per second independently of the flow through other tissues, and the circulatory system responds to the change in demand. Our human view of ourselves is that the heart has primacy over all other organs. For example, heart transplants are a much more emotive subject than liver or kidney transplants. Yet this central position of the heart is an illusion because, in the circulatory system, the blood vessels and heart RESPOND to the demands made by the tissues—the tissues have primacy. The response of the circulatory system to increased demand must be to increase the flow rate.

◇ Why is the response to increased demand an increased flow rate?

◆ To satisfy increased demand, the supply must be increased. This is achieved by an increase in the flow rate.

The mammalian circulatory system as a 'demand-led' system is considered further in Section 5.4. We must now examine the structure of the circulatory system and the means by which blood is supplied to the tissues. Remember that, at any instant, only a small percentage of the total blood volume is actually involved in exchange of materials with the tissues—some 6% in humans.

Summary of Section 5.1

1 The circulatory system maintains effective diffusion into and out of tissues, and transports nutrients, gases, hormones, cells and the waste products of metabolism.

2 Fick's equation describes the diffusion of a substance from a region of higher concentration to one of lower concentration.

3 Not all animals have a circulatory system. Effective diffusion can occur without a circulatory system if the distance between all cells and the environment is less than 1 mm.

4 There are two main types of circulatory system. In open systems the blood drains back into the heart from the body cavity. In closed systems the blood is contained within blood vessels that form a continuous loop.

5 The mammalian circulatory system is demand-led: the heart and blood vessels *respond* to demands made by the tissues.

Now attempt Question 1, on p. 164.

5.2 BLOOD VESSELS AND BLOOD FLOW

The rate at which blood flows through tissues sets limits on the rate of cellular metabolism, as you saw in Section 5.1.2. The primary factors governing the flow rate of blood through a blood vessel are the pressure gradient along the vessel (Δp) and the resistance to flow (R).

The pressure in the system is the result of the heart contracting and applying a force to the blood. Because fluid is virtually incompressible, the potential energy of the blood is increased and it flows through the blood vessels. As it does, so the potential energy will be converted into kinetic energy as a result of the friction between fluid molecules and the walls of the vessels. The resistance to flow is determined by the dimensions of the blood vessels, the structure of their walls and the properties of the fluid flowing through them. Figure 5.4 shows an artery and a vein in cross-section, exposing the structure of the walls. The innermost lining of both vessels is called the endothelium. Outside this is a layer of smooth muscle, which is generally very much thicker in arteries than in veins. Covering the muscle layer is a fibrous coat which is also generally thicker in arteries. Table 5.1 (*overleaf*) provides a summary of the characteristics of the blood vessels of a mammal. This Table will be useful for reference as you work through the rest of this chapter.

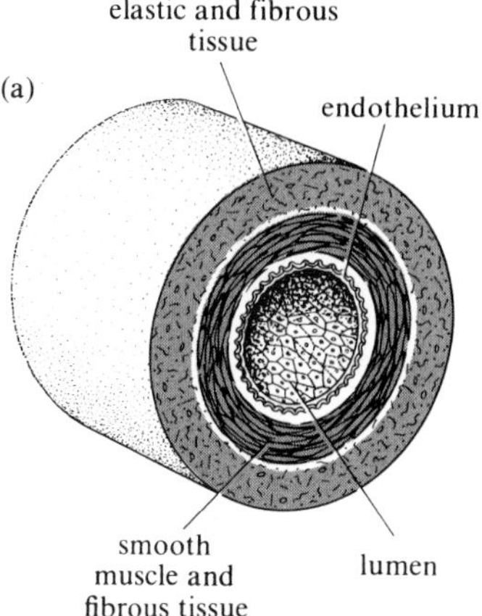

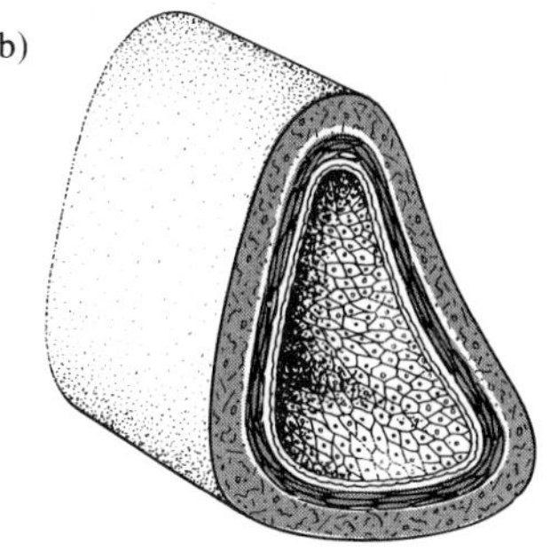

Figure 5.4 Cross-sections of (a) a generalized artery and (b) a generalized vein.

◇ From rows 1 and 4 in Table 5.1 what do you notice about the relative thickness of the walls of arteries and veins?

◆ The walls of the arteries are about twice as thick as those of veins. This is necessary because the blood flowing through arteries is generally at a higher pressure than that flowing in veins.

◇ From row 10, what happens to the rate of blood flow around the circulation?

◆ The rate of flow decreases as blood passes from the aorta to the capillaries, but increases on the venous side.

◇ From rows 3 and 10, what is the relationship between the rate of blood flow and the diameter of the blood vessel?

◆ The rate at which the blood flows is related to the diameter of the vessel for arterial and venous vessels; that is, the flow rate decreases with a decrease in the diameter of the vessel.

◇ Both the length (l) and the radius (r) of a blood vessel affect its resistance to flow. How do you explain this?

◆ The potential energy of the blood is lowered as a result of frictional losses. There are two components of frictional loss: friction between the fluid and the wall, and friction between the fluid molecules themselves. Since the size of the frictional force will be related to the surface area in contact with the fluid, both radius and length will affect the resistance.

The resistance increases markedly as the vessel radius decreases. In fact resistance is inversely proportional to the fourth power of the radius

$$R \propto 1/r^4 \quad (5.2)$$

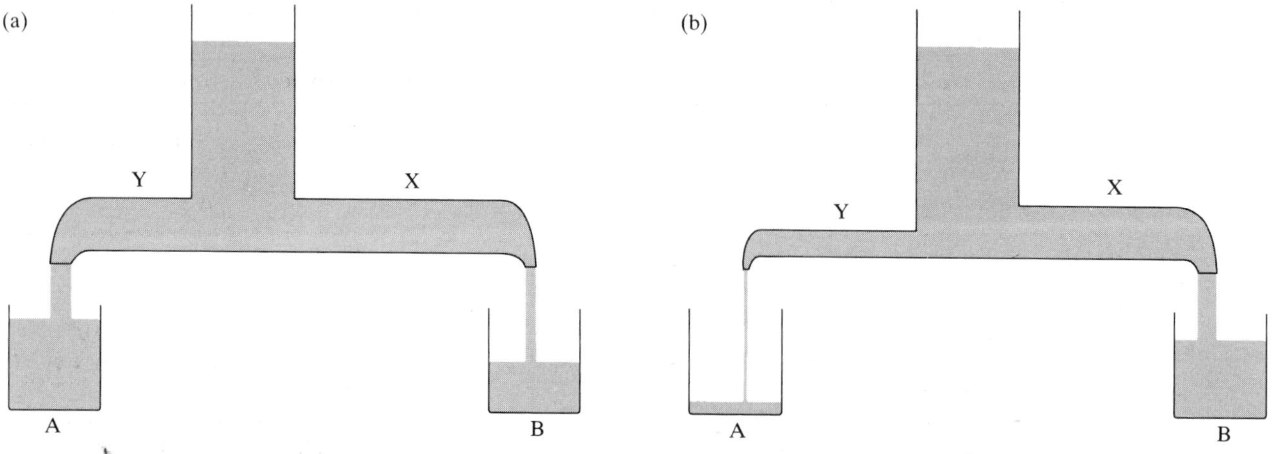

Figure 5.5 The effect of length and radius on resistance. (a) The flow rate is proportional to the length: X is twice the length of Y, so the rate of flow into A is twice that into B. (b) The flow rate is proportional to the fourth power of the radius: X is twice the radius of Y, so the rate of flow into B is sixteen times that into A.

The effect of radius on resistance can be seen in Figure 5.5. Doubling the radius of the vessel decreases its resistance by 16 times and therefore provides a 16-fold increase in flow.

There is greater friction between molecules in some fluids than in others. In treacle there is so much friction between molecules that it flows very sluggishly at room temperature. It is said to have a high viscosity(η). Water has a low viscosity.

Poiseuille's Law
A relationship that describes the flow of a fluid down a tube with rigid walls.

◇ You should now be able to write down a relationship of proportionality that describes the flow rate, J, of fluid down a tube with rigid walls.

◆ $J \propto 1/R$

So, from Equation 5.2, $J \propto r^4$. Also $J \propto 1/\eta$ and $J \propto 1/l$ and $J \propto \Delta p$, the difference in pressure along the blood vessel. Combining all these relationships for J gives

$$J \propto \frac{\Delta p r^4}{\eta l} \qquad (5.3)$$

Doesn't take into account that blood vessels stretch & store energy ie not rigid & blood contains cells etc ie isn't simple fluid.

In practice, this relationship is often simplified by combining the factors affecting resistance, the radius, viscosity and length ($r^4/\eta l$) into a single variable representing resistance to flow. So we get

$$J = \Delta p / R \qquad (5.4)$$

This equation is usually called **Poiseuille's law**, after a nineteenth century physiologist who sought to define the factors controlling the flow of blood by investigating the flow of liquid through capillary tubes. It is analogous to Ohm's law, which relates potential difference, current and electrical resistance

potential difference = current × resistance

From this relationship you can see that potential difference increases as resistance increases. In the circulatory system the greatest drop in the blood pressure occurs when the blood flows through the vessels with the highest resistance.

Poiseuille's law is of considerable use in biology when considering flow through tubes, but it must be used with caution.

Table 5.1 Characteristics of the blood vessels of a mammal. Asterisks denote values for one part of the circulation (the intestine) of a dog. Other figures are whole-body values for a human adult

	Aorta	Artery	Arteriole	Sphincter	Capillary	Venule	Vein	Vena Cava
1 Relative cross-sectional area								
2 Composition of walls—tissue type: endothelium elastic muscle collagen								
3 Diameter of lumen	2.5 cm	0.4 cm	30 μm	35 μm	8 μm	20 μm	0.5 cm	3 cm
4 Thickness of walls	2 mm	1 mm	20 μm	30 μm	1 μm	2 μm	0.5 mm	1.5 mm
*5 Number of vessels	1	2 400	40 000	—	1 200 000	80 000	2 400	1
*6 Total cross-sectional area/cm^2	0.8	5	125	—	600	570	30	1.2
*7 Length of vessel (approx)/cm	40	5	0.2	—	0.1	0.2	5	40
*8 Total volume/cm^3		190		—	60		680	
9 Blood pressure/ kPa	13.3	12.0	8.0	—	4.0	2.7	2.0	1.3
10 Rate of blood flow/cm s^{-1}	40	40–10	10–0.1	—	less than 0.1	less than 0.3	0.3–5	5–20

◇ In circulatory physiology, why should we need to be cautious about our use of Poiseuille's law?

◆ Because Poiseuille's law assumes that the walls of the tubes are rigid. You will know from personal experience that the reason that you can feel your pulse is that the blood vessel in your wrist is expanding and contracting in synchrony with the beating of your heart.

In fact, the elasticity of the walls of blood vessels is an important feature of their structure. The heart contracts rhythmically. At each contraction the walls of the arteries expand, storing potential energy (Figure 5.6). As the heart muscle relaxes after the contraction the walls of the arteries contract again, imparting their stored energy to the blood and evening out the flow of blood through the system.

Another reason for using Poiseuille's law with care is that blood is not a simple fluid because it contains cells. The viscosity of the blood is not the same as the viscosity of the plasma because of the presence of the cells, and this difference is particularly noticeable when blood flows through tubes smaller than 0.5 mm in diameter. In fact the viscosity is lower than would be predicted.

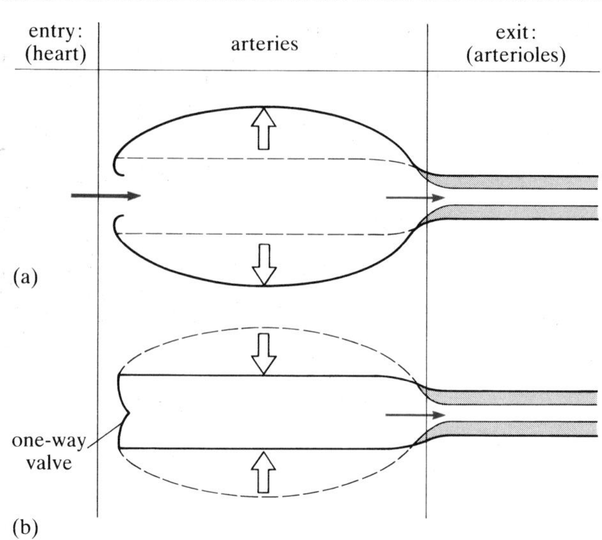

Figure 5.6 The movement of blood into and out of arteries during the cardiac cycle. The lengths of the red arrows denote relative amounts of blood. (a) When the heart contracts, less blood leaves the arteries than enters and the arterial walls stretch. (b) When the heart relaxes, the walls recoil passively, driving blood out of the arteries.

5.2.2 Blood: the transport medium

Transport media are classified according to whether they are part of an open or a closed system. The term 'blood' is usually reserved for the fluid within a closed circulation. The fluid in an open circulation performs many of the functions of blood and has a similar composition, but it is called **haemolymph**, a term derived from the Greek word haema (blood) and the Latin word lympha (clear water). Blood and haemolymph are complex fluids of variable composition. They contain a large variety of dissolved and suspended inorganic and organic substances, often with a number of different types of cell (Figure 5.7).

The storage and transport of oxygen are achieved by coloured proteins in the blood called respiratory pigments. These are able to combine reversibly with oxygen. There is a range of these respiratory pigments and they are quite different biochemically. There may be different pigments found in members of the same phylum, and there may even be more than one pigment found in the same animal. The gastropod *Buccinum* has haemoglobin in the muscles and haemocyanin in the blood. Table 5.2 lists proven pigments with some of their properties.

Haemoglobin is an iron-containing compound made up of haem, which is a **metalloporphyrin**, and a protein, globin. The haem portion of the molecule is a constant feature of all haemoglobins, but the globin varies with the species. In addition, the molecules polymerize to form units of different sizes. Vertebrate **myoglobin** (the haemoglobin found in muscle) and lamprey haemoglobin are units of similar size that appear to correspond to the smallest unit, with a relative molecular mass of 16 500 to 17 000. The haemoglobin of *Glycera*, a polychaete worm, has a relative molecular mass of 34 000; it is formed from two basic units. However, in *Arenicola*, also an annelid worm, the relative molecular mass of 3×10^6 corresponds to 180 units. The constancy of the haem part of the molecule over such a wide range of animals suggests that this respiratory pigment is, phylogenetically, a very ancient molecule.

Two of the pigments do not contain iron: **haemocyanin** and **haemerythrin**. Haemocyanin contains copper, and two atoms of copper can bind one oxygen molecule. Polymerization occurs, as in haemoglobin, and it is the means by which high pigment levels in the blood are achieved, since haemocyanin does not occur in blood cells. The basic unit has a relative molecular mass of

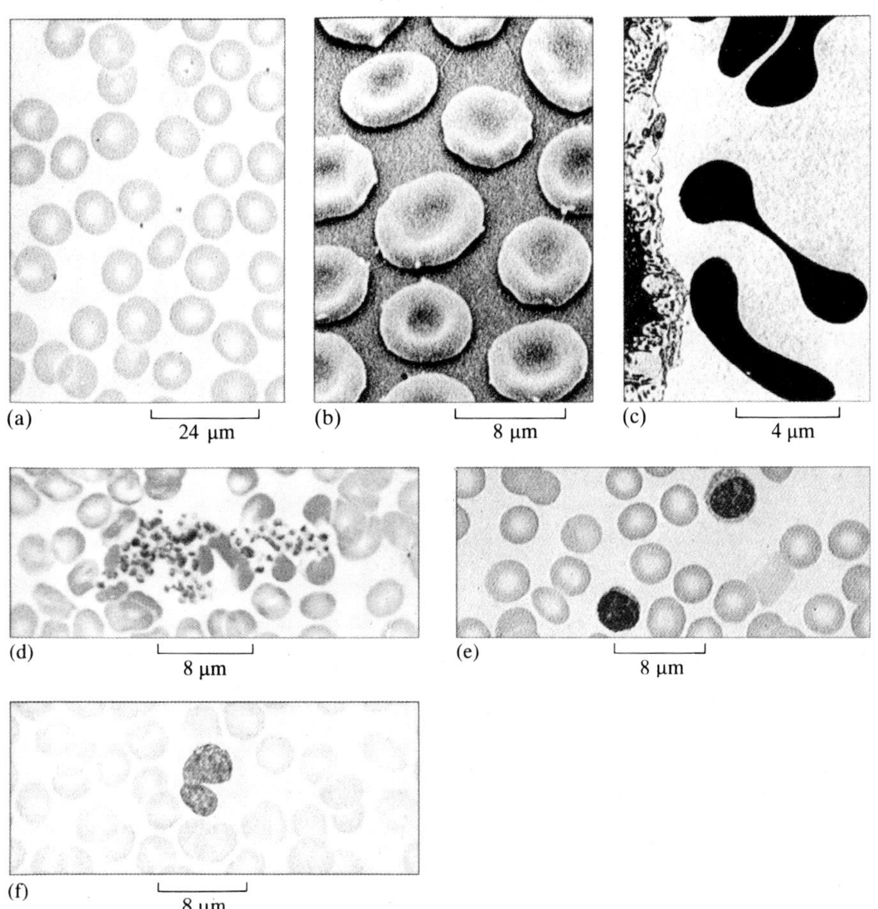

Figure 5.7 Some of the components of human blood. (a) Red cells: note that they do not have a nucleus. (b) Scanning EM photograph of red cells. (c) Section through red cells. (d) Blood platelets. (e) White cells: these are lymphocytes, concerned with the response to infection; they are much less numerous than red cells (about 1:900). (f) Monocytes: these are the largest of the white cells; they remove debris and foreign bodies from the blood.

Table 5.2 The distribution of some oxygen-carrying pigments, and their carrying capacities

Pigment	Colour	Located in	Animal	Oxygen volume (%)
haemerythrin	red	cells	annelids	2
chlorocruorin	green	plasma	annelids	9
haemocyanin	blue	plasma	crustaceans	1 to 4
			gastropod molluscs	1 to 3
			cephalopod molluscs	3 to 5
haemoglobin	red	plasma	molluscs	1 to 6
			annelids	1 to 10
		cells	amphibians	3 to 10
			reptiles	7 to 12
			fish	4 to 20
			birds	20 to 25
			mammals	15 to 30

50 000, but in the snail *Helix pomatia* polymerization produces a molecule with a relative molecular mass of 6.7×10^6. The oxygen-carrying capacity is less than that of vertebrate haemoglobin, as Table 5.2 makes clear, but it does approach that of the less efficient haemoglobins found in the plasma (non-cellular component of blood) of some invertebrates.

The oxygen-carrying role of transport fluids is familiar enough, but they do of course have other functions. You met one, transport of hormones, in Chapter 2, and you will meet others later on, acid–base regulation in Chapter 6 and excretion and osmoregulation in Chapters 11 and 12. The blood transports molecules to the tissues. We shall now consider the process of transfer into the cells of the tissues and the specialized structure of the smaller blood vessels, the capillaries.

5.2.3 The microcirculation

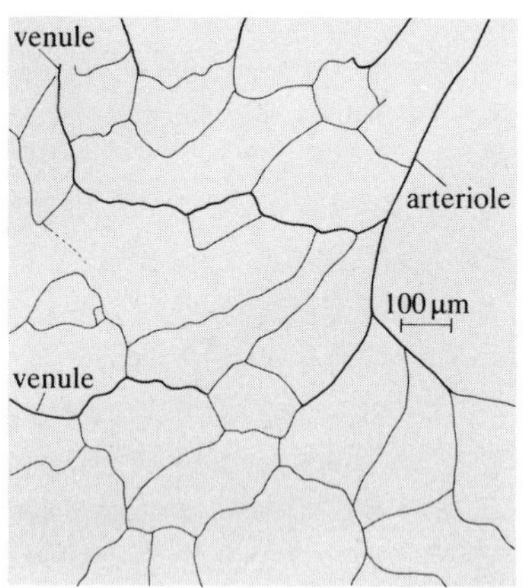

Figure 5.8 A reconstruction, from photographs, of the branching of capillaries from an arteriole in cat mesentery.

The **microcirculation** is so called because of the microscopic size of the arterioles, venules and capillaries that distribute blood throughout the tissues. It is through the walls of these vessels that solutes and macromolecules are exchanged. The size of the vessels makes them more difficult to study than the arteries and veins, so much of the current information about the microcirculation comes from work on a relatively few tissues, generally tissues that are thin or transparent. Figure 5.8 shows one part of such tissue, the mesentery of a cat: the network of capillaries is very clear, allowing the course of the individual vessels to be traced.

Notice that there are many possible pathways that the blood can take from arteriole to venule. It will generally take the path of least resistance.

◇ Look again at Table 5.1. What is the relationship between the relative amount of muscle in the walls of blood vessels and their size?

◆ The amount of muscle decreases with decreasing diameter. In capillaries, muscle is absent altogether. Note, however, that there is a small increase in the amount of muscle in the walls of the vessels immediately preceding the capillaries.

An additional ring of muscle is present at the junction between arteriole and capillary. Contraction of this band of muscle will close off the capillary and its branches, and flow will be diverted through another area of the network. Opening and closing of these **precapillary sphincters**, therefore, enables the flow to be finely controlled. It is said that all capillaries combined have a potential volume of about 140 per cent of the total blood volume.

◇ How might this difference between the potential volume of the capillaries and the blood volume be explained?

◆ Obviously, not all the capillaries can be open (or are 'patent') at any one time.

In fact, only 3–5 per cent of all capillaries are usually open for blood to flow through them, and only 5–7 per cent of the total volume of blood is contained in the capillaries.

Not only may the pathways of blood through capillary beds be altered, but also it is thought that whole capillary beds might be by-passed by vessels called **arterio-venous anastomoses**, which connect arterioles directly to venules. The question of whether such through-channels exist has generated much interest because, functioning in association with the precapillary sphincters, they would be capable of exerting a marked control of the flow of blood (and hence exchange across capillaries) within a capillary bed. Thus, if all the precapillary sphincters were to constrict, blood should be shunted to the through-channel and the capillary unit by-passed. However, such 'shunts' have been identified in only a few microcirculations, for example the skin. A more usual form of shunt is from artery to artery or vein to vein. These permit effective shunting between capillary beds but give no control *within* a bed.

As far as we know, animals with open circulations do not have the advantage of being able to operate shunts for haemolymph, although in some decapod crustaceans (e.g. the lobster) blood may be channelled to different areas within their gills.

The bypassing of capillary networks is extremely important in the functioning of the circulation because it allows blood to be moved to the areas where it is in greatest demand. For example, during running, the leg muscles require an increased supply of oxygen and nutrients. This is achieved partly by a change in ventilation and partly by an increase in blood flow to the legs at the expense of flow to other organs. In such a situation, the intestinal capillary beds often 'shut down' and because the volume of blood in the body is constant, more is available for the organs under stress (i.e. for the leg muscles).

The process of exchange at the capillaries

The capillary walls are quite thin, about 1 μm thick, and consist of a single layer of cells. Note that only endothelium is present in capillary walls. Gases diffuse readily through endothelium, but hydrophilic molecules can only enter through pores in the capillary walls. Two major exchange processes occur across the capillary wall: net diffusion and filtration.

Diffusion

Diffusion is the main process involved in the transport of respiratory gases and other dissolved substances of small relative molecular mass, such as salts, sugars and amino acids. These substances can diffuse relatively unhindered across the capillary wall in accordance with the Fick diffusion equation (Equation 5.1, Section 5.1.1). This equation, you recall, describes the diffusion of a substance from an area of higher concentration to one of lower concentration. It is quite difficult to calculate the diffusion rate of, say, oxygen from inside a capillary to inside a cell, particularly because more than one membrane has to be crossed. The rate of diffusion is increased if capillary vessels lie close to the cells; in mammals they are never more than 0.2 mm from any cell (recall that the rate of diffusion is inversely proportional to the distance to be travelled). The fact that there is flow in the capillaries means that the rate of diffusion is increased since flow steepens the gradient between the inside of the cell and the outside. In Section 5.2 the rate of flow was described as proportional to the radius to the fourth power of the vessel through which flow is taking place. Although this is not strictly accurate for the specialized type of flow that occurs in capillaries, this relationship can be used to provide an approximate estimate of flow-rate.

Filtration

The second major exchange process across capillary walls is filtration. This process is important in water movement into and out of the capillary. However, while it is indeed the second *major* process, it only accounts for around 2% of the total exchange across the capillaries. Diffusion accounts for the remaining 98%. Filtration depends on two forces that act in opposite directions. One force, the hydrostatic pressure of the blood, p, acts to 'push' water across the endothelial cell layer and out of the capillaries. The other force, osmotic pressure Π, tends to 'pull' water back into the capillary because the osmotic pressure of the blood plasma Π_p, is greater than the osmotic pressure of the interstitial or tissue fluid, Π_t, that is $\Pi_p > \Pi_t$. As most solutes can diffuse across the capillary walls freely they do not generate an osmotic gradient. Soluble proteins that cannot cross the capillary wall and are present in the interstitial fluid *do* generate an osmotic pressure of around 3.6 kPa. When the two forces (p and Π) are equal (i.e. when equilibrium is reached) the volume of fluid leaving will equal that returning and there will be no net filtration. When the forces are out of balance then either (a) more fluid is pushed out than returns, resulting in net filtration, or (b) more fluid is 'pulled' in from the interstitial space than leaves the capillaries, resulting in net absorption.

◇ If the osmotic pressure exceeds the blood pressure, will there be net filtration or net absorption?

◆ Net absorption.

Blood plasma contains about 7 g of protein per 100 cm^3 of plasma (7 grams per cent). These protein molecules (mainly albumin and globulins with a relative molecular mass greater than 600 000) are too large to pass out of the capillary readily. Interstitial fluid bathing the tissues has a much lower protein content, about 1.5 g per cent. As a result the blood plasma has a higher osmotic pressure than the fluid bathing the tissues, and water tends to pass from the interstitial fluid into the capillaries. Thus Π_p is approximately 4 kPa and Π_t is about 0.4 kPa (Figure 5.9).

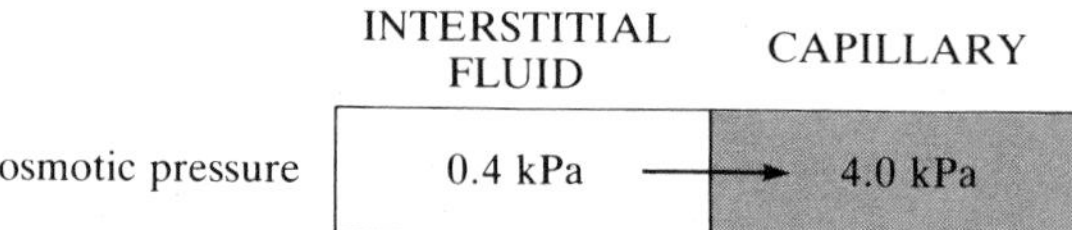

Figure 5.9

The difference, $\Pi_p - \Pi_t$ ($4 - 0.4 = 3.6$ kPa), is the force tending to move water molecules into the capillary. Note that it is difficult to measure the blood pressure in the capillary accurately.

As pressure in the capillary is generated mainly by the beating of the heart, it will tend to vary from tissue to tissue, partly depending upon the distance of the capillaries from the heart. Table 5.1 illustrates the range of pressures found within blood vessels.

◇ From Table 5.1, what happens to blood pressure around the circulation?

◆ There is a continuous drop in pressure around the circulation between blood leaving and blood entering the heart.

We see this trend clearly if we look at a capillary in the mesentery of a cat (Figure 5.10).

◇ The pressure measurements in the capillary in cat mesentery are not taken in the main (vertical) capillary, but within side branches close to the main capillary. Why?

◆ In this way, flow is not interrupted in the main capillary and more reliable pressure readings are achieved.

Hydrostatic pressure in the interstitial fluid, that is the pressure in the fluid just outside the capillary walls, is often slightly less than atmospheric pressure, by about 0.8 kPa. It must be appreciated that this is a difficult measurement to make, partly because the interstitial fluid is not really a fluid at all, but rather a matrix of gelling polysaccharides with an affinity for water. To make matters more complicated, this matrix is apparently divided into two

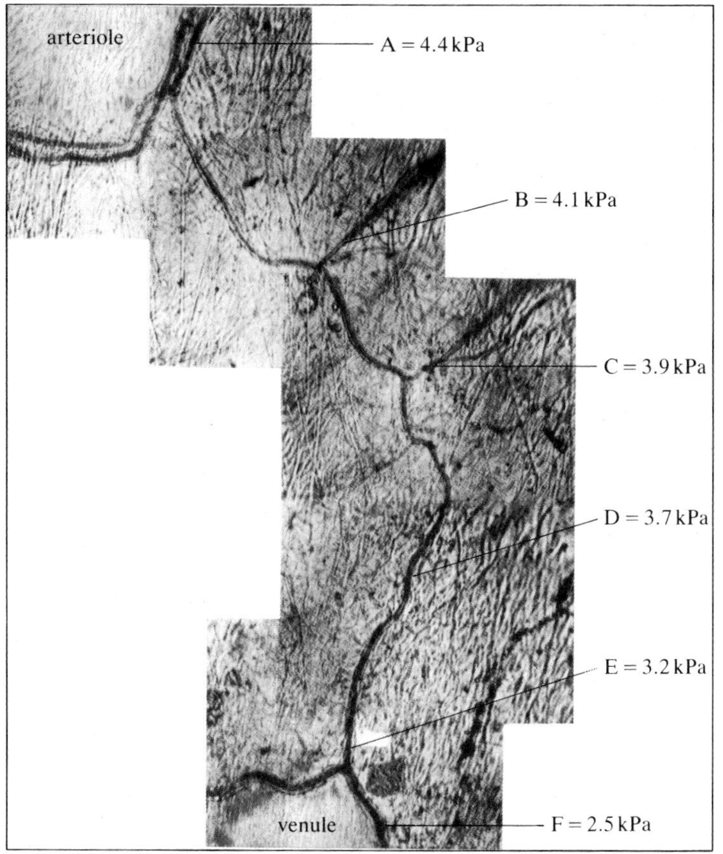

Figure 5.10 A photographic reconstruction of a capillary from arteriole to venule in the cat mesentery. Direct recordings of the mean blood pressure (expressed as kPa) were taken at points A to F. The capillaries range in diameter from 7.5 to 9 μm.

physically discrete 'pools', and the pressures recorded will thus depend very much on which pool is being recorded from. The techniques of measuring small pressures are open to criticism, and it is probable that pressures below atmospheric in the interstitial fluid are artefacts created by the measuring techniques. It is therefore safest to estimate interstitial pressure as being between −0.8 and +0.8 kPa, relative to atmospheric pressure.

Now, consider the blood pressure in the capillary in the cat mesentery in Figure 5.10.

◇ What is the blood pressure at point A in the capillary?

◆ 4.4 kPa.

◇ If we take the hydrostatic pressure of the interstitial fluid to be 0.3 kPa, what will be the difference in the total hydrostatic pressure between the two sides of the capillary wall?

◆ 4.1 kPa (Figure 5.11).

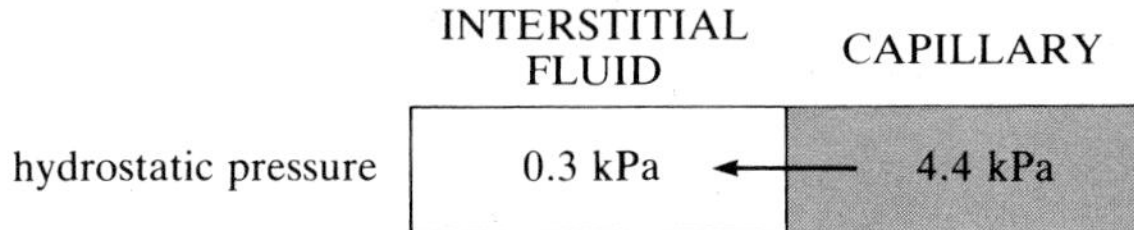

Figure 5.11

◇ What would happen to the net flow if the permeability between the compartments were to increase?

◆ Net filtration would be increased.

LYMPHATIC SYSTEM

The system of vessels that carries lymph from the tissues to the blood.

LYMPHOCYTES

A class of white blood cell responsible for specific immune defences... B-cells & T-cells

This happens when you are bitten by an insect. The irritant insect saliva causes the release of an agent, histamine, which acts to increase the permeability of the capillary walls to large molecules; the rate of filtration is therefore greatly increased, resulting in swelling around the bite.

5.2.4 The lymphatic system

As we have just seen, even a closed circulatory system is not totally closed. There is some leakage from the thin-walled blood capillaries. A secondary system of fluid-filled tubes, the **lymphatic system**, collects this fluid and returns it to the blood. The lymphatic system is usually spoken of as part of the microcirculation.

Lymphatics (lymph vessels) originate as blind-ended capillaries in the tissue spaces of most organs (Figure 5.12).

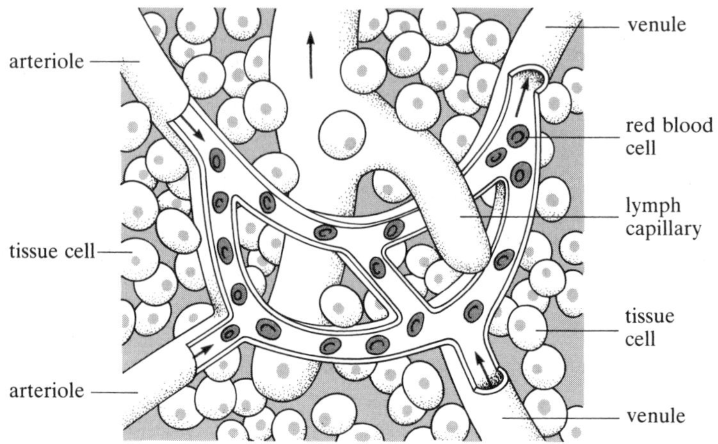

Figure 5.12 Blood and lymph capillaries: in this idealized drawing, arrows indicate the direction of blood flow.

◇ What differences can you see between blood and lymph capillaries?

◆ Lymph capillaries are blind-ended and contain no red blood cells. Lymph capillaries may be up to eight times wider than blood capillaries. This is shown clearly in Figure 5.13, a photograph of lymph and blood capillaries in rat mesentery. They also have thinner walls than blood capillaries.

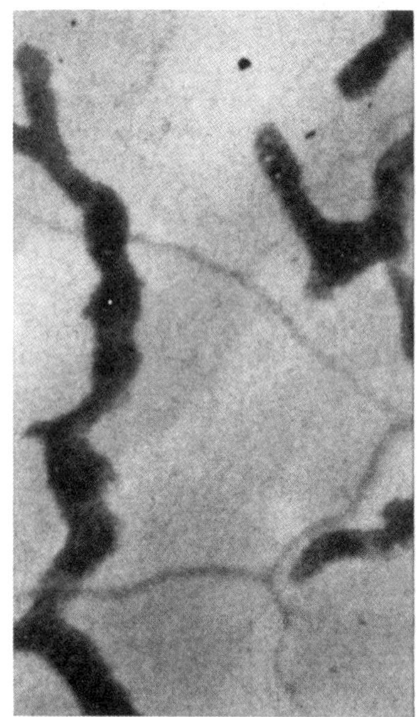

Figure 5.13 Blood and lymph capillaries in the mesentery of a rat. The lymph capillaries (lymphatics) are the large dark vessels with blind ends. The blood vessels are very much smaller. The lymphatics have been specially stained in this preparation: normally they are colourless.

There are other differences between blood and lymph capillaries. For instance, the walls of lymphatics are much more permeable than those of blood capillaries, and macromolecules such as proteins enter readily. This is particularly important because some protein does leak from the capillaries. Turn back to Table 5.1 and note the number of capillaries in a single capillary bed. This number of capillaries, each losing a small amount of protein, will together lose a substantial amount. When the amount of protein leaking from all the capillary beds in the body is considered, it obviously represents a large amount of protein. In a young cow the amount of plasma protein that leaks from the capillaries *every day* may be as much as twice the total amount of plasma protein in the circulation at any one time.

◇ What then might be the primary function of the lymphatic system?

◆ To return fluid and macromolecules to the blood.

◇ What might you expect to happen if the lymphatics 'draining' an area were to become blocked for some reason?

◆ Initially, the lymphatics would swell. Eventually, the thin walls would rupture and continued filtration would cause the interstitial fluid to increase in volume. This might result in the condition known as elephantiasis.

Under normal conditions, if fluid and macromolecules enter the lymphatics readily, why should these substances not leave just as easily? Figure 5.14 and the micrograph of lymphatic valves (Figure 5.15) give clues to the answer.

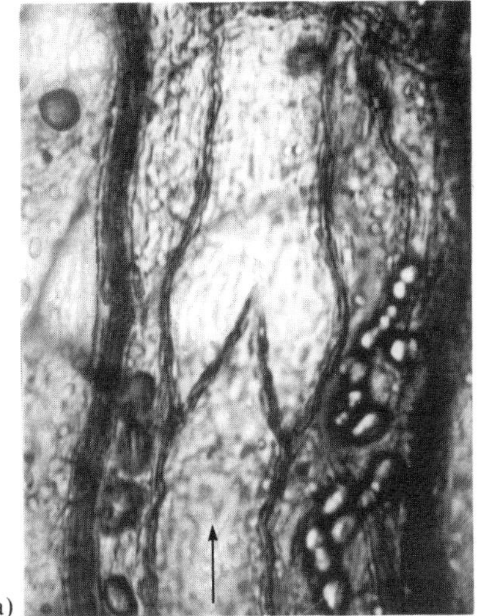

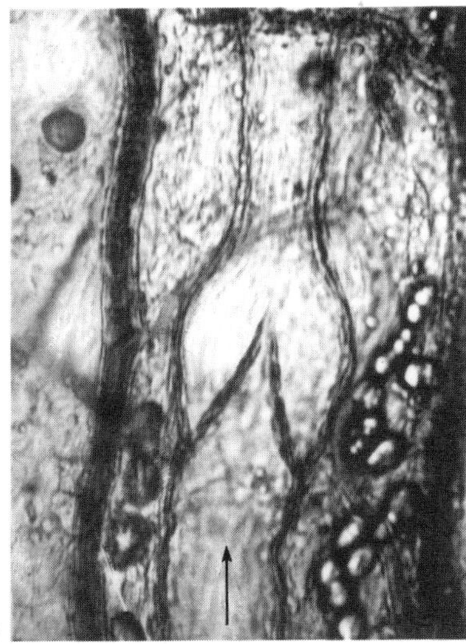

Figure 5.15 Micrographs of a valve in the lymphatic system of a rat: note the capillary alongside. (a) Partial contraction of a collecting vessel; (b) the same vessel in a relaxed state 3 s later.

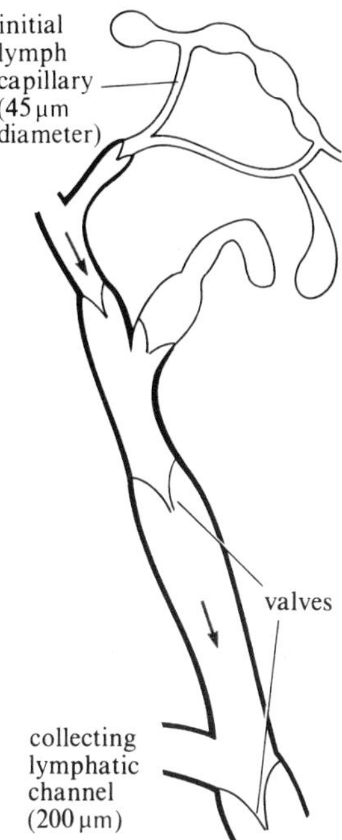

Figure 5.14 Lymph capillaries: lymph passes from blind-ended vessels to larger collecting ducts. This composite drawing was constructed from several preparations of mesentery.

Fluid entering the lymph capillaries will push the fluid already there further along the vessel. After a short distance the walls thicken (Figure 5.14) and their permeability is much decreased. The valves in the lymph capillaries must also help in preventing back-flow, as they do in veins. The lymphatics join together to form larger and larger vessels. In humans, most of the lymphatics from below the neck empty into the large thoracic duct, which in turn empties into the jugular vein in the neck.

◇ What might propel lymph up the thoracic duct to the jugular? (The thoracic duct runs from below the diaphragm to the neck.)

◆ The important factors are variations in pressure in the thorax caused by respiratory movements and the presence of valves.

In addition to its transport functions, the lymphatic system has a crucial role in the defence against foreign cells and viruses. The immune system is formed by all the physiological mechanisms that combine together to enable an organism to recognize foreign material and react to neutralize or eliminate it. In humans, the effector cells of the immune system are distributed throughout the body, though they are found particularly in the areas shown in Figure 5.16: the lymphoid tissues. The types of cell are listed in Table 5.3, with a brief indication of their function, since it is not possible to go into detail about the immune system here. The lymph nodes function as filters, removing and adding **lymphocytes** to the fluid as it passes through sinuses within the node. Lymphocytes are a class of white blood cells responsible for specific immune defences: they are stored in the node and can multiply there, though their source is the bone marrow. The walls of the sinuses are lined with **mac-**

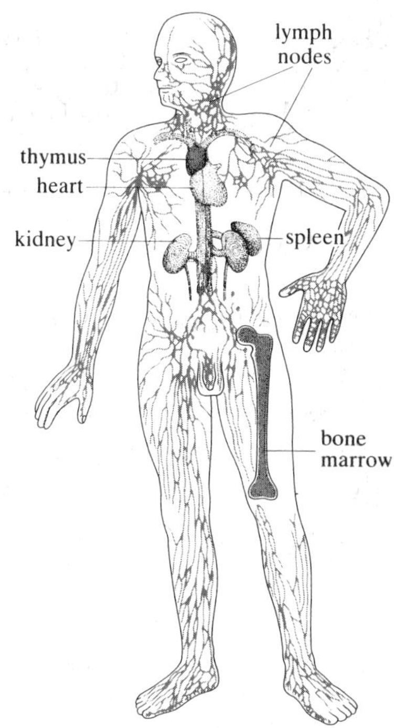

Figure 5.16 The location of the organs of the lymphatic system in a human.

Table 5.3 Major effector-cell types of the immune system

	White blood cells (leucocytes)						
	Polymorphonuclear granulocytes						
	Neutrophils	Eosinophils	Basophils	Lymphocytes	Monocytes	Plasma cells	Macrophages
percentage of total leucocytes	50–70	1–4	0.1	20–40	2–8		
primary sites of production	bone marrow	bone marrow	bone marrow	bone marrow, thymus, and lymphoid tissues	bone marrow	derived from B lymphocytes in lymphoid tissue	most are formed from monocytes
primary known function	phagocytosis; release of chemicals involved in inflammation (chemotoxins, etc.)	destruction of parasitic worms	release of histamine and other chemicals; transformed into mast cells with similar functions	B cells: production of antibodies (after transformation into plasma cells) T cells: different subgroups responsible for cell-mediated immunity, "helping" or suppressing B cells and other T cells	transformed into tissue macrophages; functions similar to macrophages	production of antibodies	phagocytosis; assist in antibody formation and T-cell sensitization; secretion of chemicals involved in inflammation, regulation of lymphocytes, and total-body response to infection or injury

rophages that remove debris from the fluid, such as dust inhaled via the lungs, bacteria and cell fragments. The **spleen**, the largest of the lymph organs, is the site of blood cell formation in the human fetus, though in the adult only lymphocytes multiply there. The spleen has a plentiful blood supply, and much of the degradation of aged red blood cells takes place there. The cell fragments are removed by macrophages.

The **thymus** produces a distinct class of lymphocytes called **T lymphocytes** or T cells. Like all lymphocytes they have their origin in the bone marrow and are swept into the thymus by the blood. Here they proliferate by successive mitotic divisions. Lymphocytes from the bone marrow that lodge in other lymphoid tissues do not form T cells upon mitosis; they form a different class called **B cells**. Figure 5.17 summarizes the routes by which the various classes of blood cells are formed.

Summary of Section 5.2

1 The rate at which blood flows through vessels is dependent upon the pressure gradient along the vessel and the resistance to flow.

2 Respiratory pigments store and transport oxygen in blood. Several pigments are known, of which the most widespread are haemoglobin, haemocyanin and haemerythrin.

3 The arterioles, venules and capillaries comprise the microcirculation. Flow through the capillaries is controlled by precapillary sphincters. Whole capillary beds may be by-passed by the circulation as it responds to changes in the demand for flow from tissues.

4 There are two major exchange processes that take place at the capillaries: net diffusion and filtration.

5 In mammals, fluid that leaks from capillaries collects in the lymphatic system. The lymphatic system has a crucial role to play in the immune system.

Now attempt Question 2, on p. 165.

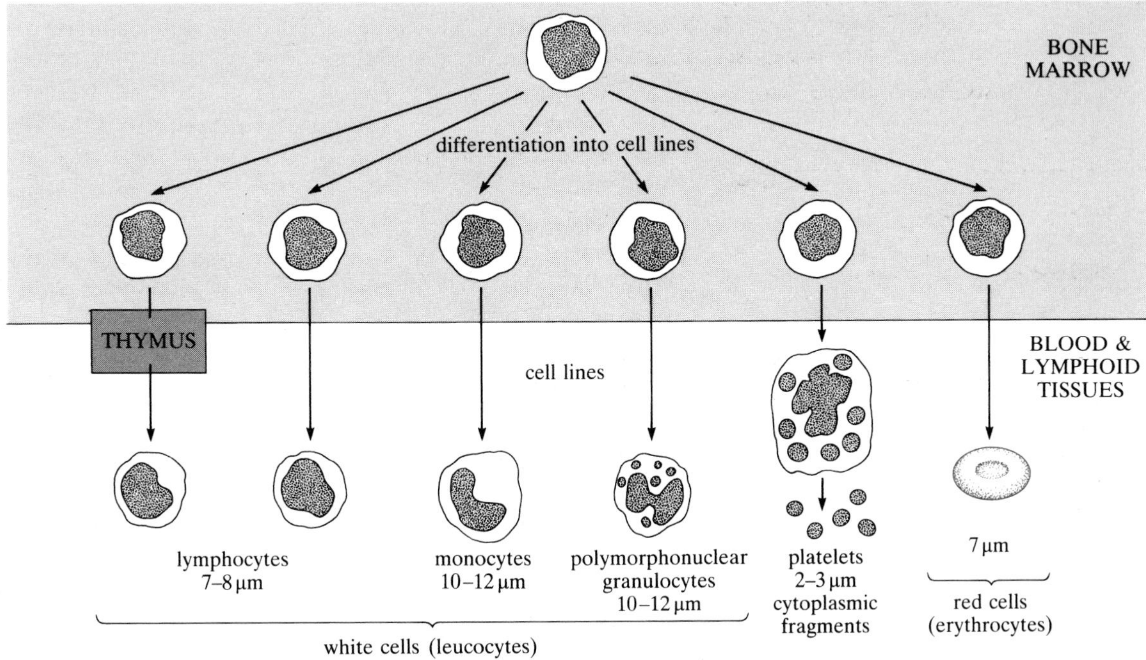

Figure 5.17 The origin of the main cells of the blood and lymph.

5.3 HEARTS AND THE CARDIAC CYCLE

In 1808 Thomas Young introduced a lecture to the Royal Society on the function of hearts and arteries thus:

> The mechanical motions, which take place in an animal body, are regulated by the same general laws as the motions of inanimate bodies . . . and it is obvious that the inquiry, in what matter and in what degree, the circulation of the blood depends on the muscular and elastic powers of the heart and of the arteries, supposing the nature of those powers to be known, must be simply a question belonging to the most refined departments of the theory of hydraulics.

The circulation of the blood in animals is indeed a question of hydraulics; there are, for example, hydraulic advantages if the flow through the blood vessels is maintained by having a large pump, sometimes with auxiliary pumps where additional flow is required. The operation of biological pumps is a fascinating area because many of us are familiar with mechanical pumps failing at regular intervals on cars, washing machines and heating systems, yet some biological pumps are regarded as failing prematurely if they do not provide service for three score years and ten. In this Section we shall examine some hearts, considering particularly their function.

The movement of viscous fluids, such as blood, through small and lengthy vessels requires considerable force, and the pressure must be maintained to overcome the resistance to flow. Muscle pressurizes the fluid by contraction, and the fluid provides the restoring force that relaxes the muscle at the end of the contraction period. When muscle is wrapped around a tube or chamber, it is possible to achieve a local reduction in volume. Figure 5.18 shows the different types of pump that operate in this manner.

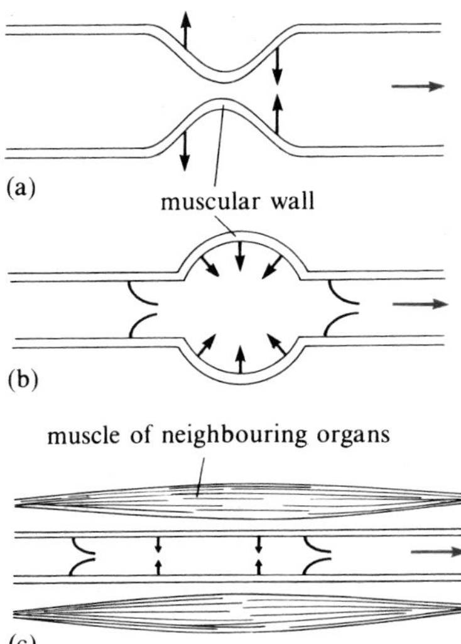

Figure 5.18 Three major types of pump that move blood around circulatory systems: (a) pulsating vessel or tubular heart; (b) chambered heart; (c) auxiliary or accessory heart.

Morphologically, hearts may be classified as pulsating vessels and **tubular hearts** (Figure 5.18a), **chambered hearts** (Figure 5.18b) and auxiliary or **accessory hearts** (Figure 5.18c). We shall briefly consider the structure and function of each in turn.

5.3.1 Pulsating vessels and tubular hearts

Blood vessels that contract with peristaltic waves (Figure 5.18a) are a widespread, primitive mechanism for the movement of the contents of the vessels. Many of the blood vessels in annelids show rhythmic peristaltic waves.

In the earthworm and other annelids the basic plan of circulation is relatively simple (Figure 5.19). Blood flows towards the anterior end of the body in a dorsal vessel situated over the digestive tract. The dorsal vessel is connected to a ventral vessel by one or more vessels passing around the gut. The ventral vessel runs beneath the gut. It carries blood towards the posterior end of the body and finally connects with a large sinus from which the dorsal vessel receives blood. In each segment of the earthworm the dorsal and ventral vessels are connected by lateral vessels, branches of which irrigate the outer layer of the body and supply the various segmental organs.

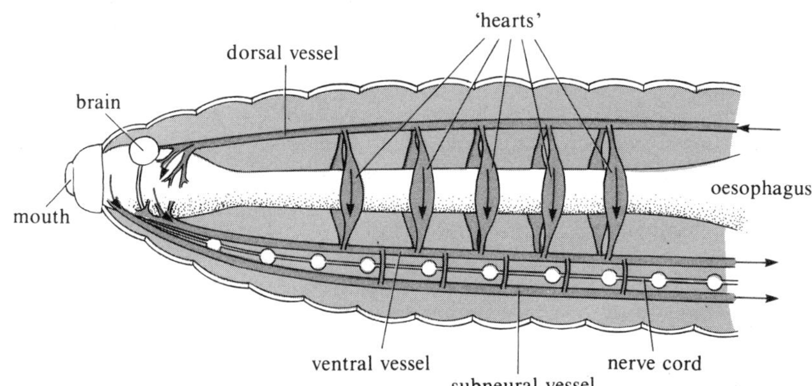

Figure 5.19 Part of the circulatory system of an earthworm. Blood vessels are shown in red; arrows mark the direction of flow.

In annelids, blood vessels typically have an internal (endothelial) lining supported by connective tissue. The outer side of the connective tissue is covered with muscle fibres, and in general the blood vessels are contractile. Contraction of the dorsal vessel is the principal means by which blood is propelled, although there are other vessels that are more conspicuously contractile and are often referred to as hearts. *Lumbricus* (the earthworm) has five pairs of these 'hearts', which function as accessory organs for blood propulsion. Both the 'hearts' and the dorsal vessel are further adapted for moving blood: endothelial folds act as simple valves.

Pulsating vessels are also found in some holothurians (sea cucumbers) and in some primitive chordates. In some vertebrates, pulsating veins aid by propelling blood (e.g. in bats' wings), but act only as accessories to the main pump, the heart.

The distinction between tubular hearts and pulsating vessels is not a very clear one. Generally, tubular hearts are more complex, having pores or ostia through which blood enters. They are usually anchored at several places and may be surrounded by a pericardial membrane.

The hearts of most arthropods are of the tubular type and are usually derived from the dorsal vessel. They may extend the length of the body, as in the insects, or may be pulsating muscular sacs in a fluid-filled pericardial sinus, as in the crustaceans.

5.3.2 Chambered hearts

Chambered hearts are characteristic of most molluscs and chordates. The chambered heart (Figure 5.18b) is potentially a more efficient pump than either a pulsating vessel or a tubular heart. In large, active animals the chambered heart is often central and can maintain a continuous circulation at high pressures. The structure and function of the mammalian heart will be considered in Section 5.3.5.

Accessory pumps

Accessory pumps are common in both vertebrates and invertebrates and are sometimes referred to as boosters. They help to propel blood through the peripheral channels that lead to vital organs. Accessory pumps are associated with the gills in the octopus. We consider this further in Section 5.3.4.

Accessory pumps often work by outside pressure (Figure 5.18c). Such a pump is found in the large veins in our own legs. The walls of these veins are relatively thin, and there are valves preventing the blood from flowing backwards. When the leg muscles contract, the veins are collapsed and the valves ensure that the blood is forced in the direction of the heart. This pumping action greatly aids the movement of blood out of the legs against the force of gravity. When a person is standing motionless the full effect of gravity is manifest and blood tends to pool in the lower legs, with the result that less blood is returned to the heart. This in turn reduces the amount of blood that the heart can pump to other tissues, including the brain, and the point can be reached where fainting occurs. This is a recognized occupational hazard of Buckingham Palace sentries who are advised to flex their leg muscles when standing stock still.

ACCESSORY PUMPS ie leg muscles contracting help to push blood back up - helped by valves to prevent back flow.

5.3.3 The cardiac cycle in lobsters

Lobsters, like all crustaceans, have an open circulation. Figure 5.20 shows the arrangement of arteries and sinuses in *Homarus vulgaris*, the common lobster found around the coasts of Britain and most of Europe. It is a large animal

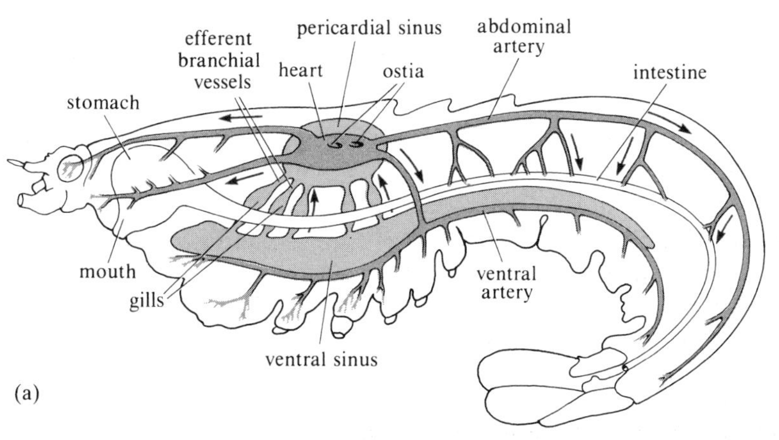

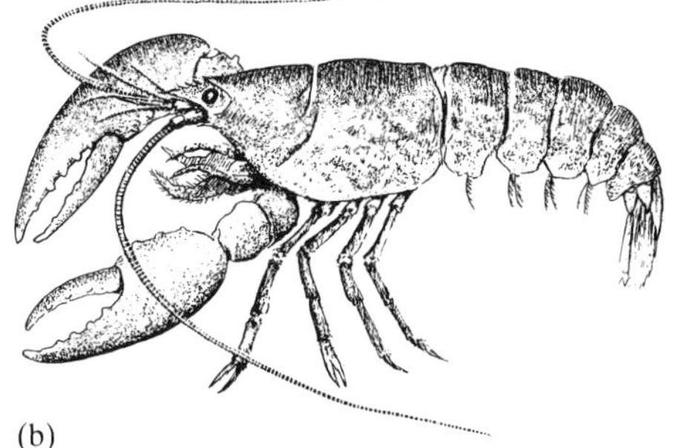

Figure 5.20 The common lobster, *Homarus vulgaris*. (a) The circulatory system. The major arteries are shown in red and the main sinuses in grey. (b) The whole animal

that can grow up to 75 cm long with a weight of around 4 kg, and the circulatory system is highly developed by comparison with other smaller crustaceans.

There is a dorsal heart with several large arteries carrying haemolymph away from it. The arteries deliver the haemolymph into large sinuses that bathe the tissues. From these the haemolymph drains into the large ventral sinus beneath the gut and through the branchial vessels to the gills. Around the heart is a large pericardial sinus that collects the haemolymph from the gills at the bases of the limbs. In the dorsal surface of the heart are four small holes, called ostia, equipped with valves. When the heart contracts, haemolymph is forced into the arteries and the valves in the ostia are kept shut by the pressure. At the end of the contraction the heart refills as haemolymph from the pericardial sinus enters through the ostia. The valves ensure a unidirectional flow.

The presence of a pulsating heart does not necessarily imply that it is the only force that drives the haemolymph around the circulation. In many crustaceans, the pool of haemolymph within the haemocoel is also moved by contractions of the muscles of the body during locomotion.

The contribution that the heart makes to the flow of haemolymph in the lobster has been investigated by measuring the pressure at various locations within the circulatory system (Figure 5.21). The phase of the heart-beat when the heart is relaxing is called **diastole** (pronounced die-ass-toll-ee); the phase of contraction is called **systole** (siss-toll-ee). Mean values of haemolymph pressure, recorded at systole, are given in red in Figure 5.21. During systole, the average haemolymph pressure in the lobster's arteries measured immediately outside the heart is 1.5 kPa.

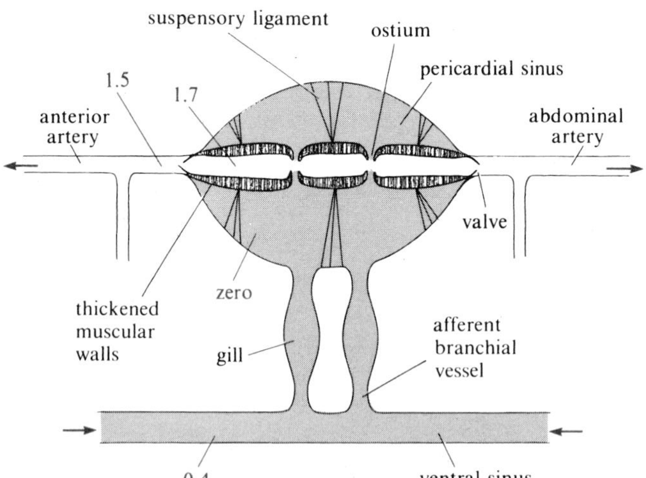

Figure 5.21 The central portion of the circulatory system of the common lobster, showing the mean systolic pressures (kPa) that have been measured.

◇ Would this value be higher, lower or unchanged in diastole?

◆ Lower, because the heart is relaxed and not pumping haemolymph forcefully.

A drop in systolic pressure (from 1.7 in the heart to 1.5 kPa in the artery) happens when haemolymph passes through closed tubes because these offer resistance to flow. Notice the large drop in pressure between the arteries and the ventral sinus. Flow occurs where there are differences in pressure within the conducting system ($J \propto \Delta p$). Contraction of the heart results in the production of a relatively high pressure and the establishment of a pressure gradient within the haemocoel. The generation of a difference in pressure results in a directional movement of haemolymph. In the lobster, as in many animals with an open circulatory system, the contractions of body muscles contribute to blood flow, but the measurements shown in Figure 5.21 show that, even when the lobster is at rest, circulation of the blood is assured by contractions of the heart.

5.3.4 The cardiac cycle in the octopus

The common octopus belongs to a group of molluscs called cephalopods. Included in this group are the cuttlefish and the squids. You have already met the squid in Chapter 1, because the large size of its axons make it a useful research animal. Cephalopods are also of interest to physiologists because they have a distinct network of arteries, veins and capillaries that form a closed circulatory system of comparable complexity to that found in vertebrates. If you were to cut away the dorsal surface of the mantle cavity of an octopus you would see the major components of the circulatory system (Figure 5.22). Because the octopus is a reasonable size it is possible to make pressure measurements in the circulatory system without impairing normal function. Measurements from four sites are shown in Figure 5.23. These traces of pressure against time were recorded simultaneously, and provide a considerable amount of information about the functioning of the system.

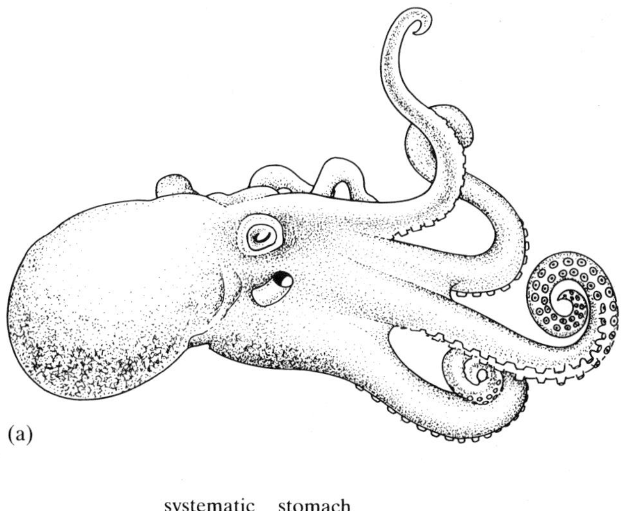

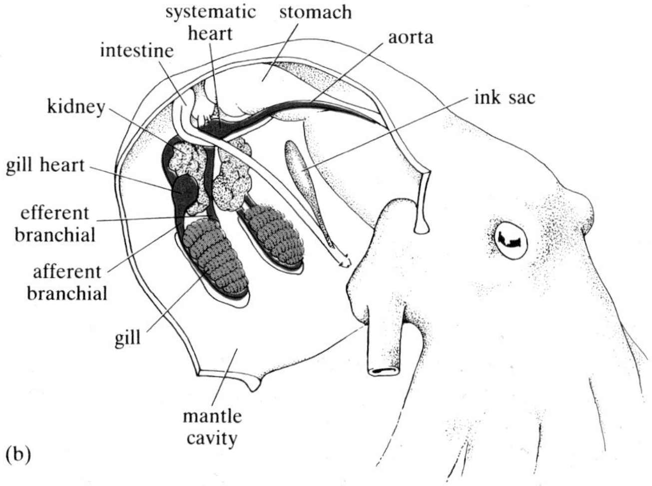

Figure 5.22 The mantle cavity of the common octopus cut open and the animal dissected to show the circulatory system and some of the other organs. The major blood vessels are illustrated in red.

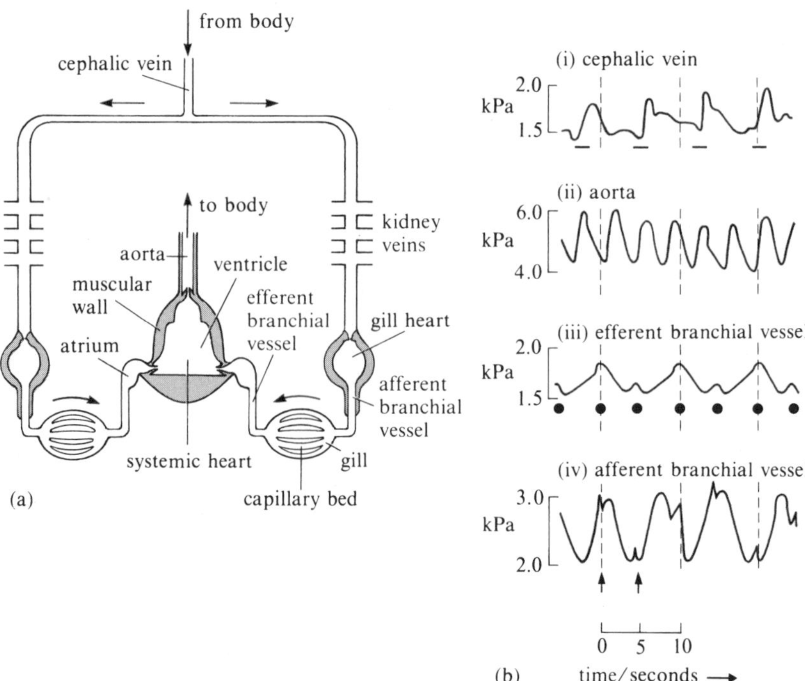

Figure 5.23 (a) A simplified diagram of the central vascular system of the octopus. Arrows indicate the direction of blood flow. Blood pressure can be measured in the intact system, and records from four different sites in the circulatory system of *Octopus dofleini* are shown in (b).

◇ First look at trace ii. What is responsible for the regular changes in blood pressure?

◆ The aorta leads from the muscular heart, and regular contractions of the heart produce regular changes in pressure.

◇ How often does the heart contract in one minute?

◆ There are five contractions in 20 seconds, therefore there are approximately 15 beats per minute.

◇ What is the lowest pressure measured in the aorta during one cycle of contraction and relaxation? Which part of the heart-beat would you expect it to correspond to? Give the same information for the peak arterial pressure.

◆ The lowest recorded pressure is approximately 4 kPa and coincides with diastole (relaxation of the heart). The highest recorded pressure is approximately 6 kPa, at systole.

◇ The difference between systolic and diastolic blood pressure is termed the pulse pressure. What is the pulse pressure in the aorta?

◆ Approximately 2 kPa.

Look at the pressure record from the efferent branchial vessel that carries blood away from the gills. The record is the sum of two separate pressure waves, one with a period of 10 seconds and a smaller one with a period of five seconds. The two waves interact because every alternate small pressure wave coincides with a large one producing a summated peak. What produces these pressure waves? First, we need to look at trace iv—the pressure changes in the afferent branchial vessel.

◇ What might be responsible for the large waves of pressure recorded in the afferent vessel?

◆ Contractions of the gill heart.

There are also regular, small rises in pressure occurring in the afferent vessel. Two of these are marked with arrows.

◇ What pressure changes in the efferent vessel occur simultaneously with the arrowed pressure changes in the afferent vessel?

◆ The small pressure waves marked with dots in trace iii.

The small pressure waves are produced by the regular contraction of the gills, but what of the large pressure wave with the 10 s period in trace iii? Water is drawn into the mantle cavity and then expelled from it in a rhythmic manner, providing a flow of oxygenated water over the gills. This periodic increase in pressure compresses blood vessels within the mantle cavity producing a pressure wave in the blood system. You can see a further effect of the respiratory water flow by looking at trace i. The cephalic vein regularly has a higher pressure just after the respiratory movements that expel water from the mantle cavity; the times of these movements are indicated by black bars below the trace.

These results show that the circulation of blood in the octopus is enhanced by several mechanisms, which you should now be able to list.

◇ List the factors responsible for pressure waves in the following vessels of the blood system of the octopus: 1, cephalic vein; 2, aorta; 3, efferent branchial vessel; 4, afferent branchial vessel.

◆ 1 *Cephalic vein*—pressure changes in the mantle cavity associated with respiratory movements.
2 *Aorta*—contraction of the main heart.
3 *Efferent branchial vessel*—contraction of the gills and respiratory movements.
4 *Afferent branchial vessel*—contraction of the gill hearts and respiratory movements. (Note that the difference in blood pressure between the cephalic vein and the afferent branchial vessels represents the boost in blood pressure caused by the contraction of the gill hearts.)

◇ What are the four factors contributing to blood flow in the octopus?

◆ Respiratory movements of the mantle cavity, and contractions of the systemic (main) heart, gill hearts and gills. Note that the flow can only be in one direction because of the presence of valves in the system.

You have now studied two invertebrate systems in some detail, one with an open circulation and one with a closed circulation. The lobster heart (Figures 5.20 and 5.21) is little more than a blood vessel with thickened, muscular walls, although valves prevent blood from leaving by the route it entered. In contrast, the octopus (Figures 5.22 and 5.23) has a systemic heart divided into three chambers: a large ventricle and two smaller atria (sing.atrium). There are valves at both the entrance and the exit to the heart and it is a more structurally specialized organ than the lobster's heart. (Note the use of 'structurally'. Some hearts found in open circulations can be quite complex in the manner in which they are controlled, e.g. the hearts of insects.)

LOBSTER HEART
– muscular blood vessel
OCTOPUS HEART
– systemic, chambered heart

The maintenance of a constant blood pressure is crucial in most vertebrates. (Blood pressure is never really constant because it varies with the cardiac cycle; what is meant here is constancy of the *average* blood pressure.) It would be disastrous for the homeostasis of the animal if the mean blood pressure fluctuated markedly because so many functions are closely connected with blood pressure. Ultrafiltration in the kidney has already been mentioned; another example is exchange across capillaries. If the blood pressure within a capillary is increased as a result of a general increase in blood pressure, then more material will leave the capillary than enters it, and toxic wastes could accumulate in the tissues.

It is hardly surprising that there is a complicated control system responsible for maintaining blood pressure at a desired level and, as in all control systems, there are receptors that monitor blood pressure (baroreceptors) and effectors, which can adjust blood pressure in either direction. The many effectors in this control system include the heart itself and the peripheral blood vessels.

Much less friction or resistance is displayed by the large sinuses of an open circulation and lower pressures suffice to circulate the fluid. Note that high-pressure open circulations do exist. An example is found in the feet of razor shells (*Ensis*) where the blood acts as a hydroskeleton (Figure 5.24). *Ensis* burrows vertically in sand. The foot is highly muscular and contains a large blood space. It acts like a piston in a cylinder. Blood pressure forces the foot out of the shell (Figure 5.24b). The end of the foot dilates to grip the sand and then the retractor muscle pulls the shell down over the foot. Thus the muscles acts against a fluid, the blood, instead of the rigid usual skeleton.

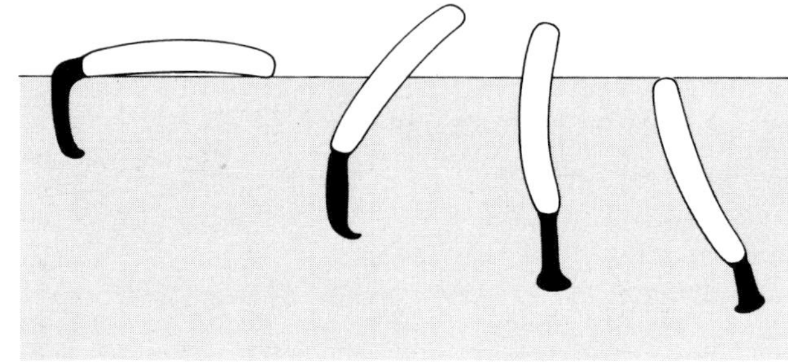

Figure 5.24 (a) The razor shell *Ensis* showing the large foot. (b) Three stages in the sequence of foot movement as *Ensis* burrows into sand.

Blood pressure is mainly, although not always exclusively, generated by a pump, the heart. A high pressure system might be expected to have a more complex and/or efficient pumping organ than a low pressure system.

5.3.5 The cardiac cycle in the mammalian heart

Figure 5.25 shows the structure of a mammalian heart. It operates as two pumps in series, with each pump consisting of two chambers: a thin-walled atrium and a thick-walled ventricle. The ventricles are the main pumping chambers, with the right ventricle pumping blood to the lungs (the **pulmonary circulation**), and the left ventricle pumping the same volume of blood around the body in the **systemic circulation**. This double circulation is shown diagrammatically in Figure 5.26. Although the flow through the pulmonary and systemic circulations must be the same, the pressures in the arterial sides are different. The systemic arterial pressure is much higher than the pulmonary. The flow through each ventricle is in one direction, the inflow and outflow being kept separate by valves. The inlet valves (from atria to ventricles) are the **tricuspid** on the right and the **mitral** on the left. The outlet valves are the **pulmonary** on the right and the **aortic** on the left.

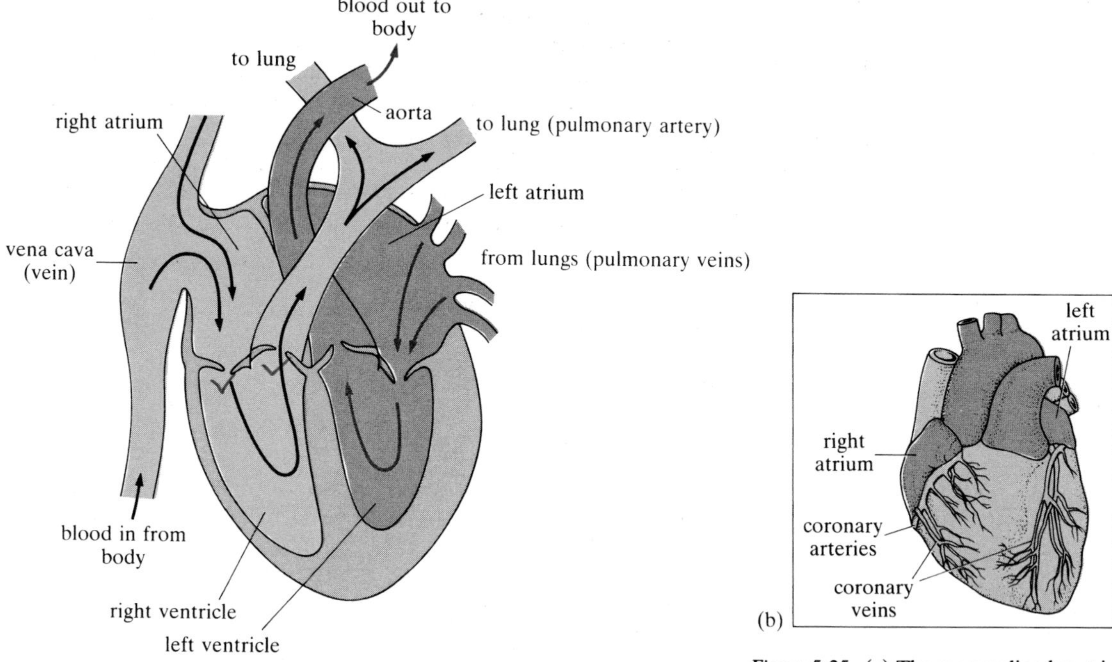

Figure 5.25 (a) The mammalian heart in section. (b) An entire mammalian heart to show the coronary blood vessels.

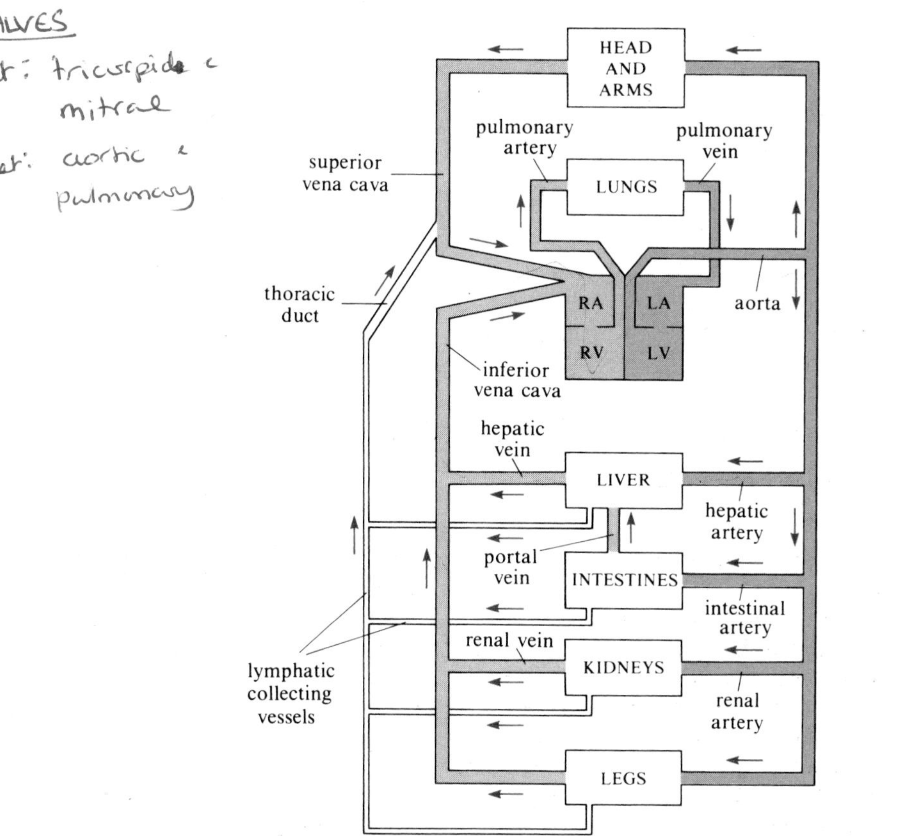

Figure 5.26 The double circulation of a mammal.

Because of the special properties of the cardiac action potential, each muscle cell in the heart will produce a contraction lasting 300 ms from a single electrical stimulus. The cells are all linked electrically; a stimulus to one will lead to contraction of them all in a defined sequence determined by their anatomical arrangement. Two groups of special cells in the right atrium generate an electrical stimulus at regular intervals. These are known as pacemaker cells, and one is in the **sino-atrial node**, the other in the **atrio-ventricular node**. As the sino-atrial node pacemaker generates action potentials more frequently than the atrio-ventricular node, it normally controls the heart rate. Excitation spreads over the atria, causing them to contract, and then passes to the ventricles via the atrio-ventricular node. The excitation is conducted to the apex of the heart (the end of the ventricles away from the valves) via a specialized bundle of muscle cells, called the bundle of His. The ventricular contraction therefore begins at the apex and, in effect, squeezes blood out of the ventricles through the outflow valves. This coordination of contraction and valve operation ensures that at each beat of the heart a volume of blood is effectively transferred from the venous to the arterial side, with an elevation of pressure. This sequence is the **cardiac cycle** (Figure 5.27).

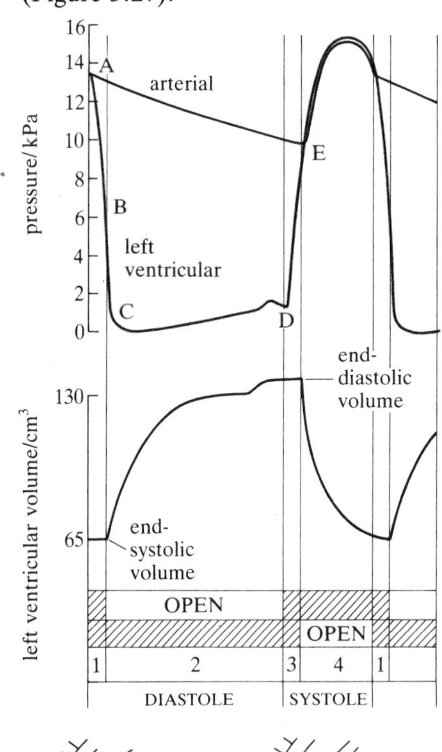

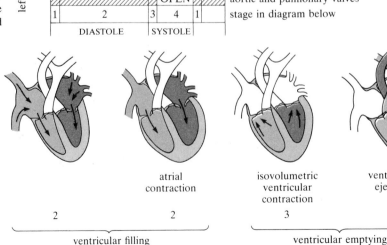

Figure 5.27 Graphs of the pressure and volume changes in the left side of the heart and aorta of a mammal during the cardiac cycle. The points A to E on the upper curve are referred to in the text. The sequence of diagrams in the lower part of the Figure shows the blood flow (*arrows*) and valve movements: the darker pink tone indicates oxygenated blood.

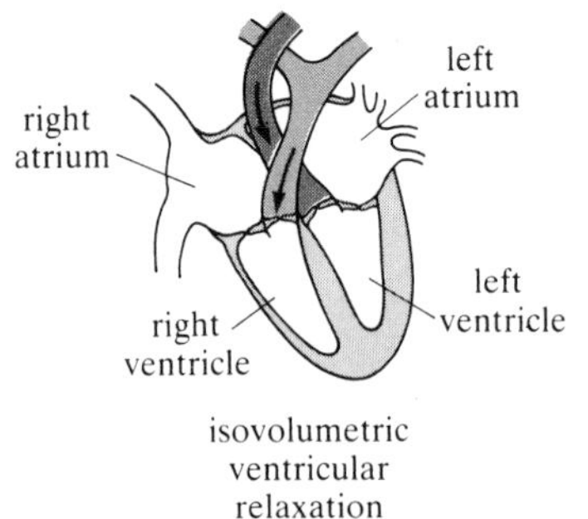

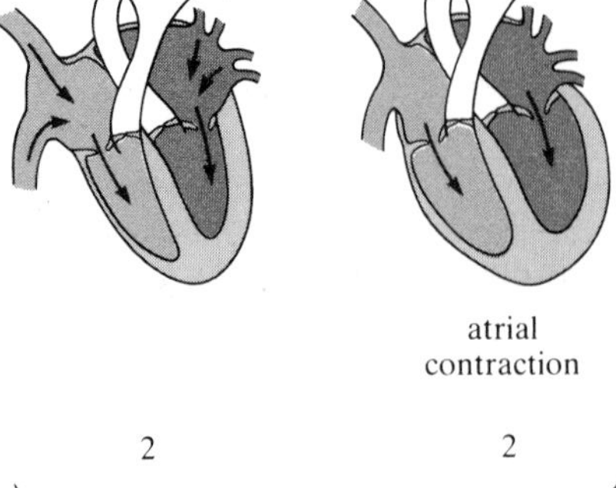
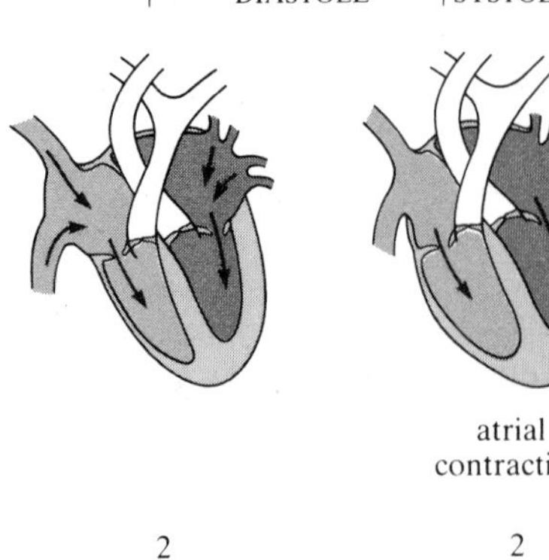

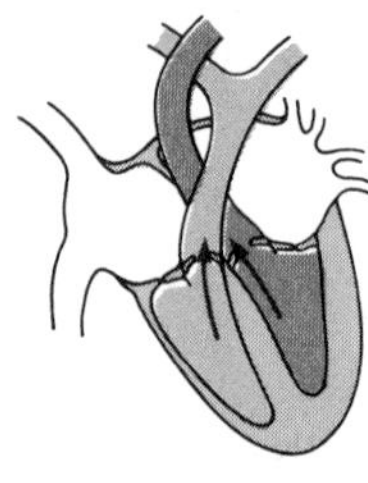

As with any repetitive event, to describe it we must break into the cycle at one point. Consider the heart just as it is relaxing after a contraction; the beginning of diastole. Valves open when there is a pressure gradient across them in the direction of allowed flow. They are closed by reversal of this pressure gradient causing momentary reverse flow which drives the leaflets of the valve together. As the ventricles relax (point A on Figure 5.27), pressure in them falls below arterial pressure, so the outflow valves (pulmonary and aortic) close. At this stage the inflow valves are also closed so pressure in the ventricle drops dramatically (point B). During the previous contraction (systole) blood continued to return to the atria, so by this stage they have distended and the atrial pressure is relatively high. As soon as ventricular pressure falls below atrial the inflow valves (tricuspid and mitral) open and the ventricle begins to fill (point C). Initially, filling of the ventricle is rapid but, as the walls stretch, filling slows down until ventricular filling is almost complete. Atrial contraction then forces a small additional volume of blood into the ventricles.

Systole begins with ventricular contraction (point D), which raises ventricular pressure, and the inlet valves are closed by momentary backflow. Continued contraction raises ventricular pressure at a high rate because both inflow and outflow valves are closed and the ventricular volume cannot change. This is the so-called **isovolumetric ventricular contraction**. Eventually, the ventricular pressure reaches the value to which the arterial pressure has fallen during diastole (point E), and the outlet valves open. Now there is a period of rapid ejection of blood, and both ventricular and arterial pressures rise rapidly. As systole ends the ventricles begin to relax and we are back to where we started at the onset of diastole.

Summary of Section 5.3

1 Morphologically, hearts can be classified as tubular, chambered or accessory. Tubular hearts are characteristic of annelids, holothurians and arthropods. Chambered hearts are characteristic of most molluscs, and the chordates. Accessory hearts are present at the bases of the gills in the octopus.

2 Experiments in which the waves of pressure produced in the blood are measured show that while the contractions of the heart produce the main pressure wave, respiratory or locomotory movements may also contribute to pressurizing the blood. This is exemplified in the octopus, where the accessory hearts also enhance the pressure in the circulatory system.

3 The mammalian heart is a highly evolved pump for controlled circulation of blood around a high pressure closed circulation. It has four chambers, two atria and two ventricles. The circulation is a double one. The pulmonary circulation is pressurized by the right ventricle and the systemic by the left. The sequence of events during contraction (systole) and relaxation (diastole) is called the cardiac cycle.

4 The electrical stimulus for contraction spreads from the pacemaker cells at the sino-atrial node and the atrio-ventricular node. The excitation is conducted to the apex of the heart by specialized cells—the bundle of His.

5 The inlet valves between the atria and ventricles are the tricuspid on the right and the mitral on the left. The outlet valves are the pulmonary on the right and the aortic on the left.

Now attempt Question 3, on p. 165.

5.4 CONTROL OF THE MAMMALIAN CARDIOVASCULAR SYSTEM

In previous sections you have read how the various arrangements of pumps and vessels ensure a flow of blood through the tissues of the body. It should also now be clear to you that for any particular organ to function effectively the flow through it needs to be matched to its metabolic activity.

Tissue metabolism is not constant, so organs need different blood flows from moment to moment. The circulation must change to cope with different conditions. This requires sophisticated regulation, which has been most intensively studied in mammals, because of the obvious relevance to medical practice. In this section we examine how the elements of the mammalian cardiovascular system work together and meet the metabolic demands of most of the tissues most of the time.

5.4.1 What are the flows required by different parts of the system?

All organs require blood flow, but their needs vary enormously, and the consequences of temporary flow reduction range from virtually harmless to lethal. Any consideration of the control of the circulation must first establish what the demands of each organ are, how much they vary, and how crucial it is that they should be continuously satisfied. Table 5.4 compares flows through different organs in the human.

Table 5.4 Typical maximum and minimum blood flow rates through different human organs and tissues: the maximum and minimum flow rates are given in $cm^3 min^{-1}$

Organ/tissue	Min rate	Max rate	Requirements
brain	750	750	The metabolic needs of the brain are constant and can be met by a flow of 0.5 $cm^3 g^{-1} min^{-1}$ for a brain of normal size. The brain is extremely intolerant of flow interruption.
heart	300	1 200	At rest, the heart needs 0.9 $cm^3 g^{-1} min^{-1}$, but if the heart has to work hard this may increase four-fold. The heart is extremely intolerant of inadequate flow.
kidney	1 200	1 200	Requires a high *constant* blood flow to maintain its function, though most flow is not nutritive; 4.0 $cm^3 g^{-1} min^{-1}$. Extremely intolerant of inadequate flow.
gut (and liver)	1 400	3 000	At rest the gut and liver, which are connected in series via the hepatic portal system, receive 1 $cm^3 g^{-1} min^{-1}$. Digestion of a meal generates a substantial increase in flow. Short term flow reduction is tolerable.
skeletal muscle	1 000	16 000	The metabolic needs of muscle vary enormously. At rest the blood flow needs to be 0.03 $cm^3 g^{-1} min^{-1}$, up to 0.6 $cm^3 g^{-1} min^{-1}$ in exercise, but this may not meet metabolic needs. Muscle can survive for a limited period on anaerobic metabolism.
skin	150	150	Skin is not metabolically very active and may be supported by 0.03 $cm^3 g^{-1} min^{-1}$, through flow may increase to 0.1 $cm^3 g^{-1} min^{-1}$ for thermoregulation.
rest	200	200	A fairly constant demand.
TOTAL	5 000	22 500	

In the adult, the total flow required ranges from 5 to 25 l $\min^{-1}$, so the control system must always meet this total demand without compromising flow to vital organs such as the brain and heart, which have an absolute requirement for continuous flow. Tissues such as skin and muscle have flows that can vary over a ten-fold range from moment to moment, yet they can tolerate short periods with no flow at all.

Control is achieved because there is a high degree of communication between different parts of the system. Tissues communicate their individual need for blood to the system, and the pump and distribution vessels must respond to the total effect of these signals to keep the total flow right. In the jargon of economics the system is **demand-led**.

5.4.2 How is the blood flow to the tissues matched to need?

In Section 5.2 you saw that flow depends on pressure and resistance. Assume for the moment that the pressure driving blood into the tissues is held constant. The flow into any organ then depends upon the resistance of its blood vessels. The major resistance in the circulation is in the arterioles, and their resistance is determined by their diameter (Section 5.2), which in turn depends on the degree of contraction of the smooth muscle in their walls. Flow is, therefore, controlled by arteriolar smooth muscle.

◇ What physiological change occurs to bring about an increase in blood flow-rate?

◆ The smooth muscle in the arterioles contracts less and the diameter of the arterioles is increased by the pressure of the blood. The increase in diameter leads to a lower resistance and hence increased flow.

The increase in the diameter of arterioles is called **vasodilation**. Decreases in flow are produced by a reduction in arteriole diameter—**vasoconstriction**. Muscles cannot actively increase their length, so it follows that except under conditions of maximal flow, there must always be some vasoconstriction. This **vasomotor tone** is then modified to control flow, so vasodilation is actually reduced vasoconstriction.

Vasomotor tone is due to a number of factors. Vascular smooth muscle tends to contract of its own accord, but this is supplemented by the action of neurotransmitters released from autonomic **sympathetic** nerves. Practically all blood vessels receive a greater or lesser sympathetic innervation, and there is constant or 'tonic' activity in the nerves, which leads to the release of noradrenalin from their terminals. This acts on receptors on the smooth muscle to cause contraction. This 'sympathetic vasoconstrictor tone' is controlled by centres in the brain, and has more powerful effects on the blood vessels of some organs than others (Figure 5.28). It generally provides a background to the operation of other factors, such as local metabolic needs.

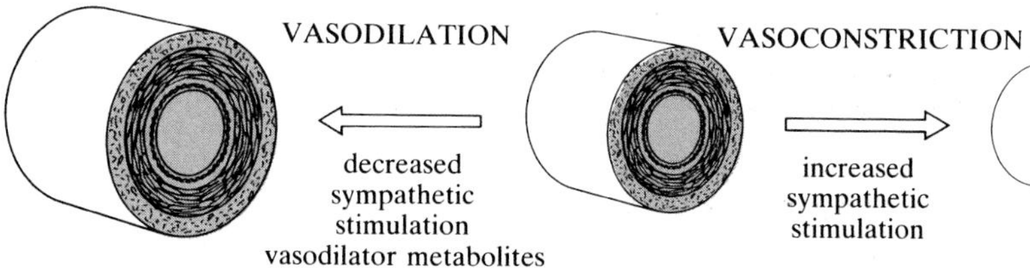

Figure 5.28 Arterioles are normally constricted to some degree by sympathetic vasoconstrictor 'tone'. Vasodilator metabolites and other factors cause a reduction in tone, which increases blood flow.

DEMAND-LED
Often applied to circ. system to emphasize role of tissues in controlling flow.
VASODILATION
Expansion of walls of arterioles and other vessels as a result of the relaxation of smooth muscle.
VASOCONSTRICTION
Narrowing of walls of arterioles and other vessels as a result of the contraction of smooth muscle.
VASOMOTOR TONE
Arterioles are slightly contracted at all times. Vasomotor tone is then modified as the flow rate is controlled.
SYMPATHETIC NERVE
In vertebrates, the sympathetic nervous system is part of the autonomic system. Symp. nerve fibres innervate organs and glands at synapses that usually use noradrenalin as the neurotransmitter.

Signals from tissues control flow by modulating vasomotor tone. Consider an example, which you may have experienced yourself. If the blood flow to your arm is cut off for a minute or two by a tourniquet, such as a pressure cuff, then its removal will be followed by a massive surge of blood, accompanied by a spreading sensation of warmth and redness of the skin. This 'reactive hyperaemia' occurs because during the period without blood flow the vasomotor tone of the arterioles has been almost completely overridden, so when they are reconnected to the arteries they are fully dilated, their resistance is very small, and blood flow very high. What causes vasodilation?

Metabolism continues in the tissues whether blood flows or not, and products of metabolism are still released by cells. Certain of these are known to act directly or indirectly on smooth muscle in arterioles, causing relaxation. They are known as vasodilator metabolites. There are many candidates, but the most likely substances are adenosine, potassium ions and hydrogen ions. During the period of no flow their concentration increases, so arterioles dilate, then as blood flow is re-established, they are washed away and flow decreases again. In fact, the flow changes so that the concentration of metabolites remains within a narrow range. This is seen very clearly if a part of the circulation is isolated and supplied with blood at a pressure that can be varied experimentally (Figure 5.29a). If this 'perfusion pressure' is increased blood flow initially rises (Figure 5.29b), but soon falls back to the original level, so resistance must have increased to match the pressure.

In this case, the metabolism of the tissues is more or less constant throughout. The blood flow changes only transiently. Within a very short time it has returned to its original value.

◇ If the flow is the same at a higher pressure, what has happened to the resistance to flow?

◆ The resistance must have increased: because flow = pressure/resistance, the arteriole diameter is reduced by contraction of its smooth muscle.

Initially, vasodilator metabolites are being produced and being washed away at a constant rate. The local concentration stays constant. When pressure rises, the initial rise in blood flow washes away extra metabolites, so their concentration falls. As a result, there is less vasodilation and the resistance to blood flow rises. As a consequence the flow falls back again and this allows the concentration of metabolites to rise. Drops in perfusion pressure have the opposite effect.

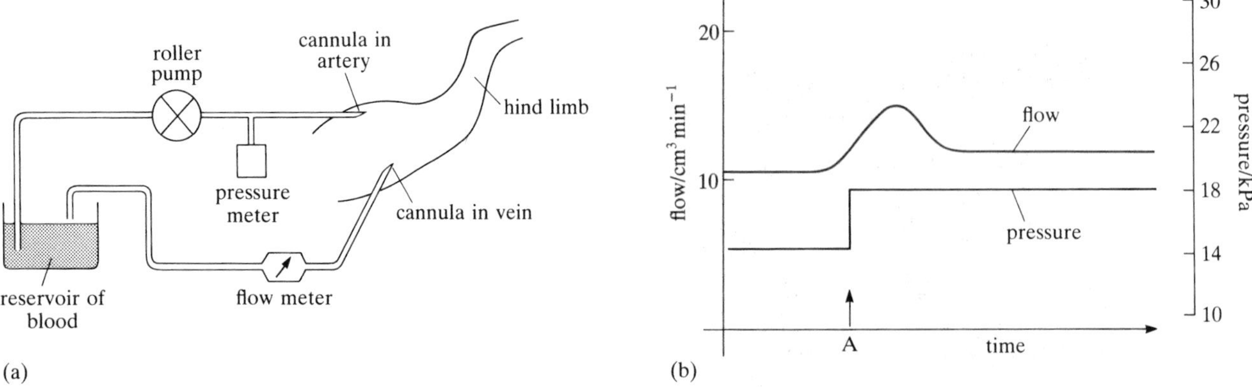

Figure 5.29 Autoregulation of blood flow. (a) A part of the circulation is isolated, and perfused with a mechanical pump. Pressure in the arteries is measured, together with flow. (b) A graph of flow against time produced from data obtained with the experimental set-up shown in (a). If the pressure is suddenly increased by turning up the pump at time A, then initially flow increases, but the removal of extra metabolites causes the arterioles to dilate less (constrict more), so resistance increases to match the pressure, and the flow returns to its previous value.

Over a fair range of pressures the flow does not, in fact, change—a phenomenon known as **autoregulation**. At very high or very low perfusion pressures, however, resistance cannot change enough to compensate.

If tissue metabolism changes but perfusion pressure remains constant, a greater or lesser amount of metabolites is produced, so again the blood flow alters and the rate at which metabolites are washed away is material to the rate at which they are produced.

At most levels of metabolic activity most organs can automatically take the blood flow they need, so long as the pressure of the blood supplying them is kept within a certain range.

In some tissues the action of the general vasodilator metabolites such as adenosine is supplemented by other substances. Often, when increases in metabolic activity are triggered by nervous or hormonal control, the trigger stimulates production of a specific vasodilator substance. The best examples are seen in exocrine glands—e.g. salivary glands. Nervous stimulation causes release of **bradykinin**, which is a potent vasodilator. In the stomach a similar function is fulfilled by **histamine**.

There is currently much interest in a substance known as **EDRF** (endothelium derived relaxing factor) which is released from endothelial cells (those lining blood vessels). EDRF acts on arteriolar smooth muscle to produce vasodilation.

Not all blood flow to organs serves just to meet their nutritional needs. Some flow is 'non-nutrient' flow. The most obvious example is in the skin. The metabolic activity of skin is very low, requiring a 'nutrient blood flow' of only 0.03 $cm^3\ min^{-1}\ g^{-1}$, yet flow may increase to 30 times this value under certain conditions.

◇ Why does the flow of blood through the skin vary so dramatically?

◆ The blood has a major role in shunting heat round the body. Heat loss at the skin is increased when blood flow increases.

This flow does not pass through capillaries, but via arterio-venous anastomoses connecting arterioles and venules (Figure 5.30). They are close to the body surface, so the blood may lose heat. Flow through arterio-venous anastomoses is controlled via the sympathetic nervous system modifying smooth muscle tone in their walls.

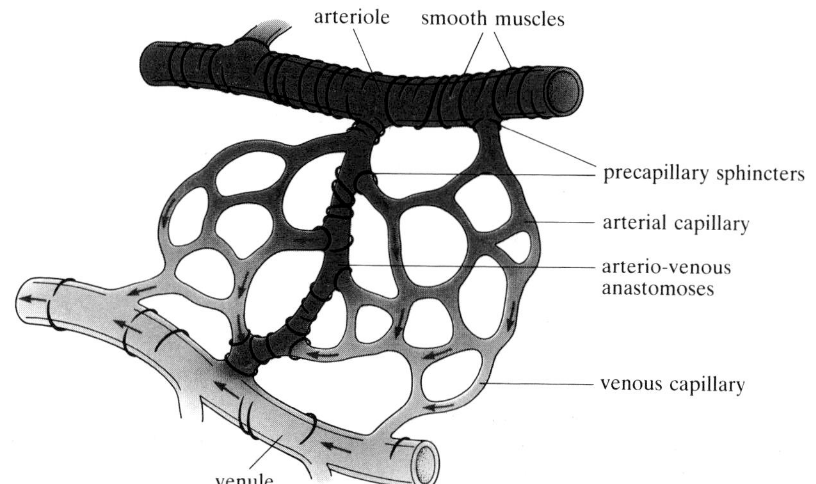

Figure 5.30 The links between arterioles and venules. The grey tone indicates the arterio-venous anastomoses.

Overall, however, the picture remains one of individual tissues looking after themselves, whether for nutritional or other reasons, and generating a total demand for blood flow that must be met by the pump if the basic requirement of adequate perfusion pressure is to be sustained.

5.4.3 How does the heart meet the total demand?

Effects of cardiac pumping and arteriolar resistance on the circulation

The heart pumps blood from the veins into the arteries, from where it leaks back against the resistance of the arterioles to the veins. The pressures in both the arteries and the veins depend upon the interaction between the active movement of blood through the heart and its passive flow through the arterioles.

Blood is pushed actively into the arteries at a total flow determined by the heart—the **cardiac output** (the total volume pumped by the heart each minute). It returns against the collective resistance of all the arterioles—the **total peripheral resistance**. The pressure in the vessels is determined by both. Increasing either will raise pressure, reducing either will make it fall. As the arteries contain little blood, and have relatively inelastic walls, small changes in cardiac output or total peripheral resistance produce relatively large changes in arterial pressure.

Blood enters the veins passively, but is actively removed by the heart. The pressure in the veins depends upon a balance between the inflow of blood from the peripheral resistance, and its removal by the heart. A fall in total peripheral resistance will tend to elevate venous pressure. A rise in cardiac output will remove more blood, and reduce venous pressure.

As the veins contain a large volume of blood, and have very elastic walls, changes in peripheral resistance and cardiac output produce relatively small changes in venous pressure. Changes in the total volume of blood circulating (blood loss caused by a wound, for example) will also alter venous pressure.

Changing either cardiac output or total peripheral resistance will affect both arterial and venous pressure.

Monitoring of these two pressures provides the signals that enable the cardiovascular system to be regulated. If the output of the heart and the total peripheral resistance are matched, then arterial and venous pressures will stay constant. If one or the other changes, both pressures change. Consider, as an example, a rise in the total demand for blood flow. This will be due to vasodilation in some organ, and expressed as a fall in total peripheral resistance. If cardiac output did not change, there would be a fall in arterial pressure; and a rise in venous pressure. So long as the heart responds to these changes by increasing its pumping, then as cardiac output rises the changes in arterial and venous pressure will be reversed, and a new equilibrium established.

Put another way, supply can only be tuned to meet demand so long as the heart is sensitive to the consequences of mismatch between cardiac output and total peripheral resistance: i.e. disturbance of arterial and venous pressures.

If the heart increases its pumping in response to rises in venous pressure and falls in arterial pressure, then supply will automatically match demand, and the circulation will be stable.

Control of cardiac output—matching supply to demand

Recall from Section 5.3.5 the sequence of events during the cardiac cycle. The total volume pumped by the heart each minute (the cardiac output) is the product of the heart rate and the volume pumped at each beat—the **stroke volume**.

$$\text{cardiac output} = \text{heart rate} \times \text{stroke volume} \tag{5.5}$$

The stroke volume itself is the difference between the amount of blood in the ventricle when it starts to contract (the **end-diastolic volume**) and the amount left when ejection ends—(the **end-systolic volume**). So,

$$\text{stroke volume} = \text{end-diastolic volume} - \text{end-systolic volume} \tag{5.6}$$

For the cardiac output to meet demand, the end-diastolic volume, the end-systolic volume and the heart rate must be matched to both venous pressure and arterial pressure.

Factors affecting filling of the heart

The ventricle fills in diastole, when the heart muscle (myocardium) is relaxed. As it fills, therefore, the elastic walls are passively stretched, and tension develops in them. Pressure within the ventricle rises by about 1 kPa. Flow ceases when the pressure in the ventricle matches the pressure in the veins supplying it. Ventricular pressure during passive stretch is related to volume by the ventricular compliance curve: the higher the pressure, the greater the end-diastolic volume.

The end-diastolic volume, therefore, depends only on the venous pressure unless there is structural change in the ventricle which alters the compliance curve. One exception to this rule occurs when the heart rate is very high, and there is too little time between the ventricular contractions for complete filling to occur. This is not a serious problem, however, until the heart rate exceeds 150 beats per minute.

End-diastolic volume depends on venous pressure.

Factors affecting ventricular contraction

End-systolic volume depends upon the extent to which the ventricles contract during systole. The electrical conduction system in the heart ensures that all cells contract on every beat (Section 5.3.5). So, the force of contraction of the ventricle as a whole can only be affected by factors acting on each cell individually. These factors may be *mechanical* or *chemical*.

Mechanical factors determine the extent to which the heart muscle relaxes during diastole, and the volume of blood that is ejected for a contraction of given force. As the ventricle fills through one valve from the atrium, and ejects through a different valve to the arteries, the forces stretching the ventricle during diastole and the factors affecting the ejection of blood can vary independently. You have just read about how the heart is stretched by filling from the veins during diastole, and how the extent of stretch depends on venous pressure. Greater stretch is achieved by an increase in the length of the individual muscle cells. Like all muscle, heart muscle, up to a limit, contracts more forcefully the more it is stretched before the contraction. A greater end-diastolic volume therefore produces a more powerful contraction.

The effect of that contraction depends upon the different mechanical factors operating during systole, when blood is being ejected into the arteries. For the left heart, the critical factor is the difficulty of driving blood into the aorta—the 'aortic impedance'. This depends in part on how easily the blood can leave the arteries via the arterioles (i.e. the total peripheral resistance), and in part on the elasticity of the arterial walls. The higher the aortic impedance, the less blood will flow into the arteries for a ventricular contraction of given force, but at the same time the higher the pressure that will be generated.

Therefore, the force of ventricular contraction depends upon the end-diastolic volume, but the effect of that contraction in ejecting blood depends upon the aortic impedance. At a constant aortic impedance, the greater the end-diastolic volume, the more blood will be ejected. If aortic impedance is increased (and therefore arterial pressure rises), less blood is ejected for a given end-diastolic volume.

So the difference between end-diastolic volume and end-systolic volume—the stroke volume—depends on two things. First, it depends upon the end-diastolic volume: the bigger the ventricle is before it contracts, the bigger will be the stroke volume. Second, it depends upon the aortic impedance: the harder it is to pump blood out of heart, and the higher the pressure, the less comes out—not really surprising!

But the end-diastolic volume, as you now know, depends on venous pressure. So, at a constant aortic impedance, if venous pressure rises, the stroke volume also rises (Figure 5.31). This is known as *Starling's law of the heart.*

It can be stated, really quite crudely, as: more in—more out!

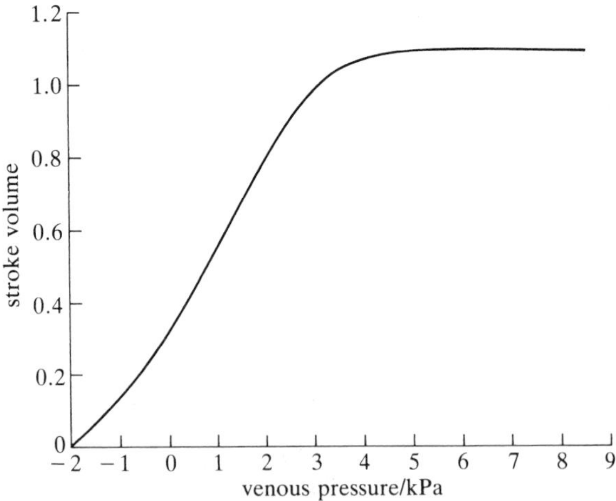

Figure 5.31 The end-diastolic volume of the heart is determined by the venous pressure. The force of contraction, which determines end-systolic volume at a constant aortic impedance, increases with end-diastolic volume. So, as venous pressure rises the difference between end-diastolic volume and end-systolic volume (stroke volume) increases.

So, by virtue of the properties of its own muscle cells, the heart is sensitive to venous pressure, its output tending to increase as venous pressure increases. A rise in arterial pressure caused by increased aortic impedance (usually due to greater total peripheral resistance), will tend to reduce stroke volume.

The response of the heart muscle itself will tend to increase stroke volume if venous pressure rises, and decrease it if arterial pressure rises.

Although the force of contraction of heart muscle always depends on its initial length, chemicals released from nerves modify the relationship. The heart is innervated by the sympathetic branch of the autonomic nervous system (Chapter 1), whose endings release noradrenalin. Noradrenalin makes heart cells beat more forcefully whatever the starting length—this is known as a change in 'contractility' of the heart. Sympathetic stimulation of the heart, therefore, reduces end-systolic volume at all end-diastolic volumes. In effect, noradrenalin steepens the relationship between venous pressure and stroke volume so that for any given venous pressure the heart pumps more (Figure 5.32).

(PARASYMPATHETIC) ACETYLCHOLINE = SLOWING OF HEART RATE

(SYMPATHETIC) NORADRENALINE = HEART CELLS BEAT MORE FORCEFULLY

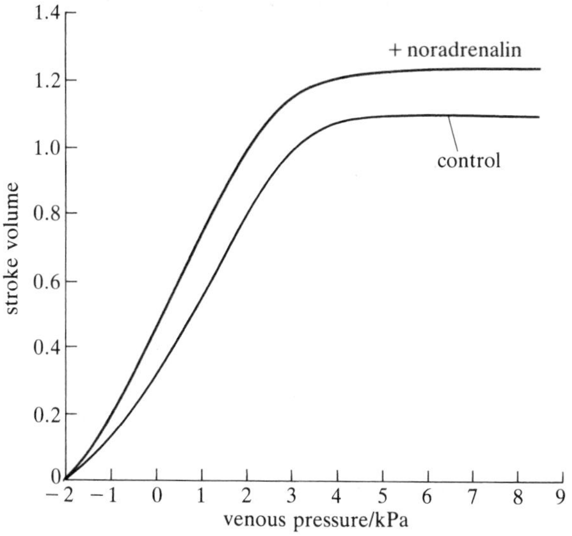

Figure 5.32 If sympathetic cells release noradrenalin, then cardiac muscle cells contract more forcefully for a given end-diastolic volume.

PARASYMPATHETIC NERVE

In vertebrates, the paras. system is part of the autonomic nervous system that is not under voluntary control. Most of the motor neurons release acetylcholine as the para. neurotransmitter.

BARORECEPTORS

Receptors in a blood vessel that are sensitive to pressure and also to the rate of change of pressure.

Factors affecting heart rate

In a normal heart the rate of beating depends only on the membrane properties of a small number of pacemaker cells in the sino-atrial node. Factors changing heart rate must act there. The major influences are nervous. The sino-atrial node is innervated by both sympathetic and parasympathetic branches of the autonomic nervous system.

Parasympathetic nerve endings release acetylcholine. Acetylcholine slows the heart beat. When the body is at rest, parasympathetic nerves are normally active, so the heart rate is low. Increases in heart rate occur by a decrease in parasympathetic activity.

Sympathetic nerves are activated to produce further increases in heart rate. The nerves release noradrenalin, which can increase heart rate up to a maximum, in a young person, of 200 beats min^{-1}. The action of noradrenalin may be supplemented by circulating adrenalin from the adrenal medulla.

The activity in both sympathetic and parasympathetic nerves is controlled by groups of cells in the brain: they are found in the medulla oblongata, which is part of the hind-brain. Their activity is controlled by many things, including sensory input from receptors in the blood vessel walls that are sensitive to pressure—the **baroreceptors**. Baroreceptors are sensitive to stretch, which is a measure of pressure. They are sited in both the arterial and venous sides of the circulation. As pressure rises the elastic walls of both arteries and veins relax so the vessels expand, and this leads to an increase of the action potential frequency in the nerves from the stretch receptors. The greatest response from the receptors occurs when pressure changes.

An increase in response from the venous receptors, located in the great veins (ascending and descending vena cavae), causes an increase in heart rate. This is often known as the Bainbridge reflex.

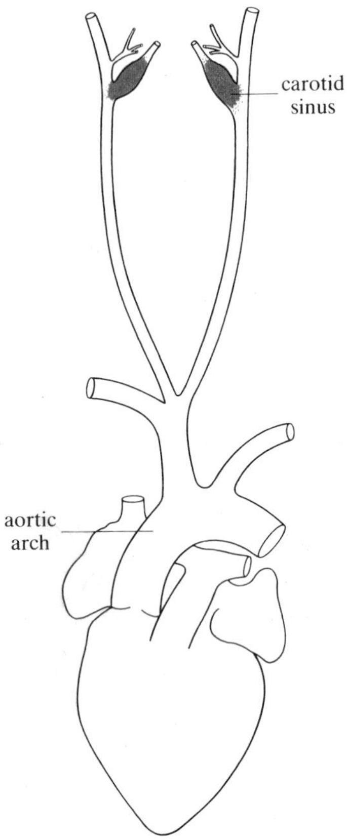

Figure 5.33 Locations of the baroreceptors in the aortic arch and carotid sinus (grey tone). As the pressure of the blood expands the walls of these vessels, stretch receptors are stimulated and send signals to the central nervous system. These signals lead to autonomic changes, which increase the heart rate and contractility, and also the flow to certain vascular beds such as the skin.

CAROTID SINUS
An enlarged area in the carotid arteries in mammals where the baroreceptors are sited.

AORTIC ARCH
Small section of the aorta that loops over the heart. Contains baroreceptors,

The arterial receptors are located principally in the **carotid sinus**, but there are also receptors in the **aortic arch** (Figure 5.33). An increase in response from these receptors *decreases* heart rate. A decrease in response increases heart rate, and contractility. So, as with the intrinsic mechanisms of the heart, a link is made between arterial and venous pressures and cardiac output.

If arterial pressure falls, the action potential frequency from the arterial baroreceptors decreases, so sympathetic activity increases, and heart rate and contractility rise. If venous pressure rises, the action potential frequency from the venous baroreceptors increases, so again sympathetic activity increases, and heart rate rises.

The response of arterial baroreceptors leads to changes in the circulation, as well as in the heart. For some parts of the circulation, notably the gut and skin, blood flow may be temporarily reduced if the system as a whole is threatened. If arterial pressure falls quickly, then there is an increase in the vasoconstriction in these tissues, so increasing one component of the total peripheral resistance, and reducing the total demand for blood. This *cannot* go on indefinitely, but it is useful as a temporary measure in control of the circulation.

The most important role of the baroreceptors, however, is to match cardiac output to arterial and venous pressure. So both stroke volume and heart rate are increased by rising venous and falling arterial pressure, and decreased by falling venous and rising arterial pressure. To put it another way, cardiac output is positively related to venous pressure, and negatively related to arterial pressure—just what is needed to match supply to demand!

5.4.4 The operating rules of the mammalian cardiovascular system

We are now in a position to state a set of rules, which can be used to make predictions of how the cardiovascular system will change under different circumstances. Each rule could be stated quantitatively, but here we shall restrict ourselves to a qualitative treatment, which will nevertheless explain most changes. In each case where two variables are related, we assume that all other variables are kept constant (though as the whole system reacts, the other variables will eventually change).

RULE 1 Total peripheral resistance is inversely related to the body's total need for blood flow.

RULE 2 Total peripheral resistance affects arterial and venous pressure.

(i) If total peripheral resistance falls, venous pressure rises. If peripheral resistance rises, venous pressure falls.

(ii) If total peripheral resistance falls, arterial pressure falls. If total peripheral resistance rises, arterial pressure rises.

RULE 3 Arterial and venous pressures affect the heart.

(i) If venous pressure rises, cardiac output rises. If venous pressure falls, cardiac output falls.

(ii) If arterial pressure rises, cardiac output falls. If arterial pressure falls, cardiac output rises.

RULE 4 The heart affects arterial and venous pressures.

(i) If cardiac output rises, venous pressure falls. If cardiac output falls, venous pressure rises.

(ii) If cardiac output rises, arterial pressure rises. If cardiac output falls, arterial pressure falls.

REGULATION

RULE 5 Arterial pressure affects total peripheral resistance.

(i) Resistance to blood flow through skin and gut may be temporarily inversely related to arterial pressure.

Rules 3 and 4 are the key to regulation, as they are reciprocal in effect. The consequence of the heart responding to changes in arterial and venous pressures is to correct the change. In other words, at any given peripheral resistance there is only one cardiac output that the system will sustain, the one that matches the tissues' demand for blood flow.

Application of the rules to two specific examples

Changes in heart rate

Some patients are fitted with a pacemaker, which allows their heart rate to be experimentally controlled. What happens if the pacemaker is used to cause a sudden increase in their heart rate?—i.e. what happens if the heart tries to increase the supply of blood with no change in the demand?

Total peripheral resistance will initially be normal, so if cardiac output begins to rise then, by rule 4, venous pressure will fall, and arterial pressure will rise. These changes will then affect the heart. By rule 3, if venous pressure falls the heart will fill less in diastole, so stroke volume will be reduced. It will be more difficult for the ventricle to eject blood into the raised arterial pressure, so stroke volume will be further reduced.

Although baroreceptors will be activated, their effects on the heart will be overridden by the artificial pacemaker, so heart rate will remain high, but stroke volume will fall. Cardiac output is the product of the two, and eventually, the rise in heart rate will be neatly balanced by the fall in stroke volume—so the cardiac output will stay the same.

The heart *cannot* drive the circulation, it can only be driven by it.

Change in tissue demand for blood

When we eat a meal, more blood needs to flow through our gut lining. The tissues of the gut produce various vasodilator metabolites, which lead to a reduction in resistance to blood flow to this organ. This produces a fall in total peripheral resistance.

By rule 2, if total peripheral resistance falls, arterial pressure will fall, and venous pressure will rise.

The rise in venous pressure will cause the heart to fill more in diastole, and the drop in arterial pressure will allow it to empty more in systole, so stroke volume will rise (rule 3). The changes in arterial and venous pressure will also be detected by baroreceptors, leading to an increase in heart rate (rule 3). Cardiac output will rise, and as it does, by rule 4, arterial and venous pressure will come back to normal.

Cardiac output rises automatically in response to a decrease in total peripheral resistance.

Now attempt Questions 4 to 9, on pp. 165–166.

5.5 EXERCISE PHYSIOLOGY

Exercise is a powerful challenge to the respiratory and cardiovascular systems. To conclude this chapter we shall examine how the mammalian cardiovascular system responds to that challenge. After discussing respiration in Chapter 6 we shall return to the subject of exercise in Chapter 7, and examine how the respiratory system responds to the challenge.

In principle, the systems we have already described should be perfectly able to cope with the demands of exercise, but in practice the magnitude and speed of the changes required cause problems. The basic mechanisms are supplemented by one or two others to 'tune' the system so that it is able to respond optimally.

5.5.1 Exercise and the cardiovascular system

The metabolic activity of skeletal muscle increases manyfold during exercise, which dramatically increases the muscle's need for oxygen, and its production of carbon dioxide. Blood flow to muscles rises by up to twenty-fold to cope with these demands, and yet the conditions in the circulation are kept within the limits that ensure adequate blood flow to organs such as the brain, heart and kidneys, a considerable feat of regulation.

The microcirculation of skeletal muscle is well adapted to widely varying blood flow. The density of capillaries is high, though at rest most do not receive blood flow, because the pre-capillary sphincters at their arteriolar end are contracted. At rest, the arterioles supplying these capillaries are also constricted by contraction of smooth muscle in their walls under the influence of the sympathetic nervous system. The total blood flow through a muscle depends on the state of contraction of both its arterioles and precapillary sphincters, which is affected by a variety of influences.

As in most tissues there is a prevailing sympathetic vasoconstrictor tone, so increases in blood flow occur when this is reduced or overridden by other influences. At the onset of exercise, the sympathetic vasoconstrictor tone is reduced. This initial vasodilation is then supplemented by the action of vasodilator metabolites, which accumulate as the metabolism of the tissues builds up. These act in part on the arterioles, but also affect precapillary sphincters, allowing more of the capillary bed to receive blood. Eventually the whole capillary bed is perfused, and the arterioles are maximally dilated, so the flow resistance through skeletal muscle falls to a minimum.

At rest, skeletal muscle receives relatively little blood, and has a high resistance to blood flow. The effect of decreasing its resistance to a very low value is to decrease the total peripheral resistance, so that in the absence of any other changes all blood would flow through muscle, leaving none for vital organs such as the brain. The exercise can only continue if the circulation as a whole is adjusted to cope with this challenge.

You read earlier that the total flow of blood to the tissues is matched to demand by the way in which the heart responds to changes in venous and arterial pressure. A variety of mechanisms ensures that if venous pressure rises and arterial pressure falls, the output of the heart is increased. Falls in total peripheral resistance tend to lower arterial and raise venous pressure, thus triggering compensatory rises in total flow through the system. These mechanisms operate during exercise, and the cardiac output rises so that the muscles get their extra blood flow without reductions in flow to the brain. The necessary change in cardiac output, however, is very large and such dramatic changes pose special problems for the system.

As exercise begins the flow of blood into the veins is greatly increased, and the movements of the muscles force it towards the heart because the veins have valves in them. This tends to produce a massive rise in venous pressure, which if unchecked would lead to the heart filling maximally in diastole. Such overfilling has adverse effects. First, the heart pumps less efficiently at high end-diastolic volumes, but second, and more important, if the venous pressure is very high, then the relationship between increases in venous pressure and cardiac output (Starling's mechanism) is lost. The heart is two pumps in series (the right and left ventricles), and the operation of Starling's mechanism is the only way by which the two ventricles can pump similar volumes of blood. Changes in heart rate and contractility affect both to the same extent, and so cannot be used for differential control. If one ventricle were to pump more than the other there would be problems in the pulmonary circulation.

Large rises in stroke volume are prevented by pre-emptive rises in heart rate just as exercise begins. The same rise in sympathetic activity that produces vasodilation also increases heart rate and contractility. Recall from the example in Section 5.4.4 that rises in heart rate without any other change in the cardiovascular system will not lead to increases in cardiac output, because as the heart tries to pump more, venous pressure falls, and by Starling's mechanism, stroke volume is reduced. In the case of exercise, however, venous pressure is tending to rise, so the increase in heart rate merely compensates for this tendency, and more blood is pumped to keep the venous pressure low enough to prevent the problems described above. The pre-emptive rise in heart rate has therefore acted to keep the system in such a state that it has all its control systems, including Starling's mechanism, in full operation, and the outputs of the two ventricles may still be matched.

The characteristic of the cardiovascular response to exercise, therefore, is of pre-emptive, sympathetically mediated changes, which prevent large changes in the variables (arterial and venous pressure) that the system uses as control signals, thus preserving them for use to control the system so that supply of blood may be precisely matched to demand.

To summarize, the sequence of events is:

1 Pre-emptive vasodilation in muscle, followed by further effects of vasodilator metabolites, cause a massive fall in total peripheral resistance as skeletal muscle demands much more blood.

2 The fall in total peripheral resistance, combined with the pumping effect of muscle contraction, tends to raise venous pressure.

3 Pre-emptive sympathetic activity increases heart rate and contractility, which compensates the tendency of stroke volume to rise, and elevates cardiac output to meet the new demand for blood.

4 Further small changes in arterial and venous pressure operate on the heart to match cardiac output precisely to demand, and keep the pressures stable, so that the brain, heart and kidneys receive the blood they need.

These mechanisms ensure that the maximum possible amount of blood flows through muscles during exercise. In extreme exercise, however, the blood flow to muscles rarely meets all their metabolic needs, mainly because as the muscles contract they compress their blood vessels, and restrict flow. This flow restriction means that at high levels of activity, muscles have to be supported in part by anaerobic metabolism and in normal individuals, blood flow to muscles is the major limiting factor on the maximum rate of delivery of oxygen to them. This limitation only applies, however, if that blood has adequate levels of oxygen, and another set of mechanisms ensures that breathing is increased sufficiently to keep the blood gases normal, despite a large turnover of oxygen and carbon dioxide.

Summary of Section 5.5

1 Exercise is a powerful challenge to the cardiovascular system since there is a very large increase in metabolic activity in a short period. Blood flow to the muscles increases as much as twenty-fold.

2 There is a pre-emptive vasodilation in muscles which reduces peripheral resistance and raises venous pressure. The heart rate increases and cardiac output rises. Large pressure changes are avoided and the brain, kidney and the heart itself receive the blood supply they require despite the increase in demand from the muscles.

OBJECTIVES FOR CHAPTER 5

Now that you have read this chapter you should be able to:

5.1 Define and use, or recognize definitions and applications of, each of the terms printed in **bold**.

5.2 Distinguish between closed and open circulatory systems. (*Question 1*)

5.3 Explain the factors that affect diffusion rates, and recall the Fick equation. (*Question 1*)

5.4 Interpret simple blood pressure records.

5.5 Predict the effect of a number of variables on the flow of blood within blood vessels. (*Question 5*)

5.6 Explain what is meant by the microcirculatory system. (*Question 2*)

5.7 Compare and contrast the structure and function of blood capillaries and the larger blood vessels. (*Questions 2 and 5*)

5.8 Describe the cell types found in blood and lymph.

5.9 Classify a heart according to morphological type. (*Question 5*)

5.10 Describe, with diagrams, the circulatory systems, including the heart, of mammals, lobster and octopus. (*Question 6*)

5.11 Explain the exchange of fluid across capillary walls and describe the ways in which the volume of interstitial fluid may be controlled. (*Questions 3, 4 and 9*)

5.12 Identify factors that influence cardiac output. (*Questions 7, 8 and 9*)

5.13 Recall and apply the operating rules of the cardiovascular system. (*Questions 7, 8 and 9*)

QUESTIONS FOR CHAPTER 5

Question 1 (*Objectives 5.2 and 5.3*) Decide whether each of the following statements is true or false.

(i) The rate at which a substance diffuses is directly proportional to the square of the distance over which diffusion takes place.

(ii) All arthropods have open circulations.

(iii) All molluscs have closed circulations.

(iv) Closed circulatory systems are often high pressure, high resistance systems.

Question 2 (*Objectives 5.6 and 5.7*) Select the appropriate number, word or phrase from the choices in brackets and insert them into the following sentences.

(i) Only . . . per cent of the circulating blood is in the capillaries at any one time. (1, 5, 50)

(ii) Gases, nutrients and waste materials are exchanged across the . . . walls of the capillaries (epithelial, interstitial, endothelial).

(iii) The capillary wall is . . . (permeable to all substances, not permeable to any substance, permeable to substances of low relative molecular mass).

Question 3 (*Objectives 5.5, 5.7 and 5.9*) Select from the following list of statements *four* that are true.

(i) The arthropod heart is typically tubular.

(ii) There are many different types of heart, but all depend upon muscular contraction for their function.

(iii) The circulation of blood in *Lumbricus* is ensured primarily by the five paired hearts.

(iv) The heart of the octopus contains two chambers.

(v) Arteries are generally of a larger inner diameter than veins.

(vi) Assuming that pressure is constant, halving the diameter of a tube will decrease the flow of a fluid within the tube to one-sixteenth of what it was.

(vii) A diameter of about 10 mm is typical of blood capillaries in mammals.

(viii) For every million mammalian red blood cells you would expect to find around 1 000 white cells.

Question 4 (*Objective 5.12*) (a) Complete Figure 5.34, which is a summary of the situation at point A in the capillary of the cat mesentery in Figure 5.10. (b) Is the net movement of water into or out of the capillary?

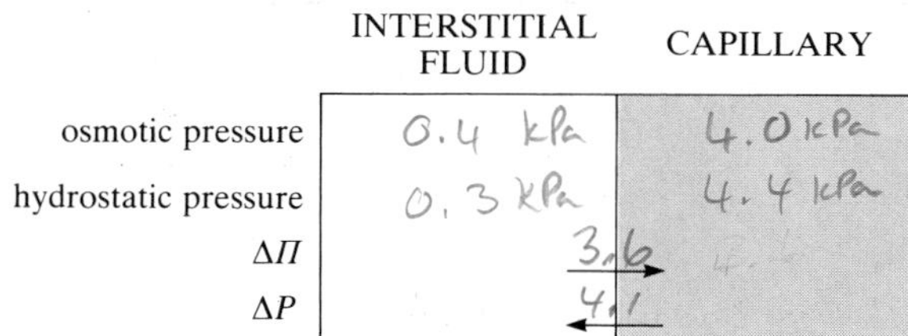

Figure 5.34 For use with Question 4.

Question 5 (*Objective 5.11*) Complete the simplified diagram of the lobster's circulatory system (Figure 5.35). Label the boxes—'gills', 'tissues' and 'ventral sinus'—as appropriate. Use arrows to show the direction of flow. Indicate in which part of the system the haemolymph is likely to be richest in oxygen.

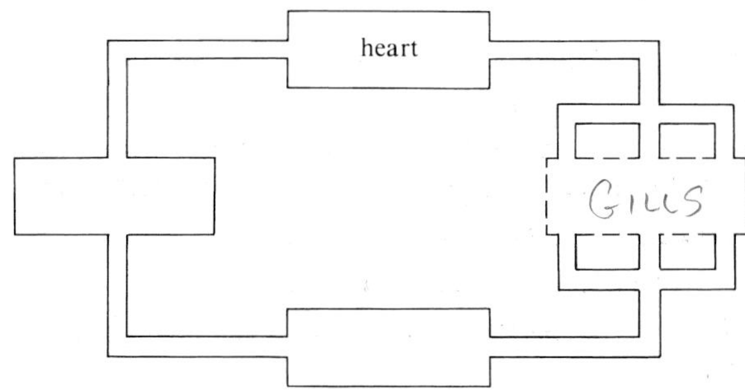

Figure 5.35 For use with Question 5.

Question 6 (*Objective 5.11*) Kwashiokor is a condition found in the tropics amongst people living on a diet that is adequate as far as energy content is concerned but very deficient in protein. One of its characteristics is swelling of tissues, particularly apparent in children where the gross swelling of the abdomen gives a pot-bellied appearance. Suggest a reason for this swelling.

Question 7 (*Objectives 5.12 and 5.13*) By applying the operating rules of the cardiovascular system, describe what would happen to the heart and circulation if you started to exercise violently.

Question 8 (*Objectives 5.12 and 5.13*) By applying the operating rules of the cardiovascular system, describe what would happen to the heart and circulation if you injure yourself and lose a litre of blood (you would have around 5 litres normally).

Question 9 (*Objectives 5.11, 5.12 and 5.13*) What would be the effect on both venous and arterial pressures of left heart failure (an inability of the left ventricle to pump all the blood that the body requires)? How would this affect the process of filtration at capillaries in the lungs and legs?

No7 - exercise violently

- arterial pressure drops
- venous pressure rises due to blood being pumped up by leg muscles
- higher stroke volume
- more blood to ventricle from ~~heart~~ lungs
- increase in pressure from heart to arteries

= speeded up heart rate and increase in blood pumped

= eventually stroke volume returns to normal but speed blood is pumped has increased.

RESPIRATION AND GAS TRANSPORT ◆ CHAPTER 6 ◆

TRACHEAE
Air-filled tubes that extend from the surface of the body of arthropods and onychophorans

6.1 INTRODUCTION

In this chapter we shall start to consider how oxygen is supplied to the tissues of various species in sufficient quantity to meet their metabolic needs. In the case of very small organisms (less than 1 mm in diameter), all the cells are close enough to the surface to exchange gases with their surroundings by simple diffusion, but above this critical size the innermost cells have to be supplied by other means. Different ways around this problem have evolved in different species so there is a wide variety of biochemical and physiological respiratory mechanisms in the animal kingdom.

First, we shall consider insects, which 'pipe' air directly into their tissues via a system of tubes (**tracheae**), so that all cells are close to the 'atmosphere'. This system produces remarkably high rates of oxygen exchange and supports tissues that can be amongst the most metabolically active in the animal kingdom. It is, however, limited to relatively small organisms. Larger animals, including all vertebrates, transfer oxygen into a circulating fluid via specialized exchange organs, and then distribute it to their tissues. We consider this process of gas transport in this chapter, and consider the organs of exchange in Chapter 7.

There is, however, a substantial problem in transporting oxygen and carbon dioxide in aqueous media—the relatively low solubility of the gases in water. The quantities of gases produced by simple dissolution in blood, for example, are much too low to meet the metabolic needs of tissues at reasonable flow rates.

In nearly all sizeable mammals, especially vertebrates, the amount of oxygen that can be carried per unit of blood is greatly increased by respiratory carriers that can reversibly bind oxygen. Such compounds are usually coloured and are called **respiratory pigments**. We examine their structure and function in Sections 6.4, 6.5 and 6.6.

The pigments also often facilitate the transport of carbon dioxide from the tissues, a process which is most completely understood in mammals, and which we consider in Section 6.7.

These processes of gas transport and exchange can, however, only be understood in the context of some basic physics of gases, which we shall discuss in Section 6.2.

6.2 THE PHYSICS OF GASES

In essence, we may consider gases to be collections of molecules moving around more or less independently within the space occupied by the gas. Their frequent collision with the walls of the container exerts pressure. The normal pressure of the atmosphere at sea-level is 101.3 kPa. If a fixed mass of

GAS LAW

The product of the pressure and volume of a gas is equal to the product of a constant and the absolute temperature.

PARTIAL PRESSURE

The component of total pressure of a gas mixture attributable to a particular gas in the mixture.

gas is confined in a smaller chamber, then collisions with the wall are more frequent, and pressure rises. So, for a fixed mass of gas, pressure is inversely proportional to volume (Boyle's law). If a fixed volume of gas is heated, the kinetic energy of the molecules increases, so the pressure rises. Pressure is directly proportional to *absolute temperature*—the temperature scale which starts at the temperature where kinetic energy is zero, i.e. 0 Kelvin (−273.15 degrees Celsius). These effects of volume and temperature are encapsulated in the **gas law**:

$$PV = RT \tag{6.1}$$

where R is the gas constant ($8.314\ \mathrm{J\ K^{-1}\ mol\ l^{-1}}$).

If two chambers containing gases at different pressures are connected, then gas will flow from the region of high pressure to that of low pressure.

In almost all cases when studying respiration we are not dealing with a single gas, but a mixture, such as air. Provided the gases do not react chemically, we may consider each gas as though it alone occupied the space, and then sum the effects of all the gases in the mixture. In other words, each gas may be considered to exert a **partial pressure**, and all the partial pressures add to give the total pressure. The partial pressure of a gas in a mixture will be the same fraction of the total pressure as the volume of that one gas is of the total gas volume, so we can calculate partial pressure thus:

◇ If dry air contains 20.9% oxygen, what is the partial pressure of oxygen (P_{O_2}) in dry air at 101.3 kPa?

◆ $P_{O_2} = \frac{20.9}{100} \times 101.3 = 21.17$ kPa.

◇ If dry air contains 0.03% carbon dioxide, what is the partial pressure of carbon dioxide?

◆ $P_{CO_2} = \frac{0.03}{100} \times 101.3 = 0.03$ kPa.

In considering respiration, the partial pressures of gases in a mixture are generally of greater significance than the total pressure. If two gas mixtures are brought into contact, then, even if there is no difference in *total* pressure, the components of the mixtures will exchange according to gradients of partial pressure. This occurs by diffusion, not by bulk flow of gases.

◇ The following gas mixtures are brought to either side of a membrane which allows gas molecules to cross. In which direction will oxygen, nitrogen and carbon dioxide diffuse?

	Gas mixture 1		Gas mixture 2
P_{O_2}	7.3 kPa	←	13.3 kPa
P_{CO_2}	7.3 kPa	→	6.3 kPa
P_{N_2}	87.6 kPa	→	81.7 kPa

◆ Oxygen will diffuse from mixture 2 to mixture 1; nitrogen and carbon dioxide will diffuse in the opposite direction.

All gas mixtures within living organisms are in contact with a large amount of water. Water molecules move into the gaseous phase as water vapour, and gas molecules move into the liquid phase both in solution and in chemical combination.

In the gaseous phase, the water vapour molecules behave as a gas and exert a partial pressure, or vapour pressure. They continually leave and re-enter the liquid phase, and eventually an equilibrium is established when molecules leave and enter at the same rate. At this equilibrium, the partial pressure of water vapour is known as the **saturated vapour pressure**. This is determined only by the tendency of water molecules to leave the liquid phase and, significantly, is independent of the total pressure of gas in the gaseous phase. So long as the gaseous phase is saturated (which is always true within the body compartments of mammals), the partial pressure of water vapour is independent of total pressure, and determined only by temperature.

For example, air at 37 °C saturated with water has a P_{H_2O} of 7.3 kPa.

◇ What is the P_{O_2} of air (20.9% O_2) at 101.3 kPa, but saturated with water vapour at 37 °C?

◆ The saturated vapour pressure of water at 37 °C is 7.3 kPa. Therefore, $P_{O_2} = (101.3 - 7.3) \times 0.209 = 19.65$ kPa.

This is of particular significance if the total pressure is reduced (say during exposure to high altitude), as the fixed vapour pressure of water becomes an increasing fraction of the total pressure.

◇ If the total pressure of air falls to 30 kPa (as it does at the top of Mt Everest), and the saturated vapour pressure of water in the lungs remains 7.3 kPa, what is the P_{O_2}, given that the composition of air has not changed?

◆ $P_{O_2} = (30 - 7.3) \times 0.209 = 4.74$ kPa.

The total pressure is reduced to 30% of the figure at sea-level, but the P_{O_2} is down to 24%.

Just as the water vapour enters the gaseous phase, so gas enters the liquid phase. As in water, an equilibrium is established where molecules of a particular gas enter and leave the liquid at the same rate. At this equilibrium, the gas is effectively exerting a partial pressure (**tension**) in the liquid, equal to its partial pressure in the gas phase. It is very important to realise that the tension of a gas in the liquid is only a measure of the tendency of the gas to leave the liquid. It is not directly a measure of quantity. Quantity is determined by how much gas enters the liquid before the equilibrium between solution and dissolution is established—the **solubility of the gas**. If carbon dioxide and oxygen at identical partial pressures are exposed to water, a much greater amount of carbon dioxide dissolves because this gas is much more soluble than oxygen. The solubility of a gas is expressed as an **absorption coefficient**, which is the volume of the gas at standard temperature and pressure (STP), 0 °C and atmospheric pressure at sea-level, i.e. 101.3 kPa, that is dissolved in a known volume of solution at equilibrium. In other words, when the partial pressure of the gas in solution is 101.3 kPa, the amount of dissolved gas per unit volume is equal to the absorption coefficient. For example, 0.049 cm^3 of oxygen at 101.3 kPa is dissolved in 1 cm^3 of water at 0 °C: the absorption coefficient of oxygen is thus 0.049. The absorption coefficient of carbon dioxide is 1.713 because, at 0 °C, 1.713 cm^3 of carbon dioxide at 101.3 kPa is dissolved in 1 cm^3 of water.

The absorption coefficients of gases can be measured at different temperatures, as shown in Table 6.1.

One important point to note from Table 6.1 is that the solubility of gases *decreases* as the temperature of the solvent increases. (You may have noticed that freshly-opened beer or coke is more fizzy when it is warm: more carbon dioxide is leaving solution.) This is the reverse of the solubility of many substances such as salts.

Table 6.1 Absorption coefficients of three gases at different temperatures (the volume in cm^3 of gas at STP that is dissolved in 1 cm^3 of water).

	Gas		
Temperature/°C	CO_2	O_2	N
0	1.713	0.048 9	0.023 9
10	1.194	0.038 0	0.019 6
20	0.879	0.031 0	0.016 4
30	0.665	0.026 1	0.013 8

The absorption coefficient of oxygen also differs in water of different salinities. Table 6.2 gives the amount of oxygen dissolved when a litre of freshwater or seawater is exposed to dry atmospheric air containing oxygen at a partial pressure of 21.16 kPa at a variety of temperatures. The value for seawater is less than that for freshwater at the same temperature because seawater contains more ions. In general, increasing the ionic content of a medium decreases the solubility of a gas in that medium. (Note that it is sometimes more convenient to express solubility as cm^3 of oxygen per *litre* of water, as here, rather than as cm^3 of oxygen per cm^3 of water.)

Table 6.2 The amount of oxygen dissolved in freshwater and in seawater (cm^3 per litre of water) at equilibrium with atmospheric air at different temperatures.

Temperature/°C	Freshwater	Seawater
0	10.29	7.97
10	8.02	6.35
20	6.57	5.31
30	5.57	4.46

The values in Table 6.2 were measured experimentally, but the same values could have been calculated by using a simple equation that relates the amount of dissolved gas to the partial pressure of that gas:

$$c = \alpha \frac{P_g}{101.3} \tag{6.2}$$

where c is the amount of the gas in solution expressed as cm^3 of gas per litre of water at STP, α is the absorption coefficient (here in cm^3 of gas per litre of solution) and P_g is the partial pressure of the gas in solution (in kPa).

This equation makes it clear that the amount of any gas per unit volume of a solution, that is the *concentration of a gas*, depends not only on the partial pressure of that gas in solution, but also on the absorption coefficient, which, as we have seen, *varies with temperature and salinity*. Consider two discrete compartments containing solutions with identical partial pressures of oxygen but different salinities. The absorption coefficients of oxygen will be different for each solution and so, from Equation 2, at equilibrium the concentration of oxygen in each solution will be different *although the partial pressures of oxygen in the solutions will be the same*.

A similar situation arises if two media of identical salinity but at different temperatures are exposed to identical partial pressures of oxygen. The colder medium will contain more oxygen, although the partial pressures of oxygen will be the same in both.

Suppose the two compartments were separated by a membrane permeable to oxygen. Although the two compartments would contain different amounts of oxygen, there would be no net movement of oxygen by diffusion from one compartment to the other. Oxygen will diffuse only if there is a *difference in partial pressure between the two compartments*; that is, if there is a *partial pressure gradient*, oxygen will diffuse down it until the partial pressures of the gas in each solution are identical. There is overwhelming evidence that all the gas movements involved in the exchange of gases in respiration in animals (and plants) can be adequately explained by diffusion.

To sum up, gases diffuse from areas of high partial pressure to areas of low partial pressure down a partial pressure gradient, which usually *but not always* means down a concentration gradient. You should note that 'gas partial pressure' is not synonymous with 'gas concentration'.

The situation is further complicated if the gas can combine chemically with any component of the liquid in which it is dissolved. Chemically combined molecules are not free to leave the liquid, so the chemical reaction must proceed as far as it can go before tension is established.

If plasma (that is, blood minus its red and white cells) is equilibrated with a gas mixture having a P_{O_2} of 13.3 kPa, then 3 $cm^3\ l^{-1}$ will dissolve. If whole blood, containing haemoglobin, which combines chemically with oxygen, is used, then content will increase to 200 $cm^3\ l^{-1}$, as the chemical reaction with haemoglobin has to proceed to completion before the P_{O_2} can rise to the same value of 13.3 kPa. When a gas enters the liquid in simple solution, the content rises in direct proportion to the partial pressure but, if chemical reaction is possible, then the dissolved gas is supplemented by the amount taken up chemically, *with no increase in partial pressure*. The amount taken up chemically, may however, depend on partial pressure and other factors in complex ways, as you will see in Section 6.5.

Now attempt Questions 1 and 2 on p. 195.

6.3 RESPIRATION IN INSECTS

In terrestrial insects, cells receive their oxygen directly by diffusion from air which is distributed throughout the body by means of a system of *tracheae*. This apparently simple system has evolved to a high degree of sophistication and we shall see how insects are able to exercise considerable control over their gas exchange and support a wide range of metabolic activities in a variety of different environments.

6.3.1 The anatomy of the tracheal system

The tracheal system arises from paired segmental invaginations of the body surface. In some insects, each branching set of tracheae remains separate from the others, but in most insects, they fuse to form a continuous, interconnected system joined by longitudinal tracheal channels (Figure 6.1).

The tracheal system opens to the exterior by spiracles, of which there are usually two pairs on the thorax and eight pairs on the abdomen. The spiracles are complex structures which prevent foreign bodies and water from entering

TRACHEOLES

Narrow subdivisions of the tracheal system that form the major sites of gas exchange in insects.

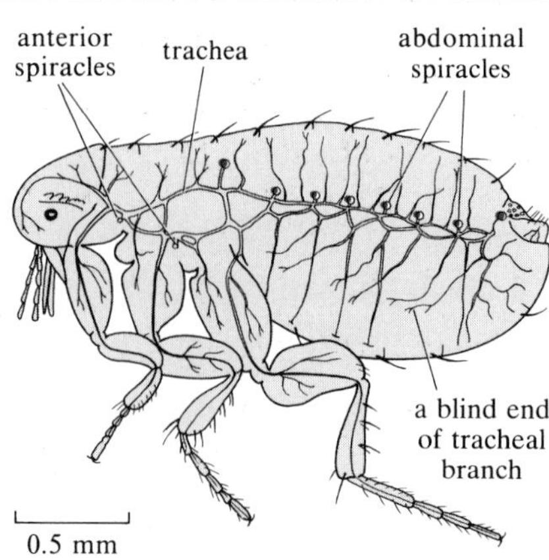

Figure 6.1 The main tracheae of the flea *Xenopsylla*, showing the location of the two anterior and eight abdominal spiracles (arrowed).

and have a mechanism that limits loss of water; also they may play a part in ventilation. The whole system is lined by ectodermal epithelium which secretes a cuticle that is hydrophobic (i.e. repels water) and is of the same nature as the exoskeleton. The cuticle lining the tracheae is often reinforced with spiral thickenings that help keep these narrow tubes open.

The tracheae become narrower as they penetrate deeper and the finest tracheal branches terminate as tracheal-end cells (sometimes called stellate cells because of their star-like shape); from each of these a number of very slender **tracheoles** extend as fine tubes reaching into various organs of the body.

Tracheoles are extraordinarily numerous (more than 1.5 million in the larva of the silkworm) and their cuticular lining is extremely thin (1 015 nm); it is not shed at the moult as is the rest of the cuticle and appears to be semipermeable and hydrophilic (i.e. water adheres to it). The blind-ended tracheoles spread out, squeezed between the cells they supply; they do not actually run into these cells but the tracheal cell membranes are intact and the tracheoles indent the cytoplasm of the other cells.

The density of tracheae and tracheoles is correlated with the metabolic activity of the tissue, in much the same way as the density of blood capillaries is in vertebrates. Not surprisingly, the flight muscle of insects (which is a very active tissue incapable of metabolizing anaerobically) is particularly well supplied with tracheae and tracheoles, but even in tissues with a more modest oxygen consumption, it seems that every cell is separated from a tracheole by no more than two or three neighbouring cells.

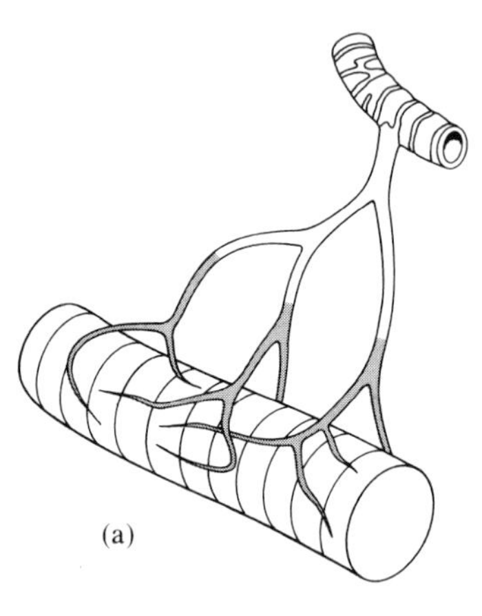

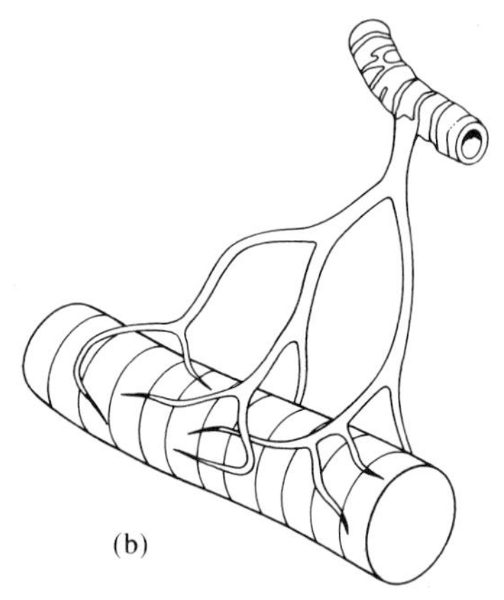

Figure 6.2 Tracheoles supplying a muscle fibre. (a) When the muscle is at rest, the ends of the tracheoles are full of fluid (shaded). (b) When the muscle has been active, air extends far into the tracheoles.

The fact that tracheoles have wettable surfaces means that they are normally at least partly filled with fluid. The proportion of the tubes that is filled when the insect is at rest varies with the tissue and the species, but that proportion of the tracheal system will be useless as a site of a gas exchange. The diffusion of oxygen in water is 10^{-6} times as fast as in air.

◇ If capillary forces attract fluids into the tracheoles, what force might partly balance such a movement by tending to draw water out of the tracheoles into the surrounding cells?

◆ The osmotic pressure exerted by the fluids around the tracheoles, just as the movement of water out of capillaries in the lungs is balanced.

When the metabolic rate of an insect increases, some fluid is drawn out of the tracheoles and the meniscus of the fluid is drawn down towards the tip of the tracheole (Figure 6.2), thus increasing the area of wall in contact with air.

There is a widespread view that the withdrawal of water is caused by the accumulation of the products of anaerobic metabolism (e.g. lactate) during oxygen shortage, which raises the osmotic pressure of the body fluid. It is not clear whether such a process could account for changes in the level of fluid in the tracheoles (ion pumps could also be involved), but the advantages of the withdrawal of fluid for gaseous diffusion are clear: during prolonged flights, insect flight muscles can increase their consumption of oxygen up to a hundred times the resting level. Tracheolar gas exchange must therefore be even more important when the metabolic rate is high.

The main part of the tracheal system of many insects consists simply of inter-connected longitudinal channels (Figure 6.3a), but in many insects there are **airsacs** associated with the major tracheae (Figure 6.3b). These airsacs are compressible and are numerous and large in active insects.

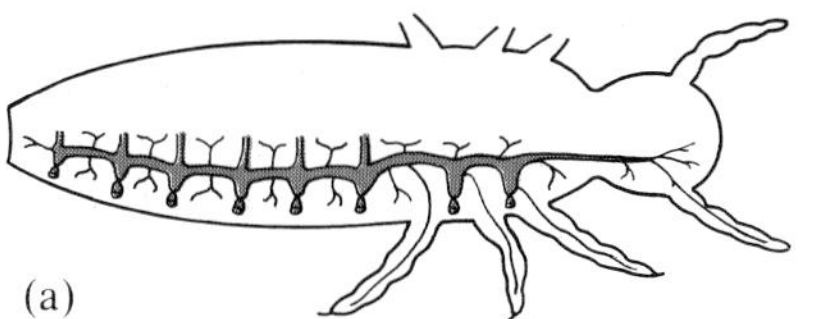

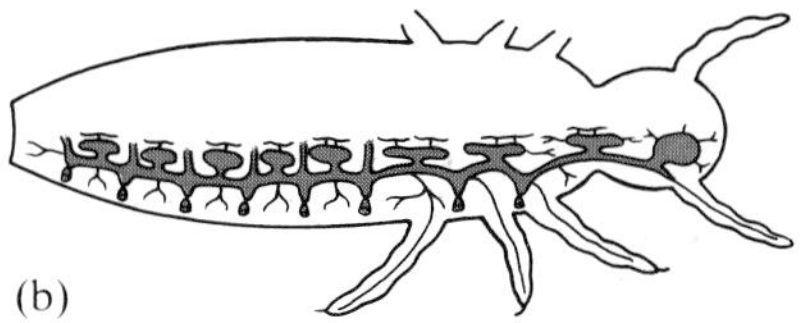

Figure 6.3 Two types of tracheal system (shown in pink on one side of the animal). In (a) there are simple tracheae with a longitudinal channel. In (b) airsacs which can be ventilated are present.

6.3.2 Ventilation of the tracheal system

The manner in which the tracheal system is ventilated varies between species. Some have no active mechanism and rely entirely on diffusion.

◇ What two major factors are likely to determine the need for active ventilation of the system?

◆ 1 The length of the tubes (which depends upon the size of the insects).
2 The rate of oxygen consumption by the tissues.

Small insects, even if they are very active (e.g. the fruit-fly, *Drosophila*), can manage simply by allowing oxygen to diffuse through the spiracles and tracheal system. Similarly, large but fairly inactive insects (e.g. some large caterpillars) do not ventilate the system actively. On the other hand, insects such as locusts maintain a flow of air through the spiracles, airsacs and primary tracheae by muscular contraction—mainly rhythmical squeezing of the airsacs and accurately timed opening and closing of various spiracles. In locusts, air enters by the anterior (thoracic) spiracles and leaves by the posterior (abdominal) ones, establishing a unidirectional flow through the airsacs and tracheal channels. Except in a few very large insects, the flow usually involves only the airsacs and larger tracheae. As far as we know, the smallest tracheae and the tracheoles exchange oxygen by diffusion only, even in the largest and most active insects. Whether a flow of air into the tracheae is maintained by muscular contraction or by diffusion, theoretical considerations suggest that diffusion could supply the necessary oxygen from the ventilated tracheae to the tissues or, in smaller insects, from the spiracles to the tissues.

This method of respiration is quite outstandingly effective. The oxygen consumption of some insect tissues, notably in their flight muscles, is the highest known for any living tissue; dragonfly flight muscles use 7.3 cm^3 of oxygen per gram of tissue per minute when fully active. This represents a metabolic rate far higher than anything achieved in mammals, yet the tracheal system delivers all the oxygen required. In this respect the insect system out-performs that of mammals, whose muscles commonly have to rely upon anaerobic metabolism because they receive insufficient blood for their needs in extreme exercise. The insect blood system has little respiratory function, except for the removal of a proportion of the carbon dioxide produced by respiration, some of which is then lost through the exoskeleton. This has a number of consequences. The insect heart is less muscular than the crustacean heart, although insects are in general more active. For the blood to supply insect flight muscles with oxygen from a central respiratory organ (lung or gill) would require a phenomenally efficient system, although of course the blood system does supply fuel (e.g. glucose) to the flight muscles. The diffuse nature of the tracheal respiratory supply involves much less energy (air is pumped rather than blood) and makes the insect very much less vulnerable to wounds.

◇ On the face of it, a tracheal system might seem superior to the respiratory system of large, warm-blooded mammals. Suggest why the tracheal system might not be appropriate for larger animals.

◆ Diffusion is adequate only over very short distances, so there would have to be ventilation right down the finest tubes in a large animal, and this is probably impractical (though science fiction authors seem unconvinced). This may be one of the main factors limiting the size of insects, but an equally important limitation is that imposed by the exoskeleton; as the animal gets larger, the exoskeleton would have to be disproportionately massive to support it.

6.3.3 Water loss during respiration in insects

One of the basic problems of life on land is how to reconcile rapid gas exchange with a minimal loss of water. Distributing air throughout the body can lead to substantial evaporation of body water. Insects have been outstandingly successful in coping with this, but even so the more active the insect, the greater its water loss.

The air in the tracheae and airsacs contains water vapour, and if there is unrestricted access to the air, the loss of water must be quite considerable. For example, if the pupa of the silkworm moth *Cecropia* is kept in an environment with 30% humidity, and the spiracles are kept open by inserting cannulae into them, within two weeks 25% of the body weight is lost and the pupa dies of dehydration. Clearly, it is important for some insects to be able to conserve water by closing their spiracles.

For insects that do not ventilate the tracheal system, that is, those that effect gas exchange by diffusion alone, the partial closure of the spiracles is the most effective way of cutting down water loss dramatically should circumstances demand.

◇ Partial closure of the spiracles, however, besides cutting down the loss of water, might be expected to have another, less favourable, effect. What is this?

◆ The inward diffusional flow of oxygen would be reduced, which might be too heavy a price to pay for conserving water.

◇ But consider the effect of constricting the spiracles on the partial pressure of oxygen in the air filling the tracheoles. What would happen if the diffusion of oxygen across the tracheal openings were reduced and the tissue continued to use oxygen at a high rate?

◆ The oxygen in the tracheoles (and eventually the tracheae) would be depleted; that is, the partial pressure of oxygen would fall and the animal's tissues might not receive enough oxygen.

◇ What effect would this have on the gradient of partial pressure across the spiracles?

◆ The gradient would steepen, which would tend to increase the rate at which oxygen diffused into the tracheae to an extent that might counteract the restriction on oxygen diffusion caused by the partial closure of the spiracles.

Note that closing the spiracles has no deleterious effect on the water vapour gradient across the tracheal openings; the air in the tracheae is well saturated even when the spiracles are open. So, the net effect is that water conservation can be ensured by constriction of the spiracles, without too serious an effect on the oxygen supply to the tissues.

There must, however, be control of the opening and closing of spiracles. This is generally achieved by their being sensitive to the carbon dioxide concentration in the tracheae. Carbon dioxide diffuses less readily than oxygen through air, as its molecular weight is higher, so it will accumulate faster if respiratory exchange is compromised.

A particularly interesting form of spiracular control is found in various insect pupae and some adults at rest. In the pupa of *Cecropia*, it has been shown that although oxygen is taken up at a fairly constant rate, carbon dioxide escapes in short bursts when the spiracles are fully open (Figure 6.4).

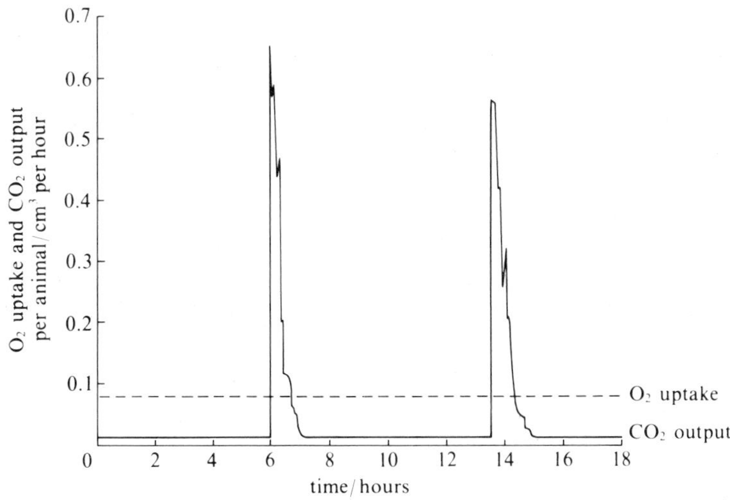

Figure 6.4 Gaseous exchange in a pupa of *Cecropia*. Note the pulses in the output of carbon dioxide in contrast to the steady rate of uptake of oxygen.

Very little water vapour is lost during the brief periods when the spiracles are open because the rate of evaporation is no greater than it would be if the spiracles were open all the time. However the loss of carbon dioxide during these brief periods is great as the spiracles open only when the internal partial pressure of carbon dioxide is high. Between periods of spiracular opening, much of the carbon dioxide produced is temporarily stored in the tissues (as bicarbonate), so keeping the P_{CO_2} within physiologically tolerable limits. When the spiracles open and the P_{CO_2} falls, most of the carbon dioxide is released from its chemically bound form and rapidly diffuses out of the tissues into the tracheae and then out of the spiracles.

What happens between bursts of carbon dioxide release is particularly interesting. Much of the carbon dioxide is stored as bicarbonate, so the volume of oxygen removed from the tracheoles is not replaced by an equivalent volume of carbon dioxide gas. This produces a subatmospheric pressure in the tracheae between bursts. The net effect is that, in the period of opening, a hundred times the amount of carbon dioxide actually produced by the metabolism in that time can escape. This escapes with only one-thirtieth of the water loss there would be if the spiracles remained open permanently. However, not all insects possess valves on the spiracles. Where water conservation is less critical (when animals live in a moist atmosphere, or where moist food is abundant) the spiracles may be unguarded.

Summary of Section 6.3

Insects have a waxy cuticle that is impermeable to water and not very permeable to oxygen. Respiratory gas exchange is therefore restricted to unique structures that directly 'pipe' gas to and from the respiring tissues without the involvement of the poorly developed circulatory system. These are the tracheae and tracheoles, which convey air directly to all parts of the tissues and are usually linked to the outside by spiracles. Most of the gas exchange with the tissues takes place via the tracheoles and the withdrawal of fluid from these fine, blind-ended tubes (possibly under the influence of increased osmotic pressure in the tissues) facilitates the movement of oxygen and carbon dioxide. Movement of gas within the tracheoles is solely by diffusion and, in small or larger inactive insects, this also accounts for movement of gas via the spiracles and along the tracheae. When the metabolic demand is greater, many insects ventilate the major tracheal channels. In those more complex species where this kind of ventilation is well developed, regular contraction of the abdominal muscles compresses large airsacs (much as in birds) particularly in flight (e.g. in bees, locusts).

Most water loss in insects occurs via the spiracles. The success of terrestrial insects in conserving water is linked with the development of mechanisms for opening and closing the spiracles, thus reducing water loss.

Now attempt Questions 3 and 4 on pp. 195–196.

6.4 OXYGEN TRANSPORT BY RESPIRATORY PIGMENTS

The size of animal that can be supported by tracheal gas exchange is limited by the length of tracheae required, the complexity of the processes required for their ventilation, and the huge water loss associated with such an enormous area for evaporation. In larger animals ventilation is restricted to exchange organs, the oxygen being transferred from there to the tissues by means of a circulating fluid which generally contains a respiratory pigment.

Respiratory pigments increase the amount of oxygen that can be carried. As we saw in Section 6.2, in human blood the dissolved oxygen is approximately 3 cm^3 per litre of blood (i.e. 0.3 vol. %), whereas the oxygen content of the same volume of blood containing red cells full of **haemoglobin** is seventy times as much—200 cm^3 per litre of blood (20 vol. %).

Simply increasing the *content* of oxygen will not, however, allow blood to act as a supplier of oxygen to the tissues. The combination of oxygen with the pigments must be *reversible* over a narrow range of conditions, so that they easily pick up oxygen at the lungs, and give it up equally readily at the tissues.

The most familiar and best understood pigment is haemoglobin, which has three important properties. Although there are other respiratory pigments in various invertebrates, as you read in Chapter 5, Section 5.2.2, we will concentrate exclusively on haemoglobin; its properties are essentially the same as those of the others.

OXYHAEMOGLOBIN
Haemoglobin with one to four oxygen molecules bound to the haem group.

6.4.1 The molecular structure of haemoglobin

DEOXYHAEMOGLOBIN
Haemoglobin in which oxygen is not combined with the haem group.

With rare exceptions, haemoglobin (Hb) ocurs in all vertebrates, but its distribution throughout the invertebrate groups is erratic and is not correlated with conventional taxonomic arrangements. Evidently this remarkably complex pigment has evolved independently in unrelated groups.

Porphyrins are a group of compounds distributed widely among animals, plants and bacteria. They associate with metals to form metalloporphyrins, which form part of a variety of compounds. For example, chlorophyll is a magnesium–porphyrin complex and cytochromes (which contain an iron–porphyrin group) play an important part in intracellular oxidation. One major group of porphyrins are the protoporphyrins, which exist in a variety of isomeric forms. When iron(II) is added to one such form, protoporphyrin IX, iron(II) protoporphyrin or haem is formed. Its full molecular structure is shown in Figure 6.5.

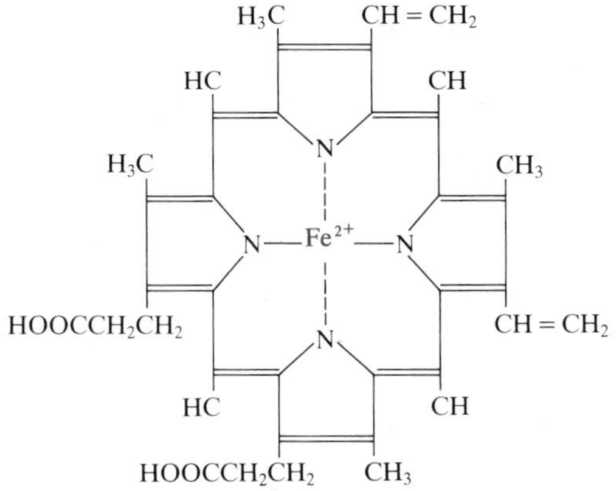

H_3C
$CH=CH_2$
HC
CH
N
H_3C
CH_3
N
Fe^{2+}
N
HOOCCH$_2$CH$_2$
$CH=CH_2$
N
HC
CH
HOOCCH$_2$CH$_2$
CH_3

Figure 6.5 The structure of the haem group of haemoglobin.

The iron(II) ion (Figure 6.5, highlighted in red) is joined to the protoporphyrin ring by bonding to four nitrogen atoms. In haemoglobin, the iron(II) ion makes two other links, one to an oxygen molecule, so forming **oxyhaemoglobin**, and the other to a histidine residue of the globin (protein) portion of the molecule at a single amino acid site. Oxygen can become bound to an isolated haem group but, unlike the binding of oxygen to haemoglobin, this leads to irreversible oxidation of the iron(II) to iron(III). Be clear about the differences between *oxidation*, which is a chemical reaction with oxygen to form an oxide and involves the removal of electrons, and *oxygenation*, which is (reversible) binding of O_2 molecules. Only when the haem group is linked to the histidine residue of an intact globin chain can the iron remain in the iron(II) state after oxygenation and thus act as an oxygen carrier. Haemoglobin which is not combined with oxygen is called **deoxyhaemoglobin**.

OXYGENATION
– Reversible binding of O_2

As far as is known, the haem groups of all haemoglobins are identical, so the wide differences that occur in the haemoglobin of different species, and even within a single species such as *Homo sapiens*, are attributable to variations in the globin portions of the molecule. A protein such as globin is an excellent vehicle for the expression of chemical variety, because changes in the type

MYOGLOBIN

A respiratory pigment with one subunit containing haem and a globular protein that binds oxygen and is found in muscle fibres. Similar to haemoglobin.

and sequence of amino acids can produce a whole family of a particular protein, though usually most variation occurs at particular sections of the polypeptide chain. Small differences in the primary structure of the globin portion can have dramatic effects on the quaternary structure and hence on the physiological properties of haemoglobin. In the disease sickle-cell anaemia, aberrant blood cells contain haemoglobin that differs from the normal form only by virtue of the substitution of *one* amino acid (valine) for the usual one (glutamic acid) at one specific point in part of the globin portion. The change in quaternary structure that results affects both the ability of the pigment to carry oxygen and the solubility of the pigment and this leads to atypical blood cells. This aberrant haemoglobin is just one example of over a hundred mutant haemoglobins that have been identified in humans, and the majority involve the substitution of a single amino acid.

We can build up a picture of vertebrate haemoglobin by first considering a 'single unit' or monomeric form called **myoglobin**. This consists of a *single* α-helical polypeptide chain—the globin moiety—in a shallow pocket of which a haem group is embedded (see Figure 6.6). Each human myoglobin molecule, with a relative molecular mass (M_r) of approximately 16 500 and precisely 153 amino acid residues, can bind only a single molecule of oxygen. The globin portions in the haemoglobin of vertebrate red blood cells are similar, though the chains are a little shorter than those of myoglobin; but the major difference is that vertebrate haemoglobins are tetramers formed by the aggregation of four polypeptide chains, each containing a discrete haem group. Two of the monomers in human haemoglobin are of a type called 'alpha' (α), each with 141 amino acids, and two are of a second type called 'beta' (β), each with 146 residues. Beta monomers are therefore slightly longer and also differ from the alpha pair in amino acid composition. As shown in Figure 6.7, the two pairs of monomers fit neatly together, and this unique quaternary structure, with two α monomers and two β monomers below (designated $\alpha_2\beta_2$), is maintained by hydrophobic links between adjacent globin chains.

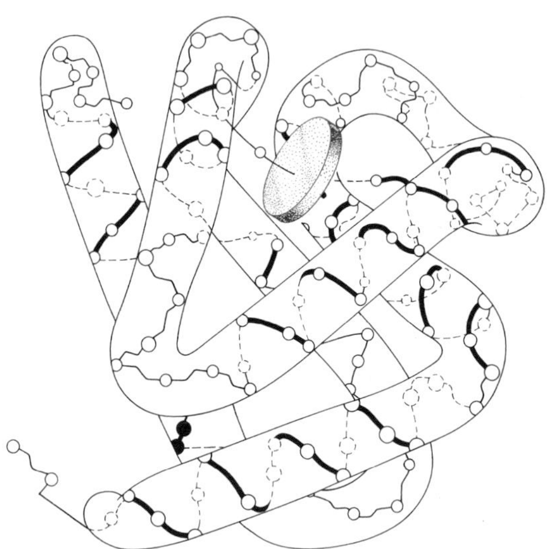

Figure 6.6 The tertiary structure of myoglobin. The haem group is shown in red.

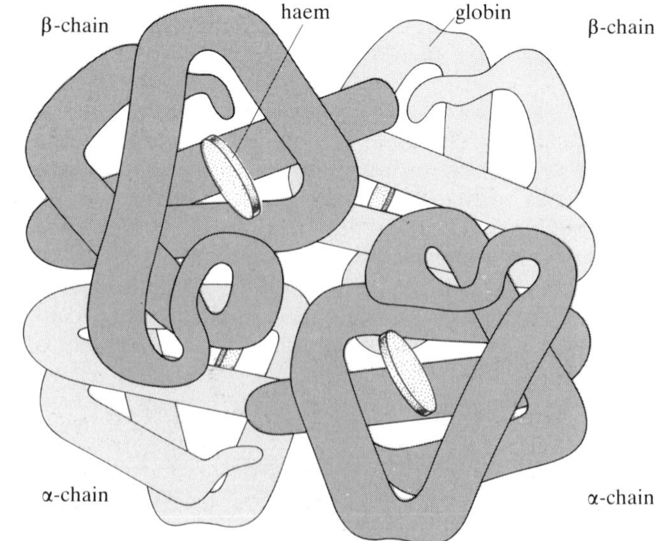

Figure 6.7 The quaternary structure of haemoglobin. The haem groups are shown in red, the globin groups in grey.

◇ Estimate the M_r of human haemoglobin. How many molecules of oxygen are likely to be bound to each haemoglobin tetramer?

◆ Each monomer of haemoglobin is roughly equivalent to the myoglobin molecule (but remember the differences in chain length and composition), so the M_r is likely to be close to $4 \times 16\,500 = 66\,000$. (In fact, the M_r is 68 000.) Each molecule of haemoglobin contains four haem groups, and so a maximum of four molecules of oxygen can become bound.

Let us concentrate on human haemoglobins for the moment. Normal individuals produce different types of haemoglobin tetramers—multiple haemoglobins—at different stages of development. The haemoglobin of a late fetus consists of two adult-type α-chains and two γ-chains, which differ from the adult β-chains in amino acid composition. This fetal haemoglobin, designated $\alpha_2\gamma_2$, gradually replaces an array of 'embryonic haemoglobins', characteristic of the first three months of embryonic life, when there are at least two additional types of globin subunits. Electrophoretic analysis of normal adult haemoglobin shows that a constant 97.5% of the total is $\alpha_2\beta_2$ but the remaining 2.5% is of a different structure ($\alpha_2\delta_2$) whose chemical composition is as yet unknown.

Besides α and β subunits, four other globin subunits are present at different stages of human life (we have mentioned only γ and δ); these aggregate in various combinations to form six different haemoglobin tetramers, of which ($\alpha_2\beta_2$) is the most common and long-lasting.

Some vertebrate species, for example, Adélie penguins, have only one type of haemoglobin, but multiple forms are more typical. Domestic fowl, for example, have three. Herrings have eight forms of haemoglobin, all of which persist throughout life, and coho salmon have 22. Geneticists, biochemists and physiologists are all very interested in multiple haemoglobins. Many workers assume that different haemoglobins play slightly different roles in gas transport and that such variety ensures that transport occurs effectively in a wide variety of environmental conditions. Nearly all of the haemoglobins from vertebrates are tetramers, but invertebrate haemoglobins are much more diverse. All contain haem groups linked to globin fractions that resemble the vertebrate type (except that the relative proportions of certain amino acids, e.g. histidine and cysteine, are different). The most distinctive feature is that these subunits often associate to form large aggregates of very high M_r. For example, the haemoglobin of *Daphnia*, a water flea, has approximately 24 subunits; one species of *Arenicola*, the lugworm, has 96 subunits (and a M_r of 2.85 million) and some polychaetes have haemoglobin with over 200 subunits. Many such pigments are not contained within red blood cells, and their very high M_r may prevent them from being lost from the circulatory system via the excretory organs.

Summary of Section 6.4

Most animals with circulatory systems possess respiratory pigments (e.g. haemoglobin) that increase the amount of oxygen that can be conveyed to the respiring tissues per unit volume of blood. Haemoglobin acts as an oxygen carrier by loosely combining with oxygen in a reversible way. All haemoglobins contain a number of subunits, each composed of a standard haem group (an iron(II) ion combined with a porphyrin molecule) and a globin (protein) portion of variable composition. Nearly all vertebrate haemoglobins are tetramers of two pairs of dissimilar subunits with a combined M_r of approximately 68 000 and are able to bind four molecules of oxygen. In adult humans, the commonest form is $\alpha_2\beta_2$, consisting of a pair of α units and a pair of β units; this differs from fetal haemoglobin ($\alpha_2\gamma_2$). Normal adults of most species have a number of different forms of haemoglobins; that is, they have multiple haemoglobins, which may have different functional properties. Haemoglobins in invertebrate species also contain monomers made up of haem groups and associated globin, but these are usually combined to form large aggregates.

Now attempt Question 5 on p. 196.

6.5 THE OXYGEN DISSOCIATION CURVE

You know already that each molecule of haemoglobin can combine with a maximum of four oxygen molecules. The degree to which oxygenation of haemoglobin occurs is determined primarily by the partial pressure of oxygen (P_{O_2}), and at a very high P_{O_2} all binding sites on the haemoglobin molecules are occupied by oxygen; that is, the pigment exists entirely in the form of oxyhaemoglobin and is fully saturated. Such a pigment remains fully saturated as long as the prevailing P_{O_2} is high, but if this falls, some of the bound oxygen becomes unstable, detaches from the haem group and is released as 'free molecular oxygen'.

One way to quantify the degree of haemoglobin oxygenation at any one P_{O_2} value is to determine the amount of oxygen carried by a particular volume of blood, and indeed this can be measured experimentally. For example, at a very high P_{O_2} when the pigment is fully saturated, human blood contains about 20 vol. %. The oxygen content at full saturation is termed the **oxygen capacity** of the blood (i.e. about 20 vol. % for human blood), and this varies between species because it depends on the number of oxygen-binding sites in a given volume of blood, that is on the 'concentration' of haem groups in the blood. Some diving mammals (e.g. seals) have a higher haemoglobin concentration than humans, and this significantly enhances the oxygen capacity of their blood (26.4 vol. %). If we compare different vertebrate species, we find that the oxygen capacity of the blood varies with the volume of red blood cells per unit volume of blood—the haematocrit—and with the concentration of haemoglobin in each red blood cell.

In making a comparison of the transport properties of blood from different species, it is often more convenient to express the amount of oxygen bound to a pigment at a particular partial pressure as a *percentage* of the amount of oxygen bound when the pigment is fully saturated—that is, as a percentage of the pigment's oxygen capacity. The **percentage saturation of haemoglobin** will vary in a predictable way over a range of partial pressures, and this can be plotted graphically, as in Figure 6.8. Such a curve is called the **oxygen dissociation curve** (or sometimes the oxygen equilibrium curve) for a respiratory pigment. When this pigment is equilibrated with oxygen at a particular partial pressure (say 12 kPa) and the P_{O_2} is then lowered to, say, 6 kPa, the number of sites at which oxygen is bound to haem is reduced, which means that the amount of oxygen carried by the pigment decreases because a small proportion of the bound oxygen has been released. If then the P_{O_2} were increased from 6 to 12 kPa, an equivalent amount of free oxygen would become bound to the haemoglobin and the percentage saturation of the pigment would return to its original value. Note that for a given drop in partial pressure of, say 1.50 kPa, the amount of oxygen unloaded will vary depending on where on the curve the change in the P_{O_2} occurs. A very steep part of the curve (for example, the portion arrowed in Figure 6.8) shows where the amount of oxygen bound to the pigment is particularly influenced by changes in the P_{O_2}.

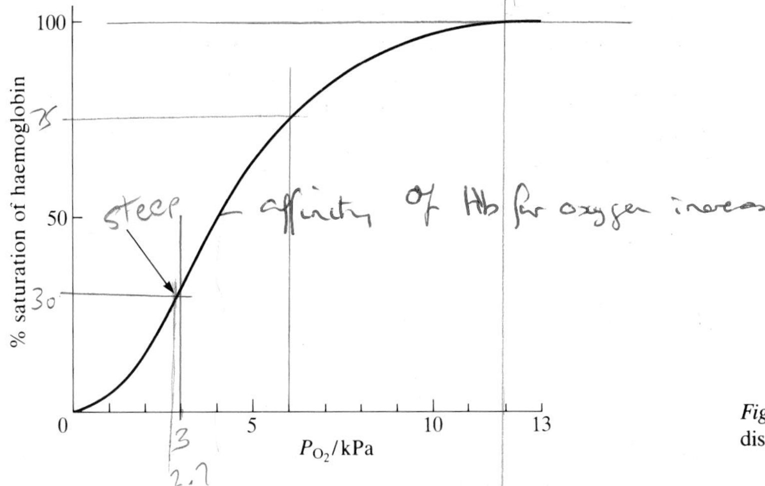

Figure 6.8 A sigmoid (S-shaped) oxygen dissociation curve for haemoglobin.

It should now be clear how haemoglobin can act as an oxygen carrier. At the relatively low partial pressures of oxygen that normally occur in the capillary beds supplying actively respiring tissues, haemoglobin tends to dissociate and release oxygen, whereas at respiratory surfaces, where the partial pressures of oxygen are usually much higher, the pigment tends to load up with oxygen.

◇ Suppose A and B in Figure 6.9 are the oxygen dissociation curves for two different pigments. Which would be the more effective oxygen transporter?

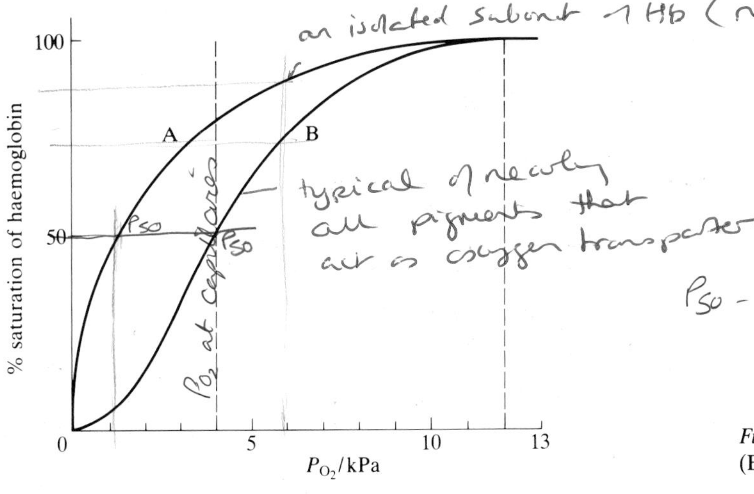

Figure 6.9 Hyperbolic (A) and sigmoid (B) dissociation curves.

◆ It is important to realize that Figure 6.9 gives no information about the *oxygen capacity* of the two pigments, even though this may have a bearing on their respective merits. But suppose (a) that the two pigments have an identical oxygen capacity and (b) that the pigments operate over a range of partial pressures between 12 kPa (at the external respiratory surface) and 4 kPa (at the capillary networks in the tissues). You might say there is little to choose between them. However, pigment A, which has a hyperbolic dissociation curve, can bind as much oxygen as pigment B, but will give up less than one-quarter of its oxygen at the P_{O_2} prevailing in the capillaries (4 kPa). Pigment B can unload almost 50% of the bound oxygen when the P_{O_2} in the capillaries falls by the same amount. Pigment A delivers the major part of its bound oxygen only when the P_{O_2} in the capillaries of the tissues falls to a very low level. In fact an S-shaped or sigmoid dissociation curve (B) is typical of nearly all pigments that act as oxygen transporters.

In molecular terms, how might an S-shaped dissociation curve arise? Look at that part of the curve (Figure 6.8) relating to low partial pressures of oxygen. At first the slope of the curve is fairly flat but then it rises more steeply up to a certain point (about 2.70 kPa). In other words, at the left end of the curve, the haemoglobin is initially reluctant to take up oxygen, but as the P_{O_2} increases, more and more oxygenation occurs for a given increase in partial pressure, which is to say, 'the appetite of the pigment for oxygen grows with the eating'. The ease with which haemoglobin takes up oxygen is normally described as the *affinity* of the haemoglobin for oxygen so here we see an increase in the affinity for oxygen as the oxygenation of the pigment increases. No equivalent increase in affinity is evident from the dissociation curves for myoglobin or for *isolated* subunits of haemoglobin (which resemble A in Figure 6.9); in both cases all the haems react to oxygen independently. This particular property is displayed only when the four subunits, each with a discrete haem group, combine to form a tetramer. Changes in affinity involve some kind of interaction between the haems of each molecule. The acquisition of a single oxygen molecule by one subunit increases the affinity of the neighbouring haems for oxygen, so that once a haemoglobin molecule takes up one oxygen molecule it is easier for it to go on and acquire the other three. Conversely, the loss of oxygen from haem in a saturated molecule of oxyhaemoglobin lowers the affinity of the remaining haems and two or three more oxygen molecules usually become detached. We now know that the binding (or release) of a single oxygen molecule to (or from) a haem influences the affinity of distant haems by changing the shape of the whole tetramer.

Nowadays, our understanding of the haemoglobin molecule is sufficient to provide a very convincing 'molecular' explanation of how cooperation between the monomers can modify the binding of oxygen at individual sites. More than 30 years of pioneering work on haemoglobin by Max Perutz in Cambridge, together with modern knowledge of allosteric effects in enzymes, led to an ingenious proposal that the haemoglobin tetramer can exist in two interchangeable forms, one called the **T (tense) structure**, the haems of which have a comparatively low affinity for oxygen, and the other the **R (relaxed) structure**, in which the haems have a high affinity for oxygen. The two pairs differ only by virtue of a slight rotation of one pair of subunits with respect to the other pair.

The uptake of oxygen by one subunit in the T structure encourages a small shift in the architecture of the whole complex to form the R structure. The transition from T to R is not induced by the binding of a set number of oxygen molecules but becomes more probable with each successive oxygen molecule that is bound. The transition involves changes in the bonds that form the contacts between the subunits, and has the effect of changing the spatial relationship between the haem iron and the plane of the porphyrin ring, so altering the binding affinity of the iron for molecular oxygen. The switch from T to R alters the binding characteristics so effectively that the last haem group has an affinity for oxygen that is much like that of a free subunit, and is one hundred times the affinity of the first haem group.

So far, it should be clear that cooperative interactions between subunits ensure that a large proportion of bound oxygen is delivered to the tissues *without the need for a massive drop in* P_{O_2} at the capillary beds in the tissues. Too low a P_{O_2} here might mean that the rate of diffusion of oxygen into mitochondria was not sufficient to allow these organelles to function optimally, for the rate of oxygen diffusion will depend on the extent of the gradient between capillaries and mitochondria (look back at Fick's equation, Equation 1 in Chapter 5, Section 5.1.1).

We shall next consider how haemoglobin is affected by factors such as the tension of carbon dioxide and the prevailing pH; but first look back at Figure 6.9. Suppose these curves to be oxygen dissociation curves for two different haemoglobins. Obviously, the two pigments differ in the rate at which they load up with oxygen as the P_{O_2} increases, although both pigments become fully saturated (or loaded) at about the same P_{O_2}. Dissociation curves that are shifted to the right in relation to the horizontal axis of the graph (like B) are produced by pigments that are reluctant to bind oxygen and therefore have a *lower overall affinity for oxygen*. Curves shifted to the left (like A) apply to pigments with a very high overall affinity for oxygen because they load up rapidly even at low tensions. (Note that high-affinity pigments give up oxygen less readily and unload their oxygen only when the P_{O_2} prevailing in the tissues is low.)

We can most conveniently express the overall affinity of a pigment by establishing the partial pressure of oxygen at which the pigment is 50% saturated. This index of oxygen affinity is termed the $\boldsymbol{P_{50}}$, and the *higher* the value of P_{50}, the more the curve extends to the right before saturation is reached and the *lower* the affinity of the pigment.

For a great many haemoglobins, an increase in the partial pressure of carbon dioxide (P_{CO_2}) shifts the oxygen dissociation curve to the right, reflecting a decrease in the affinity of the pigment for oxygen (see Figure 6.10a). This is the **Bohr shift** or Bohr effect, named after one of its discoverers. The Bohr shift is usually interpreted as the result of the increased acidity that accompanies increased partial pressures of carbon dioxide; however, it is possible that carbon dioxide might have a small direct influence on the affinity of haemoglobin for oxygen. A fall in pH alone certainly provokes the Bohr shift (see Figure 6.10b) and increased acidity may be especially important when metabolites such as lactate are produced by anaerobic metabolism.

The Bohr shift has important implications for the delivery of oxygen to the tissues. In actively respiring tissues, levels of carbon dioxide (and possibly of lactate) are high, and the Bohr shift brings about a substantial decrease in the affinity of haemoglobin for oxygen. If blood in the capillary beds of the tissues contains haemoglobin of reduced affinity, more oxygen will be unloaded from the pigment at the particular P_{O_2} that prevails. The net effect of the Bohr shift is that haemoglobin gives up a larger proportion of the bound oxygen than it would if carbon dioxide had no effect on the oxygen binding.

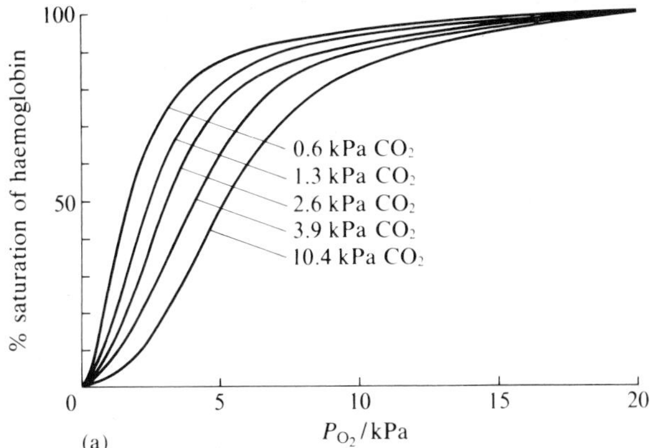

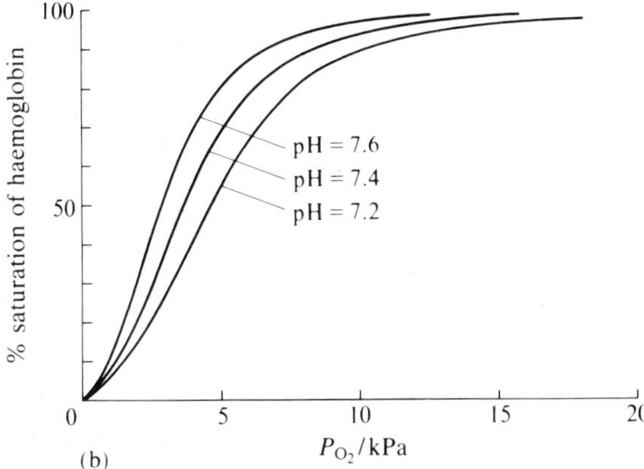

Figure 6.10 Oxygen dissociation curves at (a) various carbon dioxide tensions and (b) different pH values.

You can confirm this by making some measurements from Figure 6.11. Suppose that the P_{O_2} in arterial blood is 9.4 kPa (at a) and the P_{O_2} in the venous blood (which has obviously relinquished part of the oxygen to the tissues) is only 4.0 kPa (at v). First, assume that a pigment follows curve R and displays no Bohr shift. Read off from the vertical axis the oxygen content of arterial (a′) and venous (v′) blood.

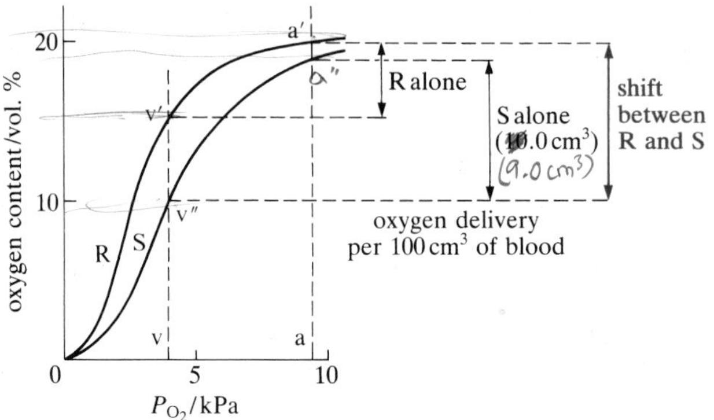

Figure 6.11 How the Bohr shift promotes the delivery of oxygen to the tissues. a = the P_{O_2} of arterial blood; v = the P_{O_2} of venous blood.

◇ How much oxygen is delivered to the tissues per 100 cm³ of blood?

◆ If the oxygen content of arterial blood is about 20 vol. % (i.e. 20 cm³ of oxygen per 100 cm³ of blood) and that of venous blood is approximately 15 vol. % (as on curve R), then each 100 cm³ of blood passing to the tissues yields about 5.0 cm³ of oxygen, which is the *oxygen turnover* to the tissues.

If the pigment followed curve S exclusively and there were no Bohr shift, the oxygen transfer to the tissues would be approximately 9.0 cm³ (= 19 − 10; see Figure 6.11).

Now suppose that the two curves relate to the same pigment and that the degree of oxygenation is measured at two *different* partial pressures of carbon dioxide, one of which (curve R) is a P_{CO_2} typical of arterial blood, and the other (curve S) is a P_{CO_2} typical of venous blood; the difference between them is attributable to the Bohr shift.

◇ What is the oxygen transfer to the tissues, given a Bohr effect of this magnitude? Remember that the P_{CO_2} at the respiratory surface will be low (see point a′), and the uptake of carbon dioxide by the capillary blood will mean that the degree of oxygenation of red blood in the tissues is represented by point v″.

◆ In these circumstances, the oxygen transfer to the tissues is 10.0 cm³ (20.0 − 10.0 = 10.0); that is the range shown in red in Figure 6.11.

This higher value for the oxygen transfer to the tissues (around an extra 10% compared with curve S alone) is a consequence of a shift in the dissociation curve that occurs when the respiratory pigment passes from the respiratory organ to the tissues or vice versa. A change in the pigment's affinity for oxygen in this way causes increased transfer of oxygen to the tissues over and above what it would be if the dissociation curve was unaffected by carbon dioxide and followed either of the curves (R or S) alone.

Observing a Bohr shift *in vitro* does not mean that the effect has any physiological significance. Knowledge of the carbon dioxide tensions and pH at the respiratory surfaces and in the tissues is obviously important, but you should note that measurements of pH in the blood refer to plasma, not to the pH within the red blood cells. The speed of the Bohr shift is often crucial too because the 'residence time' of blood cells in capillary beds is often very short, between 0.2 and 2.0 seconds. For humans, the physiological importance of the Bohr shift is well established, even when they are at rest. Venous blood has a pH (7.36) that is slightly lower than that of arterial blood (7.40) and a carbon dioxide tension (6.13 kPa) slightly higher than that of arterial blood (5.33 kPa). This is sufficient to cause an appreciable Bohr shift, and indeed there is significantly less oxygen in venous blood than would be predicted from the oxygen dissociation curve of arterial blood. During exercise, carbon dioxide tensions in the capillary beds increase, as do lactate levels and, at such times, the Bohr shift may deliver even more oxygen to the tissues.

Although we have concentrated on the effects of carbon dioxide and pH, many other factors influence the oxygen dissociation curve, including the concentrations of some common ions and that of the haemoglobin pigment itself. These effects are of greater interest to the biochemist than to the physiologist because the concentration of pigment and ionic composition of blood are normally regulated within narrow limits. One class of organic compounds—the organophosphates—occur in unusually large quantities in red blood cells. A familiar organophosphate is adenosine triphosphate (ATP), but the most abundant type in red blood cells is **2,3-diphosphoglycerate** (2,3-DPG). This is mainly formed in the cell from 1,3-DPG, which is an intermediate in the glycolytic sequence. In many red blood cells, the effect of large amounts of 2,3-DPG or of ATP appears to be to shift the dissociation curve to the right, that is to decrease the affinity of the haemoglobin for oxygen. We are beginning to understand how this might come about in molecular terms in adult human haemoglobin, because the molecular architecture of 2,3-DPG is such that it 'fits' in a convenient gap between two β chains in the T form of haemoglobin, so stabilizing the structure and decreasing the likelihood of its transformation to the R form.

It is significant that we have ended this section by talking about changes in the properties of haemoglobin that are determined by metabolic events *within* the red blood cells. It has recently become clear that such cells have a sophisticated metabolism closely linked to respiratory function.

Summary of Section 6.5

The relationship between the extent to which haemoglobin loads up with or releases oxygen and the prevailing partial pressure of oxygen is termed the oxygen dissociation curve. The degree of oxygenation is often expressed as a percentage of the amount of oxygen bound at full saturation and, when all the pigment is in the form of oxyhaemoglobin, the amount of bound oxygen represents the oxygen capacity of the blood. When haemoglobin subunits combine to form tetramers, the attachment of oxygen to one haem influences the binding properties of distant haems, thus facilitating the uptake of further oxygen molecules. This cooperative interaction between subunits is explained by a small shift in the three-dimensional shape of the haemoglobin molecule. Most dissociation curves are therefore sigmoid, which usually enables the pigment to load and unload substantial amounts of oxygen at the physiological partial pressures of oxygen that prevail at the respiratory surface and at the tissues.

The affinity of a haemoglobin for oxygen is a measure of the ease with which the pigment binds oxygen and is expressed as the P_{50}; that is, the partial pressure of oxygen at which the pigment is 50% saturated. Pigments of low affinity (high P_{50}) load up fully only at relatively high partial pressures of oxygen, and high affinity pigments unload only when the P_{O_2} in the tissues is low. Usually, a decrease in blood pH leads to a reduction in the pigment's affinity for oxygen, called the Bohr shift. An increase in carbon dioxide levels (as in respiring tissues) leads to increased acidity and a shift of the dissociation curve to the right. The result is an increase in oxygen turnover to the tissues; the Bohr shift decreases the affinity of haemoglobin for oxygen at the respiring tissues, whereas at the respiratory surface (where carbon dioxide levels are lower) the affinity of the pigment for oxygen is increased, which aids loading of the pigment with oxygen. The affinity of haemoglobin for oxygen is also influenced by important metabolites within red blood cells, especially 2,3-DPG, which shift the dissociation curve to the right as their concentrations increase.

Now attempt Question 6 on p. 196.

6.6 COMPARATIVE ASPECTS OF OXYGEN DISSOCIATION CURVES

Having dealt with some of the features of a 'typical' oxygen dissociation curve for mammalian haemoglobin, we can now compare the important features of curves from animals of different habit and habitat. We shall see that there is some correlation between the shape of an oxygen dissociation curve, the availability of oxygen and the actual demand for it in different animals.

One major variable is the affinity of different haemoglobins for oxygen. Some low-affinity pigments have dissociation curves displaced well to the right and are typically found in animals living in environments where the P_{O_2} tends to be high and constant. Such pigments can load up fully only when the prevailing level of oxygen is high, and they are especially vulnerable to a lack of oxygen. However, low-affinity pigments have the advantage of unloading a substantial portion of their bound oxygen at the tissues at a relatively high P_{O_2}. This beneficially affects the rate at which oxygen is supplied to metabolizing cells, as oxygen diffuses down a steep gradient between blood and cells.

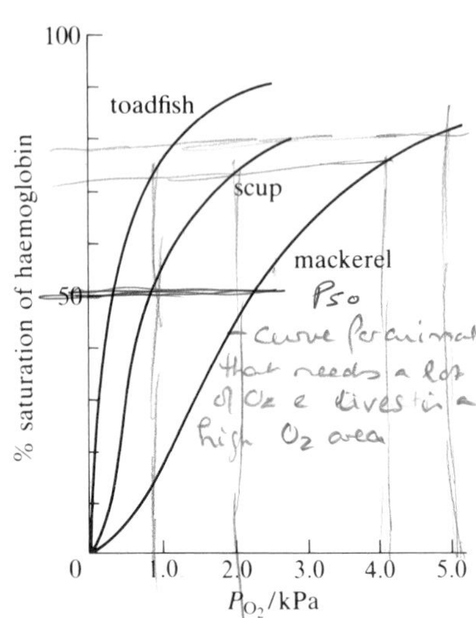

Figure 6.12 Oxygen dissociation curves for three fishes.

Consider Figure 6.12. This shows the oxygen dissociation curves for three fishes, two of which are rather inactive and live in still waters in shallow seas, the third is very active and lives in water where the P_{O_2} is high.

◇ Which of the fishes probably inhabit still waters?

◆ The toadfish and the scup, which have dissociation curves further to the left: that is, their haemoglobin has a high affinity for oxygen.

The haemoglobin of mackerel is typical of that of many active and fast-moving fish species: the dissociation curve shows the pigment is of low affinity (it is displaced well to the right), which favours a high P_{O_2} at the tissues and a high metabolic rate. The toadfish, a lethargic, bottom-dwelling teleost, has an oxygen dissociation curve that is obviously linked with its marked tolerance of oxygen deficiency—**hypoxia**—and its tendency to inhabit poorly oxygenated water. Not surprisingly, the properties of the pigment represented by the dissociation curve have an important bearing on the ability of an animal to survive in different environments. Fishes with low-affinity pigments, such as mackerel, are unlikely to move into water that is poor in oxygen, whereas fishes with high-affinity pigments, such as carp (the P_{50} is only 0.6 kPa), can exploit habitats where the P_{O_2} is low, habitats unavailable to mackerel.

Respiratory pigments from different animals also vary in their sensitivity to carbon dioxide and to temperature, and these differences, too, correlate with variations in habitat. Changes in the magnitude of the Bohr shift are particularly interesting in larval and adult amphibians, in which metamorphosis often involves a marked change from a strictly aquatic habitat to a more terrestrial one, with an accompanying switch to breathing air.

The tadpole stage of the bullfrog *Rana catesbiana* lives in stagnant ponds that vary in both P_{O_2} and P_{CO_2}. In this situation, one might expect the Bohr shift to have its customary beneficial effect on the unloading of the pigment at the tissues; but this might be more than outweighed by the deleterious influence of the Bohr shift on loading with oxygen at the gills. Indeed, in this tadpole, the pigment is practically insensitive to carbon dioxide; this is also true of teleosts living in tropical swamps, where the P_{O_2} is high and variable. Air has a lower and more constant P_{O_2} than water, and adult bullfrogs (which breathe air) have a haemoglobin that displays a moderately large Bohr shift. Their tissues derive the benefit of increased unloading of oxygen without the risk that carbon dioxide will impair oxygen uptake at the lungs. (Not surprisingly, metamorphosis involves the production of new haemoglobins of a lower affinity for oxygen and the P_{50} increases from 0.5 to about 3.3 kPa.)

The same type of correlation is evident when adult amphibians that form a series increasingly inclined to air-breathing are compared. *Necturus* (the mudpuppy), which respires aquatically across its gills and skin (though it has a simple lung), shows a small Bohr shift; *Amphiuma*, an aquatic air-breathing salamander which lacks gills, an intermediate Bohr shift and the more terrestrial *Rana catesbiana* displays the largest Bohr shift.

Changes in temperature have an important effect on the oxygen dissociation curve. Almost always, increases in temperature displace the curve to the right: that is, sensitivity to temperature (in P_{50}) is increased and the affinity of the haemoglobin is reduced at higher temperatures.

In mammals, the P_{50} increases by about 0.1 kPa for each degree Celsius rise. This has a beneficial effect during exercise, when increased production of heat by actively contracting muscles helps to unload extra oxygen.

In the blood of aquatic poikilotherms, whose body temperature varies with that of the water, the decrease in affinity accompanying increased temperature might be sufficient to prevent adequate loading of the pigment at the gills, especially in species, such as trout and mackerel, that possess low-affinity pigments (and curves positioned to the right) even at low temperatures. Problems are made worse by an increased demand for oxygen at high body temperatures (because the metabolic rate is elevated) and because oxygen is less soluble in warmer water. When the temperature increases, fishes may actively seek cooler water. In freshwaters, this may be difficult to find and mass mortalities of fishes may result. An alternative strategy for species subjected to temperature variations is the development of multiple haemoglobins of differing affinities and sensitivities. The pigment with the highest known affinity for oxygen is present in highest concentrations in the blood of fishes that live in warm waters. Other aquatic species have haemoglobins that are less sensitive to temperature than those of mammals. This is particularly true of turtles, which dive deeply and so cool rapidly, and of fish species such as tuna, which maintain high muscle temperatures and undertake extensive migrations from temperate to Atlantic waters, or in the lungfish *Neoceratodus*, which experiences large changes in ambient temperature.

Animals that are occasionally or continually exposed to low temperatures run the risk of increasing the affinity of their respiratory pigments for oxygen. Even in mammals, the effect is important, because the temperature at the extremities of the limbs may drop as low as 10 °C in cold environments, so causing a shift to the left that would restrict the unloading of oxygen to the tissues, unless the P_{O_2} in the tissues fell drastically. Many fishes inhabit ice-cold waters; for example, *Trematomus*, an Antarctic fish that lives in well-aerated water at about −1 °C. The P_{O_2} of its haemoglobin (approximately 2.8 kPa at the normal ambient temperature) is much higher than that of haemoglobin from temperate water species measured at −1.5 °C. Evidently, the haemoglobin of *Trematomus* has biochemical characteristics adapted to low temperatures; warming the blood to 4.5 °C reduces the oxygen capacity of the pigment by 30%. This effectively restricts *Trematomus* to ice-cold waters.

We have concentrated on the role haemoglobin plays in delivering oxygen to the tissues, but we should not underestimate the importance of other circulatory factors, such as blood volume, the distance between capillaries and tissues and the rate of blood flow. For example, in the family of Antarctic fishes commonly called ice-fishes, some of which reach more than 2 kilograms in weight, the blood lacks haemoglobin, and oxygen is delivered to the tissues by rapid circulation of the blood. This is achieved by a high cardiac output—the heart is five times larger than the hearts of other fishes of comparable size. In addition, ice-fishes make very effective use of the oxygen transported to the tissues by arterial blood. In these fishes the P_{O_2} at the tissues may well be unusually low, and indeed their metabolic rate is very low at these near-freezing temperatures. Once again, it is clear that the respiratory properties of blood can be understood only in terms of a wide variety of physiological and environmental data.

Summary of Section 6.6

The affinity of a particular haemoglobin for oxygen (and sometimes the oxygen capacity of the whole blood) can often be correlated with the availability of oxgyen in the environment. Where oxygen is plentiful, pigments often have a low affinity; this is especially true of pigments in active animals with a high metabolic rate, where the maintenance of a high P_{O_2} in the tissues is important. Fishes that inhabit the well-oxygenated water of swift rivers often have a low-affinity pigment which is very sensitive to increases in the level of carbon dioxide; this large Bohr shift further aids the unloading of oxygen from haemoglobin. In species that live in tropical swamps, where there is little oxygen and the level of carbon dioxide is high and variable, pigments usually have a high affinity for oxygen and a small Bohr shift. Air-breathing fishes and amphibians often have pigments of low affinity with a significant Bohr shift (carbon dioxide levels in air, unlike those in freshwaters, are remarkably steady), but during temporary shortages of oxygen (e.g. during aestivation in lungfishes) the affinity of haemoglobin for oxygen may increase.

For poikilotherms (e.g. fishes), increases in temperature raise metabolic requirements for oxygen but decrease the affinity of their haemoglobin for oxygen. In species normally subjected to wide fluctuations in body temperature, the binding of oxygen by the pigment appears to be less sensitive to temperature. The haemoglobin of an Antarctic fish (*Trematomus*) exposed continually to well-oxygenated water near −1 °C is highly sensitive to temperature but is well adapted to function at this extremely low temperature.

Now attempt Question 7 on p. 196.

6.7 TRANSPORT OF CARBON DIOXIDE

For every cm^3 of oxygen transferred to the tissues, a slightly smaller volume of carbon dioxide must return to the lungs. Carbon dioxide is more soluble in water than oxygen, but again simple solution will not transport sufficient gas. The carbon dioxide undergoes a variety of chemical reactions in blood. These are best understood in mammals, particularly in humans, where they are of great clinical significance. In this section, therefore, we will concentrate on carbon dioxide in mammalian blood.

6.7.1 Carbon dioxide in mammalian blood

In mammals, carbon dioxide is present in substantial quantities in both arterial and venous blood. There is more in venous blood, the extra being the gas transported from the tissues to the lungs. The bulk of the carbon dioxide, however, is present on both sides of the circulation, where it has a crucial role in the regulation of the acidity (pH) of blood. The transport of carbon dioxide is secondary to this function, and can only be understood in relationship to it.

Carbon dioxide reacts at several sites in the blood, and at each there is a different relationship between the partial pressure and the amount of gas taken up. We will first consider each separately, then put them together.

Carbon dioxide in plasma

Carbon dioxide dissolves in plasma and slowly reacts with water. The effective solubility is not high: at the normal arterial P_{CO_2} of 6.3 kPa, the amount of carbon dioxide dissolved is 1.2 mmol l^{-1}. There is a simple straight-line relationship between the amount dissolved and partial pressure. The combination with water produces carbonic acid (H_2CO_3), which will dissociate reversibly into hydrogen ions and bicarbonate (HCO_3^-).

$$CO_2 + H_2O \rightleftharpoons H_2CO_3 \rightleftharpoons H^+ + HCO_3^- \qquad (6.3)$$

Like all chemical equilibria, the second reaction proceeds to an extent determined by the relative concentrations of reactants and products. Plasma, however, also contains 25 mmol l^{-1} bicarbonate derived from a reaction in the red blood cell, which pushes the equilibrium to the left. The pH is determined by the **Henderson–Hasselbalch equation:**

$$\text{pH} = 6.1 + \log [HCO_3^-]/[H_2CO_3] \qquad (6.4)$$

The constant, 6.1, is the pK_a of the reaction. In normal plasma the $[H_2CO_3]$ term may be replaced by the product of P_{CO_2} and the absorption coefficient (α) (Section 6.2) so the pH is given by:

$$\text{pH} = 6.1 + \log \frac{[HCO_3^-]}{P_{CO_2} \times \alpha}$$

where P_{CO_2} is in kPa and concentration is in mmol l^{-1}.

In plasma, virtually no bicarbonate is formed from carbon dioxide as the dissociation of carbonic acid is prevented by the high concentration of bicarbonate already present. Some body fluids, such as cerebrospinal fluid (CSF) which bathes the brain and spinal cord, are like plasma, and in these the pH changes markedly as P_{CO_2} is altered, because the concentration of bicarbonate is constant.

◇ If the P_{CO_2} of a fluid such as CSF, where the bicarbonate concentration is the same as in plasma and constant in the short term, is raised to 7.3 kPa, what will be the pH? Assume an absorption coefficient for carbon dioxide in CSF at body temperature of 0.19.

◆ $$\text{pH} = 6.1 + \log \frac{25}{7.3 \times 0.19} = 7.36$$

Blood is protected from such large changes in pH by a second reaction of carbon dioxide that takes place in the red cell.

Carbon dioxide in the red blood cell

In the red blood cell, as in plasma, carbon dioxide dissolves in water and produces carbonic acid, which dissociates. The protons so produced then undergo a further reaction with haemoglobin, which acts as a buffer. This removal of product draws the reaction to the right, leading to the dissociation of more carbonic acid, until the buffering by haemoglobin is complete. Much more carbon dioxide, therefore, is taken up than by plasma, and bicarbonate is produced in quantities as a result.

The amount of bicarbonate produced is dependent on P_{CO_2}, but is much more affected by the buffering capacity of haemoglobin. This is not constant. The allosteric interactions which occur during oxygenation and deoxygenation dramatically affect the capacity of the molecule to bind hydrogen ions. Deoxyhaemoglobin binds more hydrogen ions than the oxyhaemoglobin, so it is a much better buffer and, at a given P_{CO_2}, more bicarbonate is formed in venous red blood cells than arterial.

The bicarbonate does not remain in the red blood cell. Most is exported by passive diffusion across the cell membrane into the plasma. As bicarbonate is a negatively charged ion (anion), its outward movement is compensated by the inward movement of another anion, chloride. This movement is known as the **chloride shift** (Figure 6.13).

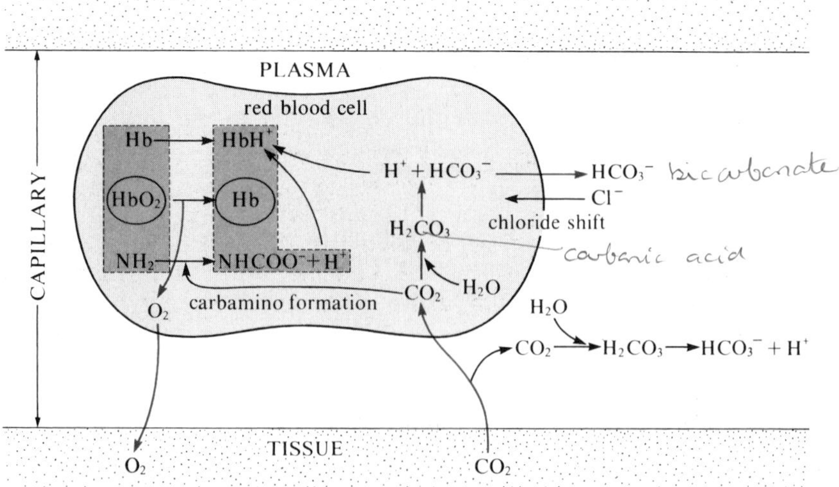

Figure 6.13 Reactions of carbon dioxide in the red blood cell.

It is this exported bicarbonate which forms the [HCO_3^-] term in the Henderson–Hasselbalch equation (Equation 6.4) for the pH of plasma. The buffering capacity of haemoglobin therefore modifies the pH of plasma indirectly by controlling the bicarbonate concentration.

In normal arterial blood,

$P_{CO_2} = 5.3\ \text{kPa}$ and $[HCO_3^-] = 25\ \text{mmol l}^{-1}$

Therefore,

$$\text{pH} = 6.1 + \log \frac{25}{5.3 \times 0.19} = 7.42$$

In normal venous blood $P_{CO_2} = 7.3\ \text{kPa}$, but more bicarbonate is formed in red blood cells, in a small part because P_{CO_2} is higher, but mainly because haemoglobin has a greater buffering capacity when deoxygenated, and draws the reaction of carbon dioxide with water to the right.

The $[HCO_3^-]$ therefore rises to $26.4\ \text{mmol l}^{-1}$. So,

$$\text{pH} = 6.1 + \log \frac{26.4}{7.3 \times 0.19} = 7.38$$

In other words, because haemoglobin has become a better buffer when relatively deoxygenated in venous blood, more bicarbonate is formed in red blood cells, which tends to counteract the fall in pH which would otherwise occur because more carbon dioxide has dissolved in the plasma.

A similar rise in P_{CO_2} in arterial blood produces a greater change in pH. In this case, if P_{CO_2} changes to 7.3 kPa, pH will fall to 7.36. This is a significant change, as most biological processes are very pH-dependent. If P_{CO_2} falls, then pH will rise. Rises in pH are potentially even more damaging than falls, as increasing alkalinity of plasma causes calcium ions to bind to plasma proteins, and the resulting fall in free calcium ion concentration in the plasma and extracellular fluid renders nerves exceptionally excitable. This will lead to uncoordinated movements ('tetany') and eventually convulsions.

The P_{CO_2} of arterial blood, which is determined by the P_{CO_2} in the lungs (see Chapter 7) critically affects its pH. Rises in P_{CO_2} produce a fall in pH (acidosis). Falls in P_{CO_2} produce a rise in pH (alkalosis).

The bicarbonate level in arterial plasma is, as we saw above, principally determined by the buffering capacity of haemoglobin. It may, however, be disturbed by other factors. Many metabolic processes produce acids (other than carbon dioxide). These so-called 'fixed acids' react with bicarbonate to produce carbon dioxide, which is breathed out via the lungs. The effect of metabolic acid production is therefore to reduce the bicarbonate concentration (metabolic acidosis). Some other metabolic processes, such as acid secretion in the stomach, tend to elevate plasma bicarbonate, producing metabolic alkalosis. So long as P_{CO_2} remains constant, these changes will alter plasma pH. As the pH depends on the ratio of bicarbonate to P_{CO_2} (see the Henderson–Hasselbalch equation, Equation 6.4), these pH disturbances may be compensated for by altering P_{CO_2} with changes in breathing, a point we will return to in Chapter 7. The important issue, however, is that the P_{CO_2} of arterial blood is a critical variable, which is intimately involved in the regulation of acid/base balance.

Chemical binding of carbon dioxide to proteins

Carbon dioxide will form **carbamino compounds** with free amino groups on proteins such as haemoglobin. Carbon dioxide bound in this way does not significantly affect pH but, as the formation is dependent on P_{CO_2}, these compounds will be formed in the tissues, and given up in the lungs, thus contributing to carbon dioxide transport. The capacity of haemoglobin to form carbamino compounds is affected to some degree by its state of oxygenation. Deoxyhaemoglobin forms them more easily.

6.7.2 The carbon dioxide dissociation curve

Just as we did in Section 6.5 for haemoglobin/oxygen binding, we may produce a carbon dioxide dissociation curve for whole blood. There are, however, a whole set of such curves, depending on the oxygenation of haemoglobin. Figure 6.14 shows two, one for normally oxygenated haemoglobin, and another for the oxygenation level of haemoglobin in normal venous blood. The difference between the two is known as the **Haldane effect**, and we now know it to be due to two factors. First, deoxygenated haemoglobin has a greater buffering capacity, so more bicarbonate is formed at a given P_{CO_2}, and second, deoxygenated haemoglobin forms carbamino compounds more easily.

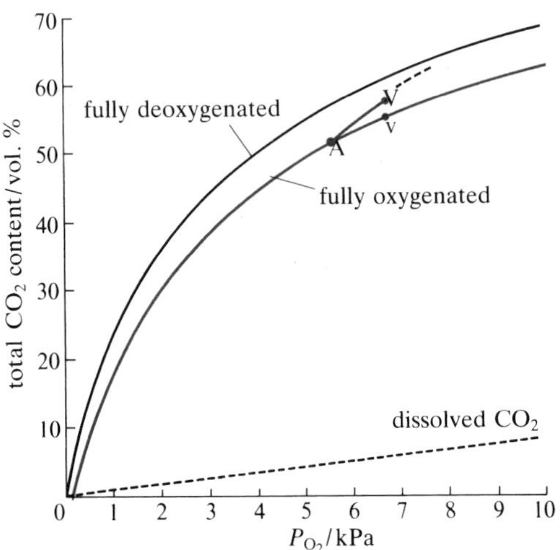

Figure 6.14 The Haldane effect, showing the total carbon dioxide content of oxygenated (red line) and deoxygenated (black line) in human blood. A is the P_{CO_2} of arterial blood, V is the P_{CO_2} of venous blood.

The Haldane effect means that more carbon dioxide can be accepted from the tissues once the haemoglobin has relinquished some of its load of oxygen. This you can see from Figure 6.14, where point A marks the partial pressure of carbon dioxide in arterial blood, and V the partial pressure of carbon dioxide in mixed venous blood. Note that mixed venous blood is never fully deoxygenated and is seldom less than 70% saturated when the subject is at rest, so V only approaches the 'fully deoxygenated' curve and never falls precisely on it. In the body, the actual carbon dioxide dissociation curve is represented by the line A–V. Its slope is steeper than either dissociation curve alone, and the Haldane effect therefore increases the amount of carbon dioxide normally accepted from the tissues of a resting person by the amount V−v. You might be surprised to see from the curves how small the change in the carbon dioxide content is between venous blood reaching the lungs (about 56 vol. %) and arterial blood leaving them (about 50 vol. %), but this is sufficient to keep pace with the rate of production in a resting human. This change in carbon dioxide content can be achieved for a drop in partial pressure of only 0.79 kPa, and without the Haldane effect the difference in P_{CO_2} between arterial and venous blood would be much greater. During exercise, the venous P_{O_2} falls appreciably, so V tends to move along the broken line without, of course, ever reaching the upper solid line. Before we leave Figure 6.14, note the reciprocal relationship between the Bohr and Haldane effects, which represent two sides of the same coin. The unloading of oxygen and loading of carbon dioxide are very closely related. An increase in the P_{CO_2} at the tissues shifts the oxygen dissociation curve to the right (the

Bohr effect), thus enhancing the delivery of oxgyen to the tissues (see Section 6.5). This unloading of oxygen increases the capacity of blood to take on carbon dioxide (the Haldane effect). Haemoglobin is essential in both.

Table 6.3 shows the breakdown of carbon dioxide content on either side of the circulation. In venous blood, the P_{CO_2} is greater, which produces a small increase in dissolved carbon dioxide and a somewhat larger increase in carbamino binding. Under the conditions of lower P_{CO_2} in the lungs, this extra dissolved carbon dioxide and carbamino carbon dioxide is given up, though together they form only a small fraction of the transported gas. The major effect is due to deoxygenation of haemoglobin. This increases its buffering capacity, leading to the formation of substantially more bicarbonate. When the haemoglobin is oxygenated in the lungs the hydrogen ions are released; they recombine with bicarbonate to produce carbon dioxide, which is breathed out.

Table 6.3 Carbon dioxide content of arterial and venous blood. Values are in mmol l^{-1}.

	Arterial blood	Venous blood	Venous minus arterial
total carbon dioxide	21.53	23.21	1.68
as dissolved carbon dioxide	1.05	1.19	0.14
as bicarbonate ions	19.51	20.85	1.34
as carbamino carbon dioxide	0.97	1.17	0.20

Overall, 8% of the carbon dioxide travels as dissolved CO_2, 12% as carbamino compounds, and 80% as bicarbonate.

Summary of Section 6.7

Carbon dioxide is a critical component of both arterial and venous blood of mammals because of its role in maintaining the pH of the body fluids within normal limits—the acid/base balance.

Carbon dioxide reacts at several sites and at each the extent of reaction is affected by different factors. In plasma, the formation of acid is limited by the presence of a high concentration of bicarbonate, so the pH is given by the Henderson–Hasselbalch equation (Equation 6.4). The amount of carbon dioxide dissolving depends directly on P_{CO_2}, but virtually none dissociates to form bicarbonate.

In the red blood cell, hydrogen ions produced as carbon dioxide dissolves react with haemoglobin, so the reaction is drawn to favour the formation of bicarbonate and, at a P_{CO_2} of 5.3 kPa, 25 mmol l^{-1} of bicarbonate is formed and exported to the plasma in exchange for chloride. The concentration of bicarbonate depends to a small extent on P_{CO_2}, but is greatly affected by changes in the buffering capacity of haemoglobin, which alters as it picks up or loses oxygen. Deoxygenated haemoglobin is the better buffer, so more bicarbonate is formed. This allows carbon dioxide to react with venous blood, and limits the change in plasma pH which occurs as the P_{CO_2} rises.

The carbon dioxide transported from tissues to lungs is only around 10% of the total content, with the difference between arterial and venous carbon dioxide content being due to a small increase in dissolved carbon dioxide in venous blood, a small increase in carbamino carbon dioxide and a somewhat larger increase in bicarbonate formed in red blood cells.

Now attempt Question 8 on pp. 196–197.

SUMMARY OF CHAPTER 6

Almost all animals need a supply of oxygen to the tissues; they acquire the oxygen from the environment, ultimately through the process of diffusion. The general body surface is frequently a useful avenue for gas exchange and may satisfy the total requirement in animals with a high ratio of surface area to volume (or a low metabolic rate). In small or flattened animals the diffusion of gases can suffice but in larger animals diffusion is supplemented by a circulatory system (Chapter 5) often with a respiratory pigment able to bind oxygen. One part of the body surface is usually specialized as the major respiratory site where the blood and respiratory medium are brought into close contact.

Respiratory pigments probably evolved separately in unrelated groups of animals, but nearly all transport oxygen in relatively high concentrations. Because haemoglobin is so effective in promoting a supply of oxygen to the tissues, animals that possess it have a substantial advantage. Nearly all vertebrate haemoglobins are tetramers; the haem component is standard but differences in the globin portion yield pigments with different properties—modifications that have adaptive significance. For example, fishes from well-aerated waters where the P_{CO_2} is low and stable have haemoglobin of low affinity for oxygen with a large Bohr shift; this aids the unloading of oxygen to the tissues without endangering the loading of haemoglobin with oxygen at the gills. Fishes exposed to high and variable levels of carbon dioxide and low partial pressures of oxygen have pigments of higher affinity, less susceptible to a Bohr shift. In general, where oxygen is readily available in the environment the shape of the oxygen dissociation curve favours high oxygen tensions for unloading oxygen at the tissues.

While oxygen is transferred to the tissues, carbon dioxide is removed. Carbon dioxide is more soluble in water than oxygen but, in mammals, simple solution is not sufficient to cope with the rate of respiration. A variety of chemical reactions takes place in the blood and much of the carbon dioxide is actually transported as bicarbonate. This has consequences for the pH of the blood. It is clear that the respiratory properties of blood can be understood only in terms of a wide variety of physiological and environmental data. In the next chapter we consider the mechanics of breathing and the changes in respiratory physiology that take place during changes in environment and exercise levels.

OBJECTIVES FOR CHAPTER 6

Now that you have completed this chapter you should be able to:

6.1 Define and use, or recognize definitions and applications of each of the terms printed in **bold** in the text.

6.2 Calculate partial pressures of gases in a gas mixture, and predict how gases will move between areas of different pressure and gases of different composition. (*Questions 1 and 2*)

6.3 Calculate the content of gas in a liquid given the tension and solubility. (*Question 2*)

6.4 Describe the structure and function of the tracheal system of insects. (*Question 3*)

6.5 Describe how some terrestrial insects limit water loss during respiration. (*Question 4*)

6.6 Describe the molecular structure of haemoglobin. (*Question 5*)

6.7 Use an oxygen haemoglobin dissociation curve to explain the transport of oxygen from lungs to tissues. (*Question 6*)

6.8 Describe the effect of pH and temperature on oxygen dissociation, and explain how they facilitate unloading of oxygen at the tissues. (*Question 6*)

6.9 Describe how the characteristics of respiratory pigments differ in animals in different habitats. (*Question 7*)

6.10 Describe the reactions of carbon dioxide in mammalian blood, and their effects on the pH of body fluids. (*Question 8*)

QUESTIONS FOR CHAPTER 6

Question 1 (*Objective 6.2*) (a) Suppose a mixture of nitrogen and oxygen has a total pressure of 200 kPa and contains 40% oxygen; what is the partial pressure of nitrogen in the mixture?

(b) Recall that physiologists often express the concentration of a gas in a solution as the volume of gas that can be extracted from solution, for example 6 cm^3 of oxygen per 100 cm^3 blood. But imagine comparing the oxygen concentrations of two samples of blood, one from a homoiothermic animal (37 °C) and another from a poikilothermic animal (20 °C). What type of adjustment could you make to account for this difference in temperature?

(c) Suppose air at 30 °C and 98.0 kPa is saturated with water vapour at a vapour pressure of 2.40 kPa. What is the partial pressure of oxygen in such saturated air?

(d) Look at the column of figures in Table 6.2 that gives the solubility of oxygen in freshwater. Compare these values with those given in Table 6.1 for the absorption coefficient of oxygen in water at the same temperatures. If both sets of figures refer to oxygen solubility at identical temperatures, why are they so different?

Question 2 (*Objectives 6.2 and 6.3*) (a) Consider a volume of distilled water at 0 °C in equilibrium with air that has a partial pressure of oxygen of 30 kPa. Given that the absorption coefficient of oxygen at this temperature is 0.0489, calculate the amount of dissolved oxygen present per litre of distilled water.

(b) Suppose that the equilibrated distilled water mentioned in (a) was brought into contact with dry atmospheric air at a total gas pressure of 101.3 kPa and a P_{CO_2} of 21.2 kPa. What is the volume of oxygen present in a litre of this dry atmospheric air?

(c) Given the answers to (a) and (b), is it correct to say that oxygen will diffuse from air into the water?

Question 3 (*Objective 6.4*) Which of statements (a)–(d) are *true* and which are *false*?

(a) The airsacs in terrestrial insects function as sites of gas exchange.

(b) Ventilation of the tracheal system is essential, because it results in the direct renewal of air within the tracheoles.

(c) In an active insect, the terminal ends of the tracheoles are normally filled with fluid.

(d) Because of the highly branched tracheal system of active insects, a unidirectional flow of air through the tracheal channels cannot be ensured and this is to the disadvantage of the animal.

Question 4 (*Objective 6.5*) (a) Suppose an insect with spiracles is placed in an atmosphere enriched with carbon dioxide. What effect would this have on the rate of water loss from the animal?

(b) Suppose that in *Cecropia* larvae the spiracles were completely sealed with wax. What would happen to the rate at which oxygen is taken up and carbon dioxide is released from the animal?

(c) Is diffusion the only mechanism responsible for the movement of gases across insect spiracles?

(d) What would you expect the total gas pressure in the main tracheae of *Cecropia* to be (i) when the spiracles were fully open, (ii) when the spiracles had been fully closed for some time?

Question 5 (*Objective 6.6*) (a) Do the subunits of myoglobin display cooperative interactions?

(b) If the subunits of haemoglobin did *not* display cooperativity, at what P_{O_2} at the tissues would the pigment be 50% saturated? (Look at Figure 6.9.)

(c) Does the cooperativity of the subunits of a haemoglobin tetramer result from the *direct* interaction of adjacent haem groups?

Question 6 (*Objectives 6.7 and 6.8*) (a) Some pigments display a *reverse* Bohr shift when carbon dioxide levels increase. Would this result in a decrease or an increase in (i) the affinity and (ii) the P_{50} of the pigment?

(b) The magnitude of the Bohr shift varies between different pigments. Which of (i)–(iv) measures the extent of the Bohr shift?
(i) A change in the oxygen capacity of the pigment per unit change of pH
(ii) A change in the pH of the pigment per unit change in P_{CO_2}
(iii) An increase in the oxygen turnover to the tissues per unit change in pH
(iv) A change in the P_{50} per unit change in pH

(c) Transfer of oxygen across the human placenta is hindered because the placental tissues impose a substantial barrier to oxygen diffusion. The P_{O_2} on the fetal side is lower than that on the maternal side. Would you expect fetal haemoglobin to have a lower or higher affinity for oxygen?

(d) The difference in the affinity of fetal haemoglobin for oxygen is the result of a difference in the level of 2,3-DPG. Would you expect a higher or lower level of 2,3-DPG in fetal red blood cells?

Question 7 (*Objective 6.9*) Suppose that an intertidal burrowing polychaete has a haemoglobin which is thought to be used solely as an oxygen store. Would your view of the pigment's function change if:

(a) The pigment had a relatively low P_{50}?

(b) The pigment had a relatively high oxygen capacity?

(c) At high tide, the animal's venous blood contained pigment that is 90% saturated?

(d) The oxygen stored in the blood would support normal aerobic respiration for only about 20 minutes?

(e) Even at low tide, the partial pressure of oxygen in the burrow was sufficiently high to ensure full loading of the pigment at the gills?

(f) At low tide, the P_{O_2} in the tissues was sufficiently low to ensure full unloading of the pigment?

Question 8 (*Objective 6.10*) (a) What would the plasma pH be if the $[HCO_3^-]$ concentration was 25 mmol l^{-1} and the P_{CO_2} 4 kPa?

(b) What would happen to a person if their blood pH was as in section (a)?

(c) Diabetics produce excess acid metabolically. If they are untreated, would their breathing increase or decrease to stabilize their plasma pH?

(d) What happens to the carbon dioxide carrying capacity of blood in tissues such as the heart which extract a much larger fraction of the oxygen from haemoglobin than other tissues?

RESPIRATORY MECHANISMS AND THE CONTROL OF BREATHING

◆ CHAPTER 7 ◆

7.1 INTRODUCTION

In this chapter we consider the organs by which animals exchange gases between the environment and their blood. We saw in Chapter 6 that respiratory pigments ensure that blood has the inherent capacity to transport oxygen and carbon dioxide at adequate rates provided it is exposed to an appropriate gaseous environment (i.e. one of higher P_{O_2} and lower P_{CO_2}).

However, exchange at a high rate is facilitated if there is a large surface area for exchange and adaptations that minimize the barrier to diffusion between blood and the 'internal environment' created within the exchange organ.

We will look first (Sections 7.2–7.4) at a familiar system, that of mammals. Here gas exchange occurs across the membranes of millions of small alveoli. Air is conducted by a series of branching airways, drawn in as the lungs are expanded during inspiration. We will look in some detail at how the air moves, and at the mechanics of expanding and contracting the lungs during breathing movements.

In Section 7.5 we consider another warm blooded group—birds. Birds have to support a very high tissue metabolic rate. The large air supply required is achieved via a system of lung ventilation which avoids many of the limitations imposed on mammals. Mammals breathe in and out via the same tubes and do not separate the inspiratory and expiratory flows.

Both mammals and birds breathe air, which is rich in oxygen. Fishes, on the other hand, though they commonly have a lower metabolic rate to support, extract oxygen from water, where it is present in much lower concentration than in air. Sophisticated gills, and complex flow patterns of blood and water maximize the extraction of oxygen. We consider these mechanisms in some detail in Section 7.6.

Whatever the organ of exchange, all animals need to regulate ventilation to match metabolic requirements. This regulation is perhaps best understood in mammals, so we restrict our consideration in Section 7.7 to the chemical control of mammalian (human) breathing as it is an excellent example of feedback control, which you read about in Chapter 4, Section 4.2.

Even humans can tolerate variation in environmental conditions, so in Section 7.8 we examine human adaptation to life at high altitude, where P_{O_2} is low.

Our capacity to cope with poor oxygen supply is pitiful compared with many other species, however, so in the final section, 7.9, we see how some species, including diving mammals, are able to tolerate low oxygen supply for very long periods.

You will see just how varied are the approaches to the problem of obtaining enough oxygen to support metabolic activity.

7.2 THE MAMMALIAN LUNG

All mammals breathe by means of lungs, which are paired structures arising from the digestive tract during development. In humans, the lungs are contained within the thoracic cavity (Figure 7.1). The dorsal (back), ventral (front), and anterior (upper) boundaries of the cavity are made up of bony structures, the vertebrae, ribs, clavicles and scapulae. The posterior (bottom) boundary is a layer of muscle separating thorax and abdomen, the diaphragm. This muscle is domed, with its fibres running from apex to edge, so when they contract the apex of the dome is drawn down into the abdomen. The volume of the thoracic cavity may therefore be changed by contraction of the diaphragm. The intercostal muscles stabilize the ribs during breathing, and may be used to expand the thorax further in heavy breathing.

The lungs themselves are lobed structures completely filling the thoracic space. Human lungs have three lobes on the right, and two on the left.

The site of gas exchange is in the **alveoli** (Figure 7.1), to which air is conducted by a branching set of airways. The wind-pipe or **trachea** carries all air into and out of both lungs. It branches into two **bronchi**, which subdivide into the lobes, segments and sub-segments, eventually feeding air to many **bronchioles**. Bronchi have cartilage in their walls; bronchioles do not, but have much more smooth muscle, so their diameter may be changed actively. After a total of twenty or so divisions, this 'tree' has around 120 000 terminal bronchioles, each supplying a bunch of millions of alveoli via respiratory bronchioles and alveolar ducts. Each alveolus is around 200 μm in diameter, and is surrounded by capillaries. The total surface area of all the alveoli is around 70 m^2, the size of a tennis court.

The blood supply to the alveolar capillaries comes via the **pulmonary circulation**, which perfuses the lung. The resistance to blood flow offered by the pulmonary capillaries is surprisingly small. The effective driving pressure for the pulmonary circuit (i.e. the mean pressure in the pulmonary artery) is much less than in the **systemic circulation** (i.e. the circulation that supplies the other main organs of the body) and hence the blood pressure in the pulmonary capillaries is only about 0.8 kPa, less than a one-quarter of that in

Figure 7.1 The location and structure of human lungs. On the left is an anterior view of the lungs and related organs. Only the major airways and bronchioles are shown. The enlargement shows a three-dimensional reconstruction of several alveolar sacs with the associated blood supply (arrows show the direction of blood flow).

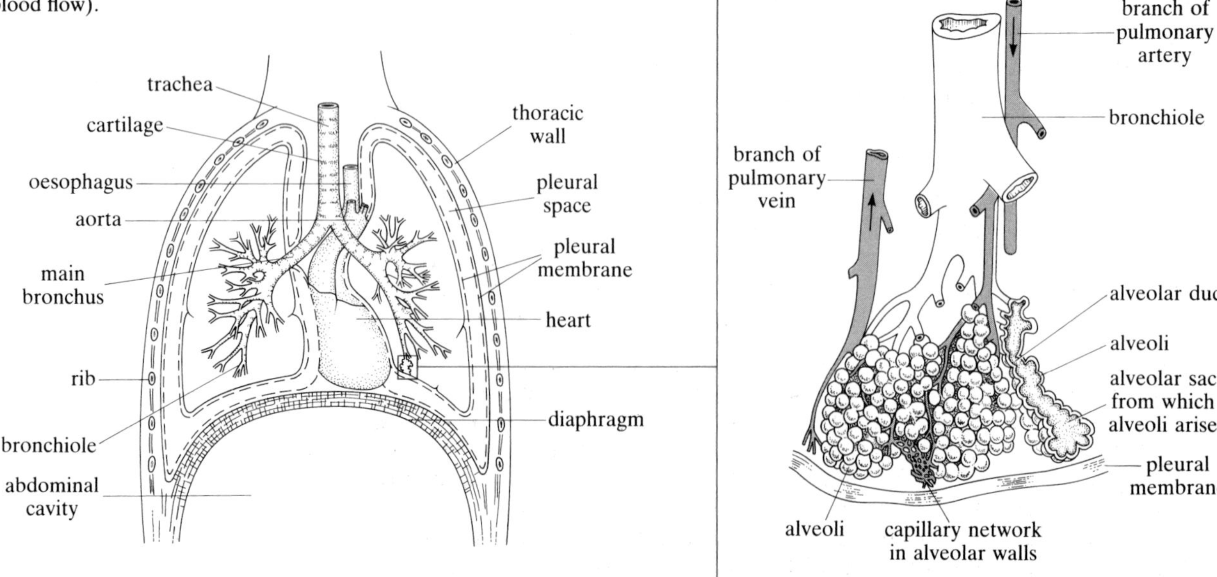

the systemic capillaries. You learnt in Chapter 5, Section 5.2, about the forces involved in determining the direction and magnitude of water movement in and out of capillaries. The thinness of the respiratory barrier means that fluid (and small dissolved molecules) are easily forced from the blood into the alveoli at a rate determined by the steepness of the gradient in hydrostatic pressure between pulmonary blood and alveolar air. The alveolar membrane is impermeable to proteins so the pulmonary blood exerts an osmotic pressure which tends to draw water back into the pulmonary capillaries. The alveolar epithelium is permanently wet; this exudation is the source of most of the water lost by evaporation as a result of breathing, which in an adult man in a temperate climate is of the order of 300 cm^3 per day. If the pulmonary capillaries were not at a low pressure, but at a pressure equivalent to that of the systemic capillaries, the hydrostatic pressure gradient across the alveolar membrane would be sufficiently great to override the osmotic pressure gradient (as happens in systemic capillaries) and would tend to flood the alveoli with fluid. This 'drowning in one's own juice', properly termed *pulmonary oedema*, often has fatal consequences in humans. It results from a change in the permeability of the alveolar epithelium, allowing proteins and other macromolecules to enter the alveoli and thus reducing the balancing effect of the blood's osmotic pressure. This change in permeability may result from damage to the epithelium, for example by noxious gases or by pathogenic organisms, and this is generally called pneumonia. Changes in permeability can also result from damage to the capillary epithelium, for example from violent allergic reactions caused by the foreign proteins introduced into the blood.

Although the pulmonary capillary bed is very extensive, the amount of blood it contains at any instant is only about 70–100 cm^3. About this volume of blood is expelled into the pulmonary artery with each contraction of the right ventricle. So at each beat, the pulmonary blood is replaced. Each red blood cell normally remains in a pulmonary capillary for less than second. During exercise, the amount of blood passing through the pulmonary circuit is much greater (i.e. there is a greater flow of blood per minute) and the residence time is as low as 0.3 seconds. Pulmonary vessels are unusually distensible and this has the advantage of accommodating large increases in blood flow during exercise without a precipitous rise in blood pressure. Capillaries in the lungs are larger in diameter than those elsewhere; this is one of several factors that lower the resistance to blood flow offered by the pulmonary capillaries.

Figure 7.2 emphasizes the thinness of the layer of tissue that separates red blood cells in the pulmonary capillaries from the alveolar air. The alveolus is lined by an alveolar membrane which lies next to the capillary endothelium (see Figure 7.2b). The total blood-to-gas distance varies between 0.2 and 0.6 μm in humans, but is nearer 4 μm in amphibians.

The lungs therefore bring gas to one side and blood to the other side of a very thin membrane of very large surface area. Gas exchange then occurs by diffusion down gradients of partial pressure. In humans, the blood returning to the lungs has a P_{O_2} of around 7 kPa, and a P_{CO_2} of around 6.3 kPa, though these values may vary considerably with changes in metabolic activity.

The gas in the alveoli has a P_{O_2} of 13.3 kPa, and a P_{CO_2} of 5.3 kPa. There are consequently steep gradients driving exchange. The important question is whether the resistance of the barrier is sufficiently low to allow adequate rates of gas exchange. Two gases have to move, and the one which moves more slowly will ultimately limit the exchange. Most of the barrier is aqueous, through which gases diffuse according to their solubility in water, and as carbon dioxide is much more soluble than oxygen, it moves 21 times as fast for a given gradient and resistance. The movement of oxygen is always,

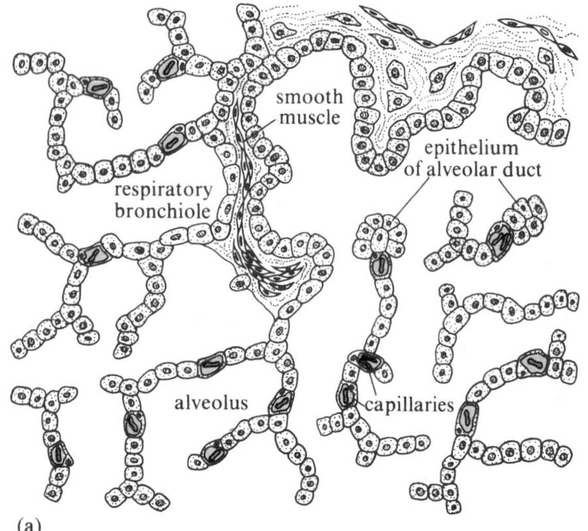

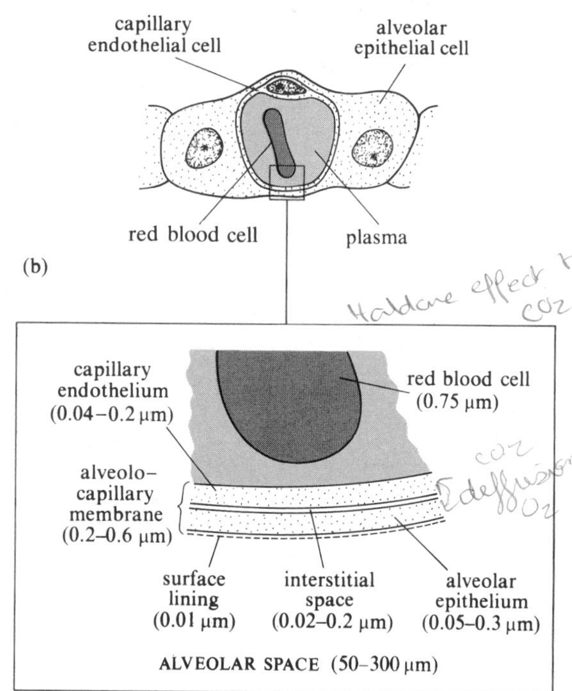

Figure 7.2 The microscopic structure of the respiratory portion of the human lung. (a) The structure of the respiratory epithelium and associated tissues. Note the presence of smooth muscle—its contraction can have a marked effect on the dimensions of the airways. (b) Progressively higher magnifications of the alveolar–capillary membrane, which separates alveolar air from capillary blood.

therefore, the limiting factor in gas exchange by diffusion. Experiments have examined the rate at which the haemoglobin in a single red blood cell acquires oxygen while it is in the pulmonary capillaries. This is possible because the pigment changes colour as it becomes oxygenated. Under average conditions oxygen loading is complete within 400 ms. As we said above, the red blood cells spend between 1 and 0.3 s in the pulmonary capillaries, more than enough time for exchange to occur, except possibly in severe exercise.

The important consequence of this is that under virtually all conditions the blood in the pulmonary capillaries comes into complete equilibrium with the alveolar gas, so the gas composition of arterial blood will be the same as that of the gas in the alveoli.

Summary of Section 7.2

The mammalian lung contains millions of very small alveoli as the sites of gas exchange—which together have a very large surface area. Air is conducted to the alveoli by a complex ‘tree’ of airways starting with the trachea and progressing through bronchi and bronchioles towards the ducts supplying the alveoli. Blood is distributed to the other side of the alveolar membrane by a vascular system which runs in series with the systemic circulation and is supplied by the right side of the heart. This pulmonary circulation is a low-resistance, low-pressure system so tissue fluid does not form except under unusual circumstances, thus ensuring that efficient gas exchange can occur.

The alveolar membrane offers a very low resistance to the diffusion of gases, though oxygen diffuses less readily than carbon dioxide. Generally oxygenation of the haemoglobin in a red cell is complete within 400 ms of arriving in the pulmonary capillaries and, except in severe exercise, the cells have more than enough time to saturate with oxygen. Thus the tensions of oxygen and carbon dioxide in the blood leaving the lungs will be very close to the P_{O_2} and P_{CO_2} in alveolar air.

7.3 LUNG VENTILATION IN MAMMALS

As blood flows continuously through the pulmonary circulation, oxygen is constantly removed from the alveolar gas and carbon dioxide added to it. Lung ventilation serves to prevent this exchange from changing the composition of the alveolar gas, so the blood leaving the lungs always has the same P_{O_2} and P_{CO_2}.

Air is moved through the conducting airways by a gradient of pressure, which is created, during inspiration, by expansion of the lung. Air flows through the airways, such as bronchi and most bronchioles, which do not expand greatly, and accumulates in the spaces which expand most, and therefore have the lowest pressure. The main accumulation of air is not, however, in the alveoli, which are relatively inelastic, and cannot expand a great deal, but in adjacent structures, the alveolar ducts, and respiratory and terminal bronchioles. Oxygen and carbon dioxide then exchange by diffusion between alveolar gas and inspired air. This diffusion step is relatively slow, preventing the composition of alveolar gas from changing greatly over the cycle of inspiration and expiration.

Expiratory flow is produced by a reverse pressure gradient generated as the lungs 'recoil' (see below) back to their original volume, though they may be compressed further by muscular effort during a forced expiration. Air flows in and out of the lung through the same airways, so the flow is said to be tidal. The volume which moves in and then out again is known as the **tidal volume**. This tidal volume is moved with each breath, and over a series of breaths the lung is said to be ventilated with the product of tidal volume and the number of breaths.

Ventilation must change to meet different circumstances, and both tidal volume and respiratory rate (the number of breaths per minute) may alter. Breathing patterns are measured using a spirometer. Subjects breathe to and from a chamber held at a fixed pressure, and changes in the chamber volume, which can easily be measured, accurately reflect the volume of air moving into and out of their lungs. Figure 7.3 shows a typical spirometer trace. The volume of air breathed in and out with each breath, the tidal volume, is around 500 cm^3 when breathing quietly, but if necessary it may be greatly increased, both by breathing in more than usual, and by breathing out more.

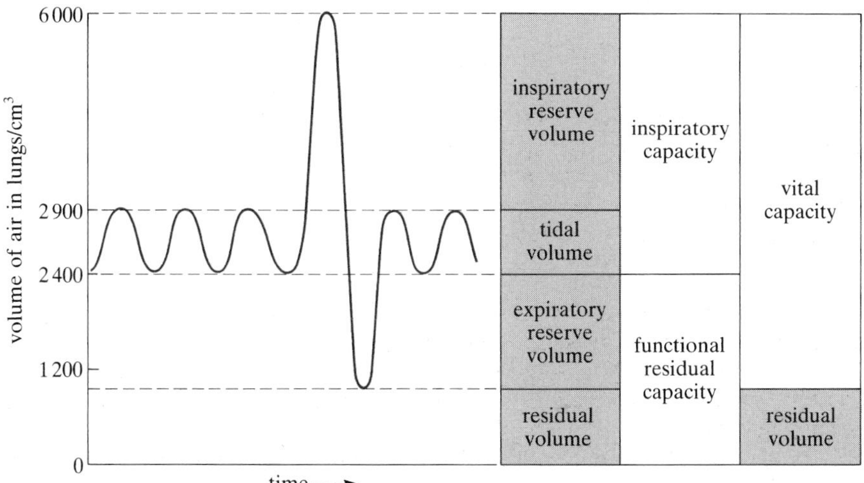

Figure 7.3 A typical spirometer trace showing lung volumes and capacities.

In each case the subject is said to be encroaching on **reserve volumes**, the inspiratory and expiratory reserve volumes, respectively. Even at maximum expiration, however, the lung cannot be completely emptied of air. There is always a **residual volume**. The biggest possible tidal volume is the sum of the 'quiet tidal' volume, the inspiratory reserve volume, and the expiratory reserve volume. This important figure is known as the **vital capacity**, the biggest breath that can be taken. The amount of air left in the lungs at the end of a quiet expiration is known as the functional residual capacity.

The pulmonary ventilation rate is the product of tidal volume and the number of breaths per minute, and, in adult humans, may range from around 6 litres per minute to over 100.

Some of the air pumped into and out of the lungs, however, is not available for gas exchange, because of the tidal pattern of flow. The last air breathed in fills the airways, and is then the first air expired, so a proportion of each tidal volume merely fills and empties the airways without ever reaching the alveoli. Lung spaces, such as the airways, which are ventilated during breathing, but do not contribute to gas exchange, are known as dead spaces. The volume of the airways is the anatomical dead space, but even in the normal lung this dead space is supplemented by some alveoli which are damaged or have insufficient blood flow to support exchange, the distributive dead space. The sum of these two is known as the **physiological dead space**.

The dead space of the lung must be ventilated before any air is available for gas exchange, so there is always 'dead space ventilation', which is the product of the dead space volume and the respiratory rate. The difference between pulmonary ventilation and dead space ventilation is the air actually available for gas exchange, the *alveolar ventilation rate*.

Changing the pattern of breathing alters the relationship between alveolar ventilation rate and pulmonary ventilation rate. Rapid shallow breathing pushes air mainly into and out of the dead space, leaving only a small fraction for alveolar ventilation. Deep, slow breathing pushes most air down into the alveoli, so relatively little is 'wasted' on ventilating the dead spaces.

◇ Consider a subject with a physiological dead space of 150 cm^3, breathing a tidal volume of 500 cm^3 16 times a minute. What is the alveolar ventilation rate?

◆ Pulmonary ventilation rate (PVR)

$$= 16 \times 0.5 \text{ litres min}^{-1} = 8 \text{ litres min}^{-1}$$

Dead space ventilation rate (DSVR)

$$= 16 \times 0.15 \text{ litres min}^{-1} = 2.4 \text{ litres min}^{-1}$$

(the dead space is always fully ventilated with each breath)

Alveolar ventilation rate

$$= \text{PVR} - \text{DSVR} = 8 - 2.4 \text{ litres min}^{-1} = 5.6 \text{ litres min}^{-1}$$

◇ Take the same subject, now breathing 250 cm^3 32 times a minute. What is the alveolar ventilation rate?

◆ PVR $= 32 \times 0.25$ litres $min^{-1} = 8$ litres min^{-1}
DSVR $= 32 \times 0.15$ litres $min^{-1} = 4.8$ litres min^{-1}
AVR $= 8 - 4.8$ litres $min^{-1} = 3.2$ litres min^{-1}

Therefore, all other things being equal, the most efficient pattern of breathing is as slow and deep as possible. We do not, however, breathe as slowly and deeply as possible, because that pattern of breathing is hard work. The mechanics of lung ventilation are such that an intermediate rate and depth is the best compromise between maximizing alveolar ventilation and minimizing effort.

Summary of Section 7.3

The composition of the alveolar gas is kept constant by ventilation of the lungs. Air is drawn through the airways by expanding the lungs, and it accumulates next to the alveoli, from where oxygen diffuses into, and carbon dioxide diffuses out of, the alveolar gas.

Air moves back and forth through the same airways—in a 'tidal' way, and with each breath the tidal volume is inhaled and exhaled. Tidal volume may be increased by utilizing inspiratory and expiratory reserves up to a limit of the biggest possible breath—the vital capacity.

Not all the air which ventilates the lung is useful, however, as some enters and leaves dead spaces, such as the airways, without coming into contact with the blood.

Shallow rapid breathing 'wastes' more ventilation in the dead spaces than slow deep breathing, but the latter is hard work, so an intermediate rate and depth is usual.

7.4 THE MECHANICS OF BREATHING

Contraction of the diaphragm and other muscles of breathing stretches the lungs and draws air through the airways. The lungs, thorax and diaphragm form a mechanical system to which the muscular forces are applied.

If the lungs are removed from the thorax they shrink to a much smaller volume, reducing in both diameter and length. Within the chest, they are stretched, and prevented from collapse by attachment to the inner surface of the thorax. This attachment is by a thin layer of fluid, the pleural fluid, whose surface tension holds the surfaces together, but allows them to slide over one another. The tendency of the lungs to collapse pulls the diaphragm up, and the chest wall in, and in the absence of any muscular effort, the whole system comes to an equilibrium with the lung volume equal to the functional residual capacity (Figure 7.4). In humans this capacity is about 3 litres.

Breathing movements displace this equilibrium by application of muscular effort but, once the effort is removed, the system will return passively to its equilibrium position. In quiet breathing, muscular effort of the diaphragm stretches the lungs during inspiration, but expiration is a passive process of 'elastic recoil'. In a forced expiration, the equilibrium is displaced by contractions of the abdominal muscles pushing up the diaphragm. The volume of air in the lungs is reduced below the functional residual capacity but the lungs are never emptied of air. A residual volume remains—about 1.5 litres in humans.

The work done in quiet breathing is, therefore, against the effort necessary to stretch the lungs. The tendency of the lungs to collapse depends in part on the

SURFACTANT

Detergent-like material lining the lungs which stabilizes their structure by reducing surface tension as lung volume decreases

Law of Laplace

$P = \frac{2T}{r}$: the pressure across a thin-walled tube is related to the wall tension divided by the inner radius of the tube.

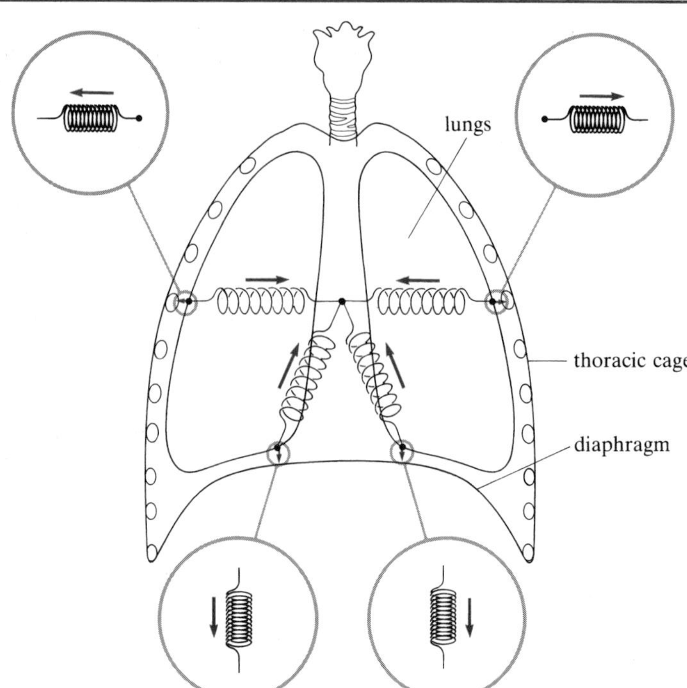

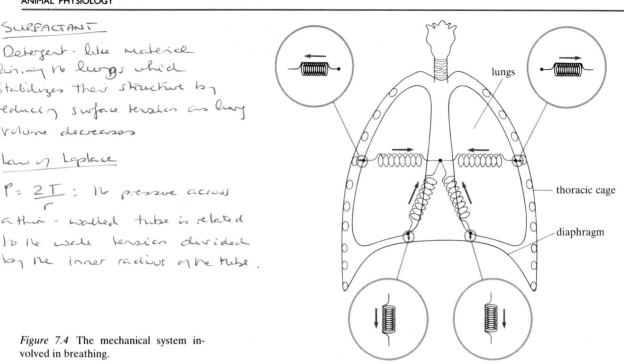

Figure 7.4 The mechanical system involved in breathing.

large amount of elastic tissue in the walls of the airways, and in part upon the surface tension forces in the film of fluid lining airways and alveoli. This surface tension is greatly modified by the secretion of a detergent-like mixture known as **surfactant**. Surfactant reduces the surface tension when the lung volume is small, but its effects become progressively less as the lung expands, so at large lung volumes it becomes more and more difficult to produce further expansion. This is one of the reasons why deep breaths are hard work.

Surfactant also serves to stabilize the lung structure. The alveoli can be considered as a set of interconnected bubbles, because they are lined with fluid. The **law of Laplace** states that the pressure in a bubble is twice the surface tension of the fluid divided by the radius of the bubble.

◇ If you were to blow two bubbles with the same soap solution, but one twice the diameter of the other, which would have the greater pressure inside it?

◆ The law of Laplace states $P = 2T/r$. As r increases, so P will reduce. The bigger bubble will have half the pressure of the small.

If large and small bubbles are connected, air will move from small to large, and the small bubble will collapse (Figure 7.5). In the lung, some alveoli are inevitably larger than others, so if this rule applied, then there would be a reduction in the number of alveoli as the smaller collapsed into the larger. This is prevented by surfactant lowering the pressure in small bubbles to the same as in the large, so allowing them to coexist. If surfactant is absent, as it occasionally is in newborn babies, then the alveoli do indeed collapse into larger spaces, and breathing becomes very difficult.

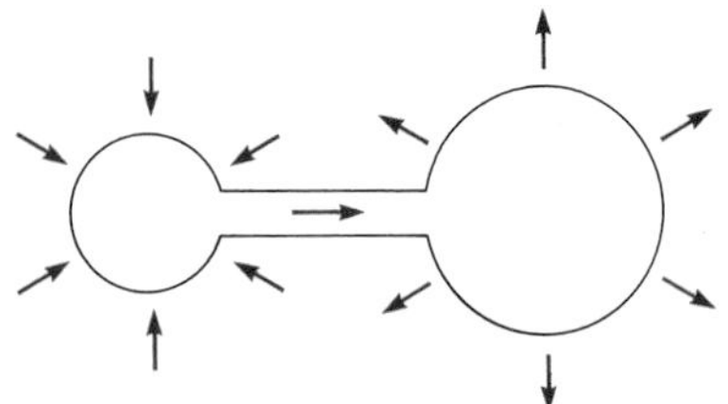

Figure 7.5 What happens if two bubbles of unequal size are connected. Arrows indicate direction of movement.

The stretching of the lung generates the pressure gradient which causes flow of air through the airways. As you have read in Chapter 5, Section 5.2, the flow of fluids through tubes depends upon the resistance, which itself depends on the nature of the fluid, the pattern of flow, and the diameter of the tube.

All other things being equal, resistance increases greatly as tubes get narrower, and it becomes more difficult to produce flow through them. The airways of the lung can have very small diameters, and therefore have a very high resistance, but they form a branching structure and, at each branch, although the tubes get narrower, they increase in number, so there are more paths for air flow. At each branch, the increase in the number of tubes more than compensates for the increase in their individual resistance. It becomes progressively easier to move air the closer it gets to the alveoli. During inspiration and most of expiration, therefore, the major resistance is in the uppermost part of the airways, the trachea, throat and nose. This is why we breathe through our mouths when ventilation is increased, for example during exercise. In the final stages of a forced expiration, however, the small airways are compressed, and become sufficiently narrow to impede and eventually prevent air flow. Overall, in the normal lung, airway resistance hardly affects the effort required to breathe, though if the airways are narrowed by disease, such as bronchitis or asthma, ventilation, and particularly expiration, can become very difficult indeed.

Summary of Section 7.4

Breathing requires muscular effort to expand the lungs and draw air through the airways. The lungs have an inherent tendency to collapse, which is resisted in the body by the thorax and diaphragm, to which they are attached by a layer of pleural fluid. In the absence of muscular effort, the lung volume is that at the end of a quiet expiration—the functional residual capacity. Application of muscular force will lead to inspiration or expiration from this volume, but return to it will be automatic if the force is removed—'elastic recoil'. Inspiration is normally active and quiet, though not forced: expiration is passive.

The effort required to breathe in depends on the ease of stretching the lungs, which depends in part on elastic tissue in them, but also on surface tension forces, though these are reduced at low lung volumes by a detergent-like material called surfactant. Surfactant also stabilizes the lungs by allowing alveoli of different size to coexist with the same internal pressure.

As air is moved through the airways, force has to be exerted to overcome flow resistance. Because of the branching nature of the airways for most of inspiration and expiration, the flow resistance is small and movement is largely in the biggest airways, but, at the end of expiration, particularly if forced, compression of the small airways makes it difficult to move air through them. This problem is greatly exacerbated in bronchitis and asthma.

Now attempt Questions 1, 2 and 3 on pp. 239–240.

7.5 THE RESPIRATORY SYSTEM OF BIRDS

The avian respiratory system is much more 'efficient' than that of mammals. Birds have a very high rate of metabolism, especially during flight, and yet their lungs are compact and semi-rigid. They occupy only about one-tenth of the volume taken up by the lungs of a similarly-sized mammal, but the structures concerned with gas exchange—the air-capillaries—are present at a greater density than the alveoli in mammalian lungs. The anatomy of the respiratory system in birds is complex, and until recently this has hindered our understanding of the pathways of air movement. The tracheae and bronchi are unexceptional but the lungs are connected with large, thin-walled **airsacs**,

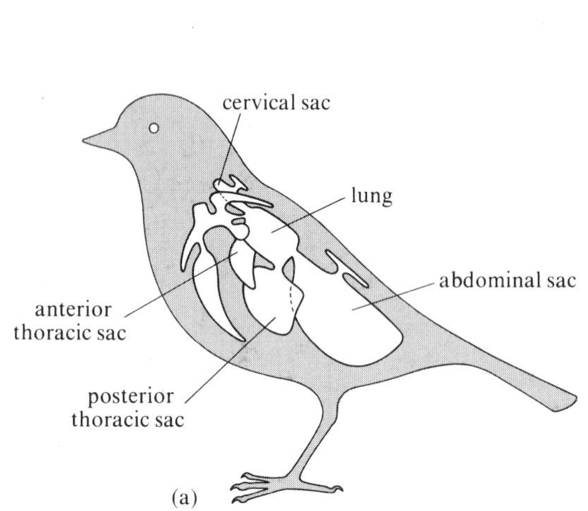

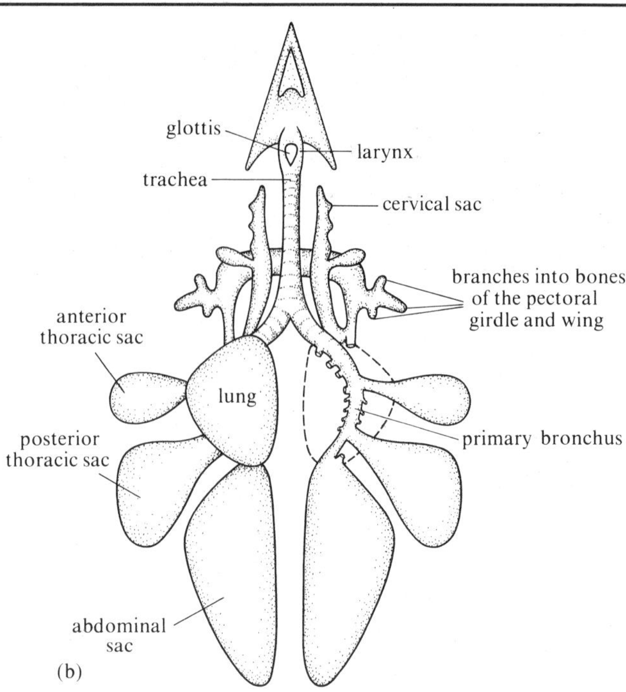

Figure 7.6 The respiratory system and airsacs of a bird, seen in lateral view (a) and dorsal view (b). The airsacs occupy much of the body of the bird and even extend into some of the larger bones. The trachea divides into left and right primary bronchi, which run through the lungs. Narrower abdominal airways, the bronchioles, branch off from each primary bronchus, but for clarity these complex branches are not shown.

some of which lie among the viscera of the abdominal cavity (Figures 7.6a and 7.6b) and others penetrate hollow bones. These airsacs seem to function as a bellows system; their blood supply is meagre and they are not directly involved in gas exchange.

The primary bronchus passes through the lung, where it branches first to form a series of anterior secondary bronchi and subsequently a series of posterior secondary bronchi (see Figure 7.7). The two sets of secondary bronchi are joined together by numerous approximately parallel parabronchi (or tertiary bronchi, as they are sometimes called). These fine tubes are the sites of gas exchange: from their walls arise an enormous number of fine air-capillaries which branch repeatedly to form thin, coiled, blind-ended tubes. These air-capillaries entwine with blood capillaries and blood passing through the parabronchial walls thus becomes oxygenated (Figure 7.8). Note therefore that there are no sac-like alveoli in this lung; the sites of gas exchange are the blind-ended air-capillaries. Some impression of the extent of the air-capillary network comes from marvellous pictures of the parabronchi obtained using scanning electron microscopy (Figure 7.9).

There has been much debate about how the lungs in birds are ventilated. Experimental manipulation in such a complex system may well upset the normal pattern of respiration, and only recently has it been possible to implant minute flow-detectors within airways and so determine the direction and relative velocity of air movements over long periods, even in unrestrained flying birds. It is certain that ventilation is not tidal. In mammals and amphibians, the alveoli are blind-ended and contain a mixture of 'fresh' and 'stale' air. In birds, the parabronchi are open at both ends and air passes through.

At inspiration, movement of the sternum (breastbone) and ribs leads to an increase in the volume of the visceral cavity and hence of the airsacs, especially the thoracic and abdominal sacs. Fresh air is thus sucked through the lungs into the sacs; the lungs themselves scarcely change in volume. It seems likely that respiratory exchange at the air-capillaries occurs during

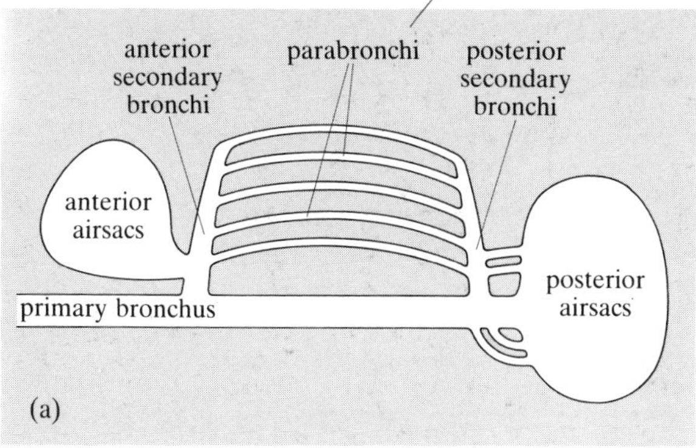

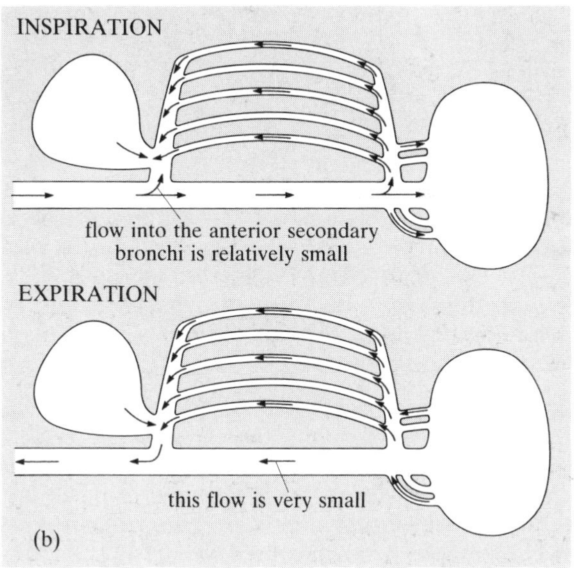

Figure 7.7 (a) The anatomy of the avian respiratory system. For simplicity, the anterior and posterior groups of airsacs are each shown as a single sac. There are a series of anterior secondary bronchi and another series of posterior secondary bronchi, all joined by parabronchi; but here we show only one of each. (b) The probable pattern of air flow during inspiration and expiration (red arrows).

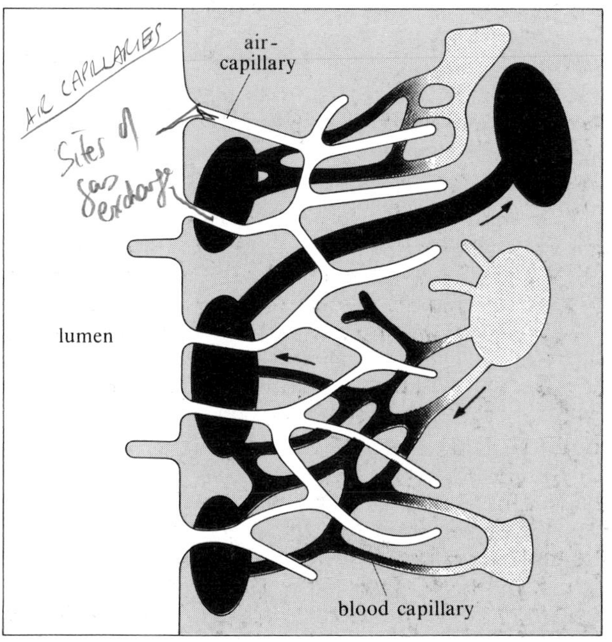

Figure 7.8 A schematic vertical section through the lumen (a cavity) and wall of a single parabronchus. The clear, fine tubes are air capillaries and arise from depressions in the parabronchial wall. The air capillaries lie close to blood vessels and, as blood passes through the parabronchial walls in the direction shown, it becomes more oxygenated.

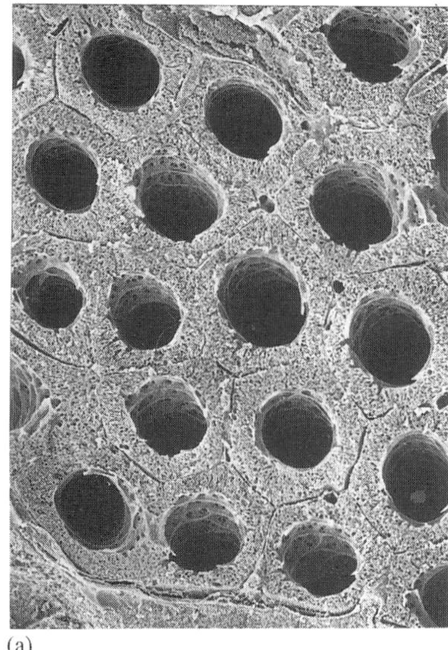

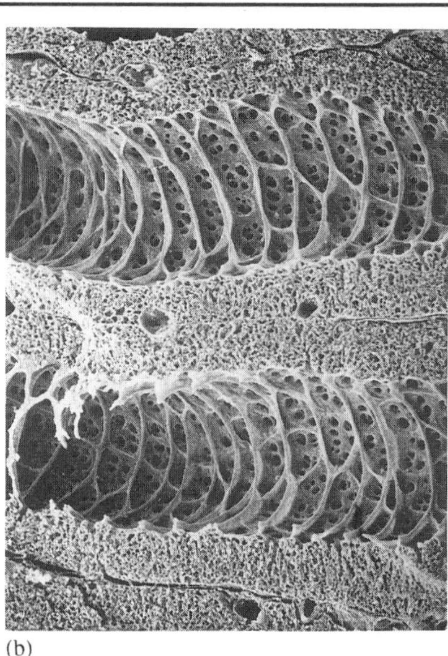

(a) (b)

Figure 7.9 Parabronchi in (a) transverse and (b) longitudinal section as revealed by scanning electron microscopy. (b) is at a higher magnification and shows only two parallel parabronchi with the air capillaries arising from the pit-like depressions.

inspiration *and* expiration because there are streams of air through the lungs twice in each respiratory cycle. The flow of air is in one direction during both phases of breathing; that is, it is unidirectional, as suggested in Figure 7.7b. This might aid the loading of the blood with oxygen, but there is no convincing evidence that blood flows in an opposite direction; that is, there is no evidence of counter-current exchange as there is in fishes (see Section 7.6).

It is remarkable that the airways apparently do not contain valves to direct the flow of air, so the mechanism controlling flow and resistance in the various tubes is obscure. Oxygen probably reaches the air-capillaries by diffusion from the airstream, which is actively moved through the bronchi and parabronchi. Whatever the mechanism, the system is extremely efficient. A high percentage of oxygen is extracted from the inspired air; this is often referred to as the 'oxygen utilization' of the animal.

The respiratory system of birds may also have secondary functions. One of these is thermoregulation, because respiratory organs are important sites of heat loss through evaporation. Ventilation is often increased when cooling is a priority, as in hot environments or during flight. In pigeons, much of the extra heat produced during flight is dissipated by panting which increases the loss of heat through evaporation (ventilation increases to about twenty times the resting level). Hyperventilation 'blows-off' more carbon dioxide from the blood and this elevates the pH of the blood. Many animals avoid the dangerous consequences of alkalosis by rapid, shallow breathing, which allows heat loss through evaporation but not excessive exchange of respiratory gases (see the discussion of dead space in Section 7.3). Many birds pant in hot environments, and the complexities of their respiratory systems may enable increases in evaporative heat loss without alkalosis. Some birds are able to direct (or shunt) the extra volumes of ventilated air away from the lungs, possibly by moving it directly into and out of the airsacs, where gas exchange is very limited but large amounts of body heat are lost via evaporation. Thus, ventilation through the parabronchi is not excessive and alkalosis is avoided. Ostriches, however, do suffer alkalosis when the ambient temperature is high, sometimes severely; but they seem to be a genuine exception for reasons linked to their large size and flightless habit.

There is still much to learn about birds' respiratory systems, particularly about the pattern of air-flow and the functions of the airsacs. Surprisingly, adequate respiratory exchange can occur if the airsacs are destroyed, which suggests they may have other functions as yet unknown. Beside their role in thermoregulation, airsacs play a crucial part in bird song by promoting a flow of air over resonating vocal chambers. Certainly, the respiratory system is well adapted to the habit of flight; one reason is that some of the smaller air spaces (e.g. those that replace the marrow of some of the major bones; see Figure 7.6b) make the bird lighter by replacing denser tissue.

COUNTER - CURRENT EXCHANGE SYSTEM

A means of generating a large gradient for passive exchange of substances between two compartments by accumulating a series of small gradients between two fluid streams travelling in opposite directions.

Summary of Section 7.5

Birds have a highly efficient respiratory system with a complicated and poorly understood ventilatory mechanism. The trachea divides into primary bronchi which run through each lung and pass into large, paired, abdominal airsacs. In the lungs, a great many small hollow tubes (parabronchi) arise from branches of the primary bronchi, and the wall of each parabronchus is penetrated by large numbers of fine, blind-ended air-capillaries (analogous to mammalian alveoli). The air-capillaries are the sites of gas exchange and lie adjacent to the pulmonary blood capillaries, which penetrate all parts of the parabronchial wall. Inspiration of air into the lungs depends on a suction-pump mechanism that enlarges the main airsacs and so sucks air along the conducting tubes. Whatever the precise mechanism of ventilation, it seems certain that (a) air always flows through the lung parabronchi (which are open at both ends) in one direction only (from the posterior to the anterior end), and (b) that this unidirectional flow of air occurs during both inspiration and expiration. This facilitates a very efficient transfer of oxygen and carbon dioxide between 'fresh' air and the adjacent pulmonary blood. The respiratory system has important additional functions, especially in bird song and in thermoregulation, because increased loss of water via evaporation can occur without increased ventilation of the lungs.

Now attempt Question 4 on p. 240.

7.6 GAS EXCHANGE IN FISHES

Fishes have to exchange oxygen and carbon dioxide with water not air, which they achieve via gills. Vertebrate gills are essentially lateral extensions of the pharynx lining the gill slits, which are a basic chordate feature. They have a very good blood supply with capillaries near the surface and a thin membrane separating these from the surrounding water.

In the larvae of many fishes and amphibians the respiratory structure may not be much more elaborate than this, the gills simply trailing in the water. In adult fishes, the gills are always carried in a cavity protected on the outside by flaps of skin and muscle. In teleosts a single flap on each side supported by bones covers all the gills; it is called the operculum (Figure 7.10a). The filaments are stacked symmetrically and supported and separated on skeletal gill arches (Figures 7.10b and 7.10c). This arrangement protects the gills; it enables a respiratory current of water to be drawn in through the mouth and passed over the gills even when the fish is at rest. It also allows a very large surface area to be exposed to the water. Figures 7.10c and 7.10d show some of the complexity of the subdivision of teleost gills into primary and secondary lamellae (Figure 7.10e), which gives rise to this large surface area.

larvae - gills trail

Adult fish - in cavity protected by skin & muscle

Teleosts - called operculum

The blood supply is such that the blood and the water are flowing in opposite directions, forming a **counter-current exchange system**. Counter-current exchange is a widespread physiological mechanism. It is found in heat exchange, for example in tuna fish, and is present in the kidney, where it aids

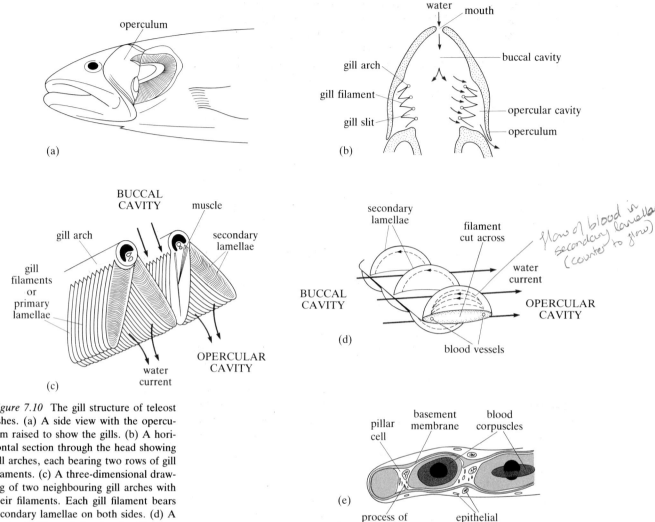

Figure 7.10 The gill structure of teleost fishes. (a) A side view with the operculum raised to show the gills. (b) A horizontal section through the head showing gill arches, each bearing two rows of gill filaments. (c) A three-dimensional drawing of two neighbouring gill arches with their filaments. Each gill filament bears secondary lamellae on both sides. (d) A single gill filament showing that the flow of blood in the secondary lamellae (broken lines) is counter to the flow of water. (e) A section through a secondary lamella, where two thin epithelial layers are separated by a series of pillar cells. Blood circulates between adjacent pillar cells.

ion concentration in the urine, as you will learn in Chapter 11. The advantage of counter-current flow is shown in Figure 7.11. If oxygen is to be transferred from water to blood (the blood capillary is shown in pink), then a counter-current flow is more effective than a concurrent one. Bear in mind that (a) the rate of diffusion of oxygen from water to blood depends largely on the gradient of partial pressure between the two compartments, and (b) the most efficient system of gas exchange is one that ensures the highest possible partial pressure of oxygen in the blood supply leaving the gill.

Suppose the partial pressure of oxygen in the ventilating water-stream (shown in white) approaching the gill capillary is 15 kPa and that the incoming blood (shown in pink) lacks oxygen. Imagine that blood and water are flowing in the *same* direction (i.e. they are *concurrent*) (Figure 7.11a). When the two streams first come into contact, there is a very steep gradient of partial

(a) concurrent

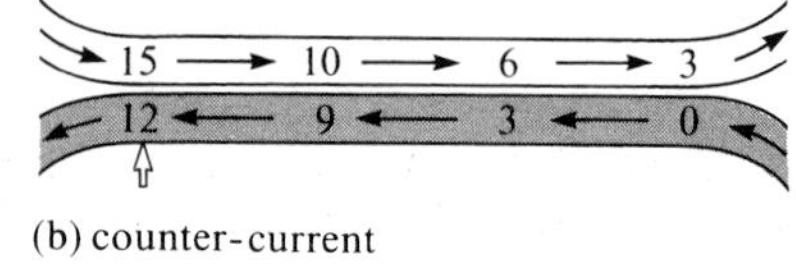

(b) counter-current

Figure 7.11 Concurrent flow (a) and counter-current flow (b).

pressures of oxygen between them (i.e. 15:0) and the flow of oxygen by diffusion is relatively high. Initially, oxygen diffuses into the blood as fast as the water loses it, so the partial pressure gradient between blood and water soon dwindles to near zero; that is, the two media equilibrate and the movement of oxygen into the blood stops. Now consider Figure 7.11b, in which the flow is *counter-current.* Blood and water now first come into contact at the right of the diagram, where the blood oxygen is zero and the water is almost completely lacking in oxygen (3 kPa). The P_{O_2} gradient from water to blood is less than that at the corresponding point in Figure 7.11a, where the two streams first meet, but is sufficiently large to promote the movement of oxygen. As blood moves along the capillary its P_{O_2} will increase as it gains oxygen, but this time the water it meets has a substantially raised P_{O_2} and so the gradient between water and blood is still significant. At all points along the capillary, the P_{O_2} gradient is sufficiently high to promote the movement of oxygen because the water encountered always has a significantly higher P_{O_2}. Although the initial rate of diffusion of oxygen from water to blood is lower in the counter-current system (Figure 7.11b), the diffusion of oxygen is maintained all the way along the length of the capillary. The net effect is that the partial pressure of oxygen in blood leaving the respiratory capillaries (see arrows) is higher if exchange is counter-current rather than concurrent and may indeed be higher than the P_{O_2} in the expired water, which is impossible with concurrent exchange.

You have probably guessed that Figure 7.11 is much simplified and that in reality many additional factors, for example the respective flow rates of the fluids and the solubility of oxygen in each medium, complicate the picture. But counter-current respiratory exchange enables teleosts to extract 80% or more of the dissolved oxygen from water. If the direction of water flow were reversed, the amount extracted would fall to about 10%. Compare this with human respiration in which inspired air contains approximately 20 cm^3 of oxygen per 100 cm^3 and expired air about 16 cm^3 of oxygen per 100 cm^3. Thus, only 20% of the available oxygen is extracted from the air.

The significance of the teleosts' high rate of extraction of oxygen will be appreciated if you consider the relatively low solubility of oxygen in water: at 15 °C, freshwater normally contains a maximum of 3.54 cm^3 of oxygen per 100 cm^3 water, and seawater has only 2.8 cm^3 of oxygen per 100 cm^3. Compare this with 20.95 vol. % of oxygen in air.

One important point is that the thickness of the membranes separating blood and water varies with the species of fish. In general, more active fishes have a thinner barrier. Thus, in mackerel, the separation is less than 1 μm, whereas in the bullhead it is 10 μm. The ratio between the total surface area of the gills and the body weight of the fish also varies because different species exhibit different degrees of subdivision of the lamellae and differences in the gross size of the gills. As you might expect, the ratio is higher in active fishes: for example, the gill area for mackerel is 1 158 mm^2 of gill per gram of body weight, whereas for the sluggish toadfish the corresponding figure is 20 mm^2 per gram of body weight.

The gills require a constant flow of water over them for normal function. In teleosts this is achieved by 'ventilation'—forced movement of water. Functionally, the important parts of the teleost respiratory system are: the buccal cavity; the gills themselves (in the normal state the tips of the filaments are closely opposed and form two 'sieves'); and the two cavities covered by opercula.

The floor of the buccal cavity can be raised or lowered, thus changing its volume. There are thin flaps of skin projecting from the upper and lower jaws in some fish, which act as passive 'oral' valves, so that when the floor of the cavity is lowered and water is drawn in through the mouth, the flaps simply trail inwards. However, when the mouth begins to close and the floor of the cavity rises, the increase in pressure causes the flaps to move so that the aperture of the mouth is reduced. If the jaws are not too far apart, the mouth will be virtually sealed as the valves meet. In the living animal, the gills are arranged in such a way that the water is forced through a number of what are virtually 'pores' or 'tubes' with diameters that may be as small as 0.025 mm. Bearing in mind the small size of the pores and the large total area of the respiratory surface (it may be ten times that of the body surface), it is apparent that the gills will offer considerable resistance to water flow. This resistance has the effect of separating the two opercular cavities (left and right) from the single buccal cavity.

Each operculum has a flap valve around its rim. This prevents water from being sucked in from the outside when the muscles move the operculum outwards.

To investigate the respiratory cycle of teleosts, G. M. Hughes made one of the earliest uses of a pressure transducer (a device by which pressure changes are transformed into electrical changes that can be displayed on an oscilloscope or recorded on paper). Using tench as experimental animals, he measured changes in pressure during ventilation by inserting very fine tubes into the buccal and opercular cavities and leading them out to the pressure transducer.

What stood out very clearly from the pressure recordings (Figure 7.12) was a big pressure difference between the buccal and opercular cavities. Thus, the

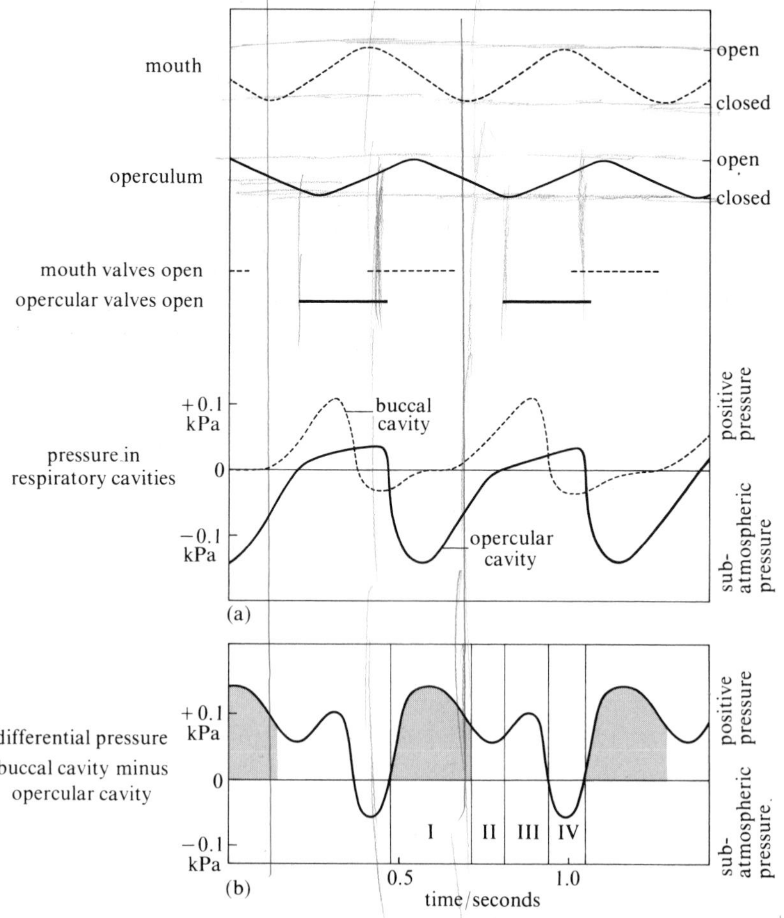

Figure 7.12 Recordings of events during two respiratory cycles in a tench. In (a) the upper lines indicate the movement of the mouth (broken line) and the operculum (continuous line). Below these are shown the periods when the month valves (broken line) and the opercular valves (continuous line) are fully open. The bottom traces in (a) show the pressures recorded in the buccal (broken line) and the opercular (continuous line) cavities. (b) shows differential pressure over the same period (i.e. pressure in the buccal cavity minus pressure in the opercular cavity). The shaded portions show the phases when water flow across the gills is produced mainly by the opercular suction pump. The four main phases of the respiratory cycle are also indicated, corresponding to the stages identified in Figure 7.13 as I–IV.

gills offer enough resistance to separate the cavities functionally as well as anatomically. It is clear that the 'buccal pump' (see Figure 7.13) is not the only part doing work in the system: in the middle of the cycle, when the pressure in the buccal cavity is falling fast, the pressure in the opercular cavity is even lower, so water is sucked across the gills.

Figure 7.13 summarizes the 'double-pump' action deduced from examination of the recordings such as those shown in Figure 7.12. In effect what is happening is that as the mouth opens and the floor of the buccal cavity begins to descend, causing a sharp fall in pressure in front of the gills, the two opercula begin to move outwards, enlarging the opercular cavities. The flaps round the edges prevent water rushing in from outside, so the opercular pressure falls even lower than the buccal pressure, thus continuing to pull water across the gills even though the mouth is opening and the buccal pressure is low. The bottom line of Figure 7.12 shows that there is an almost continual difference in pressure (and thus a flow of water) across the gills. There is a brief cross-over period when pressure is reversed, but this is only for a fraction of a second, so the inertia of the water and the resistance of the gills probably prevent any reverse flow.

Clearly, water is passed over the gills not in a series of short, sharp squirts, but in an almost continuous, relatively smooth flow—a very much more efficient process as far as gas exchange is concerned. Further work has shown that basically the same system is employed by elasmobranch fishes, though there are some differences.

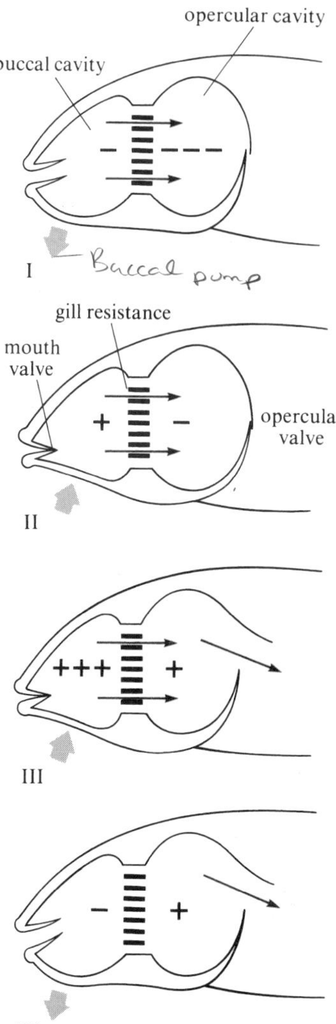

Figure 7.13 The 'double pump' theory of ventilation of the gills in teleost fishes. The magnitude of the pressures in the buccal and opercular cavities are indicated as + (above ambient pressure) and − (below ambient pressure). The grey arrows show the operation of the buccal pump; the red arrows show the direction of water movement. Only one of the opercular cavities is shown. *Stage I* The mouth is open and the buccal pump draws in water through the mouth while the opercular pump sucks water across the gills. *Stage II* The mouth is closed and the buccal pump forces water across the gills while the opercular pump sucks it across. *Stage III* The operculum is open, and the buccal pump forces water across the gills. *Stage IV* The mouth and operculum are both open, but back pressure across the gills probably does not move water.

Summary of Section 7.6

In teleosts, the gill arches bear rows of gill filaments (primary lamellae) which in turn bear a great number of secondary lamellae. This whole structure has a substantial area for gas exchange and offers considerable resistance to the flow of water. Ventilation of the gills is ensured by a buccal 'force pump' that pushes water across the gills and an opercular 'suction pump' that pulls water into the opercular cavity. Passive valves prevent any substantial backflow of water and the synchronized, alternate action of the dual pumps ensures a virtually continuous, unidirectional flow of water. The flow of blood in the very thin-walled lamellae is in a direction counter to the ventilatory flow of water; that is, it is a counter-current flow. The blood therefore always encounters water with a higher P_{O_2} and the amount of oxygen extracted from the water is therefore very high. The total area for gas exchange is greatest in those very active fishes with the highest metabolic requirements. (The gills of such fishes have water-to-blood distances similar to the air-to-blood distance in mammalian lungs.)

Now attempt Question 5 on p. 241.

7.7 CONTROL OF BREATHING IN MAMMALS

You have now read about a variety of mechanisms for getting oxygen into and carbon dioxide out of a circulating fluid. All these mechanisms have to deal with the fact that the metabolic activity of an animal is not fixed, it can vary enormously, and oxygen need and carbon dioxide production vary with it. The exchange of gases at lungs or gills must therefore be regulated to match metabolic needs. This is a complex problem with different solutions in different species. We will consider just one system—that operating in mammals, particularly humans—which for obvious reasons has been very well studied. The control of mammalian breathing involves the generation in the brain of a regular rhythm of inspiration and expiration and its continuous modulation by chemical influences with the result that the rate of gas exchange in the lung matches the metabolic need for oxygen and production of carbon dioxide.

7.7.1 The respiratory rhythm

If the nerves to the diaphragm (the phrenic nerves) are cut, together with nerves to the other muscles of respiration, then breathing stops. The muscles themselves are not intrinsically rhythmic, and depend on rhythmic patterns of activity in their nerves. If the spinal cord is cut above the level of outflow of the phrenic nerves, breathing stops, so this rhythmic pattern must originate in the brain. Early in the last century it was shown by removal ('ablation') of parts of the brain in animals that apparently normal breathing remained with only two parts of the brain intact. Both were in the hindbrain, the **pons** and **medulla**. Even with only the medulla, breathing was more or less rhythmic, though the tidal volume was very variable.

At the same time, peripheral nerves were implicated in the respiratory rhythm—the **vagi**, which carry afferent (sensory) information from the lungs to the hind brain. If these nerves are cut in the neck, with the pons and medulla intact, breathing is still rhythmic, but becomes deeper and slower. If the upper part of the pons is removed, then breathing stops in the full inspiration part of the cycle, with occasional expiratory gasps—a pattern known as apneusis. Impulses travelling up the vagi are able to reduce inspiration if the whole system is working, and are actually necessary to terminate inspiration if some structure in the upper pons is absent.

The generation of the respiratory rhythm therefore depends upon the interaction of central nervous mechanisms and sensory information from the lungs. We still do not understand exactly how this is achieved, though there are several theories.

All theories assume that there is some basic rhythm generator which is present in the medulla, with sensory information from the lungs and signals from receptors sensitive to P_{O_2} and P_{CO_2}, being fed in via other groups of neurons in the pons. The basic rhythm generator probably consists of two groups of neurons. One group is spontaneously active, and they stimulate inspiratory movements via the phrenic nerves as well as stimulating a second group of neurons. This second group, after a delay, send the impulses back to the first as inhibitory stimuli. The spontaneous activity therefore initially generates inspiration, but as the impulses arrive back via the second centre they switch off the spontaneous activity, and so inspiration stops, and the lungs empty automatically by recoil (see Section 7.2). As soon as the activity of the first group of neurons is inhibited, they stop stimulating the second, which in turn, after a delay, stop inhibiting the spontaneous activity of the first, so inspiration begins again.

This basic rhythm generator is then modulated by structures in the pons, which act to make it regular and responsive to both the mechanical state of the lungs and the concentration of blood gases. Information from the lungs comes from stretch receptors in the airways, and travels to the pons via the vagi. As the lungs expand, the airway receptors are stretched, and send impulses up the vagi, which tend to inhibit inspiration. This tendency for inspiration to inhibit inspiration is known as the **Hering–Breuer reflex**. This is why breathing becomes deeper if the vagi are cut, although the rhythm is not lost, as this is generated in the brain. This sensory feedback from the lungs ensures that inspiratory effort is matched to the difficulty of inflating the lung. Under some conditions, say when wearing a mask with a mouthpiece, or when a pregnancy or tight corsets impede the downward movement of the diaphragm, more effort is required to inspire. If the initial stimulus to breathe in is not inhibited by Hering–Breuer reflexes, because less air has moved into the lungs, then inspiratory effort will continue. These mechanisms will generate a smooth rhythm of breathing, but cannot alone ensure that the right amount of air is breathed to meet metabolic demands. This requires chemical control.

HERING - BREUER REFLEX
A breathing reflex, mediated via the vagi, in which lung inflation ~~further~~ inhibits further inspiration during that breathing cycle.

HYPOXIA
A deficiency of oxygen supply to the tissues

HYPERCAPNIA
An increase in arterial P_{CO_2}

HYPOCAPNIA
A decrease in arterial P_{CO_2}.

7.7.2 Chemical control of respiration

So long as the alveoli are ventilated, the tissues will receive blood containing adequate oxygen and carbon dioxide.

Basic rhythm generator in PONS

◇ What effect will increasing lung ventilation have on P_{O_2} and P_{CO_2}?

◆ More air will be breathed in, so alveolar, and thus arterial P_{O_2} will rise. More carbon dioxide will be 'blown off' so alveolar and arterial P_{CO_2} will fall.

– Stretch receptors in airway
– travel to pons via vagi

P_{O_2} and P_{CO_2} cannot be changed independently by changes in breathing alone. If some other disturbance causes P_{O_2} to fall and P_{CO_2} to rise, then both changes can be corrected by increasing breathing. If, on the other hand, P_{CO_2} falls with no change in P_{O_2}, then any attempt to control the P_{CO_2} by breathing more will disturb P_{O_2} from its normal value, and merely replace one problem with another. These potential conflicts are minimized by complex control mechanisms.

Changes in both P_{O_2} and P_{CO_2} can have damaging effects. A fall in P_{O_2} in the blood produces a condition known as **hypoxia**. Mild hypoxia is not a serious problem, because of the shape of the oxygen–haemoglobin dissociation curve. Look back at Chapter 6, Section 6.5, and you will see that a change in P_{O_2} from its normal value of 13.3 kPa to 9 kPa will hardly affect the saturation of haemoglobin, and therefore the oxygen content of the blood. It is only when P_{O_2} falls so that it reaches the steep part of the curve that oxygen supply to the tissues is compromised. Increases in P_{O_2} are generally harmless until it becomes very high indeed, which is impossible when breathing air at normal pressures.

On the other hand, small disturbances of P_{CO_2} are significant. A rise in the P_{CO_2} of arterial blood is known as **hypercapnia**. A fall is **hypocapnia**. Recall from Chapter 6, Section 6.7.1, that the P_{CO_2} is a critical factor in determining the pH of blood. The Henderson–Hasselbalch equation shows that, if P_{CO_2} increases, the blood becomes acid. If it falls, the blood becomes alkaline.

Small changes in P_{CO_2} produce relatively large changes in the pH of arterial blood and these changes in pH can have serious consequences.

◇ Would you expect the respiratory control systems to be more sensitive to changes in P_{O_2} or P_{CO_2}?

◆ Small changes in P_{CO_2} are potentially more damaging, so it would be advantageous if the system were more sensitive to those.

Figure 7.14 shows the respiratory responses to breathing gas mixtures which push alveolar P_{O_2} down, or P_{CO_2} up. Falls in P_{O_2} have very little effect on breathing until they are severe, but rises in P_{CO_2} immediately stimulate extra breathing up to the limits of ventilation. Figure 7.14 also shows the effects on breathing of changes in the pH of blood. Recall from Chapter 6, Section 6.7.1, that changes in P_{CO_2} due to altered ventilation are a disturbing influence on blood pH so long as $[HCO_3^-]$ is constant, but should $[HCO_3^-]$ change, say because of metabolic acid production, changes in P_{CO_2} can *correct* the disturbance of pH. So, P_{CO_2} sometimes alters from normal, compensating for a *metabolic* acidosis or alkalosis. It is clear that the brain centres controlling breathing receive information about the P_{O_2} and P_{CO_2} of blood. These must be detected by a sensory system. Sense organs which detect changes in chemistry are called **chemoreceptors**. There are chemoreceptors for both P_{O_2} and P_{CO_2}.

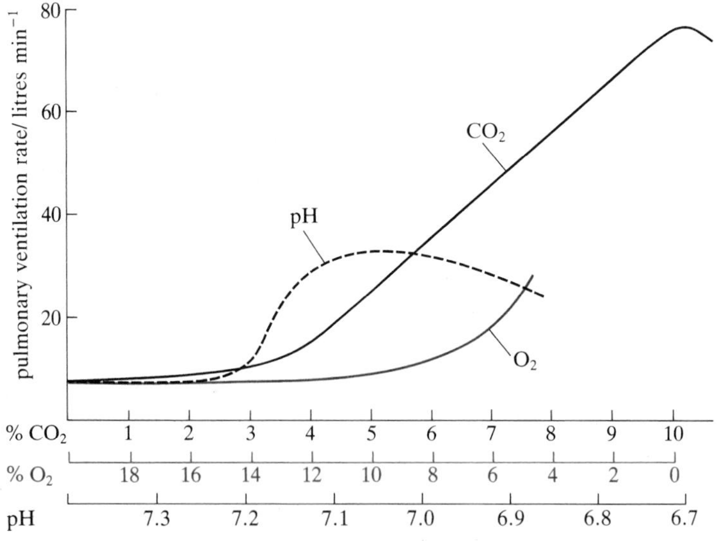

Figure 7.14 Changes in breathing following changes in percentages of inspired oxygen and carbon dioxide and pH of arterial blood.

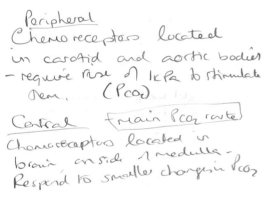

Chemoreceptors sensitive to changes in P_{CO_2} are present at several sites. The peripheral chemoreceptors are located in the carotid and aortic bodies, as shown in Figure 7.15. Peripheral chemoreceptors are relatively insensitive to changes in P_{CO_2}, and require a change in partial pressure of around 1 kPa to stimulate them. Central chemoreceptors are located in the brain, on the side of the medulla. They are not the same cells that generate the respiratory rhythm. They respond to much smaller changes in P_{CO_2} than the peripheral chemoreceptors, and therefore under most circumstances they are the main route by which carbon dioxide affects breathing.

If the central chemoreceptors are stimulated, ventilation is increased, and very small changes in P_{CO_2} will lead to large changes in breathing. For many

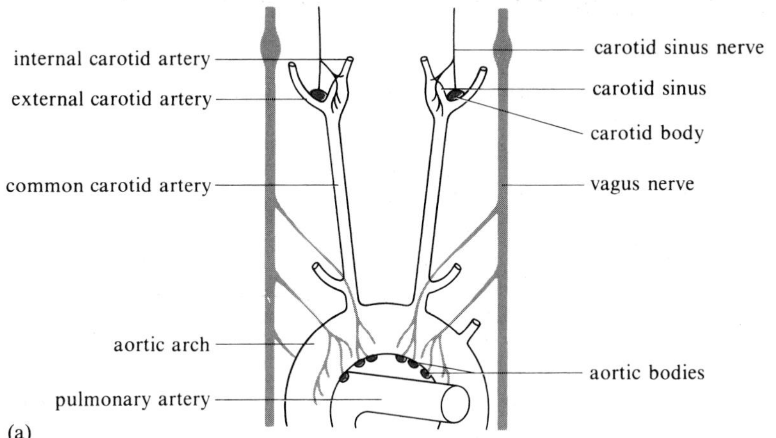

years, however, it was not clear exactly to what the central chemoreceptors responded—was it changes in P_{CO_2} directly, or some consequence of these changes?

The central chemoreceptors are not exposed directly to blood. They are bathed by the fluid which surrounds the brain—the cerebrospinal fluid (CSF). Cerebrospinal fluid is isolated from blood by specialized cells lining the brain blood vessels, which generate the **blood–brain barrier** (Figure 7.15b). Most substances cannot cross this barrier, so in many respects the composition of the CSF is separately controlled by the cells which secrete it—the choroid plexus cells.

Carbon dioxide can, however, cross the blood–brain barrier freely. If blood P_{CO_2} rises, then CSF P_{CO_2} will as well, though there may be a delay of a few seconds. However, just as in plasma, the dissolved carbon dioxide in CSF forms hydrogen ions to an extent determined by the P_{CO_2} and the concentration of bicarbonate, according to the Henderson–Hasselbalch equation (which you have seen before in Chapter 6, Section 6.7.1):

$$\mathrm{pH} = 6.1 + \log \frac{[\mathrm{HCO_3}^-]}{P_{CO_2} \times \alpha} \qquad (7.1)$$

Bicarbonate cannot cross the blood–brain barrier and its concentration in CSF is controlled by the choroid plexus cells which secrete it into the CSF. It does not change in the short term if P_{CO_2} changes. This means that small changes in P_{CO_2} produce large changes in the pH of the CSF.

For this reason many investigators considered it likely that the central chemoreceptors did not respond to P_{CO_2} directly, but to the effect of increases in it—namely to acidification of the CSF. This was shown to be the case by experiments where the pH of the CSF was changed independently of the P_{CO_2}. If the P_{CO_2} changed at a constant pH, no change in breathing occurred, but, if pH changed at constant P_{CO_2}, breathing was changed.

The central chemoreceptors are therefore just like little 'pH meters' sensing the effect changes in P_{CO_2} have on the acidity of CSF. Because the pH of CSF changes considerably for small changes in P_{CO_2}, they appear to be very sensitive to P_{CO_2}.

◇ What is the effect on breathing of an increase in metabolic production of carbon dioxide?

◆ If carbon dioxide is produced faster than it is eliminated via the lungs, P_{CO_2} in blood will rise. This rise will be translated to the CSF, where pH will fall. This will stimulate the central chemoreceptors to increase breathing.

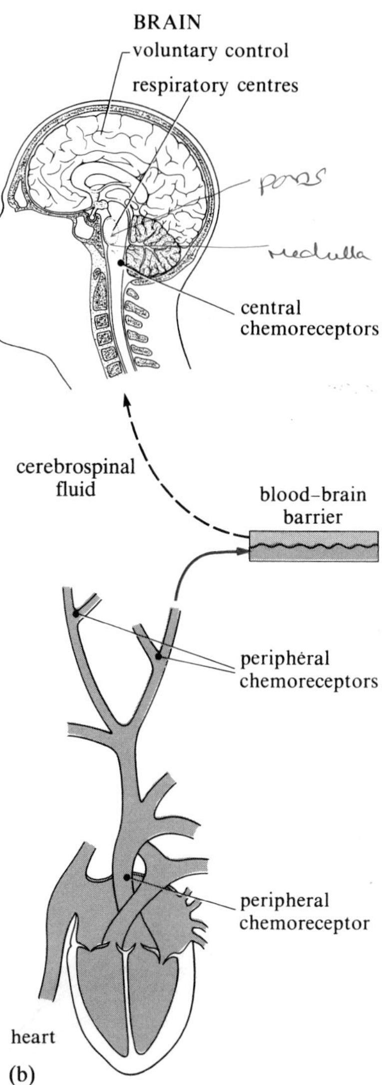

Figure 7.15 (a) The location and nerve supply of the chemoreceptors (shown in red) in the carotid and aortic bodies. Also shown is the nerve supply to the carotid sinus which contains receptors that respond to stretching. (b) The link between the central and peripheral chemoreceptors across the blood–brain barrier.

◇ What is the effect on P_{CO_2} of increasing breathing?

◆ P_{CO_2} will fall, and, as this fall is reflected in CSF P_{CO_2}, the pH of the CSF will return to normal.

This sequence of events is the way in which breathing is normally regulated (Figure 7.16). Any change in metabolism will alter production of carbon dioxide and initially, therefore, change blood P_{CO_2}, then CSF P_{CO_2}, then CSF pH. This change, when detected by the central chemoreceptors, will trigger a change in breathing which will return pH of the CSF to normal by changing blood P_{CO_2}. This is classic **negative feedback** control. Controlling breathing via the pH of the CSF does, however, have important consequences.

◇ What will happen to breathing if CSF $[HCO_3^-]$ is decreased, but blood P_{CO_2} does not change?

◆ The CSF pH depends on the ratio of $[HCO_3^-]$ to P_{CO_2}, which will decrease, therefore pH will decrease. This will trigger increased breathing despite the fact that the P_{CO_2} has not changed. The P_{CO_2} will *fall* until the pH of the CSF is restored to normal.

It is, therefore, the *pH of the CSF* which is regulated at a constant value. The ventilation rate and P_{CO_2} required depend upon the CSF bicarbonate concentration which can be changed by the choroid plexus cells. The choroid plexus cells secrete CSF continuously. The rate in a goat has been measured as $0.16\ cm^3\ min^{-1}$. In humans, the rate will be similar, and the total amount of CSF is $130–150\ cm^3$. The central chemoreceptors can, therefore, regulate at a variety of arterial P_{CO_2} values, depending on the CSF $[HCO_3^-]$, and thus the regulatory system can be 'set' to run at different levels. This ability to adjust the system is critical for the resolution of conflicts which arise when one set of chemoreceptors stimulates different changes in breathing from another, which happens when blood pH changes and the pH of the CSF does not, as we shall see in Section 7.7.4.

Figure 7.16 Pathways across blood–brain barrier which lead to changes in the pH of the CSF.

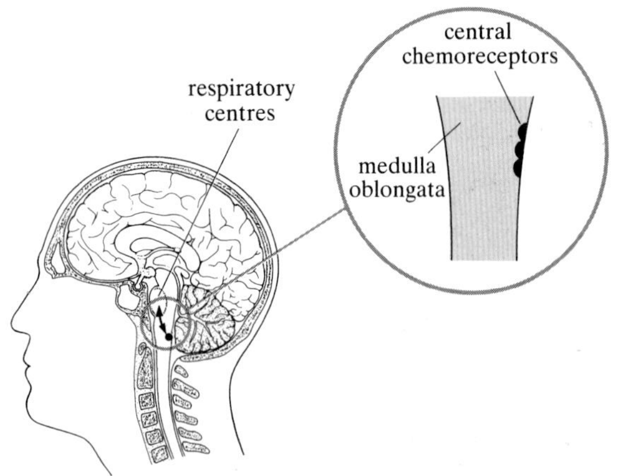

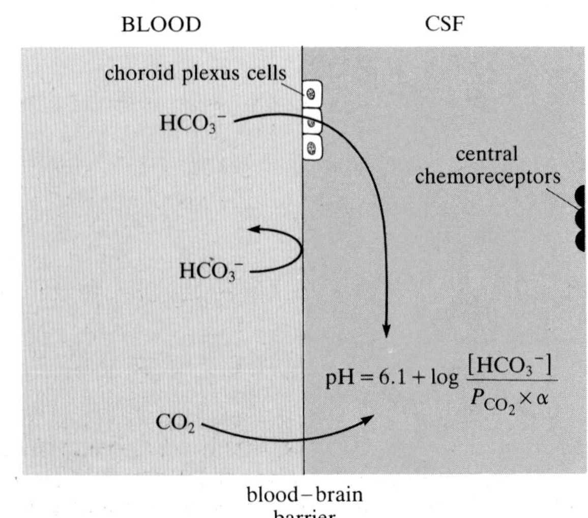

7.7.3 Effects of blood pH on respiration

Look back at Chapter 6, Section 6.7.1, where we considered the role of arterial P_{CO_2} in the determination of blood pH. You will see that, because the Henderson–Hasselbalch equation shows that pH is determined by the ratio of bicarbonate concentration and P_{CO_2}, changes in pH produced by alterations of bicarbonate concentration (metabolic acidosis and alkalosis) may be compensated for by changing P_{CO_2}. We can now see that P_{CO_2} may be changed by altering breathing. The chemical control mechanisms of breathing are therefore sensitive to changes in the pH of blood even if P_{CO_2} is constant.

◇ If blood $[HCO_3^-]$ falls, because of metabolic production of acid, what will happen to plasma pH?

◆ It will fall.

◇ What effect will this have on the peripheral chemoreceptors?

◆ They will be stimulated, and breathing increased.

◇ What effect does a change in plasma $[HCO_3^-]$ have on CSF $[HCO_3^-]$?

◆ None, in the short term, as bicarbonate cannot cross the blood–brain barrier.

This generates an immediate problem. A fall in blood pH due to low $[HCO_3^-]$ stimulates breathing, which lowers P_{CO_2}, and brings the pH back towards normal. However, CSF $[HCO_3^-]$ has not changed, so this lowering of P_{CO_2} *raises* CSF pH, a signal which will cause a *reduction* in ventilation by reducing stimulation of the central chemoreceptors. To put it dramatically, a 'fight' has developed between peripheral chemoreceptors trying to change breathing in order to correct the pH of blood, and the central chemoreceptors trying to prevent change in order to stabilize the pH of CSF.

The immediate effect of this competition is that breathing does not increase enough to stabilize blood pH, but does increase a little, so CSF pH is raised.

At this point the choroid plexus cells come into play. As you saw in Section 7.7.2, they have a crucial property—they respond to persisting (for more than a minute or two) changes in CSF pH by altering CSF $[HCO_3^-]$ concentration. If CSF pH is increased, because P_{CO_2} has been lowered, they lower CSF $[HCO_3^-]$ over an hour or two.

◇ What effect will lowering CSF $[HCO_3^-]$ have on the elevation of pH produced by persisting falls in arterial P_{CO_2}?

◆ As we saw in an earlier question in Section 7.7.2, falls in CSF $[HCO_3^-]$ at constant P_{CO_2} will lower CSF pH. As it was already high, this change will bring it back to normal.

◇ What effect will restoring CSF pH have on the central chemoreceptors?

◆ They will no longer be inhibited, so they will cease to 'fight' with the peripheral chemoreceptors.

What has happened is that the central chemoreceptors have been 'reset' by the choroid plexus activity so that the lower P_{CO_2} appropriate for the control of blood pH becomes the normal value about which they regulate.

The peripheral chemoreceptors are sensitive to the pH of blood. If it falls, breathing is stimulated. In fact, this is the mechanism of their response to rises in P_{CO_2}.

◇ What effect will a rise in P_{CO_2} have on the pH of arterial plasma?

◆ As pH depends on the ratio of $[HCO_3^-]$ to P_{CO_2}, the effect will depend on the extent to which this ratio is changed. Look back to Chapter 6, Section 6.7.1, and you will see that a rise in P_{CO_2} produces a small increase in $[HCO_3^-]$ which does not totally offset the rise in P_{CO_2}, so pH will fall. This fall will be less than that in the CSF, where $[HCO_3^-]$ is fixed.

The smaller fall in pH in blood than in CSF, for a given change in P_{CO_2}, explains why the normal control of breathing is via the central chemoreceptors. Just as in CSF, however, the level of bicarbonate in blood can change with no change in P_{CO_2}—in metabolic acidosis and alkalosis.

Under normal circumstances, therefore, breathing is regulated via the central chemoreceptors with the result that arterial P_{CO_2} is kept constant—P_{CO_2} is the principal controlling influence. If, however, there is a pressing need to change P_{CO_2}, then the dominance of the central chemoreceptors can be over-ridden, over a period of hours, and they respond to a different P_{CO_2} as 'normal'.

Another very good reason to increase breathing and possibly lower P_{CO_2} is if arterial P_{O_2} falls, i.e. during hypoxia. We shall see in Section 7.8 that mammals have chemoreceptors sensitive to low P_{O_2}, and, in Section 7.9, how their stimulation can also precipitate a conflict with the central chemoreceptors which has to be resolved by the choroid plexus.

Summary of Section 7.7.3

The peripheral chemoreceptors are sensitive to the pH of blood. Small changes in pH which follow disturbance of P_{CO_2} are detected in the CSF, and corrected by changing ventilation. The central chemoreceptors, however, do not detect changes in bicarbonate concentration in blood, as bicarbonate cannot cross the blood–brain barrier.

These changes alter ventilation via the peripheral chemoreceptors. If blood pH falls, breathing is stimulated, so P_{CO_2} falls as well, tending to restore pH. The extent of changes in P_{CO_2} is however, limited by the fact that although the pH of blood will be corrected by the change in ventilation, the CSF pH will be changed because its bicarbonate concentration was not disturbed. The central chemoreceptors will therefore tend to 'fight' the peripheral, and bring ventilation back to its previous value. This is an example of competition between the sets of receptors, which we will see is a very serious problem during acute exposure to low P_{O_2}.

7.7.4 The response to the lack of oxygen

The central chemoreceptors do not increase ventilation in response to low P_{O_2}. The response is mediated via the peripheral chemoreceptors. Arterial P_{O_2} has to fall considerably before the carotid and aortic bodies respond. Their low sensitivity means that P_{O_2} is not involved in the minute to minute control of ventilation.

The mechanism by which the cells in the carotid and aortic bodies sense low P_{O_2} is not well understood, but the receptors may be stimulated when they have insufficient oxygen for their own needs. As they have a very high blood flow, however, the oxygen content of blood must fall considerably before their oxygen supply is compromised. If blood flow is reduced they become more sensitive. What is clear, however, is that the receptors do not 'adapt', i.e. they continue to respond so long as P_{O_2} is low, which is of some significance during exposure to high altitude and in some diseases which affect oxygen transfer across the lungs.

7.7.5 Respiration during exercise

In Chapter 5, Section 5.5, you read about the changes in the cardiovascular system that occurred during exercise. Now we can add the respiratory changes. The rate of ventilation of the lungs increases up to fivefold during exercise in response to increased metabolic activity. Breathing is controlled via the central chemoreceptors, which are sensitive to the pH of the CSF (Section 7.7.2). If metabolism increases, production of carbon dioxide in the body briefly exceeds the rate of excretion via the lungs, and the P_{CO_2} in the arterial blood will tend to rise. The P_{CO_2} in the CSF rises in line with that of blood, and as it does, the pH of CSF falls, as its bicarbonate concentration is fixed and pH depends on the ratio of bicarbonate to P_{CO_2}. This fall in pH is sensed by the central chemoreceptors, which stimulate an increase in breathing, which then increases the excretion rate of carbon dioxide until it matches the production rate, and the P_{CO_2} returns to normal.

Just as in the cardiovascular system (Chapter 5, Section 5.4), this mechanism is sufficient to produce an increase in breathing during exercise, but there are problems which stem from the rapidity and magnitude of the changes required. These can be seen if we examine the time course of respiratory responses to exercise. Figure 7.17 shows that at the onset there is a sudden jump in the rate of ventilation of the lungs, which then gradually increases to a plateau. At the onset of exercise, therefore, there is a 'dynamic' phase when exercise is constant, but breathing is changing. Once exercise has been going a minute or two, both breathing and working settle into a constant 'static' phase.

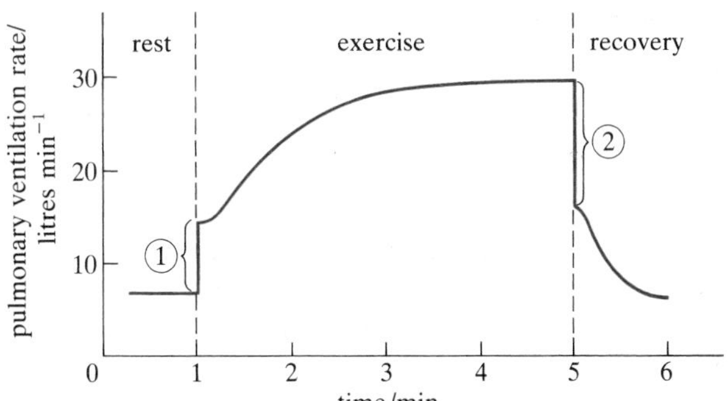

Figure 7.17 Schematic representation of changes in total ventilation per minute during exercise. The figure shows (1) the abrupt increase at the onset and (2) the abrupt larger decrease at the end of exercise.

Consider first of all the 'static' phase. We have read how respiratory control mechanisms regulate blood P_{CO_2}, so we must ask whether they are effective once exercise is established. Figure 7.18 shows the rate of lung ventilation, arterial P_{O_2} and P_{CO_2} after a few minutes at different rates of exercise. For light to moderate exercise (up to the vertical dashed line), there is a straight-line relationship between the rate of working and the rate of breathing, which suggests good regulation. This is clearly demonstrated by the fact that arterial P_{CO_2} remains constant at its normal value, despite great increases in carbon dioxide production. Once exercise becomes severe (beyond the dashed line), breathing increases even more than is necessary to

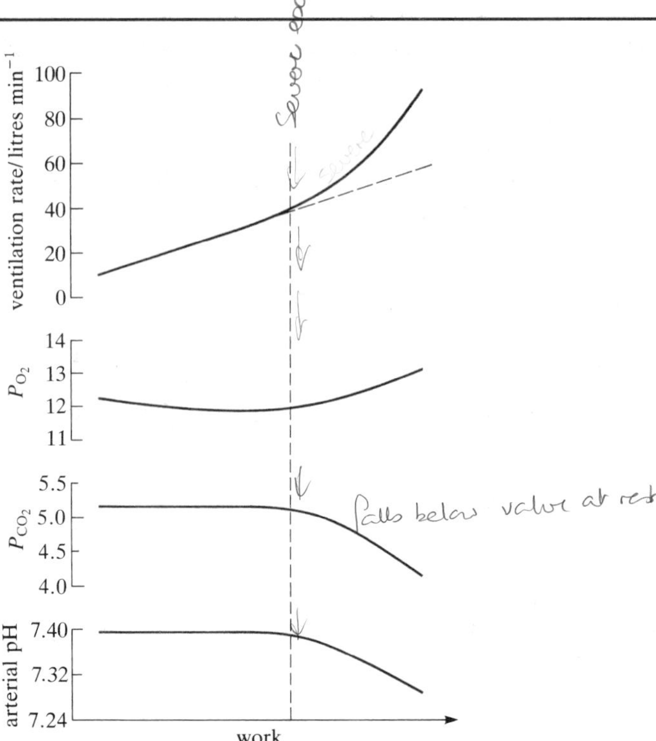

Figure 7.18 Changes in lung ventilation, blood gases and arterial pH at different levels of exercise.

stabilise P_{CO_2}, which falls below the value at rest. P_{O_2} actually rises in severe exercise, provided the subject has normal lungs. These observations show, first, that in light to moderate exercise, regulation is very good, and the mechanisms described in Section 7.7.3 fulfil their function well. In severe exercise, breathing is further driven by some other factors. Two have been identified. First, rising body temperature. As we get hotter, breathing is increased, increasing heat loss from the lungs. Second, as blood flow to exercising muscle is compromised in severe exercise (as we have discussed above), the resulting anaerobic metabolism produces ions such as lactate which tend to lower blood pH, another stimulus to respiration via the peripheral chemoreceptors.

Overall, however, in the static phase of exercise, the control systems work well. If breathing is right once it has settled down, however, it follows that it must have been wrong in the preceding dynamic phase, when exercise was at the same rate, but lung ventilation was lower. Some factor prevents breathing rising immediately to the appropriate level. In part this is because the carbon dioxide takes time to accumulate in the blood, but it also is rather slow to cross the blood–brain barrier into the CSF and stimulate the central chemoreceptors. For a period, therefore, less oxygen is being taken in than is used, and *oxygen debt* accumulates, which is paid off by continuing increased breathing after exercise has ended.

This oxygen debt is reduced by the operation of another mechanism, which, like cardiovascular changes in exercise, pre-empts the normal control mechanisms. As exercise begins, signals from sensory receptors in joints and muscles travel to the respiratory centres, and produce an increase in ventilation whose magnitude depends upon the intensity of exercise. At this stage there is no chemical stimulus to extra breathing, so the effect is to increase oxygen uptake and removal of carbon dioxide in advance of changes in P_{CO_2}. This mechanism 'sets' ventilation to about the right level, and then the normal control mechanisms (discussed in Section 7.7.2) take over to match it precisely to metabolic demand. In order to minimize the oxygen debt still further, some trained athletes override this process by deliberately increasing their ventilation in advance of exercise.

Thus, in exercise, the operation of the normal control systems is aided by a specific, pre-emptive additional mechanism which prevents the slowness of response of the control systems from producing unacceptably large changes in the regulated variables.

Summary of Section 7.7

The generation of the respiratory rhythm in mammals is a product of interactions between the central nervous system and sensory information reaching the brain from the lungs. A basic rhythm generator is thought to lie in the medulla, modulated by inputs from the pons. The rhythm generator initiates inspiration. As the lungs expand, stretch receptors send impulses to the pons that tend to inhibit inspiration. This sensory feedback from the lung reduces inspiratory effort as the expansion of the lung increases. It is known as the Hering–Breuer reflex.

These mechanisms provide the control over the breathing cycle but do not ensure that the appropriate amount of air is breathed to match the metabolic need. This is under chemical control, principally from the level of carbon dioxide in the blood, which in turn determines the pH. Changes in pH are sensed by chemoreceptors—the central chemoreceptors on the side of the medulla and the peripheral chemoreceptors in the carotid and aortic bodies. The central receptors are much more sensitive to P_{CO_2} changes than the peripheral receptors. The central receptors are not in direct contact with the blood; they are bathed in CSF and carbon dioxide can cross the blood–brain barrier, entering the CSF and altering the pH. CSF is produced continually by the choroid plexus cells and the amount of bicarbonate secreted in the fluid varies. Thus the pH of the CSF is regulated.

During exercise, the rate of ventilation increases as metabolism increases. The partial pressure of carbon dioxide in the CSF rises and so the pH falls. The central receptors sense this and stimulate an increase in breathing rate until the pH of the CSF returns to normal. In severe exercise breathing is also affected by rising body temperature and an increase in anaerobic metabolism. Lactate accumulating in the blood lowers blood pH.

Low P_{O_2} causes an increase in breathing via the peripheral chemoreceptors. They are not very sensitive, so arterial P_{O_2} has to fall considerably before they are stimulated to fire.

7.8 ADAPTATION OF MAMMALS TO HIGH ALTITUDE

The total pressure of the atmosphere diminishes rapidly with increasing altitude, but the composition of air, apart from water vapour, remains the same, so P_{O_2} must fall (recall the example in Chapter 6, Section 6.2). As the P_{CO_2} of air is very low at all altitudes, an individual exposed to high altitude suffers only from the effects that result from a reduction of P_{O_2} (hypoxia) (Figure 7.19).

The effects of acute hypoxia were first observed in balloonists ascending to ever higher altitudes. Above approximately 4 000 m, they began to feel slightly light-headed, and as they ascended further, became euphoric to the point of recklessness. Eventually, at higher altitudes they began to feel sick and dizzy, until, above 7 000 m they finally became unconscious, dying if not brought down quickly. Apart from the final lethal stages, the similarities to alcoholic intoxication are very strong. That is the great danger of hypoxia because the subjects are not aware of the risk. Figure 7.20 shows the duration of consciousness following exposure to pressures equivalent to various altitudes. Modern passenger jets fly at around 10 000 m (30 000 ft), so you will see that if your aircraft suddenly depressurises you have considerably less than one minute to fit the oxygen mask that will fall in front of your face.

You may recall, however, that a few mountaineers have walked to the top of Everest (9 000 m) without the aid of oxygen. How did they survive?

There are several reasons, all associated with responses to gradually increasing, rather than acute hypoxia. First, acute hypoxia is dangerous because, as you may have guessed, the respiratory control mechanisms will not initially allow breathing to increase as necessary to cope with the hypoxia. As you will see in Section 7.8.1, it takes time for ventilation to increase. Second, a variety of other changes in the circulation and blood increase oxygen supply to the tissues over days and weeks of exposure to hypoxia. The changes in breathing, circulation and the blood which develop over time are known as **acclimatization**.

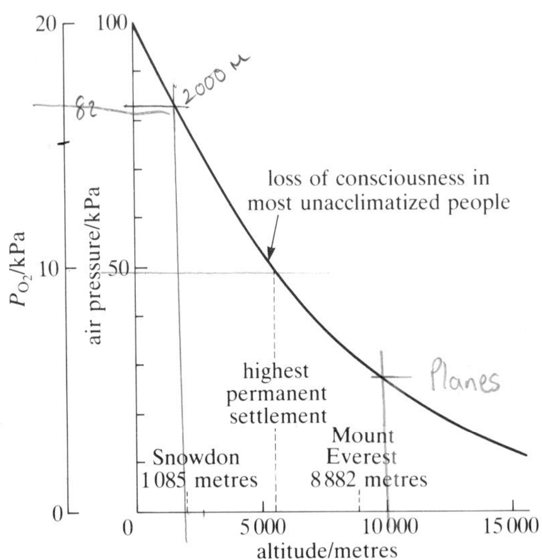

Figure 7.19 The total air pressure and the partial pressure of oxygen in the atmosphere at various altitudes.

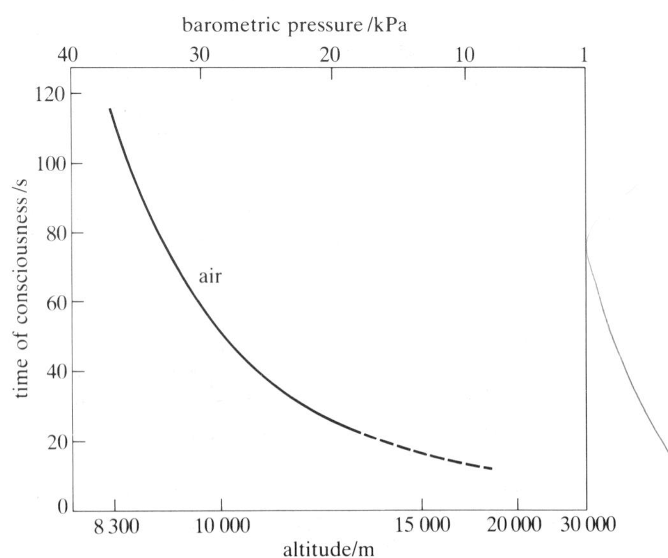

Figure 7.20 Duration of consciousness following acute exposure at different altitudes.

7.8.1 Respiratory changes

The problem with acute hypoxia is an initial failure to increase ventilation and compensate for the low P_{O_2}. This failure is not due to a failure to detect the low P_{O_2}. The peripheral chemoreceptors, though not very sensitive, respond before hypoxia is dangerous by sending impulses to the medulla which tend to increase ventilation. The problem is that arterial P_{CO_2} is not affected by exposure to hypoxia, and so these increases in ventilation reduce P_{CO_2} below its normal value (i.e. result in hypocapnia). Low P_{CO_2} renders the blood and CSF alkaline (look back at the Henderson–Hasselbalch equation in Section 7.7.2). Alkalinity of the CSF reduces stimulation of the central chemoreceptors, so their 'drive' to ventilation is reduced. This fall in central 'drive' largely offsets the rise in peripheral 'drive', so the net effect on ventilation is small—another example of conflict between the central and peripheral chemoreceptors.

The peripheral chemoreceptors triumph eventually, however, because they do not adapt. Their continued firing increases ventilation slightly, so the CSF remains slightly alkaline. Over a period of hours, the choroid plexus cells respond to this by secreting CSF which contains less bicarbonate. As the pH depends on the bicarbonate : P_{CO_2} ratio, this restores the CSF pH to normal, but at a lower P_{CO_2}. Over about 12 hours this progressive compensation allows the peripheral 'drive' to be fully expressed, and ventilation is greatly increased, so P_{O_2} in the lungs rises. The blood, however, remains alkaline for a few days, until corrected by the kidney (see Chapter 11, Section 11.3.12).

7.8.2 Circulatory changes

Another immediate and short-lived response to high altitude is an increase in cardiac output. The amount of oxygen delivered to respiring tissues depends not just on the amount of oxygen in the blood, or the P_{O_2} at the tissues, but also on the rate of blood flow to the tissues. Despite increased ventilation, the oxygen content per unit volume of arterial blood is likely to decrease at high altitudes, and this may be partly compensated for by an increase in the *amount* of blood that flows to the respiring tissues per unit time. The blood supply to the tissues is enhanced by the opening-up of hitherto constricted capillaries. Something similar happens when people exercise: more oxygen is delivered to the tissues, mainly as a result of an increase in cardiac output of up to five times the resting level. So, if a lowlander is vigorously climbing at high altitude, the physiological demands on the circulatory system are particularly marked.

Changes in the tissues also lead to a greater fraction of the blood's oxygen being given up. At sea-level, the venous blood entering the heart via the systemic veins (called mixed venous blood) has a P_{O_2} of approximately 5.3 kPa; the blood that leaves the lungs has a P_{O_2} of about 12.6 kPa.

Look at Figure 7.21. This is an oxygen dissociation curve for human blood showing the oxygen content of: (a) blood saturated with oxygen at the lungs (s); (b) blood in the main veins when the person is at rest (v); (c) blood in the main veins when the subject is exercising (e).

◇ What proportion of the oxygen bound to haemoglobin is relinquished to the respiring tissues when the subject is at rest?

◆ Only about 25% of the total load (the difference between s and v on the graph). Much of the remainder is 'reserve' capacity, which is used during exercise (the curve between v and e).

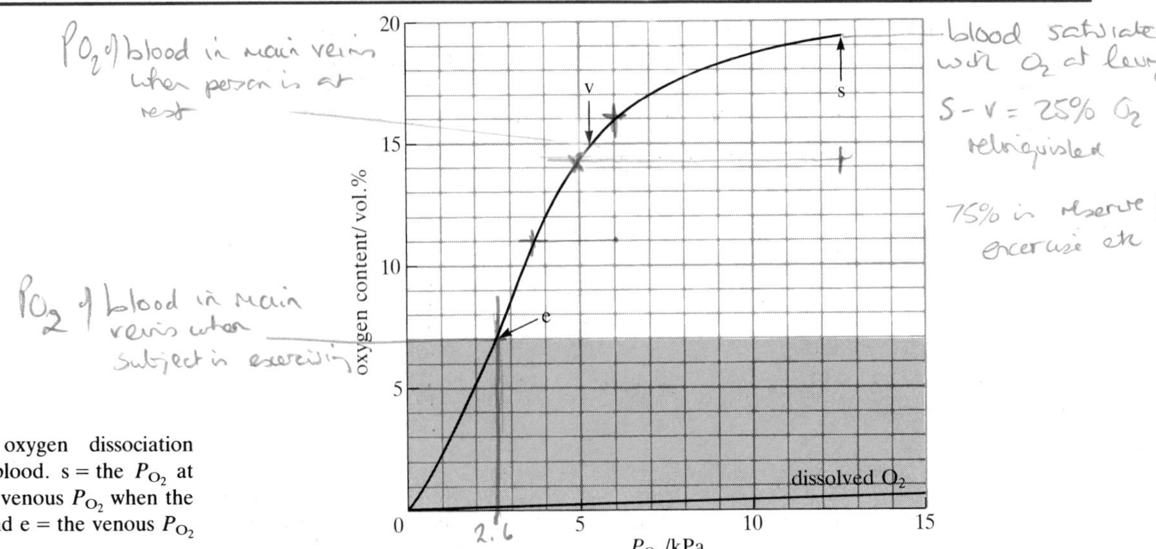

Figure 7.21 An oxygen dissociation curve for human blood. s = the P_{O_2} at saturation, v = the venous P_{O_2} when the subject is at rest and e = the venous P_{O_2} during exercise.

Because actively respiring tissues use up a great deal of oxygen, the P_{O_2} at the capillaries drops markedly during exercise, and will drop lower and lower with more intense exercise, down to a minimum of approximately 2.6 kPa (point e). This lower P_{O_2} at the tissues greatly encourages the haemoglobin to unload its oxygen, particularly because this shift in the P_{O_2} relates to a very steep portion of the dissociation curve (see Figure 7.21), where a small drop in P_{O_2} at the capillaries will result in a relatively large increase in oxygen turnover to the tissues. A drop in the P_{O_2} from 12.6 kPa at point s (pulmonary blood) to 5.3 kPa at point v (tissue capillary blood) yields about 5 vol. % turnover, whereas a relatively small decrease in venous P_{O_2}, to an exercise level of 2.6 kPa (point e), more than doubles turnover (to 13 vol. %). (The term vol. % was explained in Chapter 6, Section 6.4).

Therefore, a drop in the P_{O_2} at the tissues, even if modest, can increase the proportion of bound oxygen delivered to the tissues, and over three-quarters of the total oxygen load can be delivered to the muscles. Incidentally, this increased use of the 'venous reserve' of oxygen increases the oxygen loading at the lungs because the diffusion gradient across the pulmonary epithelium is made steeper, even if alveolar P_{O_2} is low.

Variations in the steepness of the dissociation curve also represent a great advantage for both highland residents and visitors from the lowlands. Consider arterial blood at an altitude of say 4 500 metres: it probably has a P_{O_2} of about 6 kPa and an oxygen content of approximately 16 vol. %. At sea-level, the P_{O_2} in the arteries would be at the 'normal' value of about 12.6 kPa, well up on the 'plateau' of the oxygen dissociation curve, corresponding to an oxygen content of 19 vol. %. Suppose that the oxygen demands of the tissues of lowlander and highlander are identical: say both require an oxygen turnover of 5 vol. %. You can see from Figure 7.21 that for the lowlander this would require a very low P_{O_2} in the tissue capillaries, 4.8 kPa or less, which is a drop of about 8 kPa in the P_{O_2} between arteries and capillaries. In the highlander, the same amount of oxygen would be delivered to the tissues if the P_{O_2} in the capillaries was 3.7 kPa, which represents an arterio–capillary drop in P_{O_2} of only 2.3 kPa. Simply because of the properties of haemoglobin, reflected in the shape of the oxygen dissociation curve, the highlander can deliver the same amount of oxygen to tissues as the lowlander for a much smaller drop in P_{O_2} at the capillaries. Remember, a low P_{O_2} at the capillaries adversely affects the rate of diffusion of oxygen to the tissues.

Note that even when there is a pronounced demand for oxygen and the supply is diminished—conditions equivalent to the final dash to the summit by an intrepid mountaineer—the P_{O_2} at the capillaries does not fall below a certain irreducible minimum of approximately 2.6 kPa, which means that more than one-quarter of the oxygen bound to haemoglobin remains permanently inaccessible (see the shaded portion of Figure 7.21).

The responses we have considered so far—increased ventilation and increased cardiac output linked to a widespread peripheral vasodilation—occur immediately when humans are exposed to high altitude. During the weeks or months of acclimatization to high altitudes, these initial responses are replaced by other adjustments that are less expensive in physiological terms, in particular to the haemoglobin content of the blood and the affinity of the pigment for oxygen.

When humans are first exposed to high altitudes, red blood cells stored mainly in the spleen are released, thereby increasing the haematocrit. The capacity of the blood to carry oxygen is thus increased *immediately*, but within 12 hours the rate at which red blood cells are formed also begins to increase, probably mediated by a hormone (erythropoietin) which acts on 'stem' cells in the bone marrow. The haematocrit may not reach its peak until many months after the subject is exposed to high altitude. The oxygen-carrying capacity of the blood of resident highlanders is also higher than it is in resident lowlanders. In both resident highlanders and visitors from the lowlands, the extra haemoglobin partly compensates for the reduction in the percentage saturation of the blood at high altitude. At Morococha (4 500 m), a small Bolivian village, the saturation of arterial blood is 73% compared with about 97% in lowlanders. However, the increased haematocrit of Morochan residents ensures that the oxygen content of their arterial blood is 20.7 vol. %, much the same as that of lowlanders. This increased oxygen capacity of blood at high altitude, together with the fact that the haemoglobin is operating on the steep part of the dissociation curve, means that high-altitude residents and acclimatized visitors can maintain an adequate supply of oxygen to the tissues without drastically lowering the P_{O_2} at the capillaries, and the proportion of bound oxygen utilized by respiring tissues is much the same (about 25%) in both groups. Hence, both can utilize the 'reserve' oxygen store when the demand for oxygen is increased during exercise. Indeed, the physiological adjustments described are so effective that when the work capacity of highlanders or acclimatized lowlanders is measured at high altitudes, by running to the point of exhaustion on a treadmill, it is found to be superior to that of a comparable group of lowlanders similarly tested in their own low-altitude environment!

Keep it in mind that such physiological prowess depends on a diverse range of important adjustments, and that we have considered only changes to the blood (i.e. the haematological changes). Some adjustments are at the tissue or biochemical level; for example, resident highlanders have tissues that are more closely packed with capillaries, and the important respiratory enzymes (e.g. cytochrome oxidase) are present in the mitochondria in greater amounts.

A second major adjustment people make during acclimatization to high altitude is an apparent lowering of the affinity of haemoglobin for oxygen, that is a small shift in the oxygen dissociation curve to the right. It seems that this is a result of the increased level of 2,3-DPG that occurs in red blood cells when people are exposed to high altitude; the level is elevated by as much as 10% within 24 hours of exposure to an altitude of 3 000 metres.

◇ Would you expect a decrease in the affinity of haemoglobin for oxygen at high altitude to have a beneficial effect on the uptake of oxygen by pulmonary blood or a beneficial effect on the delivery of oxygen to the tissues?

◆ The affinity of haemoglobin is of importance at two sites: (a) at the site of oxygen uptake (the lungs), where reduced affinity would be expected to have the disadvantage of reducing the saturation of pulmonary blood, and (b) at the tissues, where the delivery of oxygen might be enhanced if the affinity of the haemoglobin were reduced. You might suggest that, at a particular P_{O_2} in the tissues, a greater proportion of bound oxygen might be released if the affinity of the haemoglobin were lowered, but this gain must surely be reversed by the significant reduction in the extent of loading of the pulmonary blood. What in the end is most important is the *amount* of oxygen available to the tissues, not the P_{O_2} at which it is delivered.

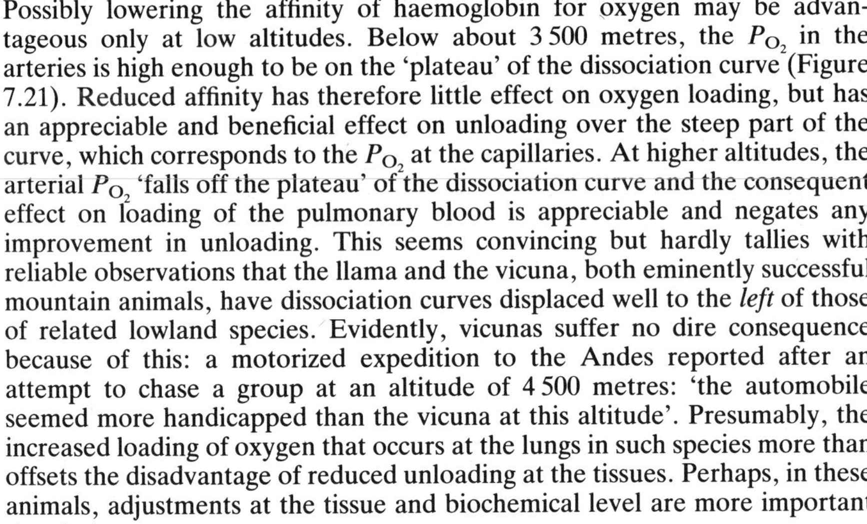

Possibly lowering the affinity of haemoglobin for oxygen may be advantageous only at low altitudes. Below about 3 500 metres, the P_{O_2} in the arteries is high enough to be on the 'plateau' of the dissociation curve (Figure 7.21). Reduced affinity has therefore little effect on oxygen loading, but has an appreciable and beneficial effect on unloading over the steep part of the curve, which corresponds to the P_{O_2} at the capillaries. At higher altitudes, the arterial P_{O_2} 'falls off the plateau' of the dissociation curve and the consequent effect on loading of the pulmonary blood is appreciable and negates any improvement in unloading. This seems convincing but hardly tallies with reliable observations that the llama and the vicuna, both eminently successful mountain animals, have dissociation curves displaced well to the *left* of those of related lowland species. Evidently, vicunas suffer no dire consequence because of this: a motorized expedition to the Andes reported after an attempt to chase a group at an altitude of 4 500 metres: 'the automobile seemed more handicapped than the vicuna at this altitude'. Presumably, the increased loading of oxygen that occurs at the lungs in such species more than offsets the disadvantage of reduced unloading at the tissues. Perhaps, in these animals, adjustments at the tissue and biochemical level are more important than in humans.

Summary of Section 7.8

At high altitude, the partial pressure of oxygen in inspired air is reduced, leading to arterial hypoxia (low P_{O_2} in blood). Hypoxia stimulates peripheral chemoreceptors, but initially the increase in the rate of breathing is limited as the resulting fall in P_{CO_2} leads to reduced respiratory drive from central chemoreceptors. This conflict is resolved on gradual exposure to altitude by the secretion of CSF containing a lower concentration of bicarbonate ions, so that it has a normal pH at reduced P_{CO_2}. The 'hypoxia drive' is then fully expressed. Despite this increase in breathing, the P_{O_2} in the arterial blood of highlanders can be up to 50% lower than that of lowlanders. However, the oxygen content of the arterial blood of acclimatized individuals is raised, because the concentration of red blood cells can more than double. Because their haemoglobin is functioning over a steeper part of the dissociation curve, highlanders deliver oxygen to the tissues with only a small drop in P_{O_2} between arteries and veins. The net effect of acclimatization is that the decline in the P_{O_2} between inspired air and the blood in the capillary beds (i.e. the P_{O_2} gradient) is much less steep than normal, which means that, even when the P_{O_2} in the inspired air is low, the P_{O_2} at the tissue capillaries is high enough to allow adequate amounts of oxygen to diffuse into mitochondria.

The dissociation curve of the blood haemoglobin of high-altitude resident humans and acclimatized visitors is shifted slightly to the right (the level of 2,3-DPG is elevated), which means that more oxygen can be unloaded at the tissues, possibly at the expense of loading up with oxygen at the lungs. At high altitudes (above 3 500 metres), the reduced affinity seriously threatens the haemoglobin saturation at the lungs and many indigenous high-altitude animals display a curve shifted to the left.

Now attempt Question 6 on p. 241.

7.9 RESPONSES TO OXYGEN DEFICIENCY

You might be tempted to believe that the role of oxygen in metabolic respiration is such a crucial one that no animal can live in the complete absence of oxygen. In fact, the great majority of animals are aerobic but many of these are surprisingly tolerant of shortage of oxygen, and indeed we shall see that, unlike humans, some can thrive for protracted periods in so-called anoxic conditions, where oxygen is absent, or is present in immeasurably small amounts.

In natural conditions, the maximum P_{O_2} in the environment is approximately 20 kPa but, particularly in water, it is commonly a good deal lower. You might think that lowered environmental partial pressures of oxygen would lead to reduced metabolic rates, and this is what happens in many oxygen-dependent animals, collectively called *oxygen conformers*.

In such animals, the consumption of oxygen varies according to the amount of oxygen in the environment over a wide range of partial pressures of oxygen (from around 1 kPa to around 20 kPa, and sometimes beyond). Many large, sluggish invertebrates are oxygen conformers, probably because the rates of oxygen diffusion into and within these animals influence the supply of oxygen to the tissues most profoundly. So, too, are many aquatic insects, nearly all parasitic worms (e.g. tapeworms), some crustaceans (some of them by no means 'primitive' or lethargic) and even a few vertebrates (e.g. toadfishes). Free-living marine animals near the surface or in shallow seas are very rarely faced with a P_{O_2} substantially below that in the air at saturation because convection currents and temperature gradients normally ensure adequate mixing of the water. The availability of oxygen in freshwater habitats may be much more limited, especially in swamps, ditches and ponds rich in pollutants or organic debris. Bacterial action may significantly deplete the supply of oxygen and raise the level of dissolved carbon dioxide; in such conditions, oxygen conformers are often slow and moribund.

In contrast, when environmental P_{O_2} changes, other animals *regulate* their consumption of oxygen, usually down to a certain *critical* P_{O_2} (or critical oxygen tension), which is usually well below 20 kPa; below this, oxygen consumption falls sharply with the falling P_{O_2} (Figure 7.22a) as the animals then behave as conformers. A good example of a regulator is the goldfish: at 15 °C, the uptake of oxygen is constant over a range of partial pressures of oxygen down to a critical P_{O_2} of 3 kPa (Figure 7.22b). In such an animal, it seems that the activities of the ventilatory system, circulatory system and blood are so modified when there is a mild shortage of oxygen that sufficient oxygen is supplied to the respiring tissues to support an unchanged metabolic rate. Other regulators include many aquatic insects, many crustaceans (including crayfish), most aquatic vertebrates and probably all terrestrial insects and vertebrates.

These concepts of 'regulation' and of 'conformity' in relation to environmental variables have a widespread application in physiology, and should remind

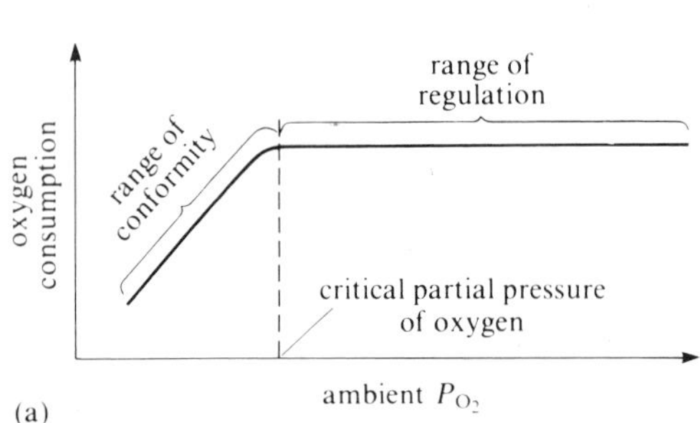

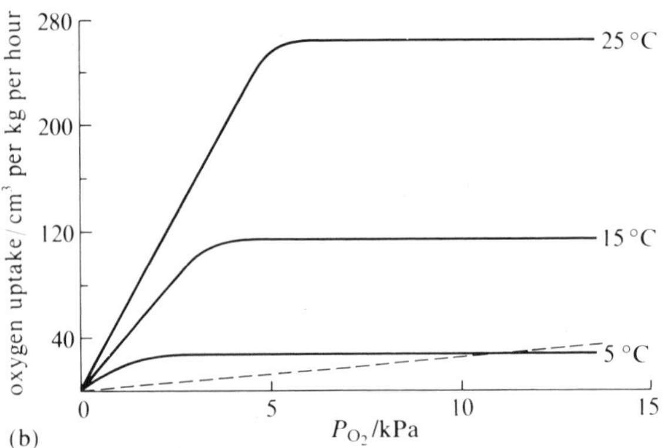

Figure 7.22 (a) An idealized view of oxygen consumption in a regulator. (b) The effect of temperature on oxygen uptake in a goldfish. The broken line shows the oxygen consumption of the toadfish (a conformer) at 20 °C.

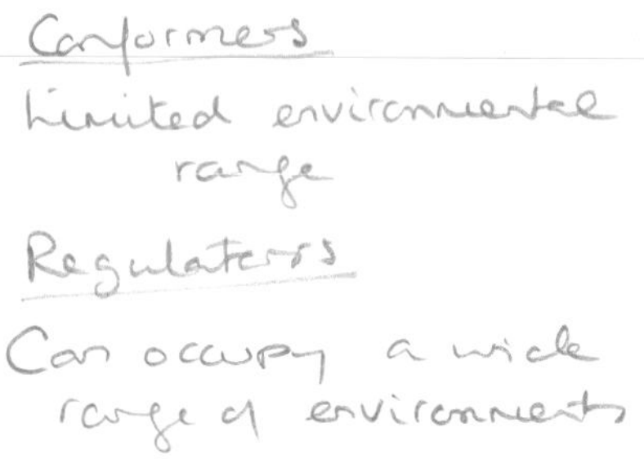

you of the different responses that animals make to changes in temperature; that is, poikilotherms are conformers and homoiotherms are regulators. For each of the major physical factors that vary in the environment (e.g. temperature, oxygen and salinity) we can recognize some animals that conform to external changes and therefore tolerate substantial internal variations. In natural conditions, such variations tend to be limited because conformers often have a limited environmental range. In contrast, regulators maintain a relatively constant internal condition by homeostatic controls, and any serious deviations (think of body temperature in mammals) have dire effects; such animals can occupy a wide range of environments. You should hardly need to be reminded that 'the real situation is more complicated', and most animals show elements of both patterns; but this division into conformers and regulators remains a useful way of distinguishing the different strategies that animals adopt under environmental stress.

Three other points of general physiological interest emerge if we look at what factors affect the value of the critical P_{O_2} in regulators.

First, an animal's response to changes in one environmental variable is normally affected by other internal and external factors. Because goldfish are poikilotherms, their level of oxygen consumption is likely to vary with the ambient temperature (see Figure 7.22b). Temperature also often influences the critical P_{O_2}, which in goldfish, trout and crayfish decreases as the ambient temperature falls. An 'internal' factor that changes the demand for oxygen in the tissues is the level of activity and, not surprisingly, the relationship between oxygen consumption and the ambient P_{O_2} is very different in stimulated, active individuals and in undisturbed, 'resting' specimens. Not only is the oxygen consumption of active fishes at a variety of partial pressures of oxygen always significantly higher than that of more sluggish ones, but the critical P_{O_2} is normally elevated by activity. The P_{O_2} at which fishes become conformers is normally lower in resting than in active individuals. Indeed, many fishes are regulators at rest, but become conformers when exercising vigorously.

A second major point is that differences between species in their response to lack of oxygen are often correlated with variations in the oxygen levels normally encountered by each species in its natural habitat. Species that live in oxygen-depleted habitats often display a greater range of regulation than related species that are found in environments that are rich in oxygen.

For example, nymphs of one species of mayfly live in well-oxygenated, swift-flowing streams whereas the nymphs of another species come from ponds thought to become depleted of oxygen at night.

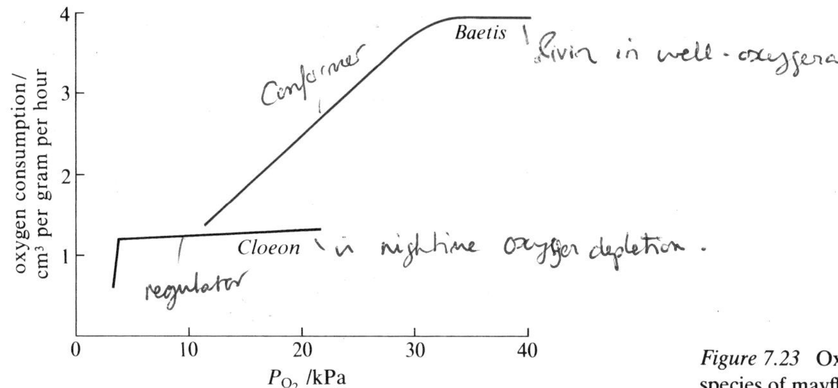

Figure 7.23 Oxygen consumption in two species of mayfly nymph.

◇ Look at Figure 7.23. Which species of mayfly occurs in which habitat?

◆ *Cloeon* is a good regulator, as one would predict of a species periodically exposed to a low P_{O_2} in ponds. *Baetis* lives in streams and has a much higher critical oxygen tension and is a conformer over most of the range of partial pressures of oxygen.

A similar logic explains why some marine crustaceans that are unlikely to be subject to oxygen depletion are conformers whereas the closely-related freshwater crayfish, which is occasionally found in ponds or slow-moving streams, is a regulator. Once again, we are reminded that the physiological properties and responses of animals are often related to the type of environment they live in.

Understandably, most animals are aerobic and intolerant of anoxia. However, some invertebrates—cnidarians, annelids, molluscs and especially nematodes and tapeworms—clearly have marked anaerobic capacities because many of them can survive anoxia for periods of one day to many weeks. Some species of sea-anemone survive for more than a day without oxygen; some annelid worms for over ten weeks. In some species, tolerance of anoxia may be important under natural conditions. This may be true of intertidal species that breathe dissolved oxygen (e.g. bivalves) and parasitic worms (e.g. tapeworms) that live in the intestine, where oxygen may be in short supply or even absent, and of some annelids that live in anaerobic freshwater muds. Among the vertebrates, carp (*Cyprinus carpio*) are able to withstand anoxia under ice in European fishponds for as long as two or three months. Some fish from freshwater swamps in Africa also appear well able to survive prolonged anoxia but most of these breathe air.

During oxygen deprivation many animals accumulate lactate by breaking down pyruvate. The breakdown of glucose to lactate yields two ATP molecules and leads to the regeneration of NAD^+, which can act as the electron acceptor necessary for one of the earlier steps in glycolysis. However, the breakdown of glucose to lactate via the glycolytic pathway cannot go on indefinitely, because the accumulation of this end-product poses acute physiological problems. Many other animals use non-glycolytic pathways, as is evident from the curious array of metabolic end-products released during metabolism in the absence of oxygen (e.g. alanine and proprionate).

Aquatic turtles certainly must depend on anaerobic glycolysis extensively to support metabolism during held-breath dives, which may last up to several days. Many turtles survive for between 6 and 33 hours when breathing pure nitrogen at 22 °C. Possibly some of these species may be able to survive indefinitely without oxygen at low temperatures. Their capacity for glycolysis is very high and unusually high levels of carbon dioxide and lactate are tolerated in the blood and tissues (the blood pH may be as low as 6.6), and for much of the dive all the energy generated appears to be derived from anaerobic glycolysis. It can come as a surprise to physiologists who are mainly concerned with rats and humans that the functioning of the heart and central nervous system of turtles appears not to be dependent on oxygen.

Most other vertebrates that dive are more dependent on oxygen, and this leads us to consider the concept of *oxygen debt.* Normally, lactate is retained during brief periods of oxygen deprivation but when the supply of oxygen to the tissues is restored, say, at the end of a dive, or (for humans) during recovery from severe exercise, the lactate can either be oxidized by being shunted into the TCA cycle (an aerobic process which yields ATP) or it can be converted back to replenish the stockpile of glycogen from which it was first derived (via the process of gluconeogenesis that utilizes ATP). Both routes require oxygen, either directly or indirectly, and so, in the recovery phase, oxygen is needed in excess of the normal 'maintenance' requirement. This elevated, post-anaerobic consumption of oxygen continues until the accumulated lactate has been dissipated. During a bout of anaerobic metabolism associated with exercise or with diving, the accumulation of lactate establishes a future requirement for oxygen; that is, an oxygen debt is incurred, which is repaid afterwards when a larger than normal amount of oxygen is made available.

Figure 7.24a shows lactate levels (red line) in the arterial blood of a seal before, during and after a dive of 18 minutes. During the dive, lactate is largely confined to the muscles that produce it but when the animal surfaces, the store is released as the circulation is restored to the muscle capillary beds.

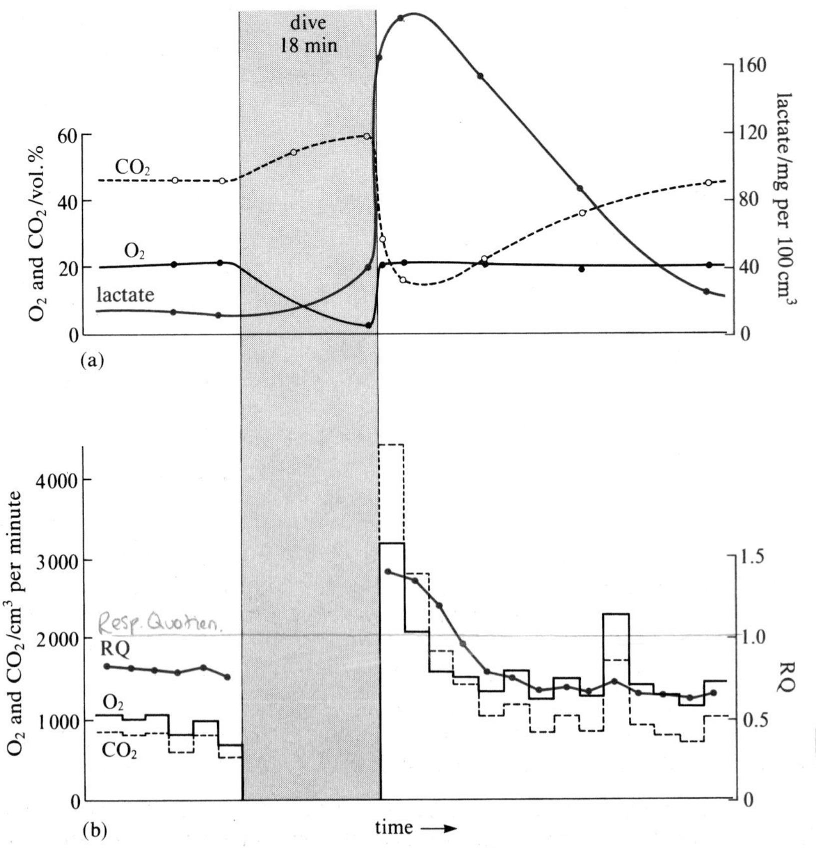

Figure 7.24 (a) The amounts of lactate (red), oxygen and carbon dioxide in the arterial blood of a diving seal. (b) The RQ (red) and the consumption of oxygen and the release of carbon dioxide over the same period, expressed as cubic centimetres of gas consumed or released per minute.

Figure 7.24a also shows the amounts of oxygen and carbon dioxide present in the blood over the same period.

Figure 7.24b shows the level of oxygen uptake (continuous line) and carbon dioxide release (dashed line) in the animal before and after the dive. These values can be used to calculate the respiratory quotient (RQ) (shown in red), which we shall discuss shortly. For the moment, note that, when the arterial level of lactate acid is high after a dive, there is a marked increase in the consumption of oxygen.

In diving mammals (unlike turtles), the heart and central nervous system cannot tolerate anoxia, and profound cardiovascular changes accompany diving. These responses ensure that a supply of oxygen is maintained to susceptible tissues, for example, heart, eyes, liver and brain. But even these tissues seem to be able to make a limited use of anaerobic mechanisms, and several enzymes are now known to be particularly important in stepping up an animal's capacity for glycolysis. One of them, lactate dehydrogenase (LDH), facilitates the conversion of pyruvate into lactate. LDH is a tetramer which can exist as various isoenzymes with varying proportions of two types of protein monomers. Of the two 'extreme' types, one (in which all the subunits are type M) can readily convert pyruvate to lactate, and this type is usually found in human skeletal muscle. The other (with all type H subunits) is found in human heart tissue and its main effect is to channel pyruvate into the TCA cycle, mainly because H_4 (all the H subunits) is inhibited by high levels of pyruvate. (H_4 also tends to promote the *reverse* reaction, i.e. lactate $\longrightarrow$ pyruvate.) Three other forms of isoenzyme exist, M_3H, M_2H_2 and M_1H_3, and the general pattern in terrestrial animals is that the M form predominates in tissue with a marked potential for anaerobic metabolism.

◇ Look at Table 7.1, which gives the proportions of the five isoenzymic forms of LDH in the brain and heart of a seal, and by way of contrast, for a terrestrial species, a sheep. What are the most striking differences between the two species?

Table 7.1 Types of isoenzyme as a percentage of the total LDH in the tissues of a seal and a sheep.

		Isoenzyme type				
		H_4	H_3M	H_2M_2	HM_3	M_4
Seal	brain	29 ± 8	40 ± 7	23 ± 6	5 ± 3	3 ± 4
	heart	33 ± 5	49 ± 5	12 ± 5	4 ± 2	2 ± 4
Sheep	brain	39 ± 4	21 ± 1	36 ± 1	4 ± 3	—
	heart	88 ± 8	12 ± 7	—	—	—

◆ In general, the diving mammal's heart and brain have higher proportions of isoenzymes that contain subunit M (normally associated with anaerobic tissue), and lower proportions of the tetramer, H_4, than the sheep's tissues.

This all suggests that although the supply of blood to the heart and brain appears to be maintained during a dive, diving mammals have a remarkable ability to tolerate a measure of anaerobic metabolism in these tissues, perhaps towards the end of a dive when oxygen tensions in the blood are low.

Several other enzymes are known to be present in increased amounts in all muscle tissue capable of anaerobic metabolism, especially enzymes such as glycogen phosphorylase and phosphofructokinase which exercise a crucial control of key reactions in the glycolytic chain. These strategically positioned enzymes are particularly important in the muscles of diving mammals, and such muscles can rapidly switch to a high rate of glycolysis and thus become independent of oxygen.

One other point relating to metabolism emerges from Figure 7.24. This shows the respiratory quotient (RQ) before and after a dive. The RQ is calculated by dividing the volume of carbon dioxide produced by the volume of oxygen consumed at standard temperature and pressure of gas. When glucose is oxidized, equal volumes of carbon dioxide and oxygen are involved, so the RQ is 1. Oxidation of lipids produces a volume of carbon dioxide that is smaller than the volume of oxygen consumed, so the RQ for lipids is about 0.7. In non-diving animals an RQ of 0.7 (see Figure 7.24) therefore suggests that *fat* is the preferred fuel. During and immediately after bouts of anaerobic metabolism, RQs are understandably abnormal and values may be unusually high when the animal surfaces and the accumulated carbon dioxide is 'blown off'. Diving mammals have the interesting habit of switching between glycogen breakdown (during anaerobic periods of the dive) and the oxidation of fatty acids during aerobic periods of migration and when swimming on the surface. So, when the animal is not diving, glycolysis is blocked and the major stores of glycogen are thus preserved or 'spared'. During aerobic activity, this turning-off of glycolysis appears to arise from an increase in the level of citrate (an important intermediate in the TCA cycle) which occurs when fatty acid oxidation becomes established (see Figure 7.25). The high level of citrate inhibits phosphofructokinase (PFK) and also pyruvate kinase, an enzyme that controls another important step in glycolysis. During diving, the oxidation of fatty acids is stopped. It is possible that lactate inhibits lipase which normally breaks down triglyceride into fatty acids and glycerol before oxidation. So diving involves a very tight metabolic control which conservatively parcels out the different types of available fuel.

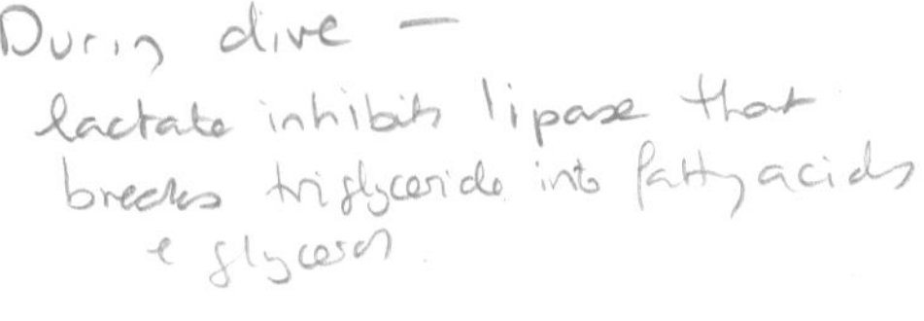

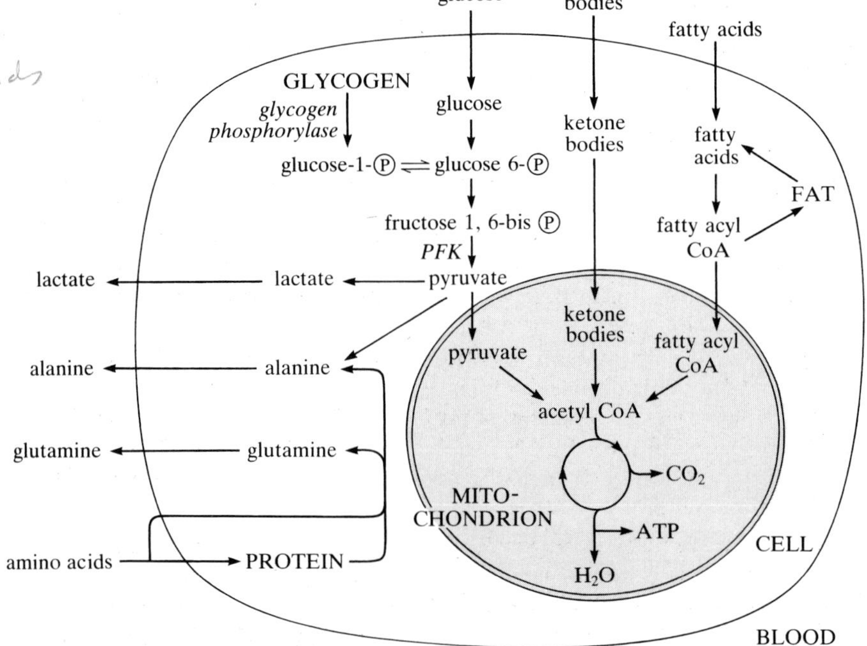

Figure 7.25 Summary of metabolic processes in a muscle cell. Exchange of metabolites between the muscle cell and the blood are also shown.

Summary of Section 7.9

The metabolic rates of animals that are oxygen conformers vary directly with the amount of oxygen in the environment over a wide range of partial pressures of oxygen. In oxygen regulators, changes in the circulatory and respiratory systems (e.g. increased ventilation) compensate for a decline in the environmental P_{O_2} and oxygen consumption remains constant. Below a critical P_{O_2}, regulators behave like conformers. Species that are naturally subject to rapid and extensive fluctuations in environmental P_{O_2} tend to be better regulators than species exposed to steady, high levels of oxygen.

Although aerobic processes yield more energy, many non-mammals can tolerate short periods without oxygen or at an extremely low P_{O_2}. Such animals often rely on anaerobic glycolysis, and in some turtles even the heart and nervous tissue can function normally without oxygen for a long time. Both turtles and mammals accumulate lactate during dives and, because much of this eventually must be converted back to glycogen by aerobic processes, an oxygen debt is incurred. Tissues of diving mammals have a huge anaerobic potential, which can be rapidly activated via key control enzymes in the glycolytic sequence, such as phosphofructokinase. In diving mammals, heart tissue (and perhaps the brain too) may tolerate elevated levels of lactate produced by the higher-than-usual amounts of LDH containing M subunits. The metabolism of some diving mammals oscillates between breaking down glycogen during anaerobic periods and breaking down triglyceride into fatty acids during aerobic periods of surface swimming.

Now attempt Question 7 on p. 241.

SUMMARY OF CHAPTER 7

The basic respiratory organs for gas exchange between blood systems and the atmosphere are lungs. In all vertebrates these are ventilated, but this is not so in invertebrates, which rely on diffusion of gases from the air. In vertebrate air-breathers the oxygen demand is high and the lungs are ventilated regularly. In mammals, amphibians and reptiles, ventilation is tidal—the same airways conduct air both to and from the alveoli. Birds have a much more complex, and probably more efficient, respiratory system in which there is a constant, unidirectional, flow of air past the gas exchange surfaces.

A major problem in aerial respiration is water loss. Many vertebrates require rapid gas exchange with minimum water loss, a requirement which is met if the respiratory surfaces are exposed to the minimum amount of air required to ensure adequate gas exchange. Invaginated lungs do keep water loss low, but they are not completely flushed out with fresh air during breathing. As a result the P_{CO_2} of the pulmonary air and the blood can increase to a high level. This may be one reason why, in air-breathers, respiration is normally regulated via the primary stimulus of carbon dioxide. By contrast most aquatic animals (fish, for example) are more sensitive to oxygen. This is probably linked to the high solubility of carbon dioxide in water. The blood of aquatic animals has a P_{CO_2} close to the ambient level and because the P_{CO_2} equilibrates with the medium so effectively, it is never substantially raised. The level of carbon dioxide is not an appropriate indicator of respiratory demand.

High altitude imposes an environmental stress on individual animals as a result of the shortage of oxygen. This hypoxia usually induces an increase in the oxygen capacity of the blood and this improves the delivery of oxygen to the tissues. The affinity of the haemoglobin for oxygen also changes as a result of a change in the intracellular level of organic phosphates. In human red blood cells, the level of ATP is low because anaerobic glycolysis is the main source of energy; the dominant phosphate is 2,3-DPG, formed from an intermediate in the glycolytic pathway. During hypoxia, the level of 2,3-DPG increases in human red blood cells and this may reduce the affinity of haemoglobin for oxygen, mainly via an allosteric effect. A shift of the oxygen dissociation curve to the right is evident within 24 h, when humans are exposed to high altitude.

OBJECTIVES FOR CHAPTER 7

Now that you have completed this chapter you should be able to:

7.1 Define and use, or recognize definitions and applications of each of the terms printed in **bold** in the text.

7.2 Describe the structure of the mammalian lung, including the airways and pulmonary circulation. (*Questions 1 and 3*)

7.3 Recall the principal lung volumes in humans, and explain factors affecting the effort required to breathe. (*Questions 1, 2 and 3*)

7.4 Describe the respiratory system of birds and how it differs from that of mammals. (*Question 4*)

7.5 Describe the operation of fish gills and explain the importance of counter-currents in gas exchange. (*Question 5*)

7.6 Explain why mammalian respiration is more sensitive to changes in P_{CO_2} than P_{O_2} and describe the function of central and peripheral chemoreceptors.

7.7 Explain why acute exposure to high altitude is more dangerous than a gradual ascent and describe the various changes associated with acclimatization to low P_{O_2}. (*Question 6*)

7.8 Distinguish between oxygen conformers and regulators. (*Question 7*)

7.9 Describe the changes in a diving mammal which allow it to tolerate long periods without breathing.

Bradycardia = (marked) slowing of heart rate

Do AC1484 Band 2
- Physiological Stress

QUESTIONS FOR CHAPTER 7

Question 1 (*Objectives 7.2* and *7.3*) Which of statements (a)–(h) are *true* and which are *false*?

(a) The trachea, the bronchi and the bronchioles are important sites of respiratory gas exchange.

(b) During normal human breathing, expiration leaves a small residual volume of about 500 cm^3 in the lungs.

(c) The lungs are emptied of gas and completely refilled with fresh air during the ventilatory cycle.

(d) The alveoli contain a mixture of gases that has a lower oxygen content than atmospheric air.

(e) The mean pressure in the pulmonary artery is substantially lower than the mean pressure in the aorta in the systemic circulation.

(f) The volume of blood passing through the pulmonary circuit per unit time is substantially lower than the volume of blood passing through the systemic circulation per unit time.

(g) During exercise, there is an alarming tendency for fluid from the plasma to enter the alveoli because of the sharply increased pressure in the pulmonary capillaries.

(h) One advantage of organs for the exchange of respiratory gases in terrestrial animals often being invaginated structures is that water loss by evaporation from the respiratory surface is thereby minimized.

Question 2 (*Objective 7.3*) During the human ventilatory cycle, lung volumes can be measured if a subject breathes into a spirometer which links the movement of air into and out of the lung with the upward and downward deflection of a writing point on a revolving drum. Rhythmic fluctuations are recorded when the subject is breathing gently, as shown in the early part of the trace in Figure 7.26. After twelve normal breathing cycles, the subject inhales to the maximum capacity, then exhales to the maximum extent, before resuming normal (resting) ventilation.

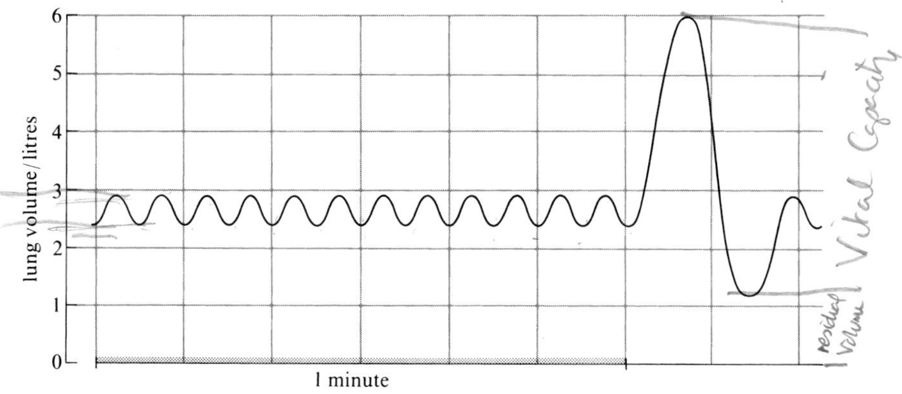

Figure 7.26 Breathing movements as recorded from a spirometer.

(a) How much air is left in the lungs of this subject at the end of normal exhalation and what term describes this?

(b) What is the tidal volume of the ventilatory cycle of this subject?

(c) What is the maximum possible tidal volume when the subject is taking forced breaths?

(d) What is the volume breathed per minute when the subject is breathing normally?

(e) Using the value calculated in (d), estimate the volume for *fresh* air moved by the lungs per minute that is available for gas exchange.

(f) Is it possible to estimate how much oxygen this subject is using?

(g) During exercise, how do you suppose the extra ventilation required might be achieved?

Question 3 (*Objectives 7.2 and 7.3*) Identify which of statements (a)–(g) referring to the ventilatory cycle in a resting human are *true* and place these correct statements in sequence, starting with inspiration.

(a) The contraction of the internal intercostal muscles now causes the ribs to be forced downwards, while contraction of the abdominal muscles forces the diaphragm upwards.

(b) Immediately after the contraction of the alveoli and exhalation of the stale air, fresh air passes into each alveolus.

(c) The thoracic cavity is now expanded by contraction of the diaphragm (which becomes less concave in side view) and the ribcage is lifted by contraction of the external intercostals.

(d) The rate of discharge of impulses from motor nerves supplying the external intercostals increases at this point. (The increase becomes more marked over the next second or so.)

(e) The lungs are now inflated by the force-pump action of the buccal cavity (the mouth), which forces the air which is at high pressure in the buccal cavity down into the lungs.

(f) The contraction of muscular elements within the lung tissue now causes the lung to recoil and consequently its volume is reduced.

(g) The relaxation of the diaphragm and external intercostal muscles, together with elastic forces that bring about a reduction in lung volume, now lead to a change in the amount of air in the lungs of about 500 cm^3.

Question 4 (*Objective 7.4*) Which of statements (a)–(f) are *true* differences between the respiratory systems of birds and mammals and which statements are *false*?

(a) Unlike the mammalian lung, the avian lung undergoes relatively little change in volume over the ventilatory cycle.

(b) In mammals, movement of the ribs in relation to the backbone helps to bring about inspiration, but birds lack ribs and ventilation is brought about by the contraction and relaxation of muscular components in the walls of the airsacs.

(c) As in birds, ventilation in mammals is brought about by air being sucked into the respiratory system; air is expelled by positive pressure.

(d) In mammals, air moves in a tidal manner into and out of the blind-ended alveoli, but, in birds, there is a constant unidirectional flow of air through the parabronchi.

(e) The surfaces of the mammalian lung are sites of evaporation and therefore of water and heat loss but the air-capillaries of the avian lung are unlikely to be sites of evaporative water loss.

(f) In birds subjected to heat stress, an increased volume of air is passed into and out of the airsacs rather than the lungs, but, in mammals, increased breathing (i.e. panting) may result in increased *pulmonary* ventilation.

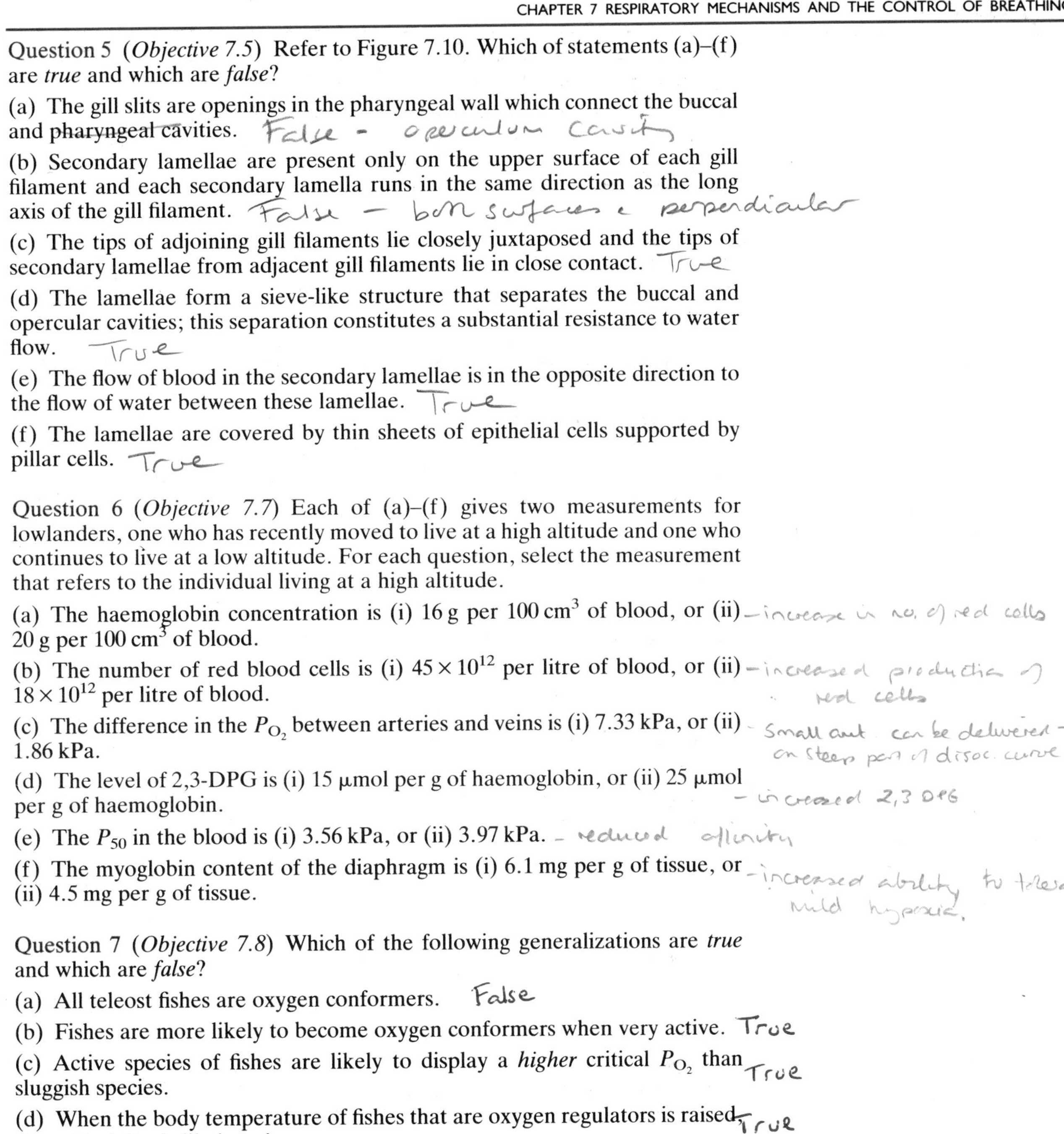

Question 5 (*Objective 7.5*) Refer to Figure 7.10. Which of statements (a)–(f) are *true* and which are *false*?

(a) The gill slits are openings in the pharyngeal wall which connect the buccal and pharyngeal cavities. False - operculum cavity

(b) Secondary lamellae are present only on the upper surface of each gill filament and each secondary lamella runs in the same direction as the long axis of the gill filament. False – both surfaces & perpendicular

(c) The tips of adjoining gill filaments lie closely juxtaposed and the tips of secondary lamellae from adjacent gill filaments lie in close contact. True

(d) The lamellae form a sieve-like structure that separates the buccal and opercular cavities; this separation constitutes a substantial resistance to water flow. True

(e) The flow of blood in the secondary lamellae is in the opposite direction to the flow of water between these lamellae. True

(f) The lamellae are covered by thin sheets of epithelial cells supported by pillar cells. True

Question 6 (*Objective 7.7*) Each of (a)–(f) gives two measurements for lowlanders, one who has recently moved to live at a high altitude and one who continues to live at a low altitude. For each question, select the measurement that refers to the individual living at a high altitude.

(a) The haemoglobin concentration is (i) 16 g per 100 cm^3 of blood, or (ii) 20 g per 100 cm^3 of blood. – increase in no. of red cells

(b) The number of red blood cells is (i) 45×10^{12} per litre of blood, or (ii) 18×10^{12} per litre of blood. – increased production of red cells

(c) The difference in the P_{O_2} between arteries and veins is (i) 7.33 kPa, or (ii) 1.86 kPa. – small amt. can be delivered – on steep part of dissoc. curve

(d) The level of 2,3-DPG is (i) 15 μmol per g of haemoglobin, or (ii) 25 μmol per g of haemoglobin. – increased 2,3 DPG

(e) The P_{50} in the blood is (i) 3.56 kPa, or (ii) 3.97 kPa. – reduced affinity

(f) The myoglobin content of the diaphragm is (i) 6.1 mg per g of tissue, or (ii) 4.5 mg per g of tissue. – increased ability to tolerate mild hypoxia.

Question 7 (*Objective 7.8*) Which of the following generalizations are *true* and which are *false*?

(a) All teleost fishes are oxygen conformers. False

(b) Fishes are more likely to become oxygen conformers when very active. True

(c) Active species of fishes are likely to display a *higher* critical P_{O_2} than sluggish species. True

(d) When the body temperature of fishes that are oxygen regulators is raised, the critical P_{O_2} is likely to increase. True

(e) Fishes that inhabit stagnant pools and slow-moving streams are less likely to be oxygen regulators than fishes that always inhabit fast-flowing streams. False

(f) When fishes are acclimatized to an environment with a low P_{O_2}, the critical P_{O_2} falls. True

FEEDING AND DIGESTION ◆ CHAPTER 8 ◆

8.1 FEEDING AND DIGESTION

All animals are heterotrophs, obtaining the carbon for their cellular components by taking in ready-made organic molecules from the environment. Unlike autotrophs (plants and chemosynthetic bacteria), they have limited ability to synthesize such molecules. Some heterotrophs, including many parasites and marine invertebrates, are **saprozoic**, absorbing relatively complex organic compounds through their body surfaces. In evolutionary terms, absorption of organic compounds from the environment may have preceded the methods of food intake now seen in animals.

Some of an animal's food provides the raw material that fuels cellular metabolism. The cellular metabolism provides the energy used in locomotion, excretion, the synthesis of new materials for body maintenance and growth, and the production of reproductive cells. But contrast the relative uniformity of cellular biochemical events with the diversity of foods utilized by different animals. The usual sources of food for animals are live plants, live animals, dead or decaying plants and animals, and micro-organisms such as bacteria and fungi. Within these general categories is a great variety of different types of food, and equally diverse types of feeding mechanism are found in the animals that make use of them. From your general knowledge, you would probably be able to list the feeding habits of a number of animals that reflect this diversity.

Animals that feed on relatively hard, resistant, food material have structural features such as biting jaws or teeth that crush food before it is swallowed. Other animals, such as aphids, feed on fluids and have remarkable structural features that enable them to pierce fluid-containing structures in plants and, in some cases, pump the fluid contents into their own bodies.

We start by considering the components of food and describe a variety of specialized feeding mechanisms. These demonstrate a relationship between the nature of the ingested food and the type of feeding mechanism used to capture it. We shall then look at the way in which the food is broken down and assimilated. Finally we consider the control of feeding, a subject already introduced in Chapter 4.

8.2 NUTRITIVE REQUIREMENTS

If the food of a number of quite different animals is analysed into its elementary constituents, most of it is found to consist of oxygen, carbon, hydrogen and nitrogen. These are the major elements that combine in various ways to form the three basic categories of organic material present in food: proteins, fats and carbohydrates. Besides these, other organic substances (such as vitamins), various inorganic substances (e.g. calcium and iron) and, of course, water must also be available, usually in the food.

8.2.1 Proteins

New proteins required for growth or for cell turnover are synthesized within the cells from amino acids derived largely from the protein in food. There are 20 or so amino acids found in animal proteins (although the total number of amino acids found in all types of living organism is over 2000, higher plants being a particularly rich source). These 20 amino acids are all of the L-configuration.

D-Amino acids are not found in the proteins of metazoan organisms. It was at one time believed that the D-amino acids had no importance in the cells of higher organisms, but it is now known that they are regularly present in small amounts in the tissues of at least insects, molluscs, annelids and mammals. Their role is not entirely clear, but certainly invertebrate and vertebrate animals all have a capacity to absorb D-amino acids from their food, albeit at a lower rate than the L-isomers. This is considered further in Section 8.6.3.

The precise number of the 20 naturally occurring amino acids required depends on the organism. Some species of bacteria, when provided with sufficient of a single amino acid in their growth medium, can make all the additional amino acids needed for incorporation into cell proteins. In contrast, in a rat a single amino acid is not sufficient—whatever the acid and however large the supply. If rats are maintained on a diet containing fats, carbohydrates and a single amino acid, growth is strongly retarded; all animals show marked signs of ill-health, and some die prematurely. For normal growth and development the diet must contain no less than *ten* specific amino acids. Since these ten are indispensable to normal growth in the rat, they are called **essential amino acids**. As the other amino acids can be synthesized by the animal from nitrogen-containing compounds, they are called non-essential. A list of the essential and non-essential amino acids for the rat is given in Table 8.1.

Table 8.1 The 20 naturally occurring L-amino acids classified by their ability to support growth in rats.

Essential	Non-essential
lysine	glycine
tryptophan	alanine
histidine	serine
phenylalanine	cysteine
leucine	tyrosine
isoleucine	aspartate
threonine	glutamate
methionine	proline
valine	hydroxyproline
arginine	citrulline

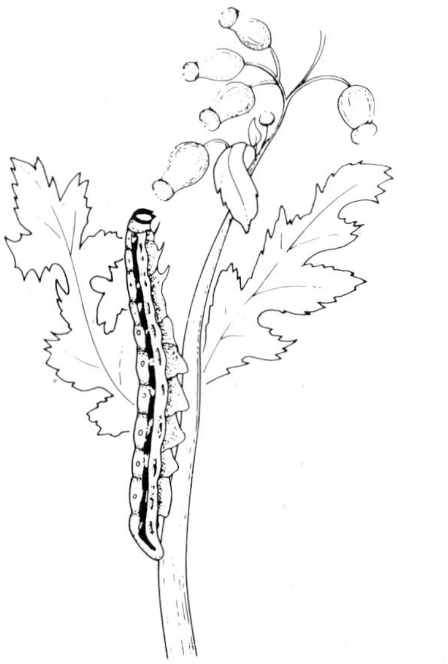

Why are some amino acids essential whereas others are not? A study of the results of the following experiments provides an explanation. These were two related investigations into the nutrition of an insect called the bollworm (Figure 8.1). In the first set of experiments, the requirements of the insect for amino acids and other nutrients were investigated, and in the second set, the metabolic fate of ingested, radioactively labelled glucose was studied.

The nutrient requirements of animals are normally determined by deletion experiments, which involve the omission (deletion) of a single nutrient from an otherwise complete diet. The complete diet of the bollworm (the control diet) consisted of 18 amino acids together with essential nutrients such as sucrose, mineral salts, water and a vitamin supply. Fat was not added because

Figure 8.1 The larva of the bollworm (*Heliothis zea*).

the bollworm can synthesize it. Deficient diets were prepared by the omission of a single amino acid from the control diet. Antimicrobial agents were included in all the prepared diets to prevent microbial growth. About 18 young bollworm larvae were fed on each diet, and tests were repeated two or four times. The nutritional adequacy of each diet was assessed by counting the number of the original larval population that reached the next stage of development (pupation).

The metabolic fate of ingested glucose can be followed by extraction and separation of radioactively labelled cellular chemicals some time after the ingestion of a meal containing radioactively labelled glucose. In a second experiment, starved larvae were fed on a test meal containing a small amount of labelled glucose. Labelled carbon from the glucose will be incorporated into the structure of newly synthesized cellular products—for example, amino acids.

Some time after feeding, the animals were killed and the amino acids of their tissues extracted by prolonged hydrolysis in hot hydrochloric acid. The amino acid content of the resulting extract was determined by using an amino acid analyser (which measured both labelled and unlabelled amino acids). The amount of radioactively labelled amino acids present in the extract was measured in a liquid scintillation counter.

Some of the results of both investigations are presented in Figure 8.2 and Table 8.2. Examine these carefully, and read the captions.

Table 8.2 Data from investigations of bollworm larvae: column 2 gives the amounts of each of the listed amino acids extracted from about 9 mg of larval tissue some time after the ingestion of [^{14}C]glucose. Column 3 gives the radioactivity of each of the amino acid fractions isolated from the same tissue. Column 4 expresses the radioactivity recorded in column 3 as a function of the amino acid content recorded in column 2; in other words, column 4 expresses the relative proportion of radioactively labelled amino acid in each of the seven listed amino acid fractions.

1 Amino acid missing from diet	2 Amino acid content/μmol	3 Radioactivity of amino acid	4 Radioactivity per micromole of amino acid
valine	0.319	none	0
leucine	0.348	none	0
isoleucine	0.218	none	0
threonine	0.233	2 049	8 794
glutamate	0.622	42 866	68 916
glycine	0.552	16 565	30 010
proline	0.260	17 262	66 392

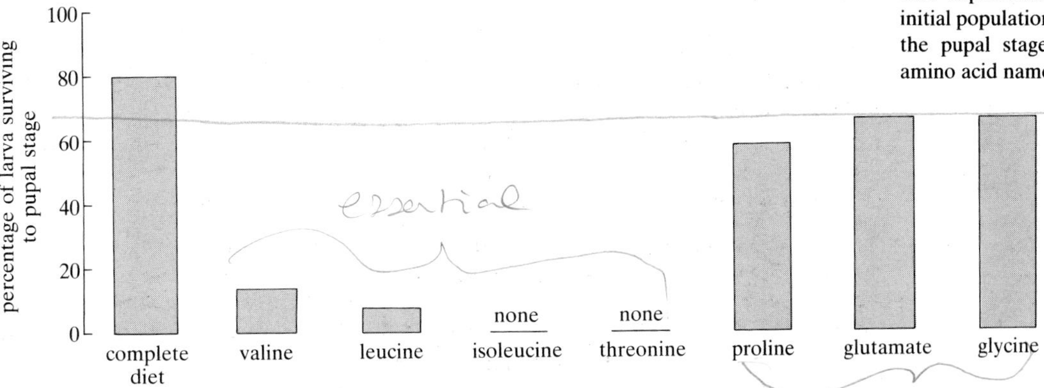

Figure 8.2 The survival of bollworm larvae fed on diets deficient in single amino acids. The height of each of the histogram bars represents the percentage of the initial population of larvae that survive to the pupal stage on a diet lacking the amino acid named below it.

◇ Which of the amino acids tested are essential?

◆ From Figure 8.2 valine, leucine, isoleucine and threonine are all essential amino acids. Survival in the absence of proline, glutamate and glycine is only slightly less than on the complete diet so these amino acids are non-essential.

◇ In what way does the radioactive labelling of the essential amino acids differ from that of the remaining amino acids?

◆ From columns three and four of the data in Table 8.2 large amounts of radioactive label have appeared in glutamate, glycine and proline, which are non-essential amino acids. There is no label in three of the essential amino acids, and only a small amount in the fourth.

◇ What explanation could you provide for this distribution, remembering that the label was originally in glucose?

◆ Non-essential amino acids can be synthesized by the animal. The label from the glucose would be expected to appear eventually in newly synthesized amino acids, as a result of metabolic interconversions. Threonine is an essential amino acid, but as a small amount of labelled carbon is incorporated into it, the larva must be able to synthesize small amounts though not enough to meet its normal needs.

The results of the investigation suggest that the extent of requirements for amino acids in the diet is partially determined by the synthetic ability of the body cells. Organisms with a marked synthetic ability require few essential amino acids. Organisms with a marked amino acid requirement are unable to effect many chemical interconversions—they have a **synthetic disability**. But the picture we have painted of the amino acid requirements of animals may not be so straightforward. For example, arginine is an essential amino acid for rats. However, rats can actually synthesize a small amount of arginine but it is not sufficient to enable optimum growth. Thus some arginine must be supplied in the diet, and it is classed as an essential amino acid. There are also cases where one amino acid may be able to satisfy partially the requirement for another. In the rat, for example, tyrosine can replace up to one-half of the phenylalanine requirement, provided sufficient is available.

8.2.2 Carbohydrates

Carbohydrates are usually the major source of energy in animals, which normally derive between 55% and 70% of the required energy directly from them.

◇ Which other compounds can provide a significant amount of energy?

◆ Fats and proteins: these two substances can both be broken down and fed into the TCA cycle.

◇ When can this happen?

◆ When the amount of carbohydrate in the diet is low—for example, when an overweight human restricts carbohydrate intake in an attempt to 'burn off' surplus fat. In certain tissues of some animals, however, this cannot happen; for example, the flight muscles of the fruit-fly *Drosophila* require carbohydrate as an energy source, and when this is exhausted, it cannot fly, even though stored fat can be used for other energy-requiring processes.

For most animals, a great variety of hexose sugars—for example, glucose, galactose, mannose and fructose—can act as sources of energy, because all these sugars are freely interconvertible and readily oxidized. The marked ability of cells to interconvert sugars means that no carbohydrates are essential in the way that some amino acids are. However, in some tissues, as we have seen, some carbohydrate (as hexose sugar), must be supplied as an energy source. But in many cases, although there are no essential carbohydrates, the growth of an animal will be better on one type of sugar than another. For instance, many sugars support the growth of young locusts. No specific sugar need be supplied in the diet, so no sugar can be called essential. Nevertheless, when identical amounts of various sugars are supplied as test diets to locusts, there are pronounced variations in the growth rate. Figure 8.3 shows that growth was at a maximum (called the optimum growth) when the dietary sugar was maltose and at a minimum when no carbohydrate was given. Other carbohydrates support intermediate or sub-optimum growth rates. A major cause of these differences in growth rates is the relative ease with which the sugars can be transported across the gut wall into the blood. Glucose is transported very easily whereas other hexoses, for example sorbose, move across the gut wall relatively slowly.

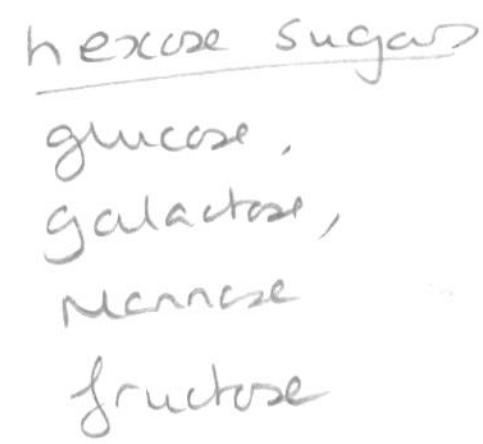

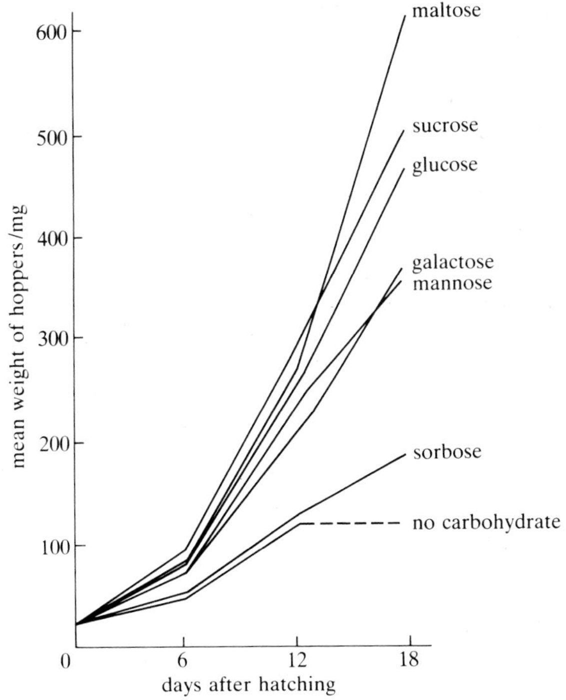

Figure 8.3 The growth of young locusts (called hoppers) on diets containing equivalent amounts of seven different carbohydrates. On day 16, all the hoppers reared in the absence of carbohydrate had died.

Thus, in the locust, glucose is not an essential constituent of the diet. In its absence, growth is maintained as long as certain other carbohydrates are substituted. However, glucose is a carbohydrate that can be more readily utilized than others, such as mannose or galactose, and in this sense glucose could be called a preferred nutrient.

8.2.3 Lipids

Lipids include the fats and several related compounds such as waxes, phospholipids and glycolipids. All animal tissues contain some fat or other lipid, present not only as stored material but also as essential components of cell membranes. Many animals can live well on little or no dietary fat, because fat can be formed from both proteins and carbohydrates.

Various important compounds are synthesized from lipid substances, for example the four-ring nucleus of the cholesterol molecule, which is found in cell membranes and which forms the skeleton of the sex and adrenocorticoid hormones of vertebrates.

However, as with amino acids, it appears that the synthetic ability of many animals is limited in respect to certain unsaturated fatty acids and more complex lipid materials such as cholesterol. In humans, cholesterol can readily be synthesized, and cholesterol in the diet is considered undesirable because it may be a major contributory factor in the development of atherosclerosis (hardening of the arteries). However, insects cannot synthesize cholesterol, and they must obtain it in their food. Rats are unable to synthesize three polyunsaturated fatty acids—linoleic, linolenic and arachidonic acid—and, if fed on a low fat diet, they lose weight, develop haematuria (presence of blood in the urine), and develop scales on their feet and tail. These three acids are sometimes called the **essential fatty acids**, although an excess of linoleic acid can partly substitute for the others. Humans, however, have the ability to synthesize much of their requirement for linoleic acid, although linolenic acid and arachidonic acid are essential fatty acids. Arachidonic acid is probably the most important, physiologically, since it plays a major role in the synthesis of prostaglandins.

8.2.4 Vitamins

Two of the best known vitamin-deficiency diseases, beri-beri and scurvy, were associated with living on a restricted diet—in famine or on long sea voyages—long before the discovery of vitamins. The term **vitamin** (from vital amine) was coined in the early part of the 20th century, and it was used to describe those accessory food factors necessary to permit animals to develop normally. Vitamins can be defined in a broad sense to include any component small molecule required for the production of a macromolecule, group-transfer molecule or coenzyme that cannot be made by the organism itself. This definition should, strictly, include the essential fatty acids, but generally the term is restricted, in normal scientific usage, to substances that are involved in the production of coenzymes and group-transfer molecules, or are such molecules themselves.

No single vitamin is a dietary requirement for all species. The need that humans have for ascorbic acid (vitamin C) is very well known, but most other mammals can synthesize it. Only some primates, the guinea-pig and a species of fruit-bat need ascorbic acid in their diet.

Vitamins may be obtained from food or as by-products of the metabolism of intestinal bacteria. Vitamin K is an essential factor in blood-clotting in humans. It is found in two forms, vitamin K_1 in plants and vitamin K_2, which is synthesized by bacteria living symbiotically in the vertebrate gut. Both forms of the vitamin can be used in the process of blood-clotting, so if no vitamin K_1 is present in the food, vitamin K_2 derived from the symbionts in the gut can be used as a source of K_1.

Vitamins were originally named by letters of the alphabet (A, B, C, D, E, K), although it is more common today to use their chemical names. There may be several types of each vitamin, for example those of the vitamin B class, all of which participate in enzymic reactions. A list of the major vitamins in humans is given in Table 8.3, for reference.

Table 8.3 Some vitamins in humans.

	Designation letter and/or name	Major sources	Metabolism	Some functions in humans	Deficiency symptoms in humans
Fat-soluble	A; carotene	Egg yolk, green or yellow vegetables and fruits	Absorbed from gut	Formation of visual pigments; maintenance of epithelial structure	Night blindness; skin lesions
	D_3; calciferol	Fish oils, liver	Absorbed from gut; little storage	Increases calcium absorption from gut; bone and tooth formation	Rickets (defective bone formation)
	E; tocopherol	Green leafy vegetables	Absorbed from gut; stored in adipose and muscle tissue	Maintains red cells	Increased fragility of red blood cells
	K; naphthoquinone	Synthesis by intestinal flora; liver	Absorbed from gut; little storage; excreted in faeces	Enables prothrombin synthesis by liver	Failure of coagulation
Water-soluble	B_1; thiamine	Brain, liver, kidney, heart; whole grains	Absorbed from gut; stored in liver, brain, kidney, heart	Formation of cocarboxylase enzyme involved in decarboxylation (Krebs cycle)	Stoppage of CH_2O metabolism at pyruvate; beriberi; neurotic heart failure
	B_2; riboflavin	Milk, eggs, liver, whole cereals	Absorbed from gut; stored in kidney, liver, heart	Flavoproteins in oxidative phosphorylation	Photophobia; fissuring of skin
	niacin	Whole grains	Absorbed from gut; distributed to all tissues	Coenzyme in hydrogen transport (NAD, NADP)	Pellagra; skin lesions; digestion disturbances, dementia
	B_{12}; cyanocobalamin	Liver, kidney, brain; bacterial synthesis in gut	Absorbed from gut; stored in liver, kidney, brain	Nucleoprotein synthesis, prevents pernicious anaemia	Pernicious anaemia; malformed erythrocytes
	folic acid (folacin, pteroylglutamic acid)	Meats	Absorbed from gut; utilized as taken in	Nucleoprotein synthesis; formation of erythrocytes	Failure of erythrocytes to mature; anaemia
	B_6; pyridoxine	Whole grains	Absorbed from gut; one-half appears in urine	Coenzyme for amino acid and fatty acid metabolism	Dermatitis; nervous disorders
	pantothenic acid	Widely distributed in foods	Absorbed from gut; stored in all tissues	Forms part of coenzyme A (CoA)	Neuromotor, cardiovascular disorders
	biotin	Egg white; synthesis by flora of gut	Absorbed from gut	Protein synthesis and transamination	Scaly dermatitis muscle pains, weakness
	C; ascorbic acid	Citrus fruits	Absorbed from gut; little storage	Vital to collagen	Scurvy—failure to form connective tissue

8.2.5 Water

Water is a fundamental constituent of all living matter. It makes up as much as 95% of the live weight of some animals' bodies and, as the principal component of blood, lymph and coelomic fluid, it is the major transport medium within the body.

◇ Water is readily lost from the body of terrestrial animals. What are three of the major ways water is lost from vertebrates?

◆ Excretion, evaporation from the respiratory surfaces and evaporation from sweat glands in some mammals.

Water must therefore be replaced when lost, and this occurs directly by drinking, from eating food with a high water content, and also to some extent from water released as a by-product of the chemical reactions going on in the body. The water balance sheet for a typical human is shown in Table 8.4. The amount of water derived from food is not much less than the amount drunk.

Table 8.4 Water gain and loss in a typical adult human: values are in cm^3 per day.

Gain		Loss	
drink	1 200	skin and lungs	900
food	1 000	sweat	50
metabolically produced	350	faeces	100
		urine	1 500
total	2 550	total	2 550

◇ List some of the chemical reactions that lead to the release of water.

◆ The synthesis of fat, protein and carbohydrate, and the oxidative breakdown of fat, protein and carbohydrate.

Aquatic animals also have a dietary requirement for water, though as you will see in Chapter 12, osmoregulation poses special problems of water balance for these animals.

8.2.6 Trace elements

There are over 90 naturally occurring elements, but not all are essential for life. In rats and chickens, only 24 are known to be absolutely necessary; in humans the number is 26. The major elements present in a 70-kg person (and similar percentages apply in living tissues generally) are shown in Table 8.5. Oxygen, carbon, hydrogen and nitrogen comprise more than 95% of the body elements, and the next six (phosphorus, calcium, potassium, sulphur, sodium and chlorine) make up 2.2%.

The additional elements listed are required in very small amounts, their combined weight in the human, for instance, totalling only 0.01% of the dry body weight. These are the **trace elements**, so called because the analytical methods available to the early investigators were unable to measure precisely the amounts in the tissues. Although present in only minute amounts, the essential trace elements are just as important as the essential amino acids and fats. For example, rats need only 1 μg of vanadium per day, but elimination of this element from the rats' diet has a deleterious effect on growth rate.

Table 8.5 Composition of human tissues.

Major elements—always present and essential	Per cent of dry weight	Trace elements—always present		
		Essential for more complex animals	Per cent dry weight	Others
oxygen	65	manganese	0.000 3	boron
carbon	18	copper	0.000 2	aluminium
hydrogen	10	iodine	0.000 04	nickel
nitrogen	3	cobalt		silicon
phosphorus	1.0	zinc		fluorine
calcium	1.5	selenium		barium
potassium	0.35	molybdenum		strontium
sulphur	0.25			chromium
sodium	0.15			tin
chlorine	0.15			lead
magnesium	0.05			titanium
iron	0.004			rubidium
				lithium
				arsenic
				bromine

Note that some of the elements are listed as essential for more complex animals, whereas others are suggested as being possibly essential. Fluorine, for example, is essential in rats for normal growth, but it is not known to be necessary in humans, although it has an important role in the prevention of tooth decay. It may be that, as with amino acids, fatty acids and vitamins, what is an essential trace element for one species is not necessarily so for another. However, our knowledge of the number of essential trace elements is incomplete and will probably increase as research continues.

Summary of Section 8.2

Nutrients that must be present in the diet of an animal are called essential, and the requirement for essential amino acids, fatty acids and vitamins is a reflection of an inability to synthesize particular substances. Animals also have a quantitative requirement for nutrients. Failure to provide appropriate amounts of nutrients produces sub-optimal growth and development.

Now attempt Questions 1 and 2, on p. 286.

8.3 FEEDING TYPES AND MECHANISMS

The chemical requirements of all animals can be defined in terms of the type and amount of different inorganic and organic nutrients required for normal growth and maintenance. All animals have successful methods for extracting sufficient of the chemical compounds they require from their environment. The physical form of the available food that satisfies this chemical need varies greatly. Animals show a great variety of feeding mechanisms for capturing and retaining food.

It is not possible to produce a classification of feeding mechanisms based on the usual taxonomic classification of animals into phyla, classes and so on, because the method of feeding of any one species may be influenced as much by the nature of the food available as by the animal's level of organization. Consequently, different types of animal living in the same habitat may obtain food by feeding mechanisms that operate on the same principle. Many marine animals (e.g. annelids, molluscs, crustaceans and whales) may filter out particles or smaller organisms from the sea. The parts of the animal concerned with the filtering process may not necessarily be structurally equivalent: the gills of bivalve molluscs are analogous to the tentacles of polychaetes as far as filter feeding is concerned, but there is no homology between these structures.

Animals can be divided into three major groups on the basis of their feeding habits. Those that feed on relatively small particles are termed **microphagous feeders**; those that feed on relatively large particles are called **macrophagous feeders**, and those that feed on fluids or dissolved substances are known as **liquid feeders**. There is often considerable overlap between the three groups, and it can be difficult to assign animals precisely to any one category.

8.3.1 Microphagous feeders

Microphagous feeders feed on particles that are small, relative to their own size, and two major groups can be distinguished—**suspension feeders** and **deposit feeders**. Suspension feeders filter their food material from the aquatic environment, and deposit feeders feed on particles that have fallen to the bottom of the water. Incidentally, because they feed on this detritus they are also called detritivores, a term like carnivore and herbivore that describes the type of food eaten, rather than describing the broad type of mechanism used or the size of the food particles.

Suspension feeders

◇ What sort of suspended food might be available to aquatic animals?

◆ The organisms that make up the plankton are the obvious food source. In addition to these living ones, there may be suspended organic material derived from live or decaying organisms.

Particulate organic matter can be present in water in surprisingly large concentrations. Planktonic organisms are very numerous at certain times of the year. In UK coastal waters in spring, when the larvae of barnacles form a major part of the plankton, the concentration of organic matter may be as high as $8\ \text{mg}\ \text{l}^{-1}$, compared with typical values of $0.2\text{–}1.7\ \text{mg}\ \text{l}^{-1}$ in the upper waters of the open sea.

There is also some organic matter dissolved in the water, and some animals can extract this. The concentrations are not high; for example sugars occur at a concentration of around $10\ \mu\text{g}\ \text{l}^{-1}$, and amino acids at around $100\ \mu\text{g}\ \text{l}^{-1}$.

Animals show a variety of methods for removing suspended food from water. Most involve setting up a current (the irrigation current) in the surrounding water by the beating of structures such as cilia or flagella. The animals extract the suspended particles from the water by means of structures that often act as filters. Secretion of mucus, which forms a sticky layer on the filtering structures, may aid filtering. Thus suspension feeders are often referred to as **filter feeders**. You should bear in mind, however, that this term is not applicable to all suspension feeders, because the ciliary structures of a number of invertebrates do not act as true filters. Suspension feeding is to a large extent non-selective; food particles are retained on the basis of their size rather than their food value.

The sponge that was introduced in Chapter 5 has a series of flagellated canals in the body wall through which water can flow. The cells that bear the flagella are called **choanocytes** (Figure 8.4). The beating flagella not only draw the water current into the interior, but they also trap the food particles, which are then ingested by the choanocytes.

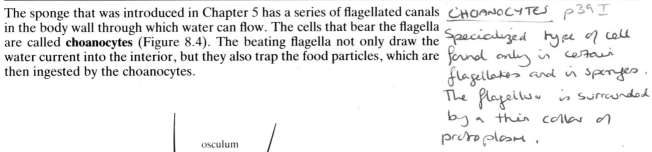

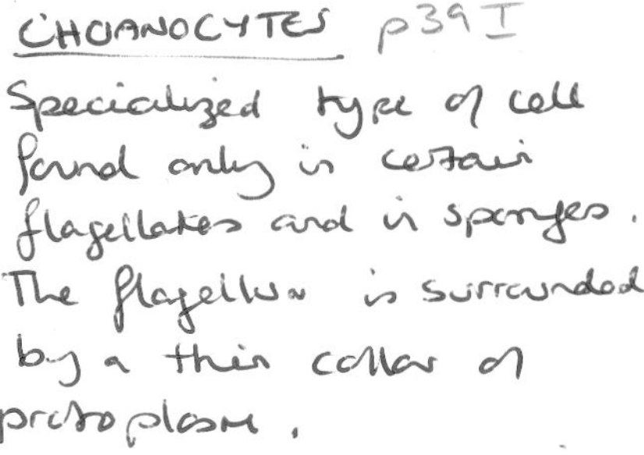

CHOANOCYTES p39 I
Specialized type of cell found only in certain flagellates and in sponges. The flagellum is surrounded by a thin collar of protoplasm.

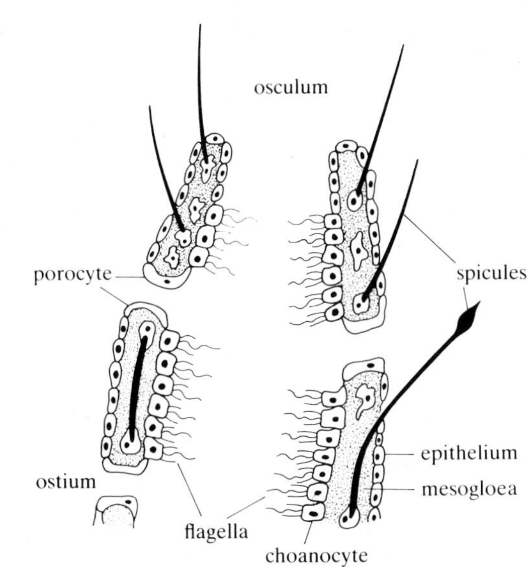

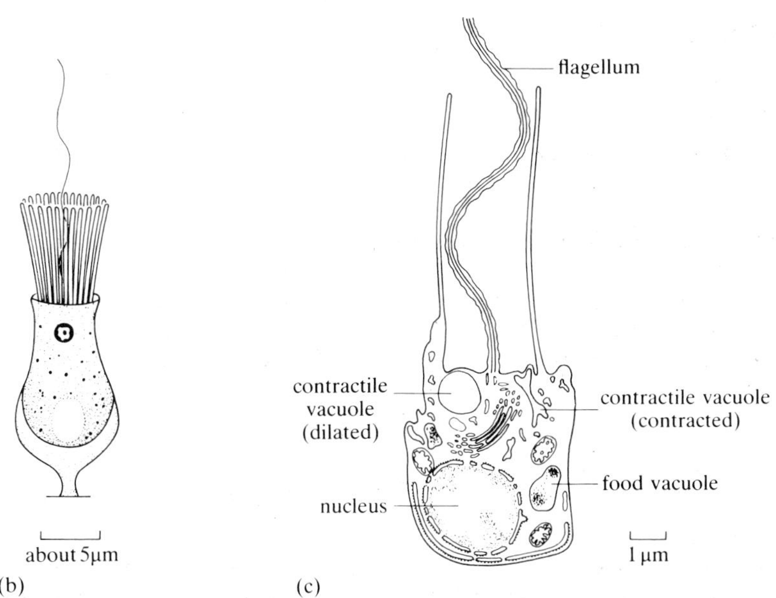

Figure 8.4 (a) Choanocytes in a canal through the body wall of a sponge. (b) A single choanocyte. (c) Section through a single choanocyte showing an ingested food vacuole.

Suspension feeding is characteristic of sedentary aquatic animals, and many show quite complicated adaptations to this way of life. One of the many beautiful polychaete worms, *Sabella*, extracts particles from the water. It lives in a vertical, blind-ended tube in the sand, and has a retractable crown of fleshy branchial tentacles (Figure 8.5). The fine branches of the tentacles (called pinnules) are covered with cilia. The continuous beat of some of these cilia moves water between the tentacles into the centre of the crown (Figure

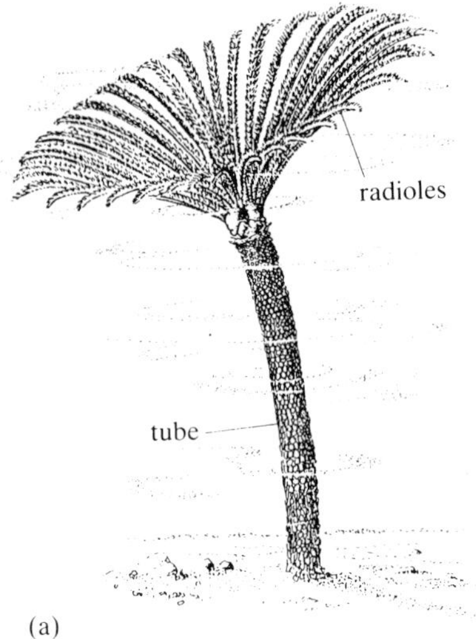

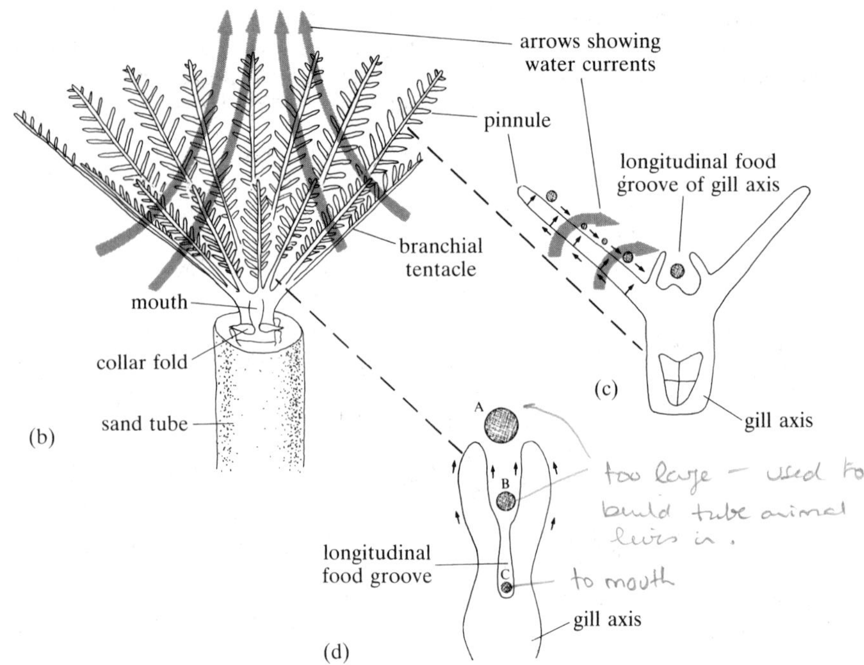

Figure 8.5 (a) The fan worm *Sabella* with its branchial crown extended. (b) The structure of the extended crown. (c) and (d) Cross-sections of a single tentacle at the level indicated, showing the movement of particles into and along the longitudinal food grooves. The small arrows indicate the direction of ciliary beat, and the large pink arrows show the direction of water currents.

8.5b). Suspended particles passing close to the tentacles within the stream of water are thrown, partly by the beat of special cilia, on the upper surface of the pinnule (Figure 8.5c). Cilia move the particles to the base of the pinnule where they enter the longitudinal food grooves. The grooves are shaped so that particles are divided into different streams based upon their size (Figure 8.5d). The stream of the smallest particles travels to the mouth. Larger particles are discarded. These large ones are not wasted since, together with secreted mucus, they are used to build the tube that the animal lives in.

Most bivalve molluscs are suspension feeders. Bivalves are encased in a pair of shell valves and are largely sedentary, often living in muddy or sandy substrata. The filtering structures are the large fleshy gills (ctenidia) that lie on either side of the body within the mantle cavity. In molluscs these were originally concerned with oxygen uptake, but in bivalve molluscs they have been greatly expanded and have a filter-feeding role. Figure 8.6a shows a simplified diagram of the internal organs of a bivalve. The gills consist of a number of filaments covered by ciliary tracts of three types (Figure 8.6c). The beating of the lateral cilia creates a current of water through the mollusc from the inhalant siphon across the gill filaments to the exhalant siphon. At one time, it was believed that a sheet of mucus produced by the gill acted as a filter and removed small particles from the water passing through the filaments, but

Figure 8.6 (a) A generalized bivalve mollusc seen from the side and with the left shell valve and ctenidium removed. Red arrows indicate water currents and the path followed by food particles; broken red arrows indicate the path of rejected particles. (b) Vertical section (at position AA in (a)) through a bivalve mollusc such as the mussel *Mytilus* to show the position and arrangement of the gills. (c) Horizontal section of one filament (at position BB in (b)).

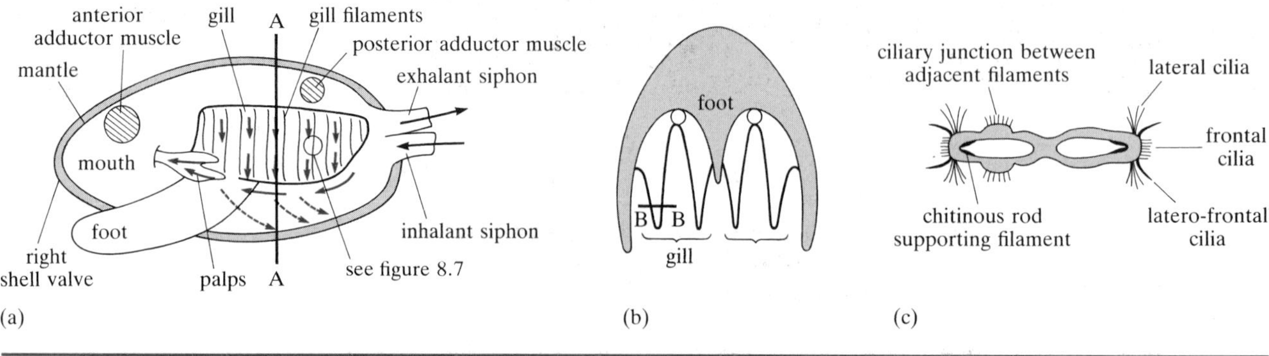

it seems more likely that interlacing latero-frontal cilia of the adjacent filaments form a net, which retains particles above a certain size. Where this happens (e.g. in bivalves such as the mussel *Mytilus edulis*), the latero-frontal cilia form compound structures known as cirri (sing. cirrus).

A scanning electron micrograph of the gill of *Mytilus* (Figure 8.7) shows how the latero-frontal cirri extend across the gap between two adjacent filaments to form a meshwork filtering arrangement. This meshwork is made finer because the component cilia (20–25 pairs) of each cirrus bend at regular intervals from its main axis to extend across the intercirral space. The maximum size of the gaps of the mesh is about 2.7×0.6 μm.

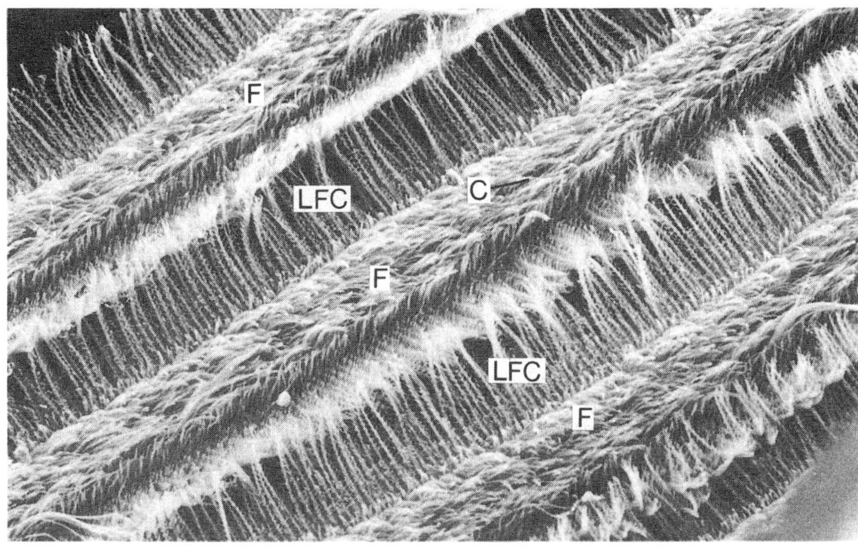

(a)

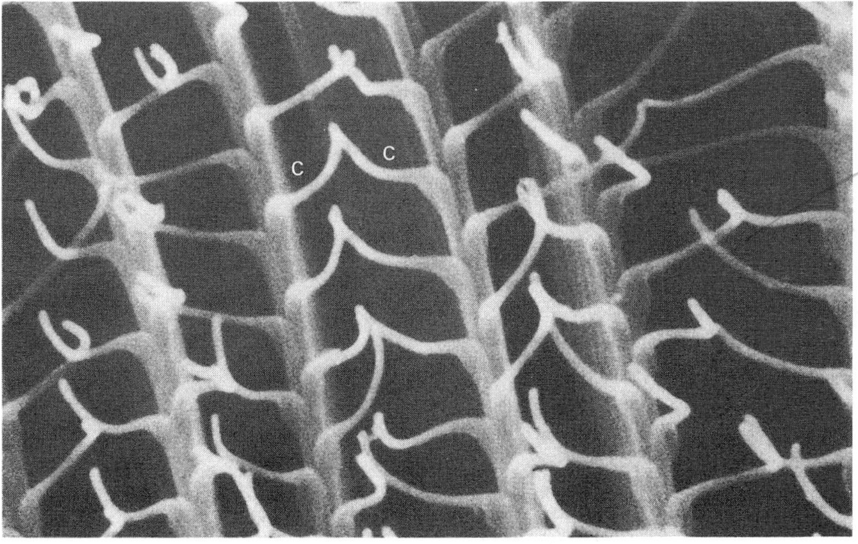

(b)

Figure 8.7 (a) Scanning electron micrograph of three gill filaments (F) of *Mytilus edulis* showing the frontal cilia (C). Note that the latero-frontal cirri (LFC) extend across the interfilamentary spaces to form a fine-mesh filter (magnification 1 100). (b) High-power view of the latero-frontal cirri in (a). Each latero-frontal cirrus consists of a double row of 20–25 pairs of cilia (C), one of each pair being on either side of the main axis of the cirrus. The effect is to form a mesh-work between the cirri and also between adjacent filaments. (magnification 8 000).

The food particles trapped by the latero-frontal cirri are thrown onto the frontal cilia, where mucus produced by the gills binds the particles into a food 'string', which is passed down the gill filament to the edges of the gills and thence to the labial palps via the pathways shown in Figure 8.6a. A proportion of the particles that reach the labial palps are later ingested via the mouth.

In the two examples of suspension feeders that we have described, *Sabella* and *Mytilus*, cilia create a water current and move food to the mouth. Mucus secretion plays a key role in binding food particles together in *Mytilus*. In *Sabella*, mucus binds together particles that are too large to be eaten and these are then passed away from the mouth and used for extending the tube that the animal lives in. How does the animal differentiate between those particles that are of the right size to eat, and those that are too large to eat but can be used for tube-building? The answer lies in a particle-sorting mechanism that forms part of the filter.

◇ What do you think are the advantages of a particle-sorting mechanism?

◆ Large, and possibly damaging particles will not enter the gut, so the sorting mechanism provides some protection. However, a more likely advantage is that it preferentially removes plankton from the water and rejects sand, so a large percentage of the material entering the gut is digestible.

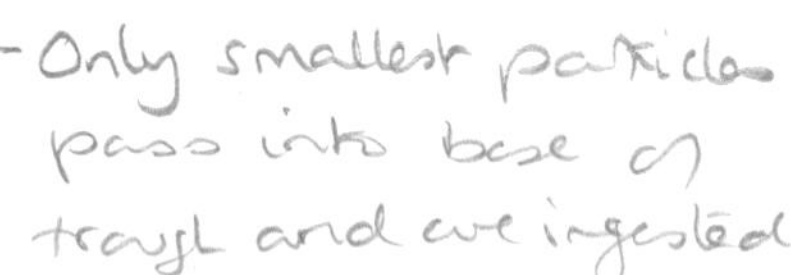

- Only smallest particles
pass into base of
trough and are ingested

Look back at the drawings of *Sabella* in Figure 8.5. The longitudinal groove is shown in cross-section (Figures 8.5c and d). This gutter-like trough, lined with cilia, leads from the individual tentacles to the crown and then the mouth. Particles too large to pass into the hollow of the trough (A) remain in the upper parts of the trough wall as shown, and on approaching the mouth, are passed to lateral, fleshy lips and are rejected. Medium-sized particles (B), while entering the trough, fail to pass to the bottom because of the lateral projections of the trough walls. Only the smallest particles (C), which pass into the base of the trough, are ingested.

Frequently, sorting is performed by the retaining or filtering surface itself, for example, the gill surface in many bivalves. Figure 8.8 illustrates how such sorting occurs. The outer surface of the gill, composed of a sheet of closely apposed gill filaments, is folded to form a series of alternating ridges and crests. All the frontal cilia (Figure 8.6c) on the outer gill face beat towards the ventral free edge of the gill, where a shallow marginal groove runs towards the mouth. Fine particles pass down the grooves of the gill surface into the food stream in the marginal groove. Coarse particles pass down the crests of the gill surface and are excluded from the marginal groove. They move anteriorly, to be passed out eventually as a rejection stream; some may even be shaken off by muscular movements of the gills.

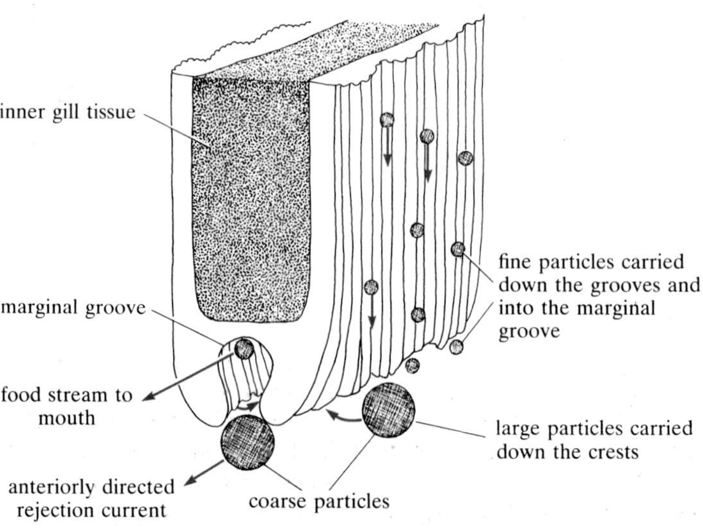

Figure 8.8 Block diagram of the free edge of a bivalve gill, showing sorting of the captured particles.

This discussion of suspension feeders has dealt only with aquatic invertebrates. Although most suspension feeders are found among animals of this type, a few vertebrates (some fish and whales) can also obtain food by filter feeding. The largest elasmobranchs are the basking and whale sharks; in these, the water that enters the mouth and flows over the gills is strained by special comb-like structures on the gills, which filter out suspended particles, mainly planktonic organisms. In one hour, a basking shark can strain plankton from 100 000 kg of water. The gill rakers of herring and mackerel serve as a sieve for catching small crustaceans. A few marine and freshwater birds are also specialized filter feeders. For example, a species of petrel has a series of hair-like lamellae that extended along the edge of the upper beak. Crustaceans are strained from the water entering the mouth. Flamingos are also filter feeders.

The largest filter feeders are the baleen whales. They have a filtering apparatus that consists of a series of horny plates hanging down from the palate (roof of the mouth). The plates bear a fringe of parallel filaments of keratin that strain crustaceans, collectively termed krill, from the water. The whale opens its mouth and takes in several tons of water, which is then squeezed out again through the baleen plates as the mouth closes. The krill are caught in the hair-like edges of the plates and swallowed. The side of the blue whale appears ridged. The ridges are a series of folds that allow a large expansion to occur when water is taken in. Thus the amount of water filtered per mouthful is increased.

The right whale (so-called because to the 19th century whalers it was the right whale to catch as it was slow and did not sink when dead) is a famous example of convergent evolution in feeding mechanisms. The lesser flamingo has a similar method of filtering small animals from the muddy water of ponds. The similarity with the baleen plates of the black right whale can be seen in Figure 8.9, but note that the flamingo feeds with its head upside down. Both mechanisms have the same function. Large volumes of water are forced through the filter mechanism, and the filter elements retain small organisms from the water. This forced-filter system can filter more water than a passive system, and whales are not like basking sharks where there is a through-flow of water from mouth to gill slits—ideal for passive filtering.

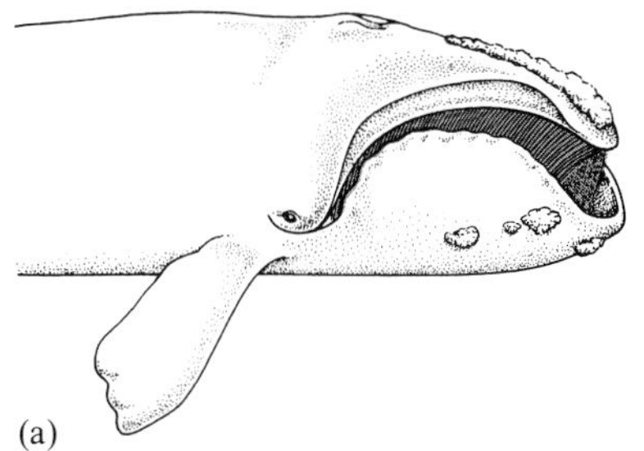
(a)

(b)

Figure 8.9 Two filter feeding mechanisms that show convergent evolution: (a) the black right whale; (b) the lesser flamingo.

Deposit feeders

Particles too dense to remain in suspension in water fall and settle, contributing to the bottom deposits. Organic debris of all types, particularly dead and decaying animals and plant detritus, is especially abundant on the surface of sediments. Some deposit feeders such as *Amphitrite* (Figure 8.10), a sessile tube-dwelling polychaete, select particles from the surface of the sediments. Others ingest the surrounding deposits indiscriminately.

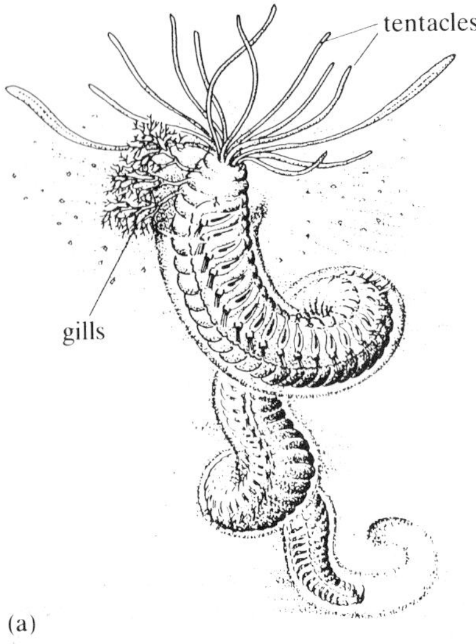

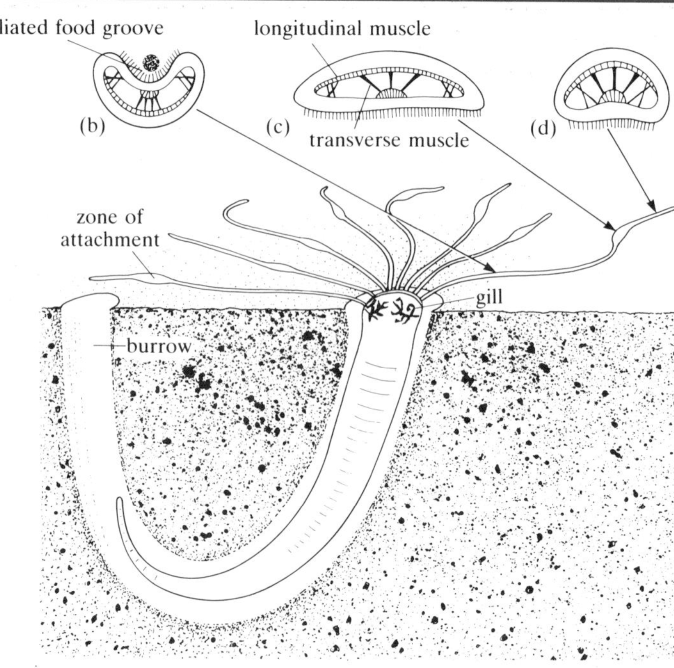

Figure 8.10 Amphitrite johnstoni, showing the structure of the tentacles. (b), (c) and (d) are cross-sections of (b) an extended tentacle, (c) the zone of attachment, and (d) the zone of extension.

Amphitrite lives embedded within the substratum, its body enclosed by a thin parchment-like tube. An anterior crown of tentacles collects food from the surface of the substratum. The tentacles of *Amphitrite* are highly mobile. In contrast, the extended tentacles of *Sabella* are passive and relatively stiff, though the crown as a whole can be rapidly withdrawn when the animal retreats into its tube. The tentacles of *Amphitrite* are ciliated on the upper surfaces (Figure 8.10b) and they convey particles to the mouth. Fine particles are passed along the tentacles in the ciliated groove (Figure 8.11a). The passage of larger particles is assisted by the squeezing of the muscular walls of the gutter towards the mouth (Figure 8.11b) and movement of the whole tentacle is required for the manipulation of very large particles (Figure 8.11c).

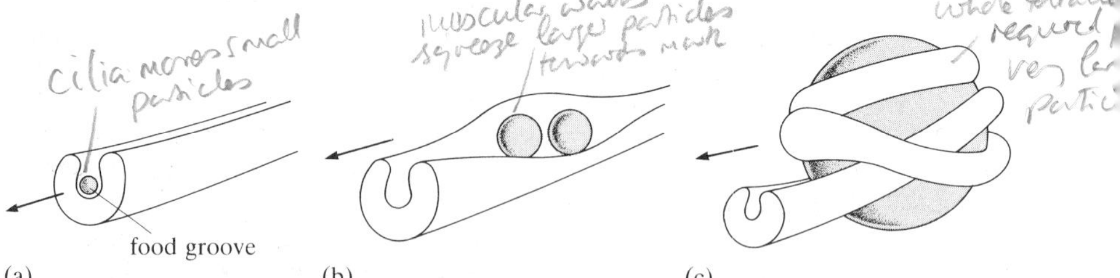

Figure 8.11 The action of the tentacles of *Amphitrite* conveying food towards the mouth.

Bivalve molluscs are not all suspension feeders—some are suspension feeders at high tide, but in shallow water as the tide falls, they may be deposit feeders. Such an animal is *Scrobicularia plana*, one of the larger British bivalves, which can grow to a diameter of about 6 cm (Figure 8.12). It is found in burrows up to 20 cm deep in mud-flats where there is considerable surface freshwater runoff. The translucent siphon can be extended over large distances, reaching up to the top of the burrow and then extending over the surface for at least 10 cm. Food particles are dragged from the surface by the siphon. Since detritus is rich in bacteria and diatoms, and is brought in on each tide, the bivalve probably never moves from its burrow.

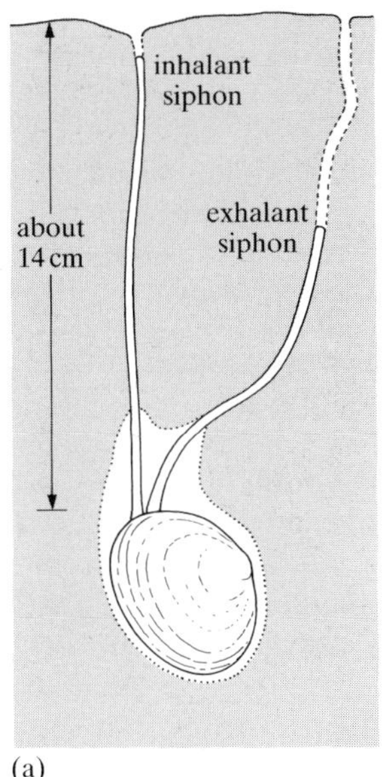

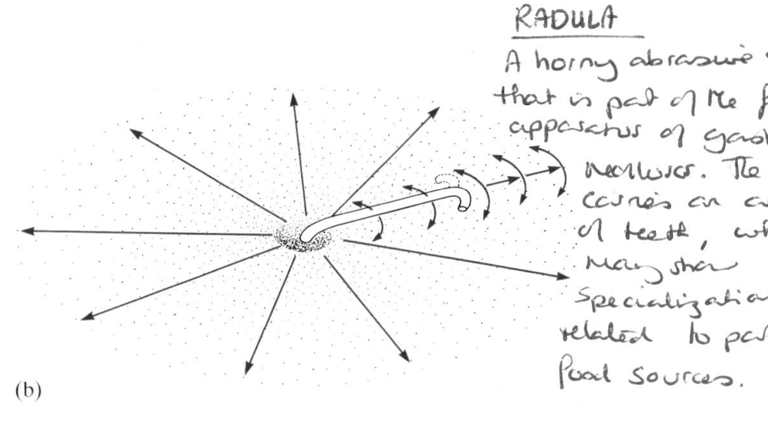

Figure 8.12 (a) *Scrobicularia plana* in its burrow in the mud. (b) A surface view showing how the inhalant siphon collects surface detritus.

8.3.2 Macrophagous feeders

Almost every animal phylum has representatives that are macrophages and, as you might imagine, there is an enormous variety of feeding mechanisms to match the variety of food sources available. There is room here to describe only a few examples to illustrate different types of mechanism:

- □ Mechanisms for scraping and boring into food
- □ Mechanisms for seizing and swallowing prey
- □ Mechanisms for seizing and masticating prey before swallowing

Mechanisms for scraping and boring into large food masses

Animals that scrape or bore into large food masses are often slow-moving, and show considerable structural and functional modifications, particularly of the buccal apparatus (the mouth parts). This is illustrated by the gastropod molluscs, a class of animals with an almost unrivalled range of food and variety of feeding mechanisms. Within the class there are grazing herbivores, carnivores that can bore holes in the shells of their prey and carnivores with teeth that can inject toxin. These gastropods have a special feeding organ, the **radula**, which is a sort of rasping tongue. It is a highly variable organ in which the shape and arrangement of the radular teeth are directly related to the type of food taken in and the way in which the food is manipulated. Figure 8.13 illustrates the structure and functioning of the radula in a gastropod such as

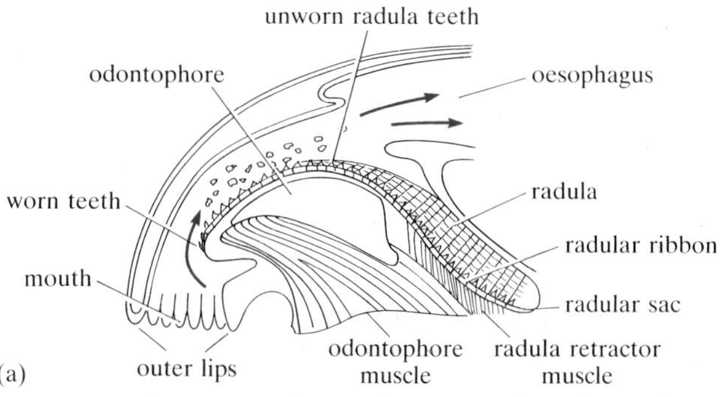

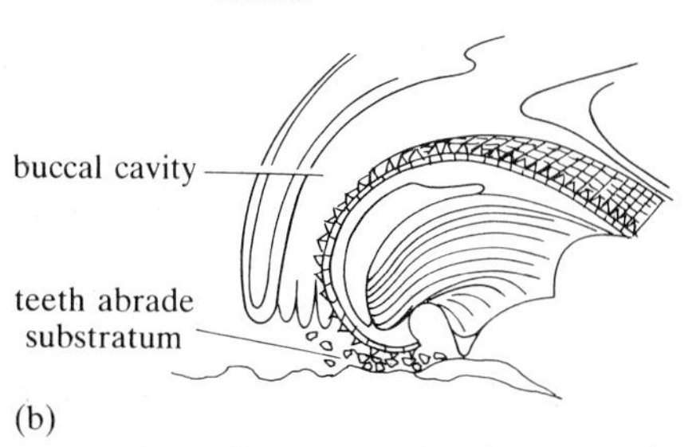

Figure 8.13 (a) Diagrammatic longitudinal section of the buccal mass of a prosobranch gastropod. (b) The radula in action.

the limpet *Patella vulgata*, which feeds on seaweeds or encrusting marine algae. The radula is a horny movable band bearing rows of teeth, and it overlies a tongue-like structure, the odontophore, which consists of a mass of muscle and cartilage. In grazing molluscs, the buccal mass containing the radula protracts forwards and downwards, contacting the substratum. The odontophore is rolled forwards so that the free surface of the radula is drawn across the expanded mouth opening. The radula teeth abrade the substratum, the fine material dislodged is passed back into the buccal cavity as the mouth closes and the odontophore is then retracted (Figure 8.13a). The particles on the radula are transferred to the ciliated lining of the roof of the mouth and then pass into the oesophagus. Note that the radula is renewed posteriorly as the teeth on the anterior part are worn away. This type of scraping, abrading radula is found in many herbivorous and omnivorous gastropod molluscs. Scraping gastropods such as snails are in one sense small-particle feeders.

◇ Bearing in mind our definition of a microphagous feeder, could gastropods such as the limpet be placed in this category rather than in the macrophagous class?

◆ The essential difference between limpets and microphagous feeders is that the grazing limpet first reduces large masses of material to fine particles before ingestion, whereas microphagous feeders ingest fine particles by filtering them from suspension or by eating fine particles that are loosely compacted in the mud or sand deposits. The distinction is a delicate one and serves once again to illustrate that there are often difficulties in deciding which feeding category an animal belongs to within a classification system. (In fact, there are some microphagous gastropods, e.g. some pteropods.)

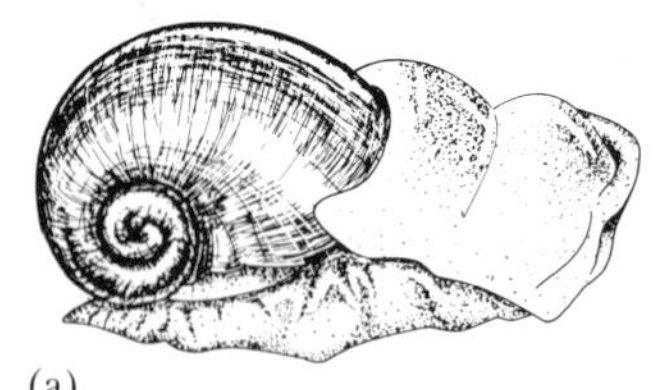

(a)

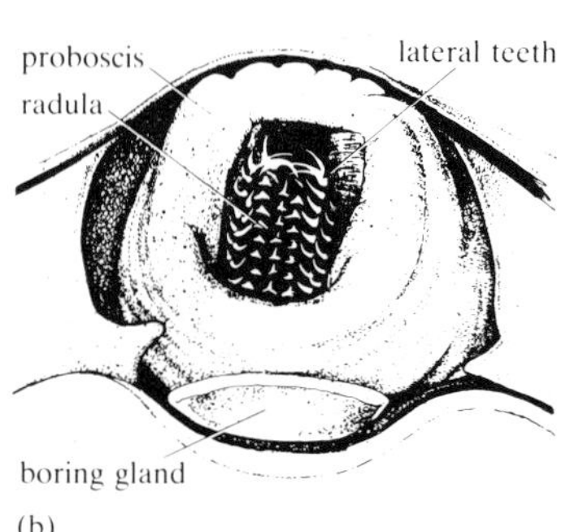

(b)

Figure 8.14 (a) A carnivorous gastropod *Natica* sp. (b) The proboscis and radula of *Natica millepunctata*.

Other gastropod molluscs, such as the carnivorous *Natica millepuncta* (Figure 8.14) have a radula, but feed in a very different way and are much more obviously macrophagous feeders. *Natica* feeds on bivalve molluscs and the radula is located in a muscular extensile searching proboscis. The blade-like cutting structures are well adapted to deal with the flesh of the prey. *Natica* drills a hole in the shell of the victim through the rasping action of the radula, which is aided by an acid secretion of an enzyme that comes from a gland on the ventral side of the proboscis.

A carnivorous gastropod you may be familiar with from UK coasts is the dog whelk *Nucella lapillus*. It feeds on barnacles and mussels.

Mechanisms for seizing prey

Carnivores generally prey on animals that are smaller than themselves or defenceless. However, there are exceptions, and one type of predator that occurs in a number of groups in the animal kingdom is the venomous one. Potent venoms enable small predators to disable or kill prey that are larger than themselves, and an interesting example of a venomous predator is the cone shell mollusc (Figure 8.15). Cone shells are macrophagous carnivores that capture relatively large prey—frequently fish. The radula teeth of the cone shell are sharp and slender, and are barbed at the tip (Figure 8.15b). Once the prey is located, the proboscis is extended towards the prey and a single radula tooth is rammed into it like a harpoon. Poison conveyed along the shaft of the radula tooth immobilizes the prey, which is then engulfed.

Some of the characteristic features of macrophagous feeders can be appreciated by comparing the method of food capture in a carnivorous mollusc such as a cone shell with that of a filter-feeding bivalve. In the bivalve, feeding

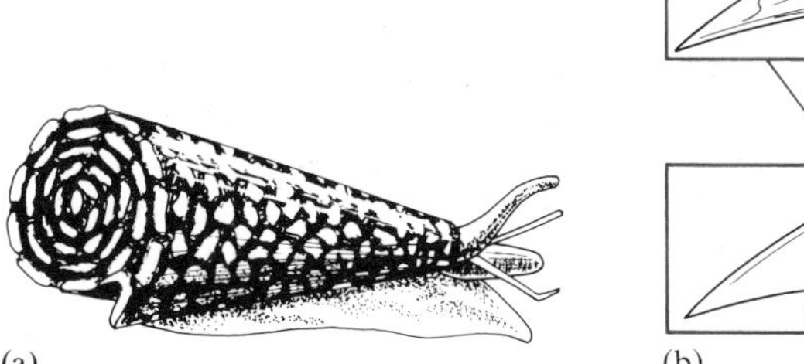

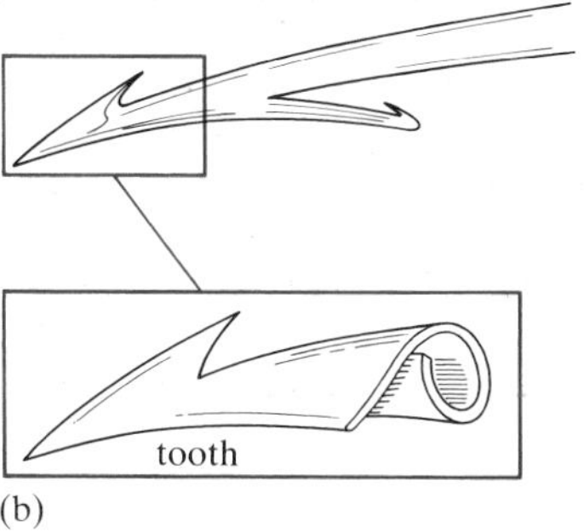

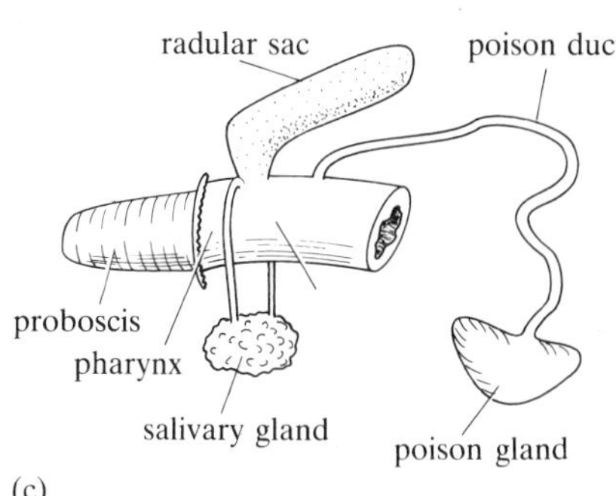

Figure 8.15 (a) A cone shell mollusc, *Conus marmoreus*. (b) The terminal part of a single tooth with an enlargement of the grooved barb. (c) Anterior part of the gut of *Conus striatus*, a cone shell mollusc similar to *C. marmoreus*.

takes place almost continuously, and selection of food is based on particle size. Feeding is almost an automatic function. But contrast these features with those found in cone shells. Information about the proximity of prey is provided by sense cells. The active prey is immobilized by a powerful nerve toxin produced by the poison glands. Prey are ingested periodically—in other words, feeding is discontinuous. The feeding of cone shells is associated with coordinated muscular responses related to catching prey.

Mechanisms for seizing and masticating before swallowing food

Mechanisms of this kind are found in a wide variety of animals, e.g. polychaetes, which have an eversible pharynx armed with teeth and jaws with which the prey is seized, and arthropods, which have chelae (claws) and mouth parts that seize, manipulate and shred the food. Many vertebrates seize and cut up or grind their food.

In most mammals, the teeth play a major role in feeding, showing considerable adaptive modification in relation to procuring food and breaking it up prior to swallowing. There are four basic types of teeth (Figure 8.16a). The incisors (at the front) cut; the canines (just behind the incisors) pierce and slash; the premolars crush; and the molars crush and grind (mastication). The number and size of these teeth vary, depending on the nature of the diet.

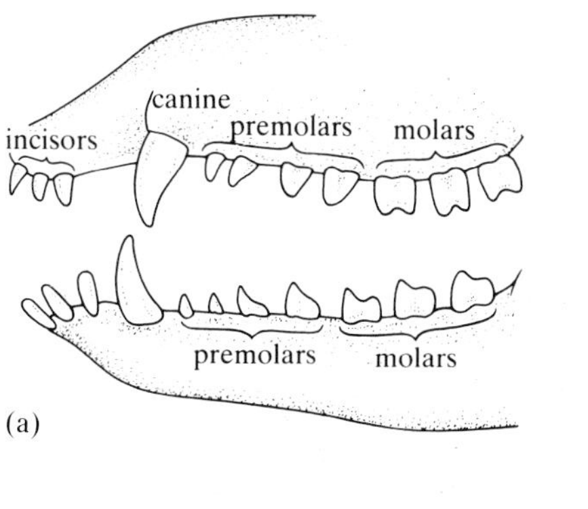

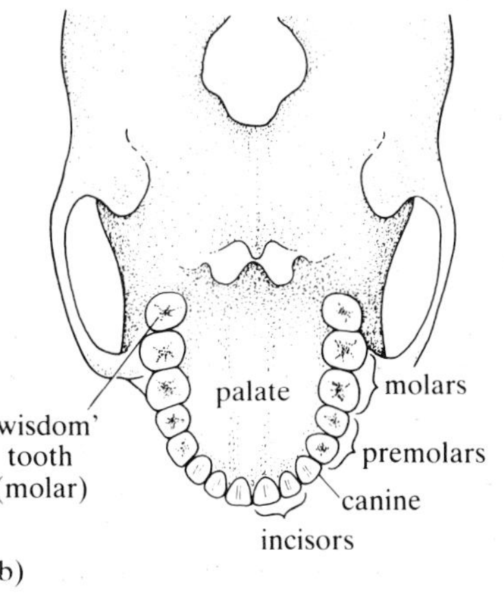

Figure 8.16 (a) The teeth of a generalized placental mammal as seen from the left side. (b) The arrangement of teeth in the upper jaw of a human.

Consider the following animals: the cat, the squirrel and the cow. There are significant differences in their dentition, related to the food that they eat. The cat is a carnivore and has enlarged canines and cutting premolars in the upper jaw, whereas the squirrel is a gnawing rodent with greatly enlarged incisors and grinding molars. The cow is a herbivore; its molars and premolars are greatly elaborated into grinding organs which break down tough grasses.

Humans, like most primates, are basically omnivorous, feeding on plants and other animals. There is little elaboration of any of the four types of teeth (Figure 8.16b). The canines are little different in shape or size from the incisors and the molars show none of the specialization seen in herbivores. In humans, as in other mammals, the mechanical disintegration of food before it is swallowed is brought about by the cutting and grinding action of the teeth, aided by movements of the muscular tongue. Food is moistened by saliva, which lubricates its passage through the anterior part of the gut. Saliva also contains a starch-splitting enzyme, though it is doubtful if this plays a significant role in carbohydrate digestion, most of which occurs in the small intestine.

8.3.3 Microphagous and macrophagous feeding contrasted

A comparison of feeding in bivalves and cone-shell molluscs indicated that in microphagous animals such as the bivalves, feeding is continuous and automatic, and that selection is on the basis of particle size. By contrast, in macrophagous animals such as the cone-shells or in mammals, selection of food is qualitative and feeding is discontinuous.

These sorts of comparisons are interesting and occasionally revealing, but they are somewhat oversimplified. Can we say that *all* microphagous feeders feed continuously? Clearly not; remember that those on the seashore must feed periodically (they stop when the tide is out), and there is evidence that bivalves cease to filter when the particle concentration in seawater is very low or very high. There is also some evidence that suspension feeders that could feed continuously do not—there are periods of almost complete inactivity. In the same way, it is unlikely that suspension feeders *invariably* select food on a quantitative basis. If some species of bivalve are fed on two species of diatom almost identical in size, one of the species is ingested, the other is rejected. This suggests that the animals are discriminating on a qualitative basis. The important point to remember is that whatever relationships or principles are constructed as a result of a comparative analysis of animal feeding habits, they must be regarded only as generalizations. In fact, all generalizations that refer to feeding probably have a number of exceptions. Any statement that attempts to generalize about feeding must be carefully qualified.

A related problem is one of classification. The immense range and diversity of animals often complicates attempts to classify any aspect of animal function in a hard and fast way. In the previous sections, the classificatory system selected was deliberately simplified. With such a system, it is impossible to classify all animals satisfactorily. Some of these difficulties are emphasized by *Scrobicularia*, which is a suspension feeder at high tide but a deposit feeder at low tide, and the grazing gastropod molluscs, which we classed as macrophagous but which could be considered as microphagous.

8.3.4 Liquid feeders

Animals such as spiders, aphids and ectoparasitic blood-sucking insects are among the most numerous of liquid feeders. Many are blood-suckers, feeding on chordates. The modifications of the mouth-parts for this role—particularly those of Hemiptera (bugs), Anoplura (sucking lice) and Siphonaptera (fleas), which feed on humans—are highly specialized.

Osmotrophs are animals that absorb food, generally dissolved organic molecules (amino acids, sugars and fatty acids), through their outer surfaces. Although the processes by which dissolved organic molecules enter the body surfaces of all osmotrophs are similar—namely, by mediated transport mechanisms—we can divide osmotrophs into two categories: those without a gut (e.g. tapeworms) and those with a gut, which are also capable of microphagous or macrophagous feeding (e.g. many marine invertebrates). Pogonophoran worms were thought to be exclusively osmotrophs, but it is now clear that, in both shallow- and deep-living forms, endosymbiotic bacteria provide organic molecules, while the worm provides carbon dioxide, oxygen and hydrogen sulphide to the bacteria.

Marine invertebrates can gain some nutrition by absorption of dissolved substances directly through their epidermal surfaces, but why animals that are perfectly capable of ingesting solid food should have the capacity to absorb dissolved substances via their epidermal tissues is an interesting question. The ability to transport substances across the cell membrane, which would have been a feature of ancestral cells, is probably a basic property of every cell, and if the environment includes nutrients, there is no reason for an evolving cell, or organism, to lose this capacity even if other sources of nutrition are present.

Summary of Section 8.3

1 Microphagous feeders consume relatively small food particles. One type is the suspension feeder, which consumes plankton and detritus suspended in an aquatic environment. The size of particles retained by suspension feeders depends on the structure of the filtering net. Most suspension feeders sort trapped particles—usually according to particle size.

2 The mechanisms that control ciliary beating are unknown but, in bivalves, neural mechanisms may be involved. In bivalves, the filtration rate appears to be independent of the particle concentration within certain limits. The concentration of organic particulate material is probably sufficient to satisfy the dietary needs of many suspension feeders.

3 Deposit feeders select particles from the substratum surface or ingest particles indiscriminately. *Amphitrite* has an anterior crown of muscular, mobile tentacles that conveys food to the mouth. It is not always possible to divide microphagous animals into either deposit or suspension feeders. *Scrobicularia*, for example, can be classed as either of these types, depending on the state of the tide.

4 Macrophagous feeders select relatively large particles, and may be herbivores, omnivores or carnivores. One type scrapes and bores; some gastropods illustrate this category very well. The various types of feeding mechanism seen in gastropods depend on modification of the molluscan radula; modifications of this are also seen in the second type of macrophage—those that seize their prey. In the cone shells, the radula teeth are poison darts, and a powerful nerve toxin paralyses the prey. The third type of macrophagous feeder seizes and masticates its food before swallowing. Numerous examples of this type are seen among vertebrates, and in mammals there are differences in dentition (teeth) related to the type of food eaten.

5 Many liquid feeders are specialized to feed on the fluids of animals and plants, but some (osmotrophs) absorb dissolved organic nutrients directly across their external body surfaces. Animals that lack a gut (endoparasitic helminths and some marine pogonophorans) are osmotrophs.

Now attempt Questions 3–5, on pp. 287–288.

8.4 DIGESTION AND THE DIGESTIVE TRACT

In previous Sections, we considered how heterotrophic organisms obtain food. Ingested food, however, can rarely be used by cells straight away. Except for some marine osmotrophs and metazoans, such as tapeworms and some pogonophorans, all animals must break down complex food molecules to their constituent building blocks. Proteins must be broken down to amino acids, polysaccharides to simple sugars, and fats to fatty acids and glycerol. These can then be utilized in metabolic and synthetic processes.

The process by which the complex food molecules are broken down into simpler constituents is called digestion, and it consists of a number of phases. In this Section we consider the digestion of the three main classes of food macromolecules, proteins, carbohydrates and fats. Digestion has evolved from an apparently simple process in lower animals, in which every cell or every cell lining the gut has the ability to carry out a wide range of digestive processes. It has become a highly organized function involving a specialized alimentary canal, where specific regions are concerned with separate phases of digestion.

Digestion is a hydrolytic process in which complex molecules are split into simpler units, this being the reverse of the condensation process by which macromolecules are formed. Two broad categories of chemical digestion are seen in animals: **intracellular digestion** (digestion within the cell) and **extracellular digestion** (digestion outside the cell, often in a specialized alimentary canal).

8.4.1 Intracellular digestion

Even without any prior knowledge of biology or anatomy, you would be able to say that, in humans, digestion is a function carried out in a specialized region of the body, the gut or alimentary canal. But consider simple animals such as the protistans. It is clear that no specialized alimentary canal is present. Food material is taken in by a process known as **endocytosis**. Figure 8.17c shows endocytosis in *Amoeba*. In *Amoeba* ingestion, digestion, absorption of the products of digestion, and elimination of waste products all occur within a single cell.

◇ Do you think that the enzymes responsible for intracellular digestion occur free in the protistan cytoplasm?

◆ No. The enzymes are contained within lysosomes; hydrolytic enzymes cannot be free within the cell. When the lysosome fuses with the ingested food vacuole, a digestive vacuole is formed, in which hydrolytic enzymes can act on the ingested food and not on other cell components.

8.4.2 Extracellular digestion

More complex animals rely principally on extracellular digestion, which normally takes place in a digestive tube or sac. The cnidarians and flatworms all have sacs with a single opening, the enteron (Figure 8.18), but all other phyla have a tubular digestive system. A significant feature of extracellular digestion is that it is taking place *outside* the animal's body. If you think of a snake, for example, the animal is essentially a hollow cylinder through which food passes, with valves at each end controlling entry and exit. The gut is lined with epidermal tissue. The tubular nature of the digestive system is

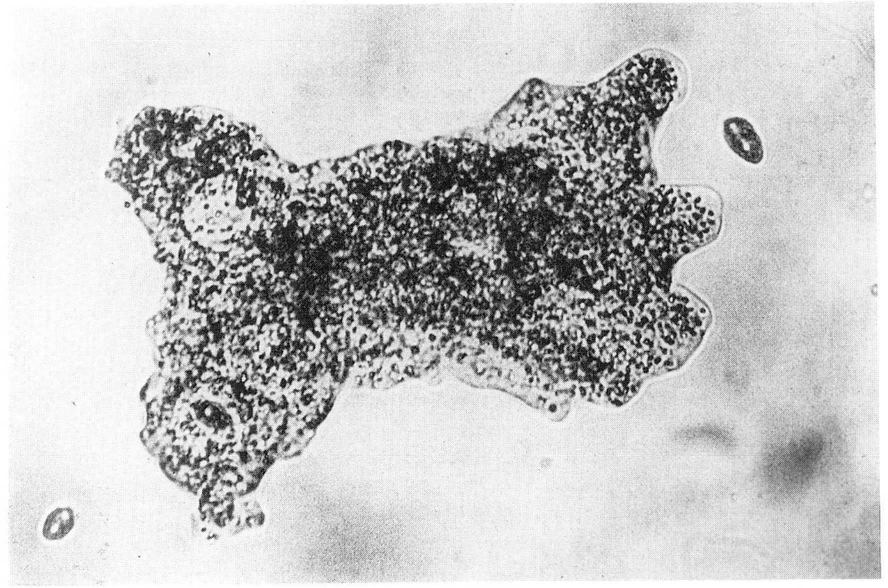

(a)

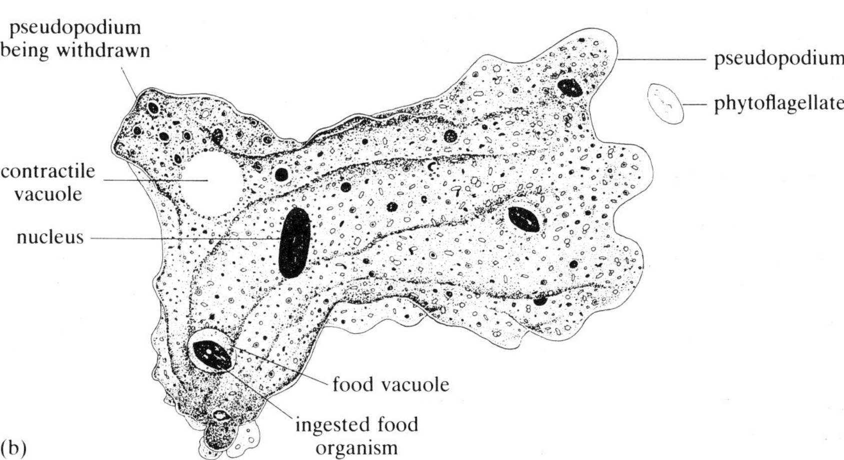

(b)

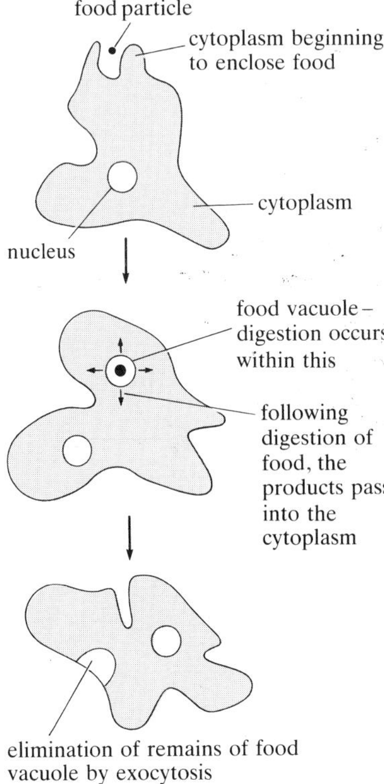

(c)

Figure 8.17 (a) A living *Amoeba.* (Magnification: ×300.) (b) Diagrammatic representation of the amoeba in (a), showing various features. (c) Endocytosis in *Amoeba.*

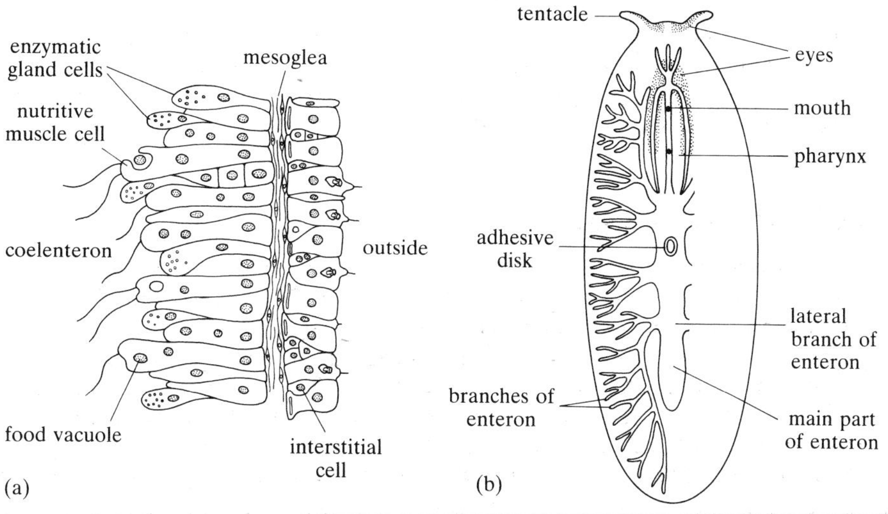

Figure 8.18 Some invertebrate digestive systems. (a) Section through the body wall of *Hydra.* The lining of the coelenteron contains cells that are phagocytic and gland cells that secrete digestive enzymes. (b) Digestive system of a flatworm.

evident in Figure 8.19, which shows examples from a range of animals. The surprising uniformity of structure is probably a reflection of the underlying uniformity of the function that digestive systems perform. For all animals, the breakdown of organic compounds into their component parts involves very much the same chemistry. An advantage of this tubular structure is that the chemistry can be split into different component parts. For example, reactions that occur optimally in an acidic environment can be partitioned off from those that require an alkaline environment. This division of the digestive tract into regions with different chemistry is reflected in the structure, but the structure also reflects a functional division. Four main areas of the digestive tract can be distinguished on the basis of their function:

- □ The receiving region
- □ The conducting and storing region
- □ The digesting and absorbing region
- □ The water-absorbing and voiding region

They are indicated by shading on Figure 8.19. We consider each region in turn.

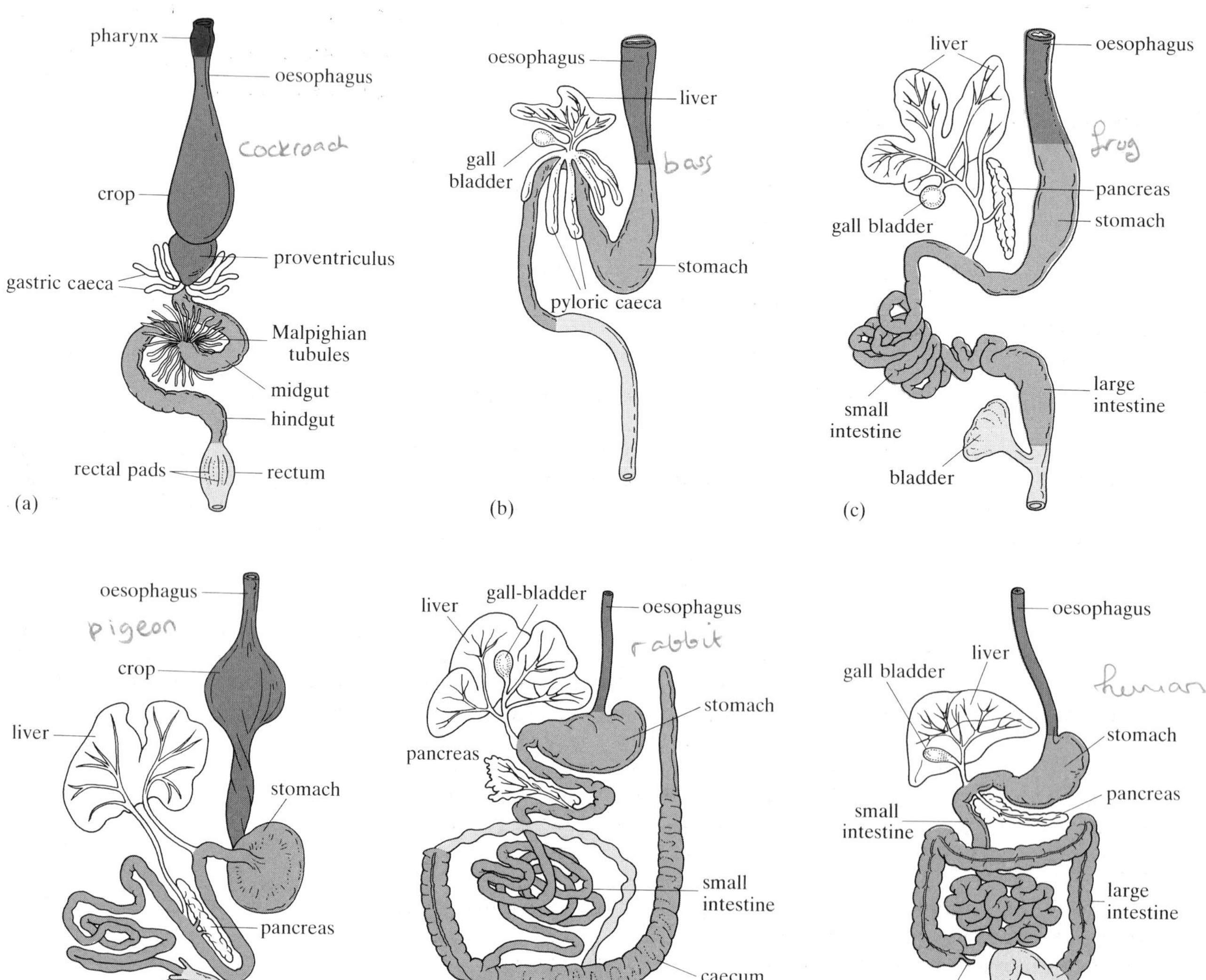

Figure 8.19 Digestive systems (a) cockroach, (b) bass, (c) frog, (d) pigeon, (e) rabbit, (f) human. Different shading is used to indicate the different regions of the system as described in the text.

The receiving region

At the start of the alimentary canal are the organs concerned with feeding and swallowing. Most animals have salivary glands in this region. These secrete enzymes that initiate digestion. In some predators, the saliva contains highly toxic enzymes that can disable or kill prey. In snakes, it is thought that the complex injection apparatus in the teeth evolved initially as a feeding aid. Injecting digestive enzymes into the bloodstream of the prey while it is still alive means that digestion gets going rapidly—particularly important if the prey is to be swallowed whole. In the gut it would take some time for digestive enzymes to get right inside a large prey animal.

The development of saliva as a toxin is a specialized feature. The primary function of the saliva is to lubricate the food, assisting swallowing.

The conducting and storing region

In chordates and some invertebrates there is a region, called the oesophagus, that conducts food from the pharynx to the stomach. There may be an expanded section of the gut, called the crop, that acts as a storage area for food, prior to digestion. Leeches, which take large blood meals at infrequent intervals, have a large crop. Birds too have a crop. Parent birds may regurgitate food from their crops to feed their nestlings.

The digesting and absorbing region

The region in which digestion takes place can be divided into two, the **stomach** and the **intestine**. Mechanical mixing of food with digestive enzymes occurs in the stomach. The contents are churned around and may be further broken down—a process known as **trituration**. Some birds have a strong, muscular gizzard in which the food is broken down by grinding. Grains of sand or small pebbles are swallowed by the bird and retained in the stomach. The movements of the muscular walls of the stomach mix seeds and pebbles and the seeds are broken down by being ground between pebbles.

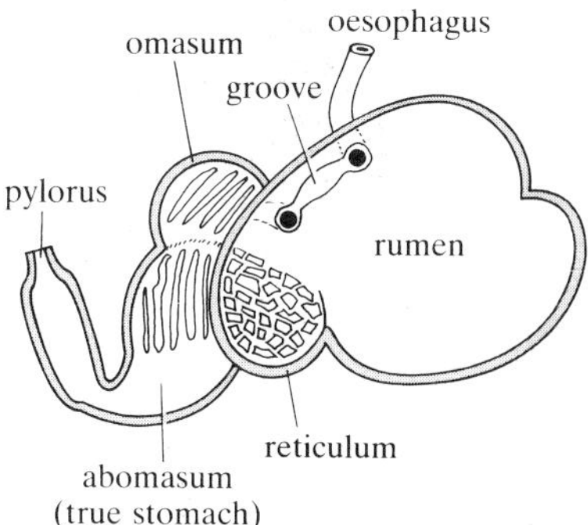

Figure 8.20 A schematic drawing of the stomach of a sheep showing the four chambers, a characteristic feature of ruminant mammals.

The stomach of ruminants (cattle, sheep, deer, giraffe, etc) is specialized in that it is multi-chambered (Figure 8.20). Ruminants are herbivores that feed on grasses and leaves, both of which have a high content of cellulose. This cannot be broken down by most animals because they lack the enzymes necessary to break the β-1,4 linkages in cellulose. However, ruminants are able to use cellulose as a food source because they have a symbiotic relationship with micro-organisms—bacteria, yeast and protistans—that live within the stomach. Some of the micro-organisms can break down cellulose, and the products become available to the ruminant.

In ruminants, food swallowed for the first time passes from the oesophagus into the rumen and reticulum. These chambers are lined with a glandular epithelium, which is unlike that of the normal mammalian stomach. The rumen is inhabited by symbiotic organisms, particularly bacteria. These ferment the cellulose material into fatty acids, methane and carbon dioxide; the gases are eructed (burped!) off, and the fatty acids are absorbed by the epithelium of the rumen. The fermenting mass, along with undigested fibrous matter, is periodically regurgitated into the mouth (the regurgitated mass is called the cud), and is mixed with saliva and thoroughly chewed so that the fibres are broken up, before being reswallowed. When food is swallowed the second time, it passes via a groove in the rumen (Figure 8.20) to the omasum, by-passing the reticulum. The food is churned by strong muscular contractions in the omasum, and is then passed into the abomasum or true stomach, where the digestive juices continue the chemical breakdown begun by fermentation in the rumen. The micro-organisms may also pass into the abomasum and they are digested: this provides the host with a source of amino nitrogen, and many vitamins, particularly those of the B group. Food then passes from the abomasum to the intestine, where digestion and absorption continue in the way described for the human gut.

Some mammals that are not ruminants nevertheless can digest cellulose. Lagomorphs (hares and rabbits) have a simple stomach, but there is a large caecum and appendix (Figure 8.21), which contain cellulose-splitting bacteria. Ingested plant material is broken down in the caecum and appendix by the bacteria; however, since the caecum is at the end of the absorptive region of the gut, the small intestine, the products of bacterial digestion cannot be utilized. The rabbit has a simple solution: the digested food material passes out in the form of soft pellets, which are immediately re-ingested. The breakdown products of cellulose can then be absorbed when the pellets reach the small intestine and new hard pellets are produced in the colon; these are not ingested.

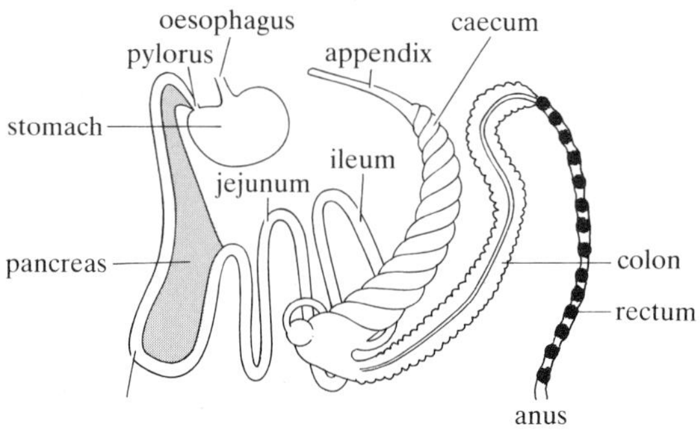

Figure 8.21 The digestive tract of the rabbit.

In many invertebrates there is no real equivalent of the mammalian stomach. In the crayfish (Figure 8.22a) there is a region lined with chitin that precedes the intestine. It is called the proventriculus and it is divided into two parts. The larger anterior chamber is the gastric mill in which food is broken up and sorted. This chamber is normally called the cardiac region, by analogy (false) with the naming of the regions of the mammalian stomach, but you should be aware that the proventriculus is not equivalent to the stomach in function. When food enters the cardiac region it is broken up by an array of teeth and ossicles, and attacked by enzymes secreted by the hepatopancreas (Figure 8.22b). The complexity of the arrangement of teeth can be seen in Figure 8.23 and Figure 8.22b. The teeth are moved by the indirect action of strong gastric

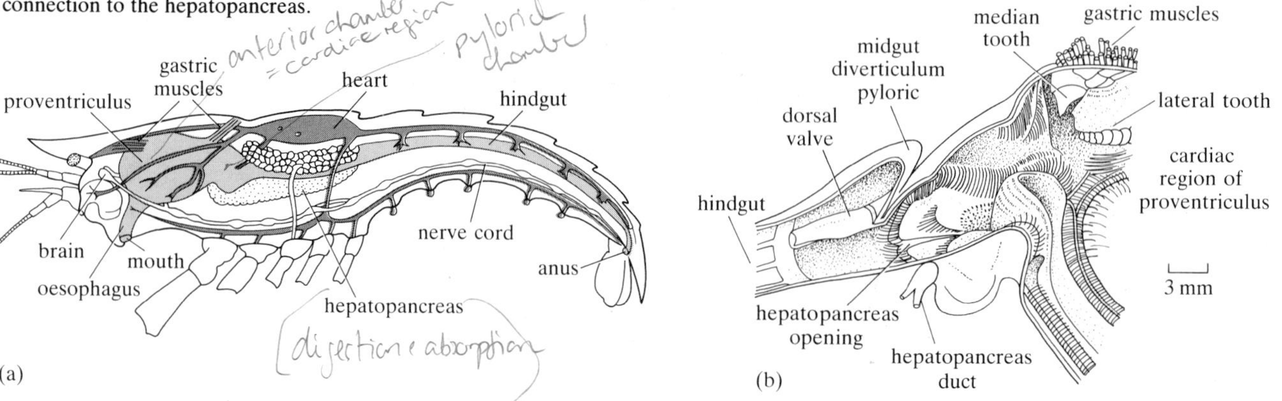

Figure 8.22 (a) Generalized diagram of the internal anatomy of a crayfish showing the main internal organs and the position of the proventriculus. (b) Section through the proventriculus to show the complex sorting apparatus and the connection to the hepatopancreas.

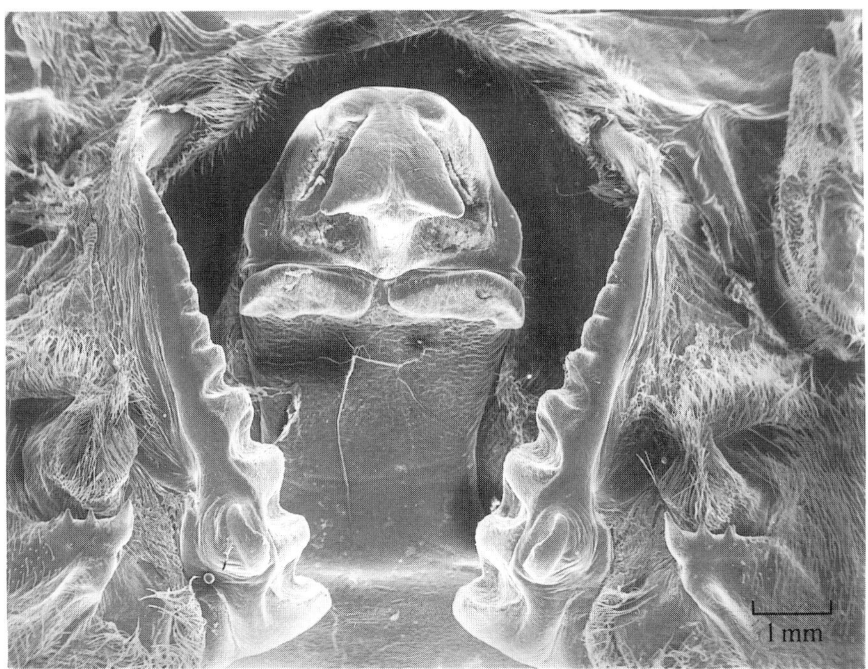

Figure 8.23 Scanning electron micrograph of proventricular teeth in the crayfish.

muscles attached to the wall of the proventriculus. The muscles distort the chamber wall, so that food is ground between the lateral teeth and the median tooth.

The smaller posterior region of the proventriculus is called the pyloric chamber; it is separated from the cardiac chamber by valves. Large indigestible pieces of food can pass through the pyloric chamber and mid-gut along the dorsal pyloric valve into the hind-gut without coming into contact with the mid-gut epithelium. Digestible particles and fluid are squeezed through fine filters into the mid-gut region and into the tubes of the hepatopancreas, where digestion and absorption takes place. The hepatopancreas is a versatile organ, being both a secretor of enzymes and the main absorbing organ. Only a small proportion of digested food is absorbed in the mid-gut.

In vertebrates, digestion initiated in the stomach is continued in the intestine. Many of the digestive enzymes enter the intestine via a duct from the pancreas, where they are produced. There is also a small duct that brings to the intestine a complex solution secreted in the liver. The intestine is the principal absorption region in vertebrates, and there are a number of structural adaptations that facilitate absorption. The surface area over which absorption can take place is greatly increased by folding of the intestinal wall and, in vertebrates other than fish, finger-like extensions called villi. The cells that line the intestine also have finger-like extensions of the cell wall called microvilli. The process of absorption will be considered in more detail in Section 8.6.

The water-absorbing and voiding region

The final section of the alimentary canal consolidates the indigestible remains of the food into faeces before it is passed out via the anus. In terrestrial animals considerable water absorption takes place in this region. In many insects the faeces are dry as a result of water absorption from the rectum (Chapter 11).

8.4.3 Digestion in humans

The human diet usually contains both plant and animal material. The gut (Figure 8.24) is relatively non-specialized by comparison with, for example, herbivorous vertebrates whose diet is dominated by one type of food. Food is hydrolysed by the co-ordinated secretion of digestive juices. To give you an idea of the range and complexity of the digestive juices in humans, they are listed in Table 8.6, together with their complement of enzymes.

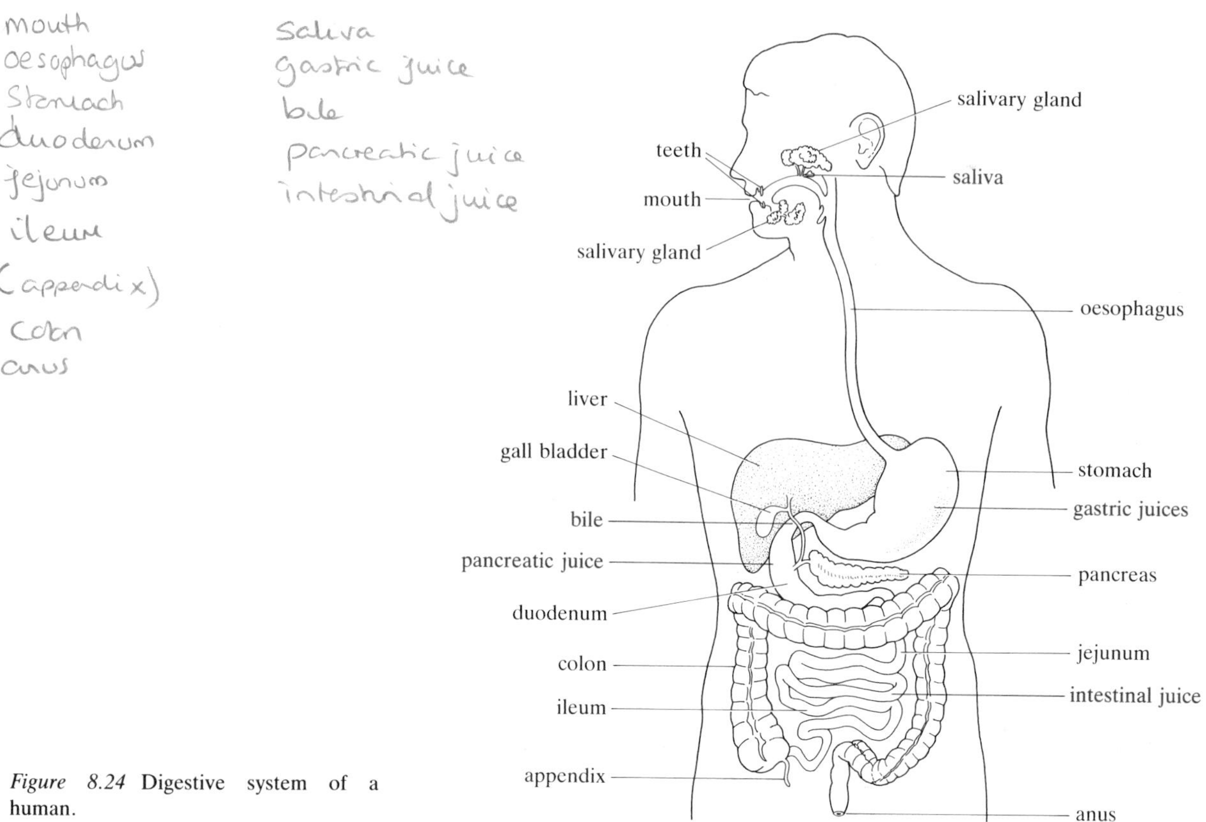

Figure 8.24 Digestive system of a human.

Summary of Section 8.4

1 Two forms of digestion are found in animals; intracellular (inside the cell) and extracellular (outside). Intracellular digestion is more characteristic of primitive animals, and there is an evolutionary trend towards extracellular digestion in more advanced types. This enables a greater division of labour between the cells of the organism.

2 Digestion of the main types of food molecule—protein, carbohydrate and fat—is a progressive, step-by-step process in which larger molecules are fragmented by enzymes into smaller and smaller units.

3 As animals become more complex, it is possible to define four distinct functional regions in the alimentary canal: a receiving region (the mouth), a conduction and storage region, a region of digestion and absorption (the intestine) and a region of compaction and consolidation of the faeces (the rectum) where considerable water absorption takes place.

Now attempt Questions 6 and 7, on pp. 288–289.

Table 8.6 Composition of the digestive juices in an adult human.

Saliva
daily volume secreted: 1–1.5 litres
pH: 6.3–6.8
total solutes: 0.5% (enzyme, salts, dry mucus (mucin))
enzyme: amylase.

Pepsin only found in vertebrates, never invertebrates

Gastric juice
daily volume: 1–3.1 litres
pH: 1–1.5
inorganic solutes: 0.1–0.5% (salts and HCl)
organic solutes: 0.1% (enzymes, mucin, haemopoietic factor)
enzymes: pepsin (+ rennin in infants for milk digestion).

Pancreatic juice
daily volume: 400–800 ml
pH: 7.1–8.2
inorganic solutes: 1% (salts including 0.5–0.7% bicarbonate)
organic solutes: 0.6% (mostly enzymes)
enzymes or precursors: trypsinogen, chymotrypsinogen, procarboxypeptidase, aminopeptidase, lipase, amylase, disaccharases, dipeptidases and nuclease.

Bile
daily volume: 500–1000 ml
pH: 7.0–8.0
inorganic solutes: 0.7–0.8% (salts)
organic solutes: fatty acids 0.02–0.14%, neutral fats 0.01–0.30%, phosphatides 0.05–0.06%, cholesterol 0.05–0.17%, bile acids 0.20–1.80%, mucin and bile pigments 0.5%.
enzymes: none.

Intestinal juice (formerly called succus entericus)
daily volume: 500–1000 ml
pH: 7.0–8.5
inorganic solutes: 1% (salts)
organic solutes: 0.6% (mucin and enzymes)
enzymes or precursors: *traces* of: enterokinase (enteropeptidases), amylase, maltase, lactase, sucrase, lipase, esterases, nucleases, nucleotidases, nucleosidases, phosphatases.

DIGESTIVE ENZYMES – show group specificity

8.5 DIGESTIVE ENZYMES

Digestive enzymes do not have the narrow specificity often characteristic of other enzymes; for example, carbohydrate-digesting enzymes can hydrolyse both plant and animal polysaccharides. They show group (rather than absolute) specificity. The same name is given to enzymes performing a similar digestive function in different animals. For example, trypsin is an enzyme that hydrolyses peptides in the vertebrate intestine but the human enzyme will not necessarily be chemically identical with the corresponding one from a fish. Furthermore, the temperature and pH optima of enzymes acting on similar substances in different animals may not be identical. Trypsin from the vertebrate pancreas has a pH optimum of 7.5–8.0, but in the silkworm and in the cricket, the pH optimum of trypsin extends from pH 6.2 to 9.0.

Same name - same function but can be chemically different between animals

Digestive enzymes, particularly those that act on proteins, are secreted in an inactive form, referred to as a zymogen.

◇ What is a possible reason for this?

◆ It avoids auto-digestion of the gland that produces the enzyme.

Enzyme activation normally takes place where digestion occurs. In many vertebrates, as you will see, activation of the enzyme that initiates the first stage of protein digestion occurs when the inactive precursor, pepsinogen, is converted into the active form pepsin, by hydrochloric acid in the stomach. The relative molecular mass (M_r) of pepsinogen is 42 000, and the conversion seems to be the removal of a substance of M_r 38 000 leaving pepsin, which has an M_r of 4 000.

8.5.1 The mechanism of digestive enzyme action

Digestion consists of the enzymic hydrolysis of macromolecules in food, the three major classes of digestive enzymes being:

- □ Proteases, which hydrolyse the peptide bonds in proteins
- □ Carbohydrases, which hydrolyse the glycosidic bonds in carbohydrates
- □ Lipases, which hydrolyse ester bonds in fats

Other enzymes, which we will not consider here but which may play a small though important role in digestion, are nucleases, nucleotidases and nucleosidases, which break down nucleic acids to their constituent building blocks.

Proteases

Food contains a great deal of protein. Two groups of proteases are responsible for chopping up the component polypeptide chains into free amino acids. **Endopeptidases** break peptide bonds well within the protein molecule, dividing the chain up into sections. Their action increases the number of chain ends at which **exopeptidases** can act. Exopeptidases attack terminal peptide bonds, until only free amino acids remain.

Some proteases are specific for particular peptide links. For example, trypsin, which is an endopeptidase, attacks peptide bonds only where the carboxyl group is part of lysine or arginine.

The conditions under which the protease operates may also be specific. In mammals the endopeptidase pepsin initiates protein breakdown in the stomach. It has a low optimum pH. HCl is secreted into the stomach, and provides an acid environment in which pepsin can function.

◇ What other role does HCl have in the breakdown of proteins by pepsin?

◆ It activates pepsinogen, as you read earlier.

Carbohydrases

Some ingested food, whether plant or animal, may contain free glucose, but most dietary carbohydrate is in a much more complex form, often consisting of numerous hexose sugar units, such as glucose, linked together by a condensation reaction in which one water molecule is removed from each glycosidic bond formed. Carbohydrases break these bonds by a hydrolytic reaction that is the reverse of this condensation reaction.

Carbohydrases can also be grouped into two categories, the polysaccharases and the glycosidases. The polysaccharases, of which the amylases are the most common, act on large polysaccharide molecules such as starch (plant polysaccharide) or glycogen (animal polysaccharide) and split them into trisaccharides or disaccharides. These, together with other carbohydrates actually ingested as trisaccharides (e.g. raffinose, in plants) or disaccharides (e.g. sucrose, in sugar cane, and lactose, in milk) are hydrolysed to monosaccharides by glycosidases such as maltase, sucrase or lactase. The digestion of carbohydrates from complex polysaccharides to monosaccharides occurs, like protein digestion, in a series of steps. Whereas protein digestion often occurs initially in an acid medium, complete hydrolysis of carbohydrates normally occurs under alkaline conditions.

◇ Could carbohydrate digestion occur in the mammalian stomach?

◆ No. The mammalian stomach contents are highly acidic and carbohydrate hydrolysis requires alkaline conditions.

The enzymes that break down carbohydrates are shown in Table 8.7. Few animals will require all of them. The occurrence of digestive enzymes is correlated with the nature of the food ingested. Amylase and maltases are of universal occurrence, and sucrases are much more common than lactases and trehalases.

Table 8.7

Polysaccharides $(C_6H_{10}O_5)_x$	Disaccharides $C_{12}H_{22}O_{11}$	Monosaccharides $C_nH_{2n}O_n$ hexoses $C_6H_{12}O_6$	pentoses $C_5H_{10}O_5$
glycogen (animals) starch (plants) *amylases* →	maltose *maltases* →	glucose	ribose ribulose
cellulose (plants and animals) *cellulases* →	cellobiose →	glucose	
	trehalose *trehalase* (insects and (some plants) →	glucose	
	lactose (mammals) *lactase* →	galactose, glucose	
	sucrose (plants) *sucrase* →	fructose, glucose	

Glucose can exist in two forms, called D-glucose and L-glucose. All the naturally occurring monosaccharides are of the D-form, but two crystalline isomers of D-glucose exist, α-D-glucose and β-D-glucose (Figure 8.25—the OH (hydroxy) groups at carbon atoms 2, 3 and 4 have been left out for clarity). By convention, α-sugars are drawn with the C-1 OH shown pointing downwards, and β-sugars are drawn with it pointing upwards. Glucose in starch, glycogen, sucrose and lactose is in the α-D form, but in cellulose it is in the β-D form, and this makes cellulose difficult to break down for most animals.

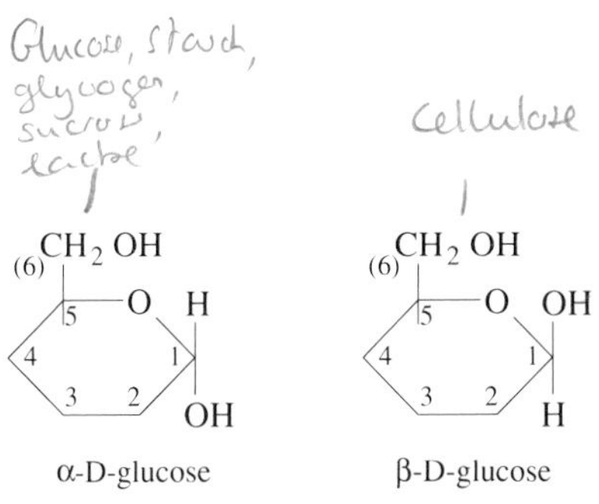

Figure 8.25 The two isomers of D-glucose.

Cellulose is an important structural component of plants, and also a major component of the diet of herbivorous animals. It is broken down by cellulase, but few animals are able to secrete this enzyme. In mammalian herbivores, for instance, it comes from symbiotic micro-organisms in the gut of the host animal. In termites, cellulase is released into the lumen of the intestine by the symbionts, where it digests the wood extracellularly. However, in cattle the symbionts digest the cellulose molecules intracellularly, and are then digested in turn by the host. A very few animals produce their own cellulases, notably the shipworm *Teredo* (a wood-boring bivalve), an isopod *Limnoria*, the silverfish (an insect) and some tilapias (cichlid fish).

Lipases

Fats are not water-soluble, and most undergo emulsification before enzymic attack. The churning of the intestinal contents mixes the fats with bile acids and lecithin, which act in the same way as detergents by lowering the surface tension between fat and water. Lipases can then hydrolyse the fatty components in food, a process that is progressive, as is the breakdown of proteins and carbohydrates. Lipases have even less specificity than proteases and polysaccharases; a single lipase can catalyse all the many stages in the breakdown of a fat. Small amounts of phospholipids, steroids and free fatty acids are found in the diet, but most fats utilized by animals are esters of fatty acids and higher alcohols, often the trihydroxy alcohol glycerol, in which case the fat is termed a triglyceride. The end-products of triglyceride digestion are monoglycerides, fatty acids and glycerol.

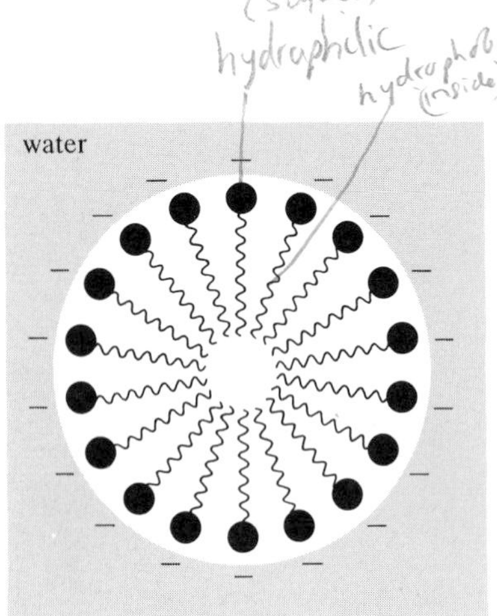

Figure 8.26 Molecules of a polar liquid in a polar solvent such as water form a spherical micelle. The hydrophobic ends tend to avoid contact with the polar solvent by grouping at the centre of the micelle.

The molecules of free fatty acids and monoglycerides have a polar hydrophilic group at one end and a non-polar hydrophobic group at the other. This means that in a polar liquid, such as water, these molecules tend to clump together, as shown in Figure 8.26. These roughly spherical clumps are called **micelles**, with the hydrophilic groups at the surface and the hydrophobic chains pointing into the middle. The micelles diffuse through the unstirred layer at the edge of the intestinal lumen (Figure 8.27) and bring the products of fat digestion to the point where they can be absorbed. The diffusion coefficient of micelles is only a seventh of that of fatty acid molecules, but the concentration inside the micelle may be a thousand times that in aqueous solution. So, although micelles take longer to diffuse than individual molecules, far more molecules move in a given time.

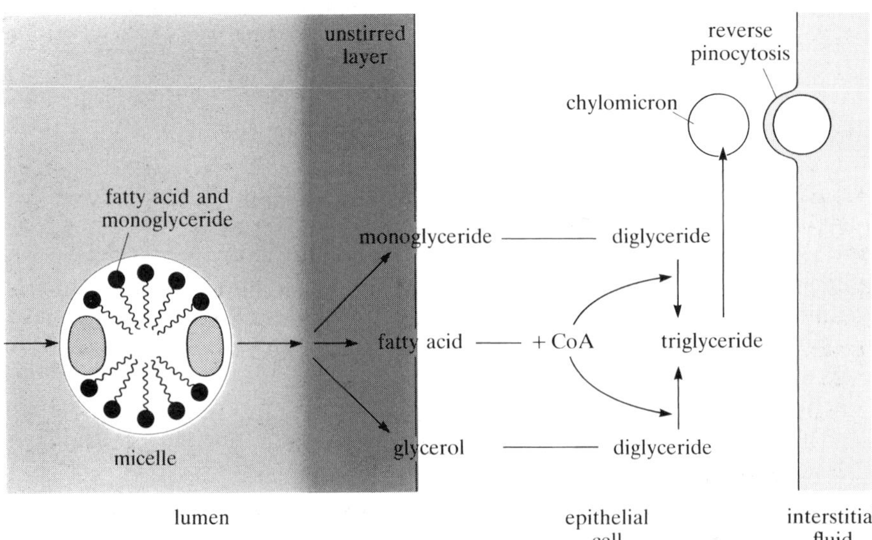

Figure 8.27 Absorption of the products of fat digestion.

8.5.2 The human liver and the secretion of bile

The liver is the largest gland in the human body. It has a wide range of functions, occupying a central position in metabolism. The role of the liver in the regulation of blood glucose levels will be described in Chapter 9. Here we consider one of its major roles in digestion—the secretion of bile.

Bile is one of the digestive juices (Table 8.6). It is a variable, and highly complex, mixture of water and solutes but it contains no enzymes. Between 700 and 1 200 cm^3 are produced each day, about 1% of this volume being the bile acids. Bile acids have several functions, particularly in the final stages of digestion and absorption of triglycerides as you read earlier.

Bile acids are also involved in the elimination of cholesterol. Cholesterol is a major component of cell membranes, but excessive cholesterol in the blood is associated with disease of the circulatory system. Atherosclerosis is a disease of the arteries. The artery walls thicken as a result of the formation of abnormal smooth muscle cells and precipitates of cholesterol. The consequence is an increase in the resistance to blood flow, and an increased risk of blood clot formation on the rough walls of the vessels (Chapter 5). The reasons for this thickening occurring are not yet understood. The level of cholesterol in the blood is kept low, under normal circumstances, by bile production in the liver. Cholesterol is removed from the blood in the liver, and converted into bile acids. These are water-soluble, and pass into the gall bladder for storage and thence into the duodenum. Cholesterol is also secreted unaltered in the bile. It is not precipitated, but remains in solution and so does not block the bile duct.

It is difficult to determine which factors control the secretion of bile. Certainly the flow of bile varies depending on whether a meal has been eaten. For example, in a fasting individual most of the bile acids are stored in the gall bladder whereas after a meal most of the bile acids have passed out of the gall bladder into the duodenum. A number of the mechanisms that control digestion also influence bile secretion, as will be discussed in Section 8.7.

Summary of Section 8.5

1 Digestive enzymes are not as specific as many other enzymes. They are often secreted in an inactive form, which prevents digestion of the secreting organ.

2 There are three major classes of digestive enzymes—proteases, carbohydrases and lipases—and a number that act on nucleic acids.

3 There are two main types of protease action. Endopeptidases attack peptide bonds within the protein molecule. Exopeptidases attack terminal peptide bonds. The presence of HCl in the stomach provides a low pH, the optimum pH for the enzyme pepsin, which initiates protein breakdown.

4 Carbohydrate hydrolysis occurs in alkaline conditions. There are two groups of carbohydrases. Polysaccharases, such as amylase, split large molecules into di- or trisaccharides. Glycosidases hydrolyse these to monosaccharides.

5 Fats are not water-soluble, so bile salts and lecithin reduce surface tension and then lipases hydrolyse the fats. There is little specificity and one lipase can catalyse many stages in the breakdown of a fat.

6 In humans the liver is the largest gland, with a wide range of functions in metabolism and digestion. Bile is one of the gastric juices and it is secreted by the liver. Bile acids are important in the digestion of triglycerides. Cholesterol is converted into bile acid in the liver, so bile is the route by which cholesterol leaves the bloodstream.

Now attempt Questions 8 and 9, on pp. 289–290.

FACILITATED DIFFUSION

Transport of a polar solute through an otherwise impermeable membrane by a process that doesn't require energy (hence also called passive transport). Transport mediated by solute binding to a membrane embedded transport (carrier) protein (or other lipid-soluble molecule). The rate of facilitated diffusion is initially faster than by free diffusion, and reaches a plateau when the transport protein becomes saturated with solute.

ACTIVE TRANSPORT

Movement of a solute against its electrochemical gradient by a process requiring metabolic energy; one type of mediated permeability.

8.6 THE ABSORPTION OF THE PRODUCTS OF DIGESTION

This Section is concerned with absorption from the alimentary canal of amino acids, sugars and fatty acids, released from their respective macromolecules by extracellular digestion. How the materials released following intracellular digestion within a lysosome pass across the lysosomal membrane into the cell cytoplasm is largely unknown. The process may involve transport mechanisms similar to those by which amino acids, sugars and fatty acids cross the luminal plasma membranes of epithelial cells lining an animal's alimentary canal.

8.6.1 The transfer of the products of extracellular digestion

In diffusion or passive transfer, the rate of movement of a transported substance (the substrate) is directly proportional to the difference in its concentration across a cell membrane. Net movement occurs down the electrochemical gradient until the concentrations are equal on either side of the membrane. In mediated transfer, the substrate moves across the membrane at a rate faster than could be accounted for by simple diffusional forces. There are two types of mediated transfer.

If the concentration of a substrate inside the cell during the transfer process never exceeds that outside, then the mediated transfer is referred to as **facilitated diffusion**. It is believed that a specific membrane component is involved in the transfer process, because the rate of transfer is at a maximum when the concentration of substrate on the outside of the cell membrane is high, and the mechanism appears to be saturated. The membrane component appears to have specific sites at which the the substrate is absorbed.

In contrast to facilitated diffusion, a second type of mediated transfer can result in a substrate being concentrated inside the cell *against* an electrochemical gradient. This form of mediated transport is called **active transport**. The active nature of the process necessitates an input of energy which involves ATP.

The types of transport process are summarized in Figure 8.28.

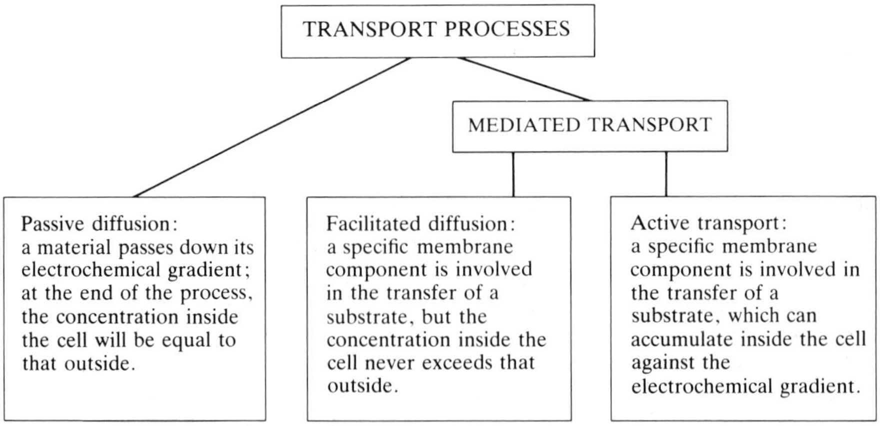

Figure 8.28 Processes for the transfer of the products of digestion across cell membranes.

The mechanism of absorption of the breakdown products of fats, at least across the cell membrane of an intestinal cell, is probably by passive (lipid-soluble) transport.

The transport of amino acids and sugar into the cells lining an animal's alimentary canal appears to be largely by processes mediated by carrier

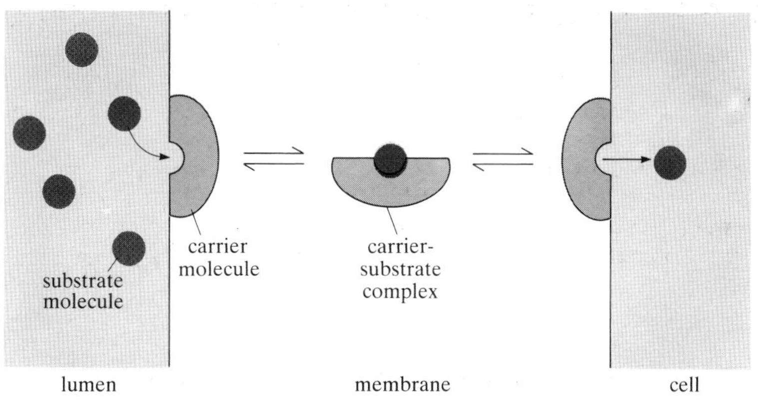

Figure 8.29 A possible transport mechanism for amino acids and sugars.

molecules (Figure 8.29). Much experimental work has been involved in attempting to isolate these carriers, and some success has been achieved, but discussion of the work is outside the scope of this book.

Amino acids and sugars can accumulate against an electrochemical gradient and this can be blocked by metabolic inhibitors such as cyanide.

AMINO ACIDS & SUGARS = ACTIVE TRANSPORT

◇ What conclusion can be drawn from the observation that cyanide blocks amino acid and sugar accumulation?

◆ The transport process is an active one.

One of the most important considerations in examining the transport processes for sugars and amino acids in nutrition is the degree of specificity of the mediated systems for a particular substrate.

8.6.2 Sugar transport

The most important mediated transport system for sugars in the alimentary canal of animals is that for hexose monosaccharides. For a molecule to be transported by this system, it must be electrically neutral and must have:

1 A six-membered carbon and oxygen ring structure of D-configuration: pentoses (e.g. xylose) and sugars of the L-configuration (e.g. L-glucose) are poorly transported by this system.

2 A free hydroxyl group on the second carbon atom of the hexose ring; 2-deoxy sugars are not usually transported.

3 A methyl or substituted methyl group on the fifth carbon atom.

These points are summarized in Figure 8.30.

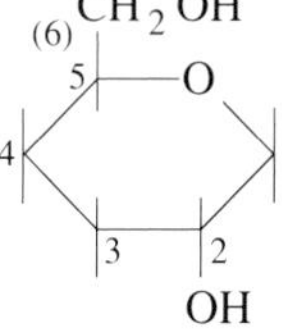

Figure 8.30 Minimum structural requirements for a sugar molecule to be transported in the intestine of vertebrates and invertebrates.

In many but not all animals, an important feature of the mediated transport of sugars is its dependence on the presence of a Na^+ gradient between the gut lumen and the cytoplasm of the epithelial cell. The Na^+ concentration in the lumen normally exceeds that inside the cell, and one hypothesis proposes that Na^+ forms a ternary complex (of three things) with the carrier molecule and the sugar (Figure 8.31). The complex forms externally, and because the Na^+ concentration is high outside and low inside the cell, it moves into the cell under the driving force of the Na^+ electrochemical gradient; once inside, the complex dissociates. Na^+ diffuses down to the basolateral membrane of the cell, where an active transport site pumps it out. The sugar diffuses through the cell to the basal surface and passes out of the cell by passive or facilitated diffusion. Thus the energy input to the transport mechanisms is indirect, occurring at the base of the cell; the pumping mechanism keeps the

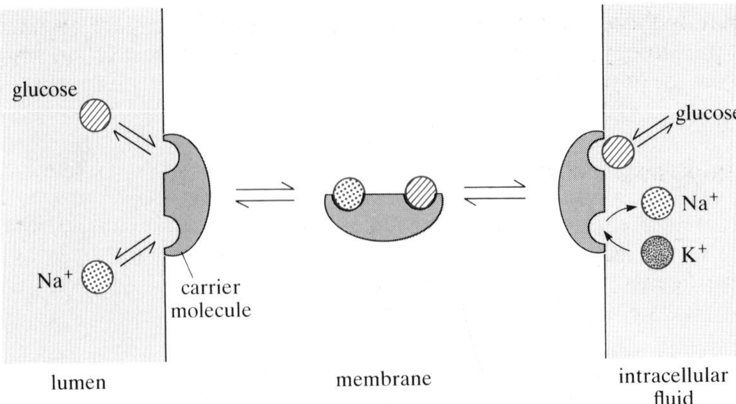

Figure 8.31 A hypothetical sugar transport system.

intracellular concentration of Na^+ low and thus maintains the electrochemical gradient between the cell and gut lumen. You should note that although the sodium gradient hypothesis for sugar transport (Figure 8.32) is backed by much evidence, it is but one of four explanations that have been advanced. None of them has been generally accepted.

Glucose and galactose are known to be transported by processes that are

- ☐ Na^+ dependent
- ☐ Unequally distributed along the length of the intestine
- ☐ Energy dependent
- ☐ Dependent on a carrier of some sort

The action of phlorrizin (phloretin 2′β-D-glucoside) is interesting because it reversibly blocks the entry of glucose and galactose into the epithelial cells of the lumen. This strongly suggests that a carrier molecule is involved. A carrier mechanism would require energy, in the form of ATP, so blocking ATP production with a metabolic inhibitor should stop glucose transport. 2,4-Dinitrophenol (DNP) is such an inhibitor, and it does indeed block glucose transfer. However, if the carrier mechanism is linked with sodium, then it could be that DNP is blocking ATP supplies to a sodium pump, rather than directly affecting a possible glucose carrier molecule.

Figure 8.32 The sodium gradient hypothesis to explain the active uptake of sugars and amino acids. The relative concentrations of the various solutes are shown by the symbol sizes (smaller symbols indicate lower concentrations).

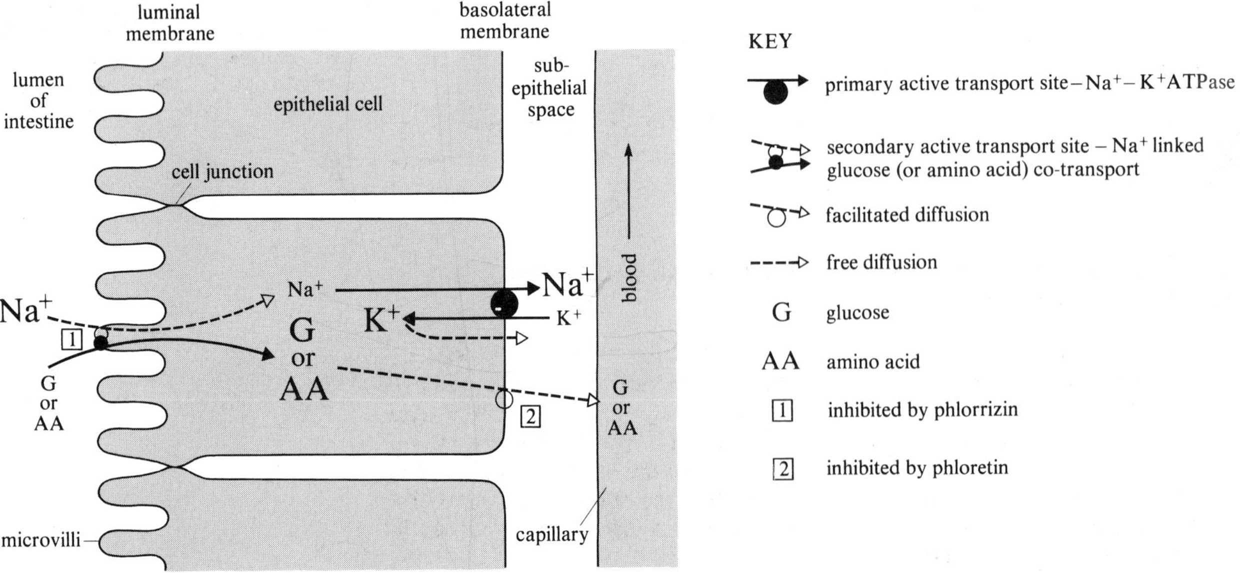

Unfortunately there is not the space to go further into this fascinating area of digestive physiology, but you should be aware that transport mechanisms are an important area of research.

8.6.3 Amino acid transport

The sodium gradient hypothesis for sugar transport may also be applicable to the amino acid transport system in the intestine of many animals. However, in a number of cases, absorption of an amino acid can also occur via mediated transport that is independent of sodium.

◇ Sugars and amino acids can interfere non-competitively with one another during transport into the cells lining the gut. Would the fact that both amino acids and sugars compete for the sodium gradient account for this?

◆ Yes, this would be a reasonable explanation. It has been verified by experiments with the mammalian intestinal transport system.

Amino acids and sugars do not compete, however, for a common transport molecule. Phlorrizin has no effect on amino acid transport, whereas it does block glucose transport.

Experimental evidence suggests that there are at least four transport processes for amino acids in the gut of vertebrates and invertebrates.

L-Amino acids have a higher affinity for transport sites than D-amino acids. This seems to be because amino acids normally have three points of attachment to the site (Figure 8.33a): the side-chain (R), the carboxyl group (COOH) and the amino group (NH_2). The spatial arrangement of the three points can be matched by only the L-configuration (Figure 8.33b): the D-configuration is able to attach itself to only two of these points (Figure 8.33c) and is thus transported much less effectively. This geometry of the transport site is perhaps not surprising when you recall that the amino acids found in animal protein occur almost exclusively in the L-configuration.

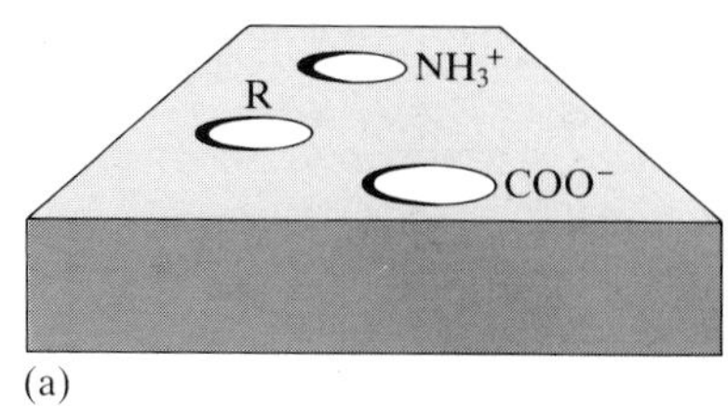

(a)

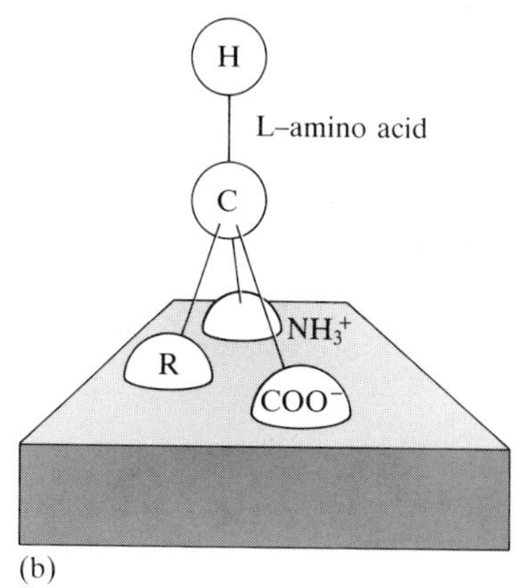

(b)

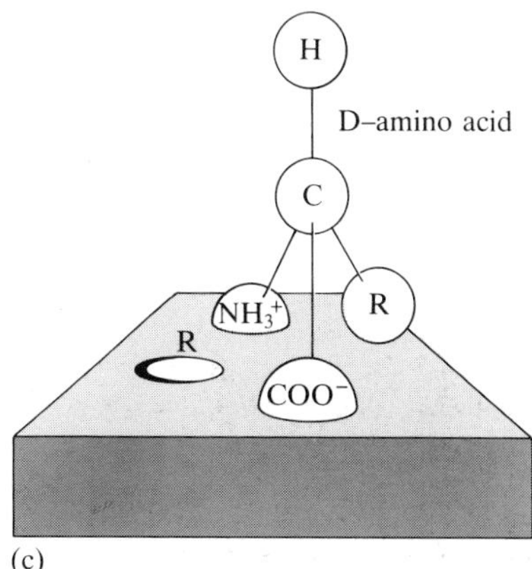

(c)

Figure 8.33 A schematic representation of the attachment of an amino acid to its carrier site. The carboxyl group is shown as COO^- and the amino group as NH_3^+.

AMINO ACIDS
- R side chain
- Carboxyl group COOH
- amino group NH2

8.6.4 Fat absorption

In Section 8.5.1 you read how free fatty acids and monoglycerides arrive at the luminal membrane as micelles. The micelles do not pass into the membrane. The fatty acids and monoglycerides are absorbed whereas the bile acids remain in the lumen, to be absorbed subsequently in the ileum. In the epithelial cells, triglycerides are synthesized and enveloped with a hydrophobic coat to form chylomicrons. The chylomicrons pass into the extracellular fluid (Figure 8.27).

8.6.5 The transport of absorbed food products from the gut cells around the body of an animal

What happens to the absorbed food products after they enter the gut epithelial cells depends on the type of circulatory system, if any, of the animal (see Chapter 5). Where a circulatory system exists, sugars and amino acids move by diffusion and mediated transfer from the gut cells into the bloodstream, then via the blood to sites in the body where they can be used in synthetic and metabolic processes. In vertebrates, these substances are initially transported via the blood to the liver, where they may be stored in various forms or metabolized.

◇ Vertebrates also have another system that can drain fluids from the intestine: what is this?

◆ The lymphatic system.

Although amino acids and sugars largely pass into the portal blood, some of these compounds, together with most of the absorbed fats, pass into the lacteals, the blind-ended lymph vessels in the centre of each intestinal villus, and reach the blood system via the thoracic duct. Chylomicrons are too large to enter capillaries and they too pass into the lacteals.

Summary of Section 8.6

1 The products of digestion (amino acids, sugars and fatty acids) enter the cells of the alimentary canal by diffusion or mediated transfer.

2 The most important transport system for sugars is that for the hexose monosaccharides, e.g. glucose. This system requires energy, and depends on a carrier molecule and the presence of sodium ions.

3 Amino acids may be transported by a similar system to that for sugar transport, but there are also other systems for their transfer that are independent of sugar transport.

4 In animals that have circulatory systems, absorbed food products are distributed around the body via the blood. In vertebrates, they pass initially to the liver where they may be stored or metabolized.

8.7 THE CONTROL OF MAMMALIAN DIGESTION

Much of the early work on the control of secretion in the mammalian digestive tract was carried out by the Russian physiologist Pavlov, who believed that the coordination and regulation of digestion were effected solely by the nervous system. Salivary secretion is indeed coordinated by nervous stimuli alone. Pavlov was able to show that salivation can be a conditioned reflex, as you saw in Chapter 4. If a bell is rung when dogs are presented with food, after a time ringing the bell will initiate salivation, even in the absence of food. Gastric secretion too is initially controlled by the nervous system. The stomach is innervated by a branch of the vagus nerve (Figure 8.34), and stimulation of the vagus near the stomach is followed by the secretion of acid, mucus and pepsinogen.

In an experiment, this effect can be mimicked by the application of the neurotransmitter acetylcholine, which is known to be released from the vagus nerve, directly to the stomach. In an experimental preparation the oesophagus can be surgically cut and both cut ends made to open separately on the surface of the skin. This preparation allows what is termed 'sham-feeding'. Food that is ingested is then diverted to the exterior via an oesophageal opening without being digested. In this preparation there *is* secretion of gastric juice, provided that the vagus is intact. Furthermore, part of the stomach can be isolated surgically and connected only by nerves and blood vessels, as was done by Pavlov in his experiments (hence the name Pavlov pouch—Figure 8.34a). Using this preparation it is possible to demonstrate that gastric juice is still released in the pouch when the vagus nerve is stimulated.

But when a different sort of pouch without a vagal (parasympathetic) nerve supply but with intact blood vessels is isolated from the rest of the stomach (a Heidenhain pouch), gastric secretion will still occur within the pouch provided that a region of the stomach known as the antrum is intact (Figure 8.34b).

◇ What does this suggest?

◆ This suggests that gastric secretion is not necessarily stimulated directly by impulses in the vagus, but that a hormone is released from the antrum into the blood.

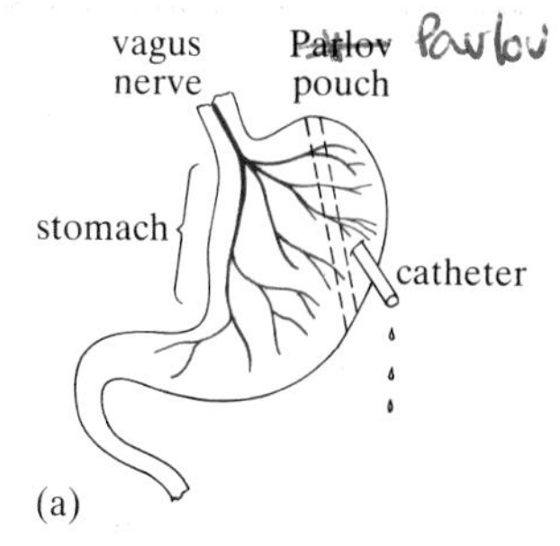

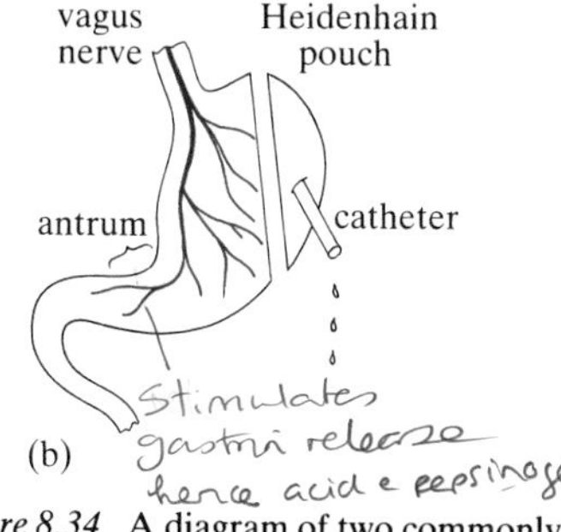

Figure 8.34 A diagram of two commonly used types of gastric pouch. (a) A Pavlov pouch. This is constructed by sewing the opposing mucosal surfaces together beneath the muscle layer in the region indicated by the two dotted lines. The vagal nerve and blood supply to the pouch are preserved intact. (b) A Heidenhain pouch. This is constructed by completely severing the mucosa and muscle of the stomach in the region indicated by the two solid lines. Separate whole stomach and body pouches are created. A blood supply continues to reach the Heidenhain pouch along the outer edge of the stomach, but the vagal innervation (*red*) is severed.

This hormone is **gastrin**. The presence of food, which causes distension of the antrum, also stimulates gastrin release. Gastrin stimulates the release of acid and pepsinogen by the cells of the stomach lining—the mucosa. The secretion of gastrin is stimulated by the introduction into the stomach of a number of substances, including ethanol, amino acids and proteins. Of these, proteins are the most effective. Gastrin release is inhibited under certain circumstances. For example, when we become frightened or depressed there may be inhibition of gastric secretion (called cephalic inhibition). There is also negative feedback from the stomach (gastric inhibition) whereby if the pH falls below 2.5 gastrin secretion is inhibited. If there is little food in the stomach then acidity will rise. When more food is taken in the stomach contents will be diluted, the acidity will fall, and gastrin secretion will be stimulated again.

So both gastrin and acetylcholine stimulate gastric secretion. There is evidence that there is a common mediator of their actions. The secretory cells are very sensitive to histamine, and the cells of the gastric mucosa are rich in the enzyme responsible for histamine synthesis—histidine decarboxylase. Anti-histamine drugs generally do not inhibit gastric secretion, which was a puzzling result when originally published in the scientific literature. However, it is now known that there are two types of histamine receptor, H-1 and H-2. H-1 receptors are found in smooth muscle, for example in the wall of the gut, whereas H-2 receptors are found in the secretory cells of the gut. H-2 receptors are not sensitive to H-1 antagonists and so gastric secretion is not inhibited by most anti-histamine drugs.

Gastrin also stimulates the production of bile by the liver. It exerts control of secretion via two routes. Firstly, it acts directly on the liver cells. Secondly, it acts indirectly through the link between stomach acidity and the hormone secretin (described later in this section) which also stimulates bile secretion.

Insulin (Chapter 9) too is a stimulator of gastric secretion, though it acts in a very complicated way. When injected into a human intravenously it causes hypoglycaemia. The hypothalamus responds to hypoglycaemia in a number of ways, including increased stimulation of the branches of the vagus nerve that innervate the gastric secretory cells. As a result, gastric secretion occurs.

◇ What would be the likely effect of injecting both insulin and a large amount of glucose?

◆ The hypoglycaemia that you would expect if insulin was injected alone would not occur since the blood glucose level would be high. You would not expect gastric secretion to be stimulated, and this is what is observed in actual experiments.

Injection of insulin provides a useful post-operative tool in medicine. One treatment for peptic ulcers is cutting the branches of the vagus nerve to the stomach. The success of the operation can be assessed by injecting insulin. There should be no increase in gastric secretion if the nerve branches have been completely severed.

A negative feedback mechanism prevents overloading of the small intestine with acid contents by inhibiting gastric secretion after acidic stomach contents have passed through into the duodenum. This feedback is thought to be hormonal, and the hypothetical hormone involved has been termed entero-gastrone. However, several hormones have been found to inhibit gastric secretion, notably **CCK-PZ** (cholecystokinin-pancreozymin), VIP (vasoactive intestinal peptide) and GIP (gastric inhibitory peptide).

Stomach function is also inhibited by somatostatin, a hormone produced in the D cells of the Islets of Langerhans (see Chapter 9). Somatostatin inhibits insulin production and exerts its inhibitory effect on the stomach partly through the hypothalamus. There may be links between this hormone and the formation of duodenal ulcers.

The secretion of enzymes by the pancreas is controlled by a number of factors. The sight or smell of food is sufficient to stimulate their release in small amounts. Both gastrin release and stomach distension also stimulate their release, but the major stimulus comes from the intestine. Bayliss and Starling were working on nervous reflexes and the alimentary system in 1902 when they discovered that acid injected into the duodenum, but not into the bloodstream, induced pancreatic secretion. They ground up cells from the intestinal mucosa with sand and produced a filtered, neutralized extract, which they then injected intravenously. This stimulated pancreatic secretion, so they suggested that a hormone was responsible and they named it **secretin**.

The pancreatic secretion has two major components, an alkaline fluid and a group of versatile enzymes. Secretin stimulates the pancreatic cells to produce the alkaline fluid, which contains a substantial concentration of bicarbonate ions. The strongest stimulus for secretin release is a fall in pH in the intestine. At pH 4.5 secretin release is below the threshold for stimulating pancreatic secretion. As the pH approaches 3.0, pancreatic secretion increases and the alkaline fluid released pushes the pH back towards normal. However, there is some doubt whether, under normal conditions, the pH ever falls this low. A possible explanation that has been put forward is that potentiation of the response to secretin occurs—that is, some other hormone acts together with secretin to enhance its effect. One possible candidate is CCK-PZ, which is released when the duodenum is perfused with phenylalanine. As well as inhibiting gastric secretion, it also enhances the secretion of bicarbonate ions in response to secretin. The response to secretin and phenylalanine together is greater than to either on its own.

There are several links between the regulation of blood glucose level and the control of gastric and duodenal secretion. For example, three gastric hormones, gastrin, secretin and CCK-PZ, may stimulate insulin production, which in turn could stimulate gastric secretion. It is not surprising that there should be these links, given that blood glucose regulation and digestion are components of the same system. However, it does highlight the need for caution when evaluating the control of a single component of a system, like gastric secretion. The importance of interactions was a theme that was developed in Chapter 4.

Summary of Section 8.7

The highly developed mammalian gut requires a precise and coordinated release of digestive enzymes at appropriate places along its length. The control of gastric and pancreatic secretion is complicated. A neurotransmitter (acetylcholine) and three hormones (gastrin, secretin and CCK-PZ) play crucial roles. The actions of acetylcholine and gastrin are mediated by histamine. Insulin stimulates gastric secretion indirectly via the hypothalamus.

Now attempt Question 10, on p. 290.

8.8 THE CONTROL OF FEEDING

Very few animals feed continuously. Even filter feeders have the ability to stop feeding, and so there must be some form of control of feeding. This control has been extensively studied in mammals, since there is an obvious application to humans.

In mammals, it has been suggested that areas in the hypothalamus control food intake. These areas have been described as the 'satiety centre' and 'feeding centre'. Brain lesioning experiments in rats have been carried out in an attempt to localize these centres and investigate their effects. Destroying the so-called satiety centre causes rats to overeat grossly, and they rapidly become obese. However, starving rats will not eat if the satiety centre is stimulated electrically. If the satiety centre is intact, but the feeding centre is destroyed rats will not eat under any circumstances. Physiologists are interested in the nature of the signals to which these two centres normally respond. However, the results of brain lesioning experiments need to be treated with extreme caution, as you read in Chapter 4.

Control of food intake is normally very accurate. A famous example of this accuracy in humans is given by one of the authors of a book on human nutrition, who asserts that he can wear a tailcoat made for him 40 years previously despite the fact that he has eaten some twenty tons of food since taking delivery of it. Not all humans are in such a satisfactory position after 40 years! Clearly, there are both long-term and short-term control mechanisms.

8.8.1 Short-term control

These are good reasons for an interest in the control of food intake in humans since many people in highly developed countries appear to have problems related to excessive consumption. Many hypotheses have been put forward to explain short-term control.

Although we often relate hunger to an empty stomach, secretions from the stomach itself play only a minor role in producing a sensation of hunger. Removal of the stomach alters the amount of food eaten per meal and there is an increase in the frequency of meals, but it does not alter the total food intake.

Protein digestion causes an increase in metabolic rate, called the specific dynamic action of protein. It has been suggested that a *decrease* in metabolic rate (and hence a decrease in temperature) as a result of depletion of reserves will produce hunger, whereas specific dynamic action produces satiety. However, it is difficult to separate the normal mechanism of temperature regulation from the effects of food intake.

The level of glucose in the blood is regulated precisely, and it has always seemed that there ought to be a link between glucose and hunger/satiety.

Glucose is a major energy source for cells. Tissues such as the brain need a continuous supply since they are unable to store glycogen (a glucose polymer). The difference between glucose levels in the arteries and in the veins (termed Δ-glucose, where Δ means change) is high during satiety and low during hunger. Note that the value of Δ-glucose tells us nothing about the *absolute* level of blood glucose. An inverse relationship between hunger and Δ-glucose has been observed in humans. Evidence for a causal connection between Δ-glucose and hunger could be obtained if reducing Δ-glucose to almost zero produced hunger. However, when humans were injected with adrenalin, which reduced Δ-glucose almost to zero, anorexia was induced instead of hunger. This observation can be explained by considering the action of glucoreceptors in the liver.

The membrane potential of the glucoreceptors responds to changes in the flow of glucose out of the liver cells; low flow rates cause depolarization, high flow rates do not. A low flow rate producing depolarization sends signals into the central nervous system, inducing a 'hunger' sequence. The effect of adrenalin is to increase the rate of glycogenolysis (the breakdown of glycogen to glucose), and the resulting large outflow of glucose from liver cells would not depolarize the glucoreceptors, although it could produce a situation where the glucose levels in the arteries and veins were the same and hence Δ-glucose was zero. The evidence from adrenalin injection casts doubt on the existence of a direct causal link between a high Δ-glucose and satiety.

8.8.2 Long-term control

Long-term control of feeding in humans is likely to be related to long-term control of fat stores. If fat stores are controlled at all, there needs to be a mechanism in the central nervous system for monitoring fat. No such mechanism has yet been found. There is some evidence that adipose cells, when depleted of fat, produce a prostaglandin that stimulates the feeding centres in the brain. There is support for this idea from evidence that injecting certain prostaglandin antagonists into the hypothalamus produces anorexia. However, injection of some other prostaglandin antagonists promotes feeding, so it is necessary to suggest that adipose cells produce a *different* prostaglandin when they are full of fat reserves. Long-term control is an expanding area of research, with its very obvious application to the human condition, but there are no satisfactory explanations of the form of the control mechanism. Many research workers believe that body weight is regulated, just as blood glucose level is, but how such regulation could work is a matter for conjecture and workers in the area of systems theory would dispute the notion that regulation of body weight is a possibility.

SUMMARY OF CHAPTER 8

The first Section examined nutritive requirements and feeding mechanisms in animals. The raw materials for cellular metabolism are supplied from an animal's diet. There is a general similarity in the major metabolic sequences of different animals, but there are variations in the extent to which different animals are able to manufacture certain chemicals. Those chemicals that must be supplied in the diet are called essential. The dietary requirements of animals can be defined quantitatively and qualitatively in terms of a number of foodstuffs.

The variety of mechanisms by which animals capture and retain food material from their environment can be seen within a single well-defined taxonomic group, such as the polychaetes. Each polychaete species has a number of structural and functional features associated with the capture and retention of a certain type of food.

This diversity of feeding habits leads to the exploitation of a variety of foodstuffs and is an example of adaptive radiation during the evolution of an animal group. Animals have been opportunists in the development of their feeding habits. If potential food material is present, existing structural and functional features may become modified and adapted, and a novel feeding habit frequently emerges. In the case of *Scrobicularia*, the gill structure enables the animal to be a suspension feeder like other lamellibranchs, but it can use its siphons in deposit feeding at low tide.

This diversity and divergence of feeding habits can be contrasted with the convergence of structural and functional features displayed by taxonomically distinct animals that have adopted identical feeding habits—for example, the feeding mechanisms employed by suspension feeders. In these animals, water movement is frequently produced by beating cilia, and particles are often trapped and transported in mucus: normally the captured particles are sorted, and a proportion of them eventually rejected. By comparing animals that feed on very different sorts of food, you should now be able to recognize that feeding involves special, adaptive structural features and each type of feeding mechanism is often associated with certain types of physiological feature or certain behavioural characteristics.

Later Sections examined digestion and absorption. Two forms of digestion occur: intracellular (inside the cell) and extracellular (outside). Intracellular digestion is characteristic of primitive animals, and there is an evolutionary trend towards extracellular digestion in more advanced types: extracellular digestion occurs in a tubular cavity within the animal, the gut, and the products of digestion are absorbed across the cells that line it. In more complex animals, there is a division of labour in the alimentary canal, and it is possible to define four distinct regions: a receiving region, a conduction and storage region, a region of digestion and absorption, and a region of compaction and consolidation of the faeces. The form and function of these regions can vary depending on the nature of the food ingested.

Digestion of the main types of food macromolecules—protein, carbohydrate and fat—whether intracellular or extracellular, occurs as a step-by-step process in which the macromolecular chains are fragmented into smaller and smaller units. The final stages of hydrolysis of protein and carbohydrate occur on or inside the intestinal cells. The means by which sugars, amino acids and fatty acids are transported into the intestinal cells was considered, and a detailed study was made of the human intestine. The complicated mammalian gut requires a precise and coordinated release of digestive enzymes at appropriate points along its length. This is achieved by nervous and hormonal control.

OBJECTIVES FOR CHAPTER 8

When you have read this chapter you should be able to:

8.1 Define and use, or recognize definitions and applications of each of the terms printed in **bold** in the text.

8.2 Explain why the essential food requirements for one animal species may differ from those of another, and interpret experimental results that involve growth responses to quantitative and qualitative variations in the supplied diet. (*Questions 1 and 2*)

8.3 Describe and give examples of two types of microphagous feeder—suspension and deposit feeders. (*Question 3*)

8.4 Interpret the results of experiments designed to assess whether microphagous feeding is, or is not, selective. (*Question 4*)

8.5 Identify which of the three types of macrophagous feeder an animal is, based on the nature of its feeding appendages or buccal apparatus. List those features of carnivorous gastropods that are associated with the macrophagous habit. (*Question 5*)

8.6 Recognize the four major regions of the digestive system, and recognize features of ruminant digestion that are different from human digestion. (*Question 6*)

8.7 Distinguish between intracellular and extracellular digestion, and summarize the advantages of extracellular digestion for more complex animals. (*Question 7*)

8.8 Explain the major properties of digestive enzymes and bile. (*Question 8*)

8.9 Describe how amino acids and sugars are transported from the gut of an animal; identify the structural requirements for amino acid and sugar transport. (*Question 9*)

8.10 Evaluate experimental evidence that demonstrates the importance of gastrin, CCK-PZ and secretin in controlling enzyme secretion. (*Question 10*)

8.11 Give a brief account of the control of feeding and digestion in humans.

QUESTIONS FOR CHAPTER 8

Question 1 (*Objective 8.2*) In a number of deletion experiments to determine the nutritional requirements of a flour-beetle, *Tribolium*, the larval stages were reared on diets lacking one of a number of amino acids. Matched groups of larvae were weighed after 15 days on each diet. The average weights of the animals on each of the diets were:

Diet	Average weight (g)	Diet	Average weight (g)
Complete diet	1.11	No hydroxyproline	1.09
No proline	1.00	No threonine	0.01
No leucine	0.03	No glycine	1.31
No phenylalanine	0.01	No serine	0.84

Which of the amino acids listed are essential?

Question 2 (*Objective 8.2*) A group of flagellate protistans use vitamin B_1 (thiamine) in metabolism. Different subgroups of flagellates have different requirements for thiamine in the diet. Subgroup 1 must be supplied with thiamine in the diet. Subgroup 2 requires only thiazole. Subgroup 3 needs to be supplied with very simple amino acids only. Subgroup 4 needs to be supplied with pyrimidine and thiazole.

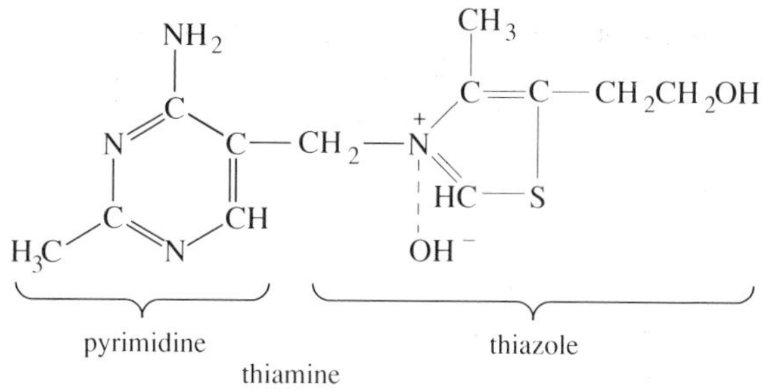

pyrimidine thiazole
thiamine

Figure 8.35 The molecular structure of thiamine.

(a) Given this information and knowing the structure of thiamine from Figure 8.35, what can you say about the synthetic ability of each of the subgroups of flagellates? Answer by selecting appropriate statements from the key below.

KEY

A It can synthesize pyrimidine.
B It can link pyrimidine and thiazole.
C It can synthesize thiazole.
D It cannot synthesize pyrimidine.
E It cannot link pyrimidine and thiazole.
F It cannot synthesize thiazole.

(b) Which of the subgroups of flagellates has the greatest 'synthetic ability'?

(c) Which of the subgroups of flagellates has the greatest 'synthetic disability'?

Questions 3 and 4 The following list contains items of information about feeding and feeding mechanisms. Study the list, and then answer the questions that follow.

1 The organism has a ring of tentacles that are used in feeding.
2 The surrounding deposit is indiscriminately consumed.
3 Feeding is more or less continuous.
4 Ciliated grooves effect a mechanical sorting on the basis of particle size.
5 The organism has special stinging cells that trap and paralyse the prey.
6 The water current passes through fine hair-like setae on the limbs.
7 Food consists of largely microscopic plankton living suspended in water.
8 Tentacles are mobile searching structures.
9 A water current is created by the beat of appendages called fans.
10 A water current is produced by ciliary beating.
11 Particles are trapped in a net-like mucus bag.
12 Only particles of a certain size are ingested.
13 Swallowing occurs periodically.
14 Sorting of the particles occurs at the palps.
15 The organism is a deposit feeder.
16 Food is passed backwards towards the mouth in ciliated grooves.
17 Secreted mucus entangles the captured particles.
18 The organism lives within a tube.
19 A proportion of the suspended particles are used for tube building.
20 The captured food is brought to the mouth by the bending movement of the tentacles.

Question 3 (*Objective 8.3*) *Sabella* and *Amphitrite* are both polychaete worms. Choose those items in the list that represent (a) similarities between the feeding mechanisms of the two worms, and (b) points that are not applicable to the feeding of either worm.

Question 4 (*Objective 8.4*) Figure 8.36 shows a feeding appendage of a species of the copepod *Calanus*. Three members of the phytoplankton are drawn to scale (*red*). Select items from the list to show what you would deduce about the feeding of *Calanus* from the information in this diagram.

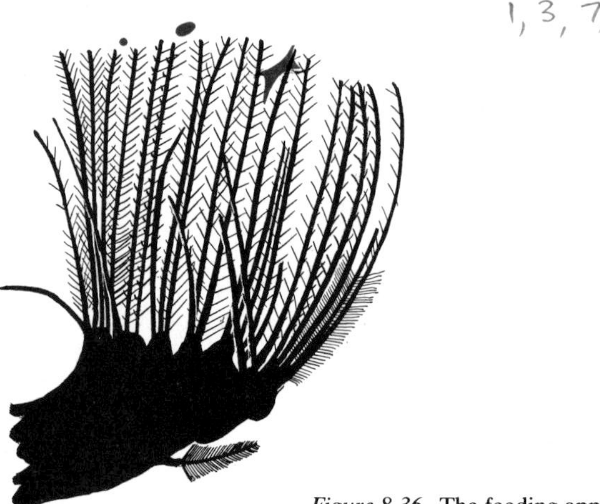

Figure 8.36 The feeding appendage of *Calanus*, a copepod.

Question 5 (*Objective 8.5*) Figures 8.37a and b show the teeth and jaws of two vertebrates. What are the predominant features of these jaws that would enable you to identify the category of macrophagous feeders these animals belong to? What type of food source are they likely to rely on?

(a)

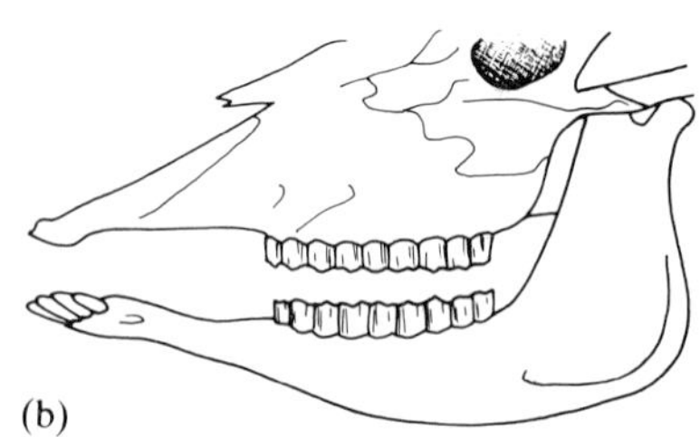

(b)

Figure 8.37 The skull, jaws and teeth of two vertebrates.

Question 6 (*Objective 8.6*) Figure 8.38 is a diagram of the digestive tract of a cockroach *Periplaneta americana* whose mouth parts are used for tearing and crushing food. Identify the regions of the digestive tract on the basis of the classification in Section 8.4.

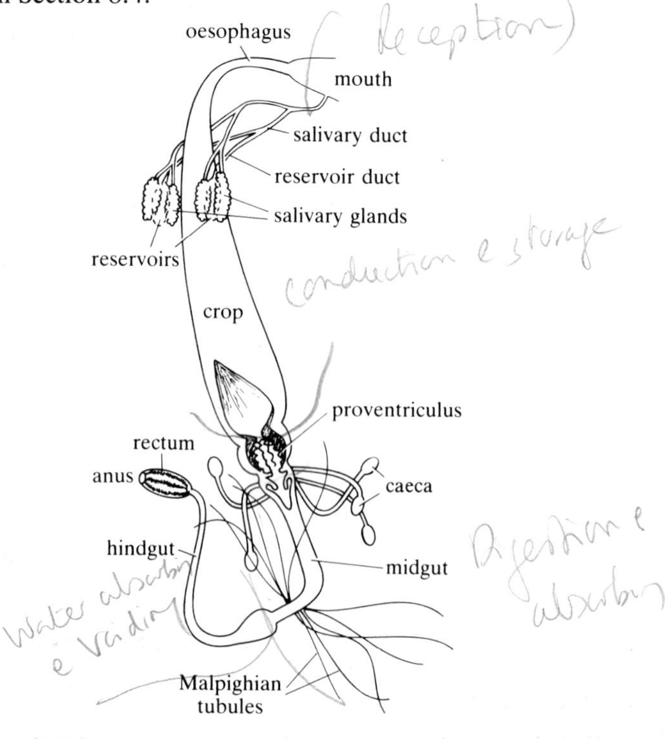

Figure 8.38 The digestive tract of the cockroach *Periplaneta americana*.

Questions 7–9 The results of three experiments on the sea-anemone *Calliactis* are described below. Using this information and that in Section 6, answer Questions 7–9. You will find it helpful to examine the diagram of another sea-anemone (Figure 8.39) to remind yourself of its anatomy.

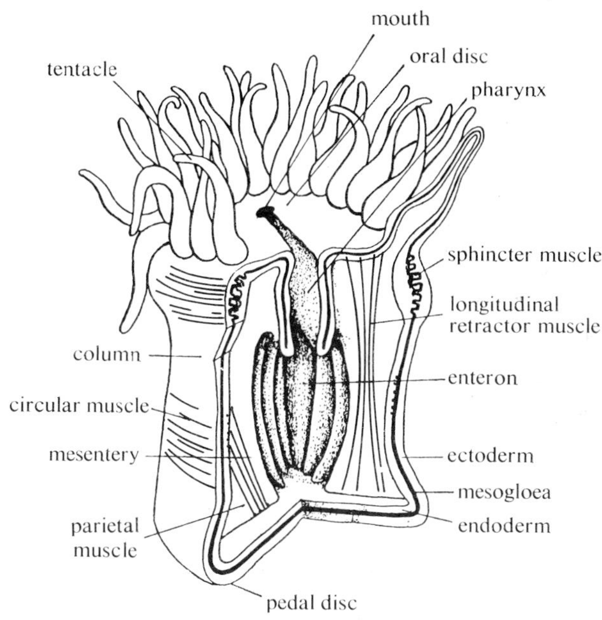

Figure 8.39 Basic anatomy of a sea-anemone.

Experiment 1 In the non-feeding anemone, the fluid-filled enteron cavity is practically free of proteases. Samples of fluid in the enteron were withdrawn from 14 specimens and in only one instance was a slight trace of protease revealed.

Experiment 2 In a series of experiments, small weighed blocks of gelatin, a protein, were fed to previously starved anemones. At selected intervals, the animals were killed, and the residual blocks recovered, weighed and analysed. These experiments revealed that, within 24 h of being fed with the gelatin blocks, the animals had digested over 98% of the block. The remnants of the gelatin blocks were completely covered with closely adhering mesenteric filaments (these line the free edge of the mesenteries, which attach to the body wall (the column) and project into the enteron).

Experiment 3 In a series of experiments, 5 g of *Calliactis* tissue, consisting of isolated mesenteries and filaments, were placed in a vessel and covered with 25 cm^3 of seawater. Samples of 1 cm^3 of the supernatant fluid were drawn at the beginning and at 1-h intervals and tested for protease activity. Very little protease was detected (on incubation of the removed samples with gelatin, 92% of the gelatin remained undigested after 24 h). Similar experiments were performed with the addition of a soluble protein, called casein, to the supernatant fluid. This resulted in the release of sufficient protease into the supernatant to digest 62% of gelatin added to the removed samples.

Question 7 (*Objective 8.7*) Does this information allow you to conclude whether digestion is primarily intracellular or extracellular in *Calliactis*? If so, what advantage is there likely to be for the animal in having this type of digestion?

Question 8 (*Objective 8.8*) Does the evidence suggest that the protease is secreted as an inactive precursor and is activated by the presence of specific chemicals in the enteron?

Question 9 (*Objective 8.8*) Why is the protease extracted from *Calliactis* unlikely to be pepsin? What proteases are likely to be present?

Question 10 (*Objective 8.10*) In the late 1930s, a controversy developed concerning the effects of the hormone secretin on pancreatic secretion. This arose from apparently contradictory results obtained by workers in an English laboratory and those obtained by some American workers. Both groups attempted to extract secretin from the duodenum of a dog and assess its physiological effect by injecting the extract into the blood system of another dog and measuring the constituents of the pancreatic juice secreted. Their techniques and results are briefly summarized.

American workers They extracted secretin from the intestinal mucosa with dilute acid. The secretin was precipitated from the extract by the addition of sodium chloride. The resulting powder, following some additional treatment, was called 'SI'. The injection of 'SI' into the circulatory system of the dog resulted in a secretion rich in enzymes and also containing significant amounts of bicarbonate.

English workers They extracted secretin from the intestine using absolute alcohol. The preparation was then adsorbed on bile salts and precipitated from aqueous solution by the addition of dilute acid. This secretin preparation was called 'Mellanby's preparation' (M). The injection of M into the circulatory system of a dog resulted in a pancreatic secretion containing a very low concentration of enzyme but large amounts of bicarbonate.

How can these results be reconciled? Which of the following explanations (i)–(iv) do you regard as most likely?

(i) The mode of action of secretin from American dogs is different from that from English dogs.

(ii) Normally, secretin evokes a pancreatic secretion rich in enzymes and in bicarbonate, but the extraction procedure used by the English workers altered its characteristics to such an extent that its enzyme-stimulating properties were inhibited.

(iii) The secretin preparation SI in fact contained *two* hormones—secretin and another (now known to be CCK-PZ). CCK-PZ stimulates the secretion of enzymes, hence the secretion resulting from SI injection contains enzymes and bicarbonate. M contains only one hormone, secretin, responsible for the production of a secretion rich in bicarbonate but poor in enzymes.

(iv) M contains *two* hormones: one of them, secretin, stimulates bicarbonate secretion, and another unnamed hormone inhibits the release of enzymes from the pancreas. SI contains only one hormone, secretin, which stimulates the release of both bicarbonate and enzymes from the pancreas.

9.1 INTRODUCTION

Active animal cells usually depend on glucose as their major 'fuel'; it is an essential substrate for cell metabolism. In vertebrates, glucose reaches the cells through their blood supply (and the concentration of glucose is generally known as blood sugar level). Figure 9.1 shows the concentration of glucose in human blood measured over a period of 24 h, during which the subject consumed three meals and slept for several hours. Using the terminology introduced in Chapter 4, glucose concentration in the blood is a *variable*.

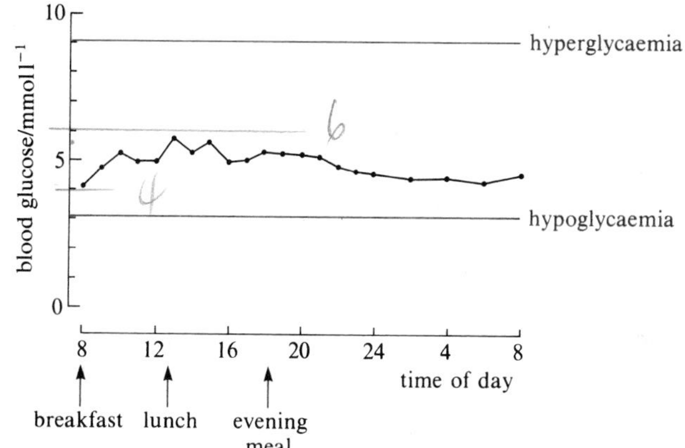

Figure 9.1 The level of glucose in human blood over a period of 24 h during which three meals were consumed and there was a period of sleep.

◇ Is blood glucose level a controlled variable or a regulated variable?

◆ Figure 9.1 shows that the level varied between about 4.5 and 6 mmol l^{-1} of blood and remained nearly constant; this strongly suggests that blood glucose is a *regulated variable*. Other evidence reviewed later in this chapter reinforces this conclusion.

There were deviations upwards after the meals, when glucose entered the blood from intestinal cells, but these were soon eliminated and the level returned to about 5 to 5.5 mmol l^{-1}. This suggests that there is a *negative feedback control system* in operation (see Chapter 4, Section 4.2). In this chapter, we try to identify parts of this system and consider how they operate in achieving *homeostasis* of blood glucose concentration in mammals.

Regulation of a variable results from *control* of other variables. In the case of blood sugar, the important variables are the input of glucose from various sources and the uptake of glucose by tissues that either make immediate use of it in metabolism or store it (usually as glycogen). The balance between input and uptake determines the level of glucose circulating in the blood; adjustments occur very rapidly. So an important question is: how are input and uptake of glucose controlled?

First, however, why is homeostasis of blood sugar level important to mammals? This can be answered by looking at what happens when regulation fails. When glucose levels rise or fall beyond the usual limits, the consequences can be severe. People with blood glucose levels below 3 mmol l^{-1}, a condition known as **hypoglycaemia**, report feelings of nausea, cold sweats and loss of concentration. If the glucose level is not raised rapidly, **coma** (unconsciousness) can result because the nervous system (with no metabolic stores of its own) is uniquely dependent on circulating glucose. If the level of glucose in the blood rises above 10 mmol l^{-1}, a state of **hyperglycaemia**, glucose begins to appear in the urine and an associated symptom is a fall in blood pH. Hyperglycaemia can also result in coma which, if unchecked, is fatal. These hypoglycaemic and hyperglycaemic states are typical of people who have problems in regulating their blood glucose level; their condition, known as **diabetes mellitus**, is considered in Section 9.6.

Homeostasis of blood glucose level is found in all vertebrates but little is known about its mechanisms except in mammals. This chapter is largely based on data from human studies because most research on this topic is connected with diabetes mellitus and is carried out on humans or animals whose blood sugar regulation is known to be similar to that of humans. The control of blood glucose level in birds, reptiles, amphibians and fishes differs from that in mammals, although the hormones present are the same or very similar. In particular, insulin seems to play a different role in different species and there is seasonal variation in fishes, reptiles and amphibians.

In Section 9.2, we consider the sources from which glucose enters the body and the blood and its use in body cells. In later sections, the parts played in control of these variables by hormones, nerves and metabolites are discussed.

9.2 SOURCES AND USE OF GLUCOSE*

The variation in the level of glucose shown in Figure 9.1 can be explained as raising of the level following absorption of food after meals and lowering of the level during moderate fasting between meals, especially overnight. It is as glucose that carbohydrate is transported round the body in the blood. Glucose is a polar molecule and its uptake by body cells is by *facilitated diffusion* promoted by glucose transporter molecules in the cell membranes. For glucose to be taken up by cells, its concentration in the blood must be higher than inside the cell and the transporter molecules must be in a state to facilitate inward diffusion across the membrane. If the intracellular glucose level exceeds that in the blood, glucose can pass out of the cell by facilitated diffusion. *Active transport* of glucose, against its concentration gradient, occurs only in cells of the intestine and kidney tubules. Glucose, produced in the gut by digestion of carbohydrates, is taken into intestinal cells by active transport and passed into the blood by facilitated diffusion (Chapter 8, Section 8.7.2). Kidney cells take up glucose from urine and pass it back into the blood; this process is discussed in Chapter 11, Section 11.3.3.

There are two important observations to be made on Figure 9.1. First, a meal may contain carbohydrate in amounts that would present the blood with about a hundred times the normal level of glucose and yet the rise in blood

*The metabolism of glucose, properties of enzymes and transport of glucose across cell membranes are explained and discussed in Chapters 4–7 of Norman Cohen (ed.) (1991) *Cell Structure, Function and Metabolism*, Hodder and Stoughton Ltd, in association with The Open University (S203, *Biology: Form and Function*, Book 2).

glucose is fairly small and transient. Second, an overnight fast of 12–14 h lowers the blood glucose level, but only by some 20%. This suggests that some mechanism must rapidly clear the glucose absorbed immediately after a meal and that glucose must be made available to the blood from a body 'store' during a fast. The needs of body cells for glucose must be satisfied, both when the body is resting and during activity which may be intense.

To understand the events after eating and during fasting and while the body is active or at rest, we must examine (in outline only) the basic biochemical processes involved. These depend on a complex series of reactions catalysed by enzymes. Four processes occur:

1 Glucose is metabolized in cells as an energy source; this process is called **glycolysis**.

2 Cells, especially in the liver and muscles, convert glucose into glycogen, a polymer of glucose units; this process is called **glycogenesis**.

3 Glycogen is a storage product and it is converted back to glucose; this process is called **glycogenolysis**.

4 Liver cells (and cells in the kidney cortex) can synthesize glucose from other substances; this process is called **gluconeogenesis**.

After considering each of these processes, we shall be in a stronger position to discuss the control of the level of glucose circulating in the blood.

The glucose which is absorbed from the gut into intestinal cells (Chapter 8, Section 8.7.2) is passed into the hepatic portal vein and transported to the liver. Liver cells take up glucose and convert it into glycogen, so reducing the level of glucose within these cells; the glycogen is stored. This process, glycogenesis, begins with phosphorylation of glucose, promoted by the enzyme hexokinase, and this is followed by isomerization and then polymerization, catalysed by the enzyme glycogen synthetase (see Figure 9.2).

To release glucose from glycogen stores, the reverse reaction, glycogenolysis, must occur. This involves phosphorylation of glycogen (at the terminal glycosidic bond), catalysed by the enzyme glycogen phosphorylase, and then separation of a phosphorylated glucose unit from the end of the polymer chain (see Figure 9.3). The glucose-1-phosphate is converted to glucose-6-phosphate and glucose is then released by activity of the enzyme glucose-6-phosphatase (Figure 9.2).

Glycogenesis and glycogenolysis involve the same biochemical reactions but in reverse directions.

◇ Are the same enzymes involved in the two processes?

◆ No. There are two pairs of 'antagonists': hexokinase and glucose-6-phosphatase, which both catalyse the glucose/glucose-6-phosphate conversions; **glycogen synthetase** and **glycogen phosphorylase** both catalyse the glycogen/glucose-1-phosphate conversions.

This should not surprise you when you recall the principles of control of metabolism through enzymes. The direction of a reaction depends on which enzyme is more active or whether one or the other is inhibited. Hexokinase is inhibited if its product is present in a high concentration.

◇ How could this affect the conversion of glucose to glycogen?

◆ The initial phosphorylation of glucose will not occur if there is a high concentration of glucose-6-phosphate in the cell.

glycogen

glycogen synthetase

glycogen phosphorylase

GLYCO-GENESIS

GLYCOGEN-OLYSIS

glucose-1-phosphate

glucose

hexokinase

glucose-6-phosphatase

glucose-6-phosphate

GLUCONEO-GENESIS

GLYCOLYSIS

pyruvate

lactate

glycerol

amino acids

LINK REACTION

acetyl CoA

fatty acids

ketone bodies

oxaloacetate

citrate

TCA cycle

Figure 9.2 Diagram to show the principal biochemical pathways associated with glucose and glycogen in mammals. Only the substances mentioned in the text are shown and some arrows cover several intermediate reactions. Enzymes are in italics.

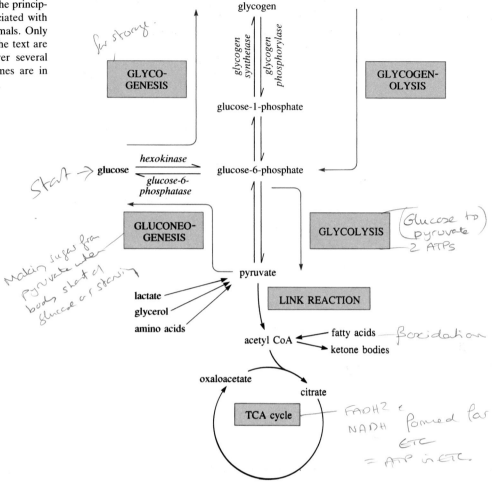

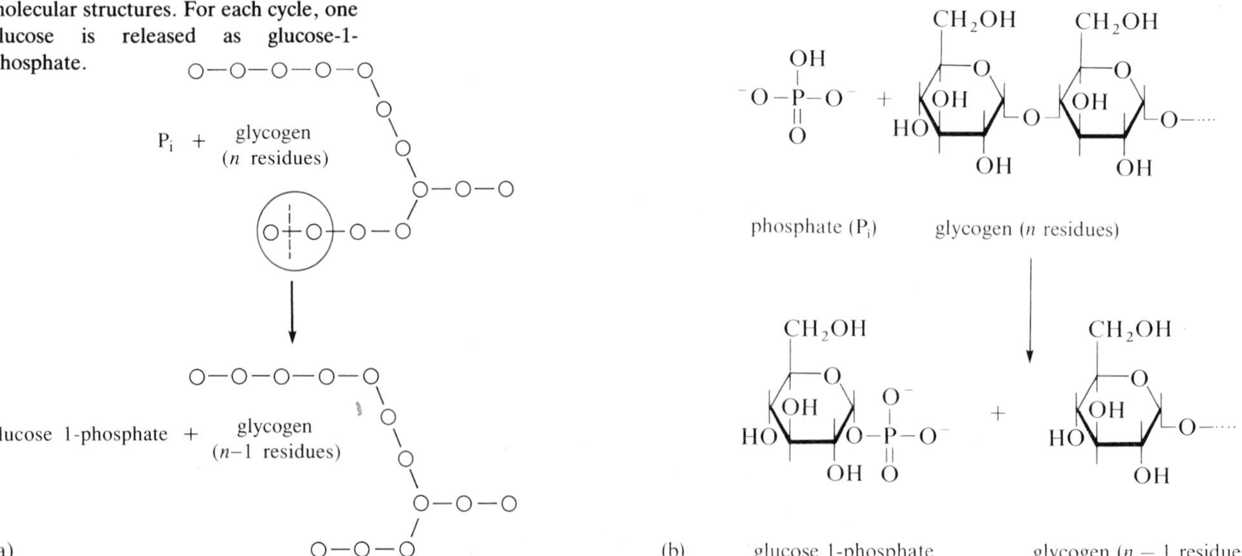

Figure 9.3 Release of glucose residues from glycogen catalysed by glycogen phosphorylase. (a) Schematic representation of the reaction. (b) Details of the molecular structures. For each cycle, one glucose is released as glucose-1-phosphate.

If glucose-6-phosphate is rapidly converted into some other product, perhaps as part of glycogenesis, more glucose can be phosphorylated through the activity of hexokinase. This phosphorylation is also the first step in glycolysis (see Figure 9.2) and the subsequent steps of this process lead through a series of reactions to the formation of two molecules of pyruvate from every molecule of glucose. In aerobic respiration, pyruvate is converted into acetyl CoA and this enters the TCA cycle (tricarboxylic acid cycle, also called the Krebs' cycle). Under anaerobic conditions, pyruvate is converted into lactate.

Pyruvate is important in gluconeogenesis because it can be converted into glucose (the reverse of glycolysis). Pyruvate can be formed from lactate, from certain amino acids and also from glycerol (derived from breakdown of lipids). The fatty acids released when storage lipids are broken down cannot be converted into glucose but can be converted into acetyl CoA; this permits fatty acids to be used in aerobic respiration. Some amino acids are converted into components of the TCA cycle, such as α-oxoglutarate and oxaloacetate; like the fatty acids, these amino acids can be used in aerobic respiration (or by an indirect pathway may be converted to glucose).

Gluconeogenesis starting from pyruvate is essentially a reversal of glycolysis but it is achieved through a different set of reactions, catalysed by different enzymes. It is not necessary to go into details here.

◇ From the information provided so far, what are the sources from which glucose enters the blood and what processes are involved in its production?

◆ Glucose enters the blood:

(a) from intestinal cells as a result of digestion of food in the gut and absorption of the products of digestion;

(b) from liver cells as a result of (i) glycogenolysis and (ii) gluconeogenesis.

Glycogenesis and glycogenolysis also occur in muscle cells, which, like liver cells, convert glucose into glycogen and store it. However, muscle cells lack the enzyme glucose-6-phosphatase.

◇ What is the significance of this deficiency?

◆ The breakdown of glycogen in muscles does not produce glucose so muscles are not a source from which glucose can be added to blood.

The glucose-6-phosphate produced by glycogenolysis in muscles is directed along the glycolytic pathway and metabolized usually via the TCA cycle, releasing energy for muscle contraction.

◇ From the information provided so far, what tissues take up glucose from the blood and what processes follow in these cells?

◆ Glucose is taken up from the blood by:

(a) liver cells, which convert it into glycogen and store this; at other times these cells convert glycogen back into glucose and release this into the blood;

(b) muscle cells, which may use it directly in glycolysis or convert it into glycogen and store this; at other times, these cells convert glycogen into glucose-6-phosphate which is then used in glycolysis;

(c) most of the cells in the body, including brain cells, which take up glucose for immediate use in metabolism; they store very little as glycogen.

The regulation of blood sugar level thus depends on:

1 Reaction to the input of glucose as a result of digestion; this involves liver cells taking up glucose and storing it as glycogen.

2 Reaction to the uptake of glucose by body cells for their normal metabolism.

3 Reaction to the uptake of glucose by muscle cells which can store it as glycogen or metabolize it immediately.

The fall in blood sugar level which follows uptake of glucose by cells other than liver cells stimulates liver cells to convert stored glycogen into glucose and to release this into the blood. Because movement of glucose through cell membranes is due to facilitated diffusion, the concentration of glucose in liver cells must be lower than that in the blood when the cells take up glucose and must be higher than that in the blood when they release it; the transporter molecules are able to operate in both directions. If liver glycogen stores are depleted, as during fasting, gluconeogenesis will take place in liver cells with subsequent release of glucose into the blood.

The reactions in liver cells clearly play a key part in control. The balance between glycogenesis and glycogenolysis is important and must be considered in more detail. The last step in glycogenesis and the first step in glycogenolysis involve the enzymes glycogen synthetase and glycogen phosphorylase, respectively (see Figure 9.2). Both these enzymes exist as two forms, *a* and *b*, one of which is active (*a*) and the other inactive (*b*). The conversion of the *b* (inactive) form into the *a* (active) form of glycogen phosphorylase stimulates the breakdown of glycogen. Similarly, the conversion of the *b* (inactive) form of glycogen synthetase to the *a* (active) form promotes synthesis of glycogen. The key to the interconversion of these *a* and *b* forms of the two enzymes is protein phosphorylation; usually this follows production of *cyclic AMP*.

◇ How might cyclic AMP be involved? (Recall Chapter 3, Section 3.3.1.)

◆ Cyclic AMP is involved in the activation of protein kinases in cells.

Cyclic AMP is a *second messenger* which amplifies and transmits stimuli between receptor molecules and effector systems. In this case, the effector system is concerned with the balance between storage and breakdown of glycogen in cells.

The inactivation ($a \rightarrow b$) of glycogen synthetase in liver is catalysed by a synthetase kinase dependent on cyclic AMP and the very same kinase activates ($b \rightarrow a$) glycogen phosphorylase. So the production of one kinase simultaneously switches off glycogen production and switches on glycogen breakdown. It will come as no surprise that the *first messengers*, the signal molecules that activate the receptors, are hormones. One of the main hormones in question is **glucagon**, and we return to it in Section 9.3.2.

Conversely, if cyclic AMP production is blocked or if the rate of its breakdown (by the enzyme cyclic AMP phosphodiesterase) is increased, or if the activation of the protein kinase by cyclic AMP is inhibited, then any one or all of these events would tend to activate the synthesis of glycogen and to prevent its breakdown. Another important hormone, **insulin**, is involved in the control of glycogen synthesis and we shall discuss it in Section 9.3.1.

Glucose level has an effect on this process of glucose–glycogen conversion. Glucose inhibits conversion of the *b* to the *a* form of glycogen phosphorylase. Figure 9.4 shows the levels of phosphorylase *a* and synthetase *a* measured in rats after a glucose meal. The level of phosphorylase *a* is seen to drop rapidly and then that of synthetase *a* rises.

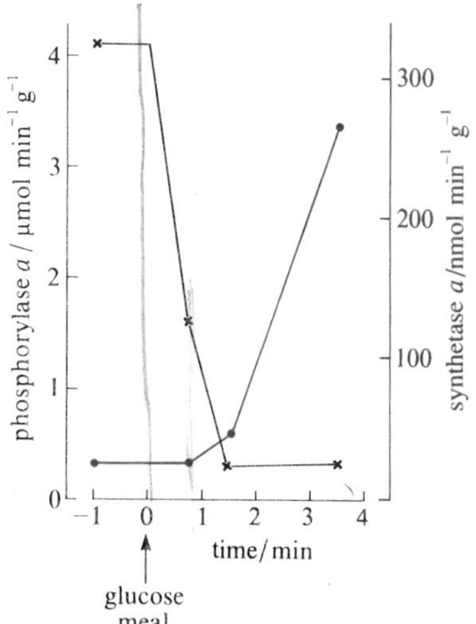

Figure 9.4 Activity of glycogen phosphorylase *a* and glycogen synthetase *a* in a rat liver after consumption of a glucose meal.

◇What will be the consequence of these changes in terms of glycogen?

◆Glycogenolysis will be inhibited and glycogenesis will be promoted, leading to build-up of glycogen stores.

It is thought that, when the amount of phosphorylase *a* exceeds a certain threshold, this inhibits synthetase *a* formation. The presence of high levels of glucose in liver causes the conversion of phosphorylase *a* to phosphorylase *b* (by blocking the reaction *b*→*a*), and so releases synthetase *a* from inhibition. The effects produced by cyclic AMP and glucose on liver enzymes are summarized in Figure 9.5.

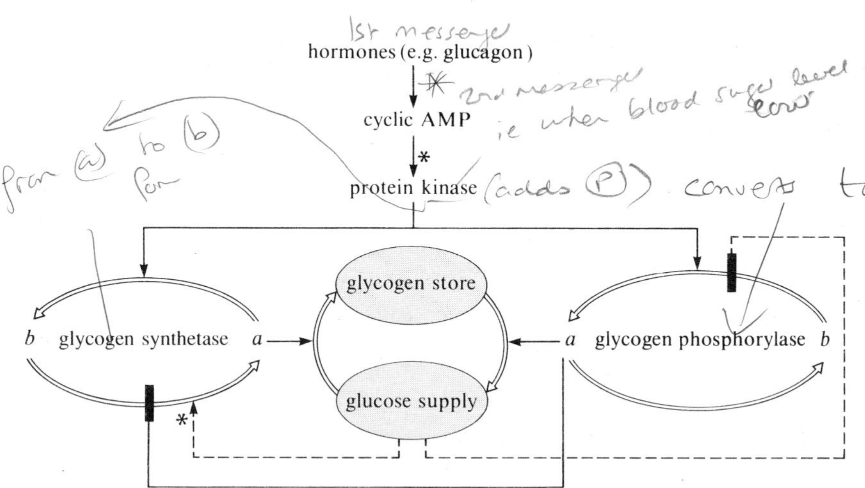

Figure 9.5 The effects of glucose supply (dashed lines) and of cyclic AMP on the interconversion of the *a* and *b* forms of glycogen synthetase and glycogen phosphorylase in liver cells. *Here and in other figures, a line ending in a thick bar indicates inhibition and a line ending in an arrow with a star beside it indicates stimulation.*

Fatty acids, as we said earlier, can be converted into acetyl CoA which can enter the TCA cycle. Oxaloacetate combines with the acetyl CoA and forms citrate (Figure 9.2). During starvation, when carbohydrate is depleted, little oxaloacetate is available and acetyl CoA accumulates; these molecules then interact with each other to produce three compounds called **ketone bodies** (acetoacetate, acetone and β-hydroxybutyrate). Mild ketosis, the condition when these substances are present at low concentrations in the blood, is normal but, in diabetes, ketone bodies build up in the blood, leading to a fruity smell in the patient's breath. Severe ketosis may lead to coma and death (Section 9.6).

In Section 9.3, we describe the release and mode of action of the main hormones concerned with regulation of blood glucose: insulin and glucagon.

Summary of Sections 9.1 and 9.2

1 Glucose levels in the blood are maintained within strict limits, rising slightly after meals and falling only slightly overnight or after a period of fasting.

2 If glucose levels rise or fall beyond these limits, the condition is called hyperglycaemia or hypoglycaemia, respectively, and there is a danger of coma.

3 Glucose enters the blood by facilitated diffusion from intestinal cells which take up, by active transport, the glucose produced by digestion of carbohydrates in the gut.

4 Glucose is used as an energy source by body cells; the first stage in metabolism is glycolysis leading to formation of pyruvate molecules.

5 Glucose is converted to glycogen (through glycogenesis) and stored in liver and muscle cells. When the level of blood glucose falls, liver glycogen is broken down to glucose (through glycogenolysis) and this passes into the blood by facilitated diffusion.

6 Fasting results in depletion of glycogen stores in liver cells and stimulates gluconeogenesis from glycerol and amino acids.

7 The enzymes responsible for the synthesis and breakdown of glycogen (glycogen synthetase and glycogen phosphorylase) both exist in an active *a* and an inactive *b* form. The conversion of these forms is sensitive to cyclic AMP and glucose. Hormones are involved in the interconversion of glucose and glycogen via cyclic AMP.

8 During starvation, acetyl CoA, formed by breakdown of fatty acids, may produce ketone bodies which accumulate in the blood.

Now attempt Question 1, on p. 320.

9.3 PANCREATIC HORMONES

Following digestion, as you read in Chapter 8, Section 8.7.5, the products are absorbed by the small intestine and carried directly to the liver, before being carried to the heart and round the rest of the circulation. Situated just below the liver and under the stomach is the *pancreas*, which plays an important role in the digestion and assimilation of food (see Chapter 8, Section 8.6). Ninety-eight per cent of the pancreatic cells produce digestive enzymes; the other 2% of the cells are arranged in discrete 'islands' among the mass of the pancreas, called the **islets of Langerhans**. These islets are composed of a number of cell types and three of these produce and secrete particular hormones.

If a glucose meal is given to an animal from which the pancreas has been removed, there is a dramatic increase in the level of glucose in the blood. When the glucose level rises above the threshold for the kidney (which will be discussed in Chapter 11), glucose appears in the urine: glucose homeostasis has been lost! This suggests that the pancreas is involved in blood sugar regulation—perhaps it secretes a hormone that normally lowers the blood glucose level. But that is not the end of the story. The islets of Langerhans produce a number of substances of which three—**insulin**, **glucagon** and **somatostatin**—are known to be important in blood glucose regulation.

By using specific histological staining techniques, it is possible to 'map' the cells within the islets. A number of cell types can be identified. The three main cell types are called A cells, which contain the hormone glucagon, B cells, which contain the hormone insulin, and D cells, which contain somatostatin. These cells are shown in Figure 9.6. In the rat, they are arranged in a distinct pattern (Figure 9.6a), with A and D cells around the

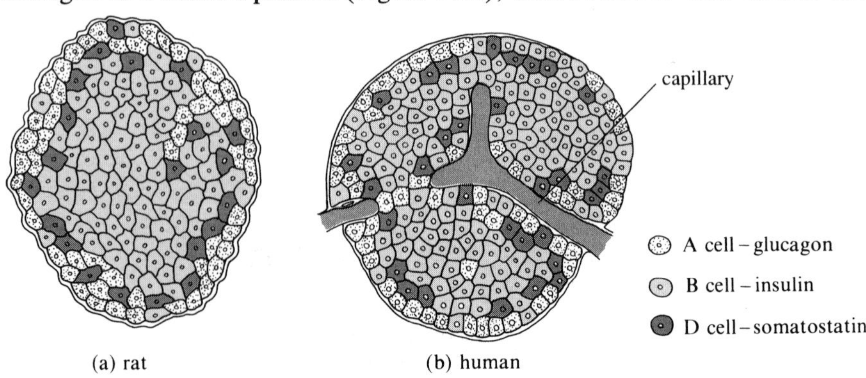

Figure 9.6 The distribution of A, B and D cells in the islets of Langerhans of (a) a rat, (b) a human.

edge of the islets while the core is made up of B cells. In fact, the exact pattern depends on the position of the islet in the pancreas. A human islet shows the same basic arrangement (Figure 9.6b), but here blood capillaries run through the tissue.

◇ What difference is there in the arrangement of A, B and D cells in the human islet compared with the rat?

◆ The A and D cells are found deeper in the human islet than in the rat; they lie along the capillaries and appear scattered throughout the islet.

To understand the role of these pancreatic hormones in blood glucose regulation, we need to look at the synthesis, release and actions of each one in turn. In Section 9.3.3 we return to the morphology of the islets to see how A, B and D cells interact with one another.

9.3.1 Insulin

B cells make up about 60% of the islet mass. They synthesize and release the hormone insulin which is a peptide of 51 amino acids but is derived from a larger parent molecule called pre-pro-insulin (110 amino acids long). This type of synthesis is quite common for hormones (Chapter 2, Section 2.2.1). On the route to secretion from the cell, pre-pro-insulin loses a hydrophobic 23 amino acid peptide and becomes pro-insulin inside the lumen of the endoplasmic reticulum. The explanation for this loss is that the hydrophobic part of pre-pro-insulin is probably the signal sequence of amino acids that directs pro-insulin through the endoplasmic reticulum (ER) membrane into the ER cisternum. Once inside the ER lumen, this signal peptide is 'clipped' off by a specific enzyme.

Figure 9.7 shows the fate of pro-insulin. It is packaged in the Golgi apparatus; then post-translational processing occurs by enzymic hydrolysis—it is clipped into two: the protein insulin and a residue known as the connecting peptide (C-peptide). Insulin is stored in granules bound to zinc and released by exocytosis. Its concentration in human blood is normally about 10^{-9} mmol l^{-1}.

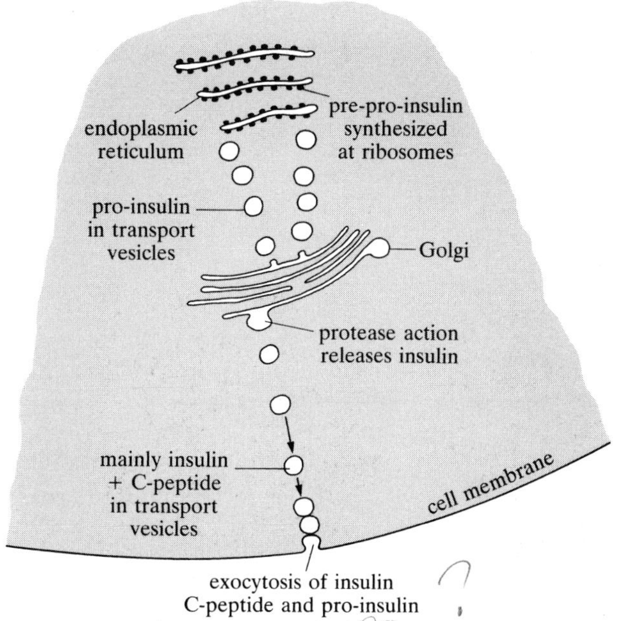

Figure 9.7 Steps in the synthesis of insulin in a B cells.

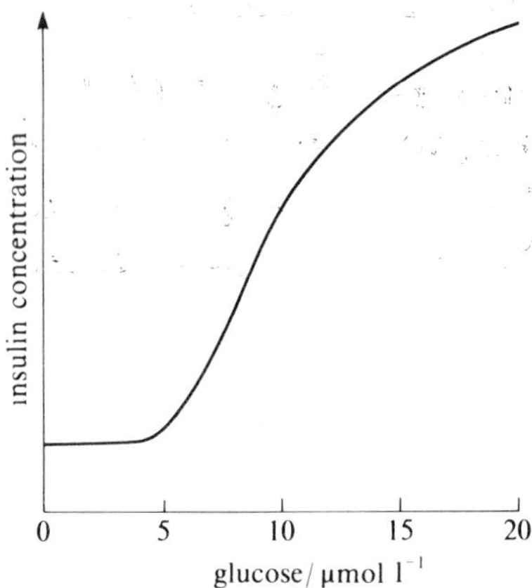

Figure 9.8 The effect of raising the external glucose concentration on the release of insulin from islets of Langerhans in an isolated pancreas.

One important stimulus for insulin secretion is an increase in the level of glucose in the blood flowing through the pancreatic islets, for example after a meal. Figure 9.8 shows the dramatic effect of raising the external glucose concentration on insulin release from islets from an isolated pancreas.

The key questions are: How does the B cell 'monitor' blood glucose levels? What then stimulates insulin release?

If islets are isolated from a pancreas and maintained under Ca^{2+}-free conditions and then exposed to glucose, insulin secretion is dramatically reduced. If a microelectrode is inserted into a normal B cell (where Ca^{2+} is present) that is then exposed to glucose, an action potential can be recorded. When Ca^{2+} is removed and the experiment is repeated, no action potential is recorded. Let us consider three more observations of the response of B cells to glucose. First, they show a rise in levels of cyclic AMP; second, they accumulate Ca^{2+} when exposed to glucose; and third, B cells contain calmodulins (Ca^{2+}-binding proteins; Chapter 3, Section 3.3.2 and Figure 3.3).

You should recall from Chapter 3, Section 3.3.2, that there are two main systems of second messengers in cells; one involves cyclic AMP and the other involves Ca^{2+}. Consider first how Ca^{2+} ions could act in these B cells. It is likely that glucose or a glucose metabolite enters the cell (via glucose transporter molecules) and then promotes the release of intracellular Ca^{2+} stores via the PIP_2 system (Chapter 3, Section 3.3.2, Figure 3.3). This Ca^{2+} is then picked up by calmodulin. The Ca^{2+}-calmodulin complex promotes fusion of the insulin-containing vesicles with the cell membrane (exocytosis), resulting in the secretion of insulin by the cell. It is also possible that extracellular glucose (and some first messengers such as acetylcholine, discussed in Section 9.4.1) could stimulate the opening of Ca^{2+} channels in the cell membrane so that extracellular Ca^{2+} ions enter the cell, bind to calmodulins and thereby promote the continued secretion of insulin. There is still uncertainty about the details.

The hormone glucagon (discussed in some detail in Section 9.3.2) promotes insulin secretion through cyclic AMP and a protein kinase. So both second-messenger systems can stimulate the secretion of insulin by B cells but they respond to different sets of first messengers, cyclic AMP to glucagon and Ca^{2+} to glucose or its metabolites. Figure 9.9 summarizes these effects of glucose and glucagon on the secretion of insulin by the B cell.

Glucose is not the only stimulus for insulin secretion. Other sugars, amino acids, ketones and a range of hormones initiate insulin secretion from the pancreas. When food is present in the stomach, nervous signals carried by the

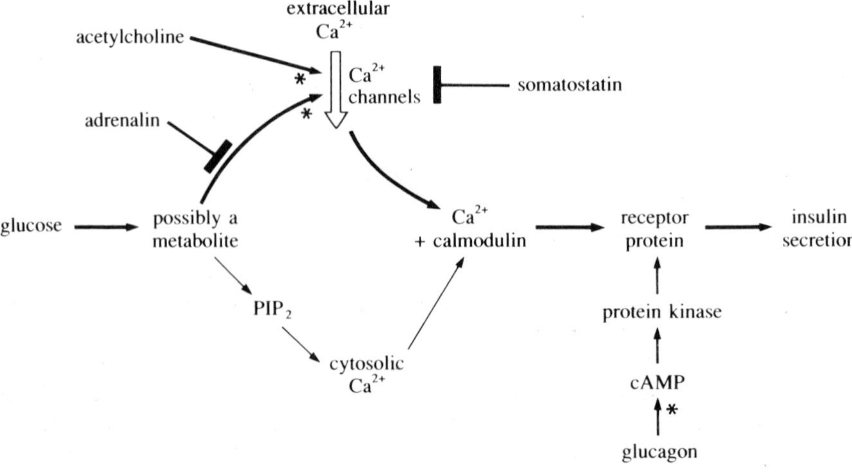

Figure 9.9 Simplified outline of the control of the secretion of insulin from a B cell by glucose, other pancreatic hormones and neurotransmitters. Note that both Ca^{2+} and cyclic AMP are involved.

vagus nerve result in stimulation of insulin release before the blood glucose level is raised.

Insulin circulates in the blood throughout the body, and **insulin receptors** (see Chapter 3, Section 3.2 and Table 3.1) have been found on three main target organs: liver cells, adipose cells and muscle tissue. Insulin binds to specific high-affinity cell-surface receptors which are glycoproteins composed of two sets of two subunits bridging the cell membrane. The inner pair of units are phosphorylated when the receptor is activated. It has been suggested that the receptor is a cytosolic tyrosine protein kinase. In 1986, investigators isolated two novel substances which appear to be second messengers. One is a diacylglycerol that is chemically different from the one usually involved with the PIP_2 system (see Chapter 3, Section 3.3.2, Figure 3.3). The other second messenger is a derivative of inositol phosphate which activates cyclic AMP phosphodiesterase (exactly how this is achieved is still not known).

◇ What effect would this have on the cells?

◆ It would lead to the degradation of cyclic AMP and hence to a change in equilibrium between the *a* and *b* forms of glycogen phosphorylase and glycogen synthetase (Figure 9.5).

◇ What would be the effect of this change?

◆ Glycogen synthetase would be activated and glycogenesis would increase.

The main effect of insulin, therefore, is to lower the level of glucose in the blood by stimulating glycogenesis in the liver and in muscles and by inhibiting gluconeogenesis in the liver. Insulin has a number of other general effects: it promotes the uptake of amino acids into cells and prevents the breakdown of lipids and protein in tissues such as adipose cells and muscle.

Generally, the actions of insulin are best thought of in terms of promoting the build-up and preventing the breakdown of lipid, protein and glycogen stores. This is achieved by quite a complex series of events, which we shall now try to unravel.

In Section 9.2, we saw how glucose in the liver is converted to glycogen, which can then be broken down again to glucose when required. Glycogen storage has been much studied. In the absence of insulin, we know that glucose on its own will activate glycogen synthetase (see Section 9.2) and, in the absence of glucose, insulin alone will activate glycogen synthetase. In combination, insulin and glucose produce a synthetase activation response greater than the sum of each separately—a synergistic effect. This suggests that there are two separate mechanisms involving insulin: it promotes glucose uptake by tissues, possibly by activating glucose transporter molecules; this alone will promote synthetase activity and the production of glycogen. The other effect of insulin appears to be more directly on the synthetase enzyme itself. The $a \rightarrow b$ conversion is dependent on a protein kinase reaction induced by cyclic AMP. Insulin prevents this reaction by promoting production of cyclic AMP phosphodiesterase and so maintains synthetase in the *a* or active form (see Figure 9.10). The result is that glycogen is deposited in the liver.

Before we leave insulin, there is one important observation to be made about its receptors. In conditions such as obesity, the level of insulin and the level of glucose in the blood are both very high. This apparent contradiction is due to the relative insensitivity of tissues to insulin. Insulin receptors are present but are very few in number, and it seems that the constant presence of the hormone actually reduces the number of its receptors. Reduced 'sensitivity' of receptors to hormones has been described for a number of hormones (see

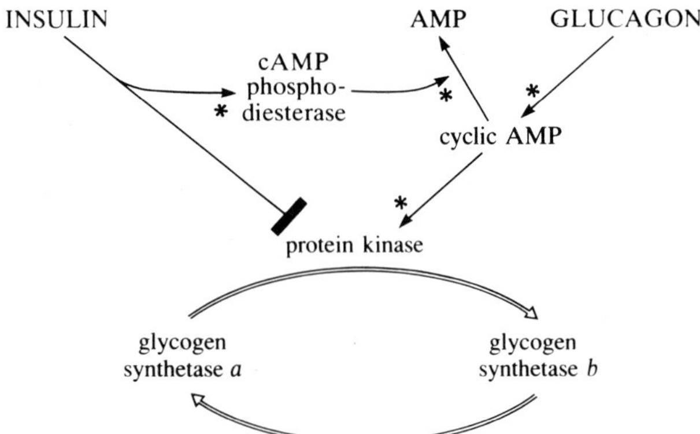

Figure 9.10 The effects of insulin and glucagon on the interconversion of the *a* and *b* forms of glycogen synthetase.

Chapter 3, Section 3.2.2). In all other ways, the hormone receptor event is normal; it is simply that there are fewer such events. When obese animals are 'slimmed' by fasting and then put on a diet to maintain a low weight, the number of insulin receptors increases again. We return to the concept of 'receptor regulation' in Section 9.6 when describing the problems of glucose regulation in some diabetics. Figure 9.11 summarizes the actions of insulin on body tissues.

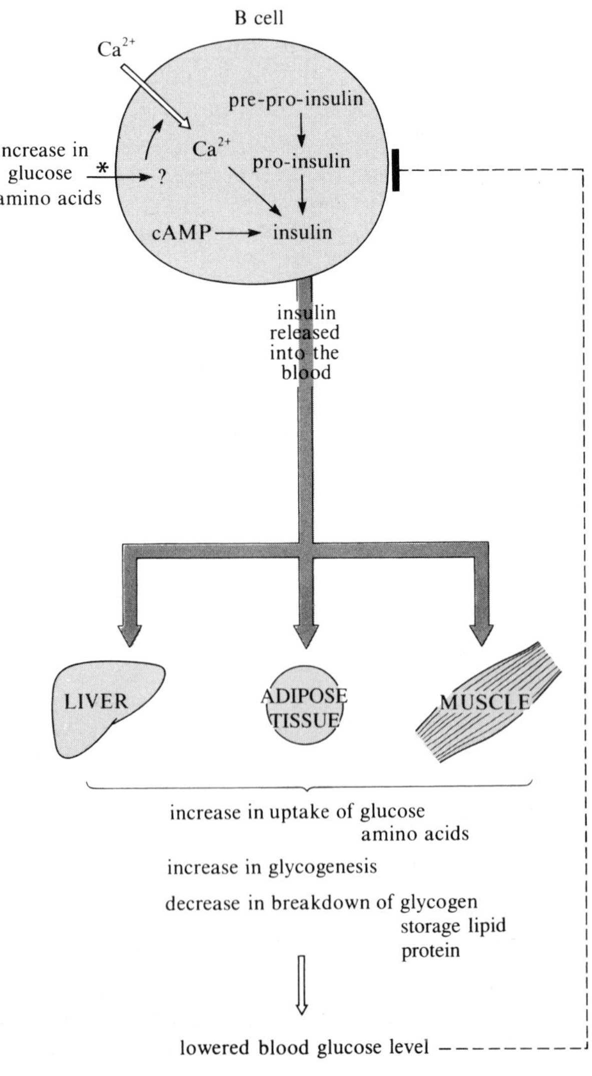

Figure 9.11 Diagram to show the role of insulin in the regulation of blood glucose. Dashed lines indicate feedback loops.

9.3.2 Glucagon

Glucagon is a hormone antagonistic to insulin in its effects, that is, it functions in ways that lead to increased levels of blood glucose during fasting or sustained exercise when glucose is in demand for metabolism in, for example, muscles. Like insulin, glucagon is a polypeptide but of somewhat smaller size—only 29 amino acids (M_r 3 500). It is synthesized in the A cells of the pancreas (which form 30% of the islet cells) from a precursor, proglucagon.

A transient fall in blood glucose in humans to less than 3.5 mmol l^{-1} for 20 min results in release of glucagon, but measurements of circulating blood glucagon over a 24-h period in healthy subjects (living on mixed diets) show little change in its level, which is about 10^{-9} mmol l^{-1}. Stimulation of a marked glucagon release from the pancreas requires a definite 'stress'—for example, starvation for a day, or cold stress, or an absolute lack of insulin.

The islets of Langerhans are arranged (Figure 9.6) such that there is considerable communication between the A, B and D cells (by way of cell junctions and extracellular fluid). Insulin release inhibits the release of glucagon, and in the absence of insulin more glucagon is released. This balance between the two ensures that glycogen is stored when there is excess blood glucose. In times of glucose demand by tissues (e.g. in exercise, or during increased metabolism in response to a cold stress), the circulating insulin level is low and release of glucagon is stimulated; this 'pulls' glucose out of store and into the blood. The two hormones—insulin and glucagon—thus both contribute to the regulation of blood sugar by control, respectively, of the uptake and output of glucose.

Little is known about the cellular system which activates the release of glucagon except that Ca^{2+} is involved. The main stimulus seems to be lack of circulating insulin but first messengers such as adrenalin (see Section 9.4.2) also promote glucagon secretion.

The major site of glucagon action is on liver cells, where it promotes the breakdown of glycogen by activating the enzyme glycogen phosphorylase and also blocks the synthesis of glycogen by deactivating the enzyme glycogen synthetase. Unlike insulin, glucagon appears to act solely via cyclic AMP; this activates a protein kinase, which in turn promotes glycogen phosphorylase activity ($b \rightarrow a$) and 'turns off' glycogen synthetase ($a \rightarrow b$). The result is that glycogen stores are broken down (Figure 9.5).

In the absence of insulin, there is breakdown of lipids and proteins liberating fatty acids, glycerol and amino acids. Glucagon stimulates the uptake of amino acids and glycerol into the liver, where they are converted to glucose by gluconeogenesis, promoted through direct effects of glucagon on liver mitochondrial enzymes. Fatty acids are metabolized by the liver to provide acetyl CoA.

Glucagon, like insulin, appears to affect the density of its receptors on liver cells. When glucagon levels are high for long periods, as during prolonged fasting, there is a desensitization of liver cells to glucagon. The physiological significance of this receptor regulation may be that it is an 'escape response' that reduces gluconeogenesis and therefore the uptake by the liver of the products of breakdown of storage lipids and structural protein.

◇ What would be the result of continual breakdown of lipids and protein?

◆ Breakdown of storage lipids liberates fatty acids, which are converted into acetyl CoA, and, if in excess, to ketone bodies (ketosis could result). Prolonged protein breakdown could result in severe muscle wastage.

Figure 9.12 summarizes the main effects of glucagon. The major stimulus for glucagon release from the pancreas is the lack of insulin. Other stimuli (such as stress, starvation and exercise) promote glucagon release from the pancreas through other pathways, as we shall see in Section 9.4.

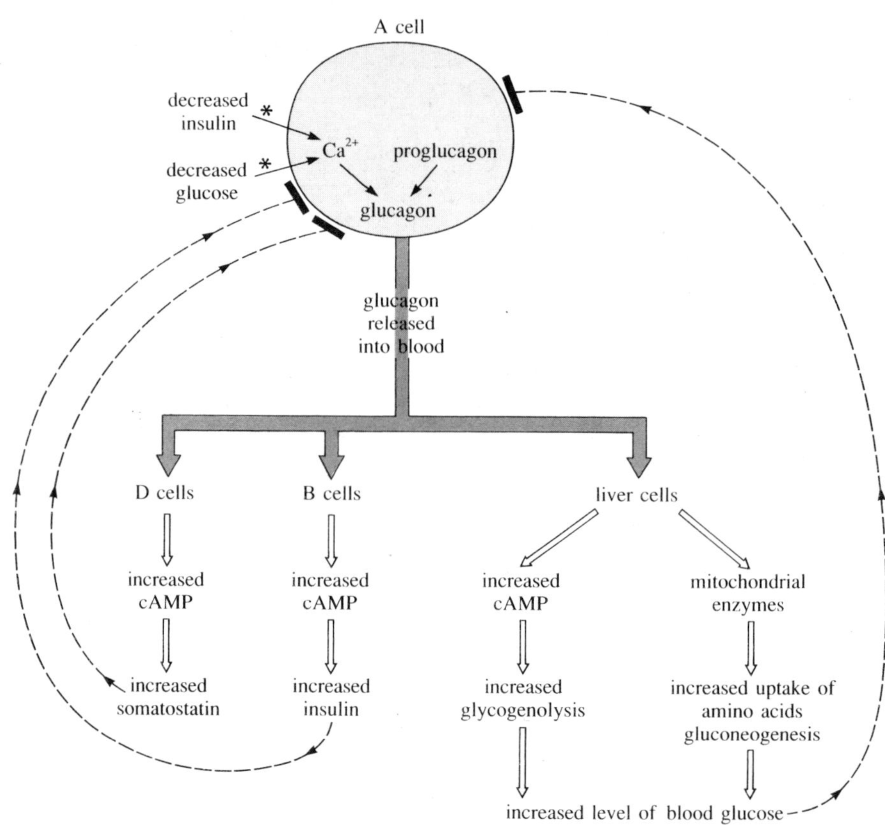

Figure 9.12 Diagram to show the role of glucagon in the regulation of blood glucose. Dashed lines indicate feedback loops.

9.3.3 Somatostatin and other pancreatic peptides

Insulin and glucagon acting in concert maintain a fairly constant blood glucose level throughout the day (Figure 9.1) and if the ratio of these two hormones in the blood were measured at any one moment, it would correlate well with the level of blood glucose.

Earlier in Section 9.3, we mentioned the presence of a third cell type, which makes up 10% of the islets of Langerhans—the D cell. Descriptions of pancreatic endocrine function before 1975 put a large question mark against the D cell. Following the isolation of somatostatin from the hypothalamus in 1975, and therefore the possibility of raising antibodies to it for the purpose of identification (Chapter 2), it soon became clear that D cells secrete somatostatin. Experiments on an isolated pancreas showed that glucagon stimulated secretion of somatostatin but insulin had no effect (Figure 9.13).

Somatostatin released from the hypothalamus is considered to be an inhibitor of the release of growth hormone (GH) from the pituitary (somatostatin is the growth hormone release-inhibiting factor). So what could its function be in the pancreas? It appears to inhibit the release of both insulin and glucagon, and the anatomical position of the D cells (Figure 9.6) seems to put them in a

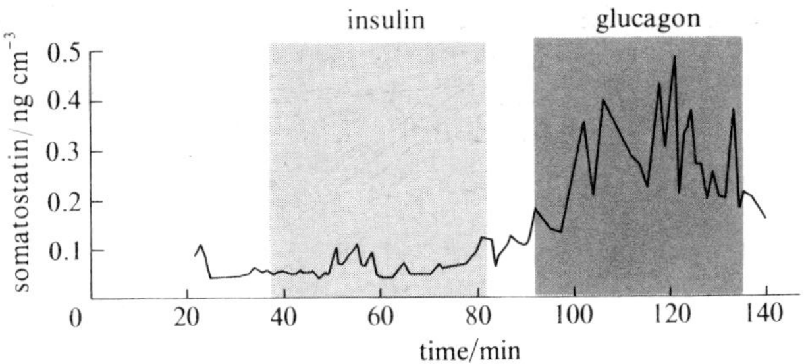

Figure 9.13 The effect on the amount of somatostatin released by an isolated pancreas of giving high doses first of insulin and then of glucagon.

perfect position to achieve this effect by releasing somatostatin within an islet. However, there is debate about the details of the blood circulation through the islets. Insulin secreted from B cells probably could affect directly the D and A cells in that islet—as we have just said, insulin has no effect on D cells but you should recall from Section 9.3.2 that insulin inhibits the secretion of glucagon from A cells. Somatostatin secreted from D cells in an islet can have an endocrine effect only on A and B cells in other islets; it inhibits secretion of hormones from both types of cell. Glucagon release from A cells stimulates insulin release through activating cyclic AMP in B cells and also promotes the secretion of somatostatin from D cells, perhaps also through cyclic AMP, although this is not certain. Here then are tight, negative feedback loops between glucagon and somatostatin, and between glucagon and insulin. The negative feedback control would mean that somatostatin release mediated by glucagon could well prevent insulin secretion mediated by glucagon. The result is a very fine control system (see Table 9.1).

Table 9.1 Actions of secretory products of islet cells on secretions of other islet cells.

Cells	Secretory product	B cell	A cell	D cell
B	insulin	–	inhibition	no effect
A	glucagon	stimulation	–	stimulation
D	somatostatin	inhibition	inhibition	–

Exactly how somatostatin achieves its inhibitory effects is not known. No receptors for somatostatin have been found on pancreatic A and B cells. However, the clue to how it might operate is that somatostatin appears to affect Ca^{2+} movements in A and B cells. Because the release of both insulin and glucagon is dependent on Ca^{2+}, any block of Ca^{2+} movement in A and B cells would inhibit insulin and glucagon release. Somatostatin thus appears to operate by blocking the Ca^{2+}-stimulated secretion of the main glucose-regulating hormones. Figure 9.14 summarizes the role of somatostatin in the regulation of blood glucose.

In addition to A, B and D cells, there are other cells within the islets that produce peptides. A number of such peptides have been identified by the use of antibodies and they are a very mixed collection indeed! Pancreatic islets, in addition to producing insulin, glucagon and somatostatin, produce thyroid-stimulating hormone-releasing factor, luteinizing hormone (or gonadotropin)-releasing factor, cholecystokinin and a small peptide designated 'pancreatic polypeptide'. Whether these peptides act locally within the islets on A, B and D cells, or interact with the rest of the pancreas that produces digestive enzymes, or whether they are released into the bloodstream and have more general actions on other tissues remains to be seen.

The effects on blood sugar regulation of the nervous system and hormones produced by tissues other than the pancreas are considered in Section 9.4.

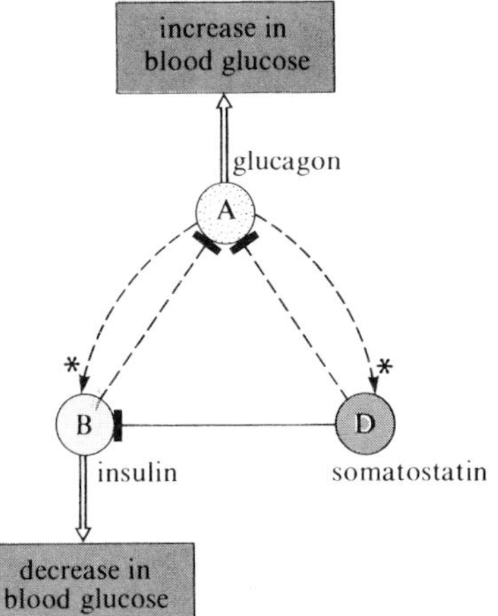

Figure 9.14 Diagram to summarize the interactions between the hormones secreted by A, B and D cells in the islets of Langerhans. Dashed lines indicate feedback loops.

Summary of Section 9.3

1 The pancreas contains discrete groups of endocrine cells called the islets of Langerhans. A number of cell types can be identified in the islet, namely the A, B and D cells, which produce and secrete the hormones glucagon, insulin and somatostatin, respectively.

2 The release of insulin is promoted by raised blood glucose level, which stimulates Ca^{2+} movement in B cells. Ca^{2+} activates a calmodulin-dependent receptor protein, and this is responsible for insulin secretion. Glucagon promotes insulin secretion through cyclic AMP acting via a protein kinase on the receptor protein. Other sugars, together with amino acids and a number of chemical messengers, can all stimulate insulin secretion.

3 Insulin promotes glucose and amino acid uptake into cells; it prevents the deactivation of glycogen synthetase by cyclic AMP and so promotes glycogen storage. Receptors for insulin have been found on many cell types, and these become desensitized in the presence of a persistently high level of insulin.

4 Glucagon release is promoted by the absence of insulin and by low blood glucose levels resulting from fasting or prolonged exercise. Glucagon raises blood glucose levels by activating glycogen phosphorylase and deactivating glycogen synthetase in liver cells; it also stimulates gluconeogenesis by increasing amino acid transport into cells and through effects on mitochondrial enzymes. Glucagon receptors also become less sensitive if glucagon levels are high.

5 Somatostatin release is stimulated by glucagon. Somatostatin inhibits the release of glucagon and insulin within the pancreas and thus regulates blood glucose levels by affecting hormone production from the A and B cells in the islets of Langerhans.

6 A number of other peptides are found in pancreatic islets, but their function in the pancreas is unknown.

Now attempt Questions 2 and 3, on pp. 320–321.

9.4 OTHER CONTROL SYSTEMS IN GLUCOSE REGULATION

So far, we have seen how glucose levels in blood are regulated as a result of biochemical events in the liver, muscles and intestine and through release of hormones from the pancreas. There have been some references already to the effects of the nervous system and hormones other than insulin and glucagon and these are described in this Section. Their mode of action is either through the control of hormone release from the pancreatic islet cells or through direct effects on tissue glycogen stores, or both.

9.4.1 The autonomic control exerted on blood glucose

The brain requires a continuous supply of glucose for continued function (some 6 g h^{-1} in humans), so it would be somewhat surprising not to identify a glucose control mechanism mediated by the nervous system. Similarly, exercise, cold stress and fasting present the body with a 'challenge' to the level of blood glucose. In the islets of Langerhans there are numerous nerve endings from the *autonomic nervous system* (ANS; see Chapter 1, Section 1.6.3 and Figure 1.23); these lie close to the A, B and D cells. However, a transplanted pancreas, or a pancreas deprived of nervous input, will still maintain blood glucose levels. These observations have, until recently, dampened any interest in the possible significance of nervous control mechanisms.

◇ What chemical messengers would you expect to find in ANS nerve endings?

◆ Neurotransmitters—for example, *acetylcholine* (from the parasympathetic system) and *noradrenalin* (from the sympathetic system).

Both acetylcholine and noradrenalin have been shown to alter the secretion of insulin and glucagon. Noradrenalin inhibits insulin release from B cells and stimulates glucagon release from A cells whereas acetylcholine increases the release of both insulin and glucagon, possibly by inhibition of somatostatin release from D cells.

What physiological significance could these observations have? First, they might represent another level of control over blood glucose, directed by the *hypothalamus* in the brain (see Chapter 2, Section 2.3.1). There are insulin receptors in the hypothalamus, and also glucose receptors; direct regulation could therefore be exerted via a feedback mechanism from the hypothalamus through the parasympathetic and sympathetic nerve fibres to the pancreas. The hypothalamus is a key area involved with temperature regulation and feeding and thus could command changes in glucose use (in response to cold) or in glucose storage (in response to fasting) by immediate control of insulin and glucagon release from the pancreas. The observation that anticipation of a meal and hunger both elicit insulin secretion probably has a basis in nerve signals from the hypothalamus to the pancreas via the parasympathetic system. If the vagus nerves are cut, the rise in insulin level stimulated by the anticipation of food is not observed.

◇ What term in Chapter 4 describes the secretion of insulin in anticipation of a meal?

◆ This is an example of *feedforward*.

Stress states are also associated with activation of the autonomic nervous system, for example exercise, trauma, surgery, burns, pain, anxiety and infection. All of these conditions can lead to a lowering of the insulin : glucagon ratio, which then results in the mobilization of the body's glucose and lipid stores. In such ways, the nervous system can alter the insulin and glucagon response, override the control through the direct response of islet cells to glucose level in the blood and so promote the metabolism of glucose to serve the particular requirements of the nervous system.

9.4.2 Adrenalin and glucocorticoids

In a less direct way, any activation of the sympathetic nervous system will also increase the release of *adrenalin* from the medulla of the adrenal glands (Chapter 2, Section 2.2.1 and Figure 2.10). Increase in circulating levels of adrenalin, as with noradrenalin, will inhibit insulin release from the pancreas and promote the release of glucagon. The release of corticotropin-releasing factor (CRF) from the hypothalamus, leading to increased secretion of adrenocorticotropic hormone (ACTH) from the anterior pituitary, is also associated with stress and the resulting rise in secretion of *glucocorticoids* (e.g. cortisol) from the adrenal cortex (Chapter 2, Section 2.3.1, Table 2.1 and Figure 2.21) affects the pancreas, leading to increased secretion of

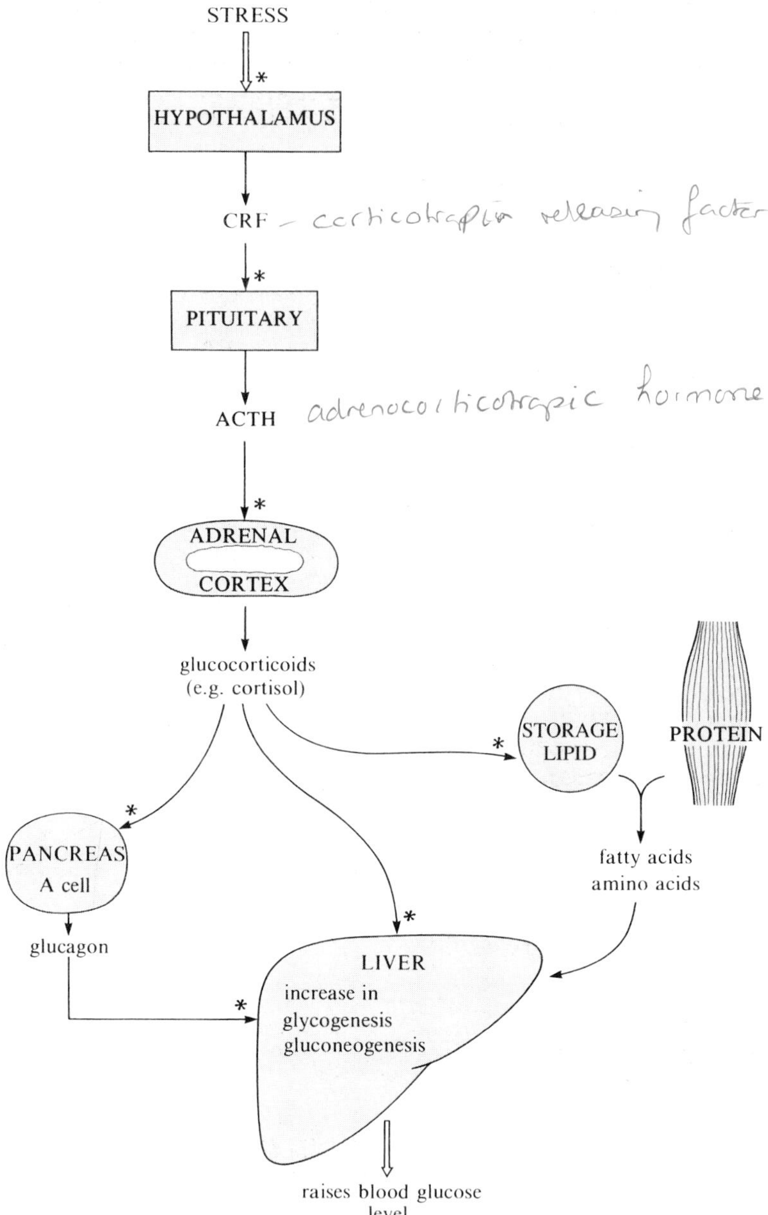

Figure 9.15 Diagram to show how stress can lead to the mobilization of glucose for metabolism at the expense of lipid and protein.

glucagon (see Figure 9.15); this results in increased levels of glucose, fatty acids and amino acids in the blood.

Adrenalin and glucocorticoids have additional effects on the liver cells. Adrenalin rapidly activates glycogen phosphorylase ($b \rightarrow a$) and promotes glycogen breakdown. In human liver cells, adrenalin works in the same way as glucagon, which, you should remember from Section 9.3.2, activates glycogen phosphorylase via cyclic AMP and a protein kinase. In rats, however, adrenalin appears to work through Ca^{2+} as the second messenger. Adrenalin also mobilizes glycogen stores in muscle and, again like glucagon, adrenalin promotes gluconeogenesis in the liver through activation of mitochondrial enzymes.

In liver cells, glucocorticoids deactivate glycogen phosphorylase ($a \rightarrow b$) and so preserve glycogen stores—the converse of the effect of glucagon—but glucocorticoids, like glucagon, promote gluconeogenesis. In other tissues, glucocorticoids promote protein breakdown, which releases amino acids. An increase in the amino acid level stimulates insulin release from the pancreas and this also maintains glycogen stores. The combined effects of glucocorticoids promote glycogen deposition in the liver (see Figure 9.15).

◇ What is the amplification mechanism induced in the tissues by glucocorticoids?

◆ You should remember from Chapter 3, Section 3.3.3 and Figure 3.5, that steroids such as glucocorticoids react with receptors in the cell cytoplasm, which then initiate transcription in the nucleus. The result is the synthesis of specific proteins in the cytoplasm. Neither cyclic AMP nor Ca^{2+} are involved as second messengers.

Summary of Section 9.4

1 Autonomic nerve fibres terminate in the pancreatic islets. Sympathetic nerves inhibit insulin and stimulate glucagon release; parasympathetic nerves stimulate both glucagon and insulin release, possibly by inhibiting somatostatin release.

2 Through the autonomic nervous system (ANS), anticipation of food promotes glucose storage; stress and exercise increase the availability of glucose.

3 Adrenalin inhibits insulin release from the pancreas and promotes glucagon secretion; it stimulates glycogen breakdown in the liver and in muscles and also promotes gluconeogenesis in the liver.

4 Glucocorticoids stimulate glucagon secretion and promote gluconeogenesis in the liver where they also inhibit glycolysis. Through promotion of protein breakdown in other cells and increase in level of amino acids in blood, they stimulate secretion of insulin.

Now attempt Question 4, on p. 321.

9.5 INTEGRATION OF CONTROL SYSTEMS IN THE HOMEOSTASIS OF BLOOD GLUCOSE

In Sections 9.2–9.4 several control systems involved in blood glucose regulation have been described, and you have studied how each system works separately. In this important Section, we try to bring the various systems together and so produce an integrated view of glucose regulation in the whole animal. You should then be able to describe the hormonal control mechanisms at work in maintaining the 24-h glucose profile that we began with in Figure 9.1. A number of organs and tissues that regulate the blood glucose level have been identified in Sections 9.3 and 9.4.

◇ What are they?

◆ 1 The hypothalamus through the autonomic nerves to the pancreas and through control of other hormones.

2 The adrenal glands (cortex and medulla) via hormones which act at the pancreas and on other target cells.

3 The islets of the pancreas through effects on target cells.

4 The liver and other hormone target tissues (mainly muscle and adipose cells).

9.5.1 Regulation of glucose and glycogen metabolism in target cells

Liver cells

The glucose concentration inside liver cells is in equilibrium with glucose concentration in the hepatic portal vein and therefore varies from about 5 mmol l^{-1} (normal state) to about 12 mmol l^{-1} following intake of a meal with high carbohydrate content. Liver cells remove glucose rapidly from the blood by facilitated diffusion and store glycogen; in feeding rats, the rate of glycogenesis is 0.5 μmol glucose converted to glycogen per g liver per minute.

Liver cells have receptors for both insulin and glucagon. The effects of these hormones are described in Section 9.3. Adrenalin has a similar effect to that of glucagon. Glucocorticoids bind to cytoplasmic receptors and deactivate glycogen phosphorylase; this activates glycogen synthetase and so increases storage of glycogen.

Muscle cells

Insulin stimulates glycogenesis in muscle cells, as in liver cells. It stimulates the transport of glucose into muscle cells by facilitated diffusion by promoting the appropriate arrangement of glucose transporter molecules in the cell membrane. The binding of insulin to receptors on muscle cells prevents breakdown of proteins and fats, thus preserving non-sugar energy stores. Adrenalin promotes glycogenolysis in muscles, resulting in increased levels of glucose-6-phosphate ready for glycolysis.

Cells in adipose tissue

Adipose tissue cells have insulin receptors and insulin promotes lipid storage, as described in Section 9.3.1. As in muscle cells, insulin promotes glucose transport into adipose cells by effects on glucose transporter molecules. The lipid in these cells is an energy store which is used when stores of glycogen in the body are low, as during starvation. You should recall from Section 9.2 that glycerol can be converted into glucose by gluconeogenesis and that fatty acids, converted into acetyl CoA, can be catabolized via the TCA cycle. If acetyl CoA accumulates, ketone bodies are formed. These can be metabolized by brain cells if glucose is very scarce but create problems by lowering blood pH.

9.5.2 Integration of glucose regulation in the blood

The one human activity that most regularly raises the blood glucose level is eating a meal. Anticipation of the meal will stimulate insulin secretion from the pancreas (Section 9.4.1) via the parasympathetic system (vagus nerves). Therefore, before glucose is actually absorbed, insulin levels are already raised in the blood. The rising level of blood glucose then triggers release of more insulin from the pancreas by acting directly within the B cells (Section 9.3.1). Insulin release will inhibit glucagon release from the A cells, so the production of glucose in the liver is prevented.

The action of insulin is relatively rapid and results in removal of excess glucose and amino acids from the blood. Glucose transport into liver, muscle and adipose cells is enhanced, along with the transport of other nutrients, and insulin maintains the enzyme glycogen synthetase in its active form and so promotes glycogen synthesis. At the same time, insulin deactivates glycogen phosphorylase and this ensures that the glycogen being formed is not immediately broken down again.

As the blood glucose level falls, insulin secretion from the pancreas slows. Lowering of the level of blood insulin releases the pancreatic A cells from inhibition, and glucagon can be released. If blood glucose is required for exercise or during the fast between meals, glucose is 'pulled out' of the liver glycogen stores and at the same time gluconeogenesis from amino acids and glycerol is promoted in the liver, all through the action of glucagon. The free fatty acids released from lipid breakdown contribute acetyl CoA, which can also be metabolized in the liver. Muscle is a one-way system for glucose: it goes in, but cannot be returned to the blood. The blood glucose level is also monitored by hypothalamic *glucoreceptors*, which activate the appropriate pancreatic cells through the autonomic nervous system.

Regulation of blood glucose level depends on maintaining a balance in liver cells between uptake of glucose from the blood and glycogenesis on the one hand and glycogenolysis, gluconeogenesis and release of glucose into the blood on the other hand. Marked increases in glucose in the blood follow digestion but fasting leads to a general shortage of glucose. There must be sufficient glucose circulating to supply the resting metabolism of cells, and the supply to the blood must be increased during muscular exercise. Stress also results in increased use of glucose. Other tissues such as adipose tissue are important in some circumstances.

The control of metabolism and uptake and release of glucose is hormonal and the two hormones concerned with 'routine' control are insulin and glucagon, secreted by B and A cells, respectively, in the islets of Langerhans.

◇ Use Table 9.2 to summarize the information for insulin and glucagon. Which of these varies more in concentration?

◆ Insulin concentration varies considerably in normal life; glucagon concentration varies much less and increases are associated with fasting and stress.

◇ Summarize the information for the other hormones and neurotransmitters listed in Table 9.2.

◆ Check your entries by referring to Table 9.3.

Table 9.2 Hormones and nerves involved in blood sugar regulation.

Hormone	Main stimulus for secretion	Effect on blood glucose	Effect on islet cells	Effect on liver cells
Insulin				
Glucagon				
Somatostatin				
Adrenalin				
Glucocorticoids				
ANS (controlled by hypothalamus): parasympathetic (acetylcholine)				
sympathetic (noradrenalin)				

Table 9.3 Hormones and nerves involved in blood sugar regulation.

Hormone	Main stimulus for secretion	Effect on blood glucose	Effect on islet cells	Effect on liver cells
Insulin	increase in blood glucose	decreases level	inhibits A (glucagon)	stimulates glycogenesis inhibits gluconeogenesis
Glucagon	lack of insulin in blood	increases level	stimulates B (insulin) stimulates D	stimulates glycogenolysis inhibits glycogenesis stimulates gluconeogenesis
Somatostatin	glucagon release	increases or decreases	inhibits A inhibits B	no direct effect
Adrenalin	'stress'	increases	stimulates A inhibits B	stimulates glycogenolysis stimulates gluconeogenesis
Glucocorticoids	CRF/ACTH	increases	stimulates A	inhibits glycogenolysis stimulates gluconeogenesis
ANS (controlled by hypothalamus):				
parasympathetic (acetylcholine)			inhibits D stimulates B	mainly as for insulin
sympathetic (noradrenalin)			stimulates A inhibits B	as for glucagon

Other hormones and neurotransmitters act either on the cells of the islets of Langerhans or directly on liver, adipose and muscle cells—or both.

There is nothing mysterious about the 'normal' level of circulating glucose. This value or 'set point' varies between individuals and simply represents the balanced insulin : glucagon ratio. Raise the glucose level, and the insulin : glucagon ratio rises; lower the glucose level, and the ratio falls. Remember, though, that this regulation by balance of the pancreatic hormones can easily be overridden by the autonomic nervous system and a new level of glucose can be set in the blood.

Blood glucose regulation is an excellent example of feedback control, both at the level of the pancreatic islet cells (e.g. insulin inhibits glucagon, glucagon stimulates insulin, somatostatin inhibits both) and also in the target tissues (e.g. low insulin levels allow glycogen, lipid and protein breakdown, these release glucose, fatty acids and amino acids into the blood, this stimulates insulin production). These control mechanisms are summarized in Figure 9.16.

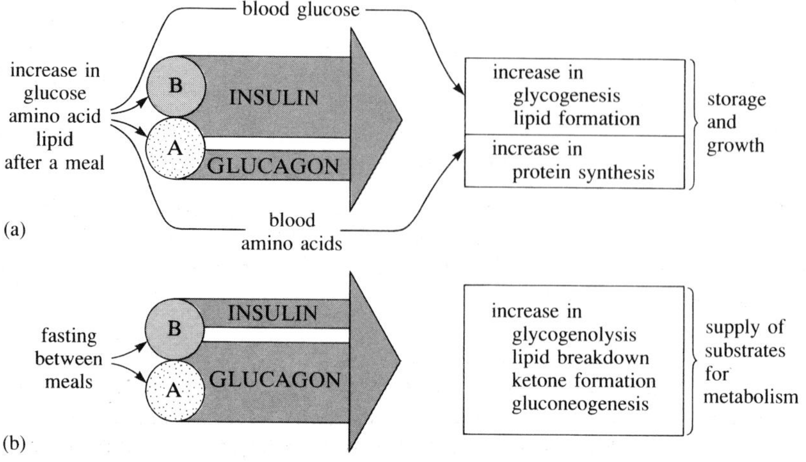

Figure 9.16 Diagrams to show the balance between the secretion of insulin and glucagon from islet cells (a) after a meal and (b) during fasting and the effects of this balance on metabolism and storage of carbohydrates, lipids and proteins.

The breakdown of normal control in humans leads to the disease diabetes mellitus, which is described in Section 9.6.

Now attempt Questions 5 and 6, on p. 321.

GLUCOSURIA

The presence of glucose in the urine; it is associated with excessive loss of water and this results in great thirst.

9.6 DIABETES—ERRORS IN THE REGULATION OF BLOOD GLUCOSE

In this section, the variety of causes underlying diabetes mellitus are discussed as illustrations of what happens when things go wrong in a homeostatically regulated system. The disease was described in a papyrus written about 1550 BC and is usually recognized by the appearance of glucose in the urine (**glycosuria**) of affected people who also suffer great thirst. The disease is known in dogs, cats, rats and mice and probably occurs in most mammals although it is only likely to be diagnosed in laboratory animals. The critical studies by Frederick Grant Banting and Charles Herbert Best, which led to the discovery of insulin in 1921, were carried out on dogs.

The regulation of blood glucose level involves an intricate network of signals, responses and feedback controls. In such a homeostatic system, any failure in one part may, therefore, be magnified throughout the network. In nearly 2% of the population of the UK, blood glucose is not properly regulated and these people suffer from diabetes mellitus. Diabetes is the sign of an error in normal glucose regulation.

The commonest symptom of diabetes is thirst. This may be so severe that an affected child can hardly stop drinking and it is associated with production of excessive amounts of urine; glucose is present in this urine so large amounts of glucose are excreted. Symptoms may come on acutely, over a few days, or gradually. In severe diabetes, the patient loses weight and children who developed the condition before the 1920s used to waste away, especially when they were put on diets low in carbohydrates and fats. Diabetes then was almost invariably fatal. Losing large amounts of blood sugar may lead to hypoglycaemia; its symptoms are slowing of speech and thought, shaking, sweating, unsteadiness, pallor, aggressive behaviour, tingling round the mouth, seeing double and finally unconsciousness. Usually the onset of hypoglycaemia is slow—and it can be relieved rapidly by taking sugar—but occasionally unconsciousness may occur quite suddenly. Nowadays, diabetes can usually be diagnosed quickly and appropriate treatment and advice given to allow patients to lead normal lives although there may be complications (discussed later) after a period of years.

You will remember from Section 9.1 that the normal glucose level in human blood can vary between 4.5 and 6 $\mathrm{mmol\,l^{-1}}$ during the day (Figure 9.1). In some diabetics, blood glucose levels can rise well above 10 $\mathrm{mmol\,l^{-1}}$ and this is associated with glycosuria.

◇ From what you have now learned about blood glucose regulation, can you provide several different explanations for persistently high glucose levels? (In other words, list the various factors that influence glucose levels, and so deduce various points in the system where malfunctions could occur.)

GLUCOSE TOLERANCE TEST

A test in which a person is given a drink containing a high dose of glucose, and the concentration of glucose is measured in blood samples taken at time zero and at intervals over the next 2 hours; the shape of the curve showing increase and then decrease in blood glucose level differs between normal subjects and those suffering from diabetes mellitus.

◆ This question should make you realize how intricate the control systems are for blood glucose. Our explanations fall into two groups: those concerned with the pancreas, and those concerned with target cells. Persistently high glucose levels could be the result of:

(a) the absence (or inactivity) of pancreatic B cells (which normally secrete insulin);
insensitivity of B cells to glucose;
overproduction by pancreatic A cells (which secrete glucagon);
overproduction by pancreatic D cells (which secrete somatostatin);

(b) lack of insulin receptors on the target cells;
an inability to store glycogen;
supersensitivity of receptors on the target cells to glucagon.

Any of the conditions listed above could result in persistently high levels of blood glucose. Diabetes is not, in fact, due to a single identifiable error in the control system but is rather the result of a number of errors.

One rapid method of identifying the error in diabetes is the **glucose tolerance test** when individuals are given glucose and the level of insulin in the blood is followed over the next few hours. This has revealed two types of diabetic (see Figure 9.17):

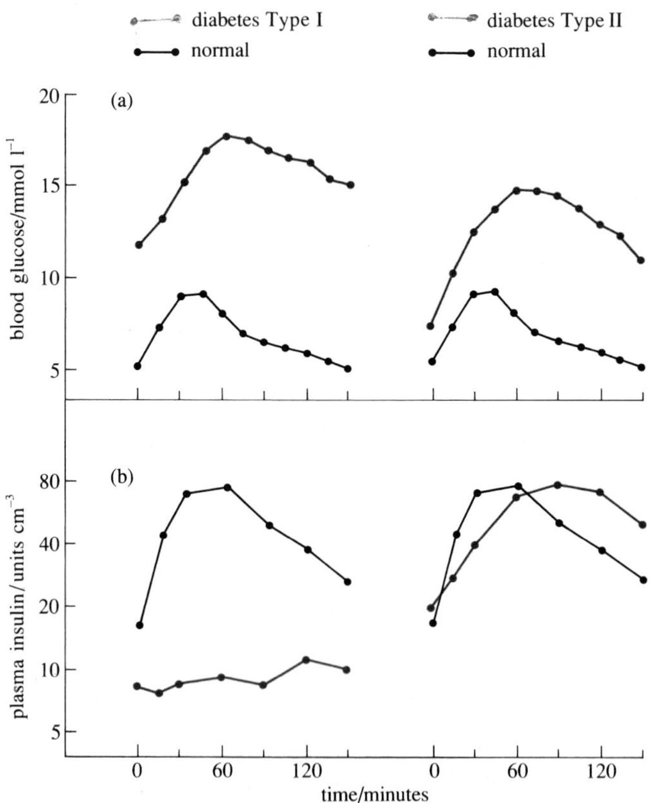

Figure 9.17 (a) Glucose tolerance curves (changes in blood glucose level) and (b) changes in plasma insulin following ingestion of a glucose meal at time 0 by normal subjects and Type I and Type II diabetics.

I A group which has a very low level of insulin or none at all; a glucose meal therefore does not stimulate insulin secretion.

II A group in which insulin levels are normal or just above or below normal; they respond to a glucose meal but the rise in blood sugar is prolonged and markedly higher than normal.

The first group of diabetics commonly consist of young people who develop the symptoms before the age of 20. These are often referred to as suffering from **Type I** or **insulin-dependent diabetes mellitus** (sometimes they are called *juvenile-onset diabetics*); they will respond to insulin treatment. The second group develop diabetes later in life, after the age of 40, and are referred to as having **Type II** or **non-insulin-dependent diabetes mellitus** (sometimes they are called *maturity-onset diabetics*). This group do not usually depend on insulin treatment and are a rather mixed group, their diabetes being the result of one of a number of possible errors in glucose regulation.

9.6.1 Type I: insulin-dependent diabetes mellitus

We consider the juvenile-onset diabetics first. Their lack of an insulin response to the glucose tolerance test (Figure 9.17) appears to be due to a massive loss of the insulin-secreting pancreatic B cells (see Table 9.4). The lack of insulin means that glucagon release is not regulated and that uptake of glucose from the blood after a meal is not promoted. Glucagon stimulates glycogen breakdown together with gluconeogenesis from the breakdown of protein and lipid so the lack of insulin results in exaggerated breakdown of glycogen and production of glucose.

Table 9.4 Percentage of A, B and D cells in islets of Langerhans of juvenile-onset diabetics and in non-diabetics (normal subjects).

	A cells	B cells	D cells
non-diabetics	29	61	10
juvenile-onset diabetics	76	–	24

◇ What will happen to body storage lipids in the absence of insulin?

◆ These lipids break down in the absence of insulin, releasing fatty acids. In liver cells, these are metabolized to form ketone bodies; ketogenesis is promoted by glucagon.

If ketone bodies build up in the blood, ketosis may result, which will lead to coma due to the increased acidity (low pH) of the blood. Excretion of excess blood glucose through the kidneys (glycosuria) is associated with considerable loss of water and this results in thirst. (The function of the mammalian kidney will be discussed in Chapter 11, Section 11.3.) Happily, the insulin-dependent diabetic's condition can be controlled by insulin treatment.

◇ How would you give insulin to a diabetic?

◆ Insulin is a protein, and therefore would be broken down in the gut if given orally. It is usually administered by injection under the skin or sometimes directly into a vein.

The subcutaneous route is commonly used because absorption is slower and the effect of the injection is prolonged. When a slow release is wanted, insulin is linked with zinc to produce a less soluble molecule or with another protein, such as protamine, which is obtained from trout. Insulin preparations are usually given as mixtures or different types are given at different times of day. It is very difficult to imitate the very complicated pattern of normal release of insulin through the day. Various devices and machines are in use or on test. Transplants of pancreatic tissue may become more successful in the future.

Diabetics treated with insulin and who are on a strict diet are normally able to exert good control over blood glucose level but small changes in such variables as temperature or level of adrenalin, or alterations in the times of meals or in physical activity, may result in hypoglycaemia or hyperglycaemia. Diabetics usually check glucose in their blood and urine daily, using various home kits. Every two months or so, the blood sugar level is usually checked by a doctor.

A discovery from the 1970s, important for diabetics and also for healthy people, is that reducing sugar can bind covalently to proteins: this **glycation of proteins** (non-enzymic glycosylation) occurs when there are high concentrations of reducing sugars, such as glucose in blood and glucose-6-phosphate in cells. The sugars bind covalently to proteins and the sugar–protein turns into an **Amadori product** in an irreversible process. It takes days to convert all the initial sugar–protein complex into Amadori product but even brief exposure to excess sugar inevitably produces some of this undesirable compound. Amadori products may become irreversibly cross-linked to other proteins, leading to structural changes in cells. Glycation and cross-linkage may be the cause of damage to myelin layers in nerves, of cataracts in eye lenses and of connective tissue problems, all frequent in diabetics—and also in elderly people. Glycation of proteins is two or three times as high in diabetics as in normal people of the same age but everyone, whether diabetic or not, accumulates glucose-damaged proteins throughout life.

Higher than normal blood sugar concentrations lead to glycation of haemoglobin and formation of an Amadori product with a different electric charge from normal haemoglobin. The life of red blood cells is known and the turnover of haemoglobin is quite rapid so measurement of the percentage of haemoglobin that is glycated can be used to estimate the average blood glucose concentration over the preceding 6 to 8 weeks. Doctors can now check glycation of haemoglobin of diabetic patients to see how well the blood sugar has been regulated over the preceding two months.

The discovery of insulin in 1921 dramatically reduced death from diabetes in insulin-dependent cases. Most commercial insulin is prepared from beef or pig pancreas, and insulin of high purity and improved long-acting insulin formulations are now available.

◇ Why should the purity of insulin matter?

◆ Impurities (e.g. other proteins) may produce an allergic reaction in the body or lead to production of antibodies which bind the insulin and alter its effects.

Genetically engineered 'human' insulin has been in use since 1982. At the time of writing (1989) there seems to be a problem in some patients because warning signs of a fall in blood glucose level, which are experienced with animal insulins, may be greatly reduced or altered when patients are switched from animal to human insulin. Sudden attacks of hypoglycaemia with no warning cause coma and may prove fatal.

The cause of the loss of B cells in young people is being investigated and there is growing evidence that viral infection may have initiated an antibody reaction that then destroyed the B cells—thus this type of diabetes is an auto-immune disease. There is evidence that an inherited predisposition may be involved and special factors, acting early in life, trigger the destruction of the B cells. A current hypothesis is that a very small virus (a retrovirus) may be involved and the initial infection may occur much earlier than the onset of the diabetic state, which often happens suddenly and is indicated by loss of weight and great thirst.

9.6.2 Type II: non-insulin-dependent diabetes mellitus

The second group of diabetics are often called *maturity-onset diabetics* because they develop problems with blood glucose regulation later in life. Their insulin levels can be normal, raised or lowered as compared with non-diabetics. These older diabetics do not develop ketosis or have comas and do not necessarily require insulin injections. This non-insulin-dependent diabetic condition is like an early ageing process of the metabolic system and this tendency may be inherited. A study of twins carried out in London showed that when one twin of a pair developed diabetes after the age of 50, in 90% of cases the other twin developed diabetes within a few years.

Some maturity-onset diabetics have high insulin and high glucose levels. These diabetics are often obese. The problem in these cases seems to be that excess glucose intake derived from the diet produces a persistently high level of insulin secretion from the pancreas.

◇ What effect would persistently high levels of insulin have on liver cells?

◆ You should remember from Section 9.3.1 that insulin receptors become reduced in number in response to a persistently high level of insulin in the blood.

There may also be a post-receptor defect, a fault in the activation of the glucose transporter molecules in muscle and fat cells. The result is peripheral resistance to insulin. By controlling the dietary intake of carbohydrate and so reversing the obesity, insulin secretion from the pancreas is reduced, and target cells such as the liver again become sensitive to insulin.

In maturity-onset diabetics who secrete low amounts of insulin, the problem may be a lack of response by B cells to glucose, or a decreased ability of B cells to synthesize and secrete insulin. Drugs, such as sulphonylureas, can be used to stimulate insulin secretion.

Development of cataracts, loss of sensation in limbs because of damaged myelin layers of nerves and various connective tissue problems are common in mature diabetics and may be the result of glycation of proteins resulting from high levels of glucose in the blood (see Section 9.6.1). Some doctors, however, believe that it is the triglycerides and free fatty acids in the diet that cause these problems. It is essential for patients to keep to specific diets, especially to the limits on carbohydrate content. Doctors look for tiny haemorrhages in the eyes (retinopathy); these can now be treated before the patient goes blind. If the nerves and arteries to the legs become affected, there may be numbness and foot ulcers may develop or tissues may become gangrenous. It is absolutely essential for older diabetics to take particular care of their feet, wearing well-fitting shoes and keeping toe-nails cut short, because, in extreme cases of infection, it may be necessary to amputate the feet; but these patients usually can learn to walk with artificial limbs and can lead fairly normal lives. All these complications are comparatively rare and tend to develop only some years after the onset of diabetes.

Summary of Section 9.6

1 Diabetes mellitus is a condition where blood glucose regulation has failed at any one of a number of control sites. Two broad types of diabetic can be identified and the differences are summarized in Table 9.5.

Table 9.5 Differences between Type I (insulin-dependent) and Type II (non-insulin-dependent) diabetics.

Features	Type I	Type II
age at onset (years)	under 20	over 40
proportion of all diabetics diagnosed	<10%	>90%
appearance of symptoms	acute	slow
ketosis	frequent	rare
obesity at onset	uncommon	common
B cells	few	variable
insulin	low blood level	variable
family history of diabetes	uncommon	common
antibody to islet B cells	present in blood	absent
treatment with insulin	almost always prescribed	seldom prescribed

2 Glycation of haemoglobin gives a useful measurement of blood sugar level in the preceding 6 to 8 weeks.

Now attempt Questions 7, 8 and 9, on p. 321.

SUMMARY OF CHAPTER 9

Glucose circulating in the blood is the main source for 'fuel' for cellular metabolism in mammals. If the concentration rises or falls beyond certain limits, the mammal suffers from hyperglycaemia or hypoglycaemia and these may result in coma and death.

Glucose, produced by digestion of carbohydrates, enters intestinal cells by active transport and is passed by facilitated diffusion into the blood. When blood glucose concentration is high, liver cells take it up by facilitated diffusion and store it as glycogen (glycogenesis). When blood glucose concentration is low, glycogen is broken down (glycogenolysis) in liver cells and glucose passes out to the blood by facilitated diffusion. If glycogen stores are depleted, liver cells convert lactate, glycerol and some amino acids into glucose by gluconeogenesis. During starvation, ketone bodies may accumulate in the blood as a result of breakdown of fatty acids and accumulation of acetyl CoA.

Body cells take up glucose by facilitated diffusion and metabolize it via glycolysis which, under aerobic conditions, is usually followed by the link reaction and TCA cycle. Many cells, including nerve cells, depend on being able to absorb glucose from the blood as they require it for metabolism.

Muscle cells can take up glucose and store it as glycogen; later they can convert this into pyruvate by glycogenolysis followed by glycolysis. Muscle cells cannot pass glucose back into the blood because they cannot convert glucose-1-phosphate into glucose.

The key to regulation of blood glucose concentration is the balance in liver cells between glucose uptake, glycogenesis, glycogenolysis, gluconeogenesis and diffusion of glucose into the blood. The key enzymes in glycogenesis and glycogenolysis are glycogen synthetase and glycogen phosphorylase, respectively. These both exist as active *a* and inactive *b* forms. The interconversion of these forms is sensitive to glucose and to the second messenger cyclic AMP; when one enzyme is active, the other is inactivated.

Two hormones, both produced in the islets of Langerhans in the pancreas, have key roles in the regulation of blood glucose and their interactions are quite complex.

The first, insulin, secreted by B cells, promotes glycogenesis in liver (and muscle) cells by conversion of synthetase into its active form; this leads to reduction in concentration of glucose in the blood. The second, glucagon, secreted by A cells, promotes glycogenolysis in liver cells by conversion of phosphorylase into its active form, which results in output of glucose into the blood. The ratio of insulin to glucagon sets the level of blood glucose.

Insulin secretion is stimulated by raised glucose levels in the blood, through Ca^{2+} movements in B cells, which involve a Ca^{2+}-calmodulin complex that promotes exocytosis. Glucagon also promotes insulin secretion through cyclic AMP acting on the receptor protein via a protein kinase. Many types of cells have insulin receptors and insulin promotes the uptake of glucose and amino acids.

Glucagon secretion is stimulated by absence of insulin and by low glucose levels in the blood, usually the result of fasting or prolonged exercise; Ca^{2+} ions are involved. Glucagon promotes gluconeogenesis by increasing amino acid transport into cells and through effects on mitochondrial enzymes. It also stimulates secretion of a third hormone, somatostatin, from D cells; somatostatin inhibits release of both insulin and glucagon.

Sympathetic nerves supplying the islets of Langerhans inhibit the release of insulin and stimulate glucagon release; parasympathetic nerves (vagi) stimulate release of both insulin and glucagon, possibly by inhibiting secretion of somatostatin. Anticipation of food stimulates insulin release and promotes glycogen storage, probably through vagal stimulation of islet B cells.

Stresses act via adrenal hormones, usually under hypothalamic and pituitary control. Adrenalin inhibits release of insulin and promotes glucagon release; it stimulates glycogenolysis and promotes gluconeogenesis. Glucocorticoids stimulate glucagon release, promote gluconeogenesis and inhibit glycogenolysis in the liver; they promote protein breakdown in other cells and the resulting increase in level of amino acids stimulates the release of insulin.

The basic regulation of blood sugar, therefore, takes place through feedback systems which control the metabolism of liver cells through release or inhibition of the pancreatic hormones insulin and glucagon. These systems are subject to further control, mainly through the autonomic nervous system and hypothalamus. As a result, homeostasis of blood sugar is normally achieved in spite of considerable variations in intake of food and in level of exercise.

Errors in blood sugar regulation in humans cause the disease diabetes mellitus. Younger Type I diabetic patients have usually suffered massive loss of pancreatic B cells; they can lead normal lives if they take insulin. In older patients, there is usually a genetic predisposition. Many symptoms resemble those of ageing and, in obese patients, insulin receptors have lost their sensitivity. Type II diabetics need to adjust their diets and pay attention to the state of their eyes and limbs.

OBJECTIVES FOR CHAPTER 9

Now that you have completed this chapter you should be able to:

9.1 Define and use, or recognize definitions and applications of each of the terms printed in **bold** in the text.

9.2 Describe the intracellular mechanisms and metabolic reactions that are involved in glucose metabolism. (*Questions 1, 2, 4 and 8*)

9.3 Describe the effects of insulin, glucagon, corticosteroids, adrenalin and glucose on the interconversion of glucose and glycogen. (*Questions 1, 4, 5, 6 and 8*)

9.4 Explain, in words or with a flow diagram, how secretion of glucagon, insulin and somatostatin is stimulated or inhibited. (*Questions 2, 3 and 7*)

9.5 Describe the effects of glucagon and insulin on protein and lipid stores. (*Questions 3, 4 and 8*)

9.6 Describe the mechanisms whereby stress or trauma can raise blood glucose levels. (*Questions 5 and 6*)

9.7 Identify the four main sites of control involved in blood glucose regulation. (*Question 5*)

9.8 Describe the signs and underlying causes of the two main types of human diabetes mellitus. (*Questions 7, 8 and 9*)

QUESTIONS FOR CHAPTER 9

Question 1 (*Objectives 9.2 and 9.3*) From (a)–(e) select two statements that you consider to be accurate.

(a) Glycogen phosphorylase and glycogen synthetase exist in two forms (*a* and *b*). In both cases, conversion from *b* to *a* is dependent on the activation of a protein kinase dependent on cyclic AMP.

(b) Fatty acids, glycerol and amino acids are all substrates for gluconeogenesis.

(c) In the absence of oxaloacetate, the oxidation of fatty acids will result in the production of ketone bodies.

(d) Stimulation of a phosphodiesterase that breaks down cyclic AMP inhibits the activation of phosphorylase.

(e) As the level of glucose is increased, liver cells show a progressive loss of glycogen synthetase *a*.

Question 2 (*Objectives 9.2 and 9.4*) What would be the effect of (a)–(e) on pancreatic insulin secretion in the presence of glucose?

(a) Inhibition of somatostatin release from the pancreas

(b) Completely removing Ca^{2+} from the pancreas

(c) Inhibiting phosphodiesterase activity in B cells

(d) The presence of food in the stomach and intestine
(e) Blocking glucose transport across B cell membranes

Question 3 (*Objectives 9.4 and 9.5*) Select two conditions from (a)–(d) under which lipid and protein stores would be depleted.
(a) Lack of B cells in the pancreas
(b) Persistently high levels of blood glucagon
(c) A meal rich in carbohydrate
(d) Blocking somatostatin release

Question 4 (*Objectives 9.2, 9.3 and 9.5*) Give examples of:
(a) Two hormones that directly raise liver glycogen levels
(b) Two hormones that stimulate glucose production in the liver
(c) Two hormones that promote protein and lipid metabolism

Question 5 (*Objectives 9.3, 9.6 and 9.7*) A classic physiological response to stress is a raised blood glucose level. By which mechanisms could stress raise blood glucose?

Question 6 (*Objectives 9.3 and 9.6*) In an experiment to investigate stress-induced hyperglycaemia, some American research workers perfused hormones singly, or as mixtures, into the vein of a dog for 5 h and then measured the increase in blood glucose concentration above the normal level. The results are summarized in Table 9.6.

Table 9.6 Increase in blood glucose concentration after various hormone treatments.

Hormone perfused	Rise in blood glucose level/ mg per 100 cm^3
adrenalin (0.05 μg min^{-1})	30
glucagon (3.5 ng min^{-1})	10
cortisol (4 μg min^{-1})	3
glucagon + adrenalin	58
cortisol + adrenalin	58
cortisol + glucagon	35
cortisol + glucagon + adrenalin	140

Briefly compare the effects of these hormones when acting singly and when acting together. What do the data indicate about the response of body tissues to these hormone mixtures?

Question 7 (*Objectives 9.4 and 9.8*) In Type I diabetics, glucagon levels are higher than in non-diabetics.
(a) What would be the effect of this raised glucagon on blood glucose level?
(b) Suggest two hormone treatments that could reduce blood glucagon level.

Question 8 (*Objectives 9.2, 9.3, 9.5 and 9.8*) Outline the events that result in ketotic coma in Type I diabetics. (To answer this question, you will need to draw on information in Sections 9.2 and 9.3 as well as Section 9.6.)

Question 9 (*Objective 9.8*) What complications can you think of that may be associated with insulin therapy in diabetes? (Think carefully about what type of insulin is given and the general effects of insulin on blood glucose.)

CONTROL MECHANISMS IN REPRODUCTION

◆CHAPTER 10◆

10.1 SYSTEMS AND CYCLES

A number of factors contribute to reproductive success. In mammals these include mechanisms that determine successful pregnancy and that maximize fecundity. Parental care is a universal feature in mammals, and it takes the form of providing the young with food and, in many species, of rearing them in a sheltered environment.

Reproductive success also requires timing, not only of gamete production, but also of the behaviour of sexually mature individuals within a species. Communication of the reproductive condition of one individual to another may involve complex behavioural displays and mass movements of groups of animals from one place to another. In many animals, especially mammals and insects, chemical signals **(pheromones)** play an important role in indicating one individual's reproductive condition to another. Complex endocrine changes underlie all these events, and the relationships between hormones, the nervous system and behaviour re-emphasize the theme we traced through Chapters 1–3.

Before looking in detail at the mechanisms involved in controlling reproduction, we introduce the major target organs—the gonads—that produce the gametes. In all mammals, there are typically two distinct sexes.

10.1.1 Female mammals

Figure 10.1 shows a generalized female reproductive system. It consists of **ovaries** and ducts that receive the eggs released from the ovaries and convey them to the site of implantation, the uterus (or womb). The **Fallopian tubes (oviducts)** also receive sperm from the uterus and convey them to the site of fertilization within the oviduct itself.

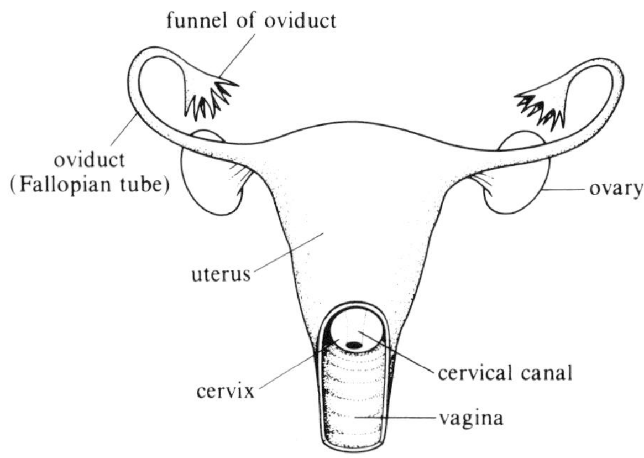

Figure 10.1 The female reproductive tract.

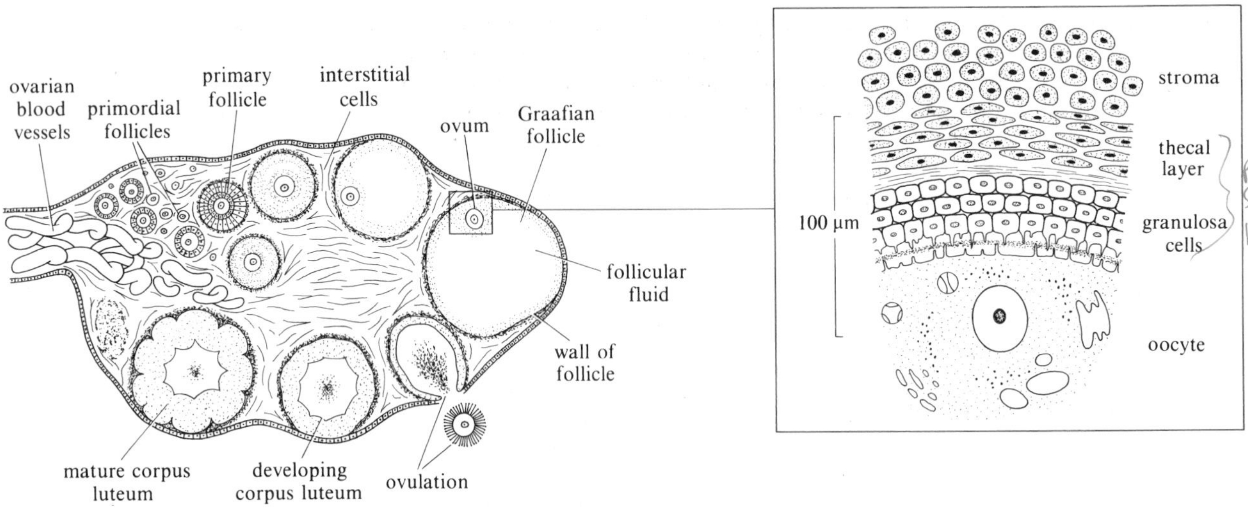

Figure 10.2 The production of follicles in the ovary. The cycle of production, ovulation and corpus luteum formation is shown clockwise around the ovary. Inset shows the wall of a mature follicle.

produce steroid hormone

PRIMORDIAL FOLLICLES
The oocyte surrounded by a layer of follicle cells, in the ovary.
GRAAFIAN FOLLICLE
A fluid-filled vesicle in a mamm. ovary containing an oocyte. The mature mammalian ovarian follicle.
INTERSTITIAL CELLS
Cells lying between follicles in the ovary and between seminiferous tubules in the testis.
GRANULOSA CELLS
Cells which surround mammalian oocyte within the ovary.
STROMAL TISSUE
Intercellular material or connective tissue component of an animal organ.
THECA
A spherical structure, made up of secretory cells and connective tissue, that surrounds an oocyte.
OVULATION
Bursting of a ripe egg (ovum) from an ovarian follicle.
BLASTOCYST
An early stage in mamm. development consisting of a thin-walled hollow sphere of cells, the thicker area on one side will become the embryo proper.

In all mammals, the ovaries are paired, and their size depends largely on the age and 'reproductive state' of the female. Within the ovaries there are **primordial follicles** (see Figure 10.2) containing the cells that give rise to the eggs (ova) and its surrounding membranes. These follicles are present in very large numbers early in development—more than can ever ripen and be shed. Primordial follicles ripen into large **Graafian follicles**, which burst and release the egg at the time of ovulation. Between the follicles lie spindle-shaped **interstitial cells**, which produce steroid hormones. Each primordial follicle contains a primary oocyte surrounded by a single layer of follicular or **granulosa cells**. As the follicle matures and the oocyte enlarges, the surrounding follicular cells enlarge and multiply. The **stromal tissue** around the follicle differentiates into a vascular inner layer made up of secretory cells and an outer layer of connective tissue. These two layers make up the **theca**. Both the secretory cells of the theca and the granulosa cells produce steroid hormones.

In most primates, including humans, usually one follicle ripens at a time, though occasionally several may ripen, leading to a number of eggs being shed together at **ovulation**. Many mammals (e.g. dog, cat, pig, ferret, rabbit) regularly produce a number of eggs from each ovary at each ovulation, and the elephant shrew sheds over 100 eggs at one time! In the nine-banded armadillo, a single egg is shed, but after fertilization the resulting zygote divides into four.

◇ How would the offspring from four eggs be related to one another as opposed to offspring from the division of one egg into four?

◆ All four offspring of the armadillo are genetically *identical*. In the case of four distinct eggs, each would be genetically different and four dissimilar individuals would be produced.

After ovulation, the egg(s) pass through the fimbria (funnel) of the Fallopian tube. Fimbriae either envelop the ovaries, or are close to them (as in Figure 10.1) and show some mobility at the time of ovulation. Fimbriae can pick up eggs lost in the body cavity or ovulated from the opposite ovary. The wall of the oviduct is lined with ciliated epithelium that beats away from the ovary, creating a 'current' towards the uterus. The walls of the oviduct are muscular, and perform a rippling motion of contraction and relaxation that helps propel the eggs down the oviduct. After fertilization in the Fallopian tube, the egg divides to form a **blastocyst**.

The oviducts lead into a **uterus**, which may be single as in primates or double as in rats and mice (see Figure 10.3). The **cervix** (or cervices in the rat or mouse) opens into the vagina, which opens externally, guarded by the **vulva**. In marsupials, for example the opossums of North and South America, there are two externally opening vaginae, two cervices and two separate uteri. The male has a forked penis capable of entering the two vaginae simultaneously. This arrangement gave rise to a superstition that copulation in the opossum is accomplished through the nostrils!

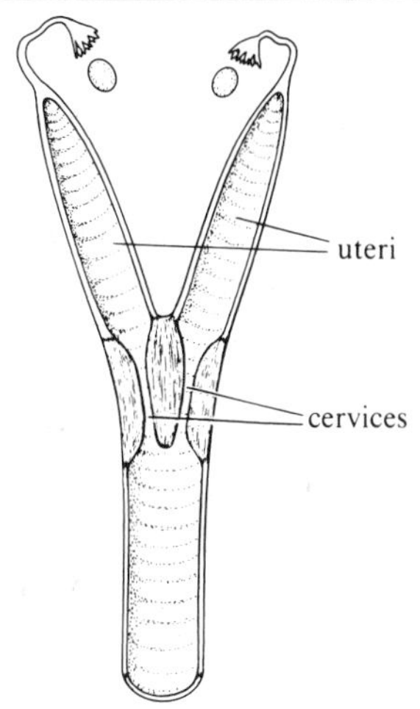

Figure 10.3 The female reproductive tract in a rat or mouse. Note the paired cervices and uteri.

Whether or not a fertilized egg implants in the uterus depends largely on the state of the epithelium that lines it—the **endometrium**. For a limited period after ovulation, the endometrium is highly receptive to anything lying on it, and it will grow around the blastocyst. It is not clear why only a few of the blastocysts implant in the species that shed many eggs at a time. In some species (e.g. rabbit and pig), it is common to find some early embryos being reabsorbed by the uterus, i.e. more blastocysts implant than are carried through to term (birth). Some species show **delayed implantation**, in which fertilized blastocysts remain unattached for various periods. The adaptive significance of delayed implantation will become clear in Section 10.3.

In women, ovulation takes place about every 28 days, from the age of 12–14 (usually) until menopause (45–50). The monthly or **menstrual cycle** is so called because the bleeding (menstruation) at the start of the cycle is a monthly occurrence. (Menstruation can be considered the end or the start of a cycle. In this chapter, the day on which menstruation starts is designated day 1 of a new cycle.) In mammals, the cycle is sometimes called the **oestrous cycle**, because many mammals show **oestrus** or 'heat' (a measurable rise in body temperature) and characteristic behaviour at about the time of ovulation. The more common term used to indicate the interval between successive ovulations is **ovarian cycle** where each ovulation is preceded by a period of increased oestrogen secretion from the ovary.

Follicular development in the human cycle takes 12–15 days. From about day 4 the endometrium begins to thicken and becomes vascularized (supplied with blood vessels). After ovulation (usually near the mid-point of the cycle), further build-up of the endometrium takes place (particularly an increase in vascularization), and in the ovary, the ruptured (empty) follicle develops into a solid secretory body called the **corpus luteum** (see Figure 10.2). If the egg is fertilized and **implantation** occurs in the uterus, the corpus luteum in the ovary persists. In some species, the corpus luteum persists throughout pregnancy, while in others, it remains only through the early stages of pregnancy. The corpus luteum will also persist if the animal is **pseudopregnant**—that is, behaving physiologically as if pregnant when actually it is not. Pseudopregnancy is common in the rat, rabbit and ferret, as we see later in Section 10.2.3.

If pregnancy does not occur, the corpus luteum breaks down—a process called **luteolysis**. In the human cycle, this occurs at about day 26 or 27. The uterine endometrium breaks down, shedding into the uterine lumen blood, epithelium, capillaries and other tissue, most of which is then voided as the menstrual flow.

The length of the ovarian cycle and its frequency or 'seasonality' depend, as does the length of breeding life, on the species. There are, in effect, three periods in reproduction: a **period of breeding potential** in the lifetime of the animal, within that period possibly a **breeding season** (although many mammals may breed continuously), and within that season one or a series of ovarian cycles. The cycle itself can be a single one **(monoestrous)**, or there can be two, or many **(polyoestrous)**. The duration of a cycle also varies between species.

PLACENTA

An organ, consisting of embryonic & maternal tissue in close union, by which the embryo of viviparous mammals is nourished.

When implantation and pregnancy occur, the corpus luteum persists in the ovary and the uterine endometrium continues to develop. A complex **placenta** (from the Greek for 'flat cake') is formed as a result of intimate apposition of fetal tissues to maternal tissues (see Figure 10.4). The placenta functions as an exchange network for fetal and maternal blood, but the two do not actually mix.

GESTATION PERIOD

Length of time from conception to birth in a viviparous mammal.

VIVIPARITY - Giving birth to free-living young, ie without an enclosed egg or cyst stage.

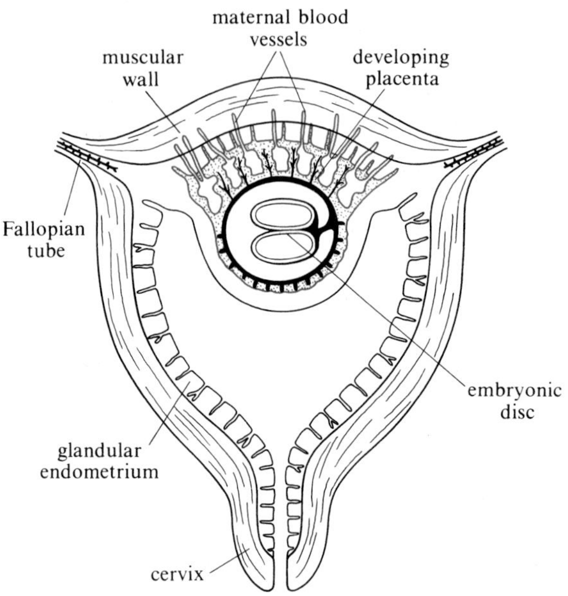

Figure 10.4 A section through the human uterus to show the commonest position of implantation and of the placenta.

PARTURITION

Act or process of birth.

NEUROENDOCRINE SYSTEM

Complex of neuronal pathways and hormones that control physiology and behaviour.

◇ Why would it be disadvantageous to allow the two blood supplies to mix?

◆ There could be immunological reactions. The mother and fetus are not genetically identical, and therefore the mother may develop antibodies against the fetus and thus reject it. This may happen in some species (including ourselves) where this 'barrier' is breached. A current or subsequent fetus may be destroyed, because blood antigens inherited from the father are distinct from those of the mother (e.g. the rhesus factor).

Respiratory gases are exchanged across the placenta together with nutrients and nitrogenous waste. The placenta is also an important source of hormones, particularly in those mammals where the corpus luteum does not necessarily persist throughout pregnancy. The precise form that the placenta takes depends on the species.

The **gestation period** (duration of pregnancy) also varies between species and depends on two factors: the size to which the fetus grows and the stage of development it reaches before **parturition** (birth). In the rat and hamster, the gestation periods are 21 and 18 days, but in the guinea-pig, it lasts 68 days. The explanation for the difference lies in the stage of development reached by the fetus at birth. Rats and hamsters have small, naked young with closed eyes and ears, and they are barely mobile. Guinea-pig young are born looking like small adults. The gestation period is remarkably consistent within a species, varying by only 5–10%. The timing of birth is intimately controlled by both maternal and fetal **neuroendocrine systems**.

Table 10.1 illustrates the variation in ovarian cycle, gestation period and number of young in mammals.

Table 10.1 The length of the ovarian cycle and the gestation period and the number of offspring for several mammals.

Species	Cycle (days)	Gestation period (days)	Offspring
giant panda	—(monoestrous, seasonal)	159	1
European hedgehog	—(monoestrous, seasonal)	34–69	5
rhesus monkey	28 (polyoestrous, seasonal)	164	1–2
chimpanzee	34 (polyoestrous, seasonal)	237	1–2
woman	28 (polyoestrous, continuous)	280	1†
Indian elephant	112 (polyoestrous)	623	1†
blue whale	—(monoestrous)	648	1–2
domestic pig	21 (polyoestrous, continuous)	113	6–16
domestic cattle	21 (polyoestrous, almost continuous)	282	1–2
domestic sheep	17 (polyoestrous, seasonal)	150	1–3
ferret	none*	42	9
domestic cat	15–21*	63	2–7
domestic rabbit	none*	31	6
laboratory rat	4–6	21	7–9
golden hamster	4	18	6
guinea-pig	16	68	3
laboratory mouse	5	19	6

* These are induced ovulators (see Section 10.3).
† Occasionally twins.

10.1.2 Male mammals

The male reproductive system consists of a pair of **testes**, paired **accessory glands** and a duct system that includes the penis (Figure 10.5). During development, the testes descend into the scrotum, and during adult life, they function as producers of sperm and hormones. Some 90% of the testicular mass is made up of the **seminiferous tubules** that produce the gametes (see Figure 10.6). These tubules are very convoluted and comprise **Sertoli cells** and **spermatogonia**, which give rise to **spermatocytes**, which in turn become **spermatids** and then sperm cells.

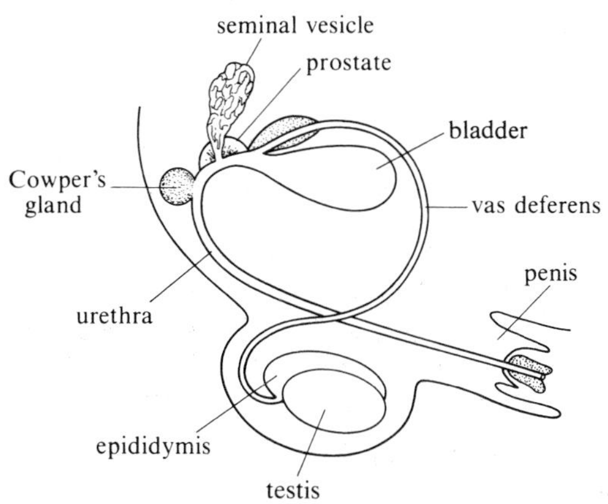

Figure 10.5 The male reproductive tract of the horse.

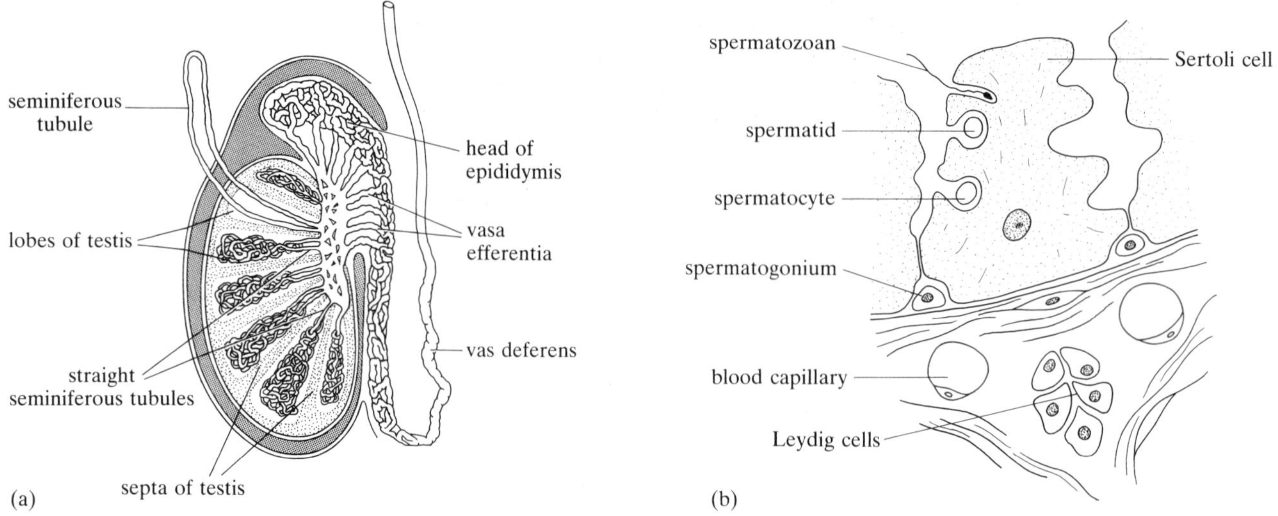

Figure 10.6 (a) A section through the testis. (b) At high power, part of the lining of seminiferous tubule and the interstitial zone.

Sertoli cells aid in the process of sperm cell formation. Many species show seasonal spermatogenesis; at other times of the year, the testes are inactive. Between the tubules are connective tissues and the **Leydig cells**; these secrete steroid hormones (particularly testosterone) and are seen in all male mammals in breeding condition.

Within an active seminiferous tubule, sperm formation (spermatogenesis) shows organization in both time and space. To go from spermatogonium, through spermatocytes and spermatids, to spermatozoa takes a constant period, called the **spermatogenic cycle**. This period is species-specific; in the rat it is 48 days, in man 64 days. Before a spermatogonium begins a new cycle it passes through a quiescent stage; this too is species-specific and is one quarter of the duration of the spermatogenic cycle. Thus, in the rat it is 12 days, in man 16 days.

In any one segment of a tubule, all epithelial cells will be at the same stage in their spermatogenic cycle, but those in adjacent segments will be either one stage ahead or one stage behind. Thus, along the length of a tubule, spermatogenesis occurs in 'waves'. Throughout the testis there are segments of seminiferous tubule at all stages of development so that, though there is a clear cycle of activity within each part of the testis, overall it is producing new spermatozoa continuously.

In most mammals (notable exceptions being marine species and the elephant, rhinoceros and the naked mole-rat), the testes descend into the scrotum. In **continuously breeding** species, they remain in the scrotum, whereas in **seasonally breeding** species, they ascend into the body cavity during the non-breeding season. The main function of the scrotum appears to be to provide the testes with an environment somewhat cooler (by 2–3 °C) than the body cavity. Additionally, the arrangement of arteries and veins to the testes functions to maintain a cooler temperature. Shepherds learned long ago that it was possible to cause temporary sterility in rams by physically tying the testes close to the body wall. Similarly, if the testes do not descend, then that individual is often sterile. This effect of high temperature on the testes appears to be due to arrested spermatogenesis and an impaired ability of the Leydig cells to secrete testosterone.

The seminiferous tubules fuse to form a single tube, the **epididymis**, which becomes a muscular **vas deferens** (Figure 10.5). The two vasa deferentia empty into the **urethra**, a tube common to the urinary and genital systems, which receives secretions (seminal fluid) from the accessory glands (prostate, Cowper's glands and seminal vesicles). The urethra runs through the penis, which is erectile. These accessory glands develop to different degrees in different mammals. No seminal vesicles are present in the cat, dog and wolf, but the prostate is well developed. There is considerable variation in the amount of seminal fluid contributed by the various accessory glands and testes, but in any species not more than 2–5% of the total ejaculate is made up of the secretion of the testes and epididymis. In man, the prostate contributes 15–30% and the seminal vesicles 40–80% of the total ejaculate. Seminal fluid has two functions: it serves as a suspending and activating medium for the sperm cells, which are not motile up to this time, and it provides the cells with a medium rich in ions, vitamins and the sugar, fructose. In many mammals, semen coagulates rapidly in the female reproductive tract. This may serve a variety of functions: buffering against the acidity of vaginal fluids, preventing the sperm from flowing out of the tract, and forming a 'plug' that prevents sperm from other males travelling up the tract.

Male mammals, like females, are active either continuously or only during a breeding season. The behaviour of male rabbits suggests that, within a breeding season, there is a cycle or rhythm of testicular function, although this rhythm (if it exists) is neither as accurately timed, nor as well defined, as the oestrous cycle of the female.

Summary of Section 10.1

1 Ripe eggs are released from the ovary and pass down the Fallopian tube (where fertilization can take place) to the vascularized uterus. If eggs are fertilized, they will implant and a placenta is then formed. If fertilization does not occur, the wall of the uterus breaks down and in the human is shed as the menstrual flow. The length of an oestrus cycle varies between species, many of which show seasonal cycles; the length of gestation also varies between species.

2 Sperm are produced in the testes continuously or during a breeding season, and accessory glands add secretions to the seminal fluid as this passes from the vas deferens into the urethra. The Leydig cells in the testes produce steroid hormones, and the Sertoli cells aid in the process of sperm formation.

Now attempt Questions 1 and 2, on p. 357.

10.2 THE CONTROL OF THE REPRODUCTIVE CYCLE

In the previous Section, we described the cyclic nature of gamete maturation and release. During the breeding season, if fertilization is successful, the cycle of ovulation is replaced by a gestation period. If the released egg is not fertilized, the ovulatory cycle is repeated. The cyclic nature of the whole process of reproduction suggests that an intricate network of control systems is involved. In this Section, the control of the ovarian (in particular, the menstrual) cycle is considered; in Sections 10.3 and 10.4, we look at the role of nerves and hormones in seasonal breeders and in the process of sex determination and maturation.

10.2.1 Releasing factors, stimulating hormones and steroids

If the pituitary of a sexually immature rat is surgically removed (an operation called **hypophysectomy**), then among other things, the gonads fail to become functional. If an identical operation is carried out on an adult rat, the gonads shrink and eventually atrophy. In both cases, when extracts of pituitary are injected into the hypophysectomized rats, gonadal function is restored. These early observations suggested a link between the pituitary and the gonads, and in 1931 it was demonstrated that the pituitary secretes two **gonadotropins**, which were later called **luteinizing hormone (LH)** and **follicle-stimulating hormone (FSH)**.

Pituitaries from both males and females contain LH and FSH. These are glycoproteins with a relative molecular mass of around 30 000, and both consist of α and β sub-units. FSH and LH are secreted from cells in the anterior pituitary.

◇ What other tropic hormones are released from the anterior pituitary?

◆ Adrenocorticotropic hormone (ACTH), thyroid-stimulating hormone (TSH) and growth hormone (GH).

The secretion of LH and FSH is stimulated by **gonadotropin releasing hormone (GnRH)**, a neurosecretory product of neurons in the hypothalamus that is carried to the anterior pituitary in blood capillaries. LH and FSH are secreted into the blood and stimulate the gonads to produce other hormones. The ovaries and testes release steroid hormones, and these have a variety of target organs and, therefore, a variety of effects. They also produce a peptide hormone, **inhibin**. In the male, inhibin is released by the Sertoli cells of the testis and selectively inhibits the secretion of FSH from the pituitary (a negative feedback relationship). Inhibin does not alter GnRH secretion from the hypothalamus. In females, inhibin is secreted by granulosa cells within the follicle and performs the same feedback function as in males. As well as effects on the pituitary, inhibin is thought to have local actions within the gonads. These might involve completion of spermatogenesis in males and of follicular development in females.

◇ What is the parent molecule from which steroids are synthesized?

◆ Cholesterol.

◇ What are the two main groups of steroid hormones?

◆ Sex steroids (oestrogens, androgens and progestins) and corticosteroids.

The route of sex steroid production from cholesterol proceeds through **pregnenolone**, which is oxidized to form **progesterone**. Conversion of either pregnenolone or progesterone will produce the **androgens** (e.g. in the testes—**testosterone** and **androstenedione**) and the **oestrogens** (e.g. in the ovary—**oestradiol** and **oestrone**). Androgens can be converted into oestrogens.

◇ What significance could this link between different sex steroid hormones have in the ovary?

◆ It means that testosterone may be present in the ovary as a precursor of oestrogens.

Similarly, small amounts of oestrogens can be found in the testis. The amount of oestrogens released in the human male is quite small, but both the boar and the stallion produce quite large amounts.

◇ In what other endocrine tissue might you expect to find sex steroids? (What other endocrine gland secretes steroids?)

◆ The cortex of the adrenal glands has steroid-containing cells (mainly producing corticosteroids), and these also produce traces of both oestrogens and androgens.

The adrenal gland and ovary receive cholesterol from the blood. The testes, however, seem to manufacture their own cholesterol (from acetate). The reasons for this difference are obscure, though it is possible that the testes are unable to gain access to the cholesterol in the blood. This blood barrier may also protect sperm in the testis from immunological attack.

Occasionally, a defect in steroid production in the adrenals leads to the production of large amounts of androgen. There is feedback from the level of androgens circulating in the blood to the pituitary, and this stops the release of gonadotropins.

◇ What sort of feedback system is this?

◆ A negative feedback system.

As the level of gonadotropins in the blood drops, so does the release of sex steroids from the gonads. Production of large amounts of androgen from the adrenal gland in a female can be disastrous.

◇ Why does the production of adrenal androgen not stop along with gonadal steroid production when gonadotropin levels drop?

◆ The adrenals are *not* under gonadotropic control but are regulated by ACTH from the pituitary.

The negative feedback system between the adrenals and the pituitary therefore breaks down because androgens are being produced instead of corticosteroids. Now the system goes wildly out of control, because the low level of corticosteroids in the blood stimulates ACTH release from the pituitary, and more ACTH causes more androgen production in the adrenals.

◇ How could this situation be quickly brought under control?

◆ The problem is the *lack* of corticosteroids. If corticosteroids are given, this will lower ACTH production (by negative feedback) from the pituitary, will prevent the release of more androgen from the adrenals, and will supply the patient with the missing corticosteroids.

The result will be that gonadotropin release from the pituitary will be resumed and normal ovarian function will be restored. The condition of producing large amounts of androgen in the adrenal glands of a female is known as the **androgenital syndrome**, and it demonstrates how catastrophic results can occur when a feedback mechanism goes slightly wrong. Such a female develops male external genitalia, but possesses the internal genitalia of *both* sexes.

Gonadotropins from the anterior pituitary therefore regulate steroid hormone production in the gonads, but this is only one part of the control system.

PROLACTIN

Hormone secreted by anterior lobe of pituitary in vertebrates, which has varied effects across species. In mammals it is responsible for the onset of lactation, but in many freshwater teleosts it is crucial in reducing diffusive ion losses via the gills.

PROLACTIN RELEASING FACTOR PRF

Hormone that promotes the secretion of prolactin.

PROLACTIN RELEASE-INHIBITING FACTOR (PIF)

Hormone that inhibits the secretion of prolactin.

◇ What stimulates gonadotropin release?

◆ The hypothalamus produces releasing (and release-inhibiting) factors that are carried to the pituitary in the blood supply.

◇ What is the releasing factor concerned with pituitary gonadotropin release?

◆ Gonadotropin releasing hormone—GnRH.

Neurosecreting neurons that produce GnRH are scattered throughout the hypothalamus. Figure 10.7 shows the position of the major nuclei (groups of neurons) in the hypothalamus. If neurons are destroyed in those areas that are shaded grey in the Figure then production of FSH and LH is diminished.

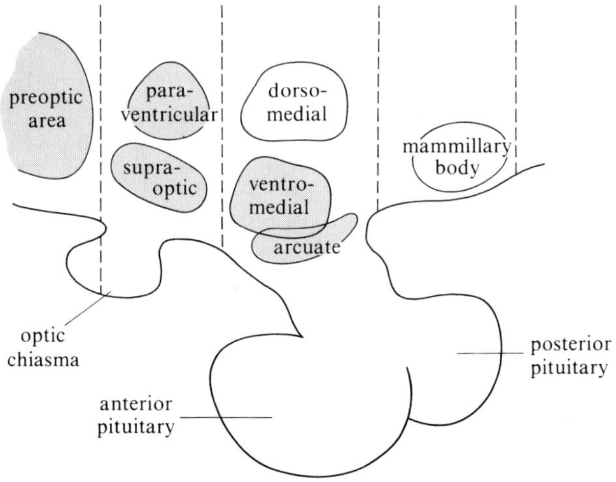

Figure 10.7 A schematic diagram of the main nuclei in the hypothalamus. Those implicated in GnRH release are shaded.

GnRH is not species-specific in structure and will cause both the release and stimulate the synthesis of FSH and LH in the anterior pituitary. The activity of GnRH is very short-lived, because it is rapidly degraded in the blood and the breakdown products are excreted by the kidneys. It was once thought that the hypothalamus also produces a gonadotropin release-inhibiting factor (GnRIF), but there is no good evidence of this, so that FSH and LH are purely under the stimulating control of GnRH. 'Blockers' (antagonists) of FSH and LH have been synthesized, however, and their usefulness in birth control is a point we return to in Section 10.5.

The control of gonadal function is thus a three-tier system between the hypothalamus, the pituitary and the gonads. The extra tier, the hypothalamus, brings with it a site for the regulation of reproduction via other areas of the brain. The importance of this relationship will become clear in Section 10.3.

Before we look at the precise control of the oestrous cycle by hormones, there is one other pituitary hormone to mention—**prolactin**. This is a peptide (relative molecular mass 25 000) produced by cells in the anterior pituitary, and its release is also under hypothalamic control. There is evidence for both a **prolactin releasing factor (PRF)** and a **prolactin release-inhibiting factor (PIF)**. We look at the role of prolactin in more detail in Section 10.2.5.

10.2.2 Fine control of the ovarian cycle

In the past 10 years, the development of the technique of radioimmunoassay has made it possible to measure accurately the levels of hormones circulating in the blood during the oestrous cycle. Figure 10.8 shows the relative changes in FSH, LH, oestradiol and progesterone in the normal human menstrual cycle. Note that both FSH and LH peak at ovulation, that the oestradiol level rises just before the FSH and LH peaks, and falls temporarily just after ovulation, and that progesterone rises after ovulation. The question is: are these changes related to one another, and if so, how?

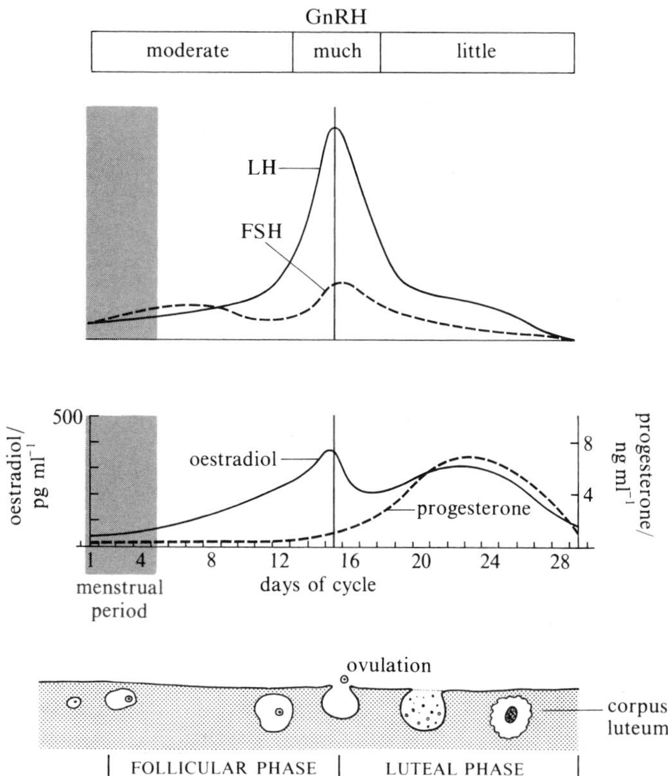

Figure 10.8 The changes in level of gonadotropins (FSH, LH) and of steroids (oestradiol and progesterone) during the menstrual cycle. The corresponding changes in the ovary are shown below.

During the follicular phase of the menstrual cycle, the low level of circulating oestrogens (from the ovary), results in an increase in the secretion of FSH from the pituitary.

◇ What sort of feedback does this illustrate?

◆ Negative feedback.

FSH stimulates follicular development in the ovaries if LH is present. Growth of the follicles results in increased secretion of oestrogens from the follicular cells. Oestrogens directly influence follicular growth and also make the ovaries more responsive to the gonadotropins LH and FSH. As the level of oestradiol increases markedly during the late follicular phase just before ovulation, there is a marked *positive* feedback effect of oestradiol on gonadotropin release, resulting in the peaks (surges) in LH and FSH seen in Figure 10.8. This combined rise in gonadotropins brings about the final maturation of the follicle and results in ovulation.

The high level of oestradiol now exerts a *negative* feedback effect on gonadotropin release (FSH and LH fall), and consequently the level of oestradiol (from the ovary) falls. The corpus luteum in the ovary (under the influence of LH) is responsible for the massive secretion of progesterone, which peaks 5–6 days after ovulation, and also for the second rise in oestradiol, which begins just after ovulation. If fertilization does not occur, the corpus luteum regresses, and consequently both progesterone and oestradiol levels fall. A new cycle then begins with menstruation.

This sequence of events is typical of the human menstrual cycle. For example, whereas progesterone appears to play no part in ovulation in the human, it is known to induce ovulation in cows, rabbits, rats and sheep. The maturing follicle in the ovary appears to be able to secrete small amounts of progesterone just before ovulation.

This pattern of hormonal changes during the menstrual cycle raises a number of questions. First, how can FSH and LH be involved with follicular growth when their levels are so consistently low during this early period of the cycle? Is it possible that the *cumulative* effect of these gonadotropins on the ovary is the important factor in follicular development?

◇ There is an alternative possible explanation. Can you think of it? (So far, we have a messenger molecule explanation, but what about changes in the ovary itself?)

◆ It is possible that receptors for LH and FSH on the follicles increase in number with time and therefore respond more strongly to the same level of gonadotropins later in the cycle. (Receptor sensitization is discussed in Chapter 3.)

Another puzzling aspect of hormonal changes during the menstrual cycle is that oestradiol appears to have both stimulatory and inhibitory effects on gonadotropin levels at different times. Oestradiol normally has a negative feedback effect on gonadotropin release, and this results in the sharp drop in FSH and LH just after ovulation. Progesterone augments the inhibitory effects of oestradiol on gonadotropin secretion. Low levels of oestradiol during the follicular phase allow FSH and LH secretion, which, in turn, promotes follicular maturation and the secretion of oestradiol from the follicle. The site of this negative feedback is on the production and release both of GnRH in the hypothalamus and of LH and FSH from the pituitary.

The target for the positive feedback effect of oestradiol just before ovulation may be either at the hypothalamus or on the pituitary itself. In primates, the pituitary seems to be involved. Here, cells that secrete FSH and LH become *more sensitive* to GnRH from the hypothalamus. The changing sensitivity of the anterior pituitary is a direct effect of rising levels of oestradiol during the follicular phase of the cycle.

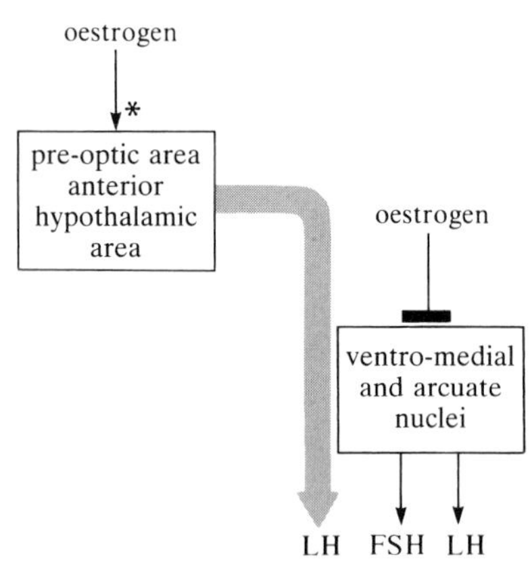

Figure 10.9 The different effects of oestrogens on the two main sites of GnRH production in the hypothalamus.

In other mammals, the main site (or sites) of the feedback actions of steroids may be different. If the GnRH-secreting neurons in the medial preoptic area of the hypothalamus in the rat are destroyed, or if their projections back to the medium eminence are severed, ovulation is prevented. If these GnRH neurons are artificially stimulated, then ovulation is initiated. Similarly, if oestrogen is implanted in this area of the hypothalamus, ovulation results. Although this area of the hypothalamus appears to be concerned with ovulation, the other area of the hypothalamus that promotes gonadotropin release (the nuclei just above the pituitary, see Figure 10.7) is involved simply with gonadal maintenance. Therefore these contradictory effects of feedback regulation of oestrogen on gonadotropins can be explained, at least in the rat, by the existence of *two* sites in the hypothalamus: one that responds by positive feedback to oestrogens (the anterior areas) and another that responds by negative feedback to oestrogens (the areas above the pituitary). This idea is summarized in Figure 10.9.

In women, unlike other mammals, it is usual for one follicle at a time to develop to the stage when it responds to the surge of LH by ovulating, although earlier in the cycle a number of follicles may have begun to develop under the influence of FSH. The sensitivity of the one follicle is probably due to its ability to secrete more oestrogen than the others. The 'fertility drugs' that became notorious and newsworthy in the late 1960s because they led to quintuplet and sextuplet births were gonadotropins. Some fertility drugs cause many follicles to mature simultaneously and provide multiple targets for fertilization.

10.2.3 The corpus luteum: maintenance and luteolysis

From Section 10.2.2 it should be clear that the complete cycle has three stages: a follicular stage, an ovulatory stage and a luteal stage. Some animals, for example primates, have all three stages. After ovulation, the ruptured follicle changes shape and becomes a corpus luteum (plural corpora lutea). Under the influence of LH, the corpus luteum starts to produce oestradiol and progesterone. If fertilization and implantation are successful, the corpus luteum persists in the ovary.

Some mammals do not show the three stages of the cycle clearly. In rats, if mating does not occur, the corpora lutea produce very little progesterone and the uterus does not undergo the changes associated with the high levels of progesterone (e.g. vascularization). If mating occurs, stimulation of the vagina and cervix by the penis is communicated via nerves to the hypothalamus, and this promotes the release of prolactin from the anterior pituitary.

◇ What mechanism is employed between the hypothalamus and the pituitary?

◆ To promote prolactin release, either the hypothalamus must secrete a PRF or there must be a reduction in the release of an inhibiting factor from the hypothalamus.

In the rat it seems that dopamine is blocked. The release of prolactin promotes cholesterol metabolism in the corpora lutea. This is not quite the full story. LH is also important in the conversion of the cholesterol into progesterone, at least for the first two weeks of gestation. After that, the placenta maintains the pregnancy by production of a **luteotropic factor** that sustains the manufacture of progesterone by the corpora lutea. The role of prolactin as a luteotropic factor is a feature of rats but is not typical of mammals in general.

In yet a third group of mammals (e.g. the cat and ferret), there is clearly only one stage of the cycle that occurs spontaneously, and that is the process of follicular development. These mammals are called induced (or reflex) ovulators because they ovulate in response to mating. We shall return to them in Section 10.3.

If pregnancy does not result, the corpus luteum undergoes luteolysis and the production of progesterone ceases (see Figure 10.8), but if the uterus is removed (hysterectomy), then the corpus luteum is maintained in some species.

◇ What can you deduce from this effect of hysterectomy?

◆ The uterus is probably producing a substance that causes the corpus luteum to break down (a **luteolytic factor**).

PROSTAGLANDINS

Group of messenger molecules synthesized from arachidonic acid. Act as local hormones, at or very near site of synthesis, and are involved in the control of a variety of cellular processes such as inflammation, regulation of the vasculature and the reproductive cycle. Prostaglandins are short-lived and are synthesized by all or most tissues in body.

This substance appears to be a **prostaglandin**, $PGF_{2\alpha}$ (see Chapter 3). Injection of $PGF_{2\alpha}$ into pregnant females causes luteolysis and subsequently abortion. $PGF_{2\alpha}$ has been used to induce abortion in women, and it appears to do so by causing contraction of the uterine muscles.

◇ If prostaglandins can cause abortion, what must happen to $PGF_{2\alpha}$ levels in pregnancy?

◆ A signal, presumably from the fetus, must suppress the synthesis and release of $PGF_{2\alpha}$ from the uterus (an anti-luteolytic signal).

So, to maintain pregnancy, the corpus luteum must produce progesterone (at least initially) and this ensures implantation of the blastocyst in the uterus and the inhibition of luteolysis. At term, or if fertilization does not occur, luteolysis must happen to initiate a new cycle.

10.2.4 Hormones and male reproduction

In the same way that the ovary is under the control of both the hypothalamus and the pituitary, so too is the testis. And just as females can be sexually active either continuously or seasonally so too can males. In most species, the breeding behaviour of the male coincides with that of the female.

The important components of the testis are the seminiferous tubules (which produce sperm) and the Leydig cells (which secrete androgen, inhibin and androgen binding protein). Both FSH and LH affect the testis. Growth of the testes is clearly under gonadotropic control. A rapid increase in testicular mass occurs at puberty and this is correlated with an increase in plasma (blood) levels of FSH and LH. Administration of LH to adult mammals that have been hypophysectomized results in growth of Leydig cells and an increase in androgen secretion. LH probably affects the synthesis of androgen at some step between cholesterol and pregnenolone. LH affects the activity of a number of enzymes in the steroid synthesis pathway, and therefore probably has a general tropic action on Leydig cell metabolism and protein synthesis. LH produces a rise in cyclic AMP levels in Leydig cells, which results in increased synthesis of steroids (see Figure 10.10).

Hormonal control of spermatogenesis is a little more complex. FSH can produce a very significant increase in the diameter of the seminiferous tubules, but both LH and FSH are essential for completion of spermatogenesis. The major site of FSH action appears to be on Sertoli cells, where membrane receptors for FSH have been identified. Sertoli cells also have receptor sites for androgens and produce androgen binding protein.

◇ Would you expect to find these androgen receptors in or on the cell?

◆ Androgens are steroids, so the principal binding sites would be on cytoplasmic proteins and not on the cell membrane.

Sertoli cells therefore have receptors for both peptide and steroid hormones.

The negative feedback control of FSH and LH production exerted by the testes appears to involve androgens, oestrogen and inhibin (see Figure 10.11). Oestrogen and androgens immediately depress gonadotropin secretion in males and, as we saw earlier (Section 10.2.1), oestrogen is a normal product of the testes.

GnRH is found in the male hypothalamus, but there is little evidence for a cycle, comparable to the ovarian cycle, in males. So what controls the release

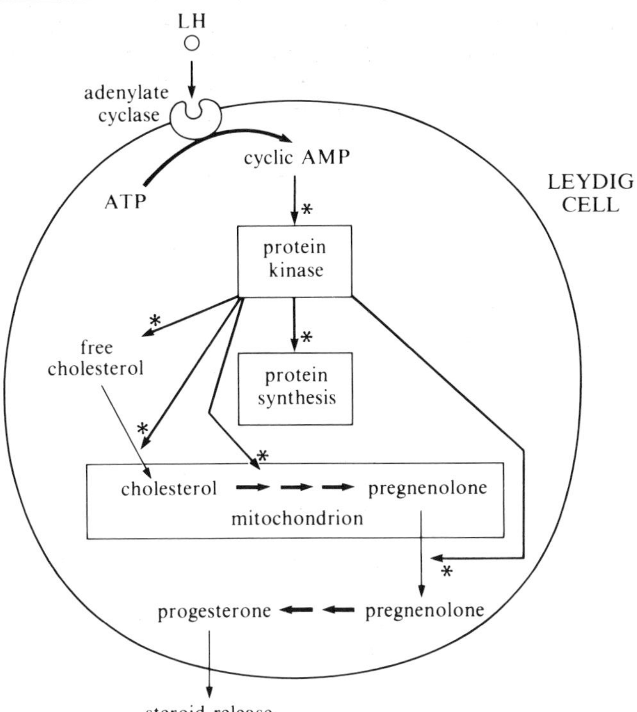

Figure 10.10 The effect of LH on Leydig cell metabolism. Cyclic AMP activation of protein kinase promotes steroid synthesis and release. As you can see, protein kinase has several effects on the route to synthesis of steroids.

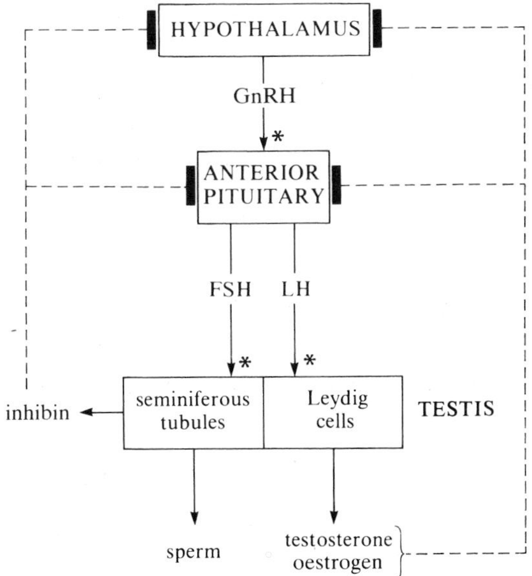

Figure 10.11 Feedback control of releasing factor and gonadotropin release is brought about possibly by testosterone, inhibin, dihydrotestosterone and oestrogen.

of GnRH in males? Unlike in females, steroids cannot cause a positive feedback effect at the hypothalamus or the pituitary in males. However, environmental changes can regulate gonadotropin levels independently of steroids. If a bull is placed close to a cow in heat, there is a rise in LH levels and consequently a rise in testosterone production from the testes. As we will see in Section 10.3, environmental effects on steroid production appear to be mediated by GnRH from the hypothalamus, and this can explain gonadal growth and regression in species that are seasonally active.

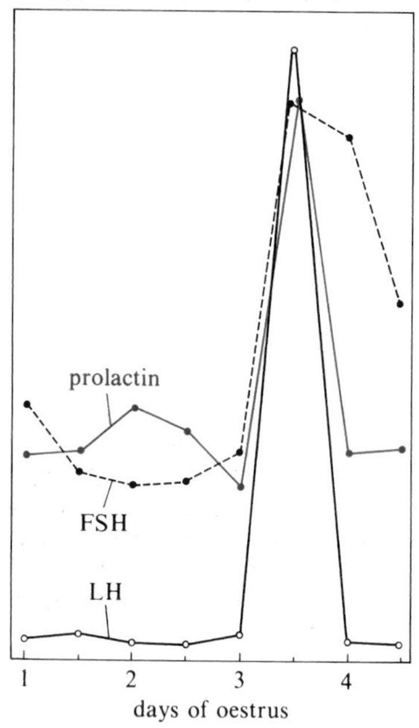

Figure 10.12 Levels of FSH, LH and prolactin during a 4-day oestrous cycle in the rat; ovulation occurs during day 3.

10.2.5 The role of prolactin

The major role of prolactin in mammals is thought to be in the regulation of milk secretion from the mammary glands. In Section 10.2.3 we saw that prolactin may also be involved in the maintenance of the corpus luteum in certain mammals.

With the development of receptor-binding techniques, prolactin binding sites have been found in the adrenals, ovaries and testes in addition to mammary tissue. High levels of prolactin in rats and in women appear to prevent ovulation by suppressing the surges of LH secretion.

◇ Where must prolactin be causing this inhibitory effect on LH release?

◆ Since LH levels rise in response to injected GnRH in rats with high prolactin levels, the site of inhibitory action of prolactin must be in the hypothalamus, or the nerve cells that produce the endogenous GnRH.

During a normal ovarian cycle, the levels of prolactin peak at the time of ovulation (Figure 10.12), probably in response to the rise in oestrogen level; if the ovary is removed, prolactin levels decrease and the mammary glands regress. The rise in prolactin may serve two functions: first, it is present ready for the luteotropic function just after ovulation, and second, it may terminate the ovulatory surge in gonadotropins, acting therefore as a fine regulator of ovulation.

There is no clear menstrual rhythm in prolactin levels, unlike the oestrus peak in the rat. The 'anti-ovulatory' effect of prolactin in women (called **lactational amenorrhoea**) can be important during the suckling period and may well be responsible for the irregular cycles that occur after a birth. In the !Kung tribe in Southern Africa, no methods of birth control are practised, and yet women have a child relatively infrequently, usually once every 4 years. The key to the relatively small number of pregnancies seems to be that children are suckled from birth to 3 years of age. Moreover, the children are allowed free access to the nipple, and the resulting high suckling frequency maintains high levels of prolactin. The dominant inhibitory factor on the ovulatory cycle may not, however, be prolactin per se. There is evidence for a direct neuroendocrine effect from the infant suckling. In western societies, breast feeding is usually of short duration and at regular times, and is therefore not a reliable method of contraception.

Some women have a persistently high level of prolactin in the blood. Associated with this is the absence of LH surges and so lack of ovulation. It appears that, in women, the absence of LH surges may be due to a decreased sensitivity of the pituitary to GnRH as well as a lack of this releasing hormone. This condition is now easy to treat, mainly because the control of prolactin secretion is better understood. The involvement of neurotransmitters in the control of pituitary function has been suspected for a very long time. In Section 10.2.1, we saw that the release of prolactin appears to be under the control of both a PRF and a PIF from the hypothalamus, probably with the PIF control dominant (i.e. prolactin release is normally inhibited). Thyrotropin releasing hormone (TRH) also releases prolactin very efficiently, but whether it is *the* PRF is uncertain. In addition, both oestradiol and progesterone in combination with oestradiol can elevate plasma prolactin levels in women.

If levels of the neurotransmitter dopamine are raised in the hypothalamus or pituitary, prolactin secretion is inhibited; if the release of dopamine is inhibited in some way, then prolactin is stimulated. Neurons that release

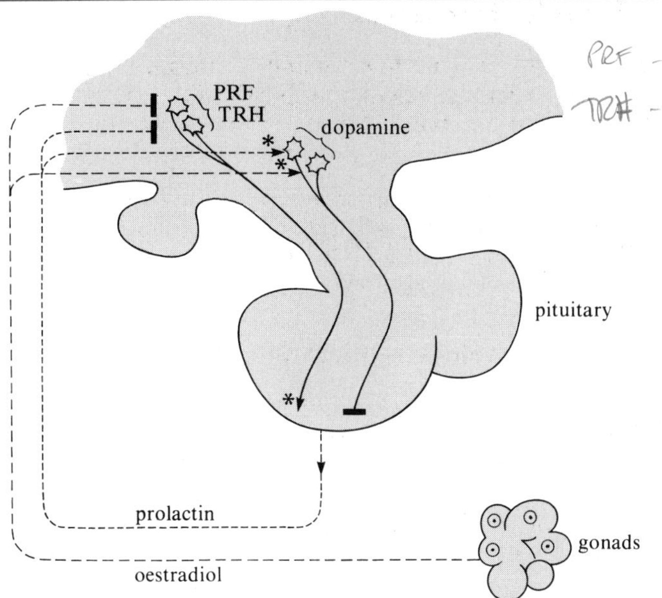

Figure 10.13 The effects of PRF (and possibly TRH) and dopamine (the PIF), released from neurosecretory cells in the hypothalamus, on the release of prolactin from the anterior pituitary. Note the feedback effects of prolactin and oestradiol on the hypothalamus.

dopamine have been identified in the hypothalamus (Figure 10.13) and dopamine appears to be an important PIF. Prolactin itself can stimulate the release of dopamine. It is possible that other neurotransmitters, namely GABA (γ-aminobutyric acid) and GAP (GnRH-associated peptide), act as PIFs, but such an effect has not been established.

◇ What would the consequence be of this feedback of prolactin onto dopamine?

◆ Because dopamine blocks prolactin release, this feedback system would be a negative one and prolactin levels would drop.

The treatment of women who are not ovulating because of a high prolactin level in the blood involves not dopamine itself but an agonist. Daily treatment with a dopamine agonist lowers prolactin levels, and ovulation and menstrual cycles return within two months.

This involvement of neurotransmitters with the secretion of hypothalamic and pituitary hormones opens up a new and exciting field. The effect of dopamine on prolactin secretion is not the only example. Noradrenalin can stimulate the release of GnRH; serotonin appears to control the rhythmic secretion of many pituitary tropins (e.g. ACTH, TSH, LH and FSH) presumably through altering the release of the appropriate releasing hormone in the hypothalamus, and dopamine can stimulate the release of GnRH. The sites of interaction of these transmitters with the neurosecretory terminals reside in the hypothalamus. This interaction of the nervous system with neurosecretory neurons could provide the mechanism for changes in the rate of release of the pituitary gonadotropins.

This close relationship between the nervous system and the endocrine system, which has been a theme in several chapters, is especially important in the timing and synchronization of reproductive cycles, shown in section 10.3.

Summary of Section 10.2

1 Two pituitary gonadotropins, FSH and LH (present in both sexes), promote gonadal steroid production and the maturation of gonadal tissue.

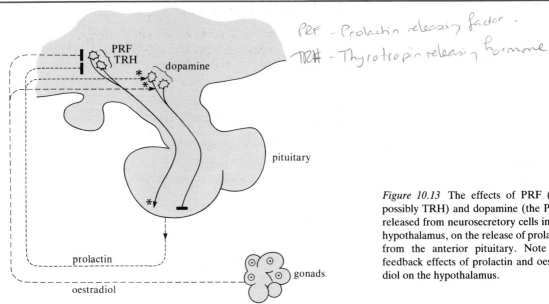

2 The ovary and testis produce steroid and peptide hormones (oestrogens, androgens, progesterone and inhibin), which, via feedback loops, control gonadotropin release. The negative feedback control of inhibin on FSH is not fully understood.

3 The hypothalamus controls gonadotropin release by a gonadotropin releasing hormone (GnRH).

4 The menstrual cycle proceeds through a number of phases: menstruation, follicular growth phase, ovulation, luteal phase.

5 FSH promotes follicular growth; LH initiates ovulation and the formation of a corpus luteum, which secretes progesterone.

6 During the follicular phase, low levels of circulating oestrogens promote (via negative feedback) secretion of LH and FSH, and rising oestrogen levels promote an LH surge and ovulation. During the luteal phase, oestrogens and progesterone suppress (via negative feedback) the secretion of FSH and LH.

7 Two separate areas of the hypothalamus have been proposed to release GnRH: one involved in the constant release of gonadotropins and the other in the intermittent release of gonadotropins (LH surge) from the anterior pituitary.

8 The corpus luteum persists during pregnancy and produces progesterone which is essential for the maintenance of a receptive uterus. Prostaglandin $F_{2\alpha}$ breaks down the corpus luteum if there is no implantation.

9 FSH and LH promote androgen production and spermatogenesis in the testes.

10 Prolactin inhibits the release of GnRH (or reduces the sensitivity of the pituitary to it), thus preventing ovulation. Prolactin is also important in the preservation of the corpus luteum, in some rodents.

11 Neurotransmitters in the hypothalamus control the production of releasing factors, and therefore pituitary hormones, or act directly on the anterior pituitary (e.g. the inhibitory effect of dopamine on prolactin release from the anterior pituitary).

Now attempt Questions 3–8, on pp. 357–358.

10.3 ENVIRONMENTAL INFLUENCES ON THE TIMING OF REPRODUCTION

In Section 10.2, we looked at the endocrine changes in the different phases of the menstrual cycle. In continuous breeders (Table 10.1), the end of one cycle is marked by luteolysis and then the breakdown of the uterine lining. New follicles ripen under the renewed influence of FSH and the cycle begins again. Most mammals, however, are not continuous breeders but show a limited number of cycles during a certain period of the year (i.e. they are seasonal breeders). So, in addition to describing the control mechanisms involved in one cycle, we have to seek other mechanisms that initiate the breeding season.

Synchronization of the reproductive state of both sexes maximizes the chances of fertilization. As the subsequent process of implantation and the length of gestation are 'fixed' in terms of duration for any given species, to improve the chances of survival of the newborn, the precise timing of fertilization in the year is crucial. Between them, mammalian species illustrate a variety of strategies resulting in the optimal timing for birth; these strategies include delayed implantation and seasonal receptiveness. Within a breeding season, species may be polyoestrous or monoestrous. The process of 'coming into season' involves environmental cues such as the change in day length, and the timing of reproductive encounters involves visual, olfactory and auditory (sight, smell and sound) cues from other individuals.

10.3.1 Light and temperature cues

The distinction between continuous and seasonal breeders is not a rigid one. Most mammals show peaks of reproductive activity. Even humans show peaks of conception in May and June in the Northern Hemisphere, and in November and December in the Southern Hemisphere. Generally, seasonal breeders are those species in which the gonads regress completely, and so become inactive, out of the breeding season.

Depending on the latitude, both day length (number of hours of light per 24-h day) and temperature vary throughout the year. In the Northern Hemisphere, day length increases from December to June and then gradually decreases. Average monthly temperature increases steadily from February to August. Depending on the size and length of gestation of a species, sexual behaviour occurs in the spring or in the autumn.

◇ What would the advantage of autumn mating be in a species (e.g. sheep) that has a long gestation period?

◆ A long gestation period, extending over the winter, means that the young are born in the following spring coincident with the increasing food supply. (If you look again at Table 10.1, you can see that, in general, larger animals have longer gestation periods, though there are exceptions.)

Figure 10.14 shows the periods of oestrus of two groups of Suffolk ewes subjected to different lighting conditions. In the control group, as the day length shortens, oestrous cycles are initiated in September. In the experimental group, with a seasonally reversed day-length cycle begun in the first year, the initiation of the second set of oestrous cycles is in May.

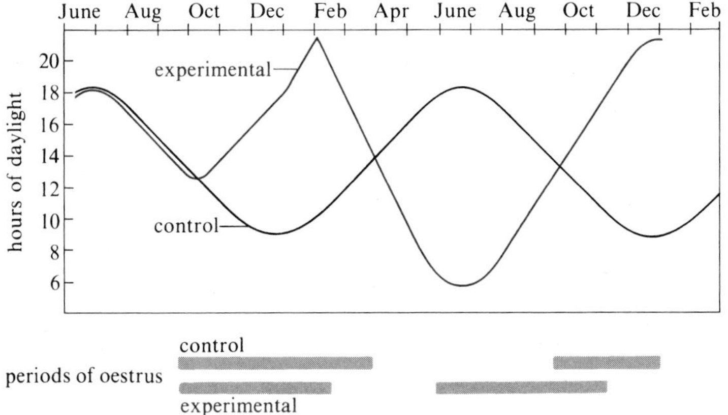

Figure 10.14 The sexual season of two groups of Suffolk ewes subjected to different lighting conditions. In each case, the sexual season (oestrus) is shown as a solid band beneath the corresponding light curve. In all cases, oestrus started 10–14 weeks after the change-over from increasing to decreasing hours of daylight.

◇ To what specific cue do the ewes appear to respond?

◆ Decreased day length; the hours of daylight have to decrease markedly before oestrous periods are initiated.

SHORT-DAY BREEDERS
A species in which breeding activity is stimulated by a DECREASE in day length

LONG-DAY BREEDERS
A species in which breeding activity is stimulated by an INCREASE in day length.

CIRCANNUAL CLOCK
An internal timing mechanism that controls one or more cyclic physiological processes with a period of about one year.

This shows that, where all the other effects are left unchanged, simply imposing a reversed light cycle can entirely rephase the breeding season. This is also true of deer. These two species are known as **short-day breeders**. Many small mammals and birds, with short gestation periods, are **long-day breeders**, coming 'into season' in the spring. So too is the horse which, at 10 months, has a very long gestation period with mating occurring in the spring. In this chapter, we use the terms short-day and long-day breeders to indicate those species in which mating coincides with a decrease in, or increase in, day length. Some species (e.g. foxes), with a wide latitudinal distribution, may come into breeding condition at different times in the year in different parts of their range (earlier in the south), but are still responding to an increase in day length.

It has recently become apparent that changes in day length are not the only cue that activates reproductive activity. In some seasonal breeders, such as deer and ground squirrels, it has been shown that seasonal cycles are under the control of an internal oscillator with a period of approximately one year, referred to as a **circannual clock** (circannual means 'about a year'). Circannual clocks are comparable to circadian clocks (circadian means 'about a day') that control many daily cycles, such as sleep/wakefulness and body temperature. Although internal clocks are dependent on external cues to maintain their accuracy, they are capable of controlling a variety of physiological and behavioural processes independently of environmental conditions.

Figure 10.15 shows some effects of day length on Soay rams. Note that the change from long days (LD) to short days (SD) stimulates FSH secretion and a growth in testis size. LH levels show episodic secretion and testosterone level gradually rises. When the rams are transferred back to a long-day schedule 18 weeks later, FSH and LH secretion falls and the testes regress. The falls in FSH and LH are not very apparent in these graphs but, had the rams not been transferred to long days, the levels of these hormones would have remained roughly the same as at the end of the SD (shaded) period. How can changes in the day length trigger gonadotropin secretion from the pituitary? There are a number of possible routes, and all inevitably lead to the hypothalamus and the level of GnRH.

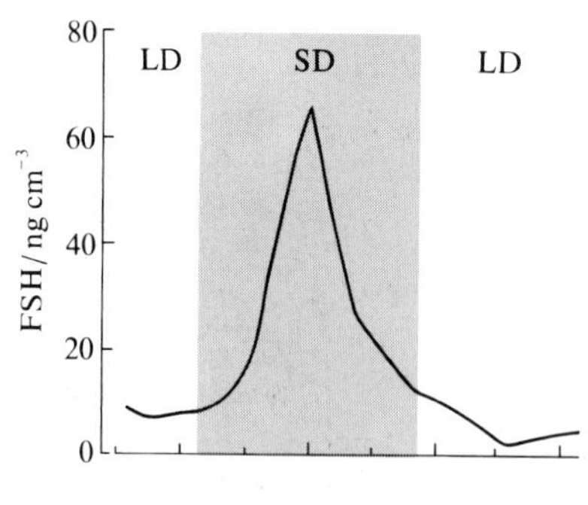

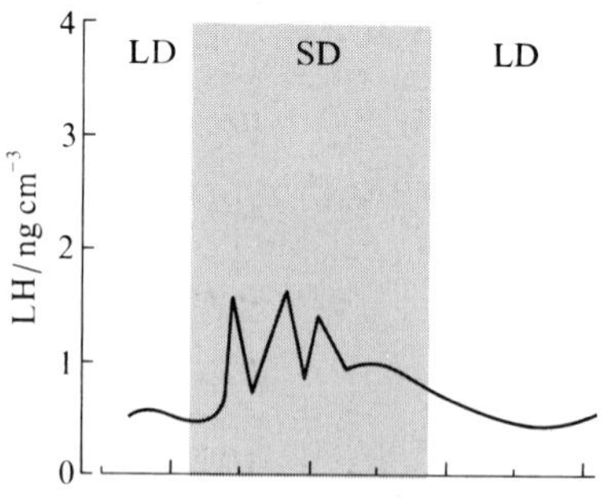

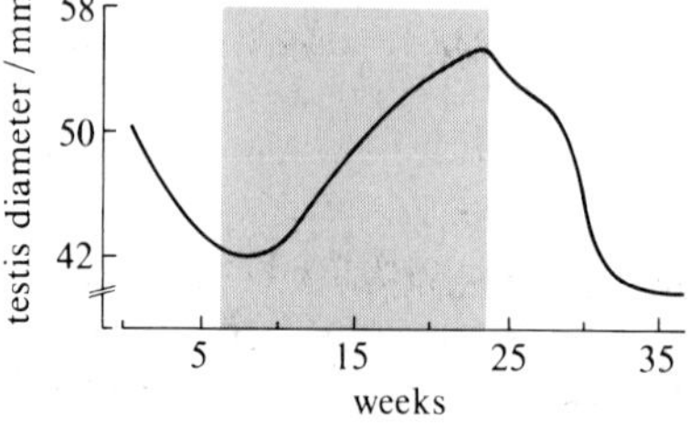

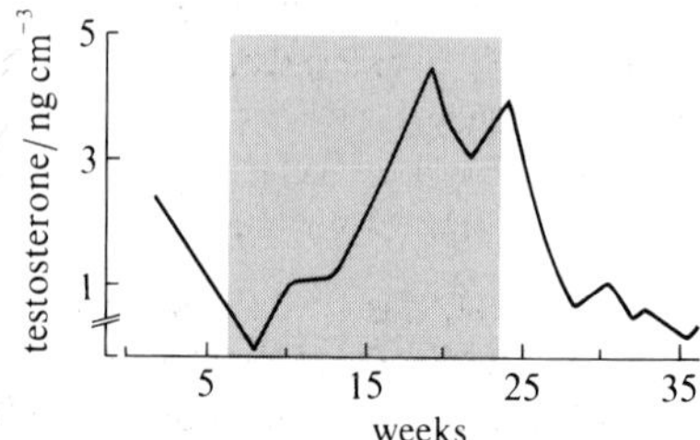

Figure 10.15 Changes in testicular size and levels of various hormones in the blood of adult Soay rams transferred from periods of exposure to long days (LD = 16 h light, 8 h dark) to short days (SD = 8 h light, 16 h dark) at week 6. Animals were maintained on short days until week 24, when they were returned to long days.

In rodents, the eyes and the **pineal gland** of the brain have been implicated in breeding cycles. If a hamster (a long-day species) on long days is blinded, there is immediate regression of the gonads, just as if it had been transferred to short days. If the pineal gland is removed, this regression does not occur. If the pineal is removed from hamsters kept under a short-day regime, the gonads grow.

◇ What does the pineal probably secrete during short days?

◆ A substance that inhibits gonadal development, most probably by inhibiting GnRH secretion.

On long days, the pineal does not exert an inhibitory effect. The pineal gland secretes an amine called **melatonin**. Melatonin sometimes causes gonadal regression; at other times it stimulates gonadotropin release. The pineal secretes melatonin mostly during the night and its removal leaves an animal insensitive to changes in day length.

Figure 10.16 shows the effects of melatonin on the weight of the testes in four rodent species. The testes of the grasshopper mouse and hamster regress when exposed to melatonin, whereas those of the house mouse and albino rat appear to be unaffected. The two melatonin-sensitive species are seasonal breeders, whereas the two melatonin-insensitive species breed continuously.

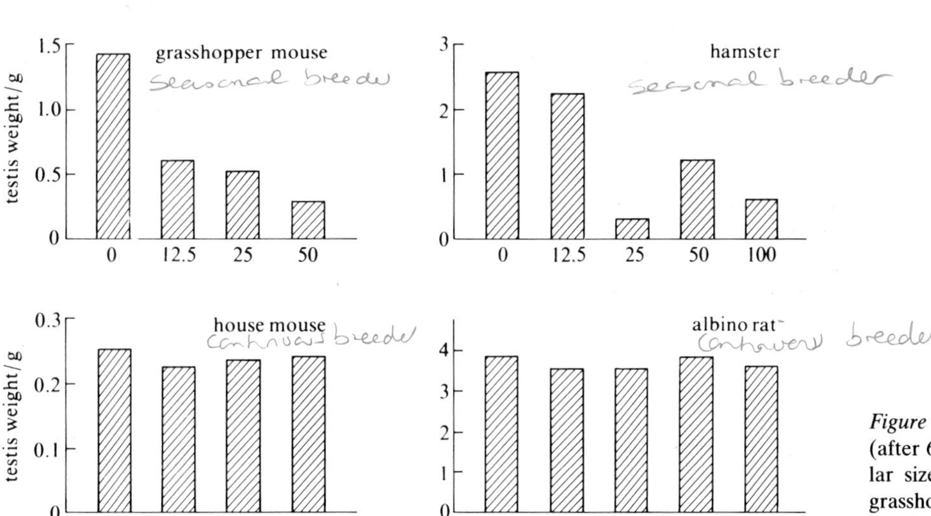

Figure 10.16 The effects of melatonin (after 60 days administration) on testicular size in four species of rodent. The grasshopper mouse and hamster are seasonal breeders; the house mouse and albino rat are continuous breeders.

Among seasonal breeders, melatonin has different effects, depending on the time of year at which they become reproductively active. In long-day breeders it acts in an inhibitory fashion, but in short-day breeders, such as deer and sheep, increased melatonin in the blood acts as an activating signal to reproduction. In some species whose reproductive activity is controlled by a circannual clock, melatonin plays a role in controlling the timing of the clock. There is increasing evidence that melatonin exerts its effect by influencing GnRH secretion from the hypothalamus, and that it does not act directly on the pituitary or the gonads.

The pineal produces melatonin on a circadian cycle, in other words there is a 24-h fluctuation in melatonin level with a peak at night. The effect of giving melatonin to an animal depends very much on *when* it is given in the day. Here, therefore, is a substance in the blood, produced by the pineal, whose effects vary depending on the time of day at which the brain is exposed to it. Because the secretion of melatonin may depend on the onset of darkness, this is clearly an important physiological mechanism by which the pineal 'tells' the brain whether it is in a 'long day' or a 'short day'.

The role of day length is probably as a 'coarse tuner', synchronizing development of the reproductive system in members of a local population, both male and female. As mentioned above, some mammals have 'built in' yearly rhythms of reproduction, for example some hibernating species. Light may act simply to adjust a circannual clock in the hypothalamus. Once this clock (which could be the production and secretion of GnRH) is set to the right time, it then continues to run without the necessity for another light cue.

Changes in day length may do more than simply prime ovulation. In some species, there is a delay between fertilization of the egg and subsequent implantation of the blastocyst—delayed implantation. This means that some species that mate in spring may delay implantation until autumn and then give birth in the following spring. Hedgehogs mate in late summer, but implantation is delayed, with the result that the young are born in the following spring. The badger, the roe deer, the mink, the polar bear and the marten can all delay implantation (for up to four months in the deer): this delay can be terminated if the day length or the temperature are experimentally changed.

Rats and mice also show delayed implantation. Females ovulate 24 hours after giving birth. If copulation occurs at this time, the subsequent gestation period is up to 40 days long instead of the normal 21. If, however, lactation is prevented, the gestation period reverts to 21 days, or if oestrogen is given during lactation, the blastocysts implant. This is interpreted as an oestrogen requirement for implantation in the rat; during lactation, this oestrogen is lost by being secreted in the milk.

The implantation mechanisms of wallabies and kangaroos are, at first sight, even more puzzling. In the wallaby, implantation will occur even if the pituitary is removed. Wallabies and kangaroos, like the rat, undergo ovulation after giving birth. The blastocyst, it seems, does not implant immediately but survives for many months. If the immature newborn, or 'joey', in the pouch dies during this time, or if it stops suckling, the blastocyst will then implant and start to develop.

In some animals, for example birds, sea turtles and bees, the female can store sperm in the reproductive tract for long periods of time (a few years in the turtle and seven years in a queen bee). One dubious case of sperm storage has been reported in mammals—in bats. Some bats copulate in autumn but do not ovulate until spring, and this was interpreted to mean that sperm were stored over winter. Closer behavioural studies have now shown that bats copulate again in the spring, and so fertilization is probably achieved with fresh sperm.

10.3.2 Cues from other individuals

While many mammals are **spontaneous ovulators**, that is they ovulate in response to a surge in gonadotropin release from the pituitary, others are **induced ovulators**. In the cat, rabbit, ferret, mink, short-tailed opossum, and some species of mice, courtship and copulation actually stimulate ovulation. Stimulation of the cervix and the vagina promotes, through the nervous system, the release of GnRH from the hypothalamus. The evidence for this role of the nervous system comes from removing the pituitary gland at various times after copulation. If the pituitary is removed up to 1 h after copulation, then no ovulation will occur. Beyond 1 h after copulation ovulation occurs even in the absence of the pituitary; by this time, the hypothalamus has initiated the release of LH and FSH.

Induced ovulators do not have an oestrous cycle comparable with those of spontaneous ovulators (Figure 10.17).

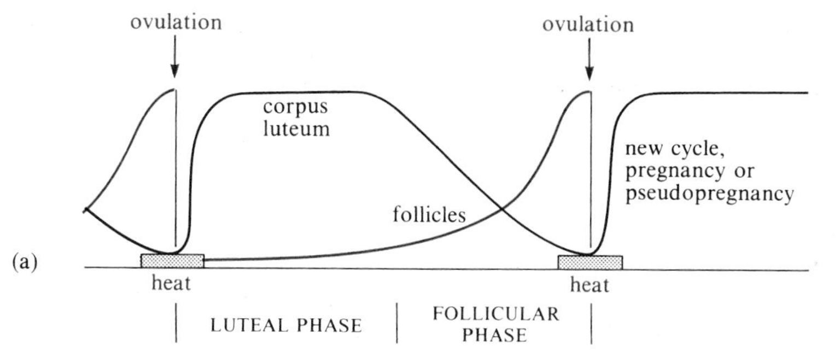

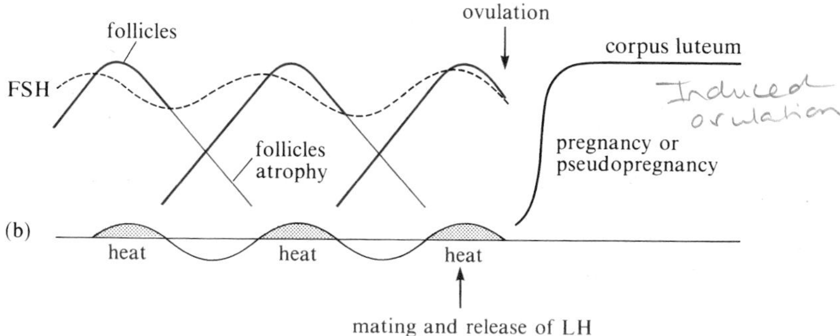

Figure 10.17 The chain of ovarian events in spontaneous and induced ovulators is basically the same. (a) Spontaneous ovulation is cyclic, following specific periods of heat in the oestrous cycle. (b) Induced ovulation occurs as a result of stimulation of the cervix during copulation. Induced ovulators experience alternating oestrous and anoestrous periods and, theoretically, are sexually receptive at all times.

◇ Why might induced ovulation offer a selective advantage over spontaneous ovulation?

◆ In induced ovulators sperm and ova are likely to meet at every mating, whereas in spontaneous ovulators many matings will be too early or too late for fertilization of the ova.

Induced ovulation is a common feature of mammals that live solitarily in field or forest habitats. Their more or less continuous state of sexual readiness provides a good chance that, at any occasional meeting, the female will be receptive. Not all solitary species are induced ovulators; in some species, seasonal 'gatherings' of individuals improve the chances of reproductive success.

Mammals show induction of other phases of the cycle in addition to the ovulatory phase. In Section 10.2.3, we saw how the luteal phase could be induced.

◇ In which species was this?

◆ In the rat; stimulation of the vagina during copulation promotes the maintenance of the corpora lutea by stimulating the release of prolactin.

The hamster and several species of mice also show such an induced luteal phase.

Physical contact is not always necessary to produce these preparatory hormonal responses. That complex displays (involving colourful plumage) or the mere sight of a finished nest may synchronize reproductive states in birds is well known, but are there parallels in mammals? Can an individual communicate its reproductive condition to another individual and thereby affect that second individual? There are a number of possible means of communication: by odour, visual displays or sound. In birds, the latter two seem to be employed, and in mammals, odours are also important.

Female rats in constant oestrus (produced by constant light exposure) induce oestrus in females in a normal cycle via an olfactory cue. If female mice are housed together, and then each is allowed access to a male, there are unexpected results.

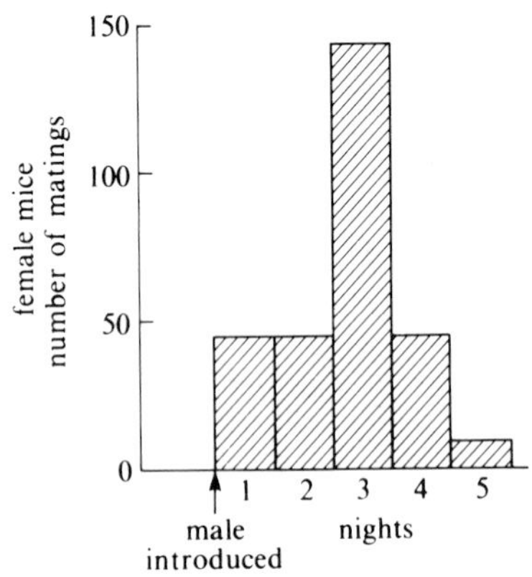

Figure 10.18 The Whitten effect: the synchronization of oestrus in groups of mice.

WHITTEN EFFECT

Synchronisation of the menstrual or oestrous cycles of a group of females, which occurs only when a male (or his odour) is present.

BRUCE EFFECT

Pregnancy failure due to the exposure of a pregnant female to an unfamiliar male.

◇ Given that the mouse oestrus cycle is 5 days long, with one day on heat, what proportion of the female mice would you expect to mate on each night?

◆ You would expect them to be on heat and therefore mate randomly, so this means a fifth of the total on each night.

But now look at Figure 10.18. Almost 50% of the females mated on the third night! This effect of females synchronizing oestrus was described first by Whitten who also ascribed it to an odour (a pheromone), and is known as the **Whitten effect**.

If a male of a different strain of mouse is put into a cage of females who have recently mated, the blastocysts do not implant successfully and the number of pregnancies is fewer than normal. In this situation, known as the **Bruce effect**, the corpora lutea of the first mating do not seem to be maintained.

◇ What could the action of the male odour be?

◆ Prolactin is luteotropic in the mouse and the pheromone blocks its release.

Pheromones are also important in the sexual behaviour of some primates. Male rhesus monkeys fitted with nasal plugs are not as interested in females they can only see, as are male monkeys that can also smell the female. If female monkeys are castrated, no mating occurs, but if they are subsequently treated with oestrogen, the mating frequencies rise sharply. It seems that oestrogen stimulates secretion of certain kinds of acid from the vagina, and this, in turn, stimulates male mating behaviour.

Summary of Section 10.3

1 Most mammals show peaks of reproductive activity during the year irrespective of whether they are seasonal or continuous breeders.

2 Light is implicated in the initiation of a breeding season; some mammals respond to decreasing day lengths while others respond to increasing day lengths.

3 The perception of day length in seasonal breeders appears to involve the pineal gland in the brain, which secretes melatonin.

4 Some mammals have the ability to delay implantation of a fertilized egg.

5 Cues from other individuals are important in the control of the reproductive cycle. These cues may be tactile, visual or olfactory. Pheromones synchronize oestrous cycles in female rats and mice, and female pheromones induce sexual activity in male monkeys.

6 Some species are induced ovulators and respond to the tactile stimulation of copulation. This stimulation is conveyed to the hypothalamus via the nervous system. A surge in LH results, and ovulation occurs.

Now attempt Questions 9–11, on p. 358.

10.4 SEX DETERMINATION: THE ACTION OF STEROIDS

In mammals, haploid gametes, produced by meiotic division from diploid cells, combine to form a diploid zygote. The genotypic sex of a mammalian fetus is determined at fertilization by the presence of an XY (male) or an XX pair of chromosomes (Figure 10.19).*

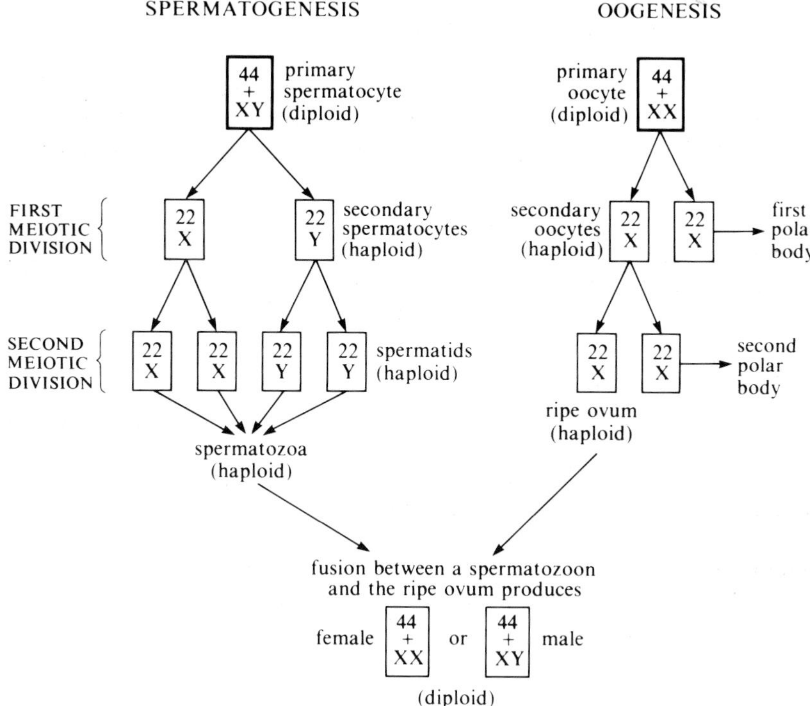

Figure 10.19 The chromosomal events in the production of spermatozoa (spermatogenesis) and of a ripe ovum (oogenesis).

But is the Y chromosome the only thing that determines 'maleness' in mammals? Certainly it appears so when clinical disorders are considered. If, due to chromosomal problems, fertilization results in a human individual with 47 (XXY) or 48 (XXXY) chromosomes, then a testis will develop, but if the individual is XO (where O represents the absence of one of the sex chromosomes), then it is phenotypically female. In recent years, however, it has become clear that genetic sex is much more complicated than simply the presence or absence of the Y chromosome, and that genes essential for the development of both male and female characteristics are located on other chromosomes.

*Spermatogenesis and oogenesis are covered in more detail in Brian Goodwin (ed.) (1991) *Development*, Hodder and Stoughton Ltd, in association with The Open University (S203, *Biology: Form and Function*, Book 5).

In the human embryo, the gonads are not recognizably of either sex until 6 weeks after fertilization. Then differentiation begins, and the secretion of steroids from the gonads determines their subsequent development. If the gonad secretes testosterone, a male reproductive system will develop. If no testosterone is present, then the female ducts and genitalia develop. The ability of the fetal testis to synthesize testosterone coincides with the differentiation of the Leydig cells (Section 10.1.2). Testosterone plays two vital roles. First, it is essential for the maturation of the seminiferous tubules and for spermatogenesis. Second, testosterone released into the fetal circulation is involved in promoting 'maleness' in other tissues. In the adult, androgen synthesis is under the control of gonadotropins (Section 10.2.4), and this appears to be true in the fetus also. The site of gonadotropin release appears to be the fetal pituitary, and gonadotropin receptors have been demonstrated on the fetal testis. Similarly, the fetal ovary can secrete oestrogens at this early stage, which may be important in brain development of the female fetus. In the female rat, the presence of ovarian oestrogen during the perinatal period is a necessary condition for normal ovulation and sexual behaviour in later life. Females whose ovarian oestrogen was blocked by the antagonist tamoxifen during late fetal and newborn life fail to show sexual behaviour or ovulatory cycles on becoming adult.

If testosterone serves as the fetal androgen, are its molecular mechanisms the same as in the adult? In adults, testosterone is thought to have few effects of its own, but it serves as a prohormone for two other steroids—**dihydrotestosterone** and oestradiol—that are active in target tissues (see Figure 10.20). Certainly the fetus demonstrates this conversion, and dihydrotestosterone is responsible for the **masculinization** of the external genitalia and male urethra. When female embryos are treated with testosterone, they develop *both* male and female genital ducts.

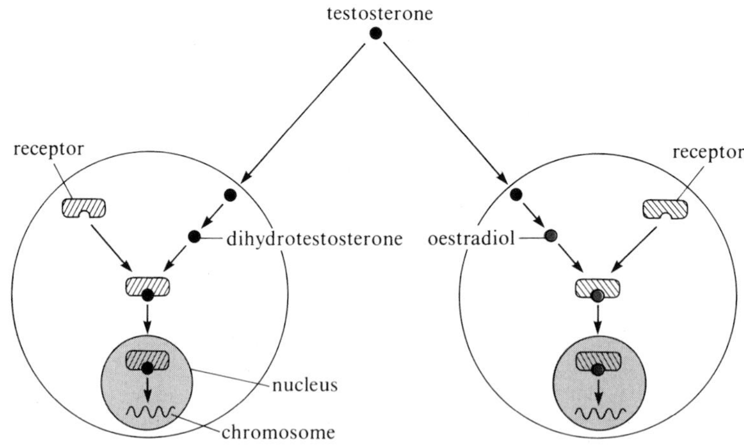

Figure 10.20 Testosterone can be converted into oestradiol or dihydrotestosterone in different target cells in the hypothalamus. These products then bind to cytoplasmic receptors and subsequently modify gene expression in the nucleus of the hypothalamic cell.

The other major physiological difference between the sexes, apart from steroid production, lies elsewhere.

◇ Where?

◆ In the hypothalamus. Whereas in females, GnRH promotes surges in gonadotropin release, in males, the levels of FSH and LH stay fairly constant.

What then determines the 'sex' of the hypothalamus? Much of the work on this has been done on rats because they are still relatively immature at birth compared with other readily available laboratory species such as guinea-pigs. Neurons within the rat hypothalamus can be distinguished at around day 15 of gestation. By day 17 GnRH is found, and LH and FSH are detectable in the fetal pituitary just before birth (day 20). However, the all-important blood circulation between the hypothalamus and pituitary does not develop until the first post-natal week.

Androgen injected into newborn female rats causes sterility and there is no ovulation (see Figure 10.21). The timing of this injection is critical; it will only work between a few days before birth until 5 days afterwards. Before this time, the pituitary is 'plastic' and releases gonadotropin according to the character of the hypothalamus. In the same way, if an anti-androgen is given to a newborn male rat, mimicking castration, the normal masculinization of the hypothalamus is prevented.

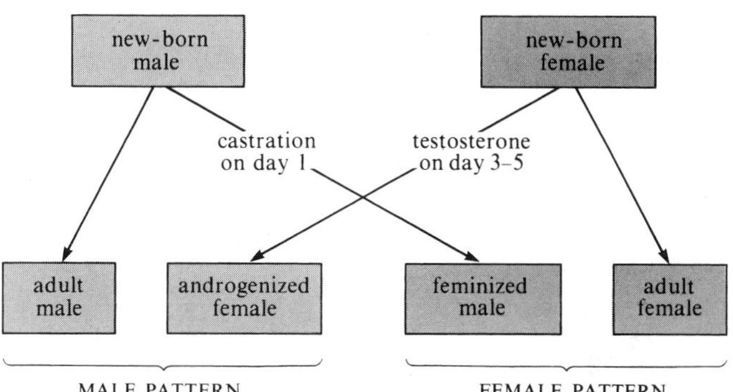

Figure 10.21 Sexual differentiation of the rat brain takes place after birth. In new-born males, testosterone secreted by the testes is converted by target cells in the brain into oestradiol, which gives rise to a permanent male pattern of brain structure. If the male is castrated at birth, however, the sexual differentiation of nerve circuits fails to take place and the brain retains a female pattern. Administration of testosterone to a new-born female rat evokes a male pattern of nerve circuits as a result of testosterone being converted intracellularly into oestradiol.

In the male, testosterone appears to be responsible for abolishing the potential for cyclic LH release (typical of females); the absence of testosterone in young females results in the hypothalamus developing the capacity to release LH in a cyclic fashion.

Evidence that the effect of testosterone is on the hypothalamus itself rather than on the pituitary stems from the discovery that female rats in which the pituitary has been replaced by one from a male will still ovulate. The pituitary is 'driven' by the 'sex' of the hypothalamus!

This short account of mammalian sex determination is, of necessity, superficial and much of it is based on the rat. The process appears to be more complicated in other mammals, for example primates. As a result of differentiation, the brain acquires certain permanent characteristics, which underlie, in some way, the sex differences in behavioural responses and neuroendocrine function. Steroids 'mould' the hypothalamus around birth, but this is not the end of the process. Reproductive competence in the rat does not appear for another 6 weeks, when steroids are produced in large amounts; these then trigger the development of secondary sexual characteristics and the changes in behaviour and physiology that are related to reproduction.

Summary of Section 10.4

In mammals sex is genetically determined (XX = female, XY = male), but on top of this, there are effects during development of steroid hormones. The actions of sex steroids are diverse. Steroids not only maintain the gonads and promote development of the gametes, but they also bring about gonadotropin release via feedback systems. In the developing mammalian fetus, sex steroids are important at three particular stages, summarized in Table 10.2.

Table 10.2 The effects of hormones on reproductive development in male and female mammals.

Stage of development	Male	Female
early in fetal life	testosterone gives male primary sexual characteristics; local factor from testes suppresses development of female genitalia	female primary sexual characteristics and ovaries develop if testosterone and local factors are absent
mid-to-late fetal life or around birth (but not in primates)	testosterone makes hypothalamus male, suppressing mechanism for cyclical LH release	mechanism for cyclical LH release develops in absence of testosterone
at puberty	increase in FSH and LH secretion leads to growth of testes and testosterone output; testosterone gives male secondary sexual characteristics	increase in FSH and LH secretion leads to growth of ovaries and higher oestrogen output; oestrogen gives female secondary sexual characteristics; cyclic LH release leads to ovulation

Now attempt Question 12, on p. 358.

10.5 CONTRACEPTION

The challenge confronting reproductive research today is a world inhabited by 5 000 million people. Each day there is a *net* population gain of 200 000 making a total gain of 70 million every year. If this increase continues at the same rate, then disaster is inevitable. Each year some 50 million abortions are performed, in some cases placing women at risk from surgical infection or causing infertility.

The ideal method of contraception should be effective, safe, simple, cheap, reversible and therefore readily acceptable. Several methods approach this ideal, but none, as yet, achieves it. Contraceptive methods have departed in the past three decades from the relatively simple 'barrier' methods refined over the centuries, to the use of steroid hormones encapsulated in the 'pill'. Hailed as a great advance in contraceptive technology in the 1960s, the pill has acquired notoriety because of side-effects associated with its use, and understandably it is now less popular. In this Section, we look at the advances in methods of contraception (and their physiological implications) and attempt to point the way to safer and more effective formulations.

In its broader sense, contraception can imply a range of interventions: prevention of gamete maturation, controlled ovulation, prevention of fertilization, prevention of implantation and, most extreme, the termination of implantation or abortion. Advances have been made in all these areas, but particularly so in the first two.

10.5.1 Steroids

Steroids exert both a positive and a negative feedback action on gonadotropin release from the pituitary. In the luteal (late) phase of the cycle, oestrogen and progesterone, in combination, inhibit LH and FSH release and prevent follicular maturation. This combination of steroids is the basis for the **'combined-type' pill**. As progesterone itself is not adequately absorbed from the gastro-intestinal tract, a synthetic **progestogen** is used. Most available preparations are standardized on a cycle length of 28 days, being either 21 days on pills and 7 days off, or 21 days on pills plus 7 days on a 'placebo' (usually in the form of lactose) so that a tablet can be taken every day without interruption. Initially, high levels of steroids were used, but lower doses of steroids are very effective, and new progestogens have been introduced. Most combined-type preparations are now very low in oestrogen.

The main site of oestrogen–progestogen action is on the hypothalamus, depressing the release of GnRH (by negative feedback) and consequently limiting the secretion of LH and FSH. Feedback inhibition may also affect the pituitary directly. FSH and LH are present in only small amounts in a woman taking the steroid pill, and the ovulatory surge of LH is abolished. Not only is ovulation prevented, but the production of oestrogens by the ovaries is suppressed. However, there are other significant changes in the ovaries and uterus as a result of steroid application. The endometrium of the uterus undergoes change, and menstruation is determined by the length of the course of tablets and not the hypothalamo-pituitary rhythm. Menstruation is shorter (in time) and lighter. The ovaries, when examined, appear to be arrested at an early stage of follicular development, and there is no evidence of fresh corpora lutea being present.

Because of the side-effects of oestrogens, a **'mini-pill'** has been designed, which contains only progestogen. A small amount is taken every single day continuously, and menstruation in about 40% of women settles down to the approximate periodicity of the normal menstrual cycle. The progestogen continuously stimulates the endometrium, but at intervals it seems that the dose is not sufficient to support further growth of the tissue, which then breaks down as a menstrual flow. In contrast to the combined pill, progestogen does not usually inhibit ovulation. Progestogen interferes with meiosis in the ovary, changes oviduct mobility, and will prevent implantation if fertilization should occur. Oestrogen levels remain close to normal, because FSH secretion is not inhibited by progesterone.

Development of a 'one-off' pill has resulted in the so-called **'morning-after' pill**. This can be either progestogen or oestrogen, and large doses are involved. The large dose of progestogen taken just after coitus stimulates the growth of the endometrium, which then breaks down as levels decrease, preventing implantation. Similarly, a series of high doses of oestrogen administered after fertilization will prevent implantation. At present, the efficacy in preventing conception by this method is not high, and large doses of oestrogen can produce nausea.

Instead of a daily dose of steroids, oestrogen and progestogen can be taken either together or separately once a month. The regime involves oral administration on about day 22–25 of each cycle, and the oestrogen is dissolved in the body storage lipids and slowly released, thus preventing ovulation. Injections and implants of steroids have also been tried. Capsules containing progestogen can be implanted just under the skin, and these are effective for periods of a year or more. Intra-uterine and intra-vaginal devices can also be used as steroid 'implants' to release progestogen slowly and inhibit ovulation.

Although the combined type (oestrogen–progestogen) pill is, at the moment, the most effective contraceptive, it is by no means ideal. Artificial intervention in any physiological control system brings with it the possibility of unwanted side-effects. The most serious risks are associated with the blood circulatory system and are related to the steroid dose, age of the taker, whether they smoke cigarettes, and the length of time on the steroid pill. There are also links between long-term use of the pill and various kinds of cancer. The pill is associated with an increased incidence of breast cancer, but with a decreased risk of ovarian and uterine cancers. In a recent report from the Royal Colleges of General Practitioners and of Obstetricians and Gynaecologists, it is recommended that women on the pill should not smoke, that the pill should not be in use continuously for more than five years, and that after the age of 35 women should look to other forms of contraception.

When oral contraceptives are taken, the rate of blood clotting can increase, and this means that there is a greater risk of a clot in a blood vessel. (Normal pregnancy also increases the dangers of blood clotting and venous thrombosis is quite a common complication of an otherwise normal pregnancy.) This danger has promoted the development of a pill with lower oestrogen levels, and it is probably safer now to take a low-oestrogen pill for 30 years than to complete one pregnancy! Long periods on the pill can lead to **amenorrhoea** (lack of menstruation) caused by over-suppression of ovarian activity. The primordial follicles do not secrete even the small amounts of oestrogen required to restart cyclic activity, but a very weak application of oestrogen will prime LH release by positive feedback.

Steroids can be used to reduce ovulation and to prevent implantation, but is it possible to use steroids to prevent sperm production? The answer, of course, is yes. Androgens, oestrogens and progestogens reduce FSH and LH secretion in males via strong inhibitory effects on the hypothalamus and pituitary. In turn, spermatogenesis is affected. Another possibility is the use of anti-androgens, which suppress spermatogenesis without any irreversible effects. The search for a useful male contraceptive continues. At the moment, nothing is as effective as **vasectomy**, a simple and relatively safe operation which involves cutting and then tying the ends of the vas deferens. Vasectomy is, however, a sterilization technique and is not necessarily reversible.

10.5.2 Novel peptides

Steroids can inhibit the release of gonadotropins either by acting on the hypothalamus (reducing GnRH) or by direct action on the pituitary. There is another way of suppressing pituitary gonadotropin release—by the use of analogues of GnRH.

◇ How would this work?

◆ The analogue could be an antagonist that would bind to the GnRH receptor in the pituitary but would not activate LH and FSH release from the pituitary.

The structure of GnRH is a sequence of 10 amino acids:

glutamate-	histidine-	tryptophan-	serine-	tyrosine-	glycine-	leucine-	arginine-	proline-	glycine-NH_2
1	2	3	4	5	6	7	8	9	10

This peptide has been synthesized and appears to be active in all vertebrates, that is, it is not species-specific. Some 300 analogues of this peptide have been synthesized in an attempt to discover which sequences are important.

Some analogues are superactive agonists, that is, they are more potent in eliciting prolonged LH and FSH release than the natural peptide. Others are antagonists to GnRH; they bind to the receptors, but do not trigger LH and FSH release, and therefore also block any naturally circulating GnRH activity. Peptides are naturally broken down by endopeptidase enzymes. If different amino acids are introduced into the peptide, this breakdown may be inhibited and consequently the potency of the peptide, whether agonist or antagonist, will be enhanced because it is not so readily destroyed.

If histidine (2) and glycine (6) in GnRH are replaced with D-phenylalanine, then a potent antagonist is produced that inhibits LH and FSH release for 6–8 hours after injection. The value of inserting a D amino acid is that these amino acids are particularly resistant to enzyme attack (the L form of an amino acid is the naturally occurring form).

Such antagonists will also suppress LH and FSH release in males. The implications of this work are enormous! A small peptide, cheaply synthesized and with long-lasting, but reversible, effects could be the new and safe method of contraception. It could be used in men and women, probably with few, if any, side-effects. While more work is needed on the effects of a long-lasting block of GnRH receptors, and on the timing and administration, it may represent a significant breakthrough.

◇ What may be the major problem of the administration of a peptide?

◆ Peptides cannot be given orally in the form of a pill because digestion would inactivate them. The only serious complication, then, might be the route of administration: injection or long term implants may be necessary.

Another new approach to contraception is to exploit the fact that receptors will become less sensitive with prolonged exposure to their chemical signals (receptor down-regulation). If GnRH is repeatedly given to a woman, ovulation will cease because the pituitary receptors to this releasing factor lose their sensitivity to it. Small amounts of a superactive agonist of GnRH can therefore be used to block ovulation. Similarly, such superactive releasing factors will block sperm production in males. In both cases the down-regulation is easily reversible, an important criterion for any contraceptive.

Summary of Section 10.5

1 The combined-type contraceptive pill is a mixture of an oestrogen and a progestogen, or progestogen alone. Steroids taken daily suppress the release of GnRH from the hypothalamus (and pituitary release of FSH and LH). These steroids therefore prevent follicular maturation and ovulation.

2 Progestogen or oestrogen taken in large doses after fertilization prevent implantation.

3 Major and minor side-effects of the steroid contraceptive pill are an increased rate of blood clotting, amenorrhoea, disruption of menstruation and cancer.

4 Steroids also affect spermatogenesis via a feedback action on FSH and LH secretion in males.

5 Analogues of the peptide GnRH may provide useful alternatives to steroids as contraceptive agents.

Now atttempt Question 13, on p. 359.

10.6 DIVERSITY IN MECHANISMS CONTROLLING REPRODUCTION

Our understanding of reproductive mechanisms in animals is largely dominated by research on a very few species, such as the mouse and the rat. Such a small sample of species provides a very impoverished picture of the diversity that exists in the way that the various aspects of reproduction, such as gonadal activity, hormonal changes and behaviour, interact with one another. In this Section, we review briefly the variation that exists between species and relate this to their ecology and natural history.

Among vertebrates, there are three major patterns of reproductive cycle (Figure 10.22):

- **Associated**—The gonads become most active and gametes mature at the time of mating.
- **Dissociated**—The gonads are most active and gametes mature after the time of mating. During the mating period, the gonads are inactive. Conception occurs between gametes that were matured previously and have since been stored.
- **Constant**—The gonads are active all the time and are continuously producing mature gametes. Mating can typically occur at any time of year.

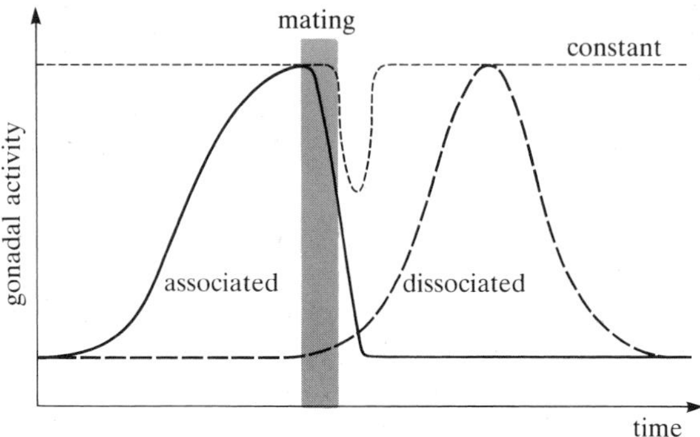

Figure 10.22 Three types of reproductive cycle, associated, dissociated and constant, found among vertebrates. Gonadal activity refers to the maturation of eggs and sperm, mating to courtship and copulation. In the associated pattern, the gonads are maximally active at the time of mating. In the dissociated pattern, gonadal activity is minimal during mating and increases after mating. In the constant pattern, the gonads are maintained continuously at or near maximum activity.

Associated breeding tends to occur in species that mate in spring and for which there is a protracted period prior to breeding, during which conditions are favourable for feeding, putting on weight and developing the gonads. In contrast, species with dissociated breeding cycles tend to mate very early in the year, often immediately after emerging from hibernation. Examples include all the amphibians found in the UK (newts, frogs and toads). In these species, individuals enter ponds in early spring with fully mature eggs and sperm that were developed in the previous autumn. During the mating season, the testes and ovaries are relatively small and are in the process of going through the early stages of gametogenesis. Eggs and sperm are developed later in the year, during a period of feeding and weight-gain in the summer, and are stored until the next spring.

The constant pattern is found in animals that live in harsh environments, such as deserts, in which breeding opportunities occur at irregular, unpredictable times of year. An example is the Australian zebra finch (*Taenatiopygia guttata*) in which breeding activity is triggered by the occasional rains that fall in the outback, causing a sudden and ephemeral flush of vegetation. Individuals are in a perpetual state of readiness to breed as soon as conditions become favourable after rainfall. Male zebra finches may have fully functional testes for as long as three years but do not breed if there is no rain.

The bias in research on reproductive mechanisms towards a few species, most of which show the associated pattern, has led to a widespread assumption that there is an association in time, and in terms of causality, between gamete production, sex hormone secretion and sexual behaviour. The recognition that other species show quite different temporal associations between these factors shows this assumption to be quite false. In particular, the paradigm that mating behaviour is activated by a seasonal increase in circulating levels of gonadal hormones has persisted despite numerous examples in which such an association clearly does not occur. For example, male ring doves (*Streptopelia risoria*) copulate during a phase in their reproductive cycle when androgen levels are relatively low. In western gulls (*Larus occidentalis*), males show androgen levels that are constant throughout the year but have a clear annual cycle of spermatogenesis and mating.

In species with dissociated reproductive cycles there is a reversal of what is assumed to be the typical association between gonadal activity, hormones and sexual behaviour, since mating occurs when the gonads are least active. It is not clear how common this pattern is, since only a limited number of species have been adequately studied, but it occurs across a range of vertebrate groups, including amphibians, reptiles and birds.

Closer scrutiny of species with associated reproductive cycles raises questions about the exact causal relationship between the various factors. If mating is associated with high hormone levels, which is cause and which is effect? Do high hormone levels stimulate mating, or are they a consequence of mating? In a reptile, the green anole lizard (*Anolis carolinensis*), male courtship stimulates pituitary gonadotropin secretion in the female, male aggression inhibits ovarian growth. In this species, ovarian development is incomplete in females that are not exposed to male sexual behaviour.

The picture becomes more complex and diverse when males and females are considered separately. It is quite possible for one sex to show an associated (A) reproductive cycle, the other a dissociated (D) one, giving four possible patterns. Examples of each of these four patterns are known: A/A, red deer, and Soay sheep; D/D, many hibernating bats, many snakes, some salamanders, and carp; A/D (female associated, male dissociated), painted turtle, tiger salamander, and pike; D/A (female dissociated, male associated), Asian musk shrew, fulmar, some skinks, and catfish. In some primates, including humans, females show clear cycles of reproductive activity which may be associated with seasonal environmental changes, but males show an essentially constant pattern. In others, such as the rhesus monkey, males also show seasonal cycles of reproductive activity.

A complete picture of the diversity in reproductive cycles awaits more detailed study of many more species, but from this brief overview it is possible to make a number of important points:

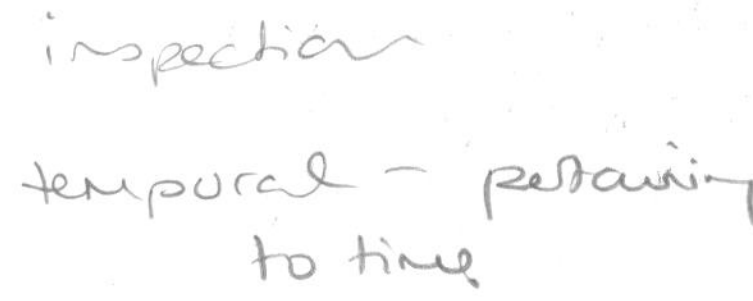

1 We cannot generalize across species about the temporal and causal relationship between gonadal activity, hormone levels and sexual behaviour. These factors are related to each other in quite different ways in different species.

2 Different patterns appear to be adaptations to different environmental factors. For example, the constant pattern is an adaptation to life in environments where conditions favourable for breeding are sporadic.

3 Within a species, males and females may show very different reproductive patterns.

SUMMARY AND CONCLUDING REMARKS

In this chapter we have concentrated on the role of hormones and the nervous system in the maturation of sperm and eggs and in the timing of the delivery of sperm to eggs.

As indicated earlier, it is not meaningful to generalize from the actions of hormones in a single mammalian species. Much of the discussion has centred around the human reproductive cycle but, on a number of occasions, descriptions have been provided of similar events in a range of species. Such comparisons in the chapter are not gratuitous, and you should, by now, have an idea of the considerable variation between mammals in terms of oestrogens, androgens and peptides from the hypothalamus and pituitary and their role in timing and inhibiting the production of gametes.

It is perhaps not surprising that we, as a species, should be so intrigued with our own reproductive physiology. The menstrual cycle is, however, unusual. Most mammals are seasonal, coming into breeding condition at a particular time of year in response to changes in light, temperature or rainfall. The result is ovulation and conception, and after a period of gestation, parturition occurs at a time that is propitious for the survival of the offspring. Should the first ovulation not result in fertilization, then the female will undergo a cycle of events that again results in ovulation. These cycles will repeat during the breeding season until fertilization is successful or until the season ends.

In one sense, therefore, the reproductive *cycle* is an unusual event and can be thought of as a fail-safe mechanism. In women, menstrual cycles occur over a period of 30–40 years, broken only by pregnancy or drastically changed by the ingestion of steroids. This 'uncoupling' of reproductive readiness from environmental cues and from other 'natural' contraceptives (such as that provided by suckling) has necessitated the tremendous efforts that are being made towards finding an acceptable form of contraception. An understanding of how natural methods of fertility operate in other species may provide us with answers.

OBJECTIVES FOR CHAPTER 10

Now that you have read this chapter, you should be able to:

10.1 Define and use, or recognize definitions and applications of each of terms printed in **bold** in the text.

10.2 Describe, in words and diagrams, the basic anatomy of the male and female reproductive systems. (*Questions 1 and 2*)

10.3 List the sequence of events in an oestrous cycle, and describe the differences between oestrous cycles typified by: apes and humans (menstruators); ferrets, cats and rabbits (induced ovulators); and rats. (*Question 1*)

10.4 Describe the major endocrine changes that control the menstrual cycle. (*Questions 3 and 4*)

10.5 Provide evidence for links between the hypothalamus, the pituitary and the gonads in the control of the menstrual cycle. (*Questions 5, 7 and 8*)

10.6 Distinguish between negative and positive feedback control of gonadotropin release. (*Questions 3 and 8*)

10.7 Describe the functions of the corpus luteum and distinguish between luteotropic and luteolytic factors. (*Question 6*)

10.8 Give two examples in which the nervous system plays a role in the timing of reproduction. (*Questions 9, 10, 11*)

10.9 Discuss the evidence for external influences on the timing of reproduction. (*Questions 9, 10, 11*)

10.10 Distinguish between three phases of mammalian development in which circulating steroids have key regulating effects. (*Question 12*)

10.11 Describe how steroids and peptides can be used as contraceptive agents. (*Question 13*)

10.12 Describe variation among vertebrates in the way that gonadal activity, hormone levels, and mating behaviour are related to one another.

QUESTIONS FOR CHAPTER 10

Question 1 (*Objectives 10.2 and 10.3*) Select three accurate statements from (a)–(g).

(a) At ovulation, follicles are released from the ovary and pass down the Fallopian tube, which is connected to the ovary and the uterus.

(b) The corpus luteum located in the uterus persists there if pregnancy occurs and aids in the development of the endometrium.

(c) The gestation period describes the time taken for a follicle to mature and ovulate.

(d) In addition to producing sperm, the testes secrete steroid hormones from the Leydig cells.

(e) Sperm are carried to the vas deferens and urethra, which also receives secretions from accessory glands.

(f) All male and a majority of female mammals are continuous, rather than seasonal, breeders.

(g) Follicular development (from a primordial to a Graafian follicle) takes about 14 days in humans.

Question 2 (*Objective 10.2*) It is not uncommon to find the Fallopian tubes partially blocked by tissue growth, which prevents the ovulated egg from reaching the uterus. Does this mean that fertilization is impossible? What complications could arise in someone with blocked Fallopian tubes?

Question 3 (*Objectives 10.4 and 10.6*) What effect would you predict from the continuous administration of progesterone to a female monkey (a menstruator) for at least 90 days at a dosage:

(a) Such that the level in the blood was at least that found normally on day 23 of the cycle (see Figure 10.8)?

(b) Such that the level in the blood was just 10% of that on day 23?

Consider the effect on the ovaries, the endometrium and the blood levels of the naturally secreted steroids.

Question 4 (*Objective 10.4*) If a male rabbit were hypophysectomized, which of the hormones FSH, LH, progesterone, and testosterone would you need to inject to maintain the entire reproductive system in a fully functional condition for (a) 2–3 weeks and (b) indefinitely?

Question 5 (*Objective 10.5*) If the gonadotropin LH were injected into a mature male mammal, which one of the following results would you expect?

(a) An increase in oestrogen

(b) An increase in testosterone

(c) A decrease in the activity of the testes

(d) A 28-day cycling of male hormones

(e) Increased sperm production

Question 6 (*Objective 10.7*) From statements (a)–(f), choose the one that accurately describes the events following fertilization of the zygote in (i) a rat and (ii) a woman.

(a) The fertilized egg implants in the uterus and begins to secrete progesterone.

(b) After implantation, the ruptured follicle forms a corpus luteum, which starts to produce progesterone under the influence of prolactin.

(c) After implantation, the corpus luteum breaks down.

(d) After implantation, the corpus luteum starts to produce progesterone under the influence of a prostaglandin.

(e) After implantation, the corpus luteum starts to produce progesterone under the influence of gonadotropin LH.

(f) After implantation, the corpus luteum starts to produce oestrogen, which maintains the endometrium.

Question 7 (*Objective 10.5*) Construct a simple flow diagram to show how prolactin secretion, initiated by suckling, can act as a contraceptive. (By flow diagram, we mean a diagram like Figure 10.11.)

Question 8 (*Objectives 10.5 and 10.6*) Construct a flow diagram to show the links between the hypothalamus, the pituitary, and the ovary in the control of the menstrual cycle.

Question 9 (*Objectives 10.8 and 10.9*) If a mature doe rabbit were hypophysectomized and soon afterwards put into a cage with a male, LH would have to be given to make fertilization possible. If you repeated the experiment, this time using mature rats, would the same action allow fertilization to take place?

Question 10 (*Objectives 10.8 and 10.9*) What sort of investigations would you perform to determine whether a mammal was an induced rather than a spontaneous ovulator? Is an induced ovulator affected by environmental factors so far as the cycle of events in its ovary is concerned?

Question 11 (*Objectives 10.8 and 10.9*) Farmers have known for many years that, in some cases, ewes may be brought into oestrus earlier than otherwise by penning them next to a ram. This also has the effect of tending to synchronize their cycles. Suggest a plausible explanation for this phenomenon.

Question 12 (*Objective 10.10*) Predict what would happen if a single dose of testosterone were given to (i) a genetically male rat early in fetal life, (ii) a genetically female rat just after it had been born, and (iii) a genetically female rat after it had become sexually mature.

Question 13 (*Objective 10.11*) Which of the following treatments should, in theory, have contraceptive effects?

(a) One injection of LH

(b) Daily oral administration of a peptide that is anti-LH

(c) Injection of progesterone as an implant that slowly releases the hormone.

(d) Daily oral administration of oestrogen and progestogen

(e) Daily oral administration of prolactin

(f) A single injection of FSH

(g) Daily injection of GnRH antagonist

HOMEOSTASIS OF BODY FLUIDS ◆ CHAPTER 11 ◆

OSMOREGULATION

Maintenance of constancy of the osmotic and ionic concentration of the body fluids of an animal.

HOMEOSTATIC MECHANISMS

The means by which the constancy of composition of, for example, body-fluid concentration is maintained in an organism.

11.1 BODY FLUIDS AND THEIR REGULATION

Water is indispensable for life on Earth; it is the universal biological solvent in which the molecular characteristics of life occur. Inorganic ions are equally important, playing a variety of roles in the metabolic processes in cells—for example, as enzyme cofactors or coenzymes, in amino acid transport, or in the depolarization of the cardiac pacemaker in the vertebrate heart. Ions and water are the principal components of the tissue fluids; indeed, tissue fluids have been described as dilute saline solutions in which sodium chloride is the predominant electrolyte. However, this is misleading, because potassium ions (K^+) rather than sodium ions (Na^+) are usually the dominant intracellular cations. Other compounds, including proteins and organic solutes such as amino acids, are often present.

In many animals, the concentrations of water and solutes within the body fluids are regulated within fairly narrow limits, and only relatively small variations in composition are permitted (as discussed in Chapter 4). One problem for animals is how to maintain the correct composition of the body fluids when this almost invariably differs from the composition of the external medium. If there is a large concentration difference across the cell membranes at the boundary with the external medium, the animal may reduce the permeability of its membranes, or it may actively move ions across them. Maintenance of this constancy is called **osmoregulation**, and the **homeostatic mechanisms** involved in the regulation of the osmotic concentrations in cells are as important as those described for the regulation of blood glucose in Chapter 9.

The homeostatic mechanisms must cope with alterations in the composition of the body fluids resulting from (i) physiological processes, such as absorption into the gut blood vessels of substances taken into the body by ingestion and modified by digestion, (ii) metabolic processes, such as the breakdown of proteins and nucleic acids in cells, and (iii) possible catastrophic events, such as haemorrhages or the intake of toxic substances. In this chapter we shall be concerned with regulation of four other groups of substances, which may be present in amounts or concentrations that are too high or low for the animal's well-being. These are:

- Water, which comprises the main bulk of body fluids and is the solvent for all the other substances;
- Inorganic ions (or salts), which are the main basis for the osmotic concentration of body fluids and also for their acidity (pH or acid–base balance);
- Nitrogenous waste substances (excretory products); some of these contribute to osmotic pressure and they may be toxic;
- Other toxic substances that dissolve in body fluids.

Most of this chapter concerns the regulation of water and ions in mammalian blood and in the haemolymph of insects. The principal organs involved are also **excretory organs**, which eliminate nitrogenous waste substances (discussed more fully in Section 11.5); the fluid produced is the **urine**.

The problems for an animal in regulating its water and solute composition will vary with the nature of its environment—clearly, the osmoregulatory problems posed in seawater will be very different from those in freshwater or on land. Here, we consider not only the movement of water and ions across an animal's exterior cell membranes, but also the role of specialized excretory organs—such as the **mammalian kidney**—which, although often thought of primarily in relation to their role in nitrogenous excretion, are of major importance in regulating the water and ionic composition of the body fluids. In most animals, it is difficult to consider osmoregulation separately from excretion, and in birds and mammals the kidney performs many of the functions shared by the skin, gills and excretory organs in fishes. Excretory organs such as the kidney can, in many respects, be considered as osmoregulatory organs in which the capability of the transporting epithelia has been enhanced by specialization and differentiation. Indeed, the **vertebrate kidney** probably evolved primarily as an osmoregulatory device, not as an organ of excretion of waste products, and it continues as this in teleosts (elasmobranchs are a special case because they produce urea). The excretory function is taken on by the vertebrate kidney only in terrestrial forms (or those that are secondarily aquatic). We shall now consider how some excretory organs function, and how excretory fluid (urine) is produced.

11.2 EXCRETORY ORGANS: THE PRINCIPLES OF THEIR FUNCTION

The two basic processes of formation of urine—**ultrafiltration** and **active transport**—are described here by examining the function of two of the best studied excretory organs, the mammalian kidney and the insect **Malpighian tubule**, but the principles involved can be applied in many instances to other excretory organs. Excretion is defined here as the removal from body fluids of unwanted substances such as excess water, salts, toxic substances and nitrogenous waste compounds.

The term kidney, or renal organ, is applied to excretory organs in a diverse array of animals. Functionally, the excretory organ of a bivalve mollusc (the kidney) may be relatively similar to that of an annelid (the nephridium), but the two organs are morphologically and embryologically very different, that is they may be functionally analogous but not homologous. It should not be forgotten that specialized organs such as the gills of teleost fish and the salt glands of birds and elasmobranchs also play an important role in excretion, which we will consider in Chapter 12.

Ultrafiltration occurs where hydrostatic or osmotic pressure differences force fluid through a membrane with pores so fine that only water and smaller molecules (but not cells or macromolecules such as protein) are able to pass through. The fluid so formed is often called the **primary urine** because it is subsequently modified during its passage through the excretory tubules to the exterior. (For vertebrates the correct terminology for primary urine is **glomerular filtrate**, for reasons which will become obvious to you later.)

Ultrafiltration is the mechanism responsible for the formation of the primary urine, but in some animals the primary urine may be formed mainly by secretion (largely active transport of ions, followed by fluid, from the body fluids into tubules of the excretory system) as in insect Malpighian tubules and some fish kidneys. A consequence of the formation of urine by ultrafiltration is that any new and perhaps toxic substance entering the animal's body fluid will be excreted provided the molecules are small enough, unless a special transport mechanism exists to reabsorb it from the filtrate. (There are, of course, other means by which animals may remove toxic substances, detoxification in the liver for example.)

Active transport can modify the composition of the ultrafiltrate in one of two ways, either by absorption from it, or by secretion into it. (Correctly speaking, because absorption occurs secondary to secretion (ultrafiltration), we should say that *re*absorption has occurred.)

◇ Active transport is a selective process. Why might this be important in urine formation in addition to ultrafiltration?

◆ Any molecule, provided that it is sufficiently small, will pass into an excretory organ with the filtrate. Hence, to prevent loss of essential components such as amino acids, sugars and vitamins, active transport is necessary to reabsorb them. Similarly, secretion enables an animal to add selectively to the ultrafiltrate.

The passive movement of substances can also be important: in the mammalian kidney, the production of a concentrated urine depends not only on active ion transport but also **passive movements** of an organic compound, urea and of course water (Section 11.3.7).

A kidney that works by filtration and absorption can process large volumes of fluid: in the mammalian kidney, up to 99% of the filtered volume can be reabsorbed so that less than 1% is excreted as urine. The kidneys of all vertebrates, (fish, amphibians, reptiles, birds and mammals), mainly work by a **filtration and reabsorption** mechanism, with the exception, as you have just read, of the secretory kidney of some teleost fish. All vertebrates can produce a urine that is **iso-osmotic** or **hypo-osmotic** to (equal to or less concentrated than) the blood, but only birds and mammals can produce a urine that is **hyperosmotic** to (more concentrated than) the body fluids. For land animals such as birds and mammals, the ability to produce a concentrated urine is of the utmost importance in controlling water loss. For birds, the removal of nitrogenous excretory products as uric acid (which is almost insoluble) further restricts water loss in the urine.

Summary of Section 11.2

1 Excretion is the removal of unwanted substances from an animal's body fluids.

2 Renal (excretory) organs in different animals may be functionally analogous but not necessarily homologous.

3 There are two basic processes by which excretory fluids (primary urine) may be formed—ultrafiltration and active transport. Transport into the excretory tubule lumen, largely active, is referred to as secretion. In insect Malpighian tubules, primary urine is formed largely by secretion whereas in vertebrate kidneys, ultrafiltration is usually responsible.

Now attempt Question 1 on p. 394.

11.3 STRUCTURE OF THE VERTEBRATE KIDNEY

The kidneys of vertebrates are paired organs that lie dorsal to the abdominal cavity. The evolution of the vertebrate kidney is an interesting but complicated story which will not concern us here. All vertebrate kidneys are composed of functional units called **nephrons**, the number, arrangement and length of which vary in the different vertebrate groups. The basic structure of the nephron is, however, the same in all vertebrate groups. To understand how a kidney functions, and in particular how the nephron contributes to the ability of the kidney to produce urine, we shall examine the workings of one of the best studied and probably most highly evolved vertebrate kidneys, the mammalian kidney.

11.3.1 The mammalian kidney

In mammals, the two kidneys lie one on either side of the vertebral column (Figure 11.1a). In adult humans each kidney weighs between 115 and 170 g

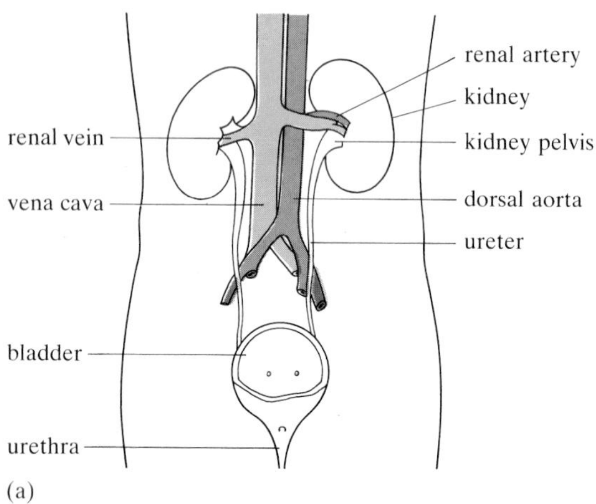

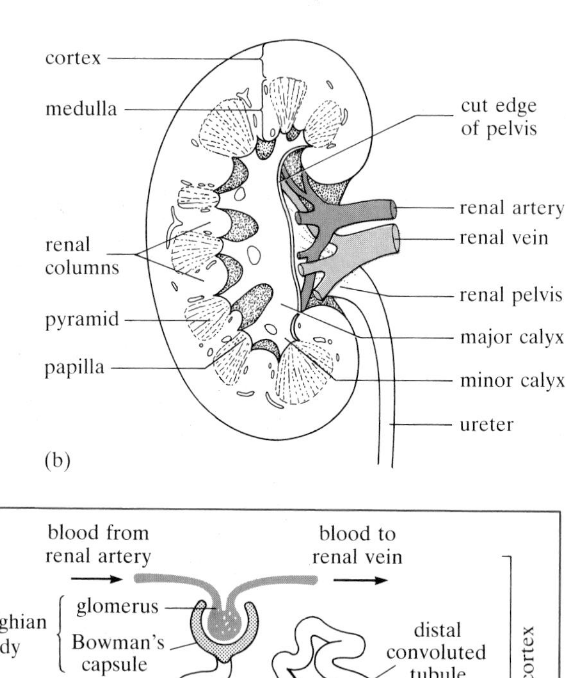

Figure 11.1 (a) The human urinary system: note the relationship of the kidney to the blood vessels and urinary ducts (ureters). (b) A bisected human kidney showing the gross structure, notably the cortex, medulla, calyces and papilla. (c) A more diagrammatic representation of a mammalian kidney showing the location of a single nephron. (d) The components of a nephron.

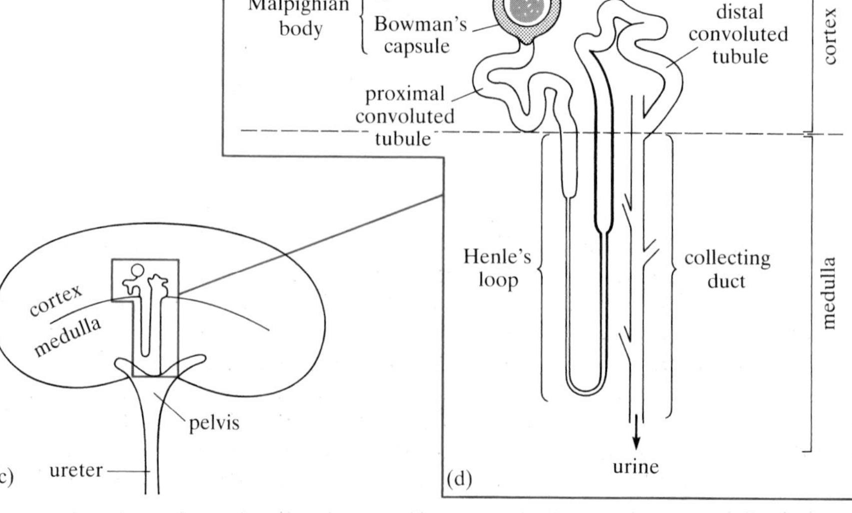

and measures 11–13 cm in length, and the right kidney is usually slightly posterior to the left (viewed from the ventral body surface). The internal arrangement can be seen in the drawing of a bisected human kidney in Figure 11.1b and in the schematic outline of a kidney in Figure 11.1c. The kidney, which has a tough outer cover called the capsule, consists of two parts: an outer region, the **cortex**, and an inner region, the **medulla**, which consists of a series of pyramids. The apices of these are known as **papillae**, each of which projects into a minor calyx (plural calyces). These in turn combine to form major calyces, which unite to form the **renal pelvis**. This leads to the ureter, which takes urine from the kidney to the **bladder**. Here the urine is stored before it is passed to the exterior. (How the bladder empties its contents to the outside is not dealt with here.)

The human kidney contains about a million nephrons, and each one consists of a number of functionally distinct parts (Figure 11.1d). At the proximal (beginning) end the tubule is closed and expanded to form a structure called **Bowman's capsule**; this almost totally surrounds a network of blood vessels, the **glomerular capillaries (glomerulus)**. Each Bowman's capsule with its glomerulus is known as **Malpighian body** (Figures 11.1d and 11.2). The distal

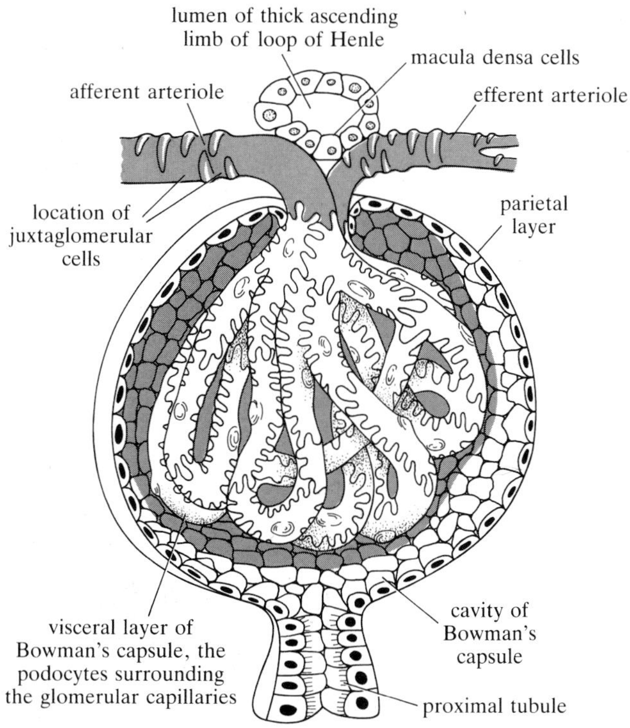

Figure 11.2 Bowman's capsule and the glomerular network or 'tuft' of capillaries (the Malpighian body).

(far) end of the tubule, the collecting tubule, joins with others to form a collecting duct that opens into the renal pelvis. The ultrastructural characteristics of cells of the various parts of the nephron show considerable variation (Figure 11.3). Note in Figure 11.2 that although the outer cells of Bowman's capsule form a flattened epithelium (the parietal layer), those on the inner side (the **podocytes**), which envelop the blood capillaries, are of a much more complicated form with extensive interdigitations at the borders of the cells. We shall return later to the role of these cells in the filtration of the blood. The **juxtaglomerular apparatus** (from the Latin juxta, meaning next to) is an important structure, primarily in regulation of sodium balance, as you will see

CORTEX = outer region
MEDULLA = inner region
PAPILLAE
Inner region of the medulla consists of a series of pyramidal structures whose apices are termed papillae.
RENAL PELVIS
Union of the major calyces of the kidney.
BOWMAN'S CAPSULE
At the proximal (beginning) end of the nephron the tubule is closed and expanded to form this structure, which almost totally surrounds a network of blood vessels, the glomerular capillaries (glomerulus).
GLOMERULAR CAPILLARIES (GLOMERULUS)
A tuft of blood capillaries from the afferent renal arterioles, which join to form efferent renal arterioles.
MALPIGHIAN BODY (of the vert. kidney)
Each consists of a glomerulus & a Bowman's capsule.
PODOCYTES
Structures than envelop the blood capillaries of the glomerulus and have extensive interdigitations at the borders of the cells.
JUXTAGLOMERULAR APPARATUS
An important structure in the regulation of fluid balance. Consists of 2 groups of cells (i) juxtaglomerular cells, which are arranged as a cuff around the afferent arterioles and (ii) the specialized macula densa cells.

For absorption - micro villi brush border like intestine

MACULA DENSA CELLS

Part of the juxtaglomerular apparatus in the vert. kidney

LOOP OF HENLE

U-shaped tubule of the nephron of kidney between the proximal & distal convoluted tubules. The movement of various ions & water in it lead to the process of counter-current multiplication in mammals.

CORTICAL NEPHRONS

Those with superficial glomeruli.

JUXTAMEDULLARY NEPHRONS

Those with glomeruli next to medulla and with loops of Henle that extend deep into the medulla.

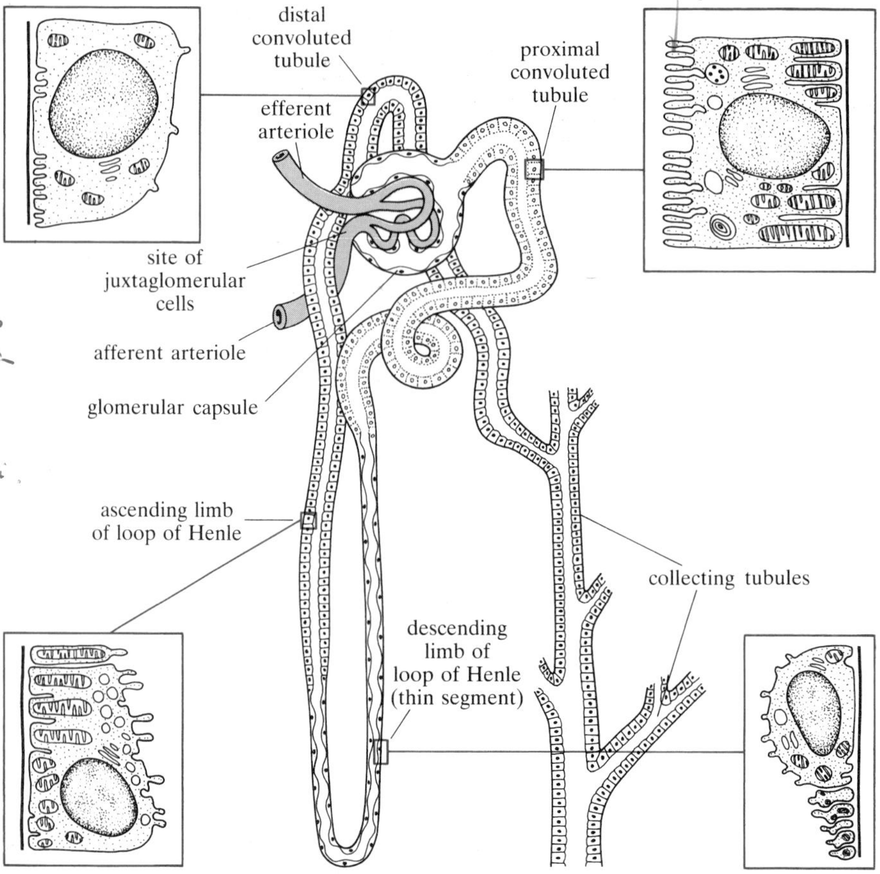

Figure 11.3 Diagrammatic representation of the ultrastructural features of different parts of the nephron.

PERITUBULAR CAPILLARIES

A network of renal capillaries, formed when the efferent renal arteriole divides; they are clumped around the cortical nephrons.

PROXIMAL CONVOLUTED TUBE

Part of the nephron next to the glomerulus, in which absorption of Na^+, Cl^- & water occurs... Structurally highly specialized as transport cells because of the presence of a microvillus brush border similar to that seen in intestinal cells. PCT

DISTAL CONVOLUTED TUBE DCT

Part of nephron after loops of Henle; in DCT there is co-transport of Na^+ & Cl^- into the DCT cells & secretion of K^+ & NH_4^+ out of DCT cells (and of H^+ by a counter transport mechanism).

in Section 11.3.11. It consists of two groups of cells: (i) cells arranged as a cuff around the afferent arteriole, the juxtaglomerular cells, and (ii) specialized cells, the **macula densa cells**, from a part of the nephron termed the thick ascending limb of the **loop of Henle**.

All the parts of the kidney labelled in Figure 11.1b and c are present in most mammals. However, there may be marked differences in the lengths of parts of nephrons, particularly of the loop of Henle, not only from one species to another but even within the same kidney. Such anatomical differences can underlie important functional differences. The glomeruli are found only in the cortex of the kidney. Some are superficial, but more commonly they occur deeper in the cortex (Figure 11.4). Those next to the medulla are known as **juxtamedullary glomeruli**.

The superficial glomeruli belong to **cortical nephrons**, and the juxtamedullary glomeruli to **juxtamedullary nephrons**. Superficial nephrons have a much shorter loop of Henle, and also the blood flow around them is different. A network of capillaries, clumped in appearance (the **peritubular capillaries**), forms around the cortical nephrons, whilst the capillaries around the juxtamedullary nephrons form two distinct populations, one forming a capillary network surrounding the **proximal and distal convoluted tubules**, and the second population proceeding directly into the medulla in the shape of a long loop (collectively these long loops are known as the **vasa recta**), parallel to the long loop of Henle.

Blood from the main abdominal artery, the dorsal aorta, passes to the kidney via the **renal artery** (Figure 11.4), which divides into a number of small

VASA RECTA Network of blood caps. from renal arteriole, which parallels the long loops of Henle & maintains the cortico-medullary gradient generated by counter-current multiplication in loop of H.

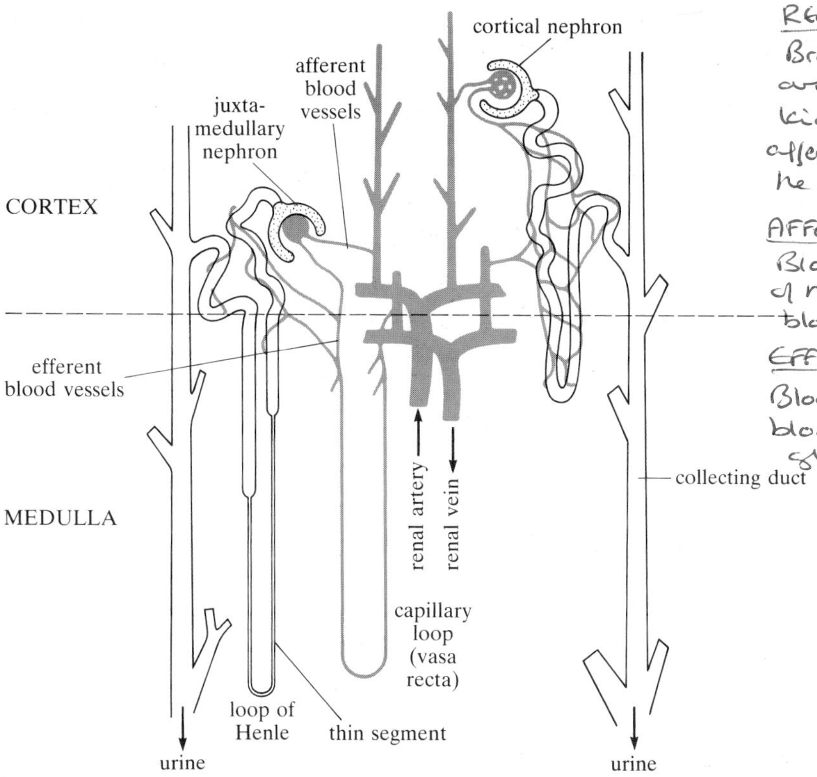

Figure 11.4 Two typical nephrons, one 'long-looped' (located close to the border of the medulla and therefore called juxtamedullary) and one 'short-looped' (or cortical). The long-looped nephron is paralleled by a loop formed by the blood capillary. The short-looped nephron is surrounded by a capillary network. Most mammalian kidneys contain a mixture of the two types, but some species have only one or the other kind.

RENAL ARTERY
Branch from the main abdominal artery, the dorsal artery, to the kidney; it divides into small afferent renal arterioles, which form the glomerular capillaries.
AFFERENT RENAL ARTERIOLES
Blood vessels formed by division of the renal artery, which supplies blood to kidney.
EFFERENT RENAL ARTERIOLES
Blood vessels formed by joining the blood capillaries that leave the glomerulus.

afferent renal arterioles before forming the glomerular capillaries. These recombine on leaving the glomerulus to form the **efferent renal arterioles**, which then divide to form the peritubular capillaries, which in turn recombine to form the **renal vein**, which enters the main abdominal vein returning blood to the heart, the posterior vena cava. This situation in which capillaries combine to form arterioles, another set of capillaries, and then veins is unusual, as you know from Chapter 5.

The glomerular filtrate (an ultrafiltrate of blood formed when blood flows through the glomerular capillaries) passes into the Bowman's capsules, and the urine is modified in composition and reduced in volume as it passes down the renal tubule into the collecting ducts. From there, it goes to the renal pelvis before entering the bladder via the ureter (Figure 11.1a). How the ultrafiltrate is initially formed and how it is modified during its course through tubules of the nephron we examine next.

11.3.2 Ultrafiltration of blood in the glomerulus: formation of the primary urine

The **effective filtration pressure** in the glomerulus (Figure 11.5) depends on three factors, the blood (**hydrostatic**) pressure in the glomerular capillaries, which forces fluid out of the blood, the **osmotic pressure** of the proteins in the glomerular capillaries, which tends to draw fluid back into the blood, and the **intracapsular pressure** (the hydrostatic pressure of the fluid in Bowman's capsule), which also tends to drive fluid back into the blood. In the rat, the glomerular capillary pressure is about 6.6 kPa. This is opposed by the osmotic pressure of the proteins in the glomerular capillaries (about 3.3 kPa) and by the intracapsular pressure (about 1.2 kPa).

RENAL VEIN
The main vein draining the kidney; it enters the main abdominal vein returning blood to the heart, the post. vena cava.
HYDROSTATIC PRESSURE
Pressure of blood in the glomerular capillaries; it is one of the three factors important in determining the effective filtration pressure of the glomerulus.
OSMOTIC PRESSURE (π)
Osmosis is the movement of water down a gradient of water potential and across a semi-perm. membrane from a region of lower to one of higher solute concentration. It occurs because solute molecules lower water potential. When the compartment into which water moves has a fixed volume, the pressure generated by osmosis defines the osmotic pressure π, of the solution in that compartment.
INTRACAPSULAR PRESSURE – Pressure within Bowman's Capsule.

GLOMERULAR FILTRATION RATE (GFR)

The rate at which plasma is filtered by the glomerulus of the kidney.

RENAL CLEARENCE

The volume of plasma that contains the amount of a substance excreted by the renal tubules in a given time. Amount excreted is the net result of 3 processes - glomerular filtration, absorption, & secretion, ...

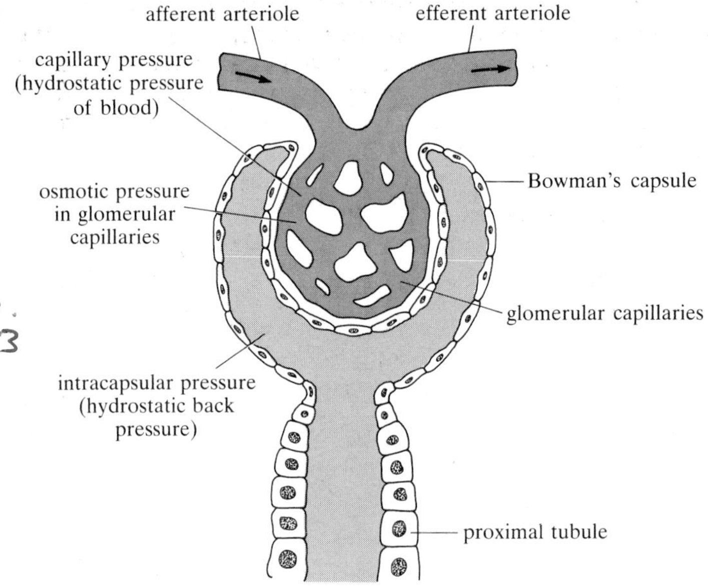

Figure 11.5 Bowman's capsule: the various forces affecting the filtration rate are indicated on the left.

◇ What will the effective filtration pressure be?

◆ 2.1 kPa $(6.6-(3.3+1.2)=2.1)$

You should realize, however, that there may be considerable differences in the effective filtration pressure along the length of the glomerular capillaries—above 2 kPa at the afferent end (the beginning of the glomerulus) (Figure 11.5) and down to 0 kPa at the efferent end (the end of the glomerulus). It is therefore very important that there is a continuous flow through the glomerular capillaries. If the blood flow rate is not maintained, the filtration pressure at the afferent end will fall. If blood pressure is too low the hydrostatic pressure may fall below the sum of the osmotic pressure and intracapsular pressure (the opposing forces).

◇ When is it likely that the blood flow rate in the glomerulus will fall?

◆ When the blood supply to the kidney decreases, for example, in extreme dehydration or during haemorrhage (blood loss).

As you will see in Section 11.3.11, there is a mechanism whereby a fall in blood pressure, and therefore a reduction in glomerular capillary pressure, can be counteracted by the kidney itself. The effective filtration pressure is responsible for determining the glomerular filtration rate. How this can be measured is examined next.

11.3.3 Glomerular filtration rate (GFR)

Glomerular filtration rate (GFR) is the rate at which plasma is filtered by the glomerulus and is usually calculated as millilitres (or cubic centimetres) per minute. It can be obtained by measuring the **renal clearance** of a substance that is freely filtered by the glomerulus, and is neither metabolized nor transported. For such a substance, the GFR equals the renal clearance.

Renal clearance is the volume of plasma that contains the amount of a substance that is excreted by the kidney in a given time. The amount excreted is the net result of three processes—glomerular filtration, absorption and secretion. If the volume of urine produced per minute is measured, then by comparing the concentration of an inert substance (one that cannot be transported or metabolized) in the urine with that in the plasma, the renal clearance can be calculated by dividing the amount of the inert substance appearing in the urine by its concentration in the plasma, and multiplying this by the rate of urine formation:

$$\text{renal clearance of substance X} = \frac{U_X}{P_X} \times V$$

where U_X is the amount of X in 1 cm^3 of urine, V is the volume of urine formed per minute (cm^3 min^{-1}), and P_X, is the amount of X in 1 cm^3 of plasma.

◇ If the concentration of X in urine is 125 mg cm^{-3}, 2.0 cm^3 of urine are formed per minute, and the concentration of X in plasma is 2.0 mg cm^{-3}, what is the renal clearance of X?

◆ $(125 \times 2.0)/2.0 = 125$ cm^3 min^{-1}

RENAL THRESHOLD
Concentration above which that substance is not reabsorbed any further by kidney tubules and begins to be excreted eg. at a certain plasma glucose conc., glucose begins to be excreted because the level of glucose in the tubular urine rises to a level above that at which all of it can be absorbed.

TRANSFER MAXIMUM Tmax
Maximum amount of a substance that can be transported across the tubular cells in a given time.

In the question above, X is an inert carbohydrate molecule, inulin, which has a relative molecular mass of 5 500, and the values are for a human kidney. Because inulin is neither reabsorbed nor secreted in the renal tubule, its renal clearance also represents a measure of GFR in humans. Inulin does not occur naturally in human blood. For clinical assessment of renal clearance in humans, the substance normally measured is creatinine, which is produced from the muscles of the body daily in relatively constant amounts.

Glomerular filtration rate can be affected by a number of factors, including the size of the individual—the bigger the body, the larger is the GFR. In women GFR is around 75 cm^3 min^{-1} and during pregnancy it may increase by up to 100%, to 150 cm^3 min^{-1}. In the course of the normal menstrual cycle in women GFR varies, being larger in the secretory than the proliferative phase. But there can also be fluctuations in GFR in any individual throughout the day, for example owing to constrictions in afferent arterioles, which would decrease glomerular plasma flow and glomerular capillary pressure.

11.3.4 Tubular transport processes

If less of a substance appears in the urine than was filtered in the glomerulus (i.e. the clearance is less than for inulin) it is likely to have undergone reabsorption, and this is true for a variety of substances, such as glucose, water and sodium chloride. Glucose is freely filtered in the glomerulus, but almost all of the glucose initially filtered is reabsorbed, at least up to a certain plasma concentration (the **renal threshold**). When this plasma glucose concentration is reached, glucose begins to be excreted because the level of glucose in the tubular urine rises to a level above that at which all of it can be reabsorbed. The maximum that can be transported across the tubular cells in a given time is called the **transfer** (or **transport**) **maximum**, T_{max}.

Examine Figure 11.6. Filtration of glucose from plasma to the nephron is a linear function of concentration, and the glucose is almost totally reabsorbed up to a plasma glucose concentration of 200 mg of glucose per 100 cm^3 of plasma. The shape of the line showing glucose reabsorption (in relation to plasma glucose concentration) represents the saturation of a transport process—that for the absorption of glucose from the urine.

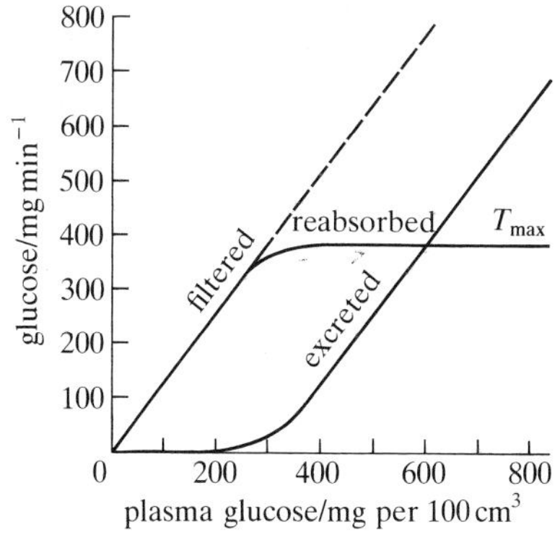

Figure 11.6 Filtration, reabsorption and excretion of glucose by a human. Below a certain limit (200 mg of glucose per 100 cm^3 of plasma), no glucose appears in the urine; this is because all the filtered glucose has been removed by tubular reabsorption. Above this limit the reabsorption mechanism is fully saturated, and although it continues to work at full extent, excess glucose is excreted. Note that 90 mg of glucose per 100 cm^3 of plasma is equivalent to 5 mmol l^{-1} of glucose in the blood.

◇ What is the approximate T_{max} for glucose in Figure 11.6?

◆ T_{max} for glucose is 375 mg min^{-1}. (At plasma concentrations above 200 mg per 100 cm^3 of plasma, glucose begins to be excreted.)

Incidentally, this has traditionally been the basis for determining the occurrence of diabetes mellitus in a subject, by testing the urine for glucose. Normally the plasma level of glucose is strictly controlled (Chapter 9), but in a diabetic subject, glucose levels cannot be kept within the normal limits and the T_{max} is exceeded in the kidney; hence glucose is excreted. However, glucose can appear in the urine for reasons unconnected with diabetes, giving too many false positives when this test is used alone as the basis for determination of diabetes, so other techniques must also be used in clinical determination of this disease.

If more of a substance appears in the urine than was filtered in the glomerulus (i.e. the clearance is greater than for inulin), then secretion into the urine has occurred in the tubules of the nephron. PAH (*para*-aminohippuric acid) is freely filtered in the glomerulus, and it is not reabsorbed anywhere in the nephron. Since the amount of PAH in the excreted urine is greater than in the glomular filtrate, the difference must represent the amount secreted into the urine by the kidney tubular cells.

The **clearance ratio** (the clearance ratio of a substance compared with the clearance of inulin) is frequently used in descriptions of processes occurring in the kidney. For a substance X, the ratio is given by

$$\frac{U_X}{P_X}V : \frac{U_I}{P_I}V$$

where I is inulin. For substances that are neither secreted nor absorbed, the inulin clearance ratio will be 1. For PAH the clearance ratio is more than 1 because $U_{PAH}V/P_{PAH}$ exceeds $U_I V/P_I$

Frequently, the concentration of a substance appearing in the excreted urine is simply expressed as a ratio of its concentration in the urine to that in the blood plasma, abbreviated as the ***U/P* ratio.**

◇ For inulin, the *U/P* ratio is 30 or 40. Why is it not 1, since it is freely filtered but is neither secreted nor reabsorbed in any part of the nephron?

◆ Because water absorption occurs in the nephron. Only in the glomerular filtrate (primary urine) will there be the same concentration in the tubular fluid as in the plasma. As soon as water starts to be reabsorbed, the concentration of inulin in the tubular fluid rises.

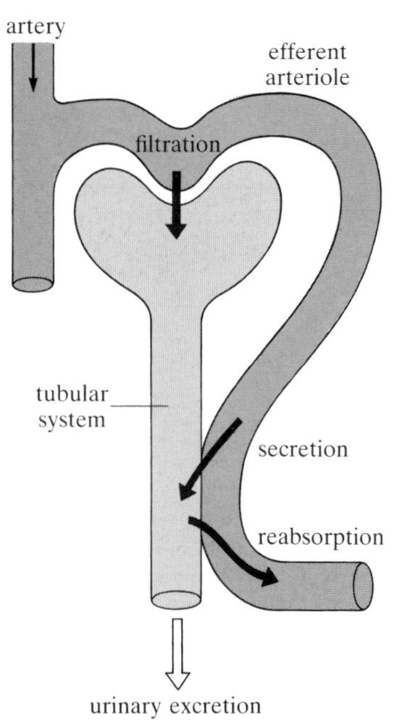

Figure 11.7 A diagram illustrating the factors that control the formation of urine: filtration, reabsorption and secretion. One or all of these may be involved in determination of whether a particular compound appears in the urine.

The possible influences on the amount of a substance appearing in the urine can be summarized as (i) filtration (exclusively), (ii) filtration and reabsorption, (iii) filtration and secretion, or (iv) filtration, secretion and reabsorption. Filtration, secretion and reabsorption are shown schematically in Figure 11.7.

Comparisons of the composition of excreted urine with that of plasma—*U/P* ratios—and the clearance technique give much information on the overall performance of the kidney, but tell us little about the sequential processing of the filtrate and its constituents along the various nephron segments.

Much information on the nature of the filtration process and of kidney function has been obtained from **micropuncture studies** of the type first carried out by Richards in 1924 on amphibian kidneys. Fine glass micropipettes were inserted into the various regions of the nephron. You might believe this to be an extremely intricate task when you recall just how small the mammalian nephron is, but the larger size of the amphibian nephron, and the superficial location of the glomerulus in Munich–Wistar rats (unlike ordinary laboratory rats), make micropuncture studies somewhat easier in these species. Analysis of the glomerular filtrate shows that it is practically free of protein, although proteins of relative molecular mass (M_r) less than 68 000 are filtered—see Table 11.1.

Table 11.1 The relation between relative molecular mass (M_r), molecular radius and the ratio of the concentration of the substance in the filtrate appearing in Bowman's capsule to its concentration in the plasma.

Substance	M_r	Molecular radius calculated from diffusion coefficient/nm	Primary urine/ plasma
water	18	0.1	1.0
urea	60	0.16	1.0
glucose	180	0.36	1.0
sucrose	342	0.44	1.0
inulin	5 200	1.48	0.98
myoglobin	17 000	1.95	0.75
egg albumin	43 500	2.85	0.22
haemoglobin	68 000	3.25	0.03
serum albumin	69 000	3.55	0.01

* If a substance is freely filtered along with water into the nephron, the ratio of its concentration in the primary urine to that in the blood plasma will be 1. As this ratio drops, it seems reasonable to assume that there is some hindrance to the substance entering the glomerulus.

◇ Which molecules in Table 11.1 are freely filtered?

◆ The molecules that are freely filtered are those of M_r up to 5 200 (inulin), but there is a progressive decrease in the efficiency with which larger molecules are filtered.

◇ What does this suggest the basis of the filtration mechanism might be?

◆ Size; from Table 11.1, the dimensions of these molecules increase with increasing relative molecular mass; however, charges on the molecule (which we have not listed here) can also be a factor.

It is believed that the glomerular capillaries have **pores**, 7.5–10.0 nm in diameter and 40–60 nm in length, occupying about 5% of the glomerular surface. Small pores, sometimes referred to as **slit pores**, are also present between the base of the interdigitations of the podocyte cells (cf. Figure 11.2).

Electron microscopy (Figure 11.8) shows that blood in the glomerular capillaries is separated from the cavity of Bowman's capsule by three layers: the capillary wall, the basement membrane and the inner layer of Bowman's capsule (podocytes). Of these, only the basement layer appears to act as a true filter because the other two layers are pierced by fenestrations (holes). Thus the molecular 'sieving' that holds back larger particles in the blood is effected by a combination of the capillary pores, the basement membrane, and the slit pores of the podocytes. Small amounts of plasma albumin are filtered from the blood but are reabsorbed in the tubules of the nephron. Just how, and where, modification of the primary urine occurs (by reabsorption or by secretion or both) in the tubules is a complicated story, and a number of

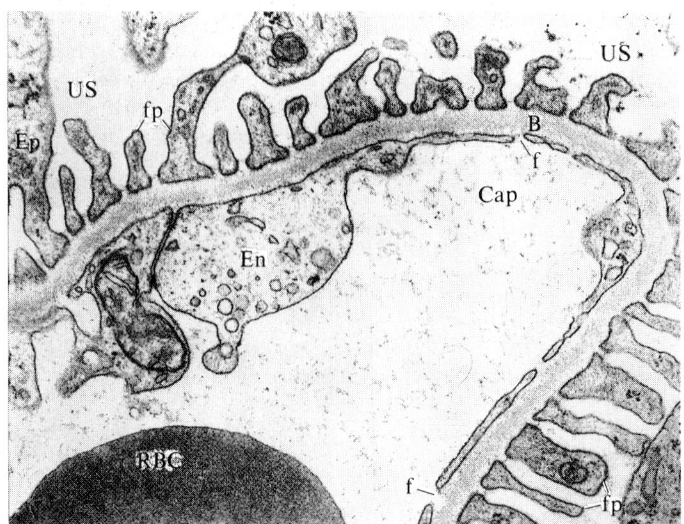

Figure 11.8 Peripheral area of a glomerular capillary from a normal rat. The capillary wall has three distinct layers: the endothelium (En) with its periodic interruptions of fenestrae (f); the basement membrane (B); and the foot processes (fp) of the epithelial cells (Ep), which are also called podocytes (because of their numerous foot processes). Cap = capillary lumen; US = urinary spaces; RBC = red blood cell. Magnification, ×32 000.

elegant techniques have been developed to examine this. Four (*in vivo* techniques) are described in Figures 11.9–11.12.

Results obtained with nephrons of the rat kidney by using various techniques, including some of those illustrated in Figure 11.9–11.12, to measure relative amounts of urea, Na^+, K^+, and water are given in Table 11.2. The values are expressed as a percentage of that in the primary urine filtered in the glomerulus. (You should remind yourself of the regions of the nephron by referring to Figure 1.11d).

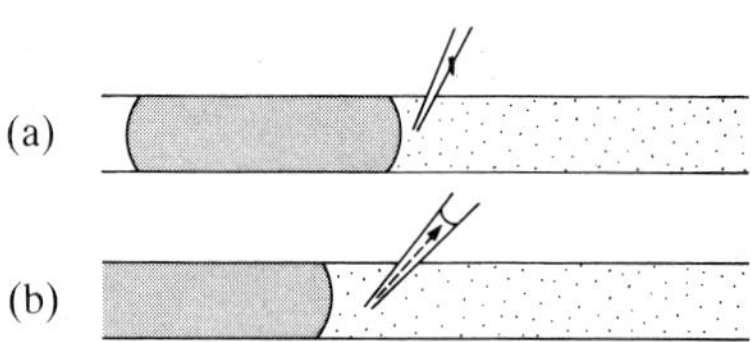

Figure 11.9 Free-flow micropuncture: (a) a micropipette is inserted into a tubule and a block of oil (grey) is injected downstream; (b) slight suction on the pipette holds the oil-block steady, and hence tubular fluid arriving at the pipette is collected into it.

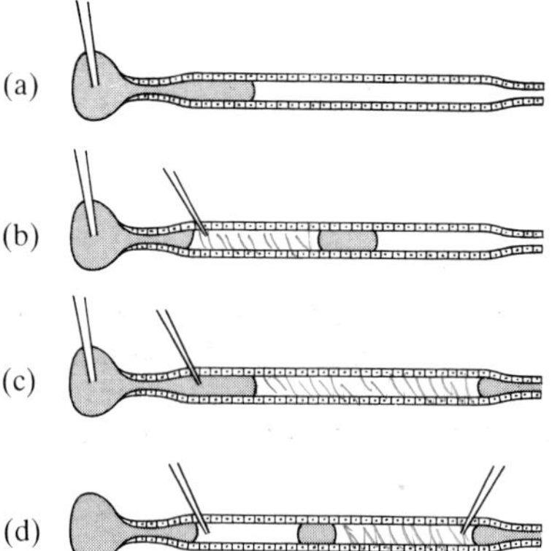

Figure 11.10 Stopped-flow microperfusion (the split-drop technique): (a) a micropipette is inserted into Bowman's capsule (left) and oil (grey) is injected; (b) a test fluid (pink) is then injected into the middle of the column of oil, forcing a droplet ahead of it; (c) The tube is full when the oil droplet reaches the far end of the tubule; (d) after about 20 min the fluid is collected in a third micropipette for analysis by the injection of a second liquid behind the oil near the glomerulus.

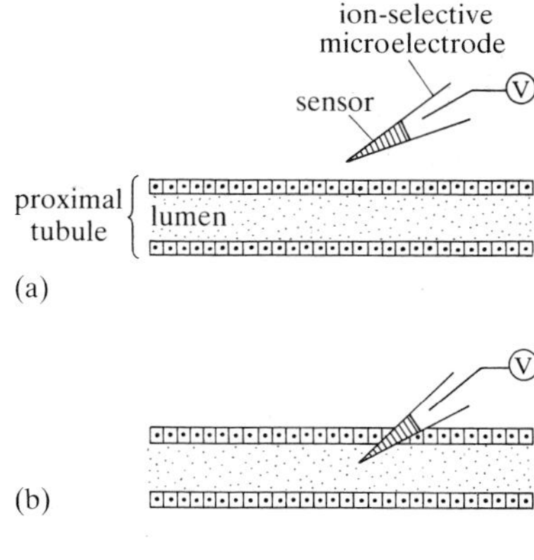

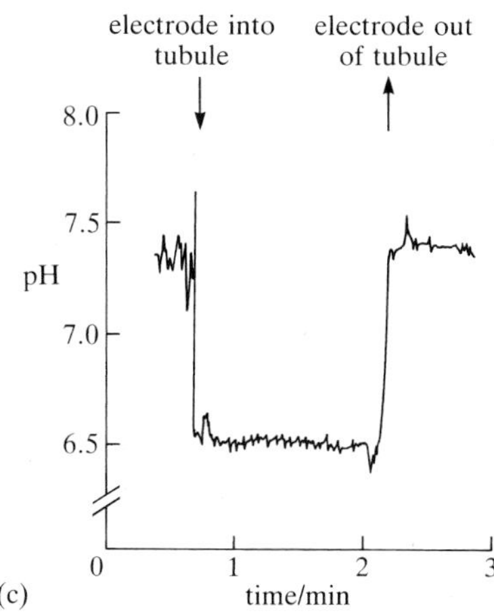

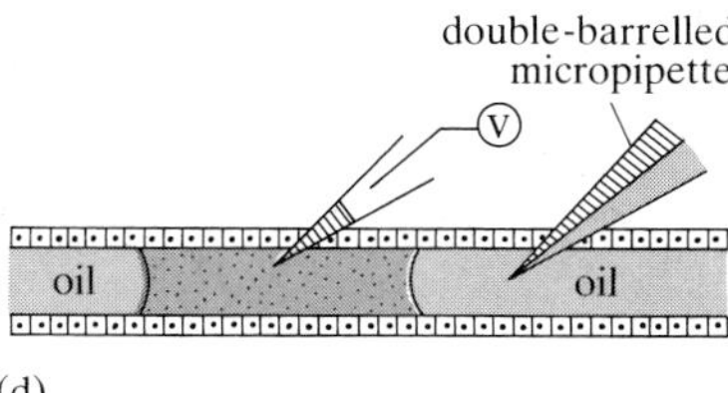

Figure 11.11 Micropuncture with ion-selective electrodes (ISEs). This is an extension of the split-drop technique and is used to measure the reabsorption of ions in, for example, the proximal tubule. It enables rapid measurements of ion concentrations and the immediate assessment of tubule function. The ISE is a very fine micropipette containing a small drop of a sensor substance in the tip, which is sensitive to a particular ion. The sensor generates an electrical potential (with respect to earth) that reflects the concentration of the ion of interest, and the potential can be measured with a sensitive voltmeter (V). (a) The ISE tip (here sensitive to H^+, i.e. pH) registers a pH of 7.4 when it is outside the tubule in the extracellular fluid bathing the kidney. (b) The ISE is placed into the proximal tubule, and this gives a direct measurement of the pH of the fluid inside, which is usually around 6.5. Proximal tubular fluid is generally more acidic than the extracellular fluid because hydrogen ions are secreted from the blood into the tubule. (c) A typical trace produced during such a manoeuvre. (d) The same tubule is punctured with a double-barrelled micropipette and a split-drop is formed as described in Figure 11.10; the change in pH in the droplet as the tubule secretes hydrogen ions is monitored continuously by the ISE. From this, the rate of H^+ secretion can be calculated and factors that may influence it can be investigated.

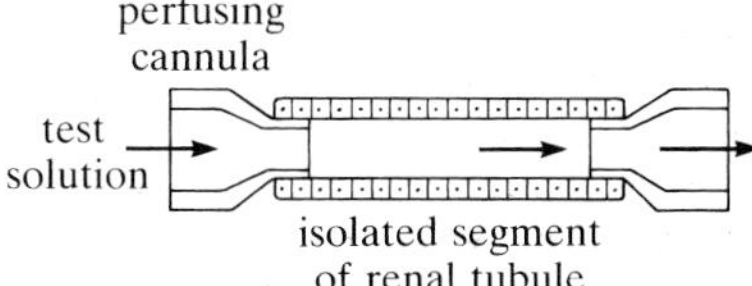

Figure 11.12 Perfusion of a cannulated segment of renal tubule: the perfusate is subjected to chemical and radiotracer analysis to determine the fluxes of ions across the tubule wall.

Table 11.2 shows that less than 1% of the water and Na^+ initially filtered is actually excreted, but about 12% of the K^+ and 26% of the urea is. Remember, however, that these values will vary, depending, for example, on the degree of dehydration of the body fluids.

Table 11.2 The relative amounts of water, Na^+, K^+, and urea in different segments of the rat kidney nephron.

	Percentage remaining relative to glomerulus				Percentage of total osmolarity		
Nephron segment	Water	Na^+	K^+	Urea	Na^+	K^+	Urea
glomerulus (=blood)	100.0	100.0	100.0	100.0	46.5	1.5	2.3
end of proximal tubule	23.0	23.0	16.4	44.7	46.5	1.0	2.3
loop of Henle (bend)	17.0	47.1	86.9	105.3	33.3	1.9	20.8
early distal tubule	17.0	9.5	3.7	133.0	35.6	0.4	24.4
end of distal tubules	6.0	2.9	26.7	77.1	23.3	6.7	30.0
end of collecting tubules	0.24	0.1	11.9	26.3	4.3	15.7	55.0

◇ In the excreted urine, will urea, K^+ or Na^+ contribute most to the total osmolarity? How does this compare with the percentages of each in the primary urine (in the glomerulus)?

◆ Urea contributes the greatest amount (55%) to the total osmolarity of the excreted urine but only 2.3% to the primary urine; K^+ actually contributes even less to the latter (1.5%). Most of the Na^+ in the primary urine is absorbed in the nephron; however, note that K^+, which contributes very little to total osmolarity of the primary urine, makes a much greater contribution to total osmolarity of the excreted urine in contrast to Na^+.

◇ Is the percentage of reabsorption of water, K^+, Na^+ and urea equal in each region of the nephron?

◆ Clearly not: 77% of Na^+, for example, disappears from the urine by the end of the proximal tubule, but the urea concentration actually increases in the bend of the loop of Henle, as do those of Na^+ and K^+.

So these data show that different amounts of Na^+, water and urea are reabsorbed from each region of the nephron. But they do not tell us how reabsorption occurs, or why the urea concentration increases in the tubular fluid. For this, we need to know the transport and permeability properties of the regions of the nephron. These measurements can be made on tubules isolated as in Figures 11.9–11.12. In particular, measurements of electrical potential differences (E_m) across tubular membranes can be very useful in allowing us to decide whether ion transport is passive or active.

◇ How can measurement of electrical potential differences (E_m) across membranes determine whether ion transport is passive or active?

◆ Unequal ion distribution across membranes can result from differential permeabilities to cations and anions (e.g. Na^+ and Cl^-), and this will lead to a membrane potential E_m that can be compared with the potential, E_N, predicted by the Nernst equation (cf. Chapter 1). If E_m is equal to E_N (for Na^+ or Cl^-), then movement of these ions could have occurred by passive means, but if E_m is different from E_N for Na^+ or Cl^-, then active transport may have occurred. In other words, $E_m = E_N + E_x$, where E_x is the component due to the active transport of ion X. (The situation is, of course, more complicated if the membrane is permeable to other ions, because these will also contribute to E_m.)

We can use information from such experiments to understand how urine is processed, and ultimately concentrated in the production of the final, excreted urine, and the next section deals with the processing of the primary urine in the various parts of the nephron. However, rather than give detailed information on values such as the measured electrical potential differences across isolated perfused tubules, or the diffusional permeability coefficients (P_d, which you met in the Fick equation in Chapter 5) for substances transported across nephron tubules, we describe these parameters in terms of the degree (low or high) of permeability of tubule segments to salts, urea and water, and whether or not there is active transport of the salts.

11.3.5 Urine concentration in mammals and birds

This section contains considerable detail and therefore requires careful reading. The important points to note are (i) the regions of the nephron in which transport of ions (especially Na^+ and Cl^-), water and urea occur, and (ii) how the counter-current multiplier in the loop of Henle acts to produce a gradient of osmolarity from cortex to medulla.

Mammals and birds are unique among vertebrates in being able to produce a urine that is hypersomotic to blood: for birds, the maximum urine concentration that can be achieved is usually about twice the plasma concentration, but for mammals such as desert rodents, the *U/P* ratio can be as high as 25. How the urine is concentrated is related to the spatial arrangement of the renal tubule, in particular the loop of Henle. We can examine what happens to the primary urine as it passes along the various regions of the nephron (Figure 11.1d)—the proximal convoluted tubule (PCT), the loop of Henle, the distal convoluted tubule (DCT), the **collecting tubules** and the **collecting ducts**.

◇ What is the most likely way that the primary urine (which consists largely of water and Na^+ and Cl^- ions) might be reduced in volume as it passes along the nephron?

◆ Water must be reabsorbed. This could be achieved by active transport of ions out of the tubules coupled osmotically to an outflow of water. (An important point to bear in mind, and which will become apparent as events occurring in the nephron are described, is that active ion transport and the consequent outflow of water need not occur at the same location.)

The concentration of urine largely depends on the movement of ions (mainly Na^+, K^+, and Cl^-), water and (as has been discovered more recently) urea. Movement of Na^+, K^+ and Cl^- across tubule membranes may be passive or active; **urea movement** is passive. Water movement is coupled in part osmotically to ion and urea movement, and is always passive (the mechanism of this coupling is beyond the scope of the present discussion).

The discussion in the remainder of Section 3 is generalized to cover 'mammalian kidneys', although much of the information comes from studies on rabbit kidneys, largely because they have more accessible tubules (experimentally) than those of many other animals; for particular animals, exceptions may exist to some of our descriptions.

11.3.6 Urine modification in the proximal convoluted tubule

Around 77% of Na^+ absorption from the primary urine occurs in the proximal tubule (see Table 11.2), and this is coupled with a proportionate amount of Cl^-, HCO_3^- and water. In addition, amino acids, sugars, inorganic phosphates and sulphates, and some proteins are transported by the PCT, mainly by active transport processes. The tubular fluid that enters the loop of Henle is iso-osmotic to the plasma—about 300 milliosmoles per litre.* However, it has been reduced in volume during its passage along the PCT.

Electron microscope studies of the proximal tubule cell show that structurally it is highly specialized for a transporting role because it has a microvillus brush border, similar to that seen in the intestinal cell. Re-examine Figure 11.3 which shows a schematic representation of the ultrastructure of tubular epithelial cells from selected regions of a nephron.

The mechanism of transport of ions and solutes by the PCT cells is complex, involving passive and active movements at the luminal and basolateral membranes. Na^+ and Cl^- are the major ionic constituents of the glomerular filtrate, and possible routes of transport of these, and hydrogen ions, in the proximal tubule are shown in Figure 11.13. Na^+ can enter the proximal cell passively down its concentration and electrical gradients, and some Na^+ entry is coupled to the movement of hydrogen ions, and a smaller amount to glucose and amino acid entry. At the basolateral cell membrane sodium is actively transported against both electrical and chemical gradients by an enzyme, Na^+/K^+ ATPase. (In the terms normally used to describe membrane transport studies there is 'primary active transport' of sodium at the basolateral membrane, and this generates the transmembrane ion gradients at the luminal membrane of the proximal tubule cell, which permit the entry of ions and solutes.)

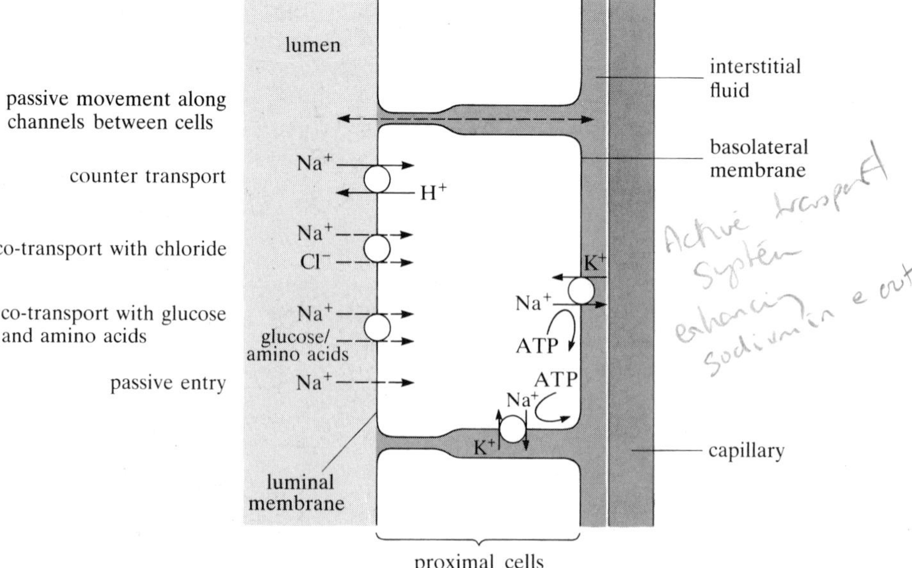

Figure 11.13 A schematic diagram showing possible routes and mechanisms of sodium transport into and out of cells of the proximal tubule. Coupled transport is indicated by the circles. There is an energy-requiring process (Na^+/K^+ ATPase) at the basolateral membrane, which actively transports Na^+ out of the cell. This is primarily responsible for establishing the electrical and concentration gradients that facilitate the passive movements of sodium across the luminal membrane.

*A milliosmole, abbreviated to mOsmol, is a thousandth of an **osmole**: this is a unit of osmotic concentration, related to molar concentration, which represents the total amount of the various osmotically active constituents dissolved in water to give a volume of one litre (see Chapter 12).

There is a difference along the length of the PCT in the capability for transport of NaCl. In the first part of the tubule lumen there is active transport, but in the last two-thirds there is a passive component to NaCl movement. Water moves across the tubular epithelium in response to hydrostatic or osmotic pressure gradients; the latter are more important in the kidney. The amount of water reabsorbed or secreted depends on the permeability of the tubule to water (which can be modified by hormonal action, as you will see, in the distal tubules). In the proximal tubule the permeability to water is relatively high, and water reabsorption appears to follow Na^+ and Cl^- reabsorption in this region passively. That water reabsorption occurs is demonstrated by the fact that the ratio of the concentration of inulin in tubular fluid to its concentration in blood plasma increases along the length of the proximal tubule.

◇ Why does this indicate that water reabsorption occurs in the proximal tubule?

◆ Because inulin cannot be transported or metabolized; an increase in inulin concentration can only result from removal of water from the tubular fluid.

The major events of water and ion and solute transport in the proximal tubule are summarized in the schematic diagram of the nephron in Figure 11.14.

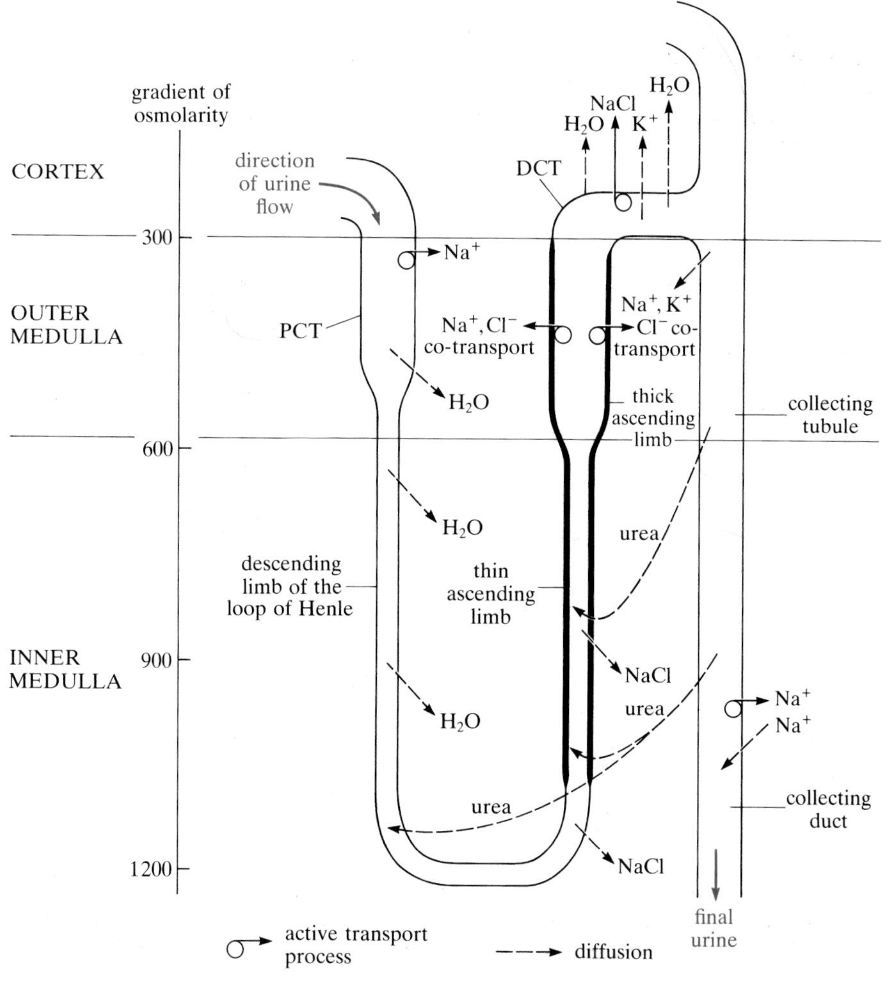

Figure 11.14 Schematic view of a mammalian nephron showing the proximal convoluted tubule (PCT), the limbs of the loop of Henle, the distal convoluted tubule (DCT), collecting tubule and collecting duct. The values on the left are the osmolarity of the interstitial tissue; note that there is an osmotic gradient, from 300 mOsmol l^{-1} in the cortex to 1 200 mOsmol l^{-1} in the inner medulla (the extent of the gradient varies between species). The fluid in the descending limb of the loop of Henle has about the same osmolarity as the surrounding tissue. Movements from the tubules of Na^+, K^+, Cl^-, urea and water occur as indicated. Relatively small amounts of salt moving from the ascending limb to the interstitial fluid will cause osmotic movement of water out of the descending limb. A small difference in the concentration between adjacent points on the two limbs leads to a large difference in concentration between the top and bottom of the loop when fluid flows around the loop. (This process, called counter-current multiplication, is explained in Figure 11.15 and the associated text.) Urea also contributes significantly to the medullary osmolarity, and there is recycling of urea from the collecting tubule through the medullary interstitium to the loop as shown.

11.3.7 Urine modification in the loop of Henle

The loop of Henle plays a crucial role in concentrating the urine. The effectiveness of the mammalian kidney is closely related to the length of the loop relative to cortical thickness, and in general, the longer the loop the greater is the ability of the kidney to concentrate the urine. Therefore, it is not surprising to find extremely long loops of Henle in desert rodents whereas in beavers, which always live close to water, the loops of Henle are relatively short.

The urine that leaves the PCT has a concentration of 300 mOsmol l^{-1}, and it passes from an iso-osmotic region to one that becomes increasingly hyper-osmotic to the body fluids. The permeability of the walls of the **descending limb** of the loop of Henle is very low to ions and urea (Table 11.3), but is very high to water, which passes out of the tubule into the more concentrated surrounding tissue, thereby concentrating the fluid in the tubule. By the time the fluid reaches the hairpin bend of the loop, its concentration can have reached 1 200 mOsmol l^{-1}, similar to that of the surrounding tissue. As fluid passes up the thin ascending limb of the loop, Na^+, K^+ and Cl^- move out of the tubule, most probably by passive means.

Table 11.3 The permeability and transport properties of parts of the mammalian nephron, indicated by a scale from zero to very high. Note also that the loop of Henle is divided into three regions, the descending limb, the thin ascending limb and the thick ascending limb. These are indicated in the highly schematic diagram of the nephron, Figure 11.14, to which you should refer as you read the account of events of urine concentration.

Segment	Active salt transport	Permeability H_2O	NaCl	Urea
Loop of Henle				
descending	nil	high	low	very low
ascending (thin)	nil	low	high	very low
ascending (thick)	high	low	low	low
Collecting ducts				
cortex/outer medulla	very low	high (if + ADH)	low	low
inner medulla	very low	high (if + ADH)	low	high

The **thin ascending limb** has a very low permeability to water (Table 11.3) so the urine becomes less concentrated by the passive loss of NaCl. The **thick ascending limb** is also impermeable to water, but there is a steep electro-chemical gradient for Na^+ between the tubule lumen and the interior of the tubule cells, which is generated by a basally located Na^+/K^+ ATPase pump, the primary active transport site. This gradient draws sodium (and potassium and chloride) ions to a transport molecule on the luminal membrane, and coupled transport (also termed co-transport) of Na^+, K^+ and Cl^- occurs across the thick ascending limb of the loop of Henle from the lumen into the tubule cells, with subsequent movement of the ions out at the basolateral membrane.

11.3.8 Urine modification in the distal convoluted tubule and the collecting tubule

The urine entering the distal convoluted tubule (DCT) is hypo-osmotic to the body fluids and it has been reduced in volume by the movements of water and ions that have occurred in the loop of Henle. In the DCT, there is co-transport of Na^+ and Cl^- at the luminal membrane, and Na^+ also enters the DCT cells by a counter transport mechanism with H^+. NaCl is actively transported out of the cell across the basolateral cell membrane by a Na^+/K^+ ATPase pump.

Bicarbonate ions are also reabsorbed (though not to the same extent as in the proximal tubules), and secretion of K^+ and $NH_4{}^+$ (and H^+ by the counter transport mechanism) occurs into the tubule lumen. Water may also leave the lumen of the DCT, although this is not linked so closely to ion movement as it is in the proximal tubule, but rather to the osmolarity of the surrounding interstitial tissue and the permeability of the DCT tubule wall, which can be modified hormonally. During its passage along the DCT, the urine normally increases gradually in osmolarity, and it is roughly iso-osmotic with blood plasma by the time it enters the final part of the nephron, the collecting tubule. In the cortical collecting tubules, more Na^+ is transported, accompanied by Cl^-, and K^+ is secreted into the tubule lumen. Again Na^+ is actively transported out of the cells by a basolateral Na^+/K^+ ATPase pump. The cortical collecting ducts have a low permeability to urea. Permeability to urea increases as the collecting ducts enter the medulla, and is very high in the inner medulla. (Ignore the details of urea movement in Figure 11.14—its significance will be explained later.)

The concentration of the urine excreted depends on what happens to the fluid as it passes along the collecting tubules and ducts; these are permeable to water, a permeability that can be modified hormonally depending on the degree of dehydration of the body fluids (see Chapter 4, and Section 11.3.11). Because the collecting ducts pass through the hyperosmotic inner medulla, water can be removed to produce a very hyperosmotic urine (provided that the duct wall is permeable).

Thus the key feature involved in the concentration of urine is the gradient of osmolarity from the cortex to the medulla. Without this, water could not be removed from the tubular urine in the loop of Henle or in the collecting ducts, nor would the diffusive ion movements occur from the thin ascending loop of Henle. Important questions that may be asked are: How does this gradient arise? What, if any, is the connection between this and the various ion movements in the loop of Henle and in the collecting tubules and ducts? Also, what is the significance of the finding that urea, as well as NaCl, contributes to the high osmolarity of the inner medulla? The answer to the first of these questions is that the generation of a large osmotic gradient is dependent on differences in solute transport and water permeability between the two limbs of the loop of Henle. These differences are affected by the close proximity of the descending and ascending limbs, which enables the loop to act as a **counter-current multiplier**.

11.3.9 Counter-current multiplication

The counter-current multiplier idea was first suggested in 1942, by Kuhn and Ryffel, as a means by which small differences in concentration of some substance between the two limbs of the loop could be magnified as fluid flowed around the system until a large gradient existed between the cortex and the medulla, with the highest concentration in the inner medulla. We can illustrate this best by considering a series of schematic drawings of the loop showing the descending and ascending parts of the loop, and introducing the characteristics of the phenomenon sequentially, though in reality the events occur simultaneously.

Figure 11.15a shows fluid at 300 mOsmol l^{-1} (iso-osmotic with plasma) in both limbs of the loop, and in the tissue around the loop, the interstitium. Flow through the loop has been stopped. The two limbs of the loop have different permeabilities to both water and solutes. The ascending limb can transfer solute to the interstitial fluid, thereby lowering the osmolarity of this limb to 200 mOsmol l^{-1} and raising that of the interstitium to 400 mOsmol l^{-1} (Figure 11.15b). The descending limb has a high permeability to water and thus water moves down its osmotic gradient from the descending limb into the interstitium, and equilibrium occurs between these two compartments (Figure 11.15c). If fluid flows through the loop (Figure 11.15d), and iso-osmotic fluid (300 mOsmol l^{-1}) enters the top of the descending limb, hyperosmotic fluid (350 mOsmol l^{-1}) is pushed round to the bottom of the ascending limb and hypo-osmotic fluid (200 mOsmol l^{-1}) moves out of the ascending limb. Osmotic equilibration between descending limb and the interstitium now begins to produce an osmotic gradient within the interstitium (Figure 11.15e)—300 mOsmol l^{-1} at the top (the cortex) and 400 mOsmol l^{-1} at the bottom (the inner medulla). If these processes are repeated, the situation depicted in Figure 11.15f is eventually reached, with a substantial gradient from cortex to medulla of 300 to 1 200 mOsmol l^{-1}.

This is the basis of the process of counter-current multiplication that *actually* occurs in the loop of Henle, with the changes in osmolarity occurring because of water and solute movements (mainly Na^+, Cl^- and urea), out of, or into, various parts of the loop of Henle, and also from the collecting duct. The key factor in this process is the co-transport of sodium, potassium and chloride from the thick ascending limb of the loop of Henle. The Na^+ and Cl^- transported from the ascending limb raise the osmolarity of the medullary

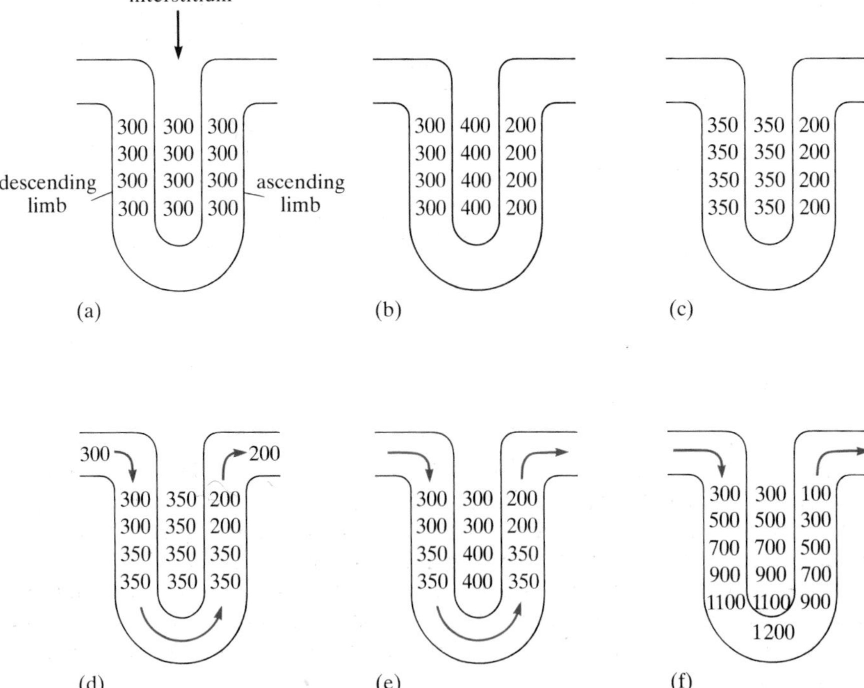

Figure 11.15 Schematic representation of six stages in the process of counter-current multiplication in the loop of Henle. In reality the process occurs continuously rather than in separate stages.

interstitium (tissue), fluid in the descending limb equilibrates osmotically with the increased osmolarity in the medulla, and a maximum horizontal gradient of 200 mOsmol l^{-1} is established between the two limbs.

So, iso-osmotic fluid from the proximal tubule enters the descending limb of the loop and becomes progressively concentrated as it flows down the descending limb (owing to water moving out) and progressively diluted as it flows up the ascending limb (due to solutes moving out). A large longitudinal gradient of osmolarity is established in the medulla, as shown in Figure 11.14. Counter-current multiplication in the loop does not actually concentrate the urine—in fact, the tubular fluid emerging from the ascending limb is hypo-osmotic to plasma (cf. Figure 11.15f). Rather, it *creates* conditions in the medulla under which a concentrated urine can be formed. The site of final concentration in the urine lies, as already mentioned, in the collecting duct. Here, tubular fluid passes through an increasingly hyperosmotic region, and the reabsorption of water, under hormonal control (see later), produces a hyperosmotic urine.

The magnitude of the longitudinal gradient in the medulla depends on:

(a) The magnitude of the transverse (horizontal) gradient (i.e. the amount of co-transport of Na^+, K^+ and Cl^- in the ascending limb);

(b) The length of the loop of Henle (the longer it is, the greater will be the gradient), and the ratio of long to short loops; in desert rodents, the kangaroo rat for example, where there are very long loops and a high ratio of long to short loops, a very concentrated urine can be generated—up to 6 000 mOsmol l^{-1}. Human loops are generally not very long and can concentrate only up to 1 200–1 400 mOsmol l^{-1}.

In addition, it has been recognized for a long time that a high protein diet increases the ability of the kidney to concentrate urine. The relevance of protein is that the major end-product of its metabolism is urea (see Section 11.5), and this, in addition to the solutes already described, plays an important role in the counter-current system. The walls of the distal tubule and early (cortical) collecting tubules have a low permeability to urea. Hence, as water is reabsorbed in these segments, the urea is progressively concentrated. In the inner medulla (Figure 11.14), urea diffuses down its concentration gradient from the collecting ducts into the surrounding interstitium, with a subsequent raising of the osmolarity of the inner medulla and thus an increase in the ability of the kidney to produce a hyperosmotic urine. Urea reabsorbed from the collecting duct can constitute up to 50% of the total solute concentration in the medulla. Although urea diffuses into the descending and ascending limbs of Henle's loop, it is recirculated to the collecting tubule and (depending on the permeability of the tubule walls), passes back into the interstitium.

11.3.10 The vasa recta

Of crucial importance in the formation of a concentrated urine is removal of the water reabsorbed from the descending limb of the loop to the medullary interstitium. If this did not occur, the medullary osmotic gradient would not exist. Also, NaCl cannot accumulate in the inner medullary tissue without limit. As you saw in Figure 11.4, the efferent blood vessels do not immediately join up to form the renal vein but form a network of capillaries around the tubules of the nephron. Most of them form the peritubular capillaries around the cortical nephrons, and NaCl entering these capillaries is carried away to the blood system.

◇ What purpose would this serve, apart from returning NaCl to the rest of the body fluids?

◆ It helps to maintain the cortico-medullary gradient by lowering the NaCl concentration in the outer medulla.

Some of the capillaries (about 2%) also form a network of capillaries called the vasa recta around the juxtamedullary nephrons, and these supply blood to the medulla, running parallel to the loop of Henle and the collecting ducts (Figure 11.3). Exactly how the vasa recta keep solutes in the medulla but dispose of the excess water extracted from the descending limb of the loop and the collecting tubules is not fully understood. It appears that, by being arranged in loops, they operate as counter-current diffusion exchangers. As blood flows down the descending limb of the vasa recta, NaCl enters by passive diffusion and water leaves by osmosis. In the ascending limbs NaCl leaves and water enters. Thus NaCl will flow out of the ascending limb of the vasa recta and into the descending limb (via the medullary interstitium). Water will flow osmotically in the opposite direction, out of the descending and into the ascending limb (Figure 11.16). Thus diffusible solutes such as urea and NaCl will be trapped in the medulla whereas excess water will be removed. For maximal concentration of the urine the blood flow through the vasa recta must be minimal, but when the flow in the vasa recta is increased solutes are washed out of the medulla and medullary osmolarity is decreased, as is the concentrating ability.

◇ What would be the consequence of the vasa recta blood vessels leaving the kidney in the medullary region rather than in the cortex (i.e. if there was no ascending limb in the vasa recta)?

◆ The osmotic gradient in the medulla would be degraded, because water drawn out of the loop of Henle would not be removed, and water would also be drawn out of the blood and into the interstitial fluid by osmosis, thereby 'washing out' the solute gradients.

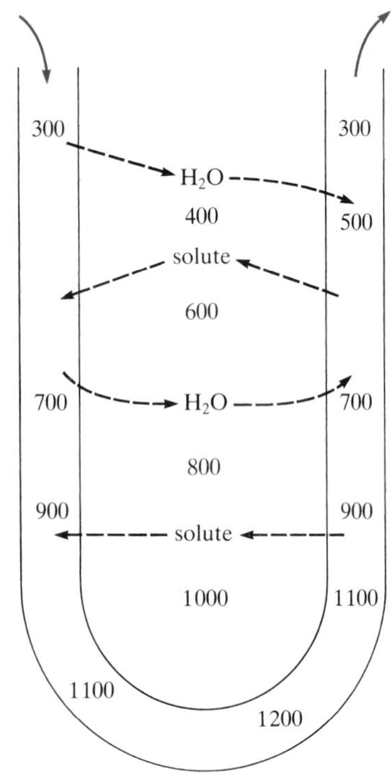

Figure 11.16 Solute and water movement in the counter-current exchange system of the vasa recta. Water effectively short-circuits the loop whereas solutes are recycled and hence retained. Numerical values are in mOsmol l^{-1}. Red lines indicate the direction of fluid flow.

The final concentration of the urine ultimately depends on the permeability of the collecting ducts, which is hormonally controlled; we examine how this control is exerted next.

11.3.11 The regulation of extracellular fluid volume and electrolyte composition: the control of kidney function

Animals can experience osmotic stress because of changes in temperature or salinity, and the ingestion of food and fluids of low, or high, salinities relative to normal plasma concentrations. Haemorrhage (loss of blood) ought not directly to affect the osmotic state of the body fluids because water and salts should be lost together. Reduction of fluid volume will cause stimulation of volume receptors which are located centrally (in the walls of the heart and the large veins that return blood to it). Peripherally the kidney will also be involved because cells among the juxtaglomerular apparatus in the kidney, the macula densa (Figure 11.2), also act in response to the fall in blood volume, although they are not strictly volume receptors. These cells monitor NaCl in the distal tubule and when haemorrhage occurs blood pressure falls, thus there is less glomerular filtration and hence less NaCl in the distal tubule. The juxtaglomerular cells also monitor the drop in blood pressure in the afferent arteriole.

In humans, the extracellular fluid volume is approximately 12 litres in a 70 kg person, and its main cationic component is sodium. The mechanisms for regulation of extracellular fluid and osmolar balance are powerful. If there is retention of sodium, water is automatically retained, thus maintaining the osmolarity of the extracellular fluid, whereas when sodium is lost, water is also lost. Therefore, maintenance of extracellular fluid volume is linked very closely to total amount of sodium in the body. Upsets in the osmotic state of the body fluids are minimized through feedback mechanisms by which the osmoregulatory organs adjust their activity to keep constant the composition of the body fluids. It can be shown that an increase of 1% in the osmotic pressure of plasma in a mammal may result in a reduction of urine flow of up to 90%. Much of this reduction is brought about by increasing the permeability of the collecting tubules, effected primarily by a polypeptide hormone secreted into the bloodstream by the posterior lobe of the pituitary gland (details of this were first introduced in Chapter 4).

◇ This hormone originates in the hypothalamus but is actually secreted from the pituitary. What is the name for this process of hormone secretion?

◆ Neurosecretion (Chapters 2 and 3).

Precursors of the hormone are secreted by nerves in the hypothalamus and travel down the nerve axons into the posterior lobe. This hormone is known as **antidiuretic hormone** (ADH) or **vasopressin**. The feedback mechanism by which it is released was first shown by Verney in 1947. He was trying to locate the receptors responsible for causing **antidiuresis** (reduction in urine flow) by perfusing hyperosmotic saline solution into experimental animals. He found that a small amount of hyperosmotic saline solution introduced into the carotid artery would induce antidiuresis—even though the general osmotic pressure in the bloodstream was not disturbed. The same amount of saline injected elsewhere did not have this effect. Thus it appears that there were **osmoreceptors** in the brain, and it is now known that they are located in the hypothalamus. These osmoreceptors shrink or expand, respectively, in response to decreases or increases in blood osmolarity.

◇ Is the feedback system that controls the blood osmotic pressure positive or negative?

◆ Negative.

If blood osmotic concentration rises, ADH release is stimulated and the urine flow is reduced—in humans to as little as 0.1–0.3 $cm^3\ min^{-1}$ (derived from an original filtration rate of 125 $cm^3\ min^{-1}$). How much the osmotic pressure of the urine may be raised depends on the length of the loop of Henle. Some desert rats with very long loops can live on metabolic water (that derived from the oxidation of foodstuffs), provided the diet is a low-protein one.

There are, of course, occasions when the body is overloaded with water and short of salts. In this case ADH secretion stops (there is no stimulation of the osmoreceptors because the plasma osmolarity is low), and maximum recovery of Na^+ occurs from the distal tubule, stimulated by the secretion of the hormone **aldosterone** from the adrenal cortex. This acts partly by increasing the activity of the Na^+/K^+ ATPase pump at the basolateral membrane in the distal tubule. If aldosterone secretion is accompanied by ADH secretion, water follows the sodium iso-osmotically out of the tubule and into the blood. If ADH is absent, the distal tubule seems to be fairly waterproof; Na^+ is extracted from the tubule and the water is left behind, giving a large flow of dilute filtrate into the collecting ducts—**diuresis**; this flow then passes down the collecting duct, and hence to the bladder, relatively unchanged.

Four factors influence aldosterone release. The first is a decrease in blood pressure in the afferent arteriole (e.g. as happens following haemorrhage). When blood pressure in the afferent arteriole falls, a group of cells of the juxtaglomerular apparatus, the juxtaglomerular cells, situated around the afferent glomerular arterioles (see Figures 11.2 and 11.17), are stimulated to release an enzyme called **renin**. This is released into the plasma and acts on a circulating plasma protein, angiotensinogen (produced in the liver), to convert it into a decapeptide **angiotensin I**, which is then converted into a highly active form, an octapeptide called **angiotensin II**. Its effect is to stimulate the synthesis and release of aldosterone from the adrenal cortex and thus Na^+ retention by the body. (Angiotensin II has a specific vasoconstrictor action on the glomerular arterioles thereby altering the glomerular filtration rate.) Secondly, aldosterone release is stimulated by a change in the proportion of Na^+ relative to K^+ in the plasma circulating through the adrenal gland. Thus as the concentration of Na^+ falls relative to that of K^+, or if the concentration of K^+ rises, aldosterone is released, and the Na^+ pumps in the distal tubule exchange Na^+ in the filtrate for K^+ in the plasma. Thirdly, stimulation of the renal sympathetic nerve causes increased renin secretion. Fourthly, decreased sodium concentration in the loop of Henle or the early part of the distal tubule is detected by the macula densa cells in the tubule wall. These, too, cause renin release from juxtaglomerular cells, which in turn leads to aldosterone release from the adrenal cortex and thus to increased sodium reabsorption.

Although the time taken for aldosterone to exert its effect is relatively long, this substance is important if the body-fluid volume is to be maintained.

◇ Why?

◆ A higher NaCl content in the plasma will mean that more fluid is retained.

Thus there is a system whereby the kidney acts as a homeostat on blood pressure and maintains its own GFR, in addition to the effect on salt balance.

Control of water and salt secretion by the mammalian kidney cannot be fully

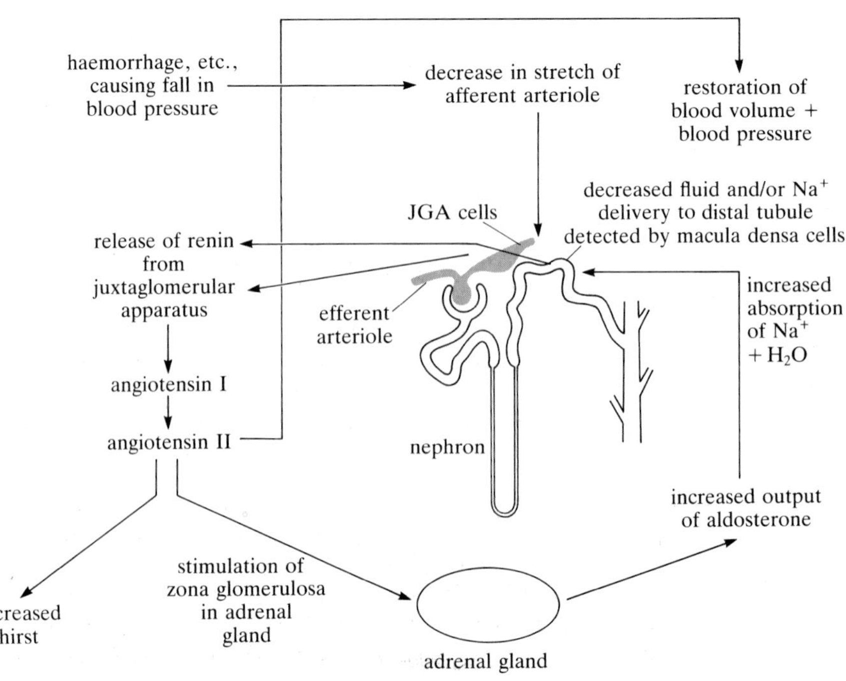

Figure 11.17 Regulation of renal function and control of salt and water excretion in a mammal by the JGA cells and the renin–angiotensin–aldosterone system.

explained on the basis of the action of ADH and aldosterone. Other factors exist that can alter sodium excretion, called **natriuretic hormones** (natriuresis means sodium excretion). These are not as yet satisfactorily identified, but may include substances such as prostaglandins. One natriuretic hormone has been postulated to promote sodium and water excretion. It is not clear what causes its release, although cerebral and renal receptors sensitive to raised levels of NaCl have been suggested to play a role. It is postulated that this hormone may inhibit the Na^+/K^+ ATPase in the tubules of the kidney, and perhaps in all cells, thereby reducing the reabsorption of Na^+ in the kidney tubules. One action of natriuretic hormone that does seem to be somewhat clearer is that of **atrial natriuretic hormone**. This, as its name suggests, is released from the walls of the atria. It is a peptide or family of peptides in which the central effect is to cause increased sodium excretion by increasing the GFR and blood flow in the inner medulla.

NATRIURETIC HORMONES
Hormones believed to promote the excretion of Na^+ ions and water; it is not clear what causes the release of the natriuretic hormone.

ATRIAL NATRIURETIC HORMONE
Hormone that is released from the walls of the atria and causes sodium excretion (natriuresis) primarily by increasing the glomerular filtration rate & possibly the inner medullary blood flow.

The kidney is not the only factor that needs to be considered in the regulation of extracellular fluid volume. The most obvious symptom of water shortage is **thirst**, which in humans can be defined as a 'conscious desire to drink'. This condition may arise after fluid loss through exercise, after eating food with a high salt content, following haemorrhage, or simply after the passage of a long time without intake of water. It is not, however, simply that the mouth is dry; it appears that there are 'thirst receptors' in a region in the hypothalamus (close to the osmoreceptors that cause ADH secretion). These thirst receptors are sensitive to the intracellular fluid volume, which is of course affected by the extracellular fluid volume. Stimulation of these receptors by, for example, intravenous hyperosmotic saline injections, induces thirst and causes fluid intake, although by what mechanism is not completely clear. Thirst may also be mediated by release of renin and angiotensin II. Renin is able to cross the blood–brain barrier and may be able to produce angiotensin II in the brain. Angiotensin is thought to act on extracellular thirst receptors (blood volume receptors) in the surface of blood vessels in the brain and cause intense thirst.

THIRST
In humans, thirst can be defined as a 'conscious desire to drink' and may arise after fluid loss eg. through exercise or haemorrhage.

11.3.12 The role of the kidney in the regulation of acid–base balance

The mammalian body produces acidic waste products daily as a result of processes such as breakdown of protein. Almost instantaneously, intracellular and extracellular buffering systems come into action. There is also a rapid excretion of carbon dioxide by the lungs, which raises the plasma pH towards normal. This is quantitatively the most important way in which acidic waste products are eliminated. The kidney represents an additional line of defence against acidosis (over-acidity of the blood) by increasing the secretion of hydrogen ions into the urine and enhancing the reabsorption of bicarbonate. If the kidney is unable to deal with these excess acids, as happens in chronic renal failure, then death will result. The mechanism of hydrogen ion excretion by the kidneys is beyond the scope of this chapter, but the importance of the kidneys in this regulatory role should be borne in mind.

Summary of Section 11.3

1 Vertebrate kidneys, including those of mammals, are paired structures that consist of functional units called nephrons, which consist of tubules with several distinct regions. At one end, each nephron expands to surround a 'tuft' of blood vessels, the glomerulus.

2 Primary urine (i.e. the glomerular filtrate in vertebrates) is an ultrafiltrate formed from blood as it passes through the glomerular capillaries. A sieving

mechanism based on size determines which substances pass from the glomerulus to Bowman's capsule, and the rate of filtration is determined by the net pressure difference across the barrier between blood in the glomerular capillaries and fluid in Bowman's capsule.

3 The glomerular filtrate is modified as it passes down the tubules of the nephron by transport out of (reabsorption), and transport into (secretion), the tubule lumen. Techniques used in mammals for following events in these tubules include stopped-flow microperfusion, micropuncture with ion-selective electrodes, cannulation of tubules, and electrical potential measurements between the tubule lumen and the bathing medium, and within tubule cells.

4 About 77% of Na^+ transport out of the glomerular filtrate occurs in the proximal convoluted tubule, followed by a proportionate amount of Cl^- and water. Much transport of amino acids and sugars out of the glomerular filtrate also occurs here.

5 Concentration of urine in mammals and birds results from counter-current multiplication in the loop of Henle. This process does not directly concentrate urine but creates the conditions in the medulla under which a concentrated urine can be formed. Counter-current multiplication is the means whereby the small osmotic concentration differences between the descending and ascending limbs of the loop of Henle, caused by the differential permeabilities to ions, water and urea between the limbs, become magnified into a large osmotic gradient in the medulla. In mammals the urine may be 25 times more concentrated than the plasma, whereas for birds the maximum urine/plasma ratio is about 2. Urea recycling plays an important role in the counter-current process, and urea can contribute up to 50% of the total osmolarity of the inner medulla.

6 Hormones determine whether the urine excreted is dilute or concentrated: if the body is short of water and plasma osmolarity increases, the concentration of NaCl in the blood rises and a negative feedback system causes the release of antidiuretic hormone (ADH). This increases the permeability of the walls of the collecting ducts to water, which is reabsorbed, and in addition the animal is stimulated to drink. If the body is short of salt, ADH secretion ceases and the reabsorption of salt is stimulated by a hormone, aldosterone, released from the adrenal cortex. If blood pressure falls, the glomerular filtration rate also falls, and this stimulates the release of renin from the juxtaglomerular apparatus. Renin eventually causes the production of a hormone, angiotensin II, that acts to increase the filtration rate, and also stimulates the release of aldosterone. There are other factors (natriuretic hormones) that can alter sodium excretion, of which atrial natriuretic hormone is perhaps the best known.

7 Thirst is an important factor in regulation of intracellular and extracellular fluid volume. In humans this can be defined as a 'conscious desire to drink', and may arise through eating food high in salt content, after fluid loss through exercise, following haemorrhage, or simply, after the passage of time without water intake. But it is not simply that the mouth is dry, it appears that there are thirst receptors in a region in the hypothalamus and stimulation of these by intravenous injections of hyperosmotic saline induces thirst and causes fluid intake. Thirst may also be mediated by release of the compounds renin and angiotensin II.

8 The kidneys play an important role in the prevention of acidosis by elimination of hydrogen ions; failure to remove these, as happens in renal failure, results rapidly in death.

Now attempt Questions 2–9, on pp. 394–395.

11.4 INSECT MALPIGHIAN TUBULES

The anatomy and physiology of insects fit them well to a terrestrial existence. They have a impervious exoskeleton, and their main nitrogenous excretory product is uric acid, which is almost insoluble in water. Insects can live in a great variety of environments, including very dry conditions where water must be conserved. Some, e.g. blood-sucking flies, absorb a great deal of water from their food and must get rid of the excess if the salt concentration in their haemolymph (the main body fluid of insects) is to remain constant. Osmoregulatory and excretory functions in insects are carried out by organs called Malpighian tubules in combination with the activities of the hind-gut.

The number of Malpighian tubules in insects varies, from two to several hundred, but their structure is basically similar. They are blind-ended ducts that arise from the gut at the point where the mid-gut becomes the hind-gut (Figure 11.18), and they are bathed in haemolymph, which circulates in the body cavity. Individual Malpighian tubules can be isolated as in Figure 11.19, a technique first developed by Ramsay in the 1950s; much of the information on how Malpighian tubules function has been gained from such preparations.

In insects, primary urine is formed largely by secretion into the tubule lumen. Fluid next passes into the hind-gut and then into the rectum. This contrasts with the mammalian kidney, in which primary urine is formed by ultrafiltra-

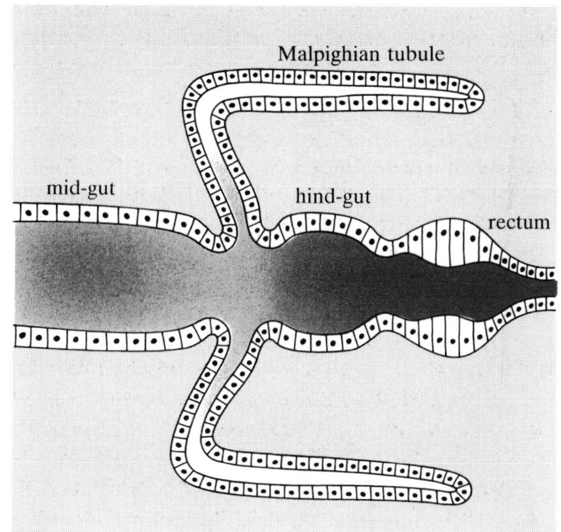

Figure 11.18 The insect excretory system. The primary urine is produced by secretion into the lumen of the Malpighian tubules and flows into the hind-gut and then into the rectum, where it is concentrated by the extraction of water. The volume of urine in the rectum decreases, showing that reabsorption of water occurs rather than secretion of solutes.

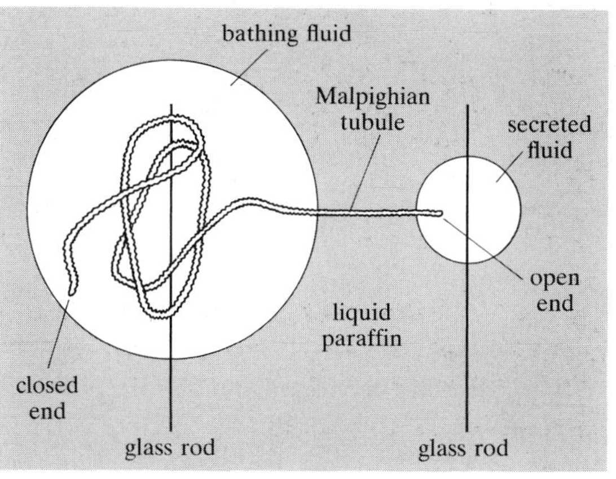

Figure 11.19 An experiment with a Malpighian tubule. The tubule is isolated and submerged in liquid paraffin, but the closed end of the tubule is bathed with a drop of fluid of known composition (usually similar to insect blood). The open end of the tubule (previously connected to the gut of the insect) is in the liquid paraffin. The fluid secreted by the tubule emerges from this end and can be collected for analysis.

Diuretic hormones (in insects) A hormone that has the opposite effect to insect ADH.

tion. In insects, the pressure of the haemolymph surrounding the Malpighian tubule is insufficient for more than a small amount of ultrafiltration to occur. In the blowfly *Calliphora*, inulin (recall from Section 11.3.3 that inulin is an inert carbohydrate molecule of M_r 5 200) appears in the fluid secreted by isolated tubules at about 4% of its concentration in the bathing solution.

◇ Since inulin is not transported by tubule cells, how can it enter the tubule lumen?

◆ Diffusion across the tubule wall would be an obvious way.

This means that the tubule wall must be relatively permeable. Compounds of M_r 400 or less freely cross the tubule wall, while compounds with large molecules, inulin for example, pass more slowly; compounds of M_r above 10 000 appear to be too large to diffuse across the tubule wall. Such permeability is important because it provides for the automatic removal of many toxic substances from the haemolymph. This is an important feature of a renal organ such as the mammalian kidney, where ultrafiltration forces out from blood passing through the glomerulus all molecules below a certain size and so automatically removes many toxic materials. The permeability of insect Malpighian tubules also enables this to happen, but to a much smaller extent; toxic compounds of M_r 5 000–10 000 would be cleared rapidly by glomeruli but only slowly by Malpighian tubules.

◇ There is another consequence of the permeability of Malpighian tubules. What is this?

◆ Apart from salts and water, essential molecules such as amino acids and sugars will pass into the tubule lumen, and these must be reabsorbed, either in a later part of the tubule, or in the hind-gut. This also happens in vertebrate nephrons.

Potassium ions appear to play a major role in urine formation in the Malpighian tubules of most insects, because if they are absent from the bathing medium around isolated tubules, the rate of fluid secretion is less than 10% of maximum. There is a positive potential across the wall of the Malpighian tubule, and this means that the movement of K^+ into the tubule lumen is against an electrochemical gradient. It seems probable that there is a K^+ pump on the luminal membrane of tubule cells. So in most insects, active transport of K^+ into the lumen of the Malpighian tubule appears to be the major driving force for primary urine formation. Most other substances follow active K^+ transport passively, driven by osmotic forces; fluid secretion in a typical Malpighian tubule cell is shown in Figure 11.20. Blood-sucking insects, e.g. *Rhodnius*, seem to be an exception in that Na^+, not K^+ is the main ion transported.

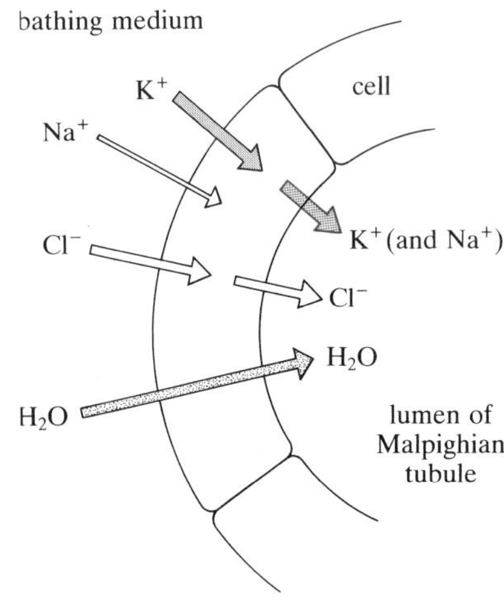

Figure 11.20 A possible mechanism for the secretion of fluid by the cells of a typical Malpighian tubule. Black arrows indicate passive processes, and the pink arrow the active transport of cations.

In the hind-gut and especially the rectum, some solutes and much of the water are reabsorbed, and nitrogenous waste, which entered the primary urine as water-soluble potassium urate, is precipitated as uric acid. How materials are transported out across the wall of the rectum is unclear, although a number of theories have been proposed. There is considerable support for the idea that active transport of solutes occurs from the lumen across the rectal wall, with water following osmotically. It has been suggested that hormones produced by neurosecretory cells in the insect brain control excretion, and thus the insect may be able to excrete a dry, or a concentrated, faeces, depending on its state of dehydration. **Diuretic hormones** and **antidiuretic hormones** (ADH) have been identified as being synthesized in neurosecretory cells of several insects. These are released into the haemolymph and act to control the activity of the Malpighian tubules and the rectum. Isolated tubules of *Rhodnius* (bug) and *Glossina* (tsetse fly) respond to their respective diuretic hormones by a thousand-fold increase in fluid secretion. This is particularly useful after a large blood meal in *Rhodnius* where the Malpighian tubules are essentially a sort of bypass system, allowing excess liquid from the meal to be eliminated very quickly.

NITROGENOUS WASTE-PRODUCTS
Compounds derived from the deamination of amino acids during metabolism.

Less is known of the action of insect ADH, but it has been postulated that in the cockroach *Periplaneta*, a single ADH hormone affects both the Malpighian tubules and rectum by reducing the passive permeability of the cells to water. It would therefore restrict water movement generated by ion transport across the Malpighian tubules, and would reduce the amount of water that leaked back across the rectal cells after it had been reabsorbed from the faeces.

◇ If insect ADH does act in this way, is it similar to mammalian ADH (vasopressin)?

◆ Yes, in that it acts to prevent water loss—antidiuresis. Mammalian ADH does this by increasing the permeability of the collecting tubules to water. (Note, however, that insect ADH is not identical chemically with vertebrate ADH.)

Summary of Section 11.4

1 In insects, osmoregulatory and excretory functions are carried out by the Malpighian tubules in combination with the rectum and hind-gut.

2 Primary urine is formed largely by secretion, but a small amount of filtration can occur from the haemolymph into the tubule lumen; compounds of M_r greater than 10 000 cannot pass through the tubule wall.

3 Active transport of K^+ into the lumen of the Malpighian tubule appears to be the major driving force for primary urine formation; most other substances follow passively. Most reabsorption of solutes and water occurs in the hind-gut.

4 There is evidence of hormonal control of excretion in insects. Antidiuretic and diuretic hormones have been identified in several insects, and these act to control the activity of the Malpighian tubules and rectum.

Now attempt Question 10 on p. 395.

11.5 NITROGEN EXCRETION

Some of the most important excretory materials are **nitrogenous waste products** resulting from the deamination of amino acids, e.g. urea (in mammals) and uric acid (in insects). Here we consider the production and properties of compounds that are most widely found as excretory products.

Most of the food eaten by animals consists of carbohydrate, fats, proteins and nucleic acids. Carbohydrates and fats are metabolized to carbon dioxide and water, which are easily removed from the body (e.g. in expired air). Nucleic acids and more importantly protein give rise to carbon dioxide, water and also nitrogen-containing products, the three most common being ammonia, urea and uric acid (Figure 11.21). We now consider the breakdown of amino acids to see how these products are formed.

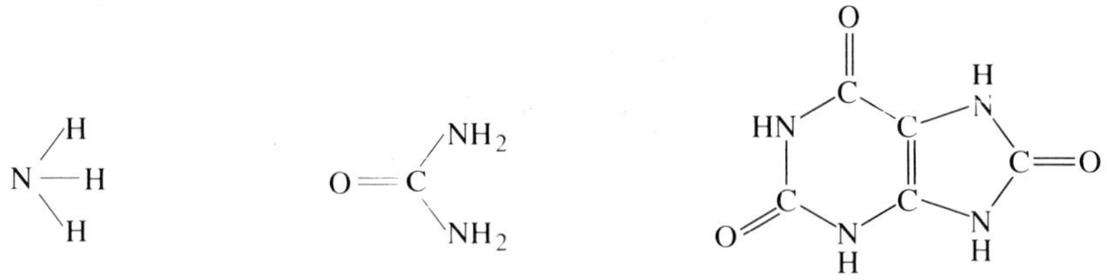

Figure 11.21 Structural formulae of three nitrogen-containing compounds.

ammonia, NH_3 urea, $CO(NH_2)_2$ uric acid, $C_5H_4O_3N_4$

When an amino acid is deaminated, the terminal NH_2 group is removed, producing a carboxylic acid and ammonia. Ammonia is very soluble but also highly toxic: a level of 1 part per 20 000 in the blood of rabbits causes immediate death, so ammonia must be disposed of rapidly as it is formed, before the level in the body fluids rises too much.

◇ How could an aquatic animal with a large permeable area in contact with the water lose ammonia to the outside?

◆ By simple diffusion, because under normal circumstances, the level of ammonia in water is negligible. Hence, the concentration gradient of ammonia will be steep even when the internal level is quite low.

Thus, all aquatic animals that 'breathe' dissolved oxygen, either through gills or the body wall (in unicells, the cell membrane), are able to dispose of much of their nitrogenous waste as ammonia, allowing it to diffuse out into the surrounding medium. In many cases, almost all the waste nitrogen is lost this way; in others, it may be about 90%. In teleost fish, some 90% diffuses out through the gills, not all as ammonia. Animals that excrete ammonia as the main end-product of amino-nitrogen metabolism are said to be **ammonotelic**.

However, in animals where water is not available in unlimited amounts and in close association with the blood, this system will not work. Elimination of nitrogenous waste is thus a major problem associated with life out of water. For example, air-breathing animals cannot generally dispose of ammonia by simple diffusion; the concentration of dissolved ammonia in the blood would have to be enormous before it began to come out of solution into the air in any quantity—the animal would be dead long before this happened. (Some woodlice appear to be an exception to this generalization!) Animals unable to get rid of ammonia by diffusion have enzyme systems that combine it with other substances to form less harmful compounds. Even in cases where water is available this may happen—for example, many fish, particulary marine teleosts, may manufacture a substance called **trimethylamine oxide**, which is largely responsible for the 'fishy' smell of fish. For quite different reasons (see Chapter 12), the cartilaginous (elasmobranch) fish combine their ammonia with carbon dioxide to produce urea.

Many vertebrates produce urea, and those that do so possess the enzyme arginase. The urea is produced by the **urea cycle** (Figure 11.22), discovered by Krebs in 1932. Animals that excrete urea are said to be **ureotelic**. In the urea cycle, NH_3 and CO_2 are condensed with phosphate to form carbamyl phosphate, which enters a synthetic pathway to form the amino acid citrulline. A second NH_3 is added from aspartic acid to form the amino acid arginine. In the presence of the enzyme arginase, arginine is decomposed into urea and ornithine, and the ornithine is then available for renewed synthesis of citrulline, and so the cycle continues.

AMMONOTELIC
Adjective applied to animals that excrete ammonia as the main end-product of amino-nitrogen metabolism. Ammonia is highly soluble but also highly toxic, so it must be disposed of rapidly as it is formed. Ammonia levels of 1 part per 20,000 in blood of rabbits cause immediate death.

TRIMETHYLAMINE OXIDE
Substance largely responsible for 'fishy' smell of fish. Found in many marine organisms & may represent a means by which ammonia can be made harmless. Also appears to prevent the destabilizing effects of urea on proteins & enables urea to build up to high concentrations in blood of fishes such as elasmo branches.

UREA CYCLE
Discovered by Krebs in 1932; many vertebrates produce urea as an end product of amino nitrogen metabolism; they have the enzyme arginase.

UREOTELIC
Animals that excrete urea as the principal nitrogenous waste product.

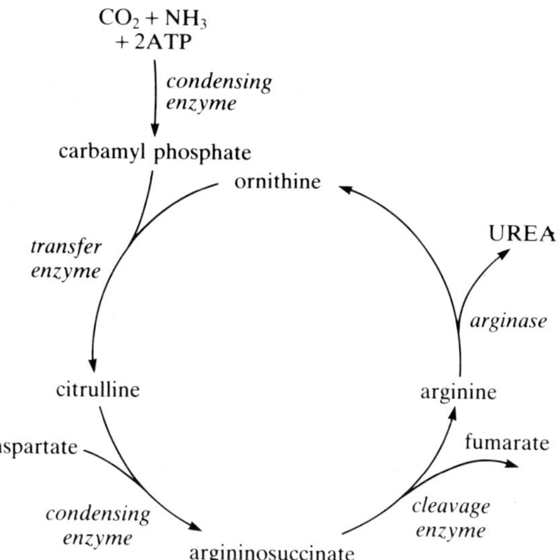

Figure 11.22 The urea cycle. Urea is synthesized from ammonia and carbon dioxide by condensation with the amino acid ornithine. Through several steps arginine is formed. This is acted on by the enzyme arginase to produce urea and ornithine: the latter re-enters the cycle.

Urea is relatively non-toxic and very soluble. It is therefore very suitable for excretion by kidneys, provided that a reasonable amount of water is available. Adult amphibians produce urea, although the eggs and young aquatic larvae produce ammonia. Figure 11.23 shows how the percentage of nitrogenous excretion in the form of ammonia or urea varies at different stages in the metamorphosis of tadpoles of *Bufo bufo* into toads. Interestingly, variations in the amount and the type of nitrogenous excretory product formed are shown within a particular group of adult animals, the lungfish, which are related to the ancestors of terrestrial vertebrates. The African lungfish *Protopterus aethiopicus* excrete 65% of their nitrogenous waste as ammonia across the gills, and the remainder is converted into urea before being excreted via the kidneys. However, in the dry season lungfish become trapped in shallow pools that dry out and when this happens the fish burrows into the mud, and secretes a cocoon of mucus around itself, within which it aestivates. Urine flow rapidly declines and eventually falls almost to zero and urea is the only nitrogenous waste product formed. It accumulates in the body reaching high levels, and is dispersed eventually (via the gills and kidneys) when the animal becomes active again in water. Urea, as we see below, can

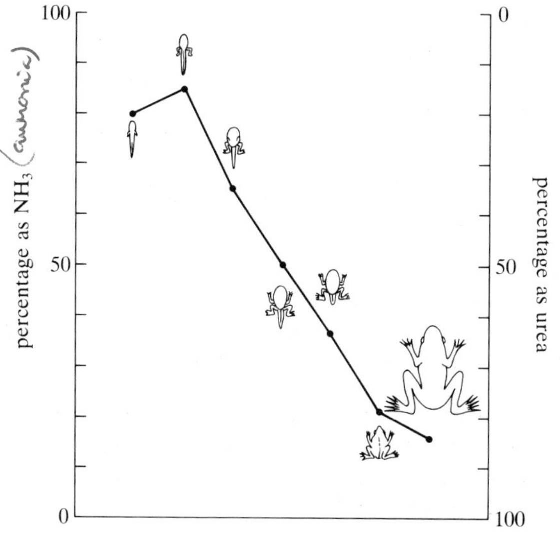

Figure 11.23 The percentage of nitrogenous excretion in the form of ammonia and urea, measured at different stages in the metamorphosis of tadpoles of *Bufo bufo* into toads.

be very harmful but fortunately the lungfish appears to be very tolerant to high levels of this substance in its blood.

Animals with shelled eggs that develop on land have special problems with regard to nitrogenous excretion. Ammonia is not suitable as the major excretory product because, as in the aestivating lungfish, the eggs are not surrounded by water. Urea is very soluble, but it cannot escape, so it would accumulate in the egg and exert a high osmotic pressure, which could damage the embryo. Thus a different excretory product is needed.

◇ What characteristics would be desirable in the excretory product of an embryo in a shelled egg?

◆ It should be non-toxic to developing tissues, and it should not exert a high osmotic pressure when its concentration in the egg rises.

The substance produced by most animals that lay shelled eggs on land is uric acid, which has both the properties suggested; such animals are said to be **uricotelic**. Uric acid does not exert a high osmotic pressure because it is rather insoluble and becomes biologically inert. The hatching embryo can leave a sludge or crystals of uric acid behind it in the shell.

Uric acid is the main excretory product of insects, terrestrial gastropod molluscs, reptiles (except some aquatic ones) and birds. The withdrawal of water from the urine causes uric acid and its salts to precipitate as a white semi-solid sludge. In some insects, uric acid is not excreted but is deposited in various parts of the body, mainly the fat bodies, and thus no water at all is required for the elimination of nitrogenous excretory products.

From what has been said, it will be apparent that in animals where the nitrogenous waste is lost by diffusion, there will be little direct connection between the processes of excretion and osmoregulation. As was pointed out at the beginning of this chapter, the vertebrate kidney appears to have evolved primarily as an osmoregulatory device rather than as an organ of excretion of waste products. This is the main role it continues to perform today in teleost fish (elasmobranchs are a special case because they produce urea), and an excretory function is taken on by the vertebrate kidney only in terrestrial forms (or those that are secondarily aquatic).

Summary of Section 11.5

1 Nitrogenous waste is an important excretory substance, and initially this often appears in the body in the form of ammonia resulting, for example, from amino acid deamination.

2 Ammonia is highly toxic and must be rapidly removed. Aquatic animals with large permeable exterior surfaces lose most ammonia by diffusion, but no such easy option exists for animals in which water must be conserved.

3 In these animals, enzyme systems exist that detoxify ammonia; uric acid is produced in animals such as insects or birds where the excretory product must be very dry, and in addition, in embryos that live in shelled eggs, uric acid can accumulate without exerting a high osmotic pressure. In terrestrial mammals, ammonia is converted into urea through the urea cycle.

Now attempt Questions 11 and 12, on pp. 395–396.

SUMMARY OF CHAPTER 11

The regulation of body fluids is often an important function of excretory (renal) organs; they rid the body of excess water, salts, nitrogenous waste products and small toxic molecules. The principles of their function were described by reference to the mammalian kidney and insect Malpighian tubule.

The two basic processes by which excretory fluids (primary urine) may be formed are ultrafiltration and active transport. Transport into the excretory tubule lumen is usually active and is referred to as secretion. In the vertebrate kidney, ultrafiltration is responsible for formation of the primary urine (glomerular filtrate). The mammalian nephron was shown to begin with a Malpighian body in which the glomerular filtrate is formed by ultrafiltration of the blood that passes through the glomerular capillaries. There is a sieving mechanism, based on size, that determines what passes from the glomerulus into Bowman's capsule, and the rate of filtration is determined by the net pressure difference across the barrier between the blood in the glomerular capillaries and the fluid in Bowman's capsule. The glomerular filtrate is modified as it passes down the nephron tubules by transport out of (reabsorption) and transport into (secretion) the tubule lumen.

Counter-current multiplication in the loop of Henle is the process whereby the small osmotic differences between the descending and ascending limbs of the loop, caused largely by differential permeabilities between the limbs (and in particular, co-transport of sodium, potassium and chloride out of the thick ascending limb), become magnified into a large osmotic gradient in the medulla. Urea recycling plays an important role in the counter-current process and contributes up to 50% of the total osmolarity of the inner medulla. Thus counter-current multiplication produces the osmotic gradient, between the cortex and medulla, responsible for much of the water removal from the primary urine in the loop of Henle; how much of the remainder is removed in the collecting ducts to produce a concentrated or dilute urine for excretion depends on the permeability of the collecting duct wall. The latter function is hormonally controlled by a negative feedback system involving antidiuretic hormone. The kidney also plays a crucial role in regulation of the acid-base balance in the body.

The excretory organs in the insects are the Malpighian tubules in combination with the hind-gut. Malpighian tubules produce a primary urine mainly by secretion, for which the driving force is active K^+ transport into the tubule lumen, though in blood-sucking insects such as *Rhodnius*, it is active Na^+ transport. Compounds of low relative molecular mass ($M_r < 10\,000$) can pass by diffusion from the haemolymph into the tubule lumen. Most reabsorption of solutes and water occurs in the hind-gut. There is evidence of hormonal control of excretion in insects; antidiuretic and diuretic hormones act to control the activity of the Malpighian tubules and rectum.

Nitrogenous waste compounds result from amino acid deamination, and the first to be formed is ammonia. This is soluble and highly toxic and must be removed rapidly. In aquatic animals with large permeable external surfaces, ammonia is lost by diffusion. No such easy option exists for animals in which water must be conserved, and these have enzyme systems that detoxify ammonia. In terrestrial mammals, ammonia is converted into urea through the urea cycle; urea is not toxic but is soluble and so exerts an osmotic effect. Uric acid is insoluble and non-toxic; it is produced by insects and birds, whose excreta may be very dry. It is also produced by embryos that live in shelled eggs where uric acid can accumulate without any osmotic consequences or other damage.

OBJECTIVES FOR CHAPTER 11

Now that you have completed this chapter you should be able to:

11.1 Define and use, or recognize definitions and applications of each of the terms printed in **bold** in the text.

11.2 Explain the two basic processes by which primary urine is formed, ultrafiltration and active transport, and discuss primary urine formation in the mammalian kidney. (*Questions 1, 3, 5, 6 and 9*)

11.3 Explain the functions of the components of the basic unit of the kidney, the nephron, and describe the differences between the two sorts of nephrons, juxtamedullary nephrons and cortical nephrons. (*Questions 2, 6 and 7*)

11.4 Explain how knowledge of differential ion and water permeabilities in the nephron, and in particular the loop of Henle, have contributed to our understanding of events that occur in the nephron. (*Question 7*)

11.5 Describe how counter-current multiplication enables a concentrated urine to be produced. (*Questions 2, 7 and 8*)

11.6 Describe the importance of the arrangement of blood vessels (named the vasa recta) around the nephron. (*Questions 5 and 8*)

11.7 Provide examples of the mechanisms of control by which the extracellular volume and the salt composition of the body fluids can be regulated. (*Questions 4, 5 and 9*)

11.8 Describe the principles of primary urine formation in the insect Malpighian tubule and discuss the role of the hind-gut in the production of excretory material. (*Question 10*)

11.9 Describe how nitrogenous waste materials are produced by animals and explain why the type of excretory product may be important in relation to the animals' environment. (*Questions 11 and 12*)

QUESTIONS FOR CHAPTER 11

Question 1 (*Objective 11.2*) Explain why, when it comes to eliminating toxic compounds of low relative molecular mass from the body, a renal organ operating by ultrafiltration will be at an advantage over one producing urine by active transport alone.

Question 2 (*Objectives 11.3 and 11.5*) Which of the items (a)–(f) are normally likely to be found in the kidney of desert rodents but not in normal (laboratory) rats?

(a) An inner medulla with an osmolarity much greater than 3 000 mOsmol l^{-1}.

(b) Active co-transport of NaCl in the thick ascending limb of the loop of Henle.

(c) A predominance of juxtamedullary nephrons surrounded by vasa recta.

(d) Extremely long loops of Henle.

(e) Ultrafiltration largely on the basis of the size of pores in the glomerulus.

(f) A collecting duct running into the pelvis of the kidney.

Question 3 (*Objective 11.2*) If a rabbit were given an intravenous injection of saline containing two non-metabolizable proteins, one of M_r 2 000, the other of M_r 100 000, how would these be excreted by the kidney, if at all?

Question 4 (*Objective 11.7*) What is the main reason for the results shown in Figure 11.24?

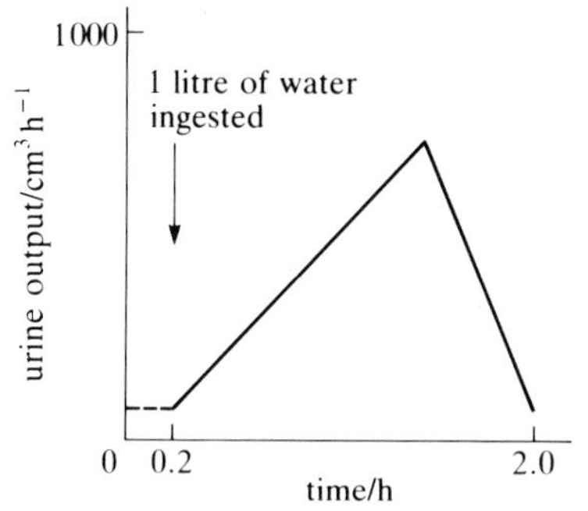

Figure 11.24 Urine output before and after a person drinks a litre of water.

Question 5 (*Objectives 11.2, 11.6 and 11.7*) Suppose the renal arteries of a mammal were clamped so as to reduce blood flow by 75%. (a) What would happen to the GFR? (b) What could restore the GFR to normal?

Question 6 (*Objectives 11.2 and 11.3*) If a human were given a glucose meal such that the plasma glucose level were elevated to 500 mg per 100 cm^3, would a urine test for glucose be positive and if so, why? (Recall the data in Figure 11.6.)

Question 7 (*Objectives 11.3, 11.4 and 11.5*) Furosemide is an inhibitor of coupled sodium chloride transport, whilst ouabain is an inhibitor of the enzyme Na^+/K^+ ATPase. Suppose that it were possible to inject these substances into the fluid around the thick ascending tubules (the peritubular fluid) in the intact kidney; what would the likely effect be on:

(a) Active NaCl transport in the loop of Henle? (Why?)

(b) The cortico-medullary osmotic gradient?

Question 8 (*Objectives 11.5 and 11.6*) Why is the counter-current mechanism in the vasa recta said to be passive while that in the loop of Henle is active?

Question 9 (*Objectives 11.2 and 11.7*) What factors determine the minimum urine flow in a mammal deprived of water?

Question 10 (*Objective 11.8*) Which of the following items are true of insect Malpighian tubules and why?

(a) Glucose in solution in the haemolymph is freely filtered to the lumen of Malpighian tubules.

(b) Haemolymph pressure is insufficient to force proteins of M_r greater than 10 000 into the lumen of Malpighian tubules.

(c) Ethacrynic acid, an inhibitor of sodium chloride co-transport, significantly inhibits primary urine formation in insects.

(d) Much water absorption from the insect excretory system occurs in the hind-gut.

(e) There is evidence of hormonal control of excretion in insects.

Question 11 (*Objective 11.9*) Are the following statements true or false? (State your reasons.)

(a) One of the main sources of ammonia in mammals is deamination of amino acids. TRUE, also from nucleic acid breakdown

(b) Ammonia is converted into uric acid in the urea cycle. FALSE, urea is produced by urea cycle.

(c) Most ammonia could be removed from the body of a terrestrial vertebrate in a similar manner to CO_2, although normally it is excreted in the form of urea. False – level of ammonia would be too high – kill animal.

(d) Uric acid is particularly suitable as a means of excreting nitrogen in embryos in shelled eggs because it does not exert a high osmotic pressure when it accumulates. TRUE.

Question 12 (*Objective 11.9*) Explain why the aquatic larvae of the newt *Triturus* can excrete nitrogenous waste products as ammonia, but the terrestrial adult excretes 87% of the excretory nitrogen as urea. When the adult returns to water to breed, urea excretion diminishes in favour of ammonia excretion.

OSMOREGULATION ◆ CHAPTER 12 ◆

12.1 OSMOREGULATION IN ANIMALS

Osmoregulation involves the regulation of water and ionic composition within an animal's body; the osmotic problems encountered depend on the animal's environment and the concentration of its body fluids (i.e. whether its body fluids are more, or less, concentrated than the external environment). This chapter considers the problems faced by animals in various aquatic and terrestrial environments, and the strategies used to overcome them. For convenience, vertebrates and invertebrates are treated separately, but in many respects, the problems posed to these animals, and the solutions to them, are similar.

12.1.1 The environment and the body fluids of animals: terminology and units of osmotic concentration

Apart from the majority of insects, spiders and terrestrial vertebrates, most animals live in aquatic environments. A metazoan animal can be visualized as consisting of a number of cells within which is the **intracellular fluid (ICF)** (see Figure 12.1). The cells are surrounded by membranes, which in turn are surrounded by an **extracellular fluid (ECF)**, and this is separated from the environment by an outer epithelial cell layer; in some animals, this may be associated with a shell-like layer. If the cell membranes of the animal were completely permeable to water and ions, and if the animal were in **osmotic equilibrium** with the environment, then the **osmotic concentration** of the ICF would equal that of the ECF, which in turn would equal that of the environment. The ICF, the ECF, and the environment would then be said to be iso-osmotic (see Chapter 11).

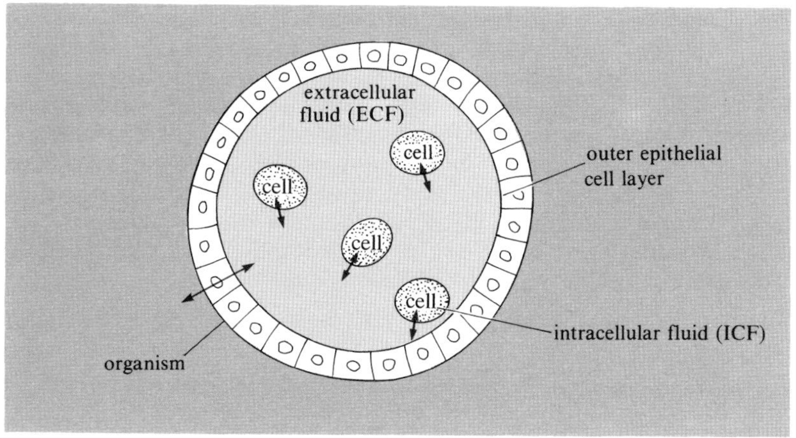

Figure 12.1 The body fluids of animals (red arrows indicate where exchanges of water and solutes may occur).

◇ Does this imply that all three would be in ionic equilibrium?

◆ Not necessarily, because osmotic concentration means the *total* concentration of osmotically active constituents in the solution, irrespective of the type of ion contributing.

body fluids), the total concentration of osmotically active constituents is usually expressed in **osmoles** per litre, that is the number of moles of solute (osmotically active particles) dissolved in a volume of water to make 1 litre in total. This is called **osmolarity**. For a non-electrolyte (e.g. sucrose or urea), the osmolarity is, in fact, equal to the molarity, but a solution of an electrolyte (e.g. NaCl) has a higher osmotic concentration because it dissociates into ions (e.g. Na^+ and Cl^-), each of which is osmotically active. The term **osmolality** is frequently used instead of osmolarity. This is not strictly the same, because it refers to an amount of solute *added* to 1 kg (1 litre) of water, but in practice in biological studies the terms are taken to be interchangeable.

Osmotic concentration is rarely measured directly, but instead is calculated from the depression of the freezing point of the solution, Δf, where the Greek letter Δ (delta) means 'change in'. (The greater the osmotic concentration, the lower is the freezing point, which is, for example, why salt melts ice on roads.) This can be converted into units of osmotic pressure, but is normally converted into concentration by defining a freezing point depression of 1.86 °C as equal to an osmotic concentration of 1 Osmol l^{-1}. In osmoregulatory studies, the term **milliosmole** (mOsmol) is often used (1 000 mOsmol = 1 Osmol). In aquatic environments, the osmotic concentration of the external medium can also be expressed in milliosmoles per litre, but more commonly it is given as salt content or **salinity** (S), in parts per thousand (‰). Normal full-strength (100%) seawater of salinity 34.3‰ has a freezing point depression of 1.86 °C and a milliosmolarity of 1 000. A salinity of less than 0.5‰ may be considered as freshwater, and anything between this and 30‰ is brackish water. A salinity above 30‰ can be considered as seawater and in some enclosed seas such as the Red Sea, where there is little freshwater input and much sun-driven evaporation, the salinity may rise to more than 40‰. Before the advent of techniques like flame photometry, which enables accurate measurements of metallic ions such as Na^+ and K^+, it was difficult to determine the precise ionic constituents of seawater. However, as about 55% of the dissolved constituents of seawater are in the form of chlorides and because chlorinity was easily determined (by titration), the concentration of seawater was often expressed as the **chlorinity** (Cl‰), and still is given as such in references to published data. Chlorinity is defined as the amount (in grams) of dissolved chlorine per kilogram of seawater. Salinity is related to chlorinity by the equation

$$S‰ = 0.3 + 1.805 \times Cl‰$$

Seawater of salinity 34.3‰ will, therefore, have a chlorinity of approximately 18.7 grams (rounded up to whole numbers this is 19 grams of chloride in 1 000 grams of seawater, i.e. 19‰), which is equivalent to 535 mmol per kilogram of seawater, or to express it slightly differently, 548 mmol per litre of seawater. In Figures and data presented later in this chapter, you will see that seawater concentrations, and those of the body fluids of animals, may be expressed either as salinity (‰), or chlorinity (mmol l^{-1}).

12.1.2 An osmotic classification of animals

Organisms living in aquatic environments fall into two categories depending on their ability to withstand a change in the osmotic concentration of the environment. **Stenohaline** (Greek *stenos*, narrow; *halos*, salt) organisms have only a limited tolerance to changes in the osmotic concentration of the external environment, and most marine invertebrates come within this category. **Euryhaline** (Greek *eurys*, wide) animals can survive a wider range of osmotic concentrations. In addition, stenohaline or euryhaline animals are also classified according to how the concentration of their body fluids varies in response to changes in the osmotic concentration of the external environment. If the osmolarity of an animal's body fluids is plotted against the osmolarity of the external medium over a range of values (the resulting graph is called a blood–medium osmotic adaptation curve or just an **adaptation curve**), and the two are found to be roughly the same (Figure 12.2), then the animal is said to be an **osmoconformer** (but not necessarily an ionic conformer); we can also say it is **poikilo-osmotic** (Greek *poikilos*, changeable).

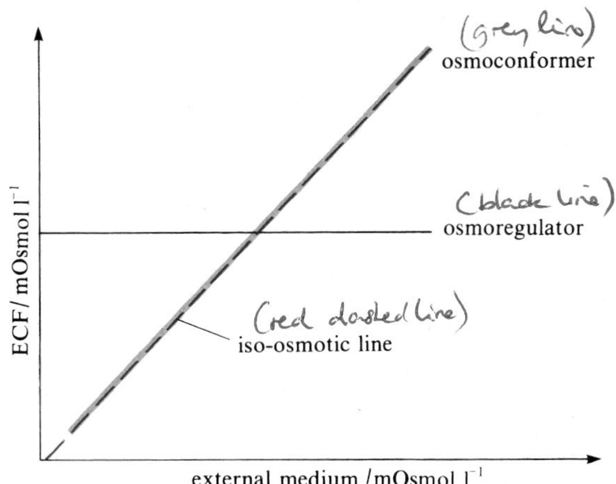

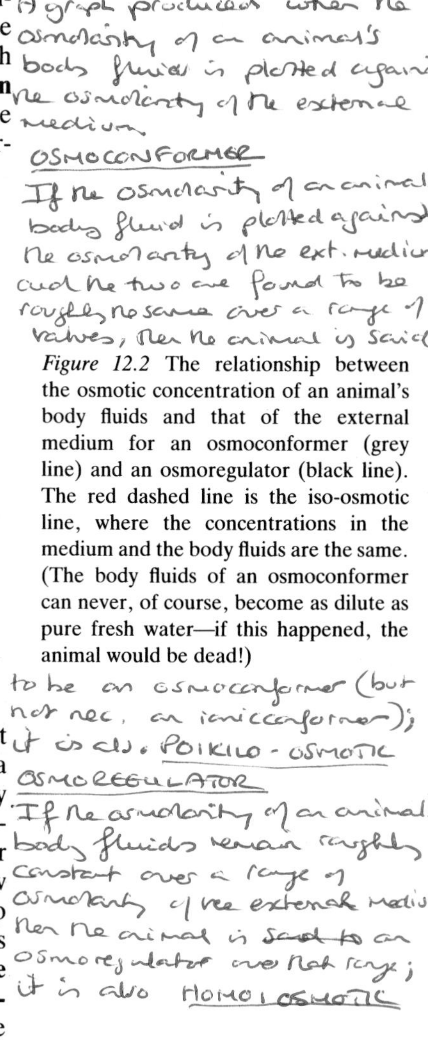

Figure 12.2 The relationship between the osmotic concentration of an animal's body fluids and that of the external medium for an osmoconformer (grey line) and an osmoregulator (black line). The red dashed line is the iso-osmotic line, where the concentrations in the medium and the body fluids are the same. (The body fluids of an osmoconformer can never, of course, become as dilute as pure fresh water—if this happened, the animal would be dead!)

If the osmolarity of the body fluids is maintained at a level different from that of the external medium, then the animal is an **osmoregulator** (Figure 12.2) a term first introduced in the previous chapter. If the concentration of the body fluids is maintained constant over the range of variation in external concentration, then the animal is also said to be **homoiosmotic** (Greek *homoios*, similar or constant). Whereas osmoregulating marine invertebrates often show considerable change in the concentration of their body fluids in response to changing external salinity, aquatic vertebrates, by contrast, show much less variation. Animals are (1) **stenohaline osmoregulators**, (2) **euryhaline osmoregulators**, (3) **stenohaline osmoconformers**, or (4) **euryhaline osmoconformers**. Few animals, however, fit one or other of these categories over the complete range of change in the concentration of the external medium to which they can be exposed: ideal osmoregulators or ideal osmoconformers as depicted in Figure 12.2 are seldom found.

◇ Re-examine the osmolarity relationships shown in Figure 12.2, and then look at the four adaptation curves in Figure 12.3. Which curve describes each of the four categories (1–4) listed above?

Figure 12.3 Adaptation curves of four osmotic categories of animals. (ΔC stands for change in the osmotic concentration.)

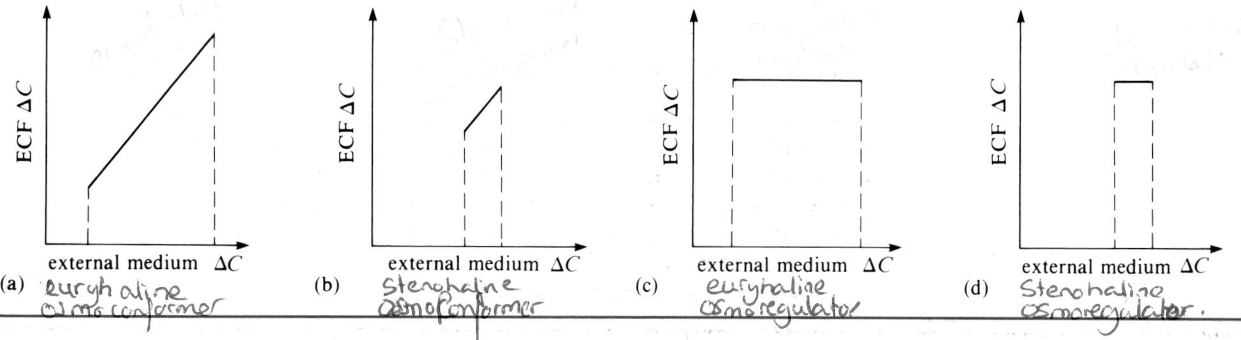

◆ Curve (a) describes category 4, a euryhaline osmoconformer; (b) describes 3, a stenohaline osmoconformer; (c) describes 2, a euryhaline osmoregulator; and (d) describes 1, a stenohaline osmoregulator.

Some animals, however, are osmoconformers over one part of the range of external medium osmolarity but osmoregulators over another part (Figure 12.4).

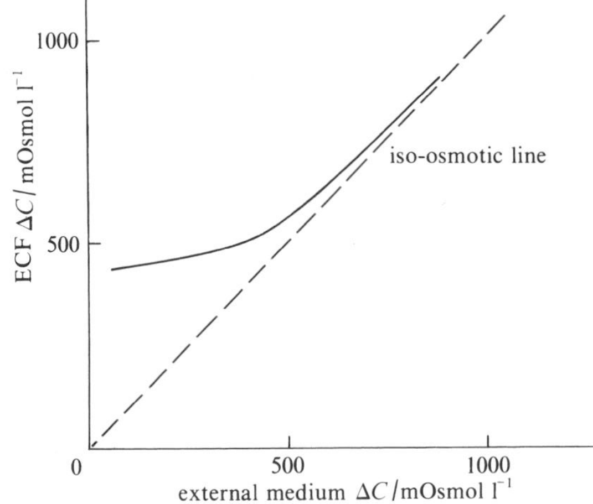

Figure 12.4 Adaptation curve for an animal that is an osmoconformer at one external salinity range but an osmoregulator at another.

◇ Identify the range of external osmotic concentrations over which each of these two terms applies in the adaptation curve shown in Figure 12.4.

◆ That part of the curve (500–900 mOsmol l^{-1} in Figure 12.4) parallel to the iso-osmotic line represents osmoconforming; that part below 500 mOsmol l^{-1} represents osmoregulating.

It is also important to realize that there are two different types of euryhaline and stenohaline osmoregulator, depending on whether the osmolarity of the body fluids is lower or higher than that of the external medium.

◇ What are these?

◆ Hyperosmotic regulators (the body fluid osmotic concentration is kept above that of the external environment) and hypo-osmotic regulators (the body fluid osmotic concentration is kept below that of the external medium).

Finally, animals may be described as **volume conformers** (these change in volume (water content) when the external osmotic concentration changes) or **volume regulators** (these show no change in volume when the external osmotic concentration changes).

12.1.3 Permeability

Biological membranes represent a barrier to the movement of many substances including water and ions, and in osmoregulation studies it is important to be able to measure membrane permeability,* and to compare the permeability of one membrane with another. There are two key terms that you will come across in this chapter, P_d, the **diffusional permeability**

*Permeability is discussed in more detail in Chapter 7 of Norman Cohen (ed.) (1991) *Cell Structure, Function and Metabolism*, Hodder and Stoughton Ltd. in association with The Open University (S203, *Biology: Form and Function*, Book 2).

coefficient, and P_{os}, the **osmotic permeability coefficient**. The former term was introduced earlier (Chapter 5) in the Fick equation for free diffusion across a cell membrane:

$$J_{oi} = P_d A(C_o - C_i)$$

where

J_{oi} is the net flux (rate of diffusion) of a substance from the *o*utside to the *i*nside of the cell,

P_d is the diffusional permeability coefficient, and its magnitude is dependent on (i) the size of the diffusing molecule, (ii) lipid solubility of the molecule, (iii) the absolute temperature, and (iv) the thickness of the membrane; it has units of centimetres per second (cm s^{-1}),

C_o is the concentration of the substance outside the membrane,

C_i is the concentration of the substance inside the membrane,

A is the surface area of the membrane.

The permeability of a membrane to solutes can be determined using radioactively labelled solute, and permeability to water can be determined using radioactively labelled water. To measure the diffusional permeability coefficient of water or a solute one places the radioactively labelled substance on one side of the membrane and measures the rate of its appearance on the other side; P_d can then be calculated using the Fick equation.

Alternatively, rather than simply measuring the exchange of radioactively labelled water, the permeability of a membrane to water can be measured in the presence of an imposed osmotic gradient. This involves a net flow of water across the membrane and results in a change in cell volume and internal pressure. Water permeability measured under an imposed osmotic pressure gradient is called **hydraulic conductance** (**L_p**) and has units of cm s^{-1} kPa^{-1}. Hydraulic conductance values can be converted into the same units as diffusional permeability (cm s^{-1}) to enable comparison between them, and values of L_p so converted are expressed as P_{os}, the osmotic permeability coefficient, to indicate that they are derived from an osmotic technique.

HYDRAULIC CONDUCTANCE (L_p)
Measure of the ease with which water moves. L_p has units of distance moved per unit pressure per unit time (m Pa^{-1} s^{-1}) and is the reciprocal of hydraulic resistance R.

DIFFUSIONAL PERMEABILITY COEFF.
A measure of the permeability of a membrane to water, or a solute; it can be obtained by measuring the isotopic flux of a solute or water across a membrane, with subsequent calculation of P_d from the Fick equation. It has units of centimetres per second (cm s^{-1}).

Summary of Section 12.1

1 Osmoregulation involves the regulation of water and ionic composition; the nature of the problem depends to a large extent on the animal's environment.

2 Osmotic concentration is a measure of the number of osmotically active particles per unit volume, usually referred to as osmolarity.

3 Aquatic organisms fall into two broad categories based on their ability to survive changes in the osmotic concentration of the environment—stenohaline animals have limited tolerance, whereas euryhaline animals can survive much greater changes. If the osmolarity of the body fluids is maintained at a level different from that of the medium, the animal is an osmoregulator. If the osmolarity of its body fluids changes to match that of the external environment, then the animal is an osmoconformer. However, many animals do not fit neatly into one of these classes; for example, some animals are osmoconformers over one external salinity range and osmoregulators over another. Efficient osmoregulators also show some degree of adjustment in their internal osmolarity in the face of large changes in external salinity.

4 Permeability of tissues (or of animals) is expressed in terms of either P_d, the diffusional permeability coefficient, or P_{os}, the osmotic permeability coefficient.

Now attempt Question 1 on p. 435.

12.2 OSMOREGULATION IN AQUATIC ENVIRONMENTS: INVERTEBRATES

It has long been argued that life on Earth began in the sea; the entire evolutionary history of the majority of marine invertebrates now living there seems to have been spent in that environment. Thus it might not be surprising to find that their enzyme systems and intracellular metabolism have evolved in such a way as to function at an osmotic concentration virtually identical with that of seawater. In Table 12.1, the concentrations of common ions in seawater are compared with their concentrations in the blood of some marine invertebrates.

Table 12.1 The composition of seawater and of the body fluids of some marine invertebrates. Body-fluid concentration is expressed as a proportion of the concentration of the seawater medium from which the animals were taken (*body fluid/medium, or B/M, ratio*).

	Na^+	K^+	Ca^{2+}	Mg^{2+}	Cl^-	SO_4^{2-}	Total $mOsmol\,l^{-1}$
seawater/$mmol\,l^{-1}$	450	10.4	10.8	55.9	573	29.5	~1 000
Arenicola (Polychaeta)	1.00	1.04	1.00	1.00	1.00	0.92	~1 000
Maia (Crustacea)	1.02	1.22	1.29	0.81	0.99	0.50	~1 000
Nephrops (Crustacea)	1.14	0.78	1.39	0.17	0.98	0.68	~1 000
Carcinus (Crustacea)	1.11	1.21	1.27	0.36	1.00	0.57	~1 000
Mytilus (Bivalvia)	0.99	1.18	1.13	0.97	0.99	1.00	~1 000
Sepia (Cephalopoda)	0.95	2.09	1.07	1.03	1.03	0.21	~1 000
Echinus (Echinoidea)	0.99	0.98	1.01	1.00	1.00	1.00	~1 000

◇ What is the relationship between the total osmotic concentration of the body fluids and that of the seawater?

◆ Clearly, these marine invertebrates are in osmotic equilibrium with seawater—they are osmoconformers—and they will not have any major problems of water regulation. Nevertheless, osmoconformers that are iso-osmotic with seawater must regulate their body water content within narrow limits because, in most cases, they must take into the body sufficient water and ions for urine production.

◇ Do they show any evidence of ionic regulation or are they all in ionic equilibrium?

◆ The sea-urchin *Echinus* appears to be in ionic equilibrium, but the other animals show differences, mostly small, between the concentrations of their body fluids and that of seawater, i.e. there is some ionic regulation.

In most instances, the concentration of K^+ is higher (except in the case of *Nephrops*) and the concentrations of Mg^{2+} and SO_4^{2-} are lower (except in the case of *Sepia*) in the body fluid than in seawater. The small differences in, say, Na^+ concentration between the fluid and seawater might be explained by positive ions being attached to negatively charged molecules such as proteins inside the cells, but the differences in concentration of K^+, Mg^{2+} and SO_4^{2-} between the inside and outside of the animal are too great to be explained in this way.

◇ What is an obvious way in which the concentrations of these ions could be held at different values in the animal's body fluids and in the environment?

◆ Active transport.

◇ What are three important criteria that can be used to determine whether there is active ion transport?

◆ Energy dependence, saturation kinetics, departure from Nernst equilibrium.

ESTUARINE ENVIRONMENT The tidal portions of river mouths, which are characterized by high salinity differences along their lengths & salinity fluctuations correlated with changes in their tidal cycle. The salinity at high water and at the seaward end may be as high as 30‰, but at low tide it may be as low as 1 or 2‰ almost freshwater.

It is important to emphasize the capacity of marine invertebrates for ionic regulation, even though they are in osmotic equilibrium with their environment in 100% seawater. Ionic regulation is also obvious in animals inhabiting estuarine or freshwater environments.

12.2.1 Osmoregulation in estuarine and freshwater environments: invertebrates

Here we deal largely with regulation of the osmotic concentration of the blood and other body fluids, i.e. regulation of extracellular fluid composition. Regulation of the composition of fluid within cells is dealt with later.

Invertebrate animals probably evolved in the sea and, in seawater, they maintain their body fluids iso-osmotic with the medium. But what happens in the **estuarine environment**, or in freshwater? Estuaries, the lowest tidal portions of rivers, are characterized by high salinity fluctuations that may be correlated with changes in the tidal cycle (Figure 12.5); the salinity in mid-estuary at high water may be as high as 30‰, but 1 or 2‰ at low tide, almost freshwater. Does the osmolarity of invertebrate body fluids change in parallel with the change in ionic concentration of the external medium, or does it remain the same? Stenohaline marine invertebrates such as cephalopod molluscs or the sea-urchin *Echinus* can survive only small dilutions of seawater, but most marine invertebrates examined can survive dilutions of up to 20%. Indeed, many can survive the much larger dilutions often found at times in estuaries, and a few can penetrate into freshwater.

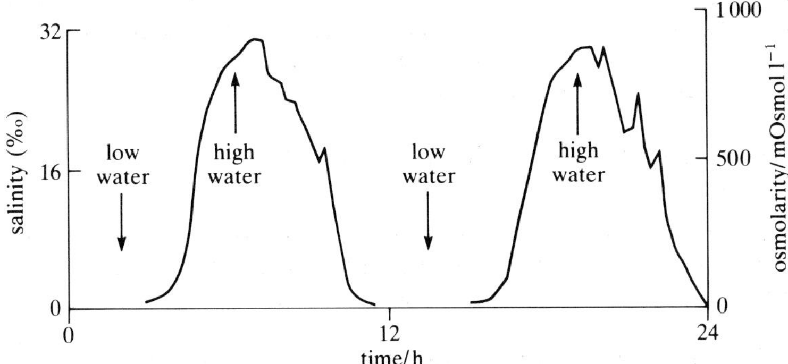

Figure 12.5 Salinity changes during the tidal cycle at a point on the shore of the Conwy Estuary, North Wales.

Osmoconformers achieve this by adjusting the osmotic concentration of their body fluids to that of the environment, whereas osmoregulators maintain the concentration of their body fluids higher, or lower, than that of the environment. Examine Figure 12.6a. The range of external medium concentration (in this case chlorinity) over which the animals can survive is indicated by the length of the adaptation curve. Figure 12.6a shows data for animals fre-

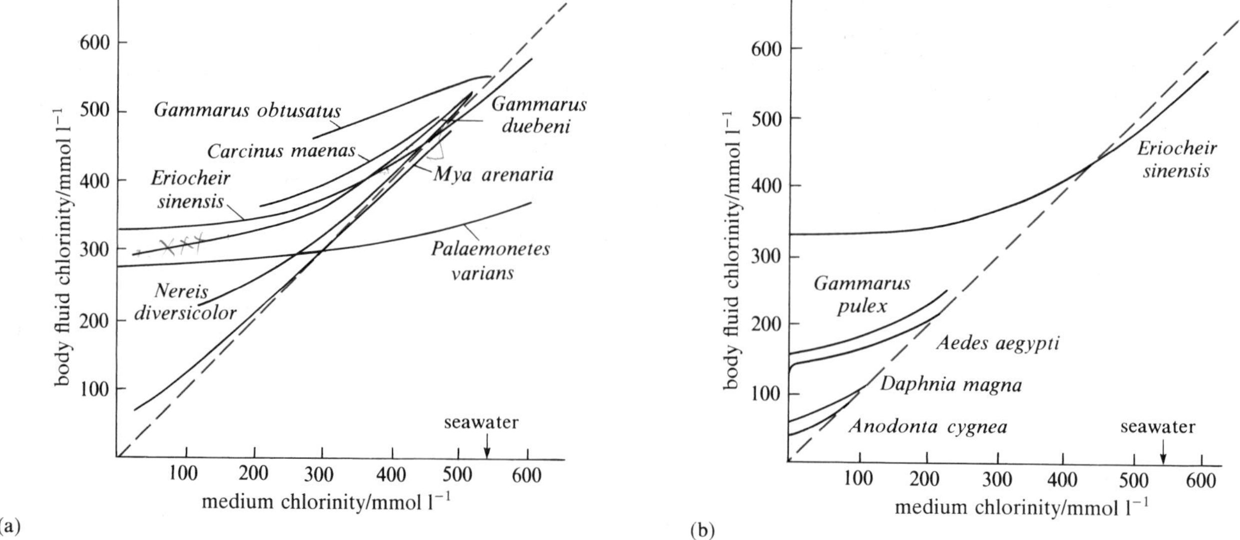

Figure 12.6 (a) The relationship between the concentration of chloride ion in the body fluids and in the medium (measured as chlorinity in mmol l^{-1}) in various brackish-water animals. (b) The same relationship in various freshwater animals. Full-strength seawater is indicated by an arrow. (Full-strength seawater of salinity 34‰ has a chlorinity of 548 mmol l^{-1}.)

quently found in brackish waters (waters of salinity varying between freshwater and seawater). The concentration of the body fluids of bivalve molluscs such as *Mya arenaria* closely follows the medium concentration.

◇ Does *Mya* exhibit osmoconformation or osmoregulation?

◆ Osmoconformation, and because this occurs over a wide range of change in medium concentration, *Mya* is a euryhaline osmoconformer.

With the exception of the shrimp *Palaemonetes varians*, the concentrations of the body fluids of the other animals shown in Figure 12.6a are higher than that of the environment at low external medium concentrations, and appear to follow the medium concentration as it approaches that of full-strength seawater.

◇ How would you classify these animals osmotically?

◆ At low external medium concentrations they are hyperosmotic regulators, but at higher concentrations they are osmoconformers. (This illustrates very clearly the danger of a rigid classification that places animals only in one category.)

Note also the degree of euryhalinity shown by some animals; the Chinese mitten crab, *Eriocheir*, the shrimp, *Palaemonetes* and the soft-shell clam, *Mya*, are truly euryhaline animals, and the habitat of *Eriocheir* and *Palaemonetes* extends from freshwater to 100% seawater, though *Palaemonetes* does not survive in freshwater for very long. However, *Eriocheir* can, and so it is also included in Figure 12.6b, which shows how the body fluids of those species normally found in freshwater vary when the external medium concentration is raised.

◇ What important point is noticeable from an examination of the concentrations of the body fluids of the animals in Figure 12.6b?

◆ All these invertebrates are hyperosmotic to the external medium at less than 40% of normal seawater chlorinity (215 mmol l^{-1}), and for stenohaline forms such as *Anodonta* or *Daphnia*, this is true for the complete range of external salinities over which they can survive.

An aquatic animal whose body fluids are more concentrated than the environment is liable to face certain difficulties, and we can examine one aspect of this using Figure 12.7. This diagram shows the results of an experiment in which the percentage change in volume of a marine polychaete, *Eudistylia vancouveri*, was measured while it was transferred from a medium at salinity 28‰ to a medium at 14‰ (distilled water plus seawater), or to freshwater. The first point to note is that the animal increases in size by osmotic swelling. The amount of swelling is proportional to the dilution of the medium; it is about twice as much in freshwater as in a medium at 14‰. Another problem (not obvious from this graph) is that salts are lost from the animal by diffusion.

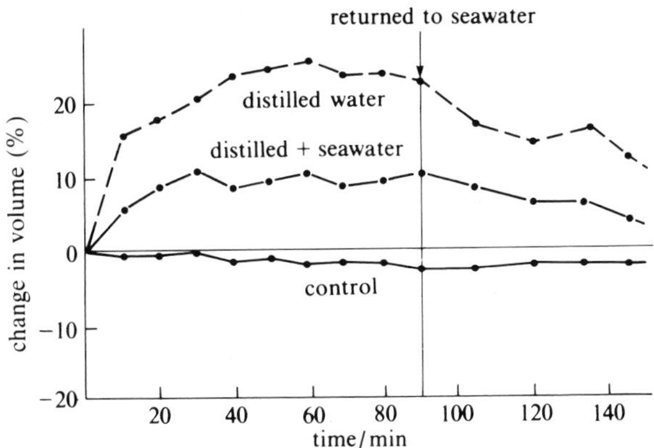

Figure 12.7 Change in volume of *Eudistylia vancouveri* as a percentage of the initial values determined in seawater at about 28‰ salinity.

These two problems, osmotic swelling and diffusive ion loss, are equally applicable whether the animal placed in dilute medium is an osmoconformer or an osmoregulator. But clearly, for the osmoconformer, the problem of continuing net water uptake is solved once its osmotic concentration is reduced to the level of the medium and the increase in fluid volume caused by osmotic water inflow is countered by the removal of this excess water, although some osmoconformers can tolerate a certain degree of swelling. However, for the hyperosmotic regulator there is potentially a danger of a continuous inflow of water and loss of salts, which if unchecked would eventually be fatal. Removal of excess water (volume regulation) can be accomplished by copious urine production, but this will aggravate ion loss because even the most dilute urine contains some ions. Ultimately, therefore, ion loss has to be made good by transport of ions into the animal's body fluids. But the extent to which this transport need occur will depend on the rates of passive exchange of water and ions between the animal's body fluids and the environment. Limitation of passive diffusion is therefore desirable in a hyperosmotic or a hypo-osmotic regulator.

12.2.2 Strategies for survival of hyperosmotic invertebrates in dilute waters

Alteration of the passive rates of exchange of water and ions

Why is it that different species have different passive rates of exchange of water and ions? The Fick equation, $J_{oi} = P_d A(C_o - C_i)$, tells us that there are three major factors relating to ion movement, alteration of which may affect passive diffusive fluxes of solute across epithelial tissues. These are changes in (i) the area over which diffusion can occur (A), (ii) the osmotic gradient, i.e. the concentration difference ($C_o - C_i$) and (iii) the permeability of the epithelial tissue (P_d). We will examine these now.

Alteration in the area over which the exchange of ions and water occurs

◇ How could altering the area over which diffusion can occur influence water influx and ion loss?

◆ Because the area over which diffusion occurs determines the ratio of permeable surface area to the volume of the body (or tissue) of the animal.

The **surface area to volume ratio** of a larger animal is lower than that for a smaller animal of the same shape. Consequently, water influx and ion loss per unit of body mass are relatively lower in the larger animal, assuming that permeability and the initial blood concentration are the same in both. (Note, however, that it is the permeable surface area, often the gill surface area rather than the total body surface area, which is most important in determining overall permeability. This makes comparisons based on whole body areas of different species difficult, a qualification that should be borne in mind in the following example.) The large spider crab *Maia* swells at a rate of 4% per hour when transferred from 100% seawater to 20% seawater, whereas the smaller *Porcellana* swells at 12% per hour on transference to 60% seawater. Clearly, the osmotic problems appear to be relatively smaller for the larger animal. Thus for a hyperosmotic species, there could be an evolutionary advantage in increasing its size, but there is not a great deal that any individual can do about this in the short term when it is transferred from seawater to dilute water. However, alterations in blood flow to permeable tissues such as gills may in effect reduce the area over which diffusion can occur.

Alteration of the osmotic gradient between the animal and the environment

The lower the osmotic gradient between the animal and the environment, the easier (other things being equal) it will be for a hyperosmotic regulator to survive in dilute waters. The explanation for this is quite simple: an animal cannot (a) produce urine free from ions, or (b) make its body surface entirely impermeable. Therefore it must transport ions into its body, and the greater the osmotic gradient with the environment, the more energy is expended.

◇ Why should this be?

◆ Transport of ions against a concentration gradient requires energy—by a process of active transport.

The osmotic concentration of the body fluids of the freshwater mussel *Anodonta* (50 mOsmol l^{-1}) is considerably lower than that of invertebrate osmoconformers living in the sea (e.g. *Mytilus* body fluids are about 1 000 mOsmol l^{-1}). If the osmotic concentration of the body fluids and cells of *Anodonta* were as high as that of *Mytilus* in seawater, its minimum energy expenditure for osmoregulation would be 50 times higher than its resting metabolic rate, which would present it with considerable difficulties! By reducing the body fluid concentration to 50 mOsmol l^{-1}, the osmotic work load is reduced by a factor of 2 500, or to 1.2% of the resting metabolic rate.

However, this conventional view of the energy cost of active transport has been disputed by some scientists. The gradients against which the transport process must operate may not be as great as suggested by simple comparison of an animal's body fluids with its environment. Ions such as Na^+ and Cl^- are not taken up directly from the external medium into blood but into cells that usually have relatively low concentrations of Na^+ and Cl^-, and in freshwater the uptake of Na^+ and Cl^- may be brought about by exchange mechanisms: Na^+ for H^+ (or $NH_4{}^+$), and Cl^- for $HCO_3{}^-$, for which energy costs might be very low.

Alteration in body surface permeability

The third factor that can influence the exchange of salts and water between an animal and the environment is body surface permeability. Reduction in permeability to water decreases the amount of water taken into the body by osmosis, and also directly reduces the rate of ion loss. Furthermore, a decrease in water intake would also indirectly decrease ion loss because less water would have to be excreted, and hence ion loss via this route would be less. Now examine Table 12.2.

Table 12.2 Body-surface permeability of marine and freshwater crustaceans.

Animal	Habitat	P_{os} (relative values)
Pugettia producta	seawater	0.33
Maia verrucosa	seawater	0.10
Carcinus maenas	40% seawater	0.042
Eriocheir sinensis	freshwater	0.003
Austropotamobius pallipes	freshwater	0.009

◇ What differences in the osmotic permeability of crustaceans in different environments are shown by these values?

◆ The osmotic permeabilities of the two freshwater species are much lower than for the two in seawater: *Carcinus* in 40% seawater is intermediate.

These are observations on *different* species; the next question is whether an *individual* of a euryhaline species can vary its permeability according to the medium in which it lives. Examine Figure 12.8, which illustrates the results of an experiment in which the body surface permeability of the amphiphod crustacean *Gammarus duebeni* was indirectly measured in environments of different salinity by determining the half-time for exchange of total body water. In this experiment the diffusional permeability coefficient P_d was measured when the animals were transferred to media of various concentrations. Note that there is a marked decrease (by a factor of approximately 2) in apparent permeability in 0% seawater (freshwater) compared with that in 100% seawater (it takes twice as long to exchange body water in 0% seawater).

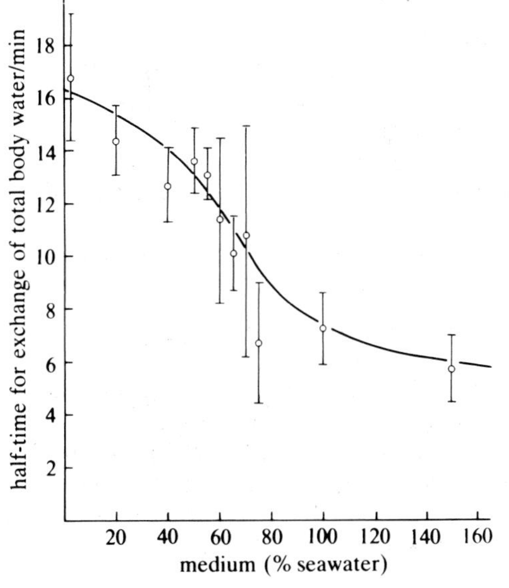

Figure 12.8 Reduction of the apparent permeability to tritiated water in relation to dilution of the medium in *Gammarus duebeni*. The vertical bars are twice the standard error.

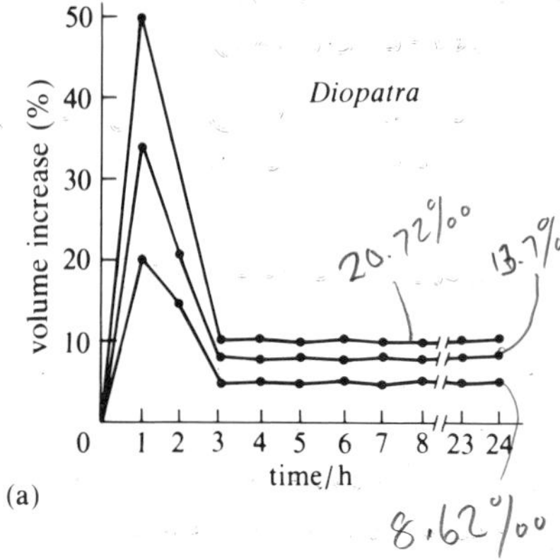

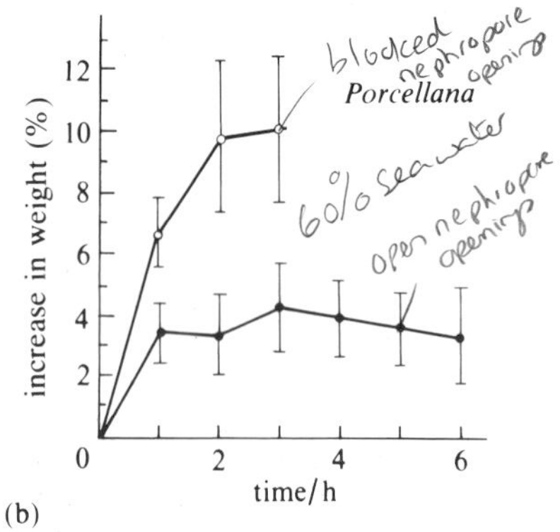

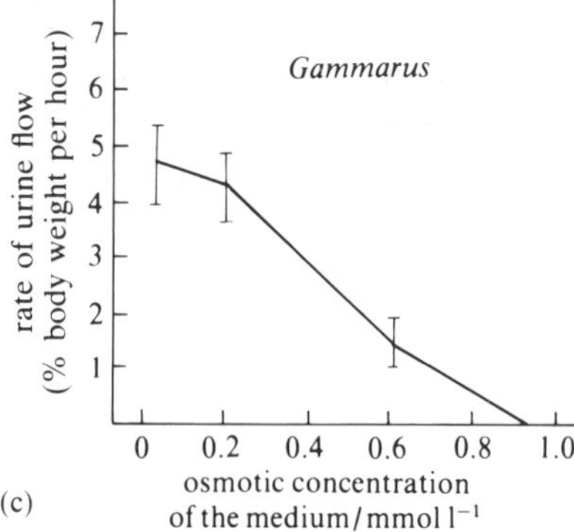

Figure 12.9 (a) Changes in volume of the polychaete annelid *Diopatra variabilis*, after sudden transfer from seawater to water of salinity 8.62‰ (top curve), 13.7‰ (middle curve) and 20.72‰ (bottom curve). (b) Weight changes of the crab *Porcellana platycheles* with blocked (open circles) or open (filled circles) nephropore openings after transference to 60% seawater. Means and 95% confidence limits for 10 crabs are shown. (c) Urine flow rate of the amphipod *Gammarus oceanicus* in media of different salinities. The vertical lines are twice the standard errors.

These results show that a euryhaline animal moving from lower to higher salinities (or vice versa) *can* change its permeability, although there is considerable variation in the abilities of individuals to do this. However, although hyperosmotic regulators have a lower permeability than osmoconformers, there is a limit to the reduction in permeability they can achieve because respiratory surfaces must be thin enough to permit diffusion of gases. Consequently ions will be lost, and water will enter across these surfaces.

Volume regulation

For a hyperosmotic regulator to survive, the influx of water, which initially causes considerable swelling, must be counter-balanced by water removal—**volume regulation**—so that osmotic equilibrium can be attained, and renal organs (Chapter 11) play an essential role in this.

Figure 12.9a shows the results of some experiments in which the polychaete worm *Diopatra variabilis* was transferred from seawater to various media of reduced salinity. The initial increase in volume is rapidly counteracted and the animal reaches a stable volume after 3 h. If the excretory pores (nephropores) of the crab *Porcellana* (Figure 12.9b) are blocked, then there is a 10% increase in weight, but this increase is much less (4%) if these pores are open; this indicates that the renal organs do in fact play a major role in volume regulation. In media of reduced salinity, the volume of urine produced by *Gammarus* is increased (Figure 12.9c).

To conserve salts, it should be advantageous to produce a urine that is hypo-osmotic relative to the body fluids. The initial fluid entering the renal organs of most animals has a similar composition to the body fluids, so the removal of ions to produce a hypo-osmotic urine requires energy. It is not always possible for marine invertebrates to produce a hypo-osmotic urine, but even in *Carcinus* in which the urine is iso-osmotic with blood, inulin, an inert marker (Chapter 11, Section 3.3), is concentrated, implying that water and other ions have been reabsorbed. Furthermore, as Table 12.3 shows, excretory organs may play a role in maintaining an ionic composition of the blood different from that of the external medium, even in species in which the urine is iso-osmotic with the blood. For example, in *Eledone*, there are differences in the blood/medium concentration ratios for all the ions listed. Where the ion is at a higher concentration in the blood, to maintain the difference, the excretory organ should conserve ions (e.g. K^+), but where the concentration is lower in the blood, a greater proportion of the ion should be excreted (e.g. SO_4^{2-}).

Table 12.3 The contribution of the excretory organ to the maintenance of differences between the ionic composition of the blood and of the medium.

		Na^+	K^+	Ca^{2+}	Mg^{2+}	Cl^-	SO_4^{2-}
Eledone (octopus)	*B*/*M* (%)	97	152	107	103	102	77
	U/*B* (%)	102	90	87	89	97	136
Carcinus (crab)	*B*/*M* (%)	110	117	108	34	104	61
	U/*B* (%)	95	78	94	390	98	224

M, *U*, and *B* are the concentrations in medium, urine and blood, respectively.

For a hyperosmotic regulator, the ability to produce a hypo-osmotic urine markedly increases ion conservation while allowing water elimination. This occurs in a number of estuarine invertebrates, including some species of nereid worms, the freshwater mussel *Anodonta*, and various gastropods and crustaceans. The freshwater crayfish *Austropotamobius* is a hyperosmotic regulator that produces a dilute urine. In this animal, the excretory organs are, as in other decapod crustaceans, the paired **antennal glands** (Figure 12.10).

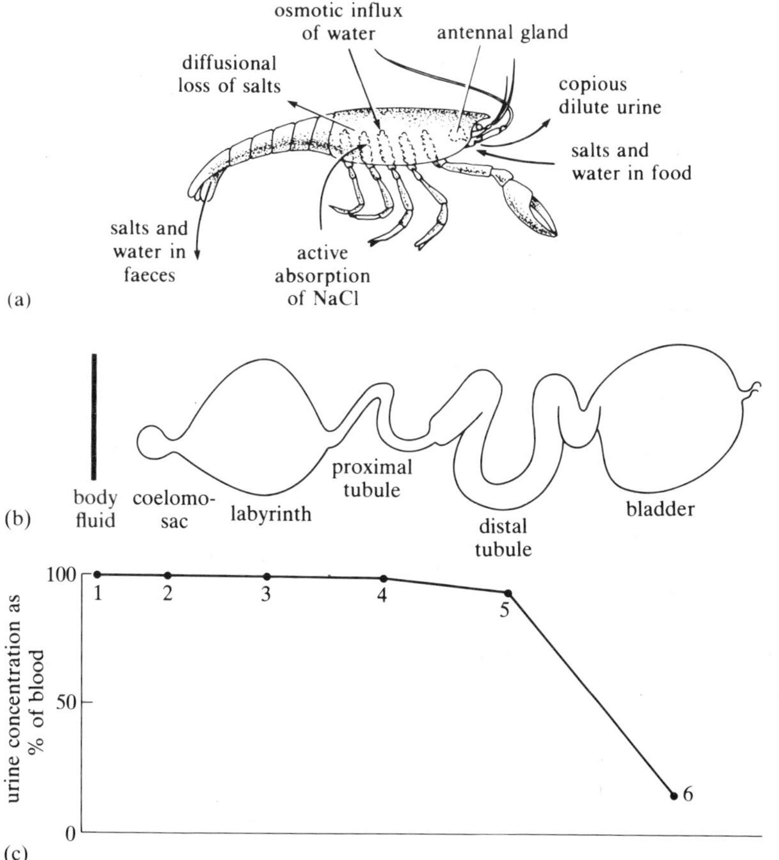

Figure 12.10 (a) A summary of the sites of salt and water exchange in the crayfish *Austropotamobius*. The antennal gland is shown in dashed outline as a saccular structure opening at the base of the second antenna. The gills, which are covered by the carapace and not visible externally, are shown at the bases of the legs. (b) An enlarged view of the antennal gland, showing its various parts. (c) The variation in the osmotic concentration of the urine, as a percentage of haemolymph (blood) concentration, as it passes through the antennal gland. The sampling points are (1) body fluid, (2) coelomosac, (3) labyrinth, (4) proximal tubule, (5) distal tubule, (6) bladder.

In freshwater crustaceans including *Austropotamobius*, there is a long tubule, the renal canal, which consists of a proximal tubule and a distal tubule. This region is absent in other crustaceans, such as *Carcinus*, which produce urine that is iso-osmotic with the blood. The concentration of urine in different parts of the antennal gland of *Austropotamobius* can be followed by micropuncture studies of the type used to examine urine production in vertebrate kidneys (Chapter 11, Section 3.4). Just as fluid enters the vertebrate Bowman's capsule by ultrafiltration, it is believed that fluid enters the coelomosac from the crustacean haemolymph by ultrafiltration. Not surprisingly, the ultrastructure of the crustacean coelomosac resembles the vertebrate glomerulus—Bowman's capsule epithelium. The coelomosac leads into a complex tubule, the labyrinth, and then into the renal canal, which ends in a bladder. The ions Cl^- and Na^+ and K^+ are absorbed from the urine in the labyrinth, but the bulk of the solute removal and water absorption occurs along the renal canal in the distal tubule and in the bladder, producing a very dilute urine.

A question arises in connection with the energy requirement for production of a dilute urine. Might it not require less energy to absorb salts from freshwater than to reabsorb them from the urine? It has been shown,

ANAL OR RENAL PAPILLAE
Structures found in some aquatic insects eg the mosquito larvae Aedes aegyptis where ion exchange occurs.

theoretically, that the energy cost of ionic regulation is less, per mole of solute transferred, by active transport in the kidney, than it is by transport across the body surface in freshwater. This is relatively easy to understand: the magnitude of the electrochemical gradient between the external medium and the body fluids of an animal is large and constant, whereas the gradients between the body fluids and the urine are initially small. Only in the final stages of urine formation is the gradient large, and only then is the amount of energy required for active transport similar to that required to pump ions in from the external medium. For the most part, the ion pumps in the renal organs extract ions from fluid that is more concentrated than freshwater. Consequently, it requires less energy to reabsorb ions from the urine than to replace them by direct absorption from freshwater.

But production of a hypo-osmotic urine will not maintain the ionic composition of the body fluids at a constant level; it will only decrease the urinary loss of ions, so ions must be gained from the environment to replace those lost. Because the concentration in the environment will be lower than in the animal, this must occur by active transport.

The active transport of ions in hyperosmotic regulators

In soft-bodied marine invertebrates, it is assumed that uptake of ions occurs over the general body surface, but in those aquatic animals in which the outer body surface is largely impermeable (e.g. arthropods, Figure 12.10a), the gills are known to be the site of active ion uptake. Sometimes these structures are considerably larger than respiratory exchange would require, and although the main reason for enlarged gills is normally related to their use in feeding (see Chapter 8), it is possible that this large size may also be partly explained by the necessity for ion exchange.

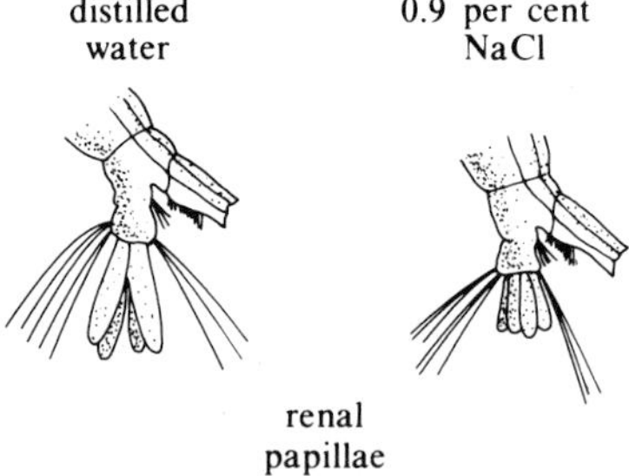

Figure 12.11 The posterior ends of *Aedes aegypti*, showing the difference in size of the renal papillae in larvae reared in distilled water and in 0.9% NaCl.

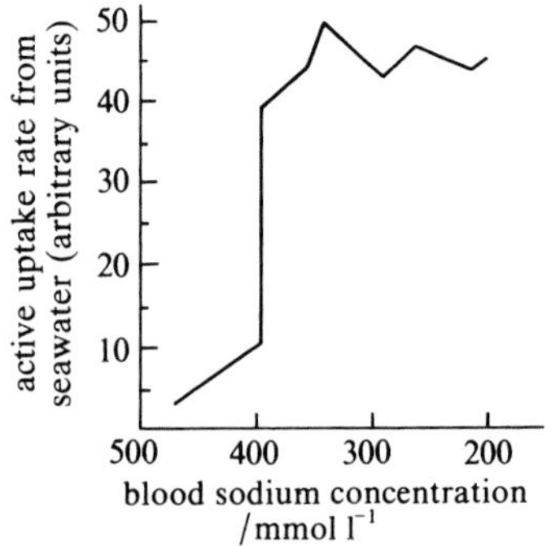

Figure 12.12 The relation between the rate of active uptake of Na^+ and the blood Na^+ concentration in *Carcinus*.

Often other body structures apart from, or in addition to the gills, may be involved in ion exchange. In the mosquito larvae *Aedes aegypti*, ion exchange appears to occur via the **anal or renal papillae** (Figure 12.11). The probable importance of the renal papillae is clear from comparison of their sizes in larvae placed in 150 mmol l^{-1} (0.9%) NaCl (in response to which the internal osmotic pressure of the body fluids will rise), and in freshwater (where the larva is a hyperosmoregulator). In 150 mmol l^{-1} NaCl, the need for active transport from the external medium is small and the renal papillae are smaller than in freshwater, where the papillae fulfil the need for active ion uptake.

In *Carcinus* in 100% seawater, the normal Na^+ concentration in the blood is slightly higher than in seawater. Although some Na^+ is lost by passive diffusion, it is replaced largely by passive influx. The amount of Na^+ uptake from the medium is very low until the blood Na^+ concentration falls below 400 mmol l^{-1} (Figure 12.12).

However, the Na^+ transport system is tending towards full saturation at an external Na^+ concentration of 300 mmol l^{-1} (this would be equivalent to approximately 40% seawater), and therefore it has the potential for rapid uptake if the blood concentration should fall for any reason.

◇ How can the active uptake system for Na^+ ions be fully saturated at an external Na^+ concentration of 300 mmol l^{-1}, and yet the amount of active uptake be low at 400 mmol l^{-1}?

◆ The most likely explanation is that the full capacity of the Na^+ transport system is not being utilized at 400 mmol l^{-1}. Those sites that are used are fully saturated, but they represent only a small proportion of the total number.

A fall of 30 mmol l^{-1} in the blood Na^+ concentration (from 400 to 370 mmol l^{-1}) causes an increase in Na^+ uptake by a factor of 13, but because the Na^+ transport system is almost saturated at 300 mmol l^{-1}, when the external Na^+ concentration falls further, it does not allow sufficient Na^+ transport into *Carcinus* to maintain a constant concentration of blood Na^+. It does, however, allow the animal to keep its blood Na^+ concentration above that of the external medium (Figure 12.13). Below a concentration of 100 mmol l^{-1} the availability of Na^+ to the transport system is too low for the crab to maintain a gradient between its blood and the medium, so an external Na^+ concentration of 100 mmol l^{-1} is about the tolerance limit for the crab.

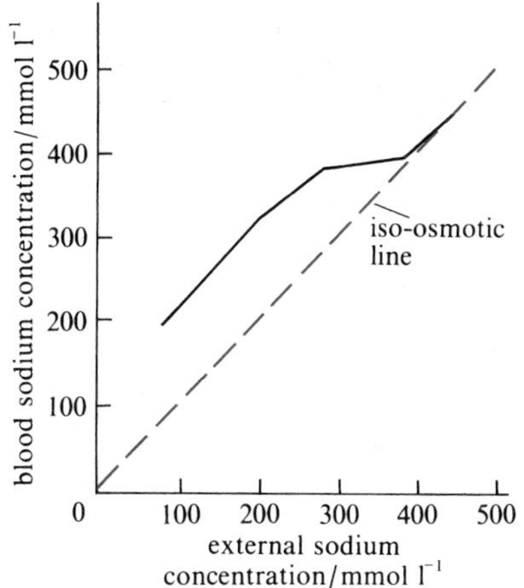

Figure 12.13 The relation between the Na^+ concentration of the blood of *Carcinus* and that of the external medium.

One especially interesting osmoregulatory feature of brackish water animals has been the evolution of ion transport systems with high affinities for their substrates, i.e. low K_t values. (The term K_t, which is analogous to K_m in enzyme kinetic studies, is a form of the Michaelis–Menten constant that represents the affinity of a membrane transport site for its substrate.)

◇ What importance would low values of K_t have to an animal inhabiting dilute media?

◆ Because a lower K_t value means a higher affinity for the substrate, a transport system with a low K_t will be better adapted to absorb ions from dilute solutions.

Consider Figure 12.14. The term J_{max}, which is analogous to v_{max} in enzyme kinetic studies, is the maximum rate of uptake of ions. The Figure shows that, for a high external concentration of Na^+ (100 mmol l^{-1}), the value of

J_{max} is almost the same whether the system has low or high affinity characteristics. When the external Na^+ concentration falls to 40 mmol l^{-1}, however, the high affinity system is still almost fully saturated and therefore working at close to the maximum rate, 10 μmol h^{-1}, whereas the low affinity system is transporting at only 5 μmol h^{-1}. Note that K_t is defined as the Na^+ concentration when the rate of ion uptake is half the maximum ($\frac{1}{2}J_{max}$).

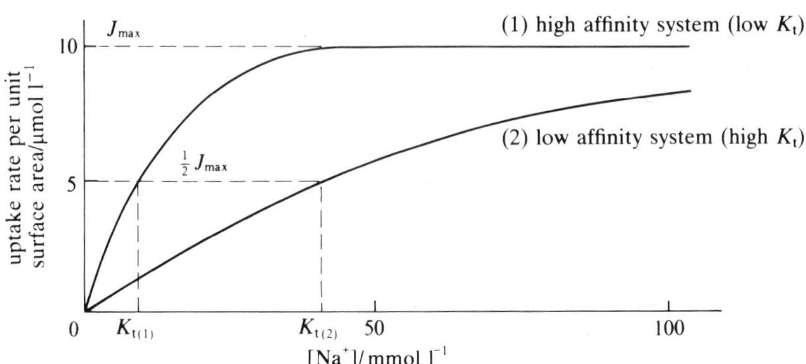

Figure 12.14 Plots of uptake against concentration of low affinity and high affinity Na^+ transport systems for the epithelial tissues of two animals. The terms K_t and J_{max} are explained in more detail in Chapter 7 of Norman Cohen (ed.) (1991) *Cell Structure, Function and Metabolism*, Hodder and Stoughton Ltd, in association with The Open University (S203, *Biology: Form and Function*, Book 2).

◇ Do the K_t values for ion transport systems in Table 12.4 confirm that there is a relationship between the K_t value of an ion transport system and the animal's environment?

◆ Clearly they do; the K_t values for the ion transport systems of animals living in freshwater are, in general, considerably lower than those in brackish water species, which in turn are considerably lower than those in seawater animals.

Table 12.4 The affinities of transport systems of various aquatic animals for Na^+ in the medium. (All are crustaceans except *Limnaea* (gastropod) and *Margaritana* (bivalve).)

Species	Habitat	K_t/mmol l^{-1}
Carcinus maenas	sea and brackish water	20
Mesidotea entomon	brackish water	9
Marinogammarus finmarchicus	sea	6–10
Mesidotea entomon	freshwater	2–3
Gammarus duebeni	brackish water	1.5–2.0
Gammarus zaddachi	brackish water	1.0–1.5
Gammarus duebeni (*celticus*)	freshwater	0.4
Gammarus pulex	freshwater	0.1–0.15
Gammarus lacustris	freshwater	0.1–0.15
Limnaea pereger	freshwater	0.25
Margaritana margaritana	freshwater	0.04

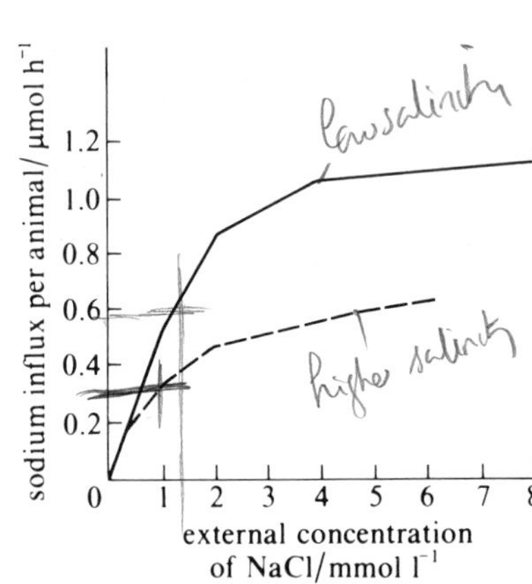

Figure 12.15 The rates of Na^+ uptake in a range of dilute media of shrimps (*Gammarus zaddachi*) previously acclimatized to a low salinity (solid line, 0.3 mmol l^{-1}) and to a higher salinity (dashed line, 10 mmol l^{-1}).

There is evidence that when animals normally found in brackish water, such as the amphipod *Gammarus zaddachi*, are transferred to freshwater and maintained there for an extended period, the capacity of the transport system for Na^+ increases, i.e. the J_{max} changes (Figure 12.15). This is an adaptive feature that will increase the ability of the animals to survive in freshwater.

◇ Estimate the K_t value for *Gammarus* in freshwater and brackish water; does this change when the animal moves from brackish water to freshwater?

◆ The K_t value in both environments is about 1 mmol l^{-1} Na^+; it does not significantly change in either environment—only the maximum rate (J_{max}) changes. It is doubtful if the K_t value for ion transport of any single species ever varies significantly, even when they are adapted to different environments.

12.2.3 Intracellular iso-osmotic regulation

Few marine invertebrates are truly homoiosmotic over the entire salinity range within which they may live; there is some variation in the osmotic concentration of their extracellular body fluids. This Section examines intracellular iso-osmotic regulation, or how the osmotic concentration of the intracellular fluid is varied to match changes in the extracellular fluid.

In most animals, the concentrations of the extracellular fluid (ECF) and the intracellular fluid (ICF) are almost iso-osmotic, and the mechanisms by which this occurs are called iso-osmotic regulation of the intracellular fluid. It is important to realize that just as the osmotic equilibrium between the ECF and the external medium does not necessarily imply ionic equilibrium, neither does the iso-osmotic regulation of the ICF. It is not just ionic regulation that may occur. Much of the intracellular osmotic pressure is due to organic solutes of low relative molecular mass, particularly amino acids, and there is considerable evidence that the concentrations of these may differ between the ICF and the ECF. The importance of iso-osmotic regulation of the ICF appears to be as a means of cell volume regulation in response to changes in ECF concentration, and obviously osmoconformers must effect larger changes for any given dilution of the medium than osmoregulators. When the osmotic concentration of the ECF of an animal is lowered on transference to a dilute medium, water uptake is likely to occur into the cells. In order to limit or prevent such water movement and cell volume change, the osmotic concentration in the cells must be reduced. Such osmotic adjustments need not necessarily involve ion pumping; 99% of the blood osmotic pressure in marine molluscs is due to inorganic ions, but only 50% of the intracellular osmotic concentration is due to these substances, the remainder is the result of the presence of organic compounds, particularly amino acids. Figure 12.16 shows the variation in the amino acid content of the muscles of the euryhaline bivalve *Mya arenaria* when it was transferred from almost full-strength seawater to media of reduced salinity.

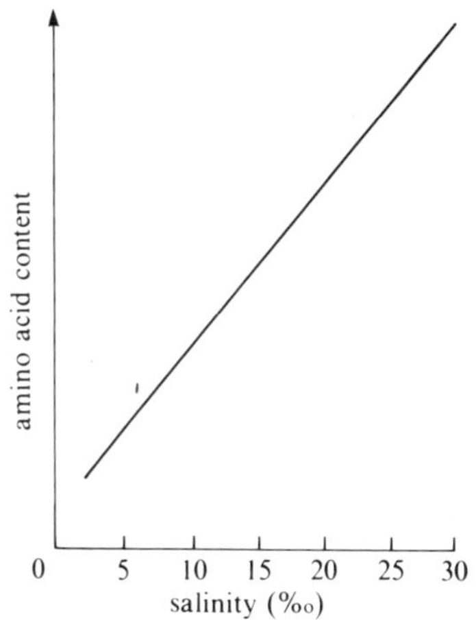

Figure 12.16 Variation in the amino acid content of muscle with salinity of the medium in the clam *Mya arenaria*. The animals were transferred from 30‰ salinity to 20‰, 15‰, 10‰, 5‰ and 2‰, and 48 h acclimatization were allowed at each salinity before the amino acid content was measured.

◇ What information can you deduce from this graph? Does it provide evidence for the regulation of amino acid content within the animal?

◆ The amino acid content of the muscle decreases as the salinity of the medium is decreased, so clearly there is some regulation of amino acid composition.

Mya is a species in which the blood remains iso-osmotic with the medium, and hence all regulation is brought about at the cellular level; as seen in Figure 12.16, the amino acid changes are dramatic, though the time-scale over which they occur is often a matter of hours and even days. Smaller changes are observed in hyperosmotic regulators, reflecting the lesser osmotic adjustments necessary at the cell level. In another experiment with *Mya*, it was shown that the amino acid concentration increases on transference back to normal seawater. An interesting question is whether the changes in amino acid content affect equally all amino acids found free within the cell (the intracellular free amino acid pool) or whether specific amino acids are involved.

◇ Examine Table 12.5, which gives the levels of most of the commonly occurring amino acids present in the intracellular free amino acid pools of the lugworm *Arenicola* and the starfish *Asterias*. Are there any changes in the level of amino acids when the animals are transferred from 100% to diluted seawater?

Table 12.5 A comparison of free amino acids in the tissues of *Arenicola* and *Asterias* adapted to seawater and to diluted seawater

	Amino acid content/mmol per 100 g			
	Arenicola		*Asterias*	
	seawater	50% seawater	seawater	60% seawater
alanine	7.45	2.20	0.17	—*
arginine	0.02	0.07	0.13	0.10
aspartate	1.38	1.46	0.11	0.05
glutamate	1.04	0.64	0.14	0.17
glycine	22.46	9.96	16.3	9.5
histidine	0.12	0.37	0	0
isoleucine	0.05	0.02	trace	trace
leucine	0.07	0.03	trace	trace
methionine	0.03	0.01	—	—
phenylalanine	0.04	0.03	trace	trace
proline	—	—	0	0
threonine	0.19	0.09	trace	trace
valine	0.06	0.02	trace	trace
asparagine	—	—	0.08	0.11
taurine	—	—	3.9	2.2
serine	—	—	trace	trace
glutamine	—	—	0.23	0.11
lysine	0.18	0.15	trace	trace
tyrosine	0.03	0.01	trace	trace

* A dash indicates that no value was recorded.

◆ Yes. Specific amino acids are involved, but there are differences between species. In *Arenicola*, changes occur mainly in alanine and glycine, whereas in *Asterias*, glycine accounts for most of the changes. In *Mytilus* (not shown in Table 12.5), taurine accounts for one quarter of the organism's total amino nitrogen in 100% seawater but is virtually absent when it is in 5% seawater.

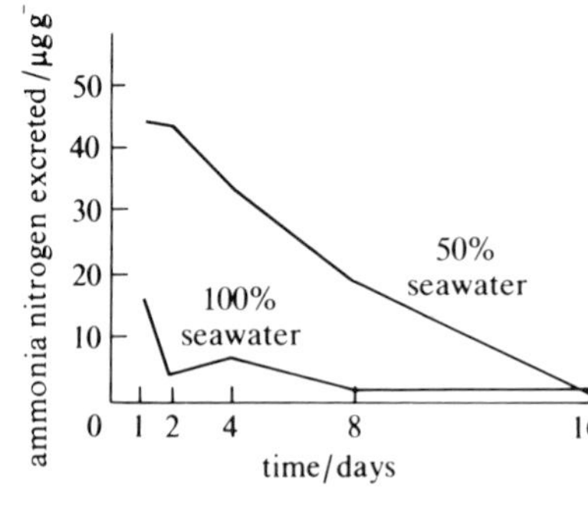

Figure 12.17 The rate of ammonia excretion in the bivalve *Macoma inconspicua* when transferred from 100% seawater to 50% seawater. The rate for animals remaining in 100% seawater is shown for comparison. Note that the excretion in 50% seawater has almost ceased at day 16, by which time the animals were fully acclimatized to the reduced salinity.

Coincident with this decrease in free amino acids, there are short-term increases in the amount of ammonia excreted by bivalves, and this is shown for *Macoma inconspicua* in Figure 12.17. Conversely, when animals are transferred back to seawater, there is a temporary decrease in ammonia excretion during the period when amino acids are being replaced in the cells.

At least two possibilities exist to explain the loss of free amino acids in dilute seawater in relation to the mechanism of regulation of intracellular amino acid content; either amino acids are incorporated into peptides or proteins, or they are broken down. The information in Figure 12.17 suggests that regulation of the intracellular amino acid concentration (when the free amino acid content decreases) principally involves the degradation of amino acids rather than their use in peptide or protein formation. Moreover, when the mitten crab *Eriocheir* is transferred from freshwater to seawater, the total amounts of alanine and proline (both in protein and in the free amino acid pool) in the muscles increase, demonstrating that the increase in amino acid content does not come from amino acid in protein.

◇ How else could the intracellular amino acid content be regulated when the animal is placed in low salinities?

◆ By transport of amino acids out of cells into the blood, and then directly (or after deamination) out of the animal. The process would happen in reverse when the animal was placed in a medium of higher salinity.

There are four principal pathways involved in the formation of free amino acids by amination of organic acids, namely:

- □ α-oxoglutarate⟶glutamate
- □ pyruvate⟶alanine
- □ oxaloacetate⟶aspartate
- □ 3-PGA (3-phosphoglycerate)⟶glycine

Regulation of amino acid levels in euryhaline animals can be achieved by influencing the cellular enzyme systems that control breakdown or synthetic pathways, and those enzyme systems that control the fate of 'reducing equivalents' (e.g. $NADH+H^+$) in cells. The most obvious way in which modification of the enzyme systems could be achieved in response to a change in salinity would be for these systems to be sensitive to changes in inorganic ion levels, and in particular to the increase or decrease in the cellular concentration of NaCl following the changes in the blood concentration of NaCl. For example, the enzyme glutamate dehydrogenase (GDH) is responsible for the amination of α-oxoglutarate to glutamate (in the presence of $NADH+H^+$):

$$NADH + H^+ + \alpha\text{-oxoglutarate} + NH_4^+ \xrightleftharpoons{\text{GDH}} NAD^+ + \text{L-glutamate} + H_2O$$

Normally, enzyme activity is directed towards the production of glutamate, and the increased cation (Na^+) concentration increases the enzyme activity and the rate of availability of $NADH+H^+$ for amino acid production. A decrease in cation concentration decreases enzyme activity and favours the breakdown of L-glutamate. Also when an animal is transferred from freshwater to seawater, Na^+ may directly inhibit enzymes responsible for amino acid breakdown, thereby decreasing amino acid breakdown. Coupled with a rise in the production of glutamate (and other amino acids), this will result in a raised level of internal amino acids.

Change in the level of free amino acids in cells need not, however, arise solely from synthetic or degradative processes. Consider the following information: when [^{14}C]glycine is injected into the haemolymph of the shore crab *Carcinus*, it is taken up into muscle cells. When the animal is transferred from 100% to 50% seawater, the level of [^{14}C]glycine in the blood rises sharply for a time before dropping, apparently owing to the glycine being taken up into the cells of the crab's hepatopancreas (a liver-like structure). When *Carcinus* is transferred from 30% to 105% seawater, the level of free amino acids in the blood falls but the concentration in the cells rises. These changes occur in addition to a drop in the level of NH_4^+ in the blood.

◇ What do you deduce from these observations?

◆ In response to a reduction in salinity, amino acids are transported out of the cells and into the blood, thus lowering the intracellular osmolarity, and the excess amino acids are transported in the blood and subsequently broken down in the hepatopancreas. When the animal is transferred back to seawater, amino acids in the blood are transported into the cells, thereby bringing about a rapid increase in intracellular osmolarity. The drop in the level of NH_4^+ in the blood indicates that amination of organic acids and synthesis of amino acids occurs *de novo*.

Before leaving the subject of intracellular iso-osmotic regulation, it is important to examine the time-scale over which adjustments in amino acid concentration may occur. The changes in *Mya* (Figure 12.16) may take hours or even days. Indeed, in the mussel *Mytilus edulis* (Figure 12.18) there appears to be a puzzling relationship between fluctuations in external osmolarity and tissue amino acid concentration (measured as ninhydrin positive substances—NPS; ninhydrin is a reagent used to detect amino compounds).

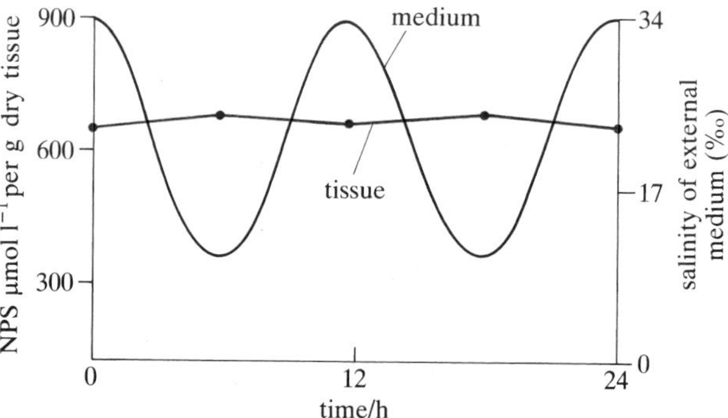

Figure 12.18 Tissue amino acid concentrations (measured as ninhydrin positive substances, NPS) in mussels, *Mytilus edulis*, exposed to fluctuating salinity.

The levels of intracellular amino acids (measured as tissue NPS) vary very little despite the large tidal fluctuations that result in salinity variations from 10‰–34‰. What then is the importance of intracellular iso-osmotic regulation? Most probably in estuaries where the salinity fluctuates relatively rapidly (Figure 12.18), osmoconformers do not become iso-osmotic with the external medium but simply tolerate some cellular swelling. However, in areas where the osmolarity is low over prolonged periods (as in stable brackish water areas such as lagoon-type environments) then alteration in intracellular amino acid content (intracellular iso-osmotic regulation) is probably the major factor in allowing colonization, especially by osmoconformers.

12.2.4 Osmoregulation in a concentrated environment: hypo-osmotic invertebrates

Most invertebrates are osmoconformers or hyper-osmoregulators: few maintain their body fluids at a lower osmolarity than the surrounding medium. However, hyporegulation is known in a few invertebrates. The shrimps *Palaemonetes* and *Leander* both maintain their body fluids hypo-osmotic to full-strength seawater. It is believed that these were originally freshwater animals that secondarily invaded the sea but then maintained their body fluids at a lower concentration than seawater.

◇ What problems of ion and water balance are these animals likely to have?

◆ The problems they face are an influx of ions and an osmotic loss of water—the reverse of those facing an invertebrate hyperosmotic regulator. They will need to expend energy (to pump out ions for example) to maintain a hypo-osmotic state.

The brine-shrimp *Artemia* is an animal that can be found in water of salinity only a few parts per thousand, but it can also survive in water containing 300 g of NaCl per litre, many times more concentrated than full-strength seawater. The relationship between the concentration of the body fluids of *Artemia* and the external medium over part of the range of concentrations it can survive is shown in Figure 12.19. Its blood is maintained hypo-osmotic to all media more concentrated than 25% seawater, and it achieves this by active regulation, not by being impermeable to water and ions. The strategies employed by invertebrate hypo-osmoregulators for survival parallel, in many respects, those used by the marine vertebrates, particularly teleost fish, and we consider such mechanisms of hyporegulation when we deal with these animals in Section 12.4.

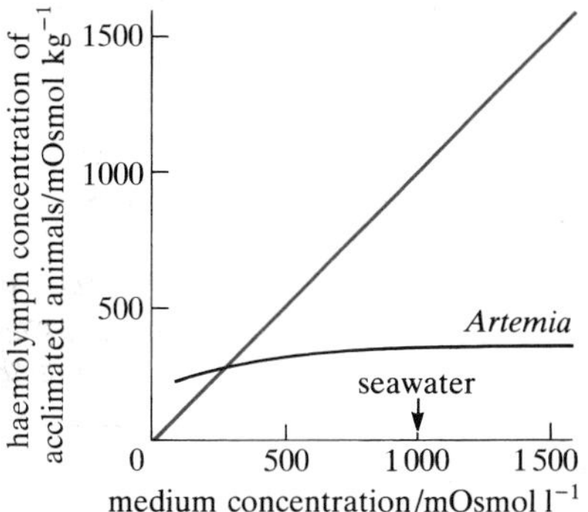

Figure 12.19 A comparison of body fluid and ambient osmolarity for the brine shrimp, *Artemia*. The red line is the iso-osmotic line.

Summary of Section 12.2

1 The body fluids of most marine invertebrates are in osmotic equilibrium with seawater, but many species show evidence of ion regulation.

2 Many invertebrates can survive in dilute waters; some of these are euryhaline animals that are also found in the sea, others are stenohaline and are restricted to dilute waters. In very dilute waters, invertebrates are normally hyperosmotic regulators, but as the external medium concentration increases, they become osmoconformers. A few may even be hypo-osmotic regulators where the external medium concentration is very high.

3 There are several strategies by which osmoregulators control the osmotic concentration of their blood and other body fluids (extracellular fluids) to ensure their survival in dilute waters. Particularly important is the alteration of factors affecting the diffusive flux of solute across epithelial tissue—the decrease in permeability and the decrease in osmotic gradient between the body fluids and the environment. However, the respiratory membranes must be permeable enough for gaseous exchange, and no animal can have body fluids as dilute as freshwater, so these factors cannot completely counteract the two major problems in dilute waters—the diffusive loss of ions and the osmotic influx of water. Excess water can be removed (volume regulation) in the urine, and although this is normally hypo-osmotic to blood, more ions are lost to the medium; ultimately, therefore, ion loss must be made good by active transport from the environment.

Measurement of Michaelis–Menten kinetic constants, K_t and J_{max}, shows that ion transport systems of animals in dilute waters are adapted to function optimally at very low external ion concentrations. When an invertebrate adapted to brackish waters becomes adapted to freshwater, the J_{max} of the transport system increases but the K_t remains similar.

4 Regulation of the intracellular osmotic concentration, iso-osmotic regulation, also occurs in invertebrate osmoregulators. But osmotic adjustment of intracellular fluid is due largely to alterations in the levels of organic compounds, especially amino acids, and a number of possibilities exist to explain how amino acid levels are elevated in seawater, but depressed in dilute waters: the degradation of amino acids in dilute waters, the synthesis of amino acids as the external salinity rises, and the transport of amino acids into, and out of, the blood to lower, or raise, intracellular osmolarity. Such changes of the intracellular osmotic concentration are likely to take place over a period of days and seem to be of most value to animals living in stable low salinity areas, such as lagoons, rather than to animals needing to make rapid alterations in intracellular osmolarity in response to rapid salinity changes, for example in estuaries.

5 Invertebrate hypo-osmotic regulators are less common than hyperosmotic regulators, but the shrimp *Palaemonetes* is one example. The osmotic problems these animals face are similar to those of vertebrate hypo-osmoregulators.

Now attempt Questions 2–5, on pp. 435–436.

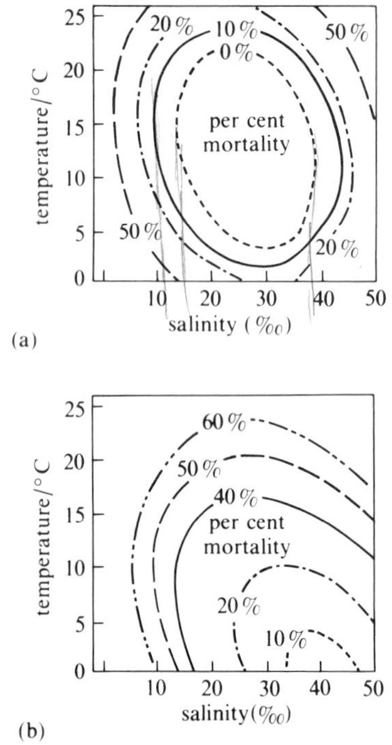

12.3 THE EFFECT OF OTHER FACTORS ON OSMOREGULATION IN ANIMALS

So far, we have considered the problems of water and ion balance that face invertebrate animals in terms of one environmental variable, salinity. Of course, the physiological effects of variation in one environmental factor should not really be considered in isolation because its effects are likely to depend on other factors, and in particular, on temperature. Examine Figure 12.20a, which shows the salinity tolerance of a shrimp as a function of temperature. Clearly, the survival of the species is affected by both temperature and salinity, and Figure 12.20b shows that in poorly oxygenated water mortality is higher, indicating that an additional factor ought to be considered in the context of the effect of salinity changes on this animal.

There are some obvious ways in which temperature can exert an influence on ion and water balance.

◇ How might changes in temperature influence passive and active ion fluxes across membranes?

◆ An increase in temperature will increase the rate of random movement of particles in solution and so will increase diffusional fluxes. Membranes will be more fluid at higher temperatures,* and thus they may be more permeable. Active ion transport depends on enzymes at some stage (e.g. for energy supply), and enzymes are temperature-sensitive. Therefore, a change in temperature may affect the rate of mediated ion transport.

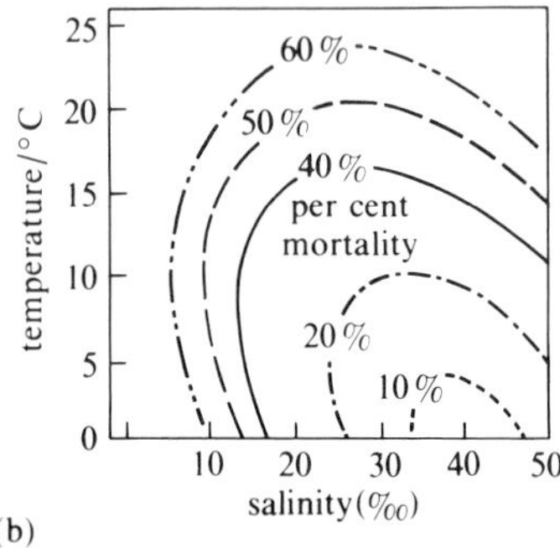

Figure 12.20 The effects of temperature and salinity on the percentage mortality of the shrimp *Crangon septemspinosa* in (a) aerated seawater and (b) poorly oxygenated seawater.

*The fluidity of membranes is discussed in detail in Chapter 7 of Norman Cohen (ed.) (1991) *Structure, Function and Metabolism*, Hodder and Stoughton Ltd, in association with The Open University (S203, *Biology: Form and Function*, Book 2).

We have mainly considered the physiological response of animals to ionic and osmotic variation, but many animals cope with osmotic stress by **behavioural strategies**, which may be as effective as physiological osmoregulation in allowing survival in environments of fluctuating salinity. For example, a marine mussel *Mytilus*, taken from seawater and dropped into freshwater, clamps its shell shut, thereby isolating itself from sudden osmotic stress. Figure 12.21 shows how the osmotic concentration of the seawater within the shell (the mantle fluid) remains above 600 mOsmol l^{-1}, even when the external salinity is as low as 3 mOsmol l^{-1}; this shows the effect of shell closure. There may also be modifications of an animal's reproductive cycle in response to salinity variations; in estuarine invertebrates, the larval phase is often shorter than in comparable marine species, and this may improve the chances of survival of a delicate stage in the life cycle as well as limit the risk of loss from the estuary to the sea.

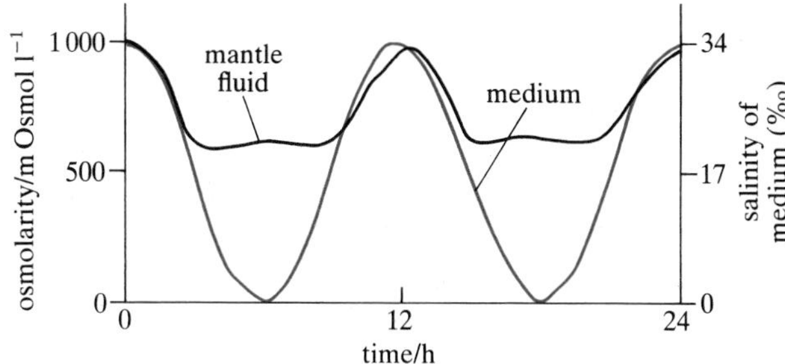

Figure 12.21 Changes in mantle fluid osmolarity in *Mytilus* exposed to fluctuating salinity.

Summary of Section 12.3

Salinity is not the only external factor that affects osmoregulation in aquatic animals. Other factors, such as temperature or oxygen levels, may influence osmoregulatory problems. Behavioural strategies, such as the shell-closing of mussels, may also help certain animals to cope with the periodic changes in salinity seen in estuaries, and with sudden osmotic shocks.

Now attempt Question 6, on p. 436.

12.4 OSMOREGULATION IN AQUATIC VERTEBRATES

This Section deals mainly with fish and to a lesser extent amphibians. Almost all vertebrates are osmoregulators and are to a large extent homoiosmotic (i.e. they maintain their body fluid osmolarity relatively constant), even when like the eel and the salmon they migrate from the sea to freshwater (or vice versa). Some, such as elasmobranchs, appear to be iso-osmotic to their environment, but they achieve this by tolerating high levels of organic compounds, largely urea, in their blood.

Table 12.6 shows the steady-state (i.e. fully acclimatized) osmotic concentration of the body fluids of some aquatic vertebrates found in the sea or in freshwater. The major solute concentrations of Na^+, K^+ and urea (or trimethylamine oxide: see Chapter 11) are also given. Fish are normally restricted to one environment, either seawater or freshwater, but there are a number of migratory teleost species. The osmotic concentrations of the body fluids of two migratory fish, the eel and the salmon, are given for both these environments.

Table 12.6 The concentrations of major solutes (in millimoles per litre) in seawater and in the blood plasma of some aquatic vertebrates.

	Habitat	Solute Na^+	K^+	urea* (TMAO)	Osmotic concentration/ mOsmol l^{-1}
seawater	—	~450	10	0	~1 000
Agnathans					
hagfish (*Myxine*)	marine	549	11	~1	1 152
lamprey (*Petromyzon*)	marine	—	—	—	317
lamprey (*Lampetra*)	freshwater	120	3	<1	270
Elasmobranchs					
ray (*Raja*)	marine	289	4	444	1 050
dogfish (*Squalus*)	marine	287	5	354	1 000
freshwater ray (*Potamotrygon*)	freshwater	150	6	<1	308
Teleosts					
goldfish (*Carassius*)	freshwater	115	4	~1	259
toadfish (*Opsanus*)	marine	160	5	~1	392
eel (*Anguilla*)	freshwater	155	3	~1	323
	marine	177	3	~1	371
salmon (*Salmo*)	freshwater	181	2	~1	340
	marine	212	3	~1	400
Crossopterygians					
coelacanth (*Latimeria*)	marine	181		355	1 181
Amphibians					
frog (*Rana*)	freshwater	92	3	~1	200
crab-eating frog (*R. cancrivora*)	brackish water (80% normal seawater)	252	14	350	830

*1 mmol per litre is osmotically insignificant. The urea values for ray, dogfish and coelacanth include trimethylamine oxide (TMAO).

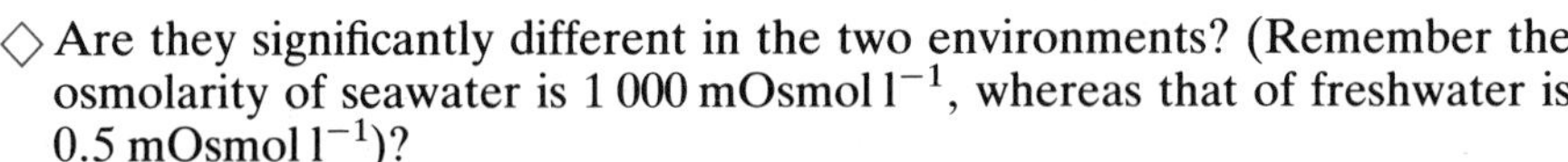
◇ Are they significantly different in the two environments? (Remember the osmolarity of seawater is 1 000 mOsmol l^{-1}, whereas that of freshwater is 0.5 mOsmol l^{-1})?

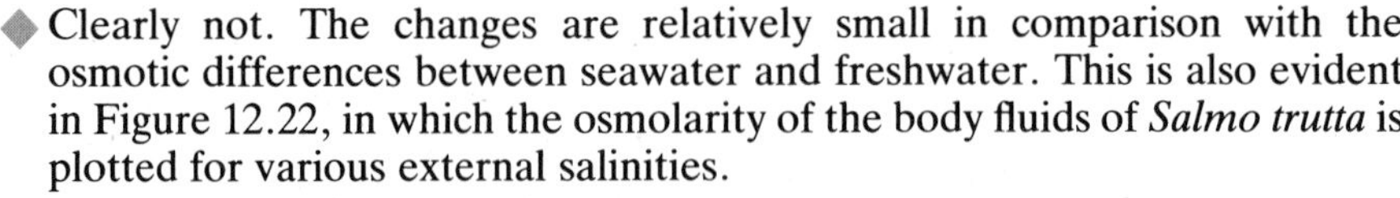
◆ Clearly not. The changes are relatively small in comparison with the osmotic differences between seawater and freshwater. This is also evident in Figure 12.22, in which the osmolarity of the body fluids of *Salmo trutta* is plotted for various external salinities.

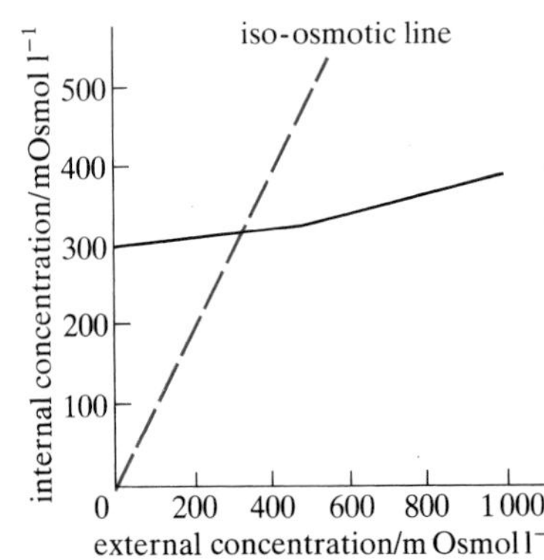

Figure 12.22 The mean plasma osmotic concentrations in the brown trout 24 h after direct transfer from freshwater to various seawater dilutions.

There is good evidence, mainly from palaeontology, but also from a study of the comparative anatomy of kidneys, that the early vertebrates, the ancestors of modern teleosts, lived in estuaries or fresh water.

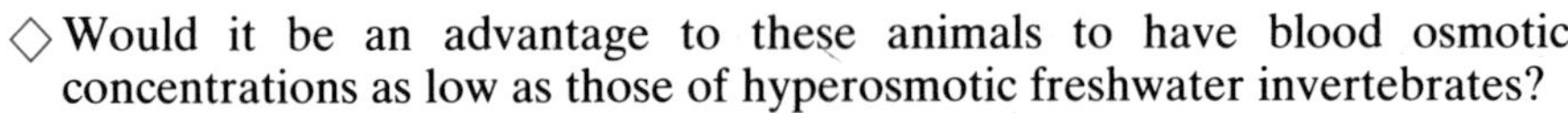
◇ Would it be an advantage to these animals to have blood osmotic concentrations as low as those of hyperosmotic freshwater invertebrates?

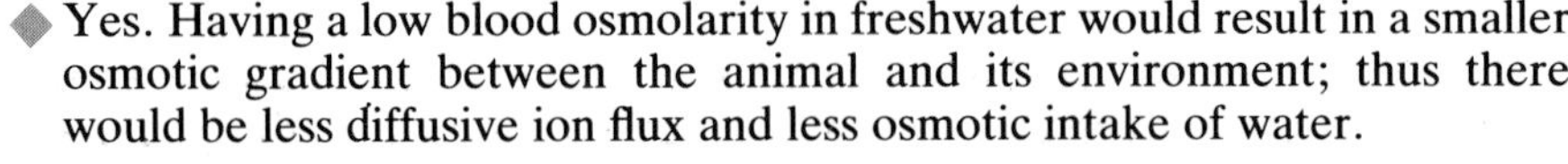
◆ Yes. Having a low blood osmolarity in freshwater would result in a smaller osmotic gradient between the animal and its environment; thus there would be less diffusive ion flux and less osmotic intake of water.

It is not surprising that the osmolarity of the blood of modern freshwater teleosts is lower (300 mOsmol l^{-1}) than that of seawater (1 000 mOsmol l^{-1}), and in all probability this lower blood osmolarity was also present in the early vertebrates. When they reinvaded the sea, the low blood osmolarity was maintained.

Table 12.6 shows that all vertebrates in freshwater are hyperosmotic to the environment but this is not true for most of those in marine environments. Teleosts are normally hypo-osmotic, as is the lamprey *Petromyzon* but elasmobranchs, the coelacanth *Latimeria* and the hagfish *Myxine* are iso-osmotic (or nearly so) to seawater, and the crab-eating frog *R. cancrivora* (found in coastal mangrove swamps in Southeast Asia) is iso-osmotic to 80% seawater. *Myxine* behaves osmotically like a marine invertebrate, and shows ionic but not osmotic regulation. But in the other vertebrates listed, the elevated blood osmolarity is maintained by the presence of high levels of blood urea and/or trimethylamine oxide: we can examine the advantage of accumulating urea by considering now the case of the elasmobranchs.

12.4.1 Osmoregulation in elasmobranchs

The ionic composition of elasmobranch body fluids is roughly similar to that of other vertebrates, except in two respects. First, the plasma Na^+ concentration is higher than in marine teleosts (Table 12.6), approaching 300 mmol l^{-1} compared with 212 mmol l^{-1} for the salmon, and 177 mmol l^{-1} for the eel in the sea. Hence the diffusional accumulation of Na^+ (and Cl^-) will be lower than that experienced by marine teleosts, and this results in the Na^+ turnover in elasmobranchs being 10% of that in teleosts. (Na^+ turnover is defined as the percentage of the total exchangeable amount of the ion that moves in or out of the fish per hour.) Secondly, some of the problems normally faced by a hypo-osmotic regulator (dehydration and ion influx) are prevented by raised levels of organic compounds in the blood—mainly urea, but also trimethylamine oxide. Urea (Chapter 11, Section 11.4) is an end-product of protein metabolism in mammals and some other vertebrates. It is a relatively non-toxic compound that permits the excretion of the amino group (NH_2) from amino acids. Urea is excreted via the kidney in mammals, but in teleost fish it passes out largely via the gills. However, in elasmobranchs, urea is retained and is present in the blood at 100 times the level in mammals. Although such levels would be toxic to mammalian cells, those of most elasmobranchs cannot function in the absence of a high urea concentration. In fact trimethylamine oxide, which is also present in elasmobranchs (see Table 12.6), protects proteins from denaturation by urea. Elasmobranchs are usually iso-osmotic or even slightly hyperosmotic to seawater. Thus they have little need to drink to obtain water because it tends to enter osmotically. But because the ionic concentration of the animal's blood is lower than that of seawater, there is still a small diffusive influx of ions, largely across the gill epithelium (the skin is relatively impermeable) and via the gut (from ingested food and the small amount of seawater that will be unavoidably ingested). Ions are also pumped in from seawater across the gill in exchange for ammonium and bicarbonate ions, which must be excreted. So there is a problem—how to excrete excess salts.

◇ What is one obvious means by which this could be achieved?

◆ Excretion through the kidney into the urine.

In fact, this process is insufficient to account for the removal of excess salts. Another possibility is that, as in marine teleosts, salts are actively secreted across the gills. Indeed, there is evidence that some elasmobranchs may be capable of such secretion. However, a special gland (the **rectal gland**), whose duct opens into the posterior intestine, can secrete a fluid that is iso-osmotic with the plasma but contains Na^+ and Cl^- at concentrations higher than in the plasma (in fact the salts in the secretion are composed almost entirely of Na^+ and Cl^-); this gland may be important in excreting excess salt. In sharks maintained in seawater where the external Na^+ concentration is 470 mmol l^{-1} the concentration of Na^+ in the rectal gland fluid is 540 mmol l^{-1}. A summary of the ionic and osmotic concentrations and the secretions in a typical marine elasmobranch are shown in Figure 12.23. Special salt-secreting glands in vertebrates are not unique to elasmobranchs—they are also found in marine reptiles and birds (see later).

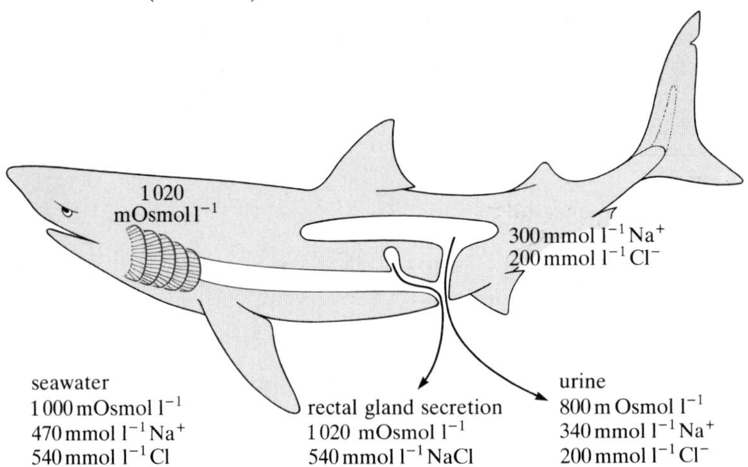

Figure 12.23 Osmotic and ionic concentrations in plasma, urine and rectal gland secretion of a 'typical' marine elasmobranch compared with those of seawater.

12.4.2 Osmoregulation in teleosts

Although teleosts are homoiosmotic, many are relatively stenohaline and cannot withstand dilutions of seawater greater than 50%. Likewise, many freshwater species are limited to fairly dilute waters. This makes even more fascinating the question of the strategies that enable **migratory species**, such as the eel and the salmon, to survive the very different osmotic problems posed in the marine and freshwater environments. This Section begins with an examination of the osmoregulatory problems posed to marine teleosts.

Teleosts in the sea

The osmotic concentration of the body fluids of **marine teleosts** (Table 12.6) is much lower than that of the environment—they are hypo-osmotic regulators.

◇ What osmotic and ionic problems will they face?

◆ A diffusive influx of ions and osmotic loss of water.

The strategies available to teleosts for overcoming these osmoregulatory difficulties will be similar to those seen in invertebrate osmoregulators.

◇ What are the four principal options?

◆ (i) Reduction in the osmotic gradient between the body fluids and the environment. (ii) Alteration in the body-surface permeability. (iii) Active transport of ions into or out of the animal as required. (iv) Alteration in the urine flow rate.

Marine teleosts in all probability maintain body fluid osmotic concentrations similar to those of their freshwater ancestors. Thus reduction in the osmotic gradient between the body fluids and the environment, option (i), would pose problems to the tissues and is not very helpful to marine teleosts. Alteration in body-surface permeability (ii) would help, because this could reduce both the diffusive influx of ions and the osmotic loss of water. In fact, osmotic and ionic fluxes occur largely via the gill epithelium (which must be permeable to effect gas exchange) and via the gut; the general body surface is relatively impermeable. Examine Table 12.7.

Table 12.7 Data for water movement and osmotic permeability constants in fish adapted to freshwater and seawater.

	Medium	Urine flow/ $cm^3\,kg^{-1}\,h^{-1}$	$[Na^+]$ urine/blood	$P_{os}/10^{-4}\,cm\,s^{-1}$
Carassius auratus goldfish	freshwater	1.4	0.08	2.08
Anguilla anguilla eel	freshwater	0.6	0.04	0.79
Platichthys flesus flounder	freshwater	1.8	0.03	0.70
Anguilla anguilla eel	seawater	0.3	0.35	0.19
Platichthys flesus flounder	seawater	0.3	0.10	0.14
Serranus sp. sea-perch	seawater	—	—	0.10

◇ Is it likely that osmotic permeability (P_{os}) is an important factor in osmoregulation in marine fish? (For this to be so, you would expect that P_{os} would be lower in fish in seawater than in those in freshwater.)

◆ Yes; the permeability (largely of the gills) is much lower in fish adapted to seawater than in those adapted to freshwater.

Why should this be? Turn back to Table 12.6 again, and compare the osmolarity of the blood of marine fish with that of seawater ($1\,000\,mOsmol\,l^{-1}$), and that of freshwater fish with that of freshwater ($<5\,mOsmol\,l^{-1}$). The osmotic gradients between the body fluids and the environment are much steeper in seawater ($600–700\,mOsmol\,l^{-1}$) than in freshwater ($300\,mOsmol\,l^{-1}$). Thus, there is a greater necessity for the osmotic permeability of fish in the sea to be lower, which is what is found. Nevertheless, as predicted, there will be an osmotic efflux of water (mainly via the gills) to the more concentrated external environment, and a diffusive influx of salts, coupled with the additional burden of intake of salts with food. Clearly, excess salts must be excreted and water gained to balance the osmotic loss.

One way to excrete salts would be via the kidney, but teleosts cannot produce a urine hyperosmotic to the body fluids. Removal of salts by a glomerular kidney of the type possessed by freshwater teleosts would therefore rid the fish of excess salt only at the expense of a very considerable water loss, which would add to the problems of osmotic water loss. In fact, marine teleosts often have lower rates of urine production (Table 12.7). Interestingly, some achieve this by having a kidney that is functionally aglomerular (in effect just as if they were without glomeruli), and in other species, the kidney is structurally aglomerular (truly without glomeruli).

◇ From your knowledge of kidney function, what will this achieve?

◆ There is no ultrafiltration of the blood but, if the rest of the kidney tubules are present, unwanted materials can be secreted into their lumens. This is what occurs.

The kidneys of marine teleosts excrete divalent ions, largely Mg^{2+} and SO_4^{2-}, with a minimal amount of water, whereas the gills are primarily responsible for excretion of Na^+ and Cl^-. This occurs against considerable concentration gradients, so active transport processes must be involved (the third of the options listed above).

Water to replace that lost osmotically is obtained by drinking seawater. Most of the water in this is absorbed following the active transport of monovalent ions across the gut wall into the fish. It appears that the osmotic permeability of the gut is higher in teleosts in seawater than in freshwater, thus facilitating solute-linked water absorption via the intestine. Divalent ions are poorly absorbed by the gut, and those that are taken in are excreted by the kidney. The high concentrations of monovalent ions taken up are secreted by the gills (excretion of Na^+ and Cl^- by this means increases by a factor of 10). Thus a net water gain is achieved, which may be offset against the losses via the kidney and gills. The exchanges of water and salts between a marine teleost and the environment are summarized in Figure 12.24.

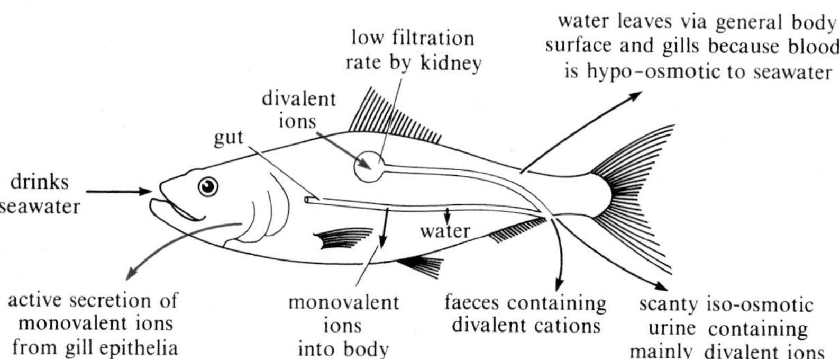

Figure 12.24 A summary of the exchange of water and ions between a marine teleost and the environment.

Teleosts in freshwater

Table 12.6 showed the osmotic concentration of the body fluids of several teleosts in the sea and in freshwater. Teleosts, in common with other vertebrates, are hyperosmotic to freshwater. Thus it is relatively easy to predict the osmotic difficulties facing **freshwater teleosts**. Unless their external membranes can be maintained impermeable, they will suffer the same problems as freshwater invertebrates.

◇ What will these be?

◆ An osmotic influx of water and a diffusive loss of ions from the body to the environment (the reverse of the problems of teleosts in the sea).

Teleosts are excellent volume regulators, so the excess water taken in by osmosis is removed, and the question that arises is how this is achieved. If the body surface is relatively impermeable, then renal loss seems to be an obvious route. Examine the urine flow rates in Table 12.7.

◇ What do the data in Table 12.7 show to be the difference in urine flow rates between teleosts adapted to freshwater and to seawater?

◆ The urine flow is considerably higher in freshwater.

But unless the animal can produce a hypo-osmotic urine, this higher flow could worsen the problem of diffusive ion loss to the medium.

◇ From Table 12.7, is a hypo-osmotic urine produced?

◆ The urine/blood (U/B) ratio for Na^+ is well below 1 for the three species listed, which suggests that unless large amounts of other cations or non-electrolytes are present, the urine is hypo-osmotic.

◇ What do the results in Table 12.7 imply, if it is assumed that primary urine produced in the glomerulus is similar in osmotic concentration to the plasma?

◆ That there is active absorption of ions by the kidney tubules.

Although the urine is hypo-osmotic to the blood, there is none the less a loss of solutes in the urine additional to the diffusive ion loss through the body surface, principally at the gills. The animal will gain some ions in its food but not sufficient to make up for these losses. Therefore, ion uptake must occur from the surrounding medium, and this is mainly via the gills. Freshwater has a salinity of less than 0.5‰ (15 mOsmol l^{-1}: the body fluids are 300 mOsmol l^{-1}), so the ions must be transported into the animal (i.e. in the reverse direction to the transport in marine teleosts) by active processes. We consider this further when the mechanisms by which migratory teleosts survive when they move from seawater to freshwater (and vice versa) are examined. The exchanges of water and salts between a freshwater teleost and its environment are summarized in Figure 12.25; note that freshwater fish do not drink significant amounts of their medium.

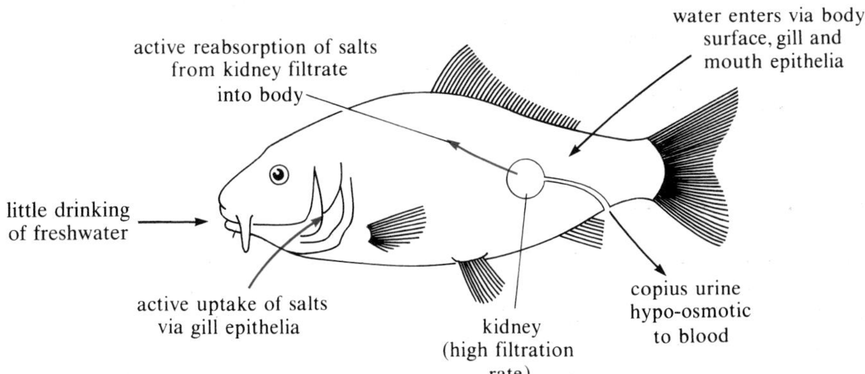

Figure 12.25 A summary of the exchange of water and ions between a freshwater teleost and the environment.

Teleosts that are subjected to changes in salinity

Some fish, such as the blenny *Xiphister* found in rock-pools on the sea-shore, are remarkably euryhaline and can survive well in dilute waters or in seawater, maintaining the ionic composition of their body fluids relatively constant. The flounder, *Platichthys flesus* (Table 12.7), is a marine fish but is also common in estuaries and lower reaches of rivers. It also maintains the composition of its body fluids at a constant level. Such an ability well suits these fish to variations in the salinity of their environment. But how they can cope with the problem of being at one time hypo-osmotic regulators and at others hyperosmotic regulators poses a number of interesting physiological questions. Similar problems exist for **migratory teleosts** like the eel, *Anguilla*

CHLORIDE CELLS
Cells found on fish gills, so called because of their histochemical similarities to the HCl*-secreting cells in the stomach. Salt secretion is not a function of all gill cells but is carried out by those specialized cells on the gill filaments. The Chloride cells have extensive basal & lateral infoldings across which ion movement is believed to occur.

sp., or the salmon, *Salmo salar*. Eels migrate from their main feeding phase in freshwater to the sea to spawn, whereas salmon move from feeding in the sea to freshwater to spawn, normally returning to their natal rivers (the rivers where they lived as fry). Why these species undertake their migratory journeys is not entirely clear.

Eels normally die in the sea once they have spawned but some salmon, notably individuals of the Atlantic species *Salmo salar*, may after spawning in freshwater return to the sea again to feed before returning once more to their natal rivers to spawn. Often at least 10% of females, but rather fewer males, return to the sea after spawning (spawning is a rather more vigorous activity for male (cock) than female (hen) fish). Since teleosts are homoiosmotic, these euryhaline teleosts will face opposite problems in each environment. Table 12.7 shows how P_{os} and urine flow rates vary in eels and flounders adapted to either freshwater or the sea.

◇ What are the problems posed to, and the strategies likely to be adopted by the blenny *Xiphister* when a pool of 100% seawater is diluted by rain to 10% seawater?

◆ The passive osmotic water efflux that occurs in 100% seawater is reversed in 10% seawater, and the diffusive influx of salts in 100% seawater becomes a diffusive efflux in 10% seawater. In 100% seawater, the blenny actually ingests relatively large amounts of the medium, and much of the water is absorbed via the gut, but in 10% seawater this would stop. In addition, urine flow is increased in 10% seawater, and the direction of active transport across the gills changes.

How all these changes are controlled is still a matter for speculation, but there is evidence from several species that endocrine mechanisms are involved. Certainly antidiuretic hormones of neurohypophysial origin are present (see Chapters 1–3), and these are believed to have a sustained antidiuretic effect in marine fish; alterations in the level of fish antidiuretic hormones (see Chapter 11) could account for part of the increased urine flow in freshwater. Neurohypophysial and adrenocorticoid hormones may be involved in the control of Na^+ transport across the gill membranes. Salt secretion does not occur from all gill cells but is carried out by specialized cells on the gill filaments, the **chloride cells** (Figure 12.26), so called because of their histochemical similarities to the cells in the stomach that secrete hydrochloric acid. The chloride cells have extensive basal and lateral tubular infoldings, across which ion movement is believed to occur.

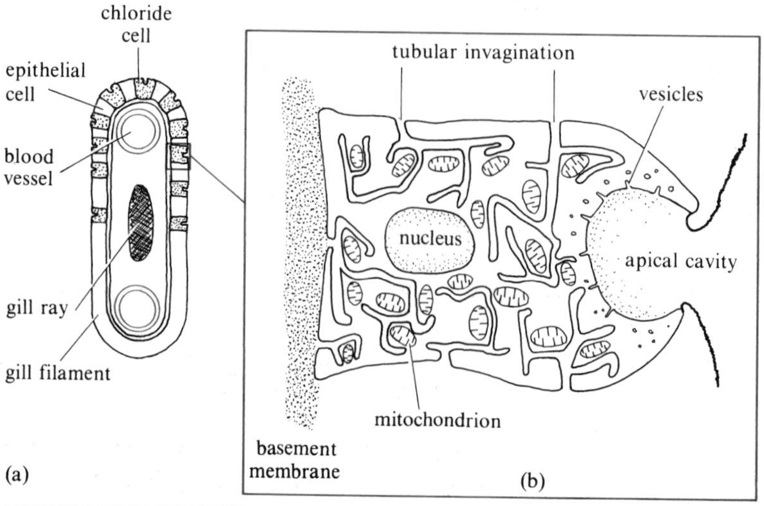

Figure 12.26 (a) Transverse section of a marine teleostean gill filament showing the location of chloride cells. (b) A chloride cell drawn from an electron micrograph. Note the presence of an apical cavity, which in some species disappears in freshwater. In the eel adapted to seawater, arginine-vasotocin (a neurohypophysial peptide hormone with antidiuretic properties) increases the blood flow to the basal lamella of the chloride cell, and this seems to facilitate an increase in secretion of Na^+; in addition, the corticosteroid, cortisol, promotes excretion of Na^+ via the gills in seawater.

In an eel adapted to seawater, a corticosteroid called cortisol promotes excretion of Na^+ by the gills. In freshwater, however, cortisol promotes *uptake* of Na^+ by the gills. Another hormone, **prolactin**, plays a crucial role in many freshwater teleosts in reducing diffusive ion losses via the gills. It was once thought that the cells that excrete Na^+ in seawater also take up Na^+ in freshwater, but it is now believed that the direction of ion transport does not change and that different cells are involved in the two environments.

◇ If hormones play a role in the activities of the chloride cells, what kind of sensory system would you expect to be present to regulate the blood hormonal level?

◆ Osmoreceptors (compare the control of the level of ADH in mammals, Chapter 11, Section 11.3).

12.4.3 Osmoregulation in amphibians

Modern amphibians are mainly aquatic or semi-aquatic (living in damp places) and their external body surface (skin) is often quite permeable. The osmoregulatory problems they encounter and the strategies used to cope with them are reminiscent of those of teleosts. Although some amphibians are found in deserts, even these cannot complete their life cycle without an aquatic phase. They breed in the small water pools formed after periodic torrential rain. The gill-breathing tadpole larvae, typical of the class, have a greatly accelerated development, which enables them to metamorphose to the adult form within a few days, whereas for the common frog *Rana temporaria*, development from egg to adult takes several months. The majority of amphibians, however, live in moist environments or dilute waters. Only a few are found in almost marine waters—for example, the crab-eating frog of Southeast Asia, *Rana cancrivora*, inhabits coastal mangrove swamps. The P_{os} for *Rana temporaria* is 3.3×10^{-4} cm s^{-1}, as opposed to 0.79×10^{-14} cm s^{-1} for the eel gill in freshwater, and 0.19×10^{-4} cm s^{-1} in seawater.

◇ How can the high skin permeability of amphibians such as *Rana* be explained?

◆ One obvious reason is the fact that a considerable portion of their respiration occurs via the skin, particularly in frogs, which, as adults, lack gills and live in damp places.

In Table 12.6, look at the osmotic concentration of the blood of the two amphibians listed.

◇ What problems will the frog face in freshwater?

◆ Clearly the same problems as a teleost hyperosmotic regulator—a diffusive loss of ions and an osmotic influx of water.

◇ What strategies is the frog likely to use to maintain itself hyperosmotic to the environment?

◆ The production of copious amounts of a urine hypo-osmotic to the body fluids and active ion transport into the animal via the skin.

◇ From an examination of Table 12.6, deduce the osmotic problems facing the euryhaline frog *Rana cancrivora*, when in full-strength seawater.

◆ Water efflux and salt influx. The presence of urea in the blood partly solves the problems facing the crab-eating frog in a way similar to that adopted by elasmobranchs.

Summary of Section 12.4

1 Aquatic vertebrates are almost invariably osmoregulators, and they are also to a considerable extent homoiosmotic in the face of changes in external salinity. In freshwater, they are hyperosmoregulators; in the sea (apart from some agnathans, the elasmobranchs and coelacanths), they are hypo-osmotic regulators.

2 Elasmobranchs show ion regulation, but the osmolarity of their body fluids is maintained roughly the same as seawater by the presence of high concentrations of organic substances (mainly urea) in the blood, which they are able to tolerate.

3 Teleosts in the sea suffer a diffusive influx of ions and an osmotic loss of water. This is reduced to a minimum by their low external osmotic permeability, but since the gills must be sufficiently permeable to effect respiration, these problems cannot be completely eliminated. Teleosts gain water and ions by drinking the external medium. Monovalent ions are transported out via the gills, divalent ions via the kidney.

4 The osmotic problems of teleosts in freshwater are the reverse of those in seawater. As in freshwater invertebrates, there is an osmotic influx of water and a diffusive efflux of ions. Excess water is eliminated via the kidneys, and ion loss is made good by transport into the animal via some gill cells. There is little drinking of the medium in freshwater.

5 Migratory teleosts and those living in the intertidal zone must cope with the osmoregulatory problems of both freshwater and the sea, and the mechanisms that enable changes in, for example, the direction of ion transport by the gill appear to be hormonal.

6 Amphibians are essentially aquatic or semi-aquatic animals, and the osmotic problems they face are similar to those of freshwater teleosts. A few amphibians can live in almost full-strength seawater to which their blood is nearly iso-osmotic, mainly because of the presence of high concentrations of urea.

Now attempt Questions 7–9, on pp. 436–437.

12.5 OSMOREGULATION ON LAND

For the most part, aquatic environments are fairly stable and, except in areas like estuaries, show few changes in parameters such as salinity or temperature, whereas on land temperature and water availability may vary greatly. The physical differences between the two environments mean that the physiological problems for aquatic animals moving onto the land are immense, the greatest being that of evaporative loss of water and the subsequent danger of dehydration. So why move to land? The colonization of land must have enabled aquatic animals to escape from competition for resources and from predators and to exploit an array of ecological niches. Successful large-scale evolution of terrestrial groups has taken place only in insects, arachnids, reptiles, birds and mammals; these animals can be found even in deserts.

In a terrestrial environment, the bulk of water and solute exchange occurs not through the body surfaces (which are largely impermeable) but via a balance between food and fluid ingested and materials excreted. Renal organs (kidneys and Malpighian tubules) play a crucial role in the control of fluid balance in vertebrates and insects, and reduction in respiratory water loss is very important in mammals and birds living in arid environments. Here, we do not consider moist-skinned animals such as annelids and most amphibians, which behave 'osmotically speaking' like freshwater animals.

Most terrestrial animals overcome the problems of salt loss by salt intake with their food. However, **evaporative water loss** is less easy to cope with and is additional to urinary and faecal water loss, all of which must be compensated for by ingestion and by the production of metabolic water. Moreover, evaporative water loss is related to the temperature and humidity (saturation deficit) of the air, and the rate of loss will be determined by the gradient of water vapour pressure between the body and the surrounding air.

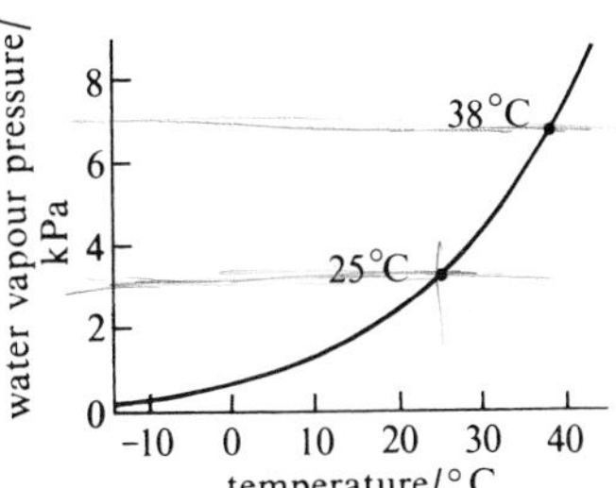

Figure 12.27 The water vapour pressure over a free water surface increases rapidly with temperature. At the body temperature of mammals (37–38 °C), it is roughly twice as high as at room temperature (25 °C), and at higher temperatures, the rise is increasingly steep.

◇ Examine Figure 12.27. What is the increase in water vapour pressure for an increase in temperature from 25 °C to 38 °C?

◆ 3.6 kPa; i.e. it almost doubles.

Thus higher body temperatures are associated with higher vapour pressure differences between the animal and the air (if it is cooler), and these cause increased rates of evaporation. Temperature is not the only factor that affects the rate of evaporation of water from the surface; air movement can result in an evaporating surface being cooled by water loss, as you may have noticed if you stand in a breeze after a swim. The rate of evaporative water loss from several land animals is shown in Table 12.8.

Table 12.8 The evaporation of water from the body surfaces of various animals at room temperature. The data indicate order of magnitude; exact figures vary with experimental conditions. All data are micrograms of water evaporated per hour from 1 cm^2 body surface at a saturation deficit of 0.13 kPa.

Animal	Water evaporated
earthworm	400
frog	300
salamander (amphibian)	600
garden snail, active	870
garden snail, inactive	39
human (not sweating)	48
rat	46
iguana (lizard)	10
mealworm (beetle larva)	6

Note the extremely high loss rates from all except those animals we class as fully terrestrial—human, rat, lizard and mealworm. An inactive snail has a low rate of water loss because in this state it has retreated into its shell and its mucous covering has dried out and formed an impermeable barrier, but when active, a snail's body is covered in a layer of slime from which water is rapidly lost. An impermeable outer covering is present in all fully terrestrial animals; we examine examples from two groups, insects and terrestrial vertebrates (birds and mammals).

12.5.1 Insects

If numbers are an indication of evolutionary success, then insects are extremely successful, with several million species. One of the major reasons for their success is their almost impermeable exoskeleton, which is covered with a thin layer of wax. The waxes are not randomly arranged but are present in organized stacks and this greatly reduces water loss. The organization is disrupted if the animal's temperature is raised above a transition level, which is species-specific. In a balance sheet of water loss and gain for an insect, the individual components are:

Water loss	Water gain
Evaporation	Drinking
from body surface	Uptake via body surface
from respiratory tracheae	from air
Faeces	from water
Urine	Water in food
Secretion	Oxidation (metabolic) water

The magnitude of the problem of water loss depends on the nature of the insect's environment; the stick insect *Carausius* is a herbivore that feeds on fresh plant material and has access to ample water supplies. In contrast, the mealworm *Tenebrio*, which feeds on flour, has little or no water available to it. The mealworm larva has an acute problem of water conservation; its blood osmotic concentration is extremely high, and the animal even has the ability to take up some water from humid air. **Respiratory water loss** from the tracheae (Chapter 7) is kept to a minimum through control of the opening of the **spiracles**; indeed in some insects the spiracles open only intermittently, allowing bursts of water vapour and carbon dioxide to be lost. During the period of apparent closure, however, there is in fact some entry of oxygen, owing to a very slight degree of opening that is not great enough to permit water loss.

The extent of the removal of water from the faeces and urine (by the rectal gland) is truly remarkable in mealworms—indeed, it is almost total, leaving a dry pellet of faeces and uric acid with an osmotic pressure of 13 000 mOsmol l^{-1}, twice that of the urine of a desert rat! Little water is gained in the food, though some may be obtained through the body surfaces. The most important source of water is from oxidation of food. For glucose, the oxidation reaction is:

$$\underset{(180\ g)}{C_6H_{12}O_6} + \underset{(192\ g)}{6O_2} \longrightarrow \underset{(264\ g)}{6CO_2} + \underset{(108\ g)}{6H_2O}$$

◇ Calculate from this equation how much metabolic water the oxidation of 1 g of glucose will yield.

◆ 0.6 g (180 g of glucose yields 108 g of water; therefore 1 g of glucose yields 108/180 g = 0.6 g).

The oxidation of 1 g of protein gives 0.5 g of water, but the oxidation of 1 g of fat gives more water: 1.07 g.

12.5.2 Terrestrial vertebrates: birds and mammals

It is not possible to discuss osmoregulation in terrestrial vertebrates without considering the role of the kidneys: mammalian kidney function was described in Chapter 11. There is an additional water balance problem particularly pertinent to birds and mammals—the use of water as a means of cooling in hot environments.

◇ Name two ways this happens in humans and other mammals.

◆ Your choice could include: by sweating, by panting, or by spreading saliva over the body surface in rats.

In mammals such as humans, the balance between the loss and gain of water and ions is affected by loss through the kidneys, the skin and respiration, and gain from food and fluids ingested. (There will also be water loss in the faeces, which is usually small in carnivores and species eating dried seeds, although it may be somewhat more significant in species, such as cows, that have a very fibrous diet and hence produce large amounts of faecal material.) Most intriguing from the point of view of the physiologist is how desert vertebrates, and especially animals such as rodents, survive in exceptionally arid environments in which there is virtually no access to free water. The kangaroo rat *Dipodomys spectabilis*, which is found in North American deserts, can remain in water balance even when fed on dried foods; a balance sheet for an experimental animal fed on dry barley for one month is shown in Table 12.9: the water loss does not exceed water gain. One reason why the water loss is so small is that urinary excretion is relatively low because the animal produces an extremely hyperosmotic urine.

Table 12.9 Water metabolism during a period in which a kangaroo rat consumes and metabolizes 100 g barley (usually about 4 weeks). Air temperature, 25 °C: relative humidity, 20%.

Water gains	cm^3	Water losses	cm^3
oxidation water	54.0	urine	13.5
absorbed water	6.0	faeces	2.6
		evaporation	43.9
total water gain	60.0	total water loss	60.0

◇ What structural feature would you expect to find in the kidneys of the kangaroo rat (refer back to Chapter 11)?

◆ Extremely long loops of Henle should be present because these desert animals must produce a very concentrated urine.

The major route of water loss is by evaporation (Table 12.9), and this can be kept to a minimum in the desert by behavioural mechanisms—the animals spend much of their time in underground burrows where the air humidity is higher than outside. This reduces respiratory water loss because the inhaled air contains a reasonably high water content. An equally important means of reducing respiratory evaporation in animals such as the camel or the kangaroo rat is achieved by exhalation of air at a lower temperature than the body core. This is effective because saturated air at 38 °C contains 46 mg of water per litre, whereas at 25 °C it contains only 23 mg per litre, so reducing the air temperature by 13 °C potentially reduces water loss by half. This indeed happens in the kangaroo rat, and lung air, which is at core temperature (38 °C), is cooled as it passes out through the nose by a simple heat-exchange mechanism (Figure 12.28). The extent of the cooling depends on the temperature and humidity of the inhaled air. In fact, this heat exchange in the nasal passages occurs in all mammals and birds, but it is more significant where the nasal passages are long, small and narrow, and the surface is large, as in small rodents. In animals with much wider nasal passages heat exchange within the nose is incomplete, as in humans.

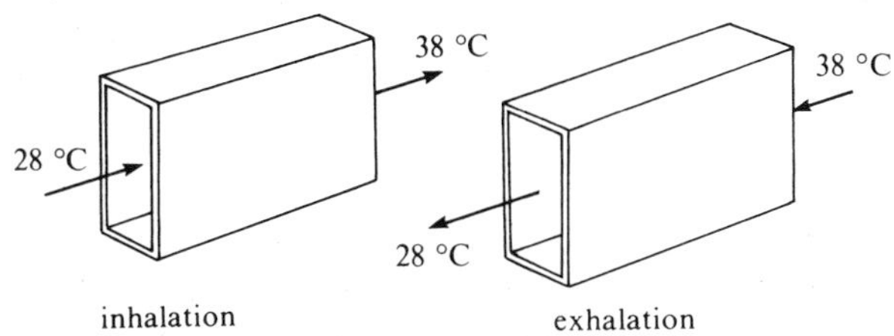

Figure 12.28 A simple model of heat exchange in the nasal passages. Ambient air is 28 °C and saturated, and body temperature is 38 °C. As inhaled air flows through the passages (left), it gains heat and water vapour and is saturated and at 38 °C before it reaches the lungs. On exhalation (right), the air flows over the cool walls and gives up heat, and water recondenses. As heat and water exchange approaches completion, the temperature of the (saturated) exhaled air approaches 28 °C.

Another adaptation that reduces respiratory water loss is shown by the camel, which has a **hygroscopic** (water-absorbing) **mucus** in the nasal passages; this enables the camel to exhale air that is less than saturated. A camel with free access to water can protect itself by sweating in the hot desert environment, but when water is not available the camel will become dehydrated after a considerable time. It can then respond by allowing its body temperature to rise in an attempt to conserve its remaining water. This may result in body temperature of 41.5 °C instead of the normal 38 °C. Such a rise in body temperature means that the camel loses less fluid and salt because it sweats less. In most other mammals, such a temperature rise would cause problems, especially to the brain. The heat-exchange mechanism in the camel's nose apparently cools the blood flowing to the brain and so prevents overheating.

12.5.3 Marine air-breathing vertebrates

Marine mammals, birds and reptiles are essentially terrestrial animals that have secondarily invaded the sea. They remain air-breathing and are physiologically isolated from their environment. However, they have only seawater to drink, and their food will often have a high salt content.

◇ Why will it be advantageous for these animals to eat teleosts rather than sharks or marine invertebrates?

◆ Because the osmolarity of the body fluids of teleosts is much lower than that of marine invertebrates, and hence less salt is ingested.

Nevertheless, there is still likely to be an intake of excess ions, and these must be excreted if balance is to be achieved. For marine birds, this is believed to be achieved via a special salt-secreting structure, the **nasal gland (salt gland)**. Table 12.10 shows the results of an experiment in which salt loss was measured as Na^+ secreted from the nasal gland and the cloaca (urinary excretion) of a gull over a period of 175 min.

Table 12.10 The nasal and cloacal excretion by a black-backed gull during intervals up to 175 min following the ingestion of seawater in an amount nearly one-tenth of its body mass.

	Nasal excretion			Cloacal excretion		
Time/ min	Volume/ cm^3	Na^+ concentration/ mmol l^{-1}	Na^+ amount/ mmol	Volume/ cm^3	Na^+ concentration/ mmol l^{-1}	Na^+ amount/ mmol
15	2.2	798	1.7	5.8	38	0.28
40	10.9	756	8.2	14.6	71	1.04
70	14.2	780	11.1	25.0	80	2.00
100	16.1	776	12.5	12.5	61	0.76
130	6.8	799	5.4	6.2	33	0.21
160	4.1	800	3.3	7.3	10	0.07
175	2.0	780	1.5	3.8	12	0.05
totals	56.3		43.7	75.2		4.41

◇ Does this support the suggestion that salt excretion occurs mainly via the nasal gland in the black-backed gull?

◆ Yes; clearly, Na^+ secretion via the nasal gland is by far in excess of cloacal Na^+ excretion.

The gland is not continuously active: it is activated by the nervous system, in response to a salt load.

It is not known if marine mammals have salt-secreting glands similar to those of birds, but certainly it would not appear to be necessary because whales and seals, unlike birds, can actually produce a urine more concentrated than seawater (Table 12.11). (This also illustrates the poor excretory performance of an animal that normally lives on land—the human!)

Table 12.11 The effect on the water balance of ingesting 1 litre of seawater in a human and in a whale.

	Seawater consumed		Urine produced		Water balance, gain or loss/cm^3
	Volume/ cm^3	Cl^- concentration/ $mmol\ l^{-1}$	Volume/ cm^3	Cl^- concentration/ $mmol\ l^{-1}$	
Human	1 000	535	1 350	400	−350
Whale	1 000	535	650	820	+350

Summary of Section 12.5

1 Terrestrial animals that do not live in moist environments (mainly insects and vertebrates) have a relatively impermeable outer body surface, and their renal organs play a crucial role in the control of fluid balance.

2 Insects without access to water can excrete urine and faeces as a dry pellet, and sufficient water can be obtained from the oxidation of food.

3 Desert mammals can produce an extremely concentrated urine, and for them, also, metabolism of food represents an important source of water. The major route of water loss is by evaporation from the respiratory surfaces, and this can be reduced by heat-exchange mechanisms.

4 Marine air-breathing vertebrates are essentially terrestrial animals that have moved back into the sea but remain physiologically isolated from their environment. They eat food with a high salt content and may also drink seawater. Many marine birds excrete excess salt via special salt-secreting nasal glands, but marine mammals are able to produce a sufficiently concentrated urine to cope with the influx of ions.

Now attempt Questions 10 and 11, on p.437.

SUMMARY OF CHAPTER 12

This chapter has dealt with osmoregulation in marine, estuarine, freshwater, and terrestrial environments. The problems posed to animals in these very different environments were examined by considering in turn the osmotic strategies for survival adopted by invertebrates and then vertebrates. In aquatic environments the animal's body fluids may be iso-osmotic, hyper-osmotic or hypo-osmotic to the external medium. Most invertebrates in seawater are osmoconformers, but show a considerable amount of ionic regulation. In dilute waters, invertebrates are almost always hyperosmotic to the medium. This is achieved by two means:

First, regulation of the extracellular fluid concentration, which comprises: (a) alterations in the passive rates of exchange of water and ions—chiefly a decrease in osmotic permeability and in the osmotic concentration of the body fluids resulting in a decrease in the osmotic gradient between the animal and the environment; (b) volume regulation—production of copious amounts of dilute urine eliminates excess water; and active transport of ions from the environment makes good the losses of salts.

Second, iso-osmotic regulation of the intracellular fluid, which is primarily effected by alterations in the level of organic compounds, notably amino acids, keeping the intracellular fluid iso-osmotic with the extracellular fluid without the necessity for any change in cell water content.

Most vertebrates are homoiosmotic regulators, though the hagfish and elasmobranchs are iso-osmotic or slightly hyperosmotic to the medium when in seawater. The hagfish behaves like marine invertebrates and the osmotic activity of its blood is determined largely by inorganic ions. By contrast, elasmobranchs and the crab-eating frog tolerate high levels of blood urea and have a proportionately lower inorganic ion content. Teleost fish face contrasting problems, being hypo-osmotic to seawater and hyperosmotic to freshwater. The problems posed and the strategies used for survival were discussed, particular attention being given to migratory teleosts, such as the salmon and the eel. In freshwater, there is a diffusive flux of ions to the exterior and an osmotic influx of water. As with invertebrates, this is counteracted by production of copious amounts of dilute urine, and ions are transported in from the exterior via the gills (the skin being relatively impermeable). In seawater, hypo-osmotic fish must counteract a diffusive influx of ions and an osmotic loss of water (hypo-osmotic regulation is seen in only a few invertebrates). The osmotic gradient between the fish and the environment is greater in the sea than in freshwater and its osmotic permeability is less (this contrasts with most marine invertebrates, which are osmoconformers and more permeable in the sea than in freshwater). Teleosts in the sea drink seawater, which is absorbed via the gut along with monovalent ions. The monovalent ions are transported out via the gills and the divalent ions by the kidneys, which produce small volumes of urine. The osmotic concentration of the urine of fish never exceeds that of the blood. Hormonal mechanisms cause the change in direction of ion transport by the gills.

Osmoregulation in fully terrestrial animals (insects, reptiles, birds and mammals) is effected largely by the excretory organs. The major problem of water loss on land is evaporative loss: a case study was given of water balance in the desert rat. Mammals, reptiles and birds that are primarily marine are essentially physiologically similar to their terrestrial counterparts, and the major problem is elimination of excess salts taken in with their food. In some birds (and reptiles), special salt-secreting glands are present, but in mammals, the kidneys alone can achieve the necessary degree of salt elimination.

OBJECTIVES FOR CHAPTER 12

Now that you have completed this chapter, you should be able to:

12.1 Define and use, or recognize definitions and applications of each of the terms printed in **bold** in the text.

12.2 Distinguish between stenohaline and euryhaline animals and between osmoconformers and osmoregulators. (*Questions 1, 2, 3, 7, 8 and 9*)

12.3 Explain how an animal in osmotic equilibrium with its environment is able to regulate its ion composition. (*Questions 2 and 3*)

12.4 Describe the major problems facing an invertebrate moving from more concentrated into dilute waters, and explain the strategies employed by hyperosmotic regulators in coping with these problems (i.e. how they regulate extracellular fluid concentration). (*Questions 3 and 4*)

12.5 Describe mechanisms of iso-osmotic intracellular regulation in invertebrates. (*Question 5*)

12.6 Explain why other factors besides variation in external salinity should be considered in relation to osmoregulation in aquatic invertebrates. (*Question 6*)

12.7 Describe the osmoregulatory strategies that enable fish and amphibians to survive the problems posed in freshwater and in the sea, and discuss the special nature of the osmoregulatory problems faced by migratory teleosts. (*Questions 7, 8 and 9*)

12.8 Describe mechanisms of water conservation in land animals, giving examples of insects and mammals. (*Question 10*)

12.9 Discuss the osmoregulatory problems faced by marine mammals. (*Question 11*)

QUESTIONS FOR CHAPTER 12

Question 1 (*Objective 12.2*) Figure 12.29 is an adaptation curve for the shrimp *Palaemonetes varians*. Is this animal stenohaline or euryhaline, and is it an osmoregulator or an osmoconformer? If it is an osmoregulator, are its body fluids hypo-osmotic or hyperosmotic to the external medium?

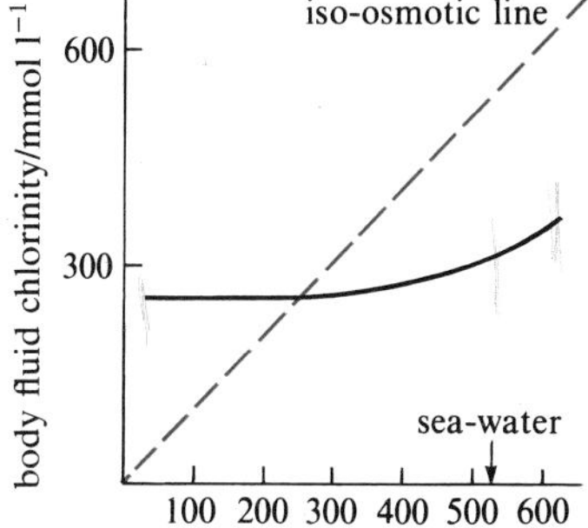

Figure 12.29 Adaptation curve for the shrimp *Palaemonetes varians*.

Question 2 (*Objectives 12.2 and 12.3*) The ratios of the concentrations of various ions in extracellular fluid to their concentrations in the medium (the *B*/*M* ratio) for the edible crab *Cancer* are: Na^+, 1.2; Ca^{2+}, 1.186; Mg^{2+}, 0.648; Cl^-, 0.974; $SO_4{}^{2-}$, 0.740. Does this provide evidence of ion regulation, and if so, which ions appear to be regulated?

Question 3 (*Objectives 12.2, 12.3 and 12.4*) Figure 12.30 is an osmotic adaptation curve for the body fluids of the polychaete *Nereis limnicola* at various external chlorinities.

(a) How does the animal respond to variations in external chlorinity from 500 mmol l^{-1} Cl^- (full-strength seawater) to 200 mmol l^{-1} Cl^- (40% seawater)?

(b) How does the animal respond to variations in external chlorinity below 200 mmol l^{-1} Cl^-, and what osmotic problems is it likely to face?

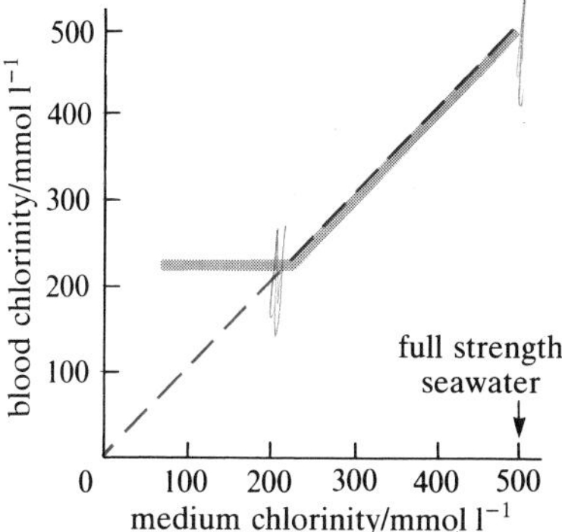

Figure 12.30 Adaptation curve for the body fluids of the polychaete *Nereis limnicola* at various external salinities.

Question 4 (*Objective 12.4*) Which three of the following items are likely to be true of the amphipod *Gammarus duebeni* on transference from full-strength seawater to 20% seawater (cf. Table 12.4 and Figure 12.6a)?

(a) The value for P_{os} decreases.

(b) The value for P_{os} increases.

(c) Larger volumes of urine are produced, becoming increasingly hypo-osmotic to the body fluids as the osmotic concentration of the external medium decreases.

(d) Less urine is produced, and what is excreted is almost as dilute as freshwater.

(e) There is active transport of ions from the medium into the animal, primarily via the gills.

(f) There is active transport of ions almost entirely from the urine back into the animal; the external surfaces of the animal become impermeable.

(g) Ninety per cent of the animals transferred die.

(h) The K_t value for Na^+ transport in *G. duebeni* acclimatized to 20% seawater will be considerably lower than that for Na^+ transport in normal full-strength seawater.

Question 5 (*Objective 12.5*) What aspect of osmoregulation is shown in the results in Figure 12.31? What interpretation can you give for them?

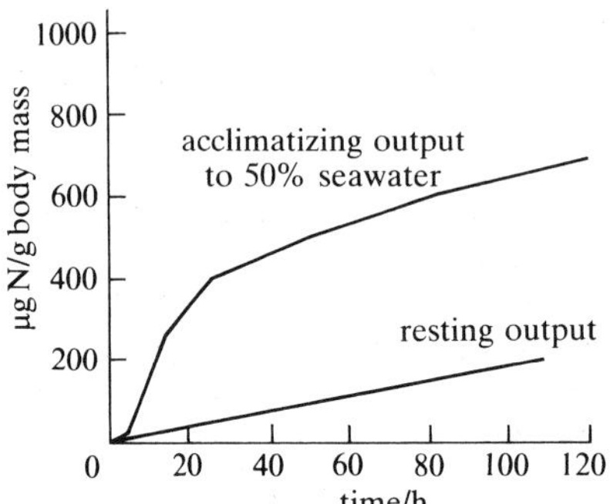

Figure 12.31 Cumulative non-protein nitrogen (N) released to the medium from *Carcinus* after transference of the animals from full-strength to 50% seawater.

Question 6 (*Objective 12.6*) Suppose two hyperosmoregulating crustaceans, acclimatized to seawater of salinity 20‰ and temperature 30 °C, were transferred to brackish water of salinity 2‰, one at a temperature of 25 °C, and the other at a temperature of 5 °C. After the animals had acclimatized to these temperatures, would the water and ionic problems of each be qualitatively and quantitatively similar?

Question 7 (*Objectives 12.2 and 12.7*) In Figure 12.32, the important sites of ion and water movement in osmoregulation in freshwater and marine teleosts are labelled 1–4 (not all sites are important in both freshwater and marine fish). Items (i) to (x) are events that may occur at these sites depending on whether the fish is osmoregulating in freshwater or in the sea.

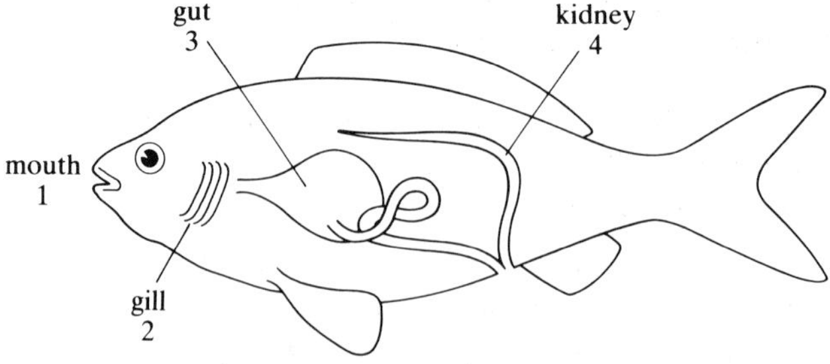

Figure 12.32 Sites of importance in water and ion regulation in teleosts in seawater or freshwater.

Out of items (i) to (x):

(a) For each of the four sites, choose one or two events that occur in a freshwater teleost.

(b) For each of the four sites, choose one or two events that occur in a marine teleost.

(i) The absorption of monovalent ions

(ii) The secretion of monovalent ions

(iii) The absorption of water into the body secondary to the absorption of monovalent ions

(iv) The drinking of the external medium

(v) The secretion of divalent ions

(vi) The absorption of divalent ions

(vii) The output of a large volume of concentrated urine

(viii) The output of a large volume of dilute urine

(ix) The output of a small volume of concentrated urine

(x) Nothing of significance to an osmoregulator occurs at this site.

Qustion 8 (*Objectives 12.2 and 12.7*) Make a list of the osmoregulatory problems posed to a teleost (e.g. the eel) when it migrates from freshwater to its spawning grounds in the sea, and discuss the strategies that will allow it to survive (Refer back to Table 12.7).

Question 9 (*Objectives 12.2 and 12.7*) Figure 12.33 is an osmotic adaptation curve that describes how the osmotic concentration of the body fluids of the frog *Rana cancrivora* changes as the external salinity alters. Is this animal an osmoconformer or an osmoregulator? How do you account for the observed concentration of the body fluids when the external concentration approaches that of full-strength seawater?

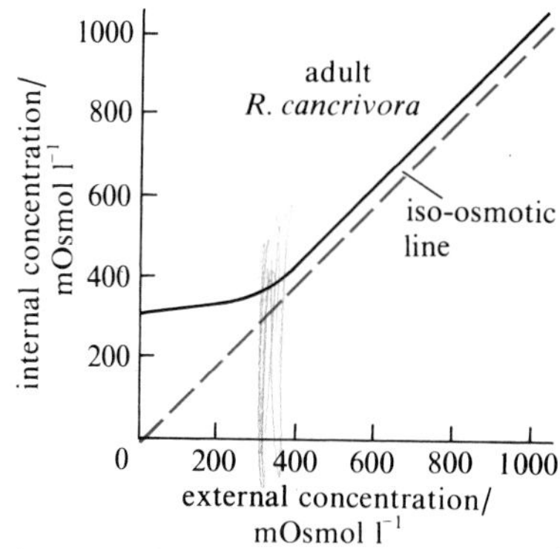

Figure 12.33 Osmotic adaptation curve for adults of the euryhaline crab-eating frog (*Rana cancrivora*) of Thailand.

Question 10 (*Objective 12.8*) The following measurements were made for water loss from various routes in two types of rat, A and B, one of which was a white laboratory rat and the other a kangaroo rat (found in deserts); oxidative water 'intake' was also measured. The values are expressed as grams of water per 420 joules of food metabolized (the joule is the unit of energy).

	Loss in faeces	Loss in urine	Evaporation	Total loss	Oxidative intake
rat A	4	5	14	23	13
rat B	0.63	3.4	3.4	8	13

Which of A and B is the laboratory rat, and which is the kangaroo rat? Give reasons for your answer, and explain any differences in the rates of water loss from the various sources in the two species.

Question 11 (*Objective 12.9*) Which of the following items are likely to be true of the harbour seal *Phoca*?

(a) There is a diffusive ion efflux to the exterior.

(b) There is a diffusive ion influx to the body fluids.

(c) There is an osmotic loss of water to the exterior.

(d) The kidneys cannot produce a concentrated urine.

(e) The seal's main food source also provides much of its water requirements.

(f) Excess salts are excreted by salt glands.

(g) The skin of the harbour seal has a relatively low osmotic permeability.

(h) The kidneys in harbour seals can excrete most of the excess salts taken in from the environment.

(i) The exterior surfaces of the seal are impermeable and permit no salt or water fluxes.

ANSWERS TO QUESTIONS

CHAPTER 1

Question 1 T, X and N are neuron cell bodies: S, W and M are dendritic (receptive) zones; V and Z are synapses: U and Y are axons. The three correct statements are therefore (iv), (v) and (vii).

Question 2 (a) An increase in permeability to Na^+ would shift the E_m towards E_{Na^+} (+45 mV), causing a membrane depolarization.

(b) An increase in permeability to Cl^- would tend to hyperpolarize the membrane because E_{Cl^-} is lower than E_m.

(c) The E_m depends upon the integrity of the membrane and its selective permeability to ions. Thus rupturing the membrane destroys the E_m.

(d) A rise in the concentration of external K^+ would depolarize the membrane because the value of E_m is dependent on E_{K^+}. In turn, E_{K^+} depends upon the difference between the concentrations of K^+ outside and inside the axon.

Question 3 The main difference between a local potential and an action potential is in the activation of voltage-dependent Na^+ channels. A local potential does not reach the E_m threshold at which Na^+ channels start to open, and so it decays with distance. During an action potential Na^+ channels open, driving the E_m towards E_{Na^+} such that the change in E_m is sufficient to generate another action potential, and so on along the axon.

Question 4 (a) Species X is probably an invertebrate because the axon has a small diameter and low speed of conduction. The axon of species Y is a giant fibre (500 μm in diameter) and is therefore from an invertebrate. Species Z is a vertebrate: the axon has a small diameter yet a high speed of conduction.

(b) Species Z has an axon of small diameter but high speed of conduction. This would suggest that the axon is insulated with a myelin sheath.

(c) Increasing the size of fibres in unmyelinated axons decreases the leakage of ions (current) across the membrane and results in a higher speed of conduction.

Question 5 At synapse A the neurotransmitter will produce a deflection of the E_m towards zero. At synapse B a large deflection of the E_m (depolarization) will result because the membrane's permeability to Na^+ alone is changed; that is, E_m will move towards E_{Na^+}.

Question 6 Acetylcholine is rapidly broken down in the synaptic cleft by the enzyme acetylcholinesterase. The enzyme splits the acetylcholine molecule into choline and acetate, which are then taken up into the presynaptic terminal where they are used to synthesize more acetylcholine.

Question 7 The correct statements are (ii)–(iv). The blood–brain barrier is selectively permeable, but not only to gases: substances such as amino acids and sugars can penetrate the brain along with certain drugs. Glial cells have a stable E_m, which can be changed in response to a number of stimuli, but action potentials have never been recorded in them.

Question 8 Sensory inputs can be coded in a number of ways:

(i) According to the frequency of action potentials (which is proportional to the duration of the stimulation);

(ii) According to changes from a baseline or resting frequency of action potential firing;

(iii) According to the way the sensory neuron is connected with other parts of the nervous system.

Question 9 Your flow diagram should look something like Figure 1.26. When you prick your finger, three distinct actions often result:

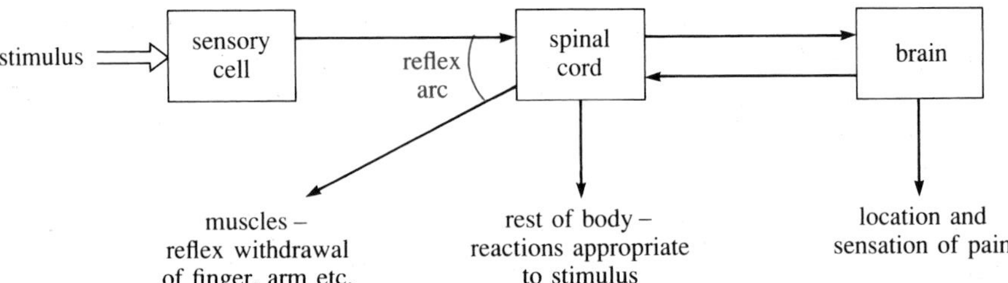

Figure 1.26 Answer to Question 9.

(i) The position of the stimulus on the body is recognized (i.e. information travels to the brain);

(ii) The finger (and sometimes the entire arm) is withdrawn from the stimulus object (i.e. information travels to the spinal cord to effect muscle movement: often the movement involves more than one set of muscles);

(iii) A sensation of pain is experienced (i.e. the brain (and often the vocal cords!) is involved).

CHAPTER 2

Question 1 (a) True, because they are both synthesized from amino acids.

(b) True, because cortisol and progesterone are both C_{21} steroids and therefore have very similar molecular structures. Growth hormone, being composed of 190 amino acids has quite a different structure.

(c) False, they also bind to binding proteins in the blood.

(d) False, LDH occurs in several different forms. Particular peptide hormones (e.g. oxytocin) show a similar polymorphism (many forms), so it is likely that growth hormone differs slightly in different species. The function of a protein is not necessarily determined by the entire amino acid sequence: only a small section of the molecule may be involved.

(e) True.

(f) True.

Question 2 The four factors are: (i) the presence of suitable receptors at target cells; (ii) the rate at which the hormone is secreted; (iii) the rate at which the hormone is broken down in the liver and at the target tissue; (iv) the rate at which the hormone is excreted via the kidney. Factors (ii)–(iv) control the concentration of the hormone.

Question 3 (a) Hormone excretion may vary over a 24-h period, so a 'normal hormone level' at, say 9 am, may be quite different from the normal level at 11 pm.

(b) Small hormones, for example steroids and very small peptides, are likely to appear in the urine.

(c) Changes in the amount of hormone excreted indicate changes in the rate at which hormones are secreted or changes in the amount of protein available in the plasma for binding to the hormone. You might also have suggested that the urine could be analysed for the products of hormone metabolism. This can also yield information about changes in hormone secretion or the rate at which they are broken down.

Question 4 (i) They possess more nerve endings per nerve. (ii) Their action potentials develop more slowly and are of longer duration. (iii) Their cytoplasm contains granules that are much larger than those seen in conventional neurons. (iv) Their products are released into the bloodstream rather than into a conventional synaptic cleft.

Question 5 (a) The writer has omitted to mention that the connection is indirect (via a portal blood system) and involves specialized (neurosecretory) neurons that secrete neurohormones.

(b) Release-inhibiting factors are also involved in some cases. Both types of neurohormone are released from a distinct region of the hypothalamus, not the whole brain.

(c) The hypothalamus is also involved.

(d) Communication via nerves usually also involves the secretion of 'hormone-like' substances (neurotransmitters), which in some cases are identical to hormones secreted by certain endocrine cells. In addition, some endocrine cells can produce electrical impulses in a fashion similar to nerves.

Question 6 (i) The level of ACTH will be decreased because of negative feedback by the elevated cortisol. (ii) The ratio of adrenalin to noradrenalin in the blood will change because high levels of corticosteroids stimulate the adrenal medulla to secrete more adrenalin. (iii) The levels of thyroid-stimulating hormone (TSH) and thyroxin (and triiodothyronine) will increase because cortisol increases the sensitivity of TSH-secreting cells to TRF. As you can see, this single change in the level of one hormone has profound effects on other parts of the endocrine system. Transplant patients must take a whole array of drugs to combat these and other side-effects, in addition to the immunosuppressant drugs.

CHAPTER 3

Question 1 A low K_d value indicates a high affinity, so nuclear receptors have a higher affinity for T_3 than T_4. Because the receptors are more efficient at binding T_3, lower doses of T_3 would be more effective at eliciting a response in the cell.

Question 2 The conclusive test would be to incubate rat liver cells with each of the substances A–D and measure changes in the level of production of cyclic AMP and protein. If any one of A–D produces an increase in both, then it must be an analogue of Z acting on the liver cell receptor. If there is no increase in either the level of cyclic AMP or protein production, this suggests that the compound does not work in the same way as Z.

Question 3 The criteria to be met by an agonist of Z are that it increases both the formation of cyclic AMP and the production of protein. The results show that both A and C are without effect, therefore they are not agonists, but they might be antagonists. Compounds B and D increase cyclic AMP levels and protein synthesis so they are both agonists, although B is relatively weak compared with D. The key findings that sort out these substances are shown in the second half of the Table. Here all have been added to liver cells in the presence of Z. In the presence of B and D, Z stimulates a higher than expected production of cyclic AMP and protein. Both B and D therefore add to the effect of Z but not in a strictly summative way. In contrast, the effect of Z is unaffected in the presence of C but abolished in the presence of A. The latter demonstrates that A is certainly an antagonist, preventing Z from acting by blocking the liver cell receptors.

So the answers are: A is an antagonist, B is a 'weak' or partial agonist, C is neither an agonist nor an antagonist, D is an agonist.

Question 4 The incorrect statements are (ii), (iii) and (vi).

Although it would be true to say that some cells may possess membrane-bound steroid receptors, the majority are cytoplasmic. Recall that steroid hormones are derived from a lipid molecule (cholesterol) and can therefore readily cross cell membranes.

The role of IP_3 is indeed to increase the cytoplasmic concentration of Ca^{2+}, but it does this by releasing Ca^{2+} from an intracellular store (often the endoplasmic reticulum) and not by opening membrane Ca^{2+} channels.

Protein kinase C is activated by diacylglycerol, which is formed, together with IP_3, as a result of receptor-stimulated breakdown of phosphatidylinositol 4,5-bisphosphate (PIP_2).

Question 5 Forskolin and a Ca^{2+} ionophore can be used to raise the intracellular concentrations of cyclic AMP and Ca^{2+}, respectively. Recall that forskolin directly activates adenylate cyclase and the ionophore allows the entry of free Ca^{2+} into the cell, thus mimicking the effect of IP_3.

In response A, forskolin elicits a maximum response but the ionophore has no effect. This indicates that response A is probably mediated by cyclic AMP. However, the effect of forskolin is reduced by the ionophore, thus increased cytoplasmic Ca^{2+} can antagonize the effect of cyclic AMP. We must conclude, therefore, that response A is initiated by an increase in intracellular cyclic AMP but regulated by changes in the concentration of cytoplasmic Ca^{2+}.

In response B, forskolin is without effect but the ionophore elicits a full response. Here we can say that response B is mediated solely by increasing the intracellular Ca^{2+} concentration. In C, both forskolin and the ionophore are effective but only to 25% of the maximum response. The maximum response is only achieved when the agents are added together. Thus response C is mediated by cyclic AMP and Ca^{2+} acting in concert, in other words, they act synergistically.

CHAPTER 4

Question 1 The disturbance to body fluids exerts *control* over the excretion of a concentrated urine and thereby serves the *regulation* of body fluids.

Question 2 The disturbance to body sodium in the form of a deficiency motivates the rat to ingest the solution of concentrated sodium chloride. The sodium ingested corrects the deficiency state. This is an example of a deviation causing action that corrects itself, therefore the system is characterized as showing negative feedback.

Question 3 Yes, there is evidence. Suppose a stimulus causes a move in membrane potential in a positive direction, i.e. membrane potential is less negative. This increases sodium permeability, which increases the flow of sodium ions into the cell. Sodium, being a positive ion, moves membrane potential still further in a positive direction, which increases sodium permeability, which allows more sodium ions to enter, and so on.

Question 4 Blood volume is regulated closely by intrinsic physiological processes and by the control over drinking exerted by blood volume. Therefore drinking in response to haemorrhage is *in response to* a disturbance in the regulated variable and is characterized as negative feedback control.

Question 5 The end served by drinking is regulation of fluid volume. The control exerted over kidney function serves the same end. During water deprivation, the kidney reabsorbs a relatively large amount of water (Section 4.4; see also Chapter 12) and thereby minimizes loss.

Question 6 Figure 4.34 shows the two components and their combination.

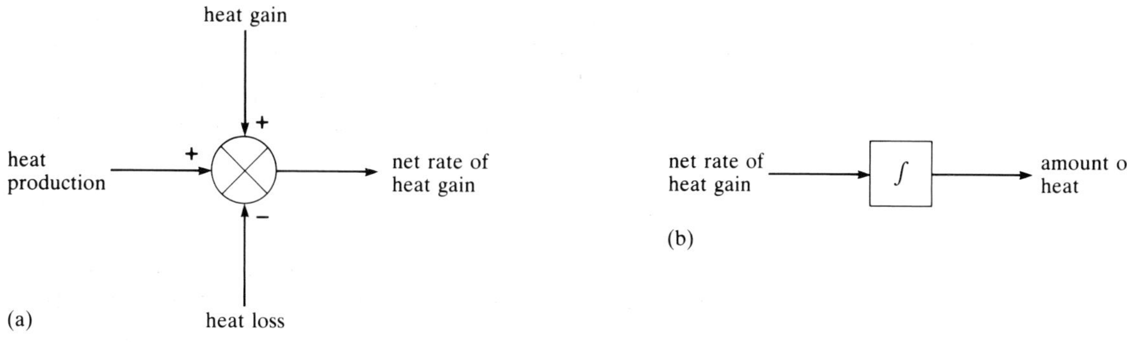

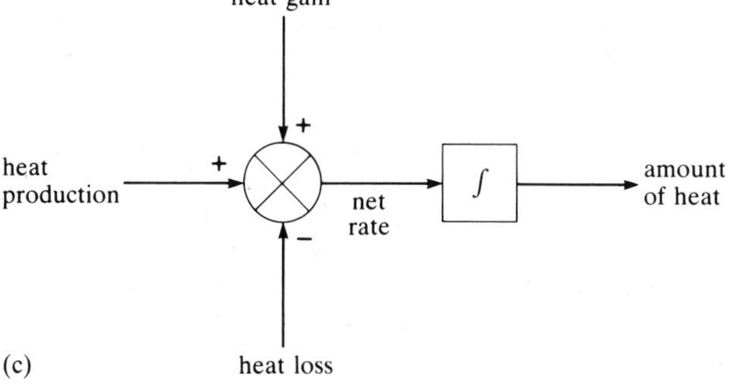

Figure 4.34 (a) The net rate of heat gain is calculated from the algebraic sum of heat gain from the environment, intrinsic heat production and heat loss to the environment. (b) The amount of heat is calculated from the integral of net rate of heat gain. (c) The combination of (a) and (b).

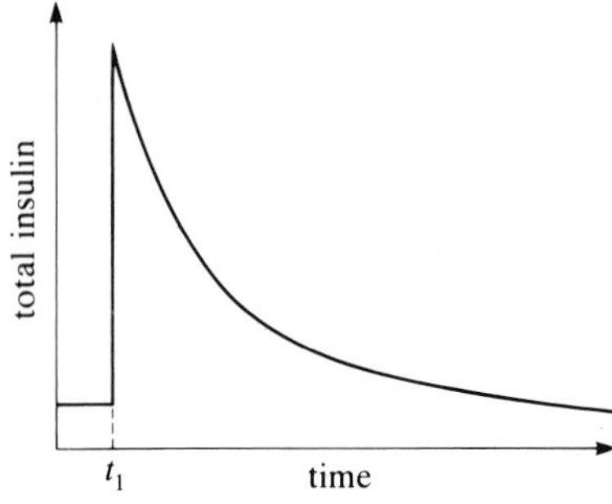

Figure 4.35 A graph showing the prediction of the model following a sudden infusion of insulin.

Question 7 Refer to Figure 4.35. At time t_1 the injection is made. Total insulin rises. Glucose transport increases, and thereby extracellular glucose concentration falls. Let us suppose that, with reference to block 4 of Figure 4.12, extracellular glucose level has fallen to the region where the rate of insulin secretion is zero. As defined by block 7, there will be a rate of destruction of insulin, and therefore 'total insulin' will fall in the manner shown in Figure 4.35. The feedback loop involving blocks 5, 6 and 7 is identical to the model shown in Figure 4.9d, and so the responses of the two systems are comparable (Figure 4.35 and Figure 4.10).

Question 8 If glucose transport into cells is at a constant value, then extracellular insulin concentration is also at a constant value, the relationship being defined by block 9. Assuming extracellular fluid volume to be constant, then total insulin must be held at a constant level over the period of observation. For this to happen, the rate of insulin secretion must be equal to the rate of insulin destruction, and so the signal leaving the summing junction (block 5) and entering the integrator (block 6) must be zero.

Question 9 It means that the frequency with which the neuron exhibits action potentials varies as a function of temperature. The frequency could go up with temperature (the 'warm neuron') or decline with increases in temperature (the 'cold neuron').

Question 10 The neuron will fire more frequently, which will increase the input to neuron 3. This will activate mechanisms of heat loss, and so body temperature will fall. It will fall to a value at which the activities in the two pathways (1⟶3 and 2⟶4) are again equal, and therefore there is no effort exerted in the interests of either losing or conserving heat.

Question 11 When the drug acts at point A, there will be a net positive signal at the output side of the summing junction, which will be the stimulus to heat loss mechanisms. Body temperature will fall to a level at which the activities in the two afferent arms are again equal, and so there is no efferent activity. When the drug acts at point B, heat production will be activated. Body temperature will rise and the signal from the warm sensor will increase. Hence, there will be a positive output from the summing junction, the stimulus to the mechanisms of heat loss. A new level of equilibrium between heat loss and heat production will eventually prevail.

Question 12 The rat would be expected to overdrink by a very large amount. It would stop drinking only when the deficit in body fluids had been corrected, by which time there would be a large amount of water in the alimentary tract that would later be absorbed.

Question 13 So long as drinking water is not available the kidney will excrete a strongly hypertonic urine. The longer the time after the injection, the greater the proportion of the extraneous load that can be eliminated by the kidney. Therefore the less is the disturbance to the body fluids, and the lower the stimulus to drinking. Of course, if the period is extended then there will come a point at which the amount drunk starts to rise as a function of deprivation time.

Question 14 Haemorrhage involves the removal of plasma, therefore the optimal solution is one that repairs both the ionic and water components of plasma, i.e. isotonic sodium chloride solution.

Question 15 The loop can be opened by sucking out blood as fast as it is ingested, a procedure that Lent and Dickinson do, in fact, describe. Another way would be to cut the neurons whose activity provides the feedback signal.

Question 16 In an experiment it is necessary to keep all conditions constant except for the one variable that is under investigation. In this experiment, variety is being compared with no variety. Therefore if the dishes are changed in the variety condition they must be changed also in the no variety condition. Suppose one did not do this and found that in the variety condition more food was ingested. One could not rule out the possible explanation that merely disturbing a rat stimulates ingestion (for example, via stress hormones). Such an experimental design would be said to be *confounded*.

Question 17 The sodium-deficient rat shows an immediate preference for a concentrated solution of sodium chloride. It does not require a learning mechanism. By contrast, rats are not able to recognize vitamins. Rather they learn to associate distinct tastes associated with a vitamin-containing meal with beneficial consequences of ingesting the meal.

Question 18 In such an experiment, the rat that has received food drinks more than the rat deprived of both food and water. Although a water-deprived rat significantly cuts down on its food intake it still ingests some food. This places a load on the kidney for excretion. There are other factors also implicated, but their discussion goes beyond our brief here.

Question 19 The rat drinks an amount of water commensurate with the disturbance that the presence of food causes to the body fluids. However, the drinking occurs before the food ingested has time to disturb body fluids. In this sense, the rat anticipates the potential disturbance, or, in other words, it is acting in a feedforward mode of control.

Question 20 We would need to be certain that it was specifically the pairing of the tone and exercise that was responsible. We would need to control against simply the passage of time being the crucial variable. A control group might be exposed to the same exercise and the same tone over the same period of time, but there would be no pairing of tone and exercise. The tone might be presented 1 h after exercise in this control group.

CHAPTER 5

Question 1 (i) False. The rate at which a substance diffuses is inversely proportional to the square of the distance diffused.
(ii) True.
(iii) False. Although cephalopods have a closed circulation, this is not true for ALL molluscs.
(iv) True.

Question 2 The correct answers are (i) 5%; (ii) endothelial; (iii) permeable to substances of low relative molecular mass.

Question 3 Statements (i), (ii), (vi) and (viii) are true.
(iii) False. Contraction of the dorsal vessel is the principal means of propelling blood in the earthworm.

(iv) False. The octopus heart has, as Figure 5.23 shows, a single ventricle and two atria.
(v) False. Arteries generally have a smaller diameter than veins.
(viii) False. Mammalian capillaries have a diameter that can be so small that it is equivalent to the diameter of a red blood cell—about 7.5 μm.

Question 4 (a) See Figure 5.36. (b) The net movement of water is *out* of the capillary.

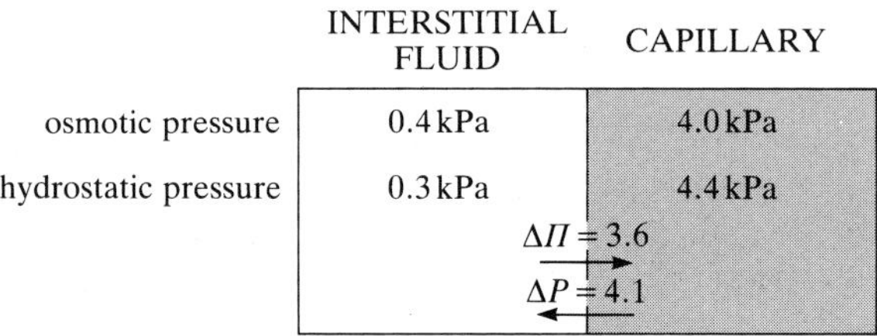

Figure 5.36 Answer to Question 4.

Question 5 See Figure 5.37.

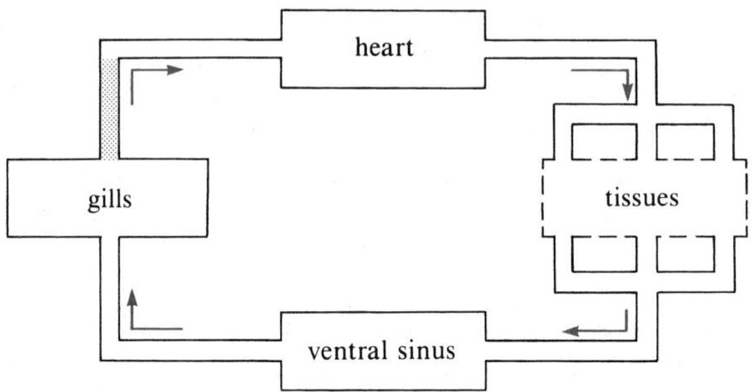

Figure 5.37 Answer to Question 5.

Question 6 Oedema is caused by the accumulation of interstitial fluid. With little or no protein in the diet there is much less protein in the blood than usual. The effective osmotic pressure of the plasma is largely due to protein macromolecules that are confined within the capillaries. With less protein, the osmotic pressure decreases and the effective filtration pressure increases. The hydrostatic pressure remains unaltered. More fluid than normal is therefore forced out of the capillaries and the excess volume is so great that the lymphatics cannot remove it fast enough and oedema results.

Question 7 The first effect of violent exercise will be a large change in the demand of the tissues for blood. Your skeletal muscles will need an enormous increase in blood flow, and their arterioles will be greatly dilated by a combination of local metabolites and other vasodilator influences. By rule 2, because the total peripheral resistance has now fallen, the arterial pressure will also fall and the venous pressure will rise. This rise means that the heart will fill more in diastole and empty more in systole, and stroke volume will rise, as predicted by rule 3. Heart rate will also increase because the fall in arterial pressure will be detected by the baroreceptors. Cardiac output will rise and hence, by rule 4, arterial and venous pressure will return to normal.

In principle, these changes could occur during the exercise, but the change in demand is so large that the circulation would be unstable during the period of adjustment. In exercise, therefore, additional nervous mechanisms elevate heart rate and contractility *in advance* of changes in tissues' demand for blood, so 'tuning' the system to respond optimally.

Question 8 At first sight the application of the rules suggests that the cardiovascular system would become highly unstable. In fact, the responses of the system are initially inappropriate, because arterial and venous pressures don't change in opposite ways. The initial effect of a temporary bleed is to reduce blood *volume*. Most blood is in the veins, so unless you have cut an artery the first change is a drop in venous pressure. At this instant, there is no change in the demand for blood. By rule 3, a fall in venous pressure leads to a fall in cardiac output, as the heart fills less in diastole. Then, by rule 4, a fall in cardiac output leads to a fall in arterial pressure. The arterial baroreceptors will detect the fall and, by rule 3, increase heart rate. By rule 5, blood flow to the skin will be reduced.

The effect of the increased heart rate is that more blood is pumped out of the veins, which reduces venous pressure. Increasing peripheral resistance helps to keep blood in the arteries, but stops it flowing to the veins, so venous pressure falls still further. Cardiac output falls again . . . leading to a further fall in blood pressure . . . which increases heart rate and peripheral resistance still more, leading to a further fall in cardiac output . . . etc. So, what you see in someone who has bled is a very rapid, feeble pulse, with dead white skin. The normal control mechanisms do not provide protection, as they tend to make the original problem, a fall in venous pressure, worse. Clearly the worst does not always happen: a separate mechanism comes into operation that protects against the continuing fall in pressure. This protection is a quite different response, constriction of the veins or 'veno-constriction', which increases venous pressure at the lower blood volume. Fluid also passes from the spaces between the body cells to increase volume in the circulation, until longer term mechanisms can restore the blood components lost in the bleeding.

Question 9 The effects of left heart failure would be (i) a fall in systemic arterial pressure, (ii) a rise in pulmonary venous pressure, and (iii) a later rise in systemic venous pressure. The rise in venous pressure would increase the hydrostatic pressure at the venous ends of the capillaries such that it exceeded the colloid osmotic pressure and fluid would not be absorbed.

CHAPTER 6

Question 1 (a) $P_{N_2} = \dfrac{\%\ N_2 \text{ in mixture}}{100} \times \text{total pressure}$

So,

$$P_{N_2} = \frac{60}{100} \times 200 = 120\,\text{kPa}.$$

(b) For each animal's blood, you should express the volume of extracted gas at a *standard temperature and pressure* (STP). The standard conditions applied are 0 °C and 101.3 kPa. The details of conversion to standard temperature and pressure need not concern you, but suppose the extraction of gas from blood was performed at 20 °C and the gas occupied 6.0 cm^3 at 100.0 kPa; then at STP the volume of gas is

$$6.0\,\text{cm}^3 \times \frac{273.15}{293}\,\text{K} \times \frac{100.0}{101.3}\,\text{kPa} = 5.52\,\text{cm}^3.$$

(c) Recall that 20.9% of air is oxygen. If the water vapour pressure is taken into account, the total pressure exerted by the gases present is 95.6 kPa, and the P_{O_2} is therefore $20.9/100 \times 95.6 = 20.0$ kPa.

(d) First, note that absorption coefficients (Table 6.1) are normally expressed as cm^3 of oxygen dissolved per cm^3 of water, whereas the solubility of oxygen in freshwater is given as cm^3 of oxygen per litre of water. Second, note that Table 6.2 refers to the amounts of gas dissolved in freshwater that is in equilibrium with *atmospheric air*, where the P_{O_2} is approximately 21 kPa. Absorption coefficients are obtained by measuring the amount of gas (e.g. oxygen) dissolved per cm^3 of water when the P_{O_2} is 101.3 kPa (i.e. a pure oxygen atmosphere). In air, with about one-fifth the P_{O_2} of an oxygen atmosphere, the amount of oxygen that dissolves in solution is about one-fifth the amount predicted by the absorption coefficient. Finally, note that Table 6.1 refers to pure water, and Table 6.2 to freshwater with an appreciable ionic content.

Question 2 (a) Recall Equation 6.2; expressing α as cm^3 of oxygen dissolved per litre of distilled water,

$$c = 48.9 \times 30/101.3 = 14.48\ cm^3.$$

(b) You know already that 20.9% of air is oxygen, therefore the amount of oxygen present per litre of this air is 209 cm^3.

(c) No, this is incorrect; you can predict oxygen diffusion only by calculating the gradient of the partial pressure of oxygen. The P_{O_2} of the water is 30 kPa, whereas that of the air is 21.2kPa. Therefore, the net diffusion of oxygen will be from water to air. This should remind you that oxygen does not always diffuse from regions of high oxygen content to regions of lower oxygen content, but it does always diffuse from regions with a high P_{O_2} to those with a lower P_{O_2}.

Question 3 (a) is false, the airsacs are compressible but are not major sites of gas exchange; they function as bellows.

(b) is false; ventilation of the tracheal system is not *essential*; many small, less active insects rely on diffusion alone and ventilation never directly renews the air in the tracheoles because they are fine, blind-ended tubes.

(c) is false; in a resting insect, the tracheoles are partly filled with fluid and during activity (or exposure to hypoxia) the fluid is withdrawn, so aiding oxygen diffusion.

(d) is false; a unidirectional flow of air is attained by some insects and it may have the advantage that it replenishes the tracheal air more effectively. Any sort of ventilation of the tracheal system is likely to be an advantage (even the tidal system shown by some insects) because it is faster than diffusion alone.

Question 4 (a) The rate at which water is lost would probably increase dramatically because the most potent stimulus for spiracular opening is carbon dioxide. Especially in a dry environment, extensive water loss would result.

(b) From Figure 6.4, when the spiracles are closed very little carbon dioxide is released. This suggests that the cuticle is highly impermeable to gases, as indeed it is. So when the spiracles are sealed, there should be very little carbon dioxide released and very little oxygen taken up because the oxygen stored in the tracheal system would be quickly exhausted and could not be replaced.

(c) No. Although in many insects it is the only mechanism, in others (e.g. *Cecropia*) a diffusional flow of oxygen inwards is supplemented by a convective flow of air; other insects ventilate the tracheal system.

(d) (i) When the spiracles are fully open, the air pressure within and outside the spiracles will be equal, that is atmospheric. (ii) Some time after the closure of the spiracles, a subatmospheric pressure would develop because oxygen would disappear from the tracheolar air, and the carbon dioxide that was released would become combined chemically to form bicarbonate; the gas remaining is mainly nitrogen which means that if all the oxygen were used up the pressure would be about four-fifths the atmospheric pressure.

Question 5 (a) No; unlike haemoglobin, myoglobin is not made up of subunits (see Figure 6.6) and neither myoglobin nor *separate* subunits of haemoglobin can interact as do the subunits in the tetramer.

(b) At about 1 kPa; note that this P_{O_2} is considerably lower than that required for 50% loading of the sigmoid dissociation curve in Figure 6.9 (4 kPa), which demonstrates the advantage of cooperativity. (Recall that the dissociation curve for isolated haemoglobin subunits resembles that labelled A in Figure 6.9.)

(c) No; Figure 6.7 shows the haem groups to be spatially separated. Although the term 'haem–haem interaction' is sometimes used, 'subunit interaction' is more accurate.

Question 6 (a) The effects of increased carbon dioxide levels are the reverse of those illustrated in Figure 6.10; that is, there is a shift to the left, which means (i) an *increase* in affinity and (ii) a *decrease* in P_{50}.

(b) The best measure is from (iv). In the Bohr shift, decreased pH does not influence the oxygen capacity of the pigments, so (i) is incorrect; (ii) is not sensible; (iii) cannot account for differences in the 'normal' level of oxygen turnover to the tissues of different pigments.

(c) A higher affinity, so the pigment can load up more fully at a lower P_{O_2}.

(d) Lower level, because a raised level of 2,3-DPG decreases the affinity of haemoglobin for oxygen. (But remember that there are *molecular* differences between fetal and maternal haemoglobins.)

Question 7 The data relate to *Arenicola* and items (a), and perhaps (b), favour a storage role. Items (c), (d), (e) and (f) suggest a limited transport function. Taking each item in turn,

(a) No; pigments with a storage function are likely to have a high affinity for oxygen.

(b) No; this might imply a storage *or* a transport role.

(c) Perhaps; this need not rule out a transport role. If the rate of blood flow to the tissue is high and the oxygen capacity is large, even a 10% fall in saturation might be useful.

(d) Yes, because exposure at low tide would obviously exceed 20 min. However, an emergency store for the brief periods between bouts of irrigation might be useful.

(e) Yes; this suggests that the P_{O_2} in the burrow never falls so low that the pigment cannot load up fully, so a transport function of alternately loading at the gills and unloading at the tissues is possible.

(f) Yes; this implies that a transport function is particularly important at low tide. In fact, reliable measurements of the P_{O_2} at the tissues are difficult to make under natural conditions.

Question 8 (a) $\text{pH} = 6.1 + \log \dfrac{[\text{HCO}_3^-]}{P_{\text{CO}_2} \times 0.19}$

Therefore $\text{pH} = 6.1 + \log \dfrac{25}{4 \times 0.19}$

$$= 6.1 + \log(32.90)$$
$$= 6.1 + 1.52$$
$$= 7.62$$

(b) At alkaline pH, free calcium concentration is reduced, nerves become excitable, and the person will suffer tetany.

(c) Diabetes produces a metabolic acidosis ($[\text{HCO}_3^-]$ falls), therefore breathing is increased to *lower* p_{CO_2} and bring the ratio ($[\text{HCO}_3^-])/p_{\text{CO}_2}$ back towards normal.

(d) As more oxygen is withdrawn from haemoglobin, its buffering capacity is increased, so more carbon dioxide can be carried.

CHAPTER 7

Question 1 (d), (e) and (h) are true. The others are false.

(a) is false; these channels only conduct air to the alveoli, which are the major sites of gas exchange.

(b) is false; after normal expiration, the lungs still contain approximately 2.5 litres of air (called the resting expiratory volume).

(c) is false; lungs that are ventilated in a tidal way can never be completely evacuated (unless the lungs fully collapse). Hence as they fill up, fresh air mixes with the stale air that remains in the lung after the previous expiration.

(d) is true; this follows from (c) above. Very little fresh air enters the alveoli on inspiration because the first air to reach the alveolar sacs is that left behind in the conducting tubes after the previous expiration. The composition of alveolar gas throughout the ventilatory cycle is therefore more stable than if the alveoli were flushed with fresh air, and this may be an important advantage.

(e) is true; (the mean blood pressure in the pulmonary artery is about 1.8 kPa; the corresponding figure in the main aorta is about 12 kPa).

(f) is false; the right side of the heart pumps per unit time the *same amount* of blood as the left side. Remember that the pulmonary and systemic circulations are in series. If a large volume of blood is to be passed through the lungs when the mean pressure in the pulmonary artery is low, then pulmonary resistance must also be low.

(g) is false; the pulmonary vessels are distensible and hence large increases in blood flow (e.g. during exercise) are *not* accompanied by large rises in pulmonary blood pressure. Filtration of fluid into the alveoli does not occur because blood pressure in the pulmonary capillaries remains very low and does not greatly exceed the osmotic pressure of the blood.

(h) True; with an invaginated structure, movement of air (which always increases loss of water by evaporation) can be kept to the minimum required for gas exchange and the respiratory surface can be close to air partly saturated with water vapour.

Question 2 (a) About 2.4 litres; this is called the resting expiratory volume.

(b) About 0.5 litres, or 500 cm^3. Obviously the amounts of air inhaled and exhaled with each breath are equal. The 'normal' tidal volume is usually stated as 500 cm^3.

(c) About 4.8 litres. During forced deep breathing, the maximum amount of air in the lungs at inspiration is 6 litres and the amount left behind after forced expiration is about 1.2 litres. The difference between them is the maximum possible tidal volume (sometimes called the *vital capacity*).

(d) 6 litres; given a tidal volume of 500 cm^3 and a respiratory rate of 12 breaths per minute, then the volume per minute is 500×12 cm^3.

(e) 4.2 litres; recall that of the 500 cm^3 inhaled per breath, only about 350 cm^3 are effective in the ventilation of the lung. The first 150 cm^3 of air taken into the respiratory spaces is stale air previously occupying the anatomical dead space. Thus the volume available for gas exchange is $350/500 \times 6 = 4.2$ litres.

(f) No; Figure 7.25 is concerned only with the total volumes of air entering and leaving the lung.

(g) When the demand for oxygen is increased during exercise, the volume per minute is increased. This can be achieved by a substantial increase in the tidal volume mainly caused by an increase in the amount of air taken in; there is a smaller increase in the amount of air forced from the lung. The ventilation rate increases; but because the entire ventilation cycle during deep breathing must take longer, this increase may be modest compared with the increase in tidal volume. (You could try to work out how large this increase is by measuring your ventilation rate before and after intense exercise.)

Question 3 The true statements, in sequence, are: (d), (c), (g). Recall the mechanics of ventilation discussed in Section 7.4.

(a) is incorrect because the question refers to the ventilatory cycle in a *resting human*. Only when expiration is an active process (as in panting) do the internal intercostals pull the ribs down and the abdominal muscles contract to force the diaphragm upwards.

(b) is incorrect; the alveoli lack muscle and cannot *contract*. The air inspired into the alveoli is a mixture of fresh and stale air.

(c) is correct.

(d) is correct; the external intercostals contract throughout inspiration and the motor nerves to these muscles fire more frequently as inspiration begins. This must therefore be the first event of this series.

(e) is incorrect; in mammals, air is sucked into the lungs; that is, a 'suction pump' operates.

(f) is incorrect; the emptying of the lungs is aided by the recoil of both the lungs and the thoracic wall but in the main this recoil is due to *elastic* not muscular tissue.

(g) is correct and is the last event in the sequence because it refers to expiration. Note that the volume of air expelled is equivalent to the tidal volume (see Figure 7.25).

Question 4 (a) True; the avian lung is semi-rigid and its ventilation is produced exclusively by airsacs (see Figure 7.6). In mammals, the lungs clearly change in volume during ventilation.

(b) False; birds have ribs, which are important for ventilation. The airsacs change in volume during ventilation but this is linked to changes in the volume of the thoracic–abdominal cavity (only mammals have a diaphragm) caused by movements of the sternum and ribs.

(c) True; in both birds and mammals air is drawn into the system by suction (i.e. by a suction pump) and is expelled by positive pressure (in both cases this is sometimes aided by muscular contractions).

(d) True; this is a major difference between mammals and birds.

(e) False; water is lost through evaporation at both sites, although in birds much evaporation takes place at the surface of the airsacs.

(f) True; ventilation of the lungs is increased during heat stress when mammals pant, but alkalosis is avoided by rapid shallow panting. (Panting mainly moves air in the dead spaces, which avoids alkalosis.) In birds, the increased ventilation appears to bypass the lungs.

Question 5 (a) False; gill slits are lateral openings in the pharyngeal wall, but they link the buccal and *opercular* cavities.

(b) False; secondary lamellae are present on both surfaces of each gill filament and the lamellae run *perpendicular* to the long axis of the gill filament.

(c) True; see Figure 7.10.

(d) True; see Figure 7.10.

(e) True; recall the importance of counter-current flow (see Figure 7.10).

(f) True; the lamellae are very thin-walled and there is an extensive flow of blood within them; the internal walls are held apart by pillar cells (see Figure 7.10).

Question 6 (a) (ii); the increase is due largely to an increase in the number of red cells, though there may be a modest increase in the amount of haemoglobin contained in each red blood cell.

(b) (i); Some reserve red blood cells (e.g. from the spleen) are added to the circulation, but most of this increase is due to an increased rate of production of red blood cells.

(c) (ii); recall that the same amount of oxygen can be delivered to the tissues of highlanders for a smaller fall in arterio–venous P_{O_2}.

(d) (ii); the concentration of 2,3-DPG increases during acclimatization.

(e) (ii); at altitude, the affinity of haemoglobin for oxygen falls; that is, there is an increase in P_{50}.

(f) (i); this should remind you that acclimatization also involves changes in the tissues (the function of myoglobin is mentioned in Section 6.4.1). Remember that the diaphragm contracts on inspiration, so hyperventilation at high altitude depends on the ability of this tissue to tolerate mild hypoxia.

(*Note* You are not expected to remember any of these data.)

Question 7 (a) False; some sluggish species, for example the toadfish, are conformers (Figure 7.22), but most active teleosts are regulators.

(b) True; oxygen consumption increases during exercise and an adequate supply of oxygen to the tissues is more difficult to ensure.

(c) True; for the reason given in (b).

(d) True; in poikilotherms, the basal rate of metabolism increases when temperatures are raised. The extra supply of oxygen to the tissues is more and more difficult to maintain when the temperature increases (Figure 7.22).

(e) False; in general, species that inhabit environments that are low in oxygen tend to be better regulators.

(f) True; the physiological changes that occur during acclimatization to an environment with low P_{O_2} appear to increase the tolerance of fishes to hypoxia.

CHAPTER 8

Question 1 The deletion of an essential amino acid results in a marked reduction of growth rate (Section 8.2). Leucine, phenylalanine and threonine are therefore essential acids. Notice that there is considerable variation in the recorded mean weights of non-essential amino acids, so the rather low value for serine is probably due to normal experimental variation.

Question 2 (a) Subgroup 1: D, E, F (needs thiamine); subgroup 2: A, B, F (needs thiazole); subgroup 3: A, B, C (needs amino acids); subgroup 4: B, D, F (needs pyrimidine and thiazole).

(b) Subgroup 3.

(c) Subgroup 1.

In these flagellates, the diversity of dietary requirements is a consequence of variation in metabolic synthetic ability. This point is discussed in Section 8.2.

Question 3 (a) Similar points: 1, 3, (12), 16, and 18. Additional comment is necessary for point 12: the particles captured by *Amphitrite* are subject to some preliminary sorting before ingestion. This has not been described in the text. Refer back to the accounts of feeding in *Amphitrite* and *Sabella* to check the other answers.

(b) Points not applicable: 2, 5, 6, 9, 11, 13, 14, and 17. You can check the answer to the question if you read the accounts of the feeding mechanisms of *Sabella* and *Amphitrite*.

Question 4 Points 3, 6, 7, and 12. *Calanus* is a suspension feeder, feeding by combing out suspended particles with its appendages.

Question 5 The predominant features of the jaws in Figure 8.37a are (1) the enlarged tooth on the upper jaw, (2) the small backwardly directed teeth posterior to the large tooth, and (3) the wide gape of the jaws. These features suggest that the animal seizes large food sources, which are swallowed whole. The jaws are those of a rattlesnake, and the anterior tooth or fang is actually modified to inject poison into the prey, and once this has been paralysed or killed, it can be swallowed. Thus, the jaws in Figure 8.37a belong to an animal that seizes and swallows its prey.

The predominant features of the jaws in Figure 8.37b are (1) the small teeth (incisors) on the lower jaw, (2) the absence of incisors on the upper jaws, (3) the gap (diastema) between the incisors and the premolars and molars at the back of the jaw (which means that the upper and lower canines and one premolar are lost), and (4) the roughness of the premolars and molars. Such an animal cannot be a predator. The nature of its incisors and molars suggest that it is a grazing animal. The jaws are in fact those of the ox. Thus, the jaws in Figure 8.37b belong to an animal that seizes and masticates its food.

Question 6 Region 1 is the mouth with extensive salivary glands, which secrete a strong amylase. The oesophagus and the crop are the regions of conduction and storage, respectively. The proventriculus is the region of internal trituration and early digestion, the cuticle forming strong teeth which crush food. The mid-gut is the site of later digestion and absorption, and proteases, carbohydrases and lipases are active here. As you might have guessed, the mid-gut caeca are a means of increasing the surface area of the absorptive region of the gut (the Malphigian tubules, as you will learn in Chapter 11, are concerned with excretion in insects). The hind-gut is concerned with the absorption of water to enable the production of very dry faeces.

Question 7 The evidence suggests that digestion is primarily extracellular. If digestion were predominantly intracellular, there would be little trace of the gelatin blocks. The fact that there is very little trace of enzyme in the enteron of the non-feeding animal or in the supernatant fluid consisting of mesenteries and filaments, but that enzymes are released when protein is present, suggests that some digestion is extracellular. Extracellular digestion would be more advantageous for this anemone, because it is a discontinuous feeder and ingests bulky food material.

Question 8 It is impossible to tell from the evidence provided whether the protease is secreted as an inactive precursor, although many proteases are released in this form; for example, pepsin is secreted in the mammalian stomach as pepsinogen.

Question 9 Pepsin is only found in the vertebrate stomach and is largely unknown in invertebrates. The main invertebrate endopeptidase is a trypsin-like enzyme. Apart from an endopeptidase, an exopeptidase is also likely to be present.

Question 10 Explanation (iii) is the most likely.

(i) Although 'racial' differences in the physiology of animals probably exist, it is extremely unlikely that the differences would be as marked as this suggestion implies.

(ii) Secretin evokes a secretion rich in bicarbonate but poor in enzymes, and hence it has no 'enzyme-stimulating' properties.

(iii) This is correct. The explanation fits in with the information on secretin and CCK-PZ action. It is now known that during the extraction of 'Mellanby's preparation' (M), secretin was absorbed on to the bile salts (see the extraction techniques) but CCK-PZ was not.

(iv) This explanation is unlikely. Secretin evokes a secretion poor in enzymes. Therefore, more than one hormone is present in the SI extract. Hormones of the intestine *promote* the release of enzymes; inhibitory gastrointestinal hormones are not known to occur (this does not prove that they do not exist!, and it is believed that GIP stops gastrin secretion).

CHAPTER 9

Question 1 The accurate statements are (c) and (d).

(a) Glycogen phosphorylase and synthetase both exist as *a* and *b* forms, but the activation of synthetase depends on the inhibition of the cyclic AMP-dependent protein kinase.

(b) Glycerol and some amino acids are substrates for gluconeogenesis by reversed glycolysis, but fatty acids are not.

(e) As the concentration of glucose rises in liver cells, the level of phosphorylase *a* drops and that of synthetase *a* rises.

Question 2 (a), (c) and (d) would all promote insulin secretion; (b) and (e) would prevent insulin secretion.

(a) Somatostatin inhibits insulin secretion by a direct effect on B cells; removing the inhibition stimulates release.

(b) Glucose stimulates insulin secretion, but the effect is via a regulatory protein that is activated by Ca^{2+}-calmodulin.

(c) Cyclic AMP promotes insulin production, and inhibiting phosphodiesterase (the enzyme that breaks down cyclic AMP) results in insulin secretion.

(d) The presence of food in the digestive system stimulates the hypothalamus and this, through the ANS and vagal stimulation, then promotes insulin secretion.

(e) Glucose must enter cells to activate Ca^{2+}-dependent secretion of insulin.

Question 3 (a) and (b) would result in the breakdown of lipid and protein.

(a) The absence of B cells implies an insulin deficiency, which means that lipid and protein are broken down and used in the liver.

(b) Glucagon promotes the breakdown of lipid and protein for gluconeogenesis.

(c) A carbohydrate-rich meal results in a plentiful supply of glucose and possibly an increase in lipid stores.

(d) Somatostatin normally inhibits insulin and glucagon secretion, and removing the inhibition would result in a rise in insulin and so enhances the protection of lipid and protein stores.

Question 4 (a) Insulin, by activation of glycogen synthetase; glucocorticoids, by deactivation of glycogen phosphorylase.

(b) Glucagon, by activation of glycogen phosphorylase and also by activation of enzymes in the TCA cycle; adrenalin, by activation of glycogen phosphorylase.

(c) Glucocorticoids (e.g. cortisol) and glucagon both promote breakdown of lipids and protein in cells.

Question 5 Stress involves the ANS and a rise in levels of glucocorticoids and of adrenalin. Glucocorticoid release from the adrenal cortex is stimulated by the action of CRF (from the hypothalamus) on the pituitary (which produces ACTH). The result is a small rise in blood glucose.

Stress via the sympathetic nervous system (through noradrenalin) inhibits insulin production directly in the pancreatic islets and stimulates glucagon release. The sympathetic system stimulates the adrenal medulla to release adrenalin, which promotes liver gluconeogenesis. Both raise the blood glucose level.

Question 6 When injected singly, the three hormones each raise blood glucose concentration. In combinations, any two together raise blood glucose by a greater amount than the sum of their individual effects (e.g. adrenalin 30 mg per 100 cm^3 and cortisol 3 mg per 100 cm^3, but adrenalin + cortisol 58 mg per 100 cm^3). When all three are injected together, the rise in blood sugar is much greater than their individual effects or their combined effects as pairs. These are examples of the synergistic actions of hormones.

Question 7 (a) Blood glucose level would rise. The effects of glucagon are shown in Figure 9.12: increase in glucagon leads to increase in the level of glucose in the blood because of glycogenolysis and gluconeogenesis in liver cells, which respond to glucagon by an increase in cyclic AMP and in mitochondrial enzymes.

(b) The raised glucagon level in Type I diabetes is due to the absence of insulin, which normally inhibits glucagon release. Therefore administration of insulin would depress glucagon levels. Somatostatin normally inhibits glucagon secretion, so giving somatostatin should depress glucagon levels.

Question 8 (a) In the absence of insulin, protein and lipids are broken down to amino acids, fatty acids and glycerol, which are metabolized mainly in the liver.

(b) In the absence of normal glucose metabolism, the liver cannot convert all the acetyl CoA produced from fatty acids via the TCA cycle.

(c) Acetyl CoA in excess will produce ketone bodies such as acetone, β-hydroxybutyrate and acetoacetate.

(d) These ketone bodies (which are strong organic acids) enter the blood and cause changes in blood pH; this can lead to coma.

(e) Excess glucose in the blood as a result of insulin deficiency causes glycosuria and water is lost from the body through the kidneys. Some excretion of excess ketone bodies increases the loss of water from the kidneys.

(f) Massive dehydration caused by the loss of water and salts from the kidneys leads to osmotic problems for cells, particularly in the brain, and coma can result.

Question 9 Insulin will lower the level of blood glucose. If too much is given, then hypoglycaemia may result, leading eventually to coma. If too little insulin is given to maintain blood glucose within the correct limits, glucose levels after a meal will rise towards the hyperglycaemic state and there is the danger of ketotic coma. Diabetics need to balance their insulin injections with their diet and physical activity very carefully.

There are other problems with insulin therapy mentioned in Section 9.6. Many commercial insulins are not of human origin but are purified from pig or beef pancreas. Although very similar to human insulin, these commercial insulins given over long periods may initiate the formation of antibodies to the insulin. These antibodies will tend to prolong the action of the insulin but reduce the peak effect of an injection. Happily, the purity of insulins is now such that antibodies are formed only to a small extent and insulin insensitivity is not a major problem in insulin-dependent diabetes. Human insulin produced by genetic engineering sounds an ideal treatment but (in 1989) there is sometimes a problem with lack of warning of the onset of hypoglycaemia and consequent danger of coma.

CHAPTER 10

Question 1 The correct statements are (d), (e) and (g).

(a) Ovulation involves the release of the ovum from a mature follicle. It is then picked up by the funnel of the Fallopian tube, which lies close to but not attached to the ovary, and is carried down into the uterus. (Section 10.1.1)

(b) The corpus luteum is a transformed ruptured follicle that persists in the ovary after ovulation. If pregnancy occurs, it remains in the ovary and secretes a steroid hormone important for the maintenance of the uterine wall. (Section 10.1.1)

(c) The gestation period is the time from implantation until birth. (Section 10.1.1)

(f) Most mammalian species are seasonal breeders, and the gonads regress out of season. (Section 10.1.2)

Question 2 Blockage of the Fallopian tubes may prevent descent of the ovum to the uterus but does not necessarily prevent sperm reaching the ovum in the tube. Fertilization can occur, and sometimes the blastocyst will implant in the tube itself. An ectopic pregnancy such as this (or in extreme cases when fertilization takes place before the ovum enters the tube) is dangerous because the fetus being in the wrong position leads to local vascularization of tissues.

Question 3 (a) This is the peak concentration of progesterone, and you would expect it to inhibit LH secretion (and therefore ovulation) and possibly reduce the effect of FSH on the follicles. The endometrium will be maintained in a 'progestational' state, and there will be no menstruation. The high level of progesterone (because of the lack of LH and thus corpus luteum formation) will depress natural progesterone levels.

(b) Low levels of progesterone would not affect the ovaries or the endometrium, and circulating levels of steroids should not be affected.

Question 4 (a) Testosterone alone will probably suffice to maintain the seminiferous tubules, allow spermatogenesis and maintain the other glands and ducts for this period. LH might be safer because it may act directly on meiosis as well as causing testosterone release.

(b) FSH is needed for the Sertoli cells, and once again it is probably better to give LH than testosterone in the long term.

Question 5 The expected result is (b). LH stimulates the Leydig cells to secrete more androgen so (c) and (a) are wrong. Spermatogenesis is under FSH control so (e) is unlikely. (d) is clearly wrong because there is little evidence for cycles of sexual activity in the human male.

Question 6 The accurate statements are (b) for rats and (e) for humans.

(a) is wrong; the implanted egg does not secrete a hormone.

(c) is wrong; after implantation the corpus luteum is retained.

(d) is wrong; prostaglandins are thought to be luteolytic and will cause abortion.

(f) is wrong; the corpus luteum produces progesterone, which maintains the endometrium of the uterus.

Question 7 Your flow diagram should look something like Figure 10.23.

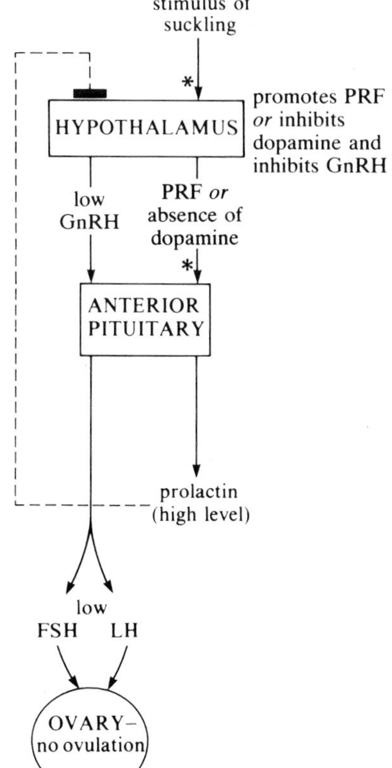

Figure 10.23 Suckling stimulates the production of prolactin from the anterior pituitary by promoting the release of PRF, inhibiting the release of dopamine from the hypothalamus, or suppressing GnRH. Prolactin in the blood can also feedback onto the hypothalamus and inhibit the release of GnRH. The consequence is low levels of the gonadotropins FSH and LH. Therefore, the follicles in the ovary do not mature, and so ovulation cannot take place.

Question 8 Your flow diagram should look something like Figure 10.24.

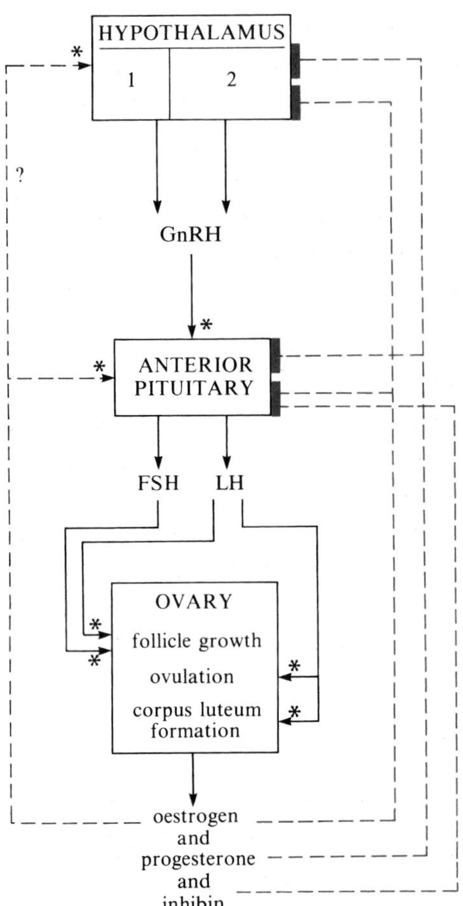

Figure 10.24 The products of the ovary (oestrogen and progesterone) have different effects at different times of the menstrual cycle. The two divisions of the hypothalamus in this diagram are (1) the pre-optic and anterior hypothamic areas and (2) the ventro-medial and arcuate nucleus.

Look at the right-hand side of the diagram first. This depicts the negative feedback effects of oestrogen and progesterone on the pituitary and hypothalamus that are operating during the follicular and luteal phases of the cycle. The left-hand side of the diagram illustrates the positive feedback effect of oestrogen that is operating just before, and in fact is responsible for, ovulation. Note that different parts of the hypothalamus release GnRH and respond differently to steroids.

Question 9 The doe is an induced ovulator, so the rabbit cannot have ovulated before hypophysectomy. In the case of the rat, the effect will depend entirely on the stage of the oestrous cycle. If the rat is in oestrus, an injection of LH will have the same effect as in the rabbit. If the rat had just ovulated, then nothing would be necessary to allow fertilization.

Question 10 Examine blood or urine for signs of cyclic steroid changes; see whether fertility is affected by anaesthesia of the genitalia or cervix or by cutting the sensory nerves from the pelvic region before mating. The ovary of an induced ovulator is affected by two separate classes of environmental effects—stimulation from the male (which induces LH release) and the normal influences of light acting on the hypothalamus, which affects the onset of reproduction in both spontaneous and induced ovulators.

Question 11 A pheromone could act through the olfactory system and the hypothalamus to advance the cycles of those ewes not yet in oestrus. Alternatively, a visual or auditory stimulus could have the same effect.

Question 12 (i) A genetically male rat starts to secrete testosterone as the gonads develop. Giving testosterone may simply speed up the development process or have no effect.

(ii) A genetically female rat at birth has ovaries and will be producing oestrogens. An injection of testosterone at this time would cause sterility, because it would abolish the ability of the hypothalamus to release LH from the 'surge' centre. An androgenized female does not ovulate.

(iii) Giving testosterone to a sexually mature female would probably prevent the release of gonadotropins from the pituitary by a negative feedback system and therefore produce a temporary halt to follicular maturation and ovulation. This would not be a permanent effect, unlike (ii).

Question 13 The correct answers are (c), (d), and (g).

(a) LH injection would initiate ovulation.

(b) Oral administration of a peptide would not work because it is liable to digestion in the gut.

(c) Progesterone at a high level through the cycle would prevent release of FSH and LH and therefore prevent follicular maturation and ovulation.

(d) This is the common formula of the contraceptive 'pill'.

(e) Prolactin is a peptide and oral administration would render it useless.

(f) FSH injection would initiate follicular maturation.

(g) If the hypothalamic control of pituitary gonadotropin release is blocked, follicular maturation and ovulation will be prevented.

CHAPTER 11

Question 1 Any toxic compound that has entered an animal's blood will, if its molecules are small enough, be automatically removed by ultrafiltration and will not be reabsorbed. Without ultrafiltration, and assuming it has a very low membrane permeability, it would not automatically be eliminated unless a special mechanism existed to transport it into the primary urine by secretion. The mammalian kidney rapidly removes toxic compounds of low molecular mass from the blood by ultrafiltration.

Question 2 The correct answers are (a), (c) and (d). The more concentrated the urine produced (i.e. the greater the *U*/*P* ratio) the greater will be the cortico-medullary gradient, and the concentration of the fluid of the inner medulla can be as high as 3 000 mOsmol l^{-1} in some desert rodents. This increased osmolarity is a function of the extremely long loop of Henle found in these animals, and these are frequently found in juxtamedullary nephrons, which are of course surrounded by the vasa recta blood vessels.

Items (b), (e) and (f) are basic features of all mammalian nephrons.

Question 3 From Table 11.1, compounds of M_r below 5 200 will be freely filtered in the glomerulus, but there is progressive hindrance to the passage of larger molecules, and there will be almost no glomerular filtration of those of M_r above 50 000. Thus, the protein of M_r 2 000 will be rapidly excreted, but that of M_r 100 000 will not, unless special mechanisms exist to secrete it to the primary urine.

Question 4 When a litre of water is ingested, the body fluids are diluted and their osmolarity falls. The absence of ADH results in collecting ducts becoming impermeable to water (no ADH will be released), and urine output rapidly rises. When the body fluids return to their normal osmolarity ADH is released and urine output decreases as a result of water absorption from the collecting ducts, whose permeability to water has increased.

Question 5 (a) The GFR would rapidly decline and become zero if the capillary blood pressure fell below that of the combined hydrostatic back-pressure (in Bowman's capsule) and the osmotic pressure of the plasma proteins in the blood (which tend to counteract the pressure forcing fluid out from the blood).

(b) The initial effect here would be a reduction in blood flow. If this then resulted in a reduction in the pressure of blood going to the kidneys it would cause the juxtaglomerular apparatus to release renin and set in motion a train of events leading to the production of the vasoconstrictor angiotensin II, and this will act to raise blood pressure (and stimulate aldosterone release).

Question 6 Yes, the urine test would be positive, because glucose is freely filtered in the glomerulus. Glucose in excess of a plasma concentration of 200 mg per 100 cm^3 cannot be transported back into the blood and will be excreted (see Figure 11.6). The T_{max} of glucose in the proximal tubule cells in humans is 375 mg min^{-1}.

Question 7 (a) Coupled transport of Na^+ and Cl^- (and K^+) occurs across the luminal membranes of the thick ascending limb of the loop of Henle. Furosemide added to the fluid around the outside of the tubules is not likely to have a direct effect on active NaCl transport. However, the Na^+/K^+ ATPase is located on the basolateral membranes so ouabain will directly inhibit it. Therefore active NaCl transport from the thick ascending limb of the loop of Henle will be inhibited by ouabain.

(b) Inhibition of active co-transport of NaCl from the thick ascending limb will in time mean that the osmolarity of the medullary interstitium will be reduced and so less water will be removed from the descending loop of Henle, with the result that the cortico-medullary osmotic gradient will eventually be reduced.

Question 8 No energy is expended in the vasa recta counter-current exchange mechanism to maintain the cortico-medullary gradient (an existing gradient). Operation of the counter-current multiplier depends largely on the co-transport of Na^+, K^+, and Cl^- in the thick ascending limb of the loop of Henle, which is driven by the primary active transport of Na^+ (the basolateral Na^+/K^+ ATPase).

Question 9 Such a mammal will be secreting the maximum amount of ADH, but the urine flow cannot be reduced to zero because urea in the glomerular filtrate exerts an osmotic pull and therefore requires water to be excreted with it. The amount of urea will therefore be one of the determining factors. Another factor will be the state of 'salt-loading'. If there is surplus Na^+ in the body, aldosterone release will be inhibited, and salt (and thus water) will be excreted. Another factor is the anatomy of the kidney of the particular animal.

Question 10 (a) True. Glucose has a relatively low relative molecular mass and can diffuse across the wall of the Malpighian tubule, though at a relatively low rate compared with the rate at which glucose is filtered from the blood to the mammalian nephron. (b) True. Haemolymph pressure is too low to permit any filtration 'under force', and while some proteins can diffuse across to the lumen of the Malpighian tubule, those of M_r 10 000 would pass to a very limited extent. (c) False. The mechanism of primary urine secretion in insects is by the active transport of K^+ with Cl^- following passively. Ethacrynic acid is unlikely to affect primary urine formation to any extent. (d) True (by cells of the rectum). (e) True. There is evidence of antidiuretic and diuretic hormones in several insects.

Question 11 (a) True. The other important source is from nucleic acid breakdown. (b) False. Ammonia is converted into uric acid in several groups of animals, but not via the urea cycle; this produces urea. (c) False. It is only theoretically possible for ammonia to be removed like carbon dioxide via respiration. In practice, the level that ammonia would need to reach in the blood for this to happen would kill the animal. (d) True.

Question 12 In newt larvae, ammonia can simply diffuse to the exterior through the skin or gills, but on land this cannot happen; ammonia could diffuse out via the lungs, but for this to happen it would have to build up to concentrations that would be toxic. On land, most ammonia is converted into urea before excretion by the kidneys, but when the animal returns to water, ammonia can once more be lost by diffusion through the skin.

CHAPTER 12

Question 1 *Palamonetes varians* is truly euryhaline: its adaptation curve stretches from freshwater to seawater. It is an osmoregulator. At low external medium concentrations, it is a hyper-osmotic regulator and at high external medium concentrations it is a hypo-osmotic regulator. In fact, this animal maintains its body-fluid composition relatively constant over a large variation in external medium concentration. It is homoiosmotic to a considerable extent.

Question 2 Yes, the B/M ratio for all the ions listed except Cl^- is substantially different from 1.00. The ions that appear to be subject to the greatest amount of regulation are Mg^{2+} and $SO_4{}^{2-}$.

Question 3 (a) The animal conforms to variation in external chlorinity greater than 200 mmol l^{-1} Cl^- (its body fluids are of the same osmolarity). (b) However, below this it is a hyper-osmoregulator. Its body fluids are more concentrated than the medium, and it faces an osmotic inflow of water and a diffusive loss of ions. It may be able to reduce its permeability and so partially counteract this, but excess water will have to be excreted and ions taken up from the environment.

Question 4 Items (a), (c) and (e) are the correct choices. The P_{os} value will decrease, which indicates a reduction in osmotic permeability—hence (b) is incorrect. As the external medium becomes more dilute, so more water will enter the animal, even though the permeability has been reduced. Hence there will be copious urine production, and clearly it will be advantageous if this is as dilute as possible; thus (c) is correct and (d) is wrong. (e) is correct: the external body surfaces of crustaceans are largely impermeable except for the gills. (f) is incorrect: active transport of ions does occur from the urine back into the animal, but a considerable portion still occurs via the gills. (g) is incorrect because these animals can live in freshwater and in the sea. (h) is incorrect. The rate of transport (J_{max}) may change, but not the affinity (K_t).

(Don't be misled by the slightly different K_t values for *G. duebeni* in Table 12.4. The freshwater animal *G. duebeni celtius* is a subspecies, and might be expected to have a somewhat different affinity for Na^+.

Question 5 The graph shows increased output of non-protein nitrogen compounds after transfer to 50% seawater. These results relate to intracellular iso-osmotic regulation of the osmotic concentration of the intracellular fluids (Section 12.2.3). Normally, a considerable part of the osmotic pressure of the intracellular fluids is caused by amino acids, and the level of these decreases when the external osmotic concentration falls. This can happen by degradation of the amino acids, and the increased release of nitrogen to the medium is a likely result of this.

Question 6 The osmoregulatory problems of the two animals would be qualitatively similar but quantitatively different. Because passive physical processes are affected by temperature, the rate of movement of ions out of the animals would be greater at the higher temperature. Membrane structure might be affected by temperature and a less fluid membrane possibly would be less permeable. But in addition, active processes such as ion transport that depend on an energy supply will certainly occur at different rates at temperatures that differ by 20 °C, and thus ion transport will be affected. Clearly, therefore, water and ionic regulation will be quantitatively different at the two temperatures.

Question 7

Freshwater teleosts		Marine teleosts	
site 1	(x)	site 1	(iv)
site 2	(i)	site 2	(ii)
site 3	(x)	site 3	(iii), (vi)
site 4	(viii)	site 4	(v), (ix)

No teleosts produce large volumes of concentrated urine (vii).

Question 8 The eel migrates from freshwater to the sea to spawn. In freshwater, it will have experienced a diffusive efflux of salts and influx of water, and in the sea these problems will be reversed. The osmotic gradient between the body fluids and the environment is greater in the sea, so the eel should reduce its osmotic permeability (to reduce water loss). In addition, urine flow should be reduced.

Transport of ions in freshwater occurs from the environment into the animal, primarily via the chloride cells on the gill, but in the sea the direction of transport must be reversed. In addition, seawater should be swallowed, and water be absorbed via the intestine, the osmotic permeability of which may be higher in the sea than in freshwater. (However, one problem is that eels' guts degenerate in the sea.) The monovalent ions are excreted by the gills, and the divalent ions by the kidney.

Question 9 *R. cancrivora* is an osmoregulator at external salinities up to about 300 mOsmol l^{-1}. At concentrations above this, it appears to be an osmoconformer. However, the ion concentration of the body fluids remains unchanged at these higher salinities, and most of the increased osmotic concentration of the body fluids is caused by a high level of urea in the blood (Table 12.6), which it can tolerate.

Question 10 Species A is the white rat; B is the kangaroo rat. Total water loss in the kangaroo rat is about a third of that in the laboratory rat for the same food intake (8 g compared with 23 g). More water reabsorption occurs from the faeces and urine in the kangaroo rat. The kangaroo rat can produce more concentrated urine because it has nephrons with extremely long loops of Henle. The evaporative losses occur mainly from the respiratory surfaces, and these are lower in the kangaroo rat because it has a more efficient nasal heat-exchange mechanism. Oxidative gains are of course similar, because the metabolic reactions are the same in both species. The point is that the oxidative intake exceeds the loss in the kangaroo rat, whereas in the laboratory rat, it is insufficient to balance losses, and extra water must be ingested to achieve water balance.

Question 11 Items (e), (h) and (i) are correct. Harbour seals are essentially physiologically similar to terrestrial mammals and they have an impermeable outer skin. Therefore, no ion or water fluxes will occur via this route, so (a), (b), (c) and (g) are incorrect. (d) is also wrong—the kidneys produce a concentrated urine ((h) is correct), but this is not nearly as concentrated as in desert mammals. (f) is unlikely, and (i) is correct.

FURTHER READING

CHAPTERS 1 TO 3

Kandel, E. R. and Sharwz, J. H. (1985) *Principles of Neuroscience*, 2nd edn, Elsevier.

Kuppfler, S. W., Nichols, J. G. and Martin, R. (1984) *From Neuron to Brain*, 2nd edn, Sinaeur.

Sheppherd, G. M. (1988) *Neurobiology*, 2nd edn, Oxford University Press.

CHAPTER 4

Booth, D. A. (1976) *Hunger Models*, Academic Press.

Gordon, C. J. and Heath, J. E. (1986) Integration and central processing in temperature regulation, *Annual Review of Physiology*, 48, 595–612.

Guyton, A. C. (1986) *A Textbook of Medical Physiology*, W. B. Saunders.

Toates, F. M. (1975) *Control Theory in Biology and Experimental Psychology*, Hutchinson Educational Limited.

Toates, F. M. (1980) *Animal Behaviour—A Systems Approach*, Wiley.

Toates, F. M. (1986) *Motivational Systems*, Cambridge University Press.

CHAPTERS 5 TO 8

Eckert, R., Randall, D. and Augustine, G. (1988) *Animal Physiology*, 3rd edn, W. H. Freeman and Co.

Hill R. W. and Wyse, G. A. (1989) *Animal Physiology*, 2nd edn, Harper & Row.

Jennett, S. (1989) *Human Physiology*, Churchill Livingstone.

Stanier, M. and Forsling, M. (1990) *Physiological Processes*, McGraw Hill.

Tortora, G. J. and Anagnostakos, N. P. (1990) *Principles of Anatomy and Physiology*, 6th edn, Harper & Row.

Vander, A. J., Sherman, J. H. and Luciano, D. S. (1986) *Human Physiology*, McGraw–Hill.

Widdicombe, J. and Davies, A. (1983) *Respiratory Physiology*, Edward Arnold.

Wood, D. W. (1983) *Principles of Animal Physiology*, 3rd edn, Edward Arnold.

CHAPTER 9

Drury, M. I. (1979) *Diabetes Mellitus*, Blackwell.

Furth, A. and Harding, J. (1989) Why sugar is bad for you, *New Scientist*, 23 September 1989, pp. 44–7.

Hardy, R. N. (1981) *Homeostasis*, 2nd edn, Studies in Biology, Edward Arnold.

Notkins, A. L. (1979) The Cause of Diabetes, *Scientific American*, **241** (5), p. 56.

Oakley, W. G., Pyke, D. A. and Taylor, K. W. (1978) *Diabetes and its Management*, 3rd edn, Blackwell.

Tattersall, R. (1986) *Diabetes: A Practical Guide for Patients on Insulin*, 2nd edn, Churchill Livingstone.

White, D. A., Middleton, B. and Baxter, M. (1984) *Hormones and Metabolic Control. A Medical Student's Guide to Control of Various Aspects of Normal and Abnormal Metabolism*, Edward Arnold.

CHAPTER 11

Green, R. (1988) in Elmslie-Smith, D., Paterson, C. R., Scratcherd, T. S. and Read, N. W. (eds.) *Textbook of Physiology*, 11th edn, Chapters 12 and 13, Churchill Livingstone.

CHAPTER 12

Lockwood, A. P. M. (1976) in Newell, R. C. (ed.) *Adaptation to Environment*, Chapter 6, Butterworths.

Rankin, J. C. and Davenport, J. (1981) *Animal Osmoregulation*, Blackies.

Schmidt-Nielsen, K. (1990) *Animal Physiology*, 4th edn, Chapters 9 and 10. Cambridge University Press.

ACKNOWLEDGEMENTS

The series *Biology: Form and Function* (for Open University Course S203) is based on and updates the material in Course S202. The present Course Team gratefully acknowledges the work of those involved in the previous Course who are not also listed as authors in this book, in particular: Ian Calvert, Lindsey Haddon, Sean Murphy and Jeff Thomas.

Grateful acknowledgement is made to the following sources for permission to reproduce material in this book.

FIGURES

Fig. 1.5: Cooper, J.R., Bloom, F.E. and Roth, R.H. *The Biochemical Basis of Neuropharmacology* (3rd edn), copyright © 1970, 1974, 1978, Oxford University Press, Inc., reprinted by permission; *Fig. 1.13(a):* Kuffler, S.W. and Nicholls, J.G. (1977) *Neuron to Brain: A Cellular Approach to the Function of the Nervous System*, Sinauer Associates, Inc.; *Fig. 2.11(a):* Krieger, Allen, Rizzo and Krieger (1971) 'Circadian periodicity in concentration of plasma, ACTH', *J. Clin. Endocrinol.*, **32**, 266–284, copyright © Williams and Wilkins, Publisher; *Fig. 2.11(b):* Krieger, D.T. (1972) 'Circadian corticosteroid periodicity: critical period for abolition by neonatal injection of corticosteroid', *Science*, **178**, 1206, copyright © AAAS; *Figs 4.3, 4.12, 4.13:* Guyton, A. (1986) *A Textbook of Medical Physiology*, W.B. Saunders, Publishers, Inc.; *Fig. 4.20:* Eikelboom, R. and Stewart, J. (1982) 'Condition of drug-induced physiological responses', *Psychol. Rev.*, **89**, 511, copyright © 1982, The Psychological Association, reprinted (or adapted) by permission; *Fig. 4.24:* Rolls, B.J., Wood, R.J. and Stevens, R.M. (1978) 'Palatability and body fluid homeostasis', *Physiol. Behav.*, **20**, 15–19; *Fig. 4.28:* Morrison, S.D. *et al.* (1967) 'Water exchange and polyuria of rats deprived of food', *Q. J. Exp. Physiol.*, **52**, 51–67; *Figs 5.5, 5.6, 5.16:* Vander, A.J., Sherman, J.H. and Luciano, D.S. (1986) *Human Physiology*, McGraw Hill Publishing Company; *Fig. 6.4:* Schneiderman, H.A. and Williams, C.M. (1955) *Biol. Bull. (Woods Hole)*, **109**, 123–145; *Figs 6.11, 6.13:* Hill, R.W. (1976) *Comparative Physiology of Animals: An Environmental Approach,* copyright © 1976, R.W. Hill, adapted with permission of Harper & Row, Publishers Inc.; *Fig. 7.2:* Eckert, R. and Randall, D. (1978) *Animal Physiology*, W.H. Freeman & Co Publishers; *Fig. 7.3:* Comroe, R.H.Jr *et al.* (1962) *The Lung* (2nd edn), Chicago Year Book Medical Publishers Inc.; *Fig. 7.7:* Hill, R.W. (1976) *Comparative Physiology of Animals: An Environmental Approach*, copyright © 1976, Richard W. Hill, adapted with permission of Harper & Row, Publishers, Inc.; *Fig. 7.12:* Hughes, G.M. (1963) *Comparative Physiology of Vertebrate Respiration*, Heinemann Educational Books; *Fig. 7.14:* Dripps, R.D. and Comroe, R.H.Jr (1947) *Am. J. Physiol.*, **149,** 43; *Fig. 7.17:* Dejours, P. in Fenn, W.O. and Rahn, H. (eds) (1964) *Handbook of Physiology,* American Physiological Society, **3(1)**; *Fig. 7.18:* Comroe, R.H.Jr (1974) *Physiology of Respiration*, The American Physiological Society; *Fig. 7.19:* Dejours, P. (1966) *Respiration*, p.187, Oxford University Press; *Fig. 7.20:* Mountcastle, V.B. (ed.) (1968) *Medical Physiology*, p.822, reproduced by permission from Lambertson, C.J. (1980) *Medical Physiology*, C.V. Mosby Co.; *Fig. 7.24(a)(b):* Scholander P.F. (1940) 'Typical results from diving experiment in a seal', permission to reproduce from Mr Scholander's estate; *Fig. 8.6(a–c):*

Jennings, J.B. (1972) *Feeding, Digestion and Assimilation in Animals* (2nd edn), Macmillan, permission to reproduce from Macmillan London and Basingstoke; *Fig. 8.9:* Milner, A. (1981) reprinted by permission from *Nature*, **289**, 347, copyright © Macmillan Magazines Ltd; *Fig. 8.12(a)(b):* Yonge, C.M. and Thompson, T.E. (1976) *Living Marine Molluscs,* p.200, Collins; *Figs 8.13, 8.15, 8.38:* Morton, J. (1971) *Guts: The Form and Function of the Digestive System*, Edward Arnold; *Fig. 8.17(a–c):* Freeman, W.H. and Bracegirdle, B. (1971) 'Amoeba and drawing of Specimen 1', *An Atlas of Invertebrate Structure*, Heinemann, copyright © W.H. Freeman and B. Bracegirdle; *Figs 8.18(a)(b), 8.19(a–f):* Eikert, R. and Randall, D. (1983) *Animal Physiology: Mechanism and Adaptations* (2nd edn), W.H. Freeman; *Fig. 8.23:* photograph of a cray fish foregut showing structure of the teeth courtesy of D.M. Holdich and T. Smith; *Fig. 8.34:* Bell, G., Emslie-Smith, D. and Paterson, C. (1976) *Textbook of Physiology*, copyright © Longman Group Ltd., p.102; *Fig. 9.4:* Horecker, B.L. and Stadtman, E.R. (1976) *Current Topics in Cellular Regulation*, p.85, Academic Press Ltd, Harcourt Brace Jovanovich Inc.; *Figs 9.6, 9.13, 9.16:* Unger, R.H., Raskin, P., Spikant, C.B. and Orci, L. (1977) *Glucagon and the A Cells, Recent Progress in Hormone Research*, **33**, copyright © Academic Press Ltd; *Fig. 9.14:* Unger, R.H. and Dobbs, R.E. (1978) Reproduced with permission from the *Ann. Rev. Physiol.*, **40**, copyright © Annual Reviews Inc.; *Figs 10.2, 10.6(a):* Johnson, M.H. and Everitt, B.J. (1980) *Essential Reproduction*, Blackwell Scientific Publications Ltd; *Fig. 10.9:* Slater, P.J.B. (1978) *Sex Hormones and Behaviour*, p.8, Edward Arnold; *Figs 10.15, 10.16:* Follett, B.K. (1978) 'Photoperiodism and seasonal breeding in birds and mammals', in Crighton *et al.* (eds), *Control of Ovulation*, pp.269–286, Butterworths; *Figs 10.20, 10.21:* Ehrhardt, A.A. (1978) 'Behaviour sequelae of prenatal hormonal exposure', in Lipton, M.A. *et al.* (eds) *Biopharmacology: A Generation of Progress*, pp.532–533, Raven Press Ltd, New York; *Figs 11.1(b), 11.3, 11.8, 11.13, 11.14, 11.16:* Emslie-Smith, D., Paterson, C.R., Scratcherd, T. and Reid, N. (1988) *Textbook of Physiology* (11th edn), Churchill Livingstone, copyright © Longman Group UK Ltd; *Figs 11.1(c)(d), 11.4:* Schmidt-Nielsen, K. (1979) *Animal Physiology*, Oxford University Press; *Fig. 11.2:* Moffatt, D.B. (1978) 'The control of water balance by the kidney', *Carolina Biology Reader Series*, copyright © Carolina Biological Supply Co., Burlington, NC; *Fig. 11.5:* Eckert, R. and Randall, D. (1978) *Animal Physiology*, W.H. Freeman & Co Ltd; *Fig. 11.7:* Lamb, J.F., Ingram, C.G., Johnston, I.A. and Pitman, R.M. (1984) *Essentials of Physiology* (2nd edn), p.297, copyright © 1980, 1984, Blackwell Scientific Publications Ltd; *Fig. 11.17:* Clegg, A.C. and Clegg, P.C. (1969) *Hormones, Cells and Organisms*, Heineman, London; *Fig. 11.19:* Madrell, S.H.P. (1971) *Philosoph. Trans. Br. Roy. Soc., Lond.*, **262**, 197–297; *Fig. 11.23:* Rankin, J.C. and Davenport, J.A. (1981) *Animal Osmoregulation*, Blackie & Sons, Ltd; *Fig. 12.6(a)(b):* Beadle (1943) 'Osmotic regulation', *Biological Review*, reproduced by permission of Cambridge University Press; *Fig. 12.7:* Hoar, W.S. (1975) *General and Comparative Physiology* (2nd edn), p.371, copyright © 1975, reprinted by permission of Prentice-Hall, Inc, Englewood Cliffs, NJ; *Figs 12.8, 12.9(a–c), 12.15, 12.20(a)(b), 12.31:* Newell, R.C. (1976) *Adaptation to Environment*, Butterworth & Co. Publishers Ltd; *Fig. 12.17:* Barrington, E.J.W. (1979) *Invertebrate Structure and Function* (2nd edn), Thomas Nelson & Sons Ltd.

TABLES

Table 12.5: Newell, R.C. (1976) *Adaptation to Environment*, Butterworth and Co. Publishers Ltd; *Tables 12.6, 12.8, 12.9, 12.10, 12.11:* Schmidt-Nielsen, K. (1990) *Animal Physiology* (4th edn), Cambridge University Press.

INDEX

Note Entries in **bold** are key terms, page numbers in *italics* refer to figures and tables.

Abbreviations in sub-entries:

ACTH adrenocorticotropic hormone
FSH follicle-stimulating hormone
LH luteinizing hormone
GnRH gonadotropin releasing hormone
TSH thyroid-stimulating hormone